Coherence and Correlation in Atomic Collisions

PHYSICS OF ATOMS AND MOLECULES

Coherence and Correlation in Atomic Collisions

Edited by

H. Kleinpoppen

Institute of Atomic Physics
University of Stirling
Stirling, Scotland

and

J. F. Williams

Department of Pure and Applied Physics
Queen's University
Belfast, Northern Ireland

Plenum Press · New York and London

Library of Congress Cataloging in Publication Data

Main entry under title:

Coherence and correlation in atomic collisions.

(Physics of atoms and molecules)
Includes index.
1. Coherence (Nuclear physics) – Congresses. 2. Angular correlations (Nuclear physics) – Congresses. 3. Collisions (Nuclear physics) – Congresses. I. Klein-poppen, Hans. II. Williams, J. F.

QC794.6.C58C66	539.7'54	79-15977

ISBN-13: 978-1-4613-2999-2   e-ISBN-13: 978-1-4613-2997-8
DOI: 10.1007/978-1-4613-2997-8

© 1980 Plenum Press, New York
A Division of Plenum Publishing Corporation
227 West 17th Street, New York, N.Y. 10011
Softcover reprint of the hardcover 1st edition 1980

To **Sir Harrie Massey**, Sec. R.S.

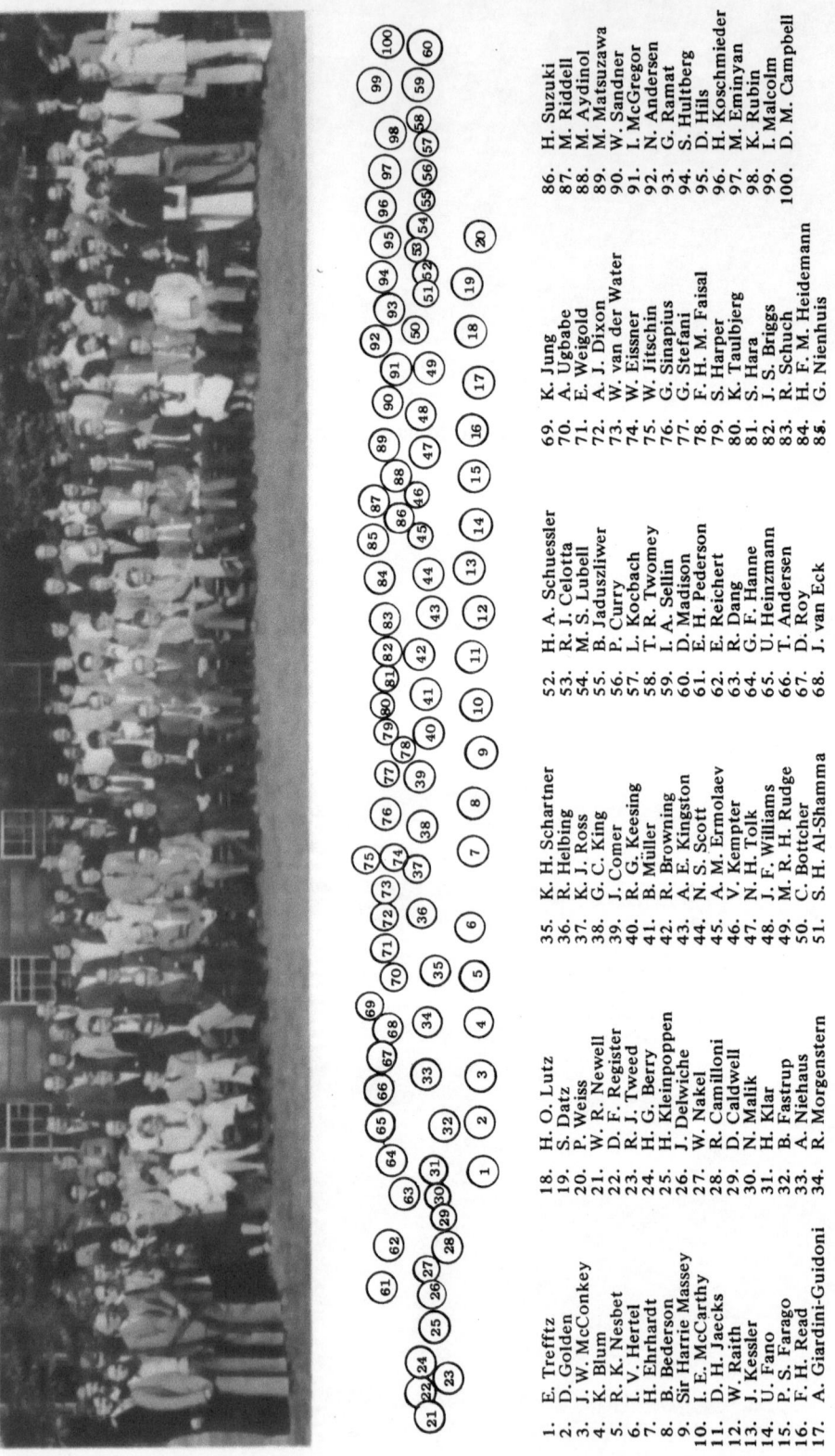

1. E. Trefftz
2. D. Golden
3. J. W. McConkey
4. K. Blum
5. R. K. Nesbet
6. I. V. Hertel
7. H. Ehrhardt
8. B. Bederson
9. Sir Harrie Massey
10. I. E. McCarthy
11. D. H. Jaecks
12. W. Raith
13. J. Kessler
14. U. Fano
15. P. S. Farago
16. F. H. Read
17. A. Giardini-Guidoni

18. H. O. Lutz
19. S. Datz
20. P. Weiss
21. W. R. Newell
22. D. F. Register
23. R. J. Tweed
24. H. G. Berry
25. H. Kleinpoppen
26. J. Delwiche
27. W. Nakel
28. R. Camilloni
29. D. Caldwell
30. N. Malik
31. H. Klar
32. B. Fastrup
33. A. Niehaus
34. R. Morgenstern

35. K. H. Schartner
36. R. Helbing
37. K. J. Ross
38. G. C. King
39. J. Comer
40. R. G. Keesing
41. B. Müller
42. R. Browning
43. A. E. Kingston
44. N. S. Scott
45. A. M. Ermolaev
46. V. Kempter
47. N. H. Tolk
48. J. F. Williams
49. M. R. H. Rudge
50. C. Bottcher
51. S. H. Al-Shamma

52. H. A. Schuessler
53. R. J. Celotta
54. M. S. Lubell
55. B. Jadus;zliwer
56. P. Curry
57. L. Kocbach
58. T. R. Twomey
59. I. A. Sellin
60. D. Madison
61. E. H. Pederson
62. E. Reichert
63. R. Dang
64. G. F. Hanne
65. U. Heinzmann
66. T. Andersen
67. D. Roy
68. J. van Eck

69. K. Jung
70. A. Ugbabe
71. E. Weigold
72. A. J. Dixon
73. W. van der Water
74. W. Eissner
75. W. Jitschin
76. G. Sinapius
77. G. Stefani
78. F. H. M. Faisal
79. S. Harper
80. S. Taulbjerg
81. S. Hara
82. J. S. Briggs
83. R. Schuch
84. H. F. M. Heidemann
85. G. Nienhuis

86. H. Suzuki
87. M. Riddell
88. M. Aydinol
89. M. Matsuzawa
90. W. Sandner
91. L. McGregor
92. N. Andersen
93. G. Ramat
94. S. Hultberg
95. D. Hils
96. H. Koschmieder
97. M. Eminyan
98. K. Rubin
99. I. Malcolm
100. D. M. Campbell

Photograph by: Photographic Unit, Dept. of Physics and Astronomy, University College, London.

The International Workshop on
COHERENCE AND CORRELATION IN ATOMIC COLLISIONS

was given financial support by

Edwards High Vacuum

I.B.M. United Kingdom Ltd.

Intertechnique Ltd.

Ortec Ltd.

Plenum Press

Spectra-Physics Ltd.

University College, London

University of Bielefeld

University of Stirling

Scientific Advisory Committee of the International Workshop

P.S. Farago, D. E. Golden, H. Kelinpoppen, J. Macek,
I.E. McCarthy, M.R.C. McDowell, W.R. Newell,
A. Niehaus, A. Scharmann, J.F. Williams

Local and Programme Committee of the International Workshop

H. Kleinpoppen (Chairman), H.O. Lutz, I. McGregor,
W.R. Newell (Secretary), J.F. Williams

Contents

Introduction

H. Kleinpoppen and J. F. Williams

It has only very recently become possible to study angular correlations and coherence effects in different areas of atomic collision processes: These investigations have provided us with an analysis of experimental data in terms of scattering amplitudes and their phases, of target parameters such as orientation, alignment, and state multipoles, and also of coherence parameters (e.g., the degree of coherence of excitation). In this way the analysis of electron–photon, ion–photon, atom–photon, or electron–ion coincidences from electron–atom, ion–atom, or atom–atom collisional excitation has led to a breakthrough such that the above quantities represent most crucial and sensitive tests for theories of atomic collision processes. Similarly, the powerful $(e, 2e)$ experiments (electron–electron coincidences from impact ionization of atoms) have attracted much attention where improved experimental studies and detailed theoretical description provide a wealth of information on either the collisional ionization process or the atomic structure of the target atom.

Interference effects, many-electron correlations, and energy and angular momentum exchange between electrons in a Coulomb field play a decisive role in the understanding of postcollision interactions. New results on coherence effects and orientation and alignment in collisional processes of ions with surfaces and crystal lattices show links to relevant interference phenomena in atomic collisions. In small-angle elastic electron–atom scattering the effect of angular coherence can be studied in a crossed beam experiment. Recent scattering experiments with laser-excited atoms reveal, through coherence effects and correlation experiments, information on scattering potentials and on alignment and orientation. Correlation and spin effects decisively influence the multiple escape of electrons near the ionization threshold. Only very recently, experiments with polarized electrons and polarized atoms revealed an interference effect in electron impact ionization of atoms. These experiments have been described by J. Kessler as belonging to "third-generation" type experiments on which experimentalists have made a start. Similar interference effects have been predicted in excitation processes with polarized electrons and polarized atoms.

Most impressive current experiments on inner-shell and bremsstrahlung processes are of topical interest in connection with x-ray–particle coincidences and anisotropies of Auger electrons and x-ray photons. New theoretical analyses of angular correlations from inner-shell excitation reveal similarities to the interpretation from relevant outer-shell processes.

The overall sophistication of collection of experimental data from atomic collisions has challenged the task of theory in atomic collision physics to the extent that particular detailed dynamical effects may now be linked to experimental observables which are not available through traditional cross-section measurements.

Together with the authors of these Proceedings we are most happy to express our dedication and admiration to Sir Harrie Massey, who has initiated, stimulated, and surveyed the field of atomic collisions over a period of almost five decades. Without his efforts we, as his followers, would not have reached the present state of advancement of most impressive and deep insight into one of the most fundamental fields of physics.

Theory of (e, 2e) Experiments

I. E. McCarthy

The (e, 2e) experiment may be analyzed in terms of a quasi-three-body model involving the incident electron, one initially bound electron, and the residual ion. If this model is valid the structure of the target and ion enters the (e, 2e) amplitude only in the form of the overlap of the target and ion wave functions. The validity of the model is verified by checking its experimental consequences. For closed-shell targets the profile of recoil momenta is very close to the square of the momentum-space orbital, which must be coupled with the target ground state to form the one-hole configuration in the ion eigenfunction. The squares of the amplitudes of the one-hole configuration in eigenfunctions of different ion states belonging to the same representation are proportional to the (e, 2e) cross section and sum to unity. For uncorrelated final states cross sections for excitation of one-hole configurations not contained in the Hartree–Fock ground state are sensitive measures of electron correlation in the target. These consequences are closely verified and establish the reaction as a very powerful structural tool for atoms and molecules. The reaction approximation for the quasi-three-body operator may be strictly tested by observing kinematic regions where the simplest approximation is adequate. The (e, 2e) reaction on the hydrogen atom provides the cleanest test of the operator. The distorted-wave impulse approximation evaluated in a factorized approximation is very good overall, but breaks down for values of the outgoing electron momenta that are too similar. The factorization of the impulse approximation is exact in an averaged eikonal approximation. In some cases experiments done with kinematics chosen to be especially sensitive to the reaction model and to eliminate the structural details indicate the validity of factorization. It is, of course, essential that the reaction theory should correctly reproduce absolute cross sections. The factorized eikonal approximation is very good in this respect in those cases where it is known to give good relative cross sections.

The $(e, 2e)$ experiment involves complete knowledge of the momenta and energies of four bodies, an incident electron (k_0), a target atom or molecule in its ground

I. E. McCarthy ● Institute for Atomic Studies, The Flinders University of South Australia, Bedford Park, S.A. 5042, Australia.

state $|g\rangle$, two outgoing electrons $(\mathbf{k}_A, \mathbf{k}_B)$, and the residual ion in an eigenstate $|f\rangle$ (which may be in the continuum). The amplitude M_f for the reaction is written

$$M_f(\mathbf{k}_0, \mathbf{k}_A, \mathbf{k}_B) = \langle \mathbf{k}_A, \mathbf{k}_B \,|\, (f\,|\,T\,|\,g) \,|\, \mathbf{k}_0 \rangle \tag{1}$$

The direct knockout approximation amounts to the assumption that T is a three-body operator, depending only on the coordinates of the center of mass of the residual ion and of two electrons 1 and 2, one of which is the incident electron and one of which is initially bound. Antisymmetry will be implicit throughout the discussion. With this assumption the state vector $|f\rangle$, which contains neither electron, commutes with T:

$$M_f = \langle \mathbf{k}_A, \mathbf{k}_B \,|\, T \,|\, (f\,|\,g)\mathbf{k}_0 \rangle \tag{2}$$

The $(e, 2e)$ amplitude now is a momentum transform of the generalized overlap function $(f\,|\,g)$, which depends only on structure properties of the target and ion. The differential cross section may be considered as a measure of this function via a profile of recoil momenta $\mathbf{q}$:

$$\mathbf{q} = \mathbf{k}_0 - \mathbf{k}_A - \mathbf{k}_B \tag{3}$$

The generalized overlap function may be calculated directly from a perturbation treatment of the one-body Green's function.[1] Such a treatment is valid for ion states $|f\rangle$ in the continuum, and has had considerable success[2] for the water molecule. However, in order to illustrate the determination of electron correlation by the $(e, 2e)$ reaction it is simpler to consider $|g\rangle$ and $|f\rangle$ as wave functions determined by a configuration interaction (CI) calculation using target Hartree–Fock functions $|\alpha\rangle$ as a basis:

$$|g\rangle = \sum_{\alpha} a_{\alpha}^{(g)} \,|\, \alpha \rangle \tag{4}$$

$$|f\rangle = \sum_{j\beta} t_{j\beta}^{(f)} C_{jr\beta} \psi_j^{\dagger}(r_2) \,|\, \beta \rangle \tag{5}$$

We will see how the $(e, 2e)$ reaction verifies that in fact the Hartree–Fock basis is the best assumption for the uncorrelated electrons. Correlations of course have no meaning if we do not specify what we mean by uncorrelated. In the CI expansions (4) and (5), correlations are expressed by the coefficients $a_{\alpha}^{(g)}$ for the target ground state and $t_{j\beta}^{(f)}$ for the final state in which the basis states consist of a single hole in an orbital j vector-coupled to a target Hartree–Fock state $|\beta\rangle$, in such a way that $|f\rangle$ belongs to the irreducible representation r of the point group of the target, whose degeneracy is n_r.

The antisymmetrized overlap function is in general

$$\langle f \,|\, g \rangle = n_r^{1/2} \sum_{j\alpha} a_\alpha^{(g)} t_{j\alpha}^{(f)} C_{jr\alpha} \psi_j \tag{6}$$

We are interested in two special cases. In the first the Hartree–Fock ground state $|0\rangle$ is an adequate description of the target ground state $|g\rangle$ so that $a_0^{(g)} = 1$:

$$\langle f \,|\, g \rangle = \sum_j t_{j0}^{(f)} \psi_j \tag{7}$$

If the Hartree–Fock basis is the best, this means that the one-body potential is such that only one orbital ψ_i contributes to the sum over j in (7). The differential cross section for the reaction is then given by a quasi-three-body model:

$$\sigma = K n_r S_i^{(f)} \,|\, \langle \mathbf{k}_A, \mathbf{k}_B \,|\, T \,|\, \psi_i \mathbf{k}_0 \rangle \,|^2 \tag{8}$$

where K is the appropriate kinematic factor.

The spectroscopic factor $S_i^{(f)}$ is defined by

$$S_i^{(f)} = [t_{i0}^{(f)}]^2 \tag{9}$$

It is essentially the probability of finding the configuration involving a single hole in the orbital i in the ion state $|f\rangle$. It obeys a sum rule derived from the normalization and closure properties of the wave functions

$$\sum_f S_i^{(f)} = 1 \tag{10}$$

The spectroscopic factor clearly is a very important quantity characteristic of electron correlations. It is determined very accurately by the $(e, 2e)$ reaction if the assumptions leading to (8) are true. The main assumption is the direct knockout assumption (2).

A direct experimental test of (2) and (8) is to verify first that momentum profiles for states belonging to the same irreducible representation r have the same shape, i.e., they are proportional to $S_i^{(f)}$, and second that the profile is correctly described by a Hartree–Fock orbital.

Figure 1 shows the $(e, 2e)$ momentum profiles for the $\frac{1}{2}^+$ states of the argon ion. They confirm the validity first of the direct knockout assumption, since all the profiles have essentially the same shape.

The verification of the Hartree–Fock orbital depends on the reaction theory being sufficiently correct in detail for (8) to give an accurate description of the momentum profile shape. We must therefore have a good model for the two-electron interaction T. Here we have used the ee Coulomb t matrix t. It is adequate for q up to about 1 a.u.

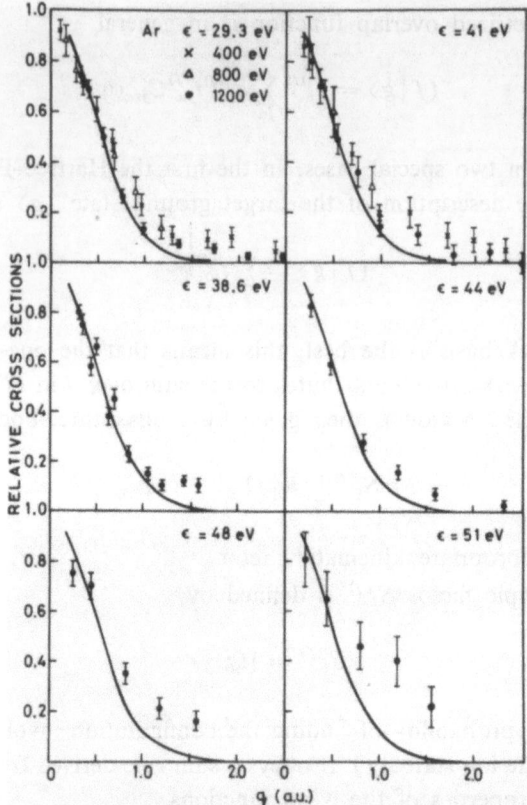

Figure 1. Momentum profiles for argon ion states for $\varepsilon_f > 29$ eV. The continuum begins at 43.6 eV. The curve is the PWIA using the $3s$ Hartree–Fock function.

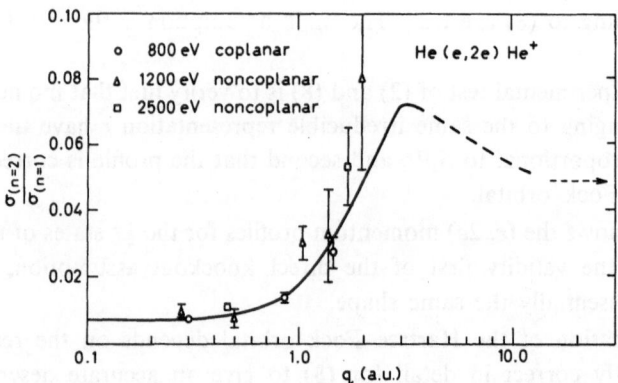

Figure 2. The ratio of summed $(e, 2e)$ cross sections for the $n = 2$ states of the helium ion to that for the ground state, plotted against the recoil momentum q. The PWIA curve is described in the text.

The second special case of the overlap function $(f \mid g)$ is the case where the final state is uncorrelated. This is exactly true for one-electron ions such as He^+ and H_2^+. In this case the ion eigenvalues ε_f are well separated and we can experimentally investigate ion states that have a large overlap with excited configurations $\mid \alpha \rangle$ of the target, but not with the ground state $\mid 0 \rangle$. An example is the $2p$ state of He^+. The $(e, 2e)$ reaction now investigates the ground-state correlation coefficients $a_\alpha^{(g)}$. Figure 2 shows the ratio of the summed $(e, 2e)$ cross sections[3] for the $2s$ and $2p$ states of He^+ to that for the $1s$ state compared with calculations using the CI expansion for He of Joachain and Vanderpoorten.[4]

Having verified the direct knockout approximation and seen that it leads to a quasi-three-body interpretation (8) of the $(e, 2e)$ reaction, we now consider the evaluation of the three-body amplitude $\langle \mathbf{k}_A, \mathbf{k}_B \mid T \mid \psi_i \mathbf{k}_0 \rangle$. The three-body equations for the problem may be written in the form due to Alt, Grassberger, and Sandhas[5] (AGS):

$$U_{ji}(E) = (1 - \delta_{ji})(E - H_0) + \sum_{k \neq j} t_k(E) G_0(E) U_{ki}(E), \qquad i, j, k = 1, \ldots, 3 \qquad (11)$$

where i, j, k label pairs, U_{ji} are the AGS transition operators, E is the total energy, H_0 is the kinetic energy plus the internal Hamiltonians of the three bodies, T_k is the t matrix for the interaction of the pair k while the third body does not interact, and $G_0 = [E^{(+)} - H_0]^{-1}$.

Suppose electron 1 is incident on the pair 1, the ion is particle 2, and the second electron is particle 3. The $(e, 2e)$ transition operator is

$$T = t_1 G_0 U_{11} + t_2 G_0 U_{21} + t_3 G_0 U_{31} \qquad (12)$$

Iterating the AGS equations (11) up to third order, using the abbreviated notation

$$ijk \cdots \equiv t_i G_0 t_j G_0 t_k \cdots$$

we have the multiple scattering series for T:

$$
\begin{aligned}
T = \quad & 2 + \; 12 + 323 \\
+ \quad & 3 + \; 13 + 232 \\
+ \quad & 23 + \; 32 + 123 + 132 \\
+ \quad & 212 + 213 + 312 + 313 \qquad (13)
\end{aligned}
$$

This series has not been calculated as such. The terms in (13) are grouped according to their relationship to the distorted-wave impulse approximation (DWIA), which is obtained by using the *ee t* matrix t_2 for T in (8) and replacing the plane waves $\mid \mathbf{k}_i \rangle$ by distorted waves $\mid \chi_i(\mathbf{k}_i) \rangle$ computed in relevant complex energy-dependent

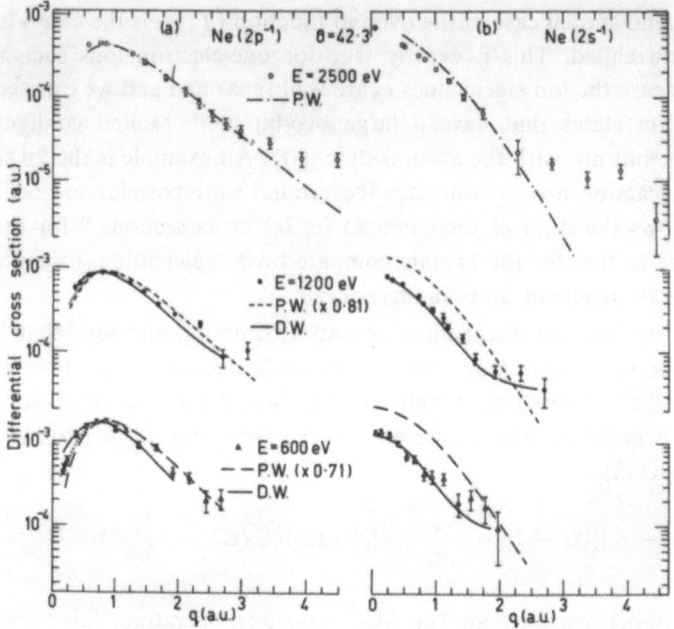

Figure 3. PWIA and DWIA momentum profiles for neon.

optical model potentials,[6] for which the t matrices are t_i, $i = 1, 3$:

$$|\chi_i(\mathbf{k}_i)\rangle = (1 + G_0 t_j)|\mathbf{k}_i\rangle, \qquad j = 3/i \tag{14}$$

$$T = (1 + t_1 G_0)(1 + t_3 G_0)t_2(1 + G_0 t_3) \tag{15}$$

It is possible to choose the observed kinematic region so that the ion is the spectator particle and the terms (15) are dominant. This is the case in symmetric geometry and related experiments.[7] If the bound electron (particle 3) is the spectator the DWIA is

$$T = (1 + t_1 G_0)(1 + t_2 G_0)t_3(1 + G_0 t_2) \tag{16}$$

and we get the heavy-particle-knockout terms, whose effects are observed as small lobes in the polar diagrams for highly asymmetric $(e, 2e)$.[8]

The first and second rows of (13) contain terms specific to (15) and (16), respectively. Row 3 contains common terms and row 4 contains excluded terms. The $(e, 2e)$ reaction has been calculated[7] in the plane-wave impulse approximation (PWIA) $T = t_2$ and in the DWIA (15).[9] Figure 3 shows the experiment and theories compared for neon. At reasonably high energies (400 eV or more) the distorted waves may be quite well represented by plane waves with wave number modified in an average potential $\bar{V}$ (the eikonal approximation).

Figure 4 shows the results of a 400-eV $(e, 2e)$ experiment on helium[10] performed in noncoplanar symmetric geometry with the angles chosen in such a way as to

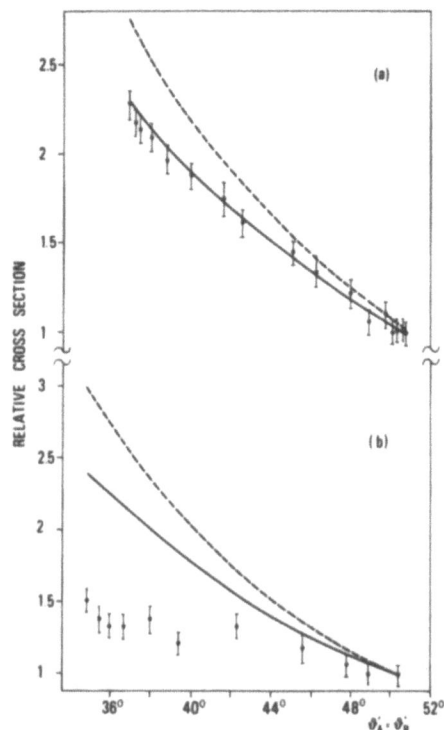

Figure 4. Relative cross sections for the helium ground state in noncoplanar symmetric geometry with $\bar{q} = 0.7$. Details are given in the text.

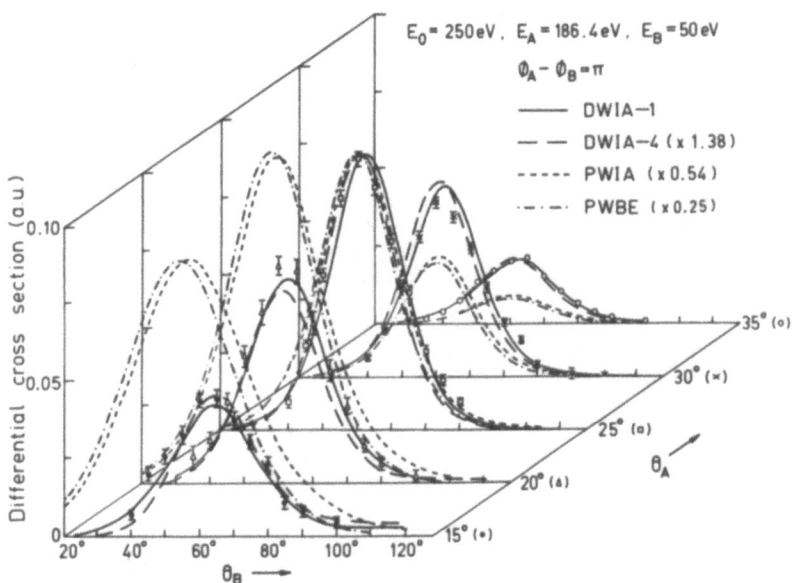

Figure 5. Coplanar $(e, 2e)$ cross sections for the hydrogen atom. The DWIA is denoted by DWIA-1. The curve DWIA-4 has $Z_A = 0$, $Z_B = 1$. The plane-wave impulse and plane-wave Born approximations (both including exchange) are denoted by PWIA and PWBE.

keep the eikonal value of the recoil momentum $\bar{q}$ (for given $\bar{V}$) constant. Figure 4(b) shows that the plane-wave impulse approximation (full line) is inadequate. The eikonal DWIA with $\bar{V} = 20$ eV is adequate [Figure 4(a), full line]. Dotted lines in both figures use the Born approximation $t = v = 1/r$. This is clearly not good enough and we should not take calculations in the plane- or distorted-wave Born approximation too seriously.

The clearest test of an $(e, 2e)$ reaction theory is the hydrogen atom, where the quasi-three-body approximation (8) is of course exact.[11] Figure 5 shows a comparison at 250 eV with the DWIA, which is quite good over a large kinematic range.

The DWIA does not obey three-body Coulomb boundary conditions. A method of circumventing this, described by Rudge,[12] is to define effective charges Z_1 and Z_3 for the pairs 1 and 3 so as to eliminate the logarithmic divergence in the phase of the three-body amplitude:

$$\frac{1 + Z_1}{k_1} + \frac{1 + Z_3}{k_3} = \frac{1}{|\mathbf{k}_1 - \mathbf{k}_3|} \tag{17}$$

This one equation does not determine Z_1 and Z_3 uniquely. For $(e, 2e)$ on the hydrogen atom no choice of Z_1 and Z_3 was found that reproduces data as well as the DWIA,

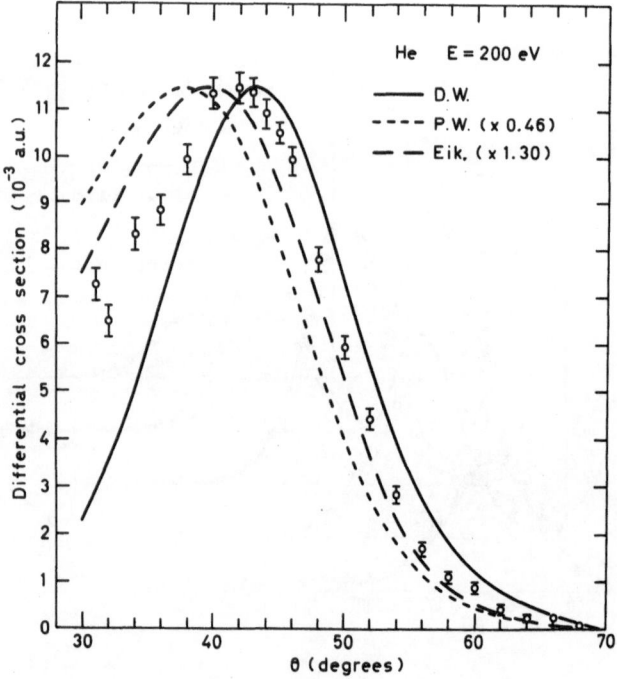

Figure 6. The coplanar symmetric $(e, 2e)$ cross sections for helium at $E = 200$ eV. The dashed line is the eikonal approximation for $\bar{V} = 10$ eV.

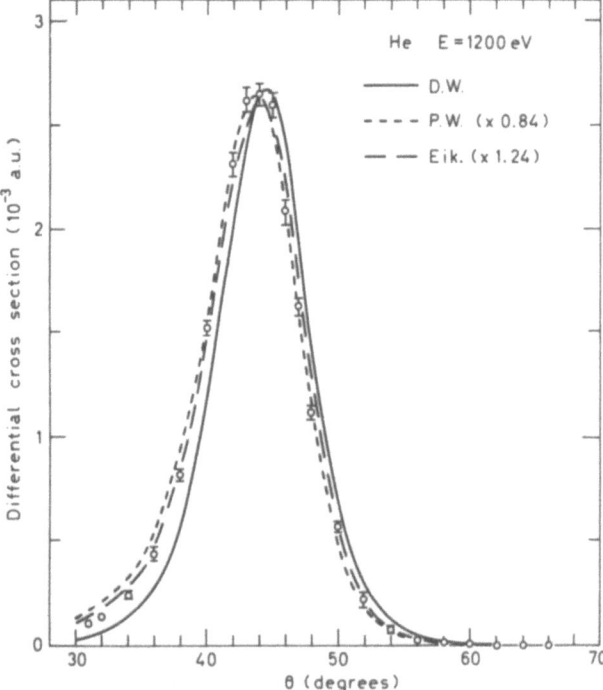

Figure 7. The same as Figure 6 but with $E = 1200$ eV.

where $Z_1 = Z_3 = 1$. This indicates that it is more important to describe the two-body quantities correctly in the interaction region than to satisfy Coulomb boundary conditions asymptotically.

A striking failure of the DWIA is the coplanar symmetric angular correlation for helium[9] at 200 eV (Figure 6). It improves with increasing energy and is quite good at 1200 eV (Figure 7).

The qualitative success of the PWIA and the considerable improvement effected by the DWIA give us grounds for believing in the convergence of the multiple-scattering series (13). Terms in this series that have not been calculated and are relevant to the neglect of three-body Coulomb effects in the final state of the DWIA are, for example, 232 and 212. Calculation of such terms is a possible direction for improving the general theoretical description of the reaction. However, for the purpose of extracting molecular structure information concerning orbitals and their correlations, PWIA in noncoplanar-symmetric geometry has already proved to be good enough in every case tried, for $E > 1200$ eV. Figure 8 shows momentum profiles determined for different orbitals of ethane[13] [cross sections for different final states $|f\rangle$ belonging to a particular orbital i are summed]. We see not only the validity of the plane-wave transform but, more significantly, the satisfaction of the spectroscopic sum rule (10), which verifies all the basic assumptions of the quasi-three-body model.

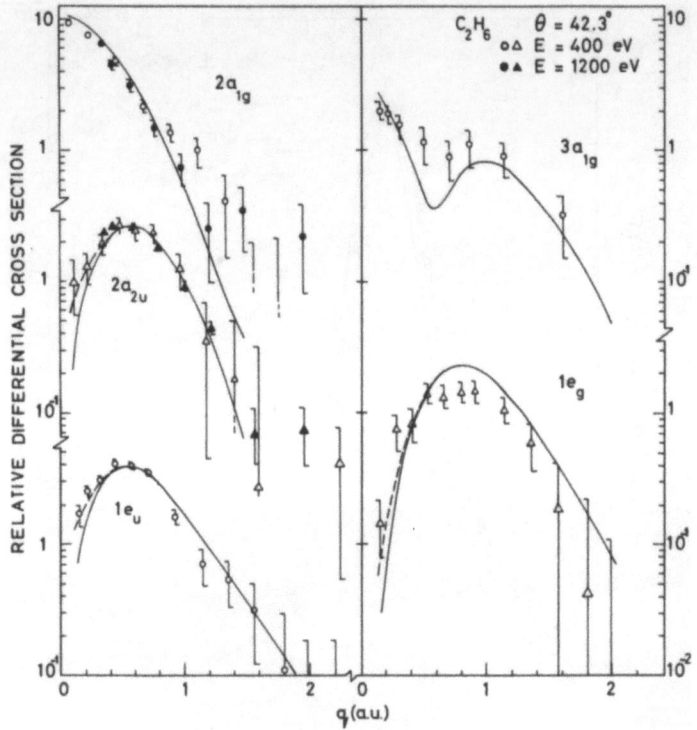

Figure 8. Noncoplanar symmetric (e, 2e) momentum profiles for ethane. The curves are the PWIA with Hartree–Fock orbitals. Relative normalizations are preserved for all orbitals. The dashed lines show the effect of folding in finite experimental angular acceptance.

ACKNOWLEDGMENT

Supported in part by the Australian Research Grants Committee.

References

1. G. R. J. Williams, I. E. McCarthy, and E. Weigold, *Chem. Phys.* **22**, 281 (1977).
2. A. J. Dixon, S. Dey, I. E. McCarthy, E. Weigold, and G. R. J. Williams, *Chem. Phys.* **21**, 81 (1977).
3. I. E. McCarthy, A. Ugbabe, E. Weigold, and P. J. O. Teubner, *Phys. Rev. Lett.* **33**, 459 (1974).
4. C. J. Joachain and R. Vanderpoorten, *Physica* **46**, 333 (1970).
5. E. Alt, P. Grassberger, and W. Sandhas, *Nucl. Phys. B* **2**, 167 (1967).
6. I. E. McCarthy, C. J. Noble, B. A. Phillips, and A. D. Turnbull, *Phys. Rev. A* **15**, 2173 (1977).
7. I. E. McCarthy and E. Weigold, *Phys. Rep.* **27C**, 275 (1976).
8. H. Ehrhardt, K. H. Hesselbacher, K. Jung, and K. Willmann, *Case Stud. At. Phys.* **2**, 159 (1971).
9. A. J. Dixon, I. E. McCarthy, C. J. Noble, and E. Weigold, *Phys. Rev. A* **17**, 597 (1978). I. Fuss, I. E. McCarthy, C. J. Noble, and E. Weigold, *Phys. Rev. A* **17**, 604 (1978).

10. R. Camilloni, A. Giardini-Guidoni, I. E. McCarthy, and G. Stefani, *Phys. Rev. A* **17**, 1634 (1978).
11. E. Weigold, C. J. Noble, S. T. Hood, and I. Fuss *J. Phys. B* **12**, 291 (1979).
12. M. R. H. Rudge, *Rev. Mod. Phys.* **40**, 564 (1968).
13. S. Dey, A. J. Dixon, I. E. McCarthy, and E. Weigold, *J. Electron. Spectrosc. Relat. Phenom.* **9**, 397 (1976).

2

Impulsive (e, 2e) Experiments: A Tool to Test Different Ionization Theories and Electronic Structure of Atoms and Molecules

A. Giardini-Guidoni, R. Camilloni, and G. Stefani

Basic (e, 2e) experiments carried out by our group are reviewed with the aim of outlining the main objectives and the results obtained. First of all the reaction mechanism has been investigated in order to find a working theory successful in interpreting experimental data. The basic theory whose validity has been investigated is the eikonal-averaged, distorted-wave impulse approximation. In this framework the (e, 2e) amplitude factorizes into parts depending on the target structure and on the two-body interaction. Through absolute value measurements and test of the energy and momentum dependence of the cross section performed on He and Ne, it has been ascertained that the approximation works well in a large kinematic range going not under 400 eV for the incident energy and not under 80° for the angle between the outgoing electrons. Dependence of the validity limits on the binding energy and symmetry of the state investigated is evident. Secondly some results on molecules are discussed in terms of information obtainable on the dynamics of the electronic structure. Finally the application of the technique to solids is considered on the basis of both preliminary data from experiments carried out on thin films and theoretical considerations of typical effects to be expected in the interaction with the lattice.

1. Introduction

Among the atomic scattering processes, those by electron impact have played an outstanding role in the atomic physics field. The most recent examples in which

A. Giardini-Guidoni ● C.N.E.N.—Divisione Nuove Attività, Centro di Frascati, C.P. 65 00044 Frascati, Rome, Italy. R. Camilloni and G. Stefani ● C.N.R. Laboratorio Metodologie Avanzate Inorganiche, Rome, Italy.

problems of fundamental physics are involved are the $(e, 2e)$ electron impact ionization processes. In these reactions the kinematics are fully determined by carefully measuring energy and momenta of the three electrons involved in the process and by detecting the two final electrons in coincidence. The first coincidence measurement was reported by Erhardt et al.[1] in 1969 for He. The aim of this experiment was to investigate the reaction mechanism at low energies. The second experimental paper on $(e, 2e)$,[2] which also appeared in 1969, dealt with the use of $(e, 2e)$ reactions as a fine probe of the single-particle character of electrons bound in atoms, molecules, and solids, in particular as a test of reliability of theoretical wave functions by comparing measured electron momentum distributions with calculated ones. That experiment was performed in Rome by a group also involved in nuclear knockout experiments, whose aim was the measurement of nuclear wave functions. Following this preliminary experiment, our group, in 1972, published the first recoil momentum profile measured for a resolved single-particle state.[3] This was the $1s$ state of solid carbon and the experiment clearly demonstrated the sensitivity of the recoil momentum profile to the shape of the orbital wave function. In 1973 the experiment of Weigold, Hood, and Teubner[4] on argon resolved for the first time the valence states. They found strong electron–electron correlation effects in the $3s$ state and demonstrated the power of the technique in extracting structure information. Since these first studies, $(e, 2e)$ experiments have been performed on the valence electrons of atoms and molecules with energy resolutions always improving.[5,6] The best resolution has been reported in the 1978 paper by Williams,[7] whose energy resolution of 0.06 eV enabled the measurement separately of the momentum distributions for the $^2p_{1/2}$ and $^2p_{3/2}$ spin–orbit-split states.

The relevance of the $(e, 2e)$ reactions as a tool to test different reaction theories and to extract structure information about atoms, molecules, and their ions is nowadays amply demonstrated.

It is outside the aim of this paper to discuss all the $(e, 2e)$ experiments performed in recent years. We will concentrate only on the work performed by our group. $(e, 2e)$ experiments performed in our laboratory are characterized by high and intermediate energies and equal energies of the final electrons. Our prime objective is to use these high-momentum-transfer reactions to get information on the electronic properties of atoms, molecules, and solids. To pursue this task it is necessary to use a good reaction theory, allowing the description of the interaction in terms of a quasi-three-body problem. For computational reasons it is necessary to factorize the cross-section expression. This is made possible by reducing the quasi-three-body problem to the simpler two-body problem by treating the interaction in the framework of the impulse approximation (IA).

Our group was deeply involved in testing the goodness of the IA and in trying to set validity limits as a function of the energy and momentum transfer. In fact only when the range of validity of the IA is settled can the experiment be a measurement of all there is to know about single-particle orbitals.

Theories developed for these reactions are amply reviewed in the paper by

McCarthy, which also appears in this book, and they will be recalled here only to be compared with experimental data. In addition to the experiments testing the IA, measurements of momentum distribution for valence orbitals in molecules have been undertaken. Recent results achieved are summarized in the following sections: in Section 2 details of the experimental technique used for gaseous targets are described. Section 3 gives a short review of the approximation used in describing the (*e, 2e*) process. Section 4 deals with comparison between data and predicted cross sections. In Section 5 the structural information obtained on molecules is outlined; in Section 6 the present status of (*e, 2e*) reactions on solid targets is recalled; and finally in Section 7 conclusions are given.

2. Experimental Method

In Figure 1 a schematic diagram of the kinematics of the (*e, 2e*) reaction is shown. A beam of electrons of energy E_0 and momentum $\mathbf{k}_0$ is incident on a gaseous target, which is assumed to be at rest. To identify the two electrons coming from the knockout process $e + \mathrm{M} \rightarrow \mathrm{M}^+ + e + e$, with energies E_A, E_B and momenta $\mathbf{k}_A$, $\mathbf{k}_B$, fast-timing coincidence technique is necessary. This gives discrimination from other events that yield electrons that have the same $\mathbf{k}_A$ and $\mathbf{k}_B$ momenta but do not reach the detectors at the same time. In our (*e, 2e*) apparatus, two rotatable electrostatic hemispherical analyzers are used to select the momenta and energy of the outgoing pair of electrons. From the energy conservation relation $\varepsilon_\lambda^f = E_0 - E_A - E_B$, where E_0 is the energy of the incident electron and E_A and E_B are the energies of the outgoing electrons, ignoring the very small ion recoil energy, a particular electronic quantum state ε_λ^f of the residual ion can be selected. When these energies are grouped into sets, depending on which orbital has been involved in the ionization, the lowest value of each set gives the threshold ionization potential relative to that orbital. The other values are due to ionization leading to a final ionic state having the same symmetry.

At sufficiently large values of E_0, E_A, and E_B the recoil momentum $\mathbf{q}$ of the residual ion, obtained through the relationship $\mathbf{q} = \mathbf{k}_0 - \mathbf{k}_A - \mathbf{k}_B$, is very nearly

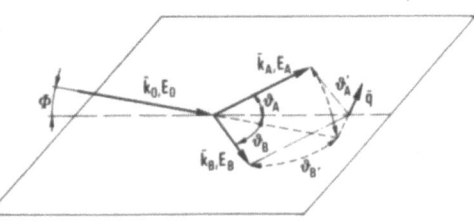

Figure 1. Schematic diagram of the kinematics of the process. Energy and momenta conservation laws: $E_A + E_B = E_0 - \varepsilon_\lambda^f$, ε_λ^f being the separation energy, and $\mathbf{q} = \mathbf{k}_0 - \mathbf{k}_A - \mathbf{k}_B$, $\mathbf{q}$ being the momentum of the recoiling system. Φ is the azimuth angle defining the direction of the incoming electrons. ϑ_A' and ϑ_B' are the scattering angles defining the directions of the outcoming electrons with respect to $\mathbf{k}_0$, while ϑ_A and ϑ_B have the same meaning but are referred to $\mathbf{k}_0$ projections in the $\Phi = 0$ plane.

equal in magnitude to the momentum of the knocked-out electron q_e; and the measure of this momentum at fixed E_A, E_B, E_0 yields the radial momentum distribution $\varrho(q)$ of the initially bound electron for randomly oriented targets. A schematic diagram showing our experimental setup and the essential structure of the electronic coincidence circuit[3] capable of 4 nsec full width at half maximum (FWHM) time resolution is given in Figure 2. In Figure 3 a sketch of the apparatus is shown. The apparatus mainly consists of a stainless steel cylindrical chamber (60 cm high and 130 cm in diameter) in which the basic components, an electron gun and two independently rotatable electron spectrometers are mounted on the bottom flange. The chamber is pumped down to $\simeq 2 \times 10^{-7}$ Torr by an 8000 liter/sec mercury diffusion pump and two baffles cooled with water and liquid nitrogen, respectively. When the gaseous jet is sent into the chamber the pressure rises to about 10^{-6} Torr. A He cryopump,

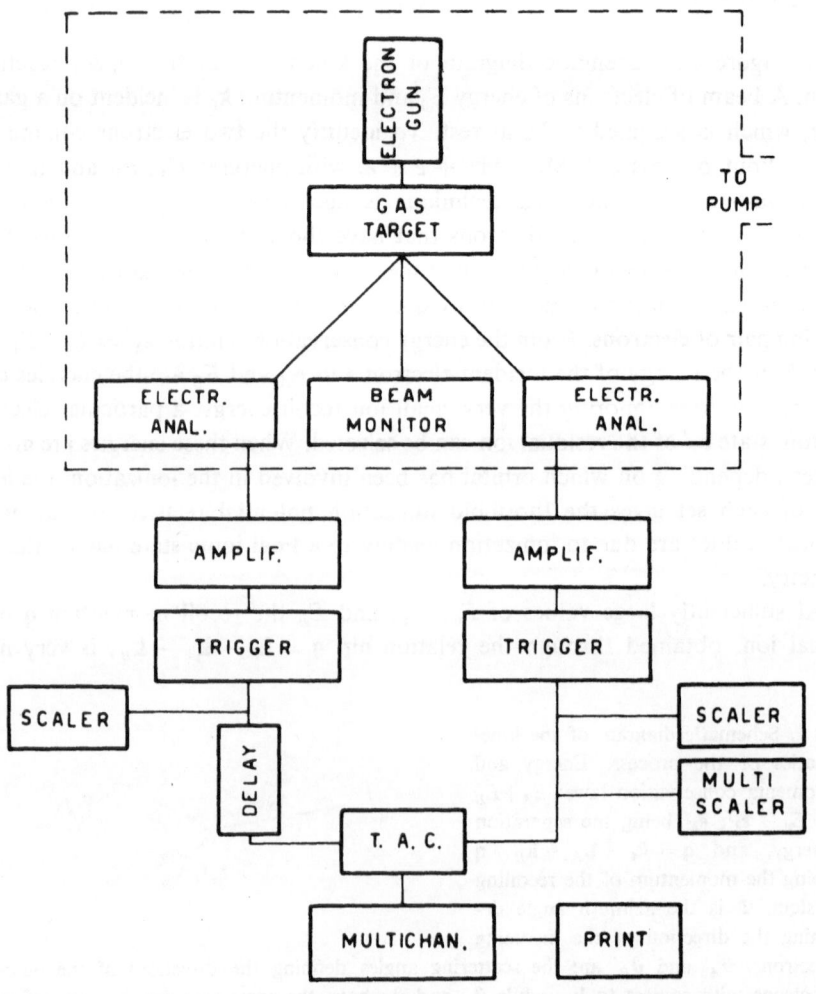

Figure 2. Schematic diagram of the experimental setup and fast electronic chain.

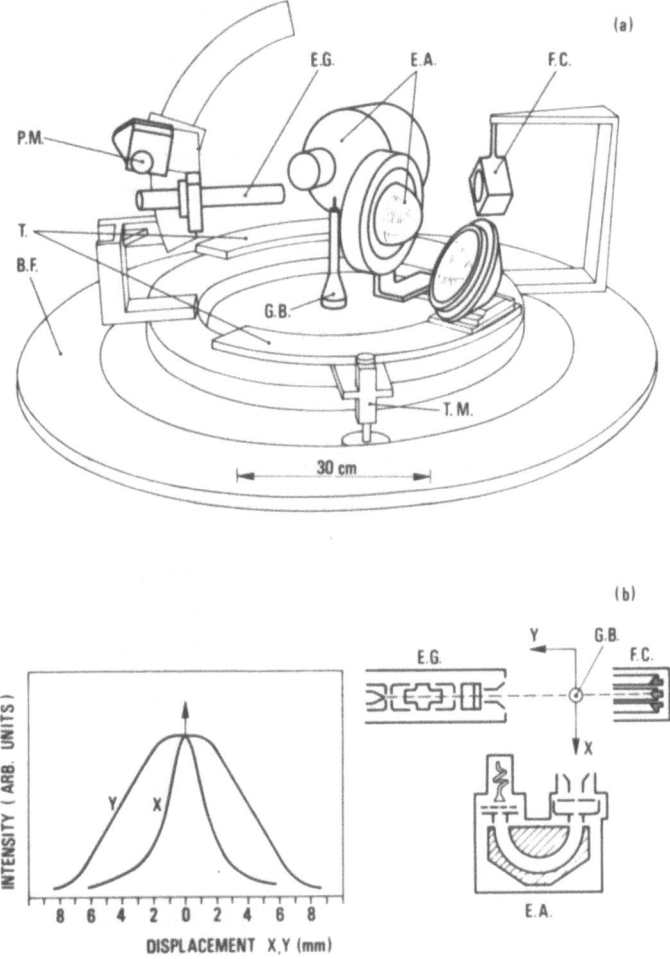

Figure 3. (a) Experimental apparatus. E. G., electron gun; E.A., electron analyzer; F.C., Faraday
cup; G.B., gaseous beam; T.M., turntable movement; T., turntables; B.F., bottom flange;
P.M., Φ movement. (b) Comparison between the gaseous beam profile (X) at the scattering
point and the electron analyzer field of view (Y).

sitting on the top flange in front of the jet, is sometimes used, giving extra pumping
speed. Three Helmholtz coils, placed perpendicularly to each other outside the
chamber, reduce the earth's magnetic field to less than 5 mG in the interaction region.
The spectrometers are further shielded by μ-metal foils. The electron beam is produced by an electron gun suitably collimated ($\Delta\vartheta$ about $\pm 1 \times 10^{-2}$ rad; Figure 3b).
The energy of the beam ranges between 150 and 4000 eV. In the same energy range
the beam spot at the target has dimensions of the order of 1 mm. The maximum
available beam intensity increases, with increasing energy of the beam, from about
10 μA to about 200 μA. The intensity and position of the electron beam are continuously tested during the measurement by five small Faraday cages arranged

together to give information on the total current, direction, and profile of the electron beam. The azimuth angle of the gun can be varied with respect to the detection plane of the electrostatic detectors from $-10°$ up to $+30°$ with a precision of $\pm 1°$. The two twin electron spectrometers, which detect the two emerging electrons at different angles, are formed by hemispherical electrostatic selectors having a retarding field at the entrance and a channeltron as electron detector at the exit. The retarding field is used to enhance the overall energy resolution of the spectrometer in such a way as to obtain a suitable compromise between energy resolution and accepted solid angle. Retarding ratios as large as 1/300 have been obtained. Under these conditions the energy resolution is still Gaussian in shape and as good as a FWHM of $\simeq 5\%$. Each spectrometer can be rotated independently around the interaction volume in a plane defined by $\phi = 0°$. The angular variation θ allowed is from $0°$ to $120°$; the precision is $\pm 0.1°$ and the solid angle accepted is about 3×10^{-4} sr. Calibrations have been tested by measuring the elastic scattering of electrons on noble gases at the energies at which sharp minima are present.[8]

The gaseous beam is obtained by allowing the gas to effuse through a Bendix multichannel array whose thickness is 0.25 mm. Each channel is 10 μm i.d. and the active area is about 50% of the total. The multichannel is sealed on a hypodermic needle 0.6 mm i.d. placed about 2 mm below the electron beam path. In this way, at the electron–gaseous beam crossing point the gas density is made to be about 20 times larger than the background gas density, thus providing a suitable scattering target. The target density can be as high as 2×10^{13} mol/cm³.

To obtain the absolute value of the cross section it is necessary to measure the coincidence rate R_C and use the relationship[8]

$$\frac{d^5\sigma}{d\Omega_A \, d\Omega_B \, dE_A} = \frac{R_C \varrho l}{I_0(\varrho l \varepsilon_A \, \Delta\Omega_A)(\varrho l \varepsilon_B \, \Delta\Omega_B) \, \Delta E} \qquad (1)$$

where I_0 is the incident electron current, ϱ is the gas beam density, l is the gas target effective length, and $\Delta\Omega_i$ is the solid angle subtended by each electron analyzer whose overall efficiency is ε_i.

The quantity $(\varrho l \varepsilon_i \, \Delta\Omega_i)$ has been experimentally determined for each analyzer at any scattering angle and any incident energy in which coincidence rates have been measured. It has been done by relating the count rate of elastically scattered electrons to the incident electron current and to the well-known absolute differential elastic cross section through the relationship

$$\frac{d^2\sigma}{d\Omega_i} = \frac{R_e}{I_0(\varrho l \varepsilon_i \, \Delta\Omega_i)} \qquad (2)$$

where R_e is the measured elastic scattering count rate.

Then the quantities ϱ and l have been independently determined by measuring the incoming gas flow and by using a gas beam scanning procedure similar to the one described by Wellenstein[8b]: the relative gas density distribution has been measured at the scattering point by performing a transverse scanning of the gas beam with a thin, well-collimated electron beam and looking for the intensity of electrons elastically scattered at 90°.

By repeating this procedure at various heights of the gas beam, the diameter and divergence at the scattering point can be determined. In the experimental conditions in which measurements on Ne have been performed, a FWHM dimension of 2 mm and a divergence angle of about ±30° have been measured. A sample convolution of the electrons and gas beams is shown in Figure 3b (curve X). The field of view of each analyzer has been measured in a similar way but on the background gas. The resultant profile Y is shown in Figure 3b, and is seen to be much larger ($\simeq 10$ mm) than the target so that corrections on the effective dimensions need not be made when the angles of the detectors are varied.

The factor ΔE appearing in expression (1) is determined by folding the two gaussian energy resolution curves and taking into account the mentioned energy conservation. In the present case the result is $\Delta E = 0.76 \, \Delta E_i$.

An analysis of all the uncertainties affecting the experimental determination of the parameters in expression (1) leads to an overall estimated uncertainty on the absolute value of the (*e*, 2*e*) cross section of a factor of 2. The largest contribution to the error comes from the uncertainty on the value of $\varrho \cdot l$. This has been confirmed by the data reproducibility achieved on former measurements of the absolute (*e*, 2*e*) cross section of He gas,[8a] where the experimental conditions (density and dimension of the gas target, incoming current) were widely varied.

3. Approximation for the Reaction Mechanism

In common with all the theoretical problems for realistic quantum systems, the (*e*, 2*e*) reaction must be reduced to a problem involving few degrees of freedom. By considering a reaction leaving the residual ion in its ground state, the simplest problem that has sufficient features of the reaction is a quasi-three-body problem, in which two electrons interact with each other through the Coulomb potential V and with the ion through optical model potentials V_1, V_2 in which unobserved channels are treated by polarization and absorption terms.

It was shown by McCarthy and Weigold,[9] with approximations amounting to closure over target states and weak coupling between channels in both two- and three-body systems, that the (*e*, 2*e*) amplitude is given in the quasi-three-body model with implied antisymmetry by

$$T(\mathbf{k}_A, \mathbf{k}_B) = \langle X_A^{(-)}(\mathbf{k}_A) X_B^{(-)}(\mathbf{k}_B)$$

$$\times \langle f | V + V \frac{1}{E^{(-)} - K_1 - K_2 - V_1 - V_2 - V} V | g \rangle X_0^{(+)}(\mathbf{k}_0) \rangle \quad (3)$$

$X_A^{(-)}$ and $X_B^{(-)}$ are distorted waves computed in the optical model potentials V_1 and V_2. $X_0^{(+)}$ is a distorted wave computed in the entrance channel optical model potential. The ground state $|g\rangle$ of the target and the final state $|f\rangle$ of the ion are functions of the many-body coordinates of the ion although these coordinates are assumed not to affect the potentials V_1 and V_2 (quasi-three-body approximation).

This approximation involves a three-body Green's function. It can be reduced to a two-electron Green's function by making a Taylor expansion of the electron–ion potentials V_1, V_2 about the electron–electron center-of-mass coordinate R. The transformation from the particle system is

$$\mathbf{r}_1 = \mathbf{R} + \tfrac{1}{2}\mathbf{r}, \qquad \mathbf{r}_2 = \mathbf{R} - \tfrac{1}{2}\mathbf{r}$$
$$\mathbf{k} = \tfrac{1}{2}(\mathbf{k}_A - \mathbf{k}_B), \qquad \mathbf{K} = \mathbf{k}_A + \mathbf{k}_B \tag{4}$$

Neglecting the gradients of the Taylor expansion with respect to $\mathbf{r}$, the distorted waves $X_A^{(-)}$ and $X_B^{(-)}$ are eigenfunctions of the operator $K_R + V_1 + V_2$, where K_R is the center-of-mass kinetic energy operator of the two electrons. The operator (3) then becomes the e–e t-matrix computed at the energy $p^2 = (\hbar^2/2m)k^2$:

$$T = \langle X_A^{(-)} X_B^{(-)} | t(p^2) | (f|g\rangle X_0^{(+)}\rangle \tag{5}$$

This is the distorted-wave impulse approximation (DWIA). The amplitude depends on the target and ion structure only through the overlap function $(f|g\rangle$ since $t(p^2)$ is independent on the internal ion coordinates. The resulting nine-dimensional integral is extremely difficult to evaluate because of the difficulty of the coordinate transformation from the relative coordinates (in which the t matrix is expressed) into particle coordinates in which the distorted waves are expressed. This difficulty is removed if the distorted waves, calculated usually from the optical model equations[10] are further approximated at intermediate energies by modified plane waves in a radially localized region.[9] Each distorted wave X_I ($I = 0, A, B$) becomes

$$X_I(r_i) = e^{\pm \nabla(r/2)} X(\mathbf{R}) \tag{6}$$

We approximate the gradient operator acting on the distorted wave by an effective propagation vector $\varkappa_I(\mathbf{R})$, where

$$\varkappa_I{}^2 = \frac{\hbar^2}{2m} [E_I + \bar{V}(\mathbf{R}) + i\bar{W}(\mathbf{R})] \tag{7}$$

The imaginary part is responsible for an attenuation factor γ in the cross section.

In order to introduce a useful eikonal approximation it is necessary to formulate the amplitude in terms of an effective radial coordinate $\bar{\mathbf{R}}$. We will expand the distorted waves about $\bar{\mathbf{R}}$, but make the assumption that the effective wave number for the two-body t matrix element in expression (5) are calculated at $\bar{\mathbf{R}}$. Numerical experience has shown that they are insensitive to small changes in the effective potential $\bar{V}(\mathbf{R}) + i\bar{W}(\mathbf{R})$. This is equivalent to an eikonal-averaged approximation if $\bar{\mathbf{R}}$ can be chosen so that[11]

$$e^{\mathbf{q} \cdot \mathbf{R}} = X_A^{(-)} \bar{\mathbf{R}} X_B^{(-)}(\bar{\mathbf{R}}) X_0^{(+)}(\bar{\mathbf{R}}) \tag{8}$$

where $\mathbf{q} = \mathbf{k}_0 - (\mathbf{k}_A + \mathbf{k}_B)$. The amplitude (5) then factorizes as follows:

$$T = \langle \mathbf{x}' \mid t_{ee}(p^2) \mid \mathbf{x} \rangle \langle \mathbf{x}_A \mathbf{x}_B \mid (f \mid g) \mathbf{x}_0 \rangle \tag{9}$$

where the distorted $\mathbf{x}_I$ expression (7) is used everywhere and $\mathbf{x}' = \frac{1}{2}(\mathbf{k}_0 + \mathbf{q})$. The factorization introduces much simplification into the use of the t matrix. Since only the absolute square of the matrix element is used, the difficult phase[12] is avoided. Furthermore, only half-off-shell matrix elements are required. The effective potentials $\bar{V}$ and $\bar{W}$ cannot be predicted *a priori* by fitting elastic scattering, where the relevant radial region is too broad to give a good approximation for X_I and must be extracted from the experiment.[13] When the potentials $\bar{V}$ and $\bar{W}$ are set equal to zero, the DWIA becomes the simpler plane-wave impulse approximation (PWIA). The explicit expression of the cross section is given by

$$\frac{d^5\sigma}{d\Omega_A \, d\Omega_B \, dE_A} = \frac{4k_Ak_B}{k_0} f_\lambda \gamma \varrho(\mathbf{q}) \tag{10}$$

where the f_λ factor is the free electron–electron cross section, which can be computed in the Born (r-matrix) approximation if the higher-order terms in the potential are dropped in expression (9). The v-matrix approximation is

$$f_v = \frac{1}{|\mathbf{x} - \mathbf{x}'|^4} + \frac{1}{|\mathbf{x} + \mathbf{x}'|^4} - \frac{1}{|\mathbf{x} - \mathbf{x}'|^2 \, |\mathbf{x} + \mathbf{x}'|^2} \cos\left[\ln \frac{|\mathbf{x} + \mathbf{x}'|^2}{|\mathbf{x} - \mathbf{x}'|^2}\right] \tag{11}$$

The t-matrix element is

$$f_t = C_0^2(\eta) \left\{ \frac{1}{|\mathbf{x} - \mathbf{x}'|^4} + \frac{1}{|\mathbf{x} + \mathbf{x}'|^4} - \frac{1}{|\mathbf{x} + \mathbf{x}'|^2 \cdot |\mathbf{x} - \mathbf{x}'|^2} \right.$$
$$\left. \times \cos\left[\eta \ln \frac{|\mathbf{x} + \mathbf{x}'|^2}{|\mathbf{x} - \mathbf{x}'|^2}\right] \right\} \tag{12}$$

where $C_0^2(\eta) = 2\pi\eta/e^{2\pi\eta} - 1$; $\eta = me^2/2\hbar^2 \varkappa'$. γ is the attenuation factor already mentioned:

$$\gamma = \exp\left[-\left(\frac{k_0}{E_0} + \frac{k_A}{E_A} + \frac{k_B}{E_B}\right) \bar{W} R_N\right] \tag{13}$$

where R_N is a normalization radius chosen so that the magnitude of the distorted waves is 1 on the scattering axis at a distance R_N before the interaction region, and $\varrho(\mathbf{q})$ is simply

$$\varrho(\mathbf{q}) = |\langle \mathbf{x}_A \mathbf{x}_B \mid (f \mid g) \mathbf{x}_0 \rangle|^2 \tag{14}$$

This is the factor that carries the structure information of the target and ion.

For closed-shell systems it is reasonable to represent the target by the Hartree–Fock approximation, in which the modulus squared of the overlap function reduces to the product of the modulus squared of the Fourier transform of the jth orbital ionized to the $\varepsilon_\lambda^{(f)}$ level and a spectroscopic factor $S_j^{(f)}$. This factor is the probability that the many-body wave function $|f\rangle$ contains the one-hole configuration, that is, the configuration with a hole in the orbital j of the ground-state Hartree–Fock wave function. It has been experimentally verified that the spectroscopic factor obeys the sum rule $\sum_f S_j^{(f)} = 1$ in all cases studied so far.[14]

The distortion effects are taken into account more accurately by using a fully distorted factorized DWIA off-shell model.[15] This model retains the factorization into a distorted-wave transform and a t-matrix element. The eikonal approximation for X_I is replaced by fully partial-wave-expanded distorted waves calculated by solving the optical model Schrödinger equation.

4. Tests of the Reaction Mechanism

A series of tests has been done in order to determine the range of validity for the approximation outlined in Section 3. The best systems to choose are the noble gases. In fact, since they are closed-shell systems, the single-particle wave functions are calculated by the Roothaan method with sufficient accuracy.[16] The $\varrho(\mathbf{q})$ factor can be easily computed and the e–e scattering factor can be extracted from the cross-section expression.

Data have been taken in three different geometries very sensitive to one of the factors belonging to the cross-section expression (10).

(a) In coplanar symmetric conditions (i.e., equal kinetic energies $E_A = E_B = E$ and equal scattering angles $\vartheta_A = \vartheta_B = \vartheta$ for the electrons emerging in the plane containing the incident beam), both the factors f_λ and $\varrho(\mathbf{q})$ have a large variation over the range of ϑ for which the cross section is measurable. This gives an overall test of the goodness of the approximations. Moreover, in this kinematics, by suitably varying ϑ, it is possible to scan $\mathbf{q}$ values parallel to $\mathbf{k}_0$ (larger scattering angles) or antiparallel to $\mathbf{k}_0$ (smaller scattering angles). The $\varrho(\mathbf{q})$-form factor is thus measured twice in the angular correlation spectrum, while the f_λ factor is continuously varying. This adds a further possibility in testing the cross section.

(b) In the noncoplanar symmetric condition (i.e., equal kinetic energies $E_A = E_B = E$ and equal scattering angles $\vartheta_A = \vartheta_B = 44°$ and ϕ variable) the f_λ factor is essentially constant as the variable azimuth ϕ is changed to vary the recoil momentum $\mathbf{q}$. The distortion effects are less emphasized in this geometry, which is particularly suitable to study $\varrho(\mathbf{q})$-form factors.

(c) Kinematic conditions are used in which the distorted-wave transform [$\varrho(\mathbf{q})$-form factor] is kept constant and the polar angles ϑ_A, ϑ_B of the two detectors and the azimuth ϕ are varied so as to vary the e–e factor. The energies E_Q and E_B

are kept equal. In this way it is possible to study the question of what is the appropriate two-body operator as well as the validity of the factorized approximation.

Moreover, the knowledge of the absolute value of the cross section gives the possibility of studying absorption effects.

The complete set of these measurements has been performed up to now only in He and Ne gases. The other noble gases have been studied in symmetric coplanar conditions. From all the measurements performed the range of validity of the factorized PWIA and of the factorized eikonal-averaged DWIA have been determined.

To start with we report and discuss data obtained for He, which has been the system most thoroughly investigated so far. In Figure 4 the measured absolute values for scattering angles $\vartheta_A = \vartheta_B = 44°$ corresponding approximately to $\phi = 0$ are reported together with the cross sections calculated in various factorized approxima-

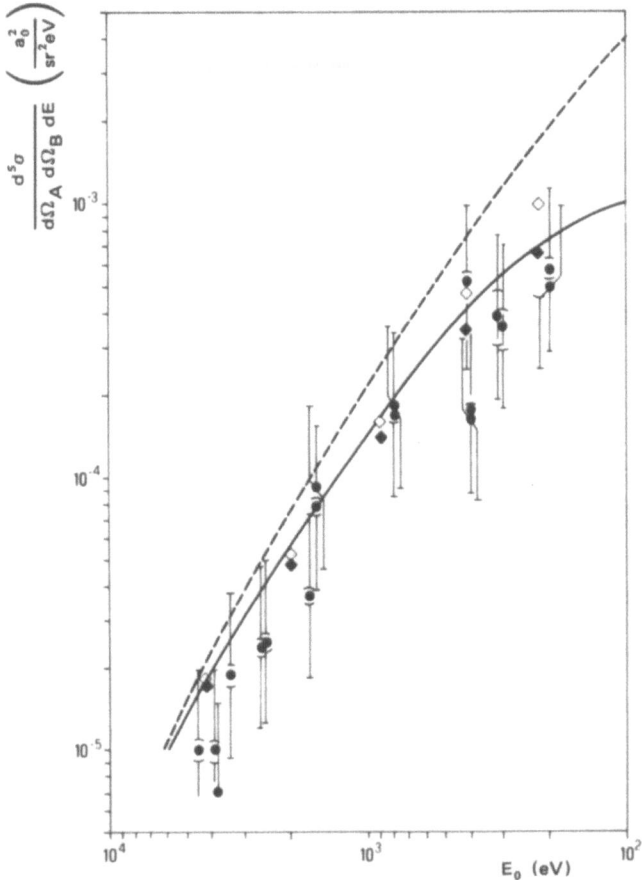

Figure 4. Absolute differential cross section for the He ground-state (e, 2e) reaction. $\vartheta_A = \vartheta_B = 44°$, $\phi = 0$. Filled circles are the experimental points. Dashed curve is the PWIA v-matrix cross section, solid curve is the PWIA t-matrix cross section. Values calculated by Geltman in Born approximation ($\diamond$) and CPBE approximation ($\blacklozenge$) are also reported.

tions by using the He $1s$ Clementi wave function[17] and an unfactorized calculation by Geltman.[18] The various calculated cross sections differ enough to indicate that the factorized Born is a worse approximation than the t matrix. The unfactorized Geltman's cross section is in good agreement with the data, although the experimental errors are too large to discriminate between Born and Coulomb projected Born with exchange (CPBE). These indications are confirmed by the data taken in different geometries. In Figure 5 the absolute measurements performed on He in symmetric coplanar conditions are reported for energies ranging from 200 to 3600 eV. It can be seen that eikonal-averaged DWIA is good in describing the process when the $\bar{V}$ potential is taken as equal to 20 eV and $\bar{W}$ is set equal to zero, from 400 eV on. At 1600 eV PWIA and DWIA become indistinguishable. The unfactorized cross section (c) is instead worse than the eikonal DWIA even at 800 eV. A confirmation of these results has been given by the measurements at fixed q, reported in Figure 6. While the f_λ factor obtained from constant q in PWIA does not fit the experimental data,

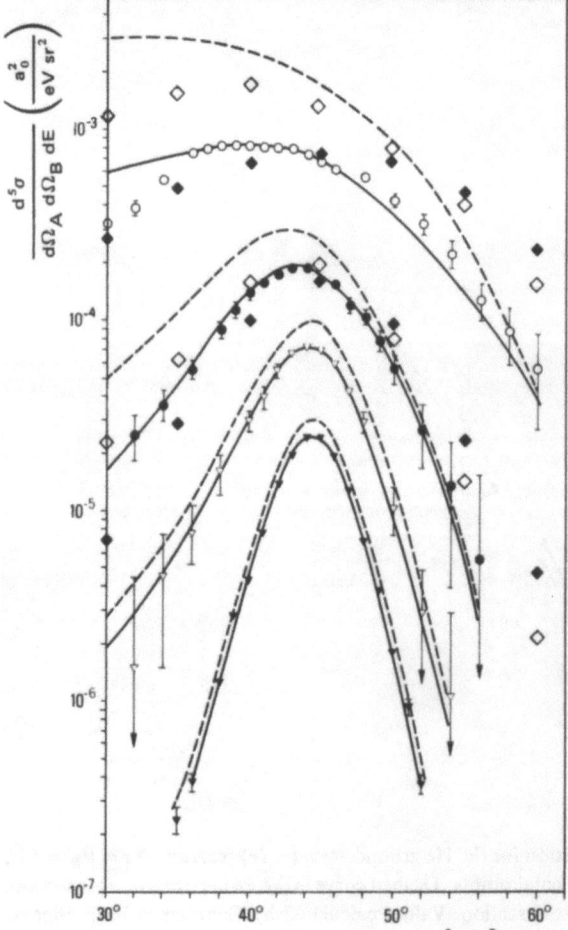

Figure 5. Coplanar symmetric $(e, 2e)$ cross section as a function of scattering angles taken at various incident energies. (⌀) 200 eV; (◆) 800 eV; (Ῡ) 1600 eV; (Ῡ) 3600 eV. Dashed curve is the PWIA; solid curve is the eikonal-averaged DWIA with $\bar{V} = 20$ eV, $\bar{W} = 0$ eV. Values calculated by Geltman in Born approximation (◇) and CPBE approximation (◆) are also reported.

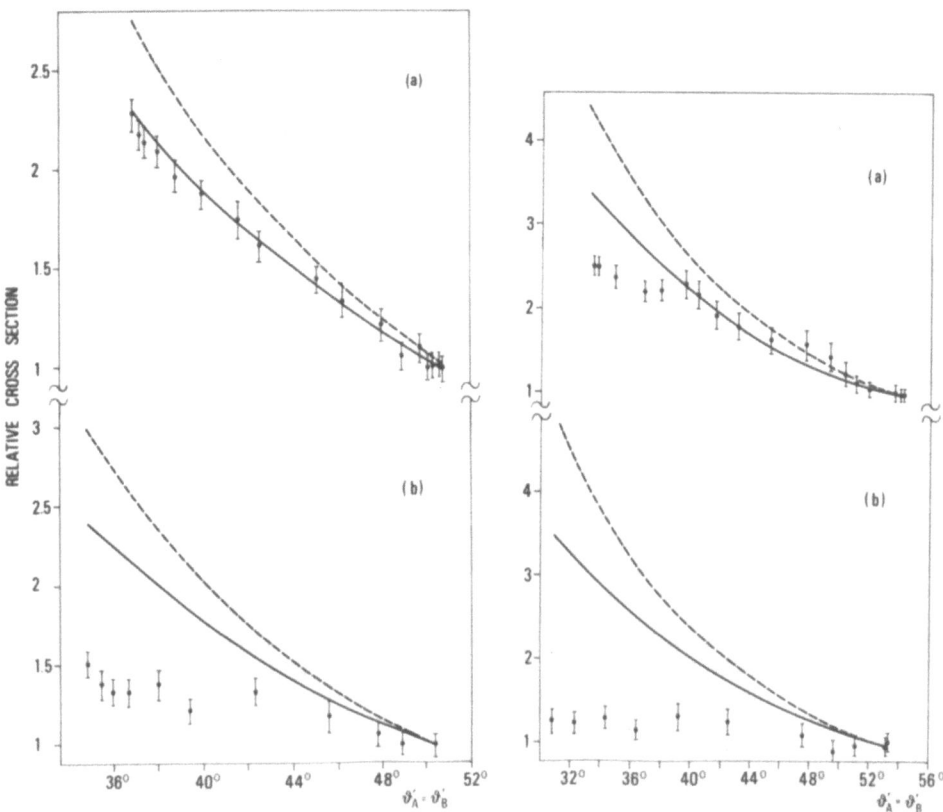

Figure 6. Relative cross section for the ejection of He $1s$ electrons in ϑ and ϕ variation so as to keep q fixed. Incident energy $E_0 = 400$ eV. In the abscissa the scattering angle ϑ' between the incident and final electrons is reported. (a) Data taken in eikonal-averaged DWIA $\bar{V} = 20$ eV, $\bar{W} = 0$ eV are compared with computed $f_v(---)$ and $f_t(\text{———})$ e–e factors. (b) Data taken in PWIA kinematic conditions ($\bar{V} = 0$, $\bar{W} = 0$) are compared with $f_v(---)$ and $f_t(\text{———})$ e–e factors. Left, $q = 7$ a.u.; right, $q = 1$ a.u.

the f_λ factors obtained from q reconstructed in eikonal-averaged DWIA fit quite well with calculated values in t-matrix approximation. From all the data reported it is evident that the angular distribution is not well fitted in the low-momentum transfer region in the range of energies up to 400 eV. This together with the failure observed for lower angles in the constant-q geometry[14] gives evidence of the role played by the momentum transfer and constitutes strong evidence that it is the impulse approximation itself, not the factorization (or eikonal approximation), that is breaking down.

In conclusion we can affirm that for He the eikonal-averaged DWIA works well from 800 eV upward in all the q momentum range. The fact that a good working theory allows for a complete reconstruction of the q momentum distribution is shown in Figure 7, where the squared Fourier transform calculated for He is compared with data ranging from 800 eV upwards.

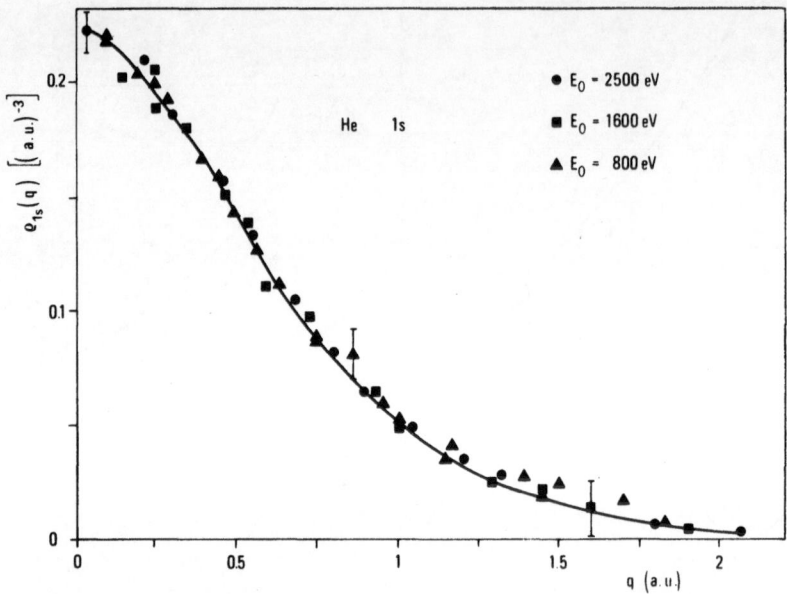

Figure 7. Experimental q distribution measured for He in coplanar symmetrical conditions for incident electron energies of 800, 1600, and 2500 eV. The solid-line curve is the squared Fourier transform of the He $1s$ Clementi wave function.[17]

The most relevant information obtained from Ne data reported in Figures 8–11 is that the validity limits for the approximations are different and depend on the subshell.

From two kinds of data obtained from the Ne $2s$ electron ejection, absolute values of the cross section (Figure 8a) at $\vartheta_A = \vartheta_B = 40°$, $\phi = 0$ and coplanar symmetric data (Figure 9b) in the energy range 200–3000 eV, two interesting features appear. Firstly it is possible to infer from the comparison of the data and values calculated in various approximations that for the Ne $2s$ state the agreement does not become satisfactory unless the imaginary part of the average eikonal potential $\bar{W}$ is taken into account. The best fit with the data is obtained in eikonal-averaged DWIA by choosing $\bar{V} = 30$ eV, $\bar{W} = 10$ eV, and $R = 0.7$ Å.[19] By using these empirical effective potentials the predicted values for the cross section are close to the ones calculated by the factorized fully distorted-wave off-shell impulse approximation.[15] Secondly a failure in the coplanar symmetric angular distribution for angles $\vartheta_A = \vartheta_B < 38°$ similar to that of the for $E_0 = 400$ eV is observed at $E_0 = 800$ eV. For that reason the measurements at constant q have been performed at the higher energy, namely, 800 eV. It appears that, utilizing the $\bar{V}$ determined from the coplanar geometry and absolute values, the eikonal-averaged DWIA is adequate to account for the data. Perhaps surprisingly, the fully distorted waves are not as good as the DWIA in predicting shapes in q-fixed measurements, as seen from the data reported in Figure 10. As the energy increases the agreement becomes better in all the angular

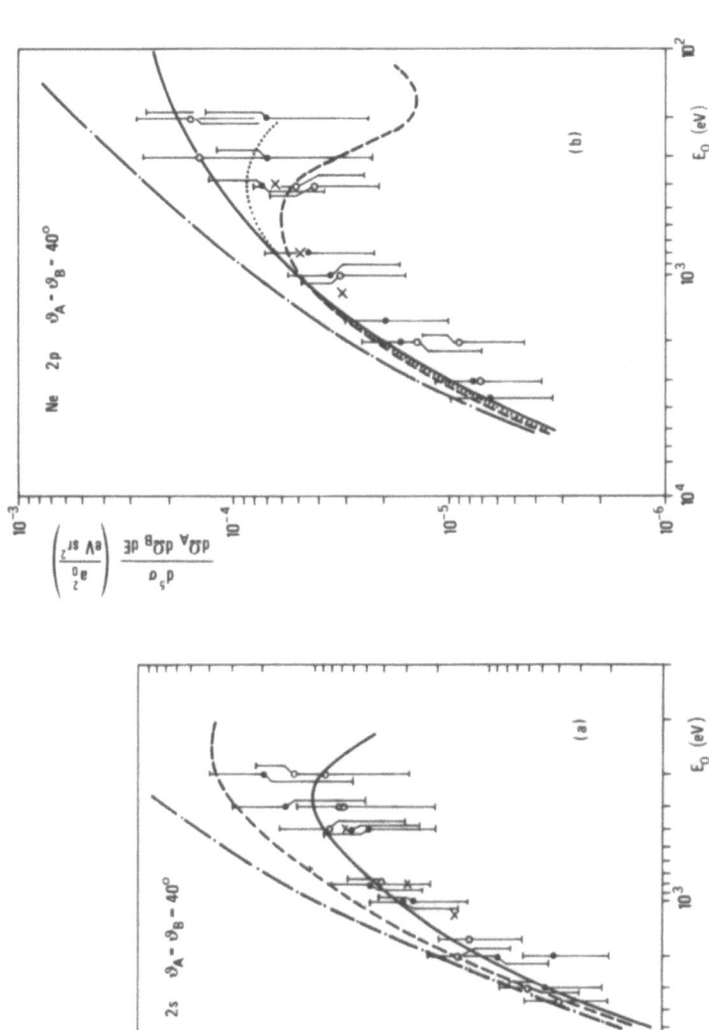

Figure 8. Absolute differential (e, 2e) cross section for Ne at various incident energies. Filled circles are the experimental points; open circles are the points normalized to He cross section. (a) Ne 2s: ---, Eikonal-averaged Born DWIA (v matrix), $\bar{V} = 30$ eV, $\bar{W} = 0$; ---, eikonal-averaged DWIA (t matrix), $\bar{V} = 30$ eV, $\bar{W} = 0$; ——, eikonal-averaged DWIA (t matrix), $\bar{V} = 30$ eV, $\bar{W} = 0$; ——, eikonal-averaged DWIA (v matrix), $\bar{V} = 30$ eV, $\bar{W} = 0$; ——, eikonal-averaged DWIA (t matrix), $\bar{V} = 30$ eV, $\bar{W} = 0$; ···, eikonal-averaged DWIA (t matrix), $\bar{V} = 10$ eV, $\bar{W} = 0$; ---, eikonal-averaged DWIA (t matrix), $\bar{V} = 0$, $\bar{W} = 0$; ×, fully distorted calculations. (b) Ne 2p: ---, Eikonal-averaged Born DWIA (v matrix), $\bar{V} = 30$ eV, $\bar{W} = 0$; ---, eikonal-averaged DWIA (t matrix), $\bar{V} = 30$ eV, $\bar{W} = 10$ eV; ×, fully distorted calculations. (b) Ne 2p: ---, DWIA (t matrix), $\bar{V} = 0$, $\bar{W} = 0$; ×, fully distorted calculations.

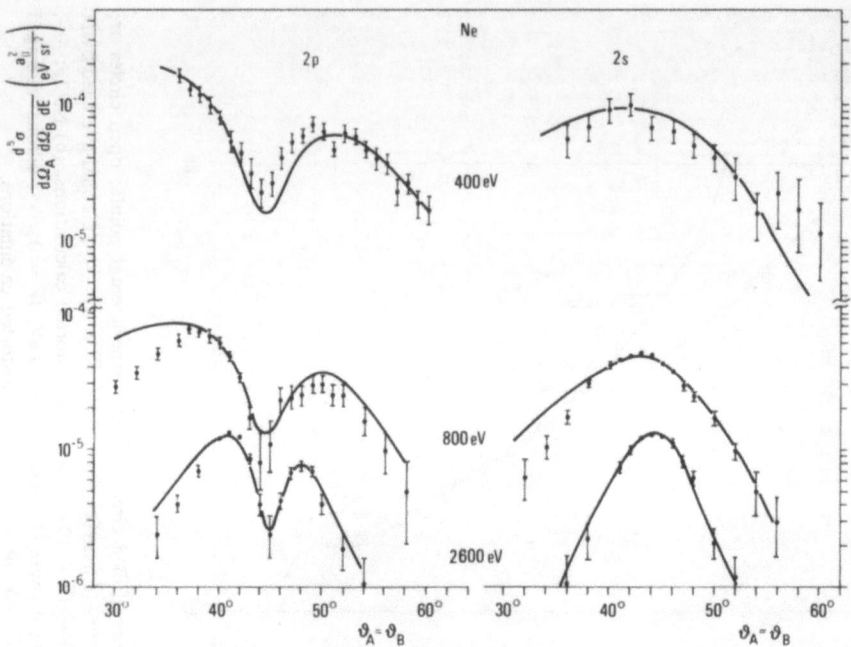

Figure 9. Coplanar symmetric $(e, 2e)$ cross section as a function of scattering angles taken at various incident energies. Left: Ne $2p$ electrons cross section; solid-line curve, eikonal-averaged DWIA with $\bar{V} = 10$ eV, $\bar{W} = 0$ eV. Right: Ne $2s$ electrons cross section; solid-line curve, eikonal-averaged DWIA with $\bar{V} = 30$ eV, $\bar{W} = 10$ eV.

range investigated, as shown in Figure 9 for energies of 2600 eV. At this energy the PWIA is almost indistinguishable from the eikonal-averaged DWIA values. The Ne $2p$ case is more complex. From the absolute values of the cross section shown in Figure 8b for $\vartheta_A = \vartheta_B = 40°$ and $\phi = 0$ it can be seen that the type of approximation that gives the best fit to the data is more difficult to determine than in the case of Ne $2s$. By applying the eikonal-averaged DWIA a reasonably good fit is obtained by choosing a value of 0 eV for the imaginary potential $\bar{W}$ and values of $\bar{V}$ ranging from 10 to 30 eV. Although these measurements cannot clarify which value of the real part of the potential to use they allow us to assess that for the $2p$ orbital the attenuation factor is negligible even at energies as low as 200 eV.

Also, the factorized fully distorted IA appears to predict the data reasonably well. From the coplanar symmetric data taken in the energy range 200–2600 eV we can assess that the eikonal-averaged DWIA is able to reproduce only the correct forward-to-backward peak ratio but is inadequate to fit the data in all the angular range. The same happens for the data taken at fixed q and reported in Figure 11. Measurements of the momentum profile in noncoplanar symmetric kinematics (Figure 12), which is less sensitive to the distortion effects, still show a discrepancy in comparison with values predicted from eikonal-averaged DWIA. Clearly it is possible to fit the q profile over the large range of q appropriate to this particular orbital only with a q-dependent effective potential or with a fully distorted-wave calculation. (More details of this q-dependent assumption are given in a paper to be published.[11]) Also in the Ne $2p$ case as soon as the energy increases the approximation becomes more realistic in explaining the results. This is clearly shown in the measurements at 1600 and 2600 eV both in coplanar and q-fixed geometry.[11] In conclusion we can affirm that for Ne $2s$ the eikonal-averaged DWIA works well from 1500 eV upwards in all the range of q momenta investigated while for Ne $2p$ the theory is working properly at energies higher than 3000 eV. For the other noble gases investigated, for which the data are not reported here for the sake of brevity, a confirmation of the different behavior of the different subshells has already been given.[13] The lower limit of the validity for a working theory, namely, the eikonal DWIA, depends on the binding energy ε_λ and on the average momentum of the initially bound electron.

5. Test of Electronic Molecular Structure

The finding of a working reaction theory, namely, the factorized PWIA, at energies larger than 2000 eV and the eikonal-averaged DWIA at intermediate energies ($\simeq 400$–2000 eV), allowed us to extract structure information on simple molecules. As pointed out in Section 2 the best way to obtain such information is by using energy

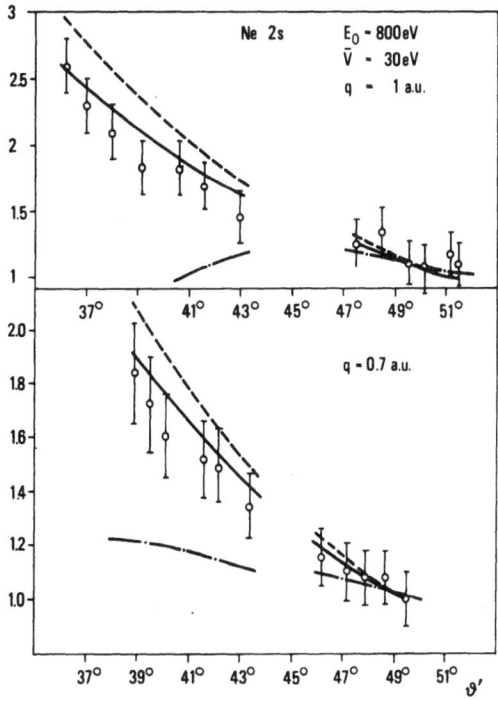

Figure 10. Relative cross section for the ejection of Ne $2s$ electrons in ϑ and ϕ variation so as to keep q fixed. Incident energy $E_0 = 800$ eV. In the abscissa the scattering angle θ' between the incident and final electrons is shown. Data taken in eikonal-averaged DWIA $\bar{V} = 30$ eV, $\bar{W} = 10$ eV are compared with computed f_v(– – –) and f_t(——) e–e factors. –·–·–, fully distorted calculations. Top: $q = 1$ a.u.; bottom: $q = 0.7$ a.u.

and angular correlation spectra in various geometries. The explicit expression of (14) is

$$\varrho(\mathbf{q}) = S_j^f \,|\, \psi_j(\mathbf{q})\,|^2$$

$$= S_j^f \sum_{1p}^{N} C_{j,p}\,|\,\Phi_p(\mathbf{q})\,|^2 + \sum_{1p}^{N}\sum_{p \neq s}^{N} C_{jp}C_{js}\{\exp[i\mathbf{q}\cdot(\mathbf{R}_p - \mathbf{R}_s)]\}\Phi_p^*(\mathbf{q})\Phi_s(\mathbf{q})$$

where $c_{j,\varkappa}$ $(\varkappa = p, s)$ are the linear combination of atomic orbitals (LCAO) coefficients of the jth molecular orbital, $\Phi_\varkappa(\mathbf{q})$ is the Fourier transform of the atomic basis set, and $\mathbf{R}_\varkappa$ are the atomic coordinates. For randomly oriented molecules $\varrho(\mathbf{q})$ has to be averaged over all the molecular orientations.[20]

Before showing a few significant examples, it is worthwhile to recall the main kinds of structural information we can obtain: (i) determination of energy orbital levels, (ii) localization and assignment of configuration interaction peaks, and (iii) testing of the calculated molecular wave functions.

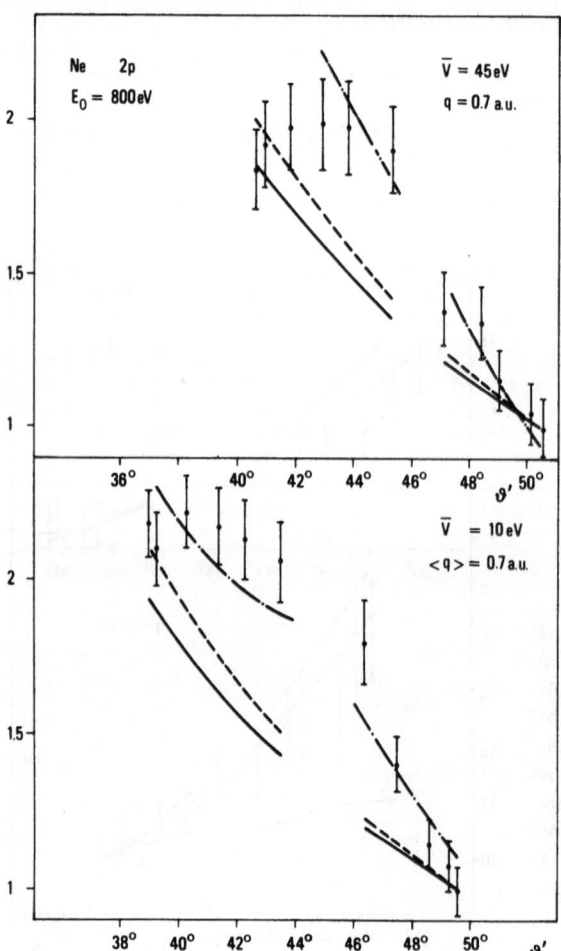

Figure 11. Relative cross section for the ejection of $2p$ electrons from Ne in the noncoplanar ϕ variation so as to keep q fixed. Incident energy $E_0 = 800$ eV. On the abscissa the scattering angle ϑ' between the incident and final electrons is shown. Top: data taken in eikonal-averaged approximation; $\bar{V} = 45$ eV kinematic conditions are compared with $f_v(---)$ and $f_t(\text{——})$ $e\text{–}e$ factors and fully distorted calculations $(-\cdot-\cdot-)$. Bottom: data taken in eikonal-averaged approximation; $\bar{V} = 10$ eV kinematic conditions are compared with $f_v(---)$ and $f_t(\text{——})$ $e\text{–}e$ factors and fully distorted calculations $(-\cdot-\cdot-)$.

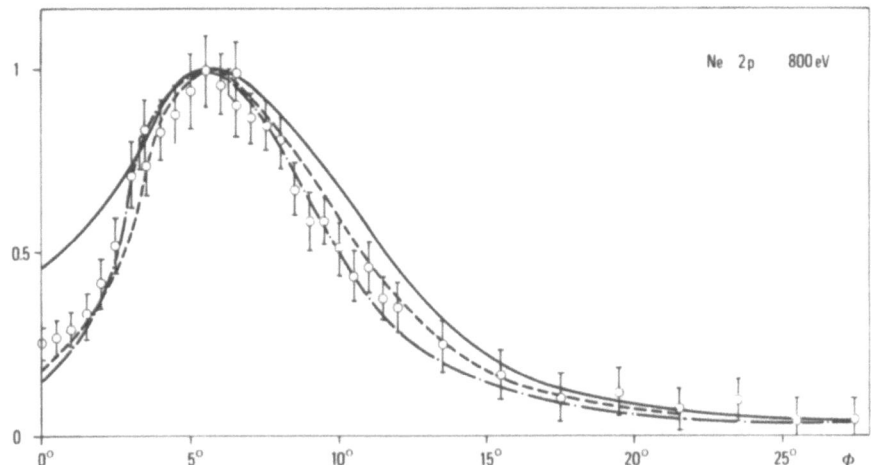

Figure 12. Relative $(e, 2e)$ cross section for Ne $2p$ at $E_0 = 800$ eV $\vartheta_A = \vartheta_B = 44°$, ϕ variable. Solid curve, eikonal-averaged DWIA $\bar{V} = 30$ eV. Dashed curve, q-dependent DWIA. Dotted dashed curve, fully distorted IA.

Many studies on molecules have been performed by other groups active in the field;[5,6,9] we report here only our recent data, which is particularly suitable for clarifying this information.

In the case of the CO_2 molecule,[21] we must underline that although the energy resolution was not enough to discriminate between all the external states, and although the internal states are a large convolution of many contributing states, it was possible to localize in energy for the first time the $2\sigma_u$ and the $3\sigma_g$ states. This was accomplished utilizing the extra degree of freedom due to the control of the momentum balance. Moreover, evidence of a configuration-interaction (CI) peak involving the $3\sigma_g$ state was found in the satellite region preceding the innermost valence states. It was also shown that the Hartree–Fock one-electron wave functions of the CO_2 molecule appear quite inadequate for predicting the electron momentum distribution for the outermost orbitals, one example of which is shown in Figure 13a. They seemed more adequate for describing the inner states, which are more atomic in character, as reported in Figure 13b.

The same features appear also in other more complex molecules such as the SF_6. Since only a preliminary paper has appeared[22] on this system we will spend more time on the description of those data. SF_6 has ten valence orbitals containing 48 electrons, and because of its very high symmetry several molecular orbital levels are degenerate. The ordering of the orbitals has been assigned recently[23] and found to be

$$(1t_{1g})^6 (1t_{2u})^6 (3t_{1u})^6 (2e_g)^4 (1t_{2g})^6 (2t_{1u})^6 (2a_{1g})^2 (1e_g)^4 (1t_{1u})^6 (1a_{1g})^2$$

The binding energies are, respectively, 15.7, 17 (for both $1t_{2u}$ and $3t_{1u}$), 18.6, 19.8, 22.9, 27, 38.3, 41.1, and 43.9 eV. In all the measured energy spectra reported in

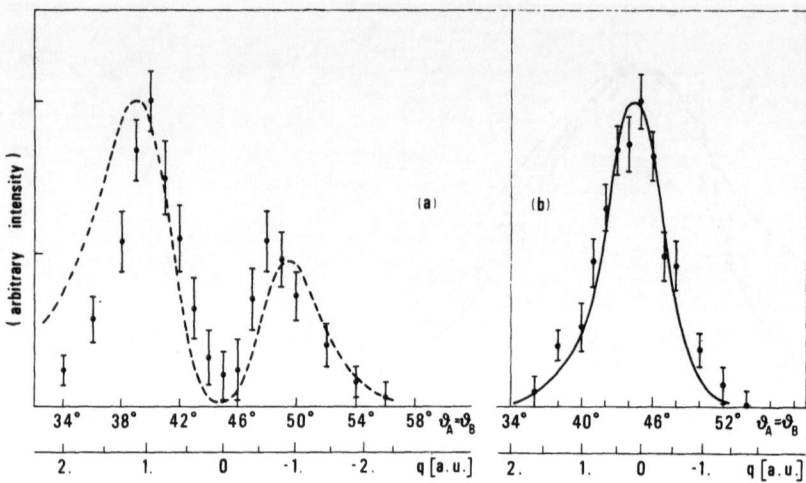

Figure 13. Relative cross section for the ejection of (a) $1\pi_g$ ($\varepsilon_\lambda = 13.5$ eV) and (b) $3\sigma_g$ ($\varepsilon_\lambda = 39$ eV) electrons of CO_2 molecule. Data taken at $E_0 = 1600$ eV in coplanar symmetric conditions are compared with values calculated in PWIA by using McLean and Yoshimine[22] wave functions.

Figure 14, the outermost external states and the innermost valence states are grouped into two large bands whose shape and intensity vary markedly as a function of the scattering angle due to the different q dependence of the various orbitals involved. Configuration-interaction peaks are clearly evident in the high-energy region of the

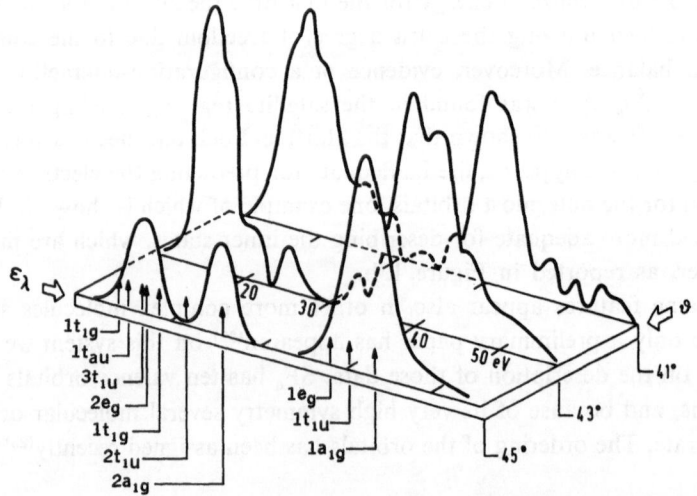

Figure 14. Measured energy spectra of SF_6 versus the separation energy. In the vertical axis, relative (e, 2e) coplanar symmetric cross sections measured at different scattering angles are shown. Incident energy $E_0 \simeq 2600$ eV. The ionization potentials[23] of the valence orbitals are indicated by arrows.

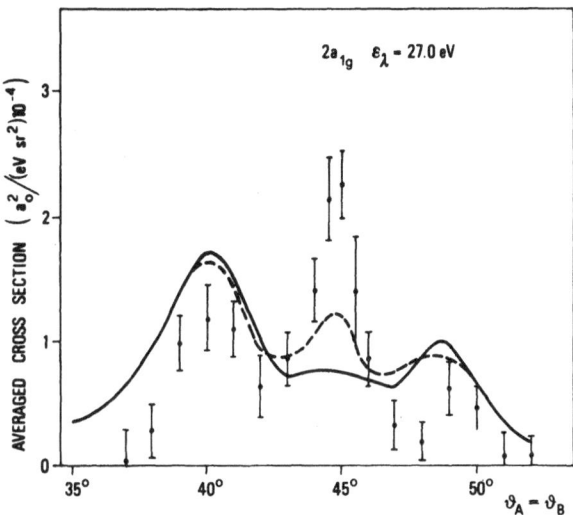

Figure 15. Absolute (e, 2e) cross section for the SF$_6$ $2a_{1g}$ orbital in symmetric coplanar conditions, averaged on the angular and energetic acceptances of the apparatus. Filled circles are experimental points. Comparison is made with cross sections calculated in PWIA by using $2a_{1g}$ wave function of Gianturco et al.[23] with sulfur d orbital excluded (– – –) and included (——).

spectrum, and they have been assigned on the basis of momentum distribution measurements.[24]

Absolute values of (e, 2e) cross sections have been measured for some of the electronic states of this system. This fact allows a better comparison with values calculated by Hartree–Fock basis sets. In Figure 15 the cross section measured for the $2a_{1g}$ state, completely resolved in energy, has been reported. Data, taken in coplanar conditions at an incoming energy of 2600 eV, are compared with values calculated in PWIA by using Hartree–Fock wave functions whose basis set differs from the presence or the absence of d orbitals.[25] While the energy eigenvalue of the $2a_{1g}$ state without d orbitals does not compare well with the measured ε_λ, its cross section is in better agreement with data. Moreover, the presence of the low-q components reflected by the high value of the cross section near $q = 0$ suggests a contribution from the 2s that is larger than the 2p fluorine orbitals in the sulfur–fluorine bond formation. A deconvolution procedure allowed us to measure the absolute cross section also for the three more internal valence states $1eg$, $1t_{1u}$, and $1a_{1g}$. Data are reported in Figure 16. When comparison is made with values calculated from the previously mentioned basis set, it can be inferred that both the $1t_{1u}$ and the $1e_g$ are quite different from the calculated values in shape and in absolute value. Cross sections calculated from that basis set, which does not take into account d orbitals, are very similar in shape and intensity. The shape and absolute value predicted for the ϕ distribution of the $1a_{1g}$ state is instead in good agreement with data when the contribution of the CI peak is taken into account. This fact confirms that orbitals more atomic in character are usually better described whichever set is chosen.

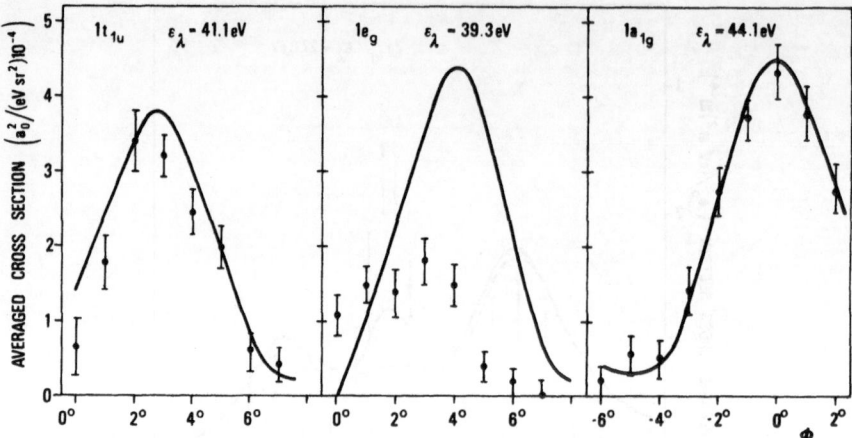

Figure 16. Absolute $(e, 2e)$ cross section for the SF_6 orbitals $1t_{1u}$ ($\varepsilon_\lambda = 41.1$ eV), $1e_g$ ($\varepsilon_\lambda = 39.3$ eV), and $1a_{1g}$ ($\varepsilon_\lambda = 44.1$ eV). They were measured at $\vartheta_A = \vartheta_B = 45°$ and ϕ variable, and averaged on the angular and energetic acceptances of the apparatus. Data are compared with cross sections calculated in PWIA by using Gianturco *et al.* wave functions with sulfur d orbital included.

This feature, i.e., that the most internal valence orbitals are slightly affected by the molecular bonding and so more easily described by Hartree–Fock wave functions was already found for the N_2 and NH_3 molecule.[26] With the aim of throwing more light on this point the study of some nitrogen compounds has recently been undertaken. Measurements on N_2O are still preliminary. For this molecule the 2π orbital cross section measured at 2600 eV in ϕ variable geometry, which is the most sensitive

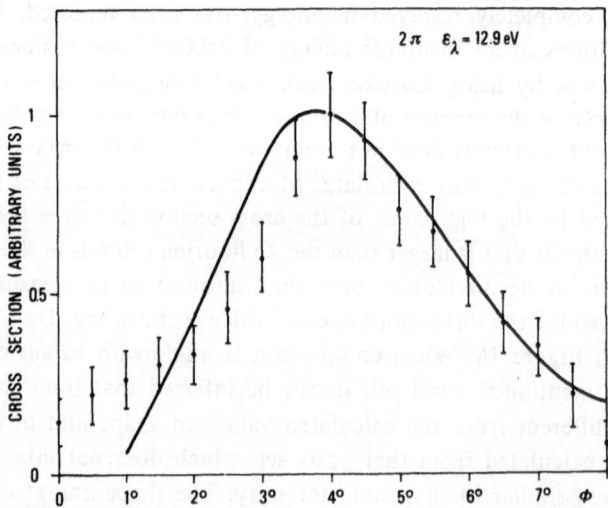

Figure 17. Relative cross section for the ejection of 2π ($\varepsilon_\lambda = 12.9$ eV) electrons of N_2O molecule. Data taken at $E_0 = 2600$ eV in out-of-plane $\vartheta_A = \vartheta_B = 45°$, ϕ variable geometry are compared with values calculated in PWIA by using McLean and Yoshimine wave functions.

to the $\varrho(\mathbf{q})$ measurements, appears to agree quite well with values calculated in PWIA by using the McLean and Yoshimine[22] wave functions (see Figure 17). The study of other orbitals is in progress and shows how sensitive is the (e, 2e) technique to the bonding problem.[33]

6. (e, 2e) Reactions with Solid Targets

The application of (e, 2e) spectroscopy to the study of solid targets is a task pursued for a long time and not yet satisfactorily achieved. Actually this experimental technique started with experiments performed on solid targets. Namely, it started with the experiment of Amaldi, Egidi, Marconero, and Pizzella,[2] which demonstrated the experimental feasibility, and with the experiment of Camilloni, Giardini Guidoni, Tiribelli, and Stefani,[3] which was the first to measure a recoil momentum profile for a resolved single-particle state. A doctoral thesis has been devoted to testing the possibility of studying bulk electronic properties of solids by (e, 2e) reactions,[27] while Levin, Neudatchin, and Smirnov[28] developed the theory of the (e, 2e) reaction in crystals in the impulse one-electron approximation. More recently the possibility of applying the (e, 2e) technique in surface-state spectroscopy has been suggested;[29] D'Andrea and Del Sole formulated a theory of the (e, 2e) reaction near solid surfaces based on Green's function formalism, this formalism allowing one to take into account the electron–solid interaction. Measurements are in progress, both at the University of Melbourne[30] and at SNAM L.R.B. of Monterotondo.[31] Performing (e, 2e) experiments with solid targets allows one to measure the dispersion curve $E(\mathbf{K})$ and the relative density of states of valence and conduction bands of the crystal, and, provided the target is a single crystal, anisotropies of the dispersion curve $E(\mathbf{K})$ are determined as well. Similar information is derived from the angular resolved photoemission spectroscopy, but we believe the interpretation of (e, 2e) results would be more reliable because the incident energy is higher and conditions for validity of the theoretical model applied are better fulfilled. Such information is believed to be particularly important in understanding solid-surface properties; namely, the problem of finding out what kind of atomic displacement gives rise to surface reconstruction could be investigated by the (e, 2e) technique. The paper by D'Andrea and Del Sole gives some examples of the sensitivity of the (e, 2e) form factor to the various models of superficial reconstruction in the case of silicon. From the experimental point of view a major difference between the (e, 2e) experiments in gases and solids lies in the respective density of the scatterer at the collision center. It means that owing to the high value of the electron impact cross section (typically $\simeq 10^{-16}$ cm^2) the incoming electron and the two outgoing ones, which must be scattered in forward directions, are unlikely to leave the solid target without further interactions with the target itself. The short penetration range of the electrons in a solid target is certainly a serious limitation in performing an experiment in kinematic conditions similar to those used for gaseous targets. But it becomes a valuable

feature in studying surfaces because the short penetration range of low-energy electrons allows one to deal with (e, 2e) events that took place in the first few layers
of the crystal. It can be achieved by using suitable kinematic conditions in which
both incoming and outgoing electrons are entering and leaving the surface with low
incident angles. This is the so-called reflection geometry. In any case the electron–
solid interaction affects the pure (e, 2e) process also in reflection geometry. The main
interaction processes are as follows.

(i) *Interaction with the Periodic Potential.* The plane waves are distorted by the
crystal potential, and Block functions should be used to describe incident and outgoing electrons. It introduces mixing between different reconstructed q's. The effect
of this distortion is minimized by increasing the incident energy.

(ii) *Interaction with Phonons.* Such interaction produces beam attenuation via
incoherent generation of optical phonons and momentum spread due to interaction
with acoustic phonons. However, owing to the high electron energy, the cross section
is small and the spread induced in the energy and momentum distributions is negligible
compared with experimental resolutions. Interaction with phonons during the (e, 2e)
process can also occur; however, this interaction is quite small at the energies usually
used for (e, 2e) reactions.

(iii) *Interaction with Plasmons.* This is by far the most relevant to the destruction
of information coming from (e, 2e) events, being the dominant scattering process
in the energy range 1–10 keV. Generation of one or more plasmons by the incoming
or outgoing electrons results in broadening of both the energy and momentum
spectrum because the two outgoing electrons are not brought out of time coincidence
while the energy balance can be identical to that for electrons coming from other
bands after a single (e, 2e) process. Because of the proper momentum of the generated
plasmons similar confusions have to be expected in the momentum distribution
spectrum.

As a result, the worst effect of electron–solid interaction is the broadening of
measured energy and momentum spectra. This effect is clearly shown by measurements of the (e, 2e) energy spectrum reported in Figure 18a. These are low-energy
resolution spectra relating to carbon and aluminum solid targets. Measurements
have been performed in transmission geometry at incident energy of about 9 keV.
Measurements on carbon 100 and 250 Å thick show a broadening of both K-shell
and L-shell peaks as the thickness is increased.[32] The effect is more evident in the
case of the thicker aluminum target whose results are reported in Figure 18b. The
energy spectrum shown relates to the two outermost shells of aluminum. For a target
thickness of 800 Å, despite a long tail of events that suffered multiple scattering, some
kind of structure in the energy spectrum is still recognizable. For a target thickness
as large as 1300 Å every structure is smeared out and the energy spectrum is reduced
to a big bump. For transmission geometry the depletion of the solid target due to
energy deposition by the electron beam in the thin film should not be forgotten.
Nevertheless, the transmission geometry on a solid target has proved to be useful

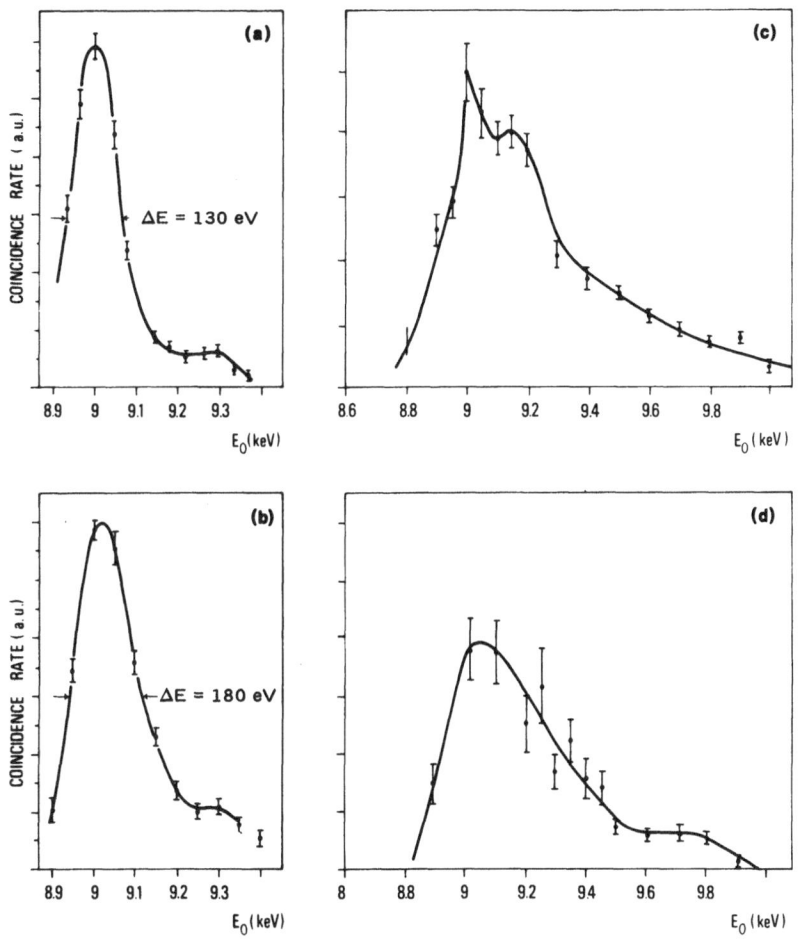

Figure 18. (e, 2e) energy spectrum taken in symmetric coplanar kinematics in solid targets. Filled circles are experimental points and the solid-line curve is an eye interpolation to the data. (a) Solid carbon 100 Å thick at $\vartheta_A = \vartheta_B = 45 \pm 0.5°$. (b) Solid carbon 250 Å thick at $\vartheta_A = \vartheta_B = 43.5 \pm 0.5°$. (c) Solid aluminium 800 Å thick at $\vartheta_A = \vartheta_B = 45°$. (d) Solid aluminium 1300 Å thick at $\vartheta_A = \vartheta_B = 45°$.

in performing momentum distribution measurements on inner shells (namely, the $1s$ orbital of carbon) and would become particularly useful in performing measurements on inner shells of heavier atoms. In that case, owing to the high binding energy and high proper momentum of the orbital involved in the reaction, to satisfy impulse-approximation requirements, very high incoming energies are required (several hundreds of keV). At such high energies the electron–solid interaction cross section are low and the previously mentioned broadening is negligible with respect to the expected energy and momentum distributions. Moreover, the (e, 2e) cross section decreases as the energy is increased and a target more dense than a gas is particularly valuable.

7. Summary and Conclusions

We have given examples of $(e, 2e)$ experiments performed for several different reasons. One of the main objectives of the $(e, 2e)$ program is to find information about the dynamic electronic structure of atoms and molecules. For this we need a good working theory in a kinematic range wide enough to enable the information to be obtained. This in turn requires a thorough investigation of the range of validity of reaction theories or in other words a detailed study of the mechanism of ionization.

The basic theory whose validity has been investigated is the distorted-wave impulse approximation (DWIA) with distorted waves approximated by plane waves computed in a constant complex potential $(\bar{V} + i\bar{W})$, where $\bar{V}$ and $\bar{W}$ are positive real numbers. This theory is the eikonal-averaged DWIA. The $(e, 2e)$ differential cross section factorizes into a spectroscopic factor, an e–e collision factor (approximated by the square modulus of a half-off-shell Coulomb t-matrix element for experimentally determined momenta), and an orbital momentum profile $\varrho(\mathbf{q})$, where $\mathbf{q}$ is the recoil momentum calculated in the average eikonal potential. For randomly oriented molecules ϱ is the rotationally and vibrationally averaged modulus squared of a Hartree–Fock molecular orbital, the characteristic orbital. The spectroscopic factor is the probability of finding the one-hole configuration (i.e., a hole in the Hartree–Fock ground state of the target) in the configuration-interaction eigenstate of the residual ion.

At present the parameters $\bar{V}$ and $\bar{W}$ must be determined empirically from momentum-profile shifts and absolute cross sections, respectively. Their values are consistent and make sense in terms of the potentials which describe elastic scattering. More accurate descriptions of data in some kinematic ranges can be obtained by allowing $\bar{V}$ to depend on the recoil momentum $\mathbf{q}$. The eikonal-averaged approximation is no longer valid in such cases, but the basic assumption of local translational invariance of the two-body collision factor is still valid. At lower energies for the $2p$ orbital of neon even the $\bar{V}(\mathbf{q})$ approximation is not accurate in detail near the maximum in $\varrho(\mathbf{q})$.

The wide range of validity of the $\bar{V}(\mathbf{q})$ eikonal approximation leads us to believe in its effectiveness as an alternative approximation to the DWIA. The breakdown of the whole approximation in many cases for $\vartheta_A + \vartheta_B < 80°$ is therefore tentatively interpreted as a breakdown of the DWIA itself, due perhaps to the fact that it underemphasizes the e–e collision contribution to the amplitude for matrix elements in which the two electrons have more similar final trajectories.

Notwithstanding the breakdown of the approximation in certain special regions, it works extremely well in noncoplanar symmetric kinematics where the e–e collision factor varies only imperceptibly. Therefore it gives an excellent basis for extracting the orbital momentum profile and the spectroscopic factor from data, certainly at incident energies (for valence states) of the order 2000 eV or higher.

This information on molecular orbitals and their configuration interaction is an extremely critical test of the structure calculations of quantum chemistry. While there are other spectroscopies that determine the energy levels of ions there is no other that (i) assigns an energy level to an irreducible representation of the point group which is done in (*e*, 2*e*) by identifying its characteristic orbital, (ii) verifies in detail the validity of an orbital calculation, or more generally a complete calculation of the target-ion overlap, which is done in (*e*, 2*e*) by comparing the details of experimental and theoretical profiles $\varrho(\mathbf{q})$, or (iii) determines spectroscopic factors that are a critical test of configuration interaction, which is done in (*e*, 2*e*) by comparing intensities for excitation of ion states belonging to the same group representation and, most importantly, verified in all cases studied by the satisfaction of the spectroscopic sum rule.

In the future, more detailed three-dimensional information on molecular orbitals in momentum space will be available from (*e*, 2*e*) reactions on oriented molecules.

It would be a very exciting step in solid-state physics to be able to obtain corresponding information about electron bands from (*e*, 2*e*). There are severe experimental difficulties that have not yet been overcome. Experiments have been reported that give some idea of the broadening of structure-related effects due to electron–solid interactions. However, it must be remembered that the first successful (*e*, 2*e*) experiment was performed on a solid and that there is hope that a refinement of techniques will fulfill the promise for solids that has already been fulfilled for targets in the gas phase.

ACKNOWLEDGMENTS

The authors gratefully acknowledge the help of Professor I. E. McCarthy in critically reading this manuscript. The authors would like to thank R. Fantoni, R. Tiribelli, and D. Vinciguerra for their assistance during various phases of the experimental part of this work.

References

1. H. Ehrhardt, M. Schulz, T. Tekaat, and K. Willmann, *Phys. Rev. Lett.* **22**, 89 (1969).
2. U. Amaldi, A. Egidi, R. Marconero, and G. Pizzella, *Rev. Sci. Instrum.* **40**, 1001 (1969).
3. R. Camilloni, A. Giardini-Guidoni, R. Tiribelli, and G. Stefani, *Phys. Rev. Lett.* **30**, 475 (1973).
4. E. Weigold, S. T. Hood, and P. J. O. Teubner, *Phys. Rev. Lett.* **30**, 475 (1973).
5. S. T. Hood, A. Hamnett, and C. E. Brion, *J. Electron Spectrosc. Relat. Phenom.* **11**, 205 (1977).
6. J. H. Moore, M. A. Coplan, T. L. Skillman, Jr., and E. D. Brooks III, *Rev. Sci. Instrum.* **49**, 463 (1978).
7. J. F. Williams, *J. Phys. B* **11**, 2015 (1978).
8. (a) G. Stefani, R. Camilloni, and A. Giardini-Guidoni, *Phys. Lett.* **64A**, 364 (1978). (b) H. Wellenstein, *Rev. Sci. Instrum.* **46**, 92 (1975).
9. I. E. McCarthy and E. Weigold, *Phys. Rep.* **27C**, 275 (1976).
10. J. B. Furness and I. E. McCarthy, *J. Phys. B* **7**, 541 (1974).
11. R. Camilloni, A. Giardini-Guidoni, I. E. McCarthy, and G. Stefani, *J. Phys. B* **12** (1979).
12. L. Hostler and R. H. Pratt, *Phys. Rev. Lett.* **10**, 469 (1963).

13. A. Giardini-Guidoni, R. Tiribelli, D. Vinciguerra, R. Camilloni, G. Stefani, and G. Missoni, *A. I. P. Conf. Proc.* **36**, 205 (1977).
14. R. Camilloni, A. Giardini-Guidoni, I. E. McCarthy, and G. Stefani, *Phys. Rev. A* **17**, 1634 (1978).
15. I. Fuss, I. E. McCarthy, C. J. Noble, and E. Weigold, *Phys. Rev. A* **17**, 604 (1978).
16. E. Clementi, D. L. Raimondi, and W. P. Reinhardt, *J. Chem. Phys.* **47**, 1300 (1967).
17. E. Clementi and C. Roetti, *At. Data Nucl. Data Tables* **14**, 177 (1974).
18. S. Geltman, *J. Phys. B* **7**, 1994 (1974); and private communication (1978).
19. R. Camilloni, G. Stefani, and A. Giardini-Guidoni, *J. Phys. B* **2** (1979).
20. V. G. Neudatchin, G. A. Novoskol'tseva, and Y. F. Smirnov, *Sov. Phys.-JETP* **28**, 54 (1969).
21. A. Giardini-Guidoni, R. Tiribelli, D. Vinciguerra, R. Camilloni, and G. Stefani, *J. Electron Spectrosc. Relat. Phenom.* **12**, 405 (1977).
22. A. D. McLean and M. Yoshimine, *IBM J. Res. Dev.* (1967).
23. N. Gelius, *J. Electron Spectrosc. Relat. Phenom.* **5**, 985 (1974).
24. R. Camilloni, A. Giardini-Guidoni, G. Stefani, R. Tiribelli, and D. Vinciguerra, Xth International Conference of the Physics of Electronic and Atomic Collisions, Book of Abstracts, 194 (1977); and *J. Chem. Phys.* **71** (1979).
25. F. A. Gianturco, C. Guidotti, U. Lamanna, and R. Moccia, *Chem. Phys. Lett.* **10**, 269 (1971).
26. R. Camilloni, G. Stefani, A. Giardini-Guidoni, R. Tiribelli, and D. Vinciguerra, *Chem. Phys. Lett.* **41**, 17 (1976).
27. G. Stefani, Thesis, Istituto Fisico Università Roma, Rome (1970).
28. V. G. Levin, N. G. Neudatchin, and Y. F. Smirnov, *Phys. Status Solidi B* **29**, 618 (1972).
29. A. D'Andrea and R. Del Sole, *Surf. Sci.* **71**, 306 (1978).
30. N. R. Avery, *A.I.P. Conf. Proc.* **36**, 195 (1977).
31. P. Ascarelli and G. Missoni, private communication (1978).
32. R. Camilloni, A. Giardini-Guidoni, G. Stefani, and R. Tiribelli, Frascati Report No. LNF-72/53, Rome (1972).
33. A. Giardini-Guidoni, R. Fantoni, R. Camilloni, G. Stefani, *Adv. in Mass Spectr.* **8** (1979).

3

Low-Energy Electron Impact Ionization with Completely Determined Kinematics

H. Ehrhardt, K. Jung, and E. Schubert

Nonresonant ionization of helium and neon atoms by electrons has been measured at collision energies between 30 and 250 eV. In a coincidence experiment scattering and ejection angles as well as energies of both outgoing electrons are determined. The results divide themselves into two energy ranges: (i) For $E_0 > 50$ eV the ionizing collision proceeds nearly like a classical binary collision in which the momentum transfer of the scattered electron is added to the momentum distribution of the target electron. The resulting angular correlation between both outgoing electrons, which may be modified by the influence of the ionic potential, has a symmetry axis close to the momentum transfer vector. Calculations of the triple differential cross section in different approximations reproduce the experimental data qualitatively and, for high impact energy, also quantitatively. (ii) For $E_0 < 50$ eV, especially for E_0 in the range below 40 eV, the exchange amplitudes have similar magnitude as the direct scattering amplitude. All scattering amplitudes interfere with each other and produce angular dependences of the coincidence cross section which are no longer symmetric about $\mathbf{K}_{0a}$ and cannot be explained by the classical model. None of the theoretical results obtained until now are in quantitative agreement with experiment. Results from resonant ionization via the aligned $3s3p^64p$ state of argon are also discussed.

1. Introduction

Today's knowledge of electron impact ionization of atoms and molecules comes from many sources. The results of many measurements of total cross sections are

H. Ehrhardt, K. Jung, and E. Schubert • Fachbereich Physik der Universität Kaiserslautern, Kaiserslautern, West Germany.

available as functions of the impact energies. Total ion currents have been measured that include singly and multiply charged ions in their ground or excited states; and also mass spectrometric separation has been used to study the energy dependence of fragments from dissociative ionization. A few experiments of this type had already been performed in the years between 1927 and 1934.[1,2] The data obtained are of great importance for practical applications, but they give little insight into the ionization process itself. More information can be obtained from the differential cross sections $d\sigma(E_0, E)/dE$ and $d^2\sigma(E_0, E, \vartheta)/dE\, d\Omega$, where E is the energy and ϑ the angle of detection of the electrons after the collision.

Because of technical difficulties, triple differential cross sections

$$d^3\sigma(E_0, E_a, \vartheta_a, \vartheta_b, \varphi_b)/dE\, d\Omega_A\, d\Omega_b$$

were not measured until 1969.[3,4] Such cross sections contain all the information that can be obtained apart from spin, since the momenta of all the particles before and after the collision are in effect determined.

Figure 1 shows the momentum $\mathbf{k}_0$ of the incoming electron and the momenta $\mathbf{k}_a$ and $\mathbf{k}_b$ of the two outgoing electrons. The three momenta $\mathbf{k}_0, \mathbf{k}_a, \mathbf{k}_b$ need not be in one plane because the remaining ion may have momentum too.

Figure 2 shows a schematic diagram of our experimental arrangement. The momenta $\mathbf{k}_0$, $\mathbf{k}_a$, and $\mathbf{k}_b$ are determined by measuring the energies E_0, E_a, and E_b and the scattering angles ϑ_a and ϑ_b. The collectors A and B can be rotated independently from each other around the scattering center. A detailed description has been given in Jung *et al.*[5]

The aim of the work of the Kaiserslautern group was to measure triple differential ionization cross sections as precisely and in as much detail as possible in the low- and intermediate-impact-energy range in order to give the theoreticians new data for the improvement of the solutions of the three-particle problem with long-range forces, and to gain insight into the ionization process and if possible to establish models for this process.

Similar measurements in the intermediate energy range have been performed by Beaty *et al.*,[6,7] who were able to determine coincidence cross sections in plane and out of plane in absolute units. Until now they have measured argon and helium

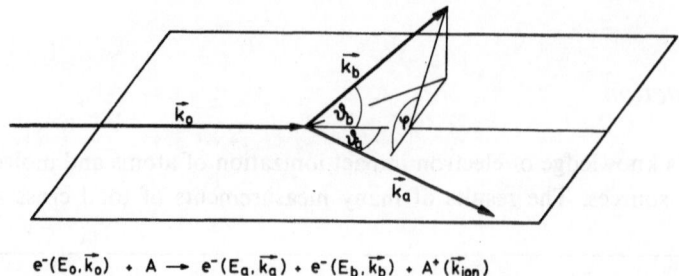

$$e^-(E_0, \vec{k}_0) + A \longrightarrow e^-(E_a, \vec{k}_a) + e^-(E_b, \vec{k}_b) + A^+(\vec{k}_{ion})$$

Figure 1. Schematic diagram of the kinematics of an ionizing electron collision with an atom.

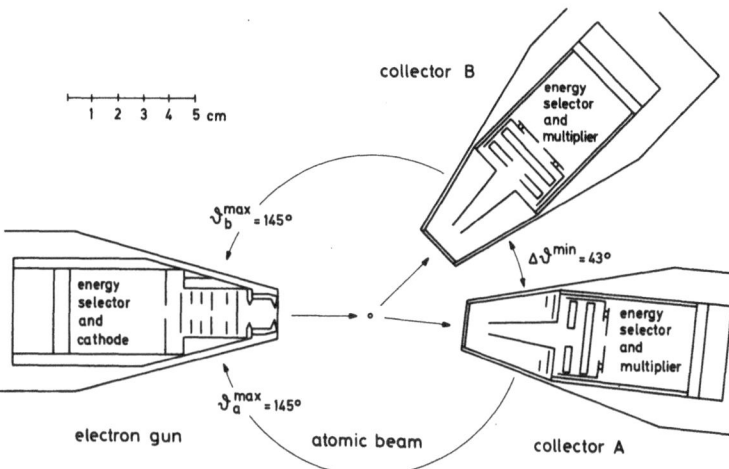

Figure 2. Experimental arrangement for the measurement of triple differential cross sections. The equipment has been described elsewhere.[5,26]

for impact energies of about 100 eV. Quite a lot of theoretical papers have accompanied the measurements, and in this way have contributed very much to our understanding.

Other groups[8–11] have performed $(e, 2e)$ measurements at 1 keV or higher energies at which ionization theory has been adequately developed, and these studies shed light on the momentum distribution of target electrons in molecules. Similarly, van der Wiel and Brion[12] and Tan and Brion[13] used the coincidence method at impact energies of several keV and at small momentum transfers to simulate photoionization processes.

Special measurements have been done by Cvejanović and Read,[14] who have studied the threshold behavior for two pairs of scattering angles by a coincidence-time-of-flight technique to test threshold theories.[15–17]

With all these activities during the last ten years knowledge of the ionization process by electrons in the energy range from threshold through the intermediate region up to very high energies has very much increased.

2. High- and Intermediate-Energy Ionization

Our first measurements[3,18] were aimed at studying the coplanar angular distributions of the triple differential cross section in an energy region where the scattering theories are now fairly well established. First we studied the ionization of the $1s$ electron of helium. Figure 3 shows a typical result at 250 eV primary energy. The cross section is plotted in a polar diagram. The direction of the incoming electron is given by the arrow from the bottom to the center. The faster (in general the scattered) electron is detected at the scattering angle ϑ_a, which is given by the arrow to

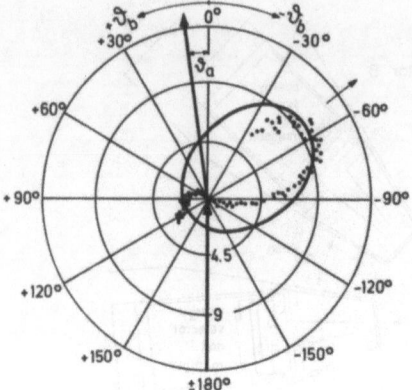

Figure 3. Angular dependence of the triple differential cross section for the electron impact ionization of helium. The scattering parameters are as follows: $E_0 = 256.5$ eV, $E_a = 212$ eV, $E_b = 20$ eV, $\vartheta_a = 8°$. The distance of the dots from the center of the polar plot is proportional to the rate of measured true coincidences. The solid-line curve shows the results of the plane-wave Born approximation of Glassgold and Ialongo.[20]

the upper left. The probability of finding the second (in general the ejected) electron at the scattering angle ϑ_b is proportional to the distance between the center and the dots. The small arrow indicates the direction of the momentum transfer $\mathbf{K}_{0a} = \mathbf{k}_0 - \mathbf{k}_a$.

The angular distribution of the slow electron displays two distinct peaks nearly opposite to each other and separated by two sharp minima. At first this form seemed to us very unlikely because the theories that were available at the time for the discussion of the coincidence cross section (namely, the binary encounter theory of Vriens[19] and the plane-wave Born approximation of Glassgold and Ialongo[20]) predicted only one peak in the direction of the momentum transfer and a flat minimum in the opposite direction, as can be seen by the solid-line curve in Figure 3.

The two theories give nearly the same mathematical expression for the angular dependence of the triple differential cross section, although they are differently deduced. The binary encounter theory is in principle a classical theory for the collision of a quasi-free electron with another electron, and the plane-wave Born approximation is a quantum mechanical theory which only takes into account the two electrons involved in the ionization process.

Our results (examples are given in Figures 3 and 4) show that for a few hundred

Figure 4. Angular correlation of the coincidence cross section. The energies of the incoming and the two outgoing electrons are the same as for the measurement shown in Figure 3. In this case the scattering angle ϑ_a of the faster electron amounts to 4°. The results of the plane–plane approximation of Veldre et al.[21] are represented by the solid-line curve.

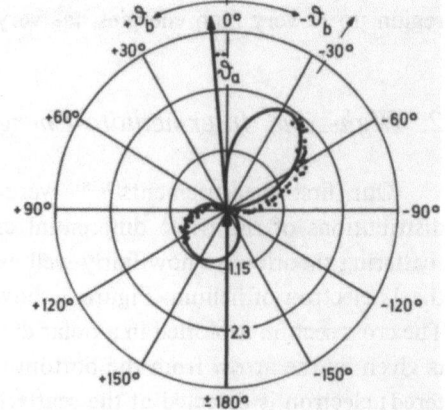

eV impact energy one should not neglect the influence of the ionic potential. The introduction of the two potential terms $2/r_1 - 1/r_{12}$ produces two scattering amplitudes with opposite sign and similar magnitudes. The amplitude caused by the electron–electron interaction $-1/r_{12}$ is forward peaked with a symmetry axis parallel to the momentum transfer direction and the amplitude caused by the electron–ion interaction $+2/r_1$ is constant with respect to the angle ϑ_b. In the directions nearly perpendicular to the momentum transfer the two amplitudes cancel each other. The cross section is zero at these points (see Figure 4).

We called the peak in the $\mathbf{K}_{0a}$ direction the binary peak because the binary interaction amplitude predominates, and the other we called the recoil peak. This nomenclature suggests that, classically speaking, electrons in the recoil peak have recoiled around the ion so as to leave the scattering region in the direction opposite to that expected.

Taking into consideration the fact that the resulting potential is not spherically symmetric because of the repulsive potential term from the scattered electron, the recoil peak may even be larger than the binary peak. This happens especially when the energy of the ejected electron is rather small.

For $E_0 = 250$ eV we found nearly symmetric angular dependences. But the symmetry axis is not parallel to the momentum transfer direction. Instead it is

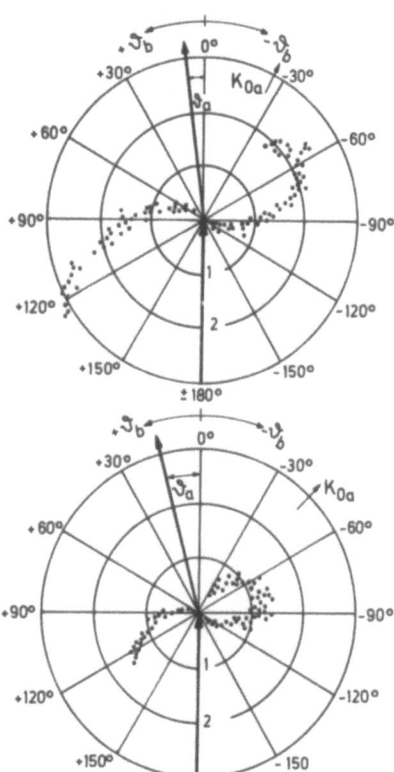

Figure 5. Angular dependences of the triple differential cross section for the electron impact ionization of helium. $E_0 = 80.5$ eV, $E_a = 53.5$ eV, $E_b = 2.5$ eV. Top: $\vartheta_a = 7°$. Bottom: $\vartheta_a = 15°$.

shifted by several degrees towards larger scattering angles because of the influence of the repulsive potential between the two outgoing electrons. For lower energies this deviation is even larger. Both peaks are deflected from the direction where the scattered electrons have been detected. Figure 5 shows the angular correlation for a collision energy of 80 eV. Both peaks have nearly symmetric shape but they are no longer opposite to each other. For even lower energies the peaks are no longer symmetric. This can be seen in the polar plots of Figure 6 at an impact energy of 50 eV.

Although the binary encounter theory seems not to be adequate to describe the angular distribution, it is a rather good tool to explain qualitatively the angular variation of the triple differential cross section in the region of the momentum transfer direction, especially if the energies of the scattered and ejected electrons are not too small.

In this theory the ionizing collision is treated as a purely electron–electron collision. The electrons that participate in the collision are treated as free electrons and their momenta can be written as

$$\mathbf{k}_0 + \mathbf{k}_e = \mathbf{k}_a + \mathbf{k}_b$$

In this equation $\mathbf{k}_e$ is the instantaneous momentum of the target atomic electron.

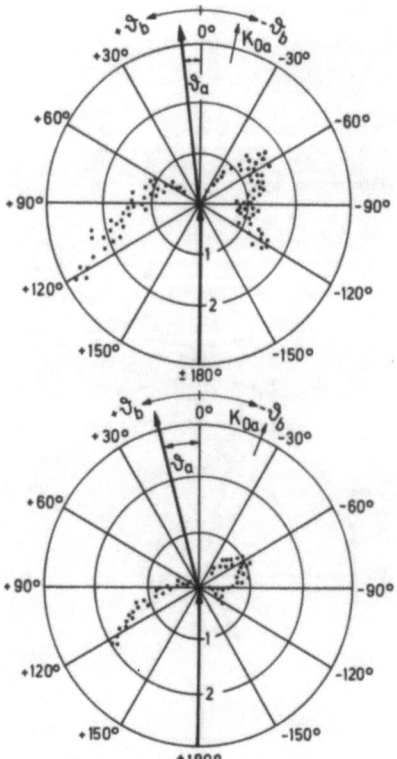

Figure 6. Same as for Figure 5 but with $E_0 = 50$ eV, $E_a = 20$ eV, $E_b = 5.5$ eV. Top: $\vartheta_a = 7°$. Bottom: $\vartheta_a = 15°$.

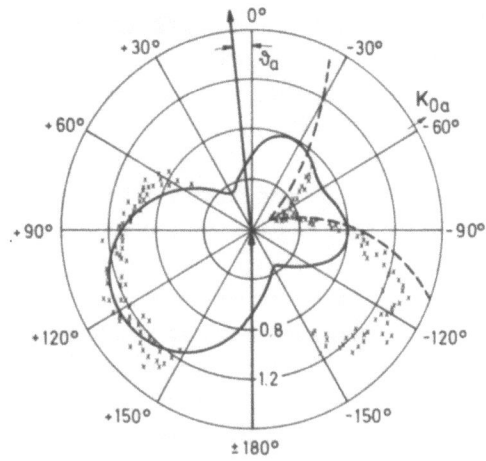

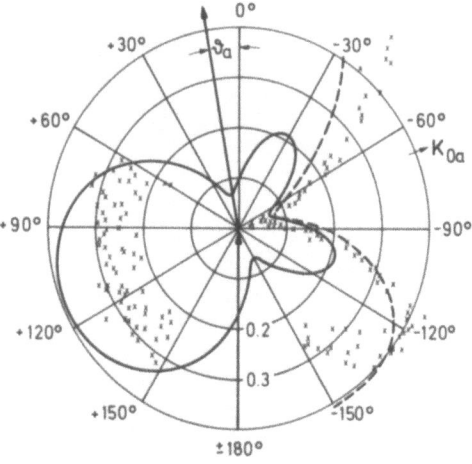

Figure 7. Triple differential cross section as a function of the scattering angle ϑ_b for the $2p$ ionization of neon. The solid-line curve shows the calculated data of the distorted-wave Born approximation of Knapp and Schulz.[28] The dashed-line curve represents the angular behavior predicted by an impulse model.[5] $E_0 = 250$ eV, $E_a = 223.5$ eV, $E_b = 5$ eV. Top: $\vartheta_a = 6°$. Bottom: $\vartheta_a = 10.5°$.

As the atom is at rest before the impact and as the ion is assumed not to change its momentum during the collision, k_e is equal to $-k_{ion}$ in this approximation. In what follows $k_{ion} = k_0 - k_a - k_b$ is called the momentum defect.

If there is no strong recoil peak, the shape of the cross section is determined only by the momentum distribution of the atomic electron. The momentum distribution of the $1s$ electron decreases monotonically with increasing momentum. As a consequence the cross section is maximal for minimal momentum defect $k_0 - k_a - k_b$. This expression is minimal if k_b has the direction of the momentum transfer $k_0 - k_a$. The cross section forms a single peak in the momentum transfer direction. This peak is rotationally symmetric. This has been predicted by all Born theories and could be proven experimentally by Beaty et al.[22]

The ionization of a $2p$ electron from neon leads to different structure in the cross section.[5] Figure 7 shows the angular correlation at 250 eV collision energy.

The ejected electron has been detected with an energy of 5 eV and the scattered electron has left the scattering center at $\vartheta_a = 6°$ and 10.5°. The momentum distribution for $2p$ electrons has a maximum at about $0.85a_0^{-1}$. If one chooses the momentum transfer in such a way that the momentum defect varies in the region of $0.85a_0^{-1}$ there is a minimum of the angular distribution in the direction of the momentum transfer and the binary peak seems to be split. In the angular region of this minimum the angular correlation looks symmetric. As for helium the neon data also display a shift of the symmetry axis to larger scattering angles by 10°–20°. The dashed line shows the results of a modified impulse model (Jung et al.[5]). The results of a distorted-wave Born approximation (Knapp and Schulz[23]) are given by the solid curve. The overall agreement of the theory is rather good, but the description by the impulse model is better for the scattering angles around $\mathbf{K}_{0a}$. This can also be seen in Figure 8, which shows the results for the same collision energy and the

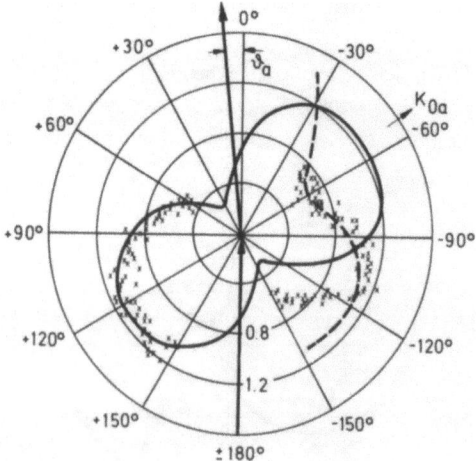

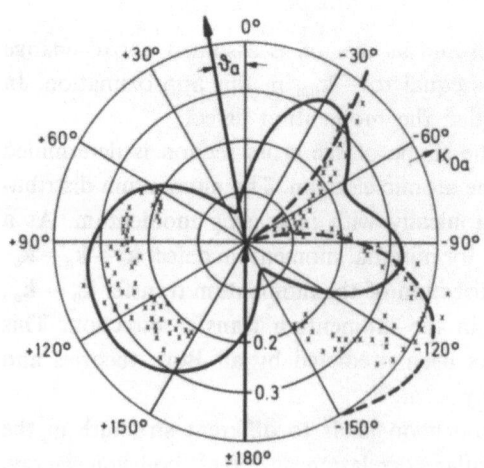

Figure 8. Same as for Figure 7 but with $E_0 = 250$ eV, $E_a = 218.5$ eV, $E_b = 10$ eV. Top: $\vartheta_a = 6°$. Bottom: $\vartheta_a = 10.4°$.

same scattering angles $\vartheta_a = 6°$ and $10.5°$. In this case the energy of the ejected electron amounts to 10 eV. We conclude between 100 and 250 eV that the relative shape of the cross section does not depend on the impact energy but on the energy of the ejected electrons and on the momentum transfer $\mathbf{K}_{0a}$, a fact that additionally has been proven by Ugbabe et al.[24] for higher collision energies.

3. Low-Energy Ionization

In the low-energy region we have made two series of measurements, one at 35 eV collision energy and the second series at 30.5 eV. In both cases helium has been used as the target gas, i.e., the excess energy above the ionization potential is 10.5 eV or 6.0 eV, respectively. This excess energy is carried away by the two outgoing electrons, which means that both electrons have rather low kinetic energies. For high and intermediate collision energies most scattering events are characterized by relatively large E_a and small values ϑ_a, i.e., practically all scattered electrons are scattered in a cone in the forward direction, whereas the ejected electrons are slow and are nearly isotropically distributed. Low-energy ionization processes are quite different. The number of ionization events is more or less equally distributed over all possible values of ϑ_a, and also the number of events is equally distributed over all values of E_a from 0 to $E_0 -$ IP. Figures 9–13 show examples of the angular correlation measurements for 35-eV and 30.5-eV impact ionization.

Figure 9 shows the results for 35 eV collision energy and 7 and 3.5 eV kinetic energy of the two ionization electrons. The scattering angles are $\vartheta_a = 36°, 60°,$ $75°, 120°$. For values of ϑ_a up to about $50°$ the measured angular distributions of the triple differential cross section are dominated by a broad peak nearly in the backward direction $(\vartheta_b \sim \pm 180°)$.

In addition, there exists a second smaller peak with its maximum approximately at $\vartheta_b = -80°$, independent of the choice of ϑ_a, and less pronounced at smaller values of ϑ_a. The full curve in each diagram represents the data of the Coulomb projected Born approximation including exchange, calculated by Geltman.[25,26]

For larger scattering angles ϑ_a the behavior of the coincidence cross section as a function of ϑ_b is completely different from that shown at smaller ϑ_a. For $75°$ the angular distribution is nearly flat without any structure. The coincidence cross section shows only a slight increase in the backward direction. In the angular region between $\vartheta_b = -20°$ and $\vartheta_b = -145°$ the agreement between theory and experiment is rather good, apart from a normalization factor (Figure 9 upper right-hand diagram and lower left). However, the constancy of the cross sections as a function of angle does not carry over into the other half of the plane, because the measured cross sections between $\vartheta_b = 110°$ and $145°$ were zero to within experimental error: i.e., no significant number of true coincidences was obtained. The theory predicts a radical departure from the constant cross section for positive ϑ_b values. The calculated data

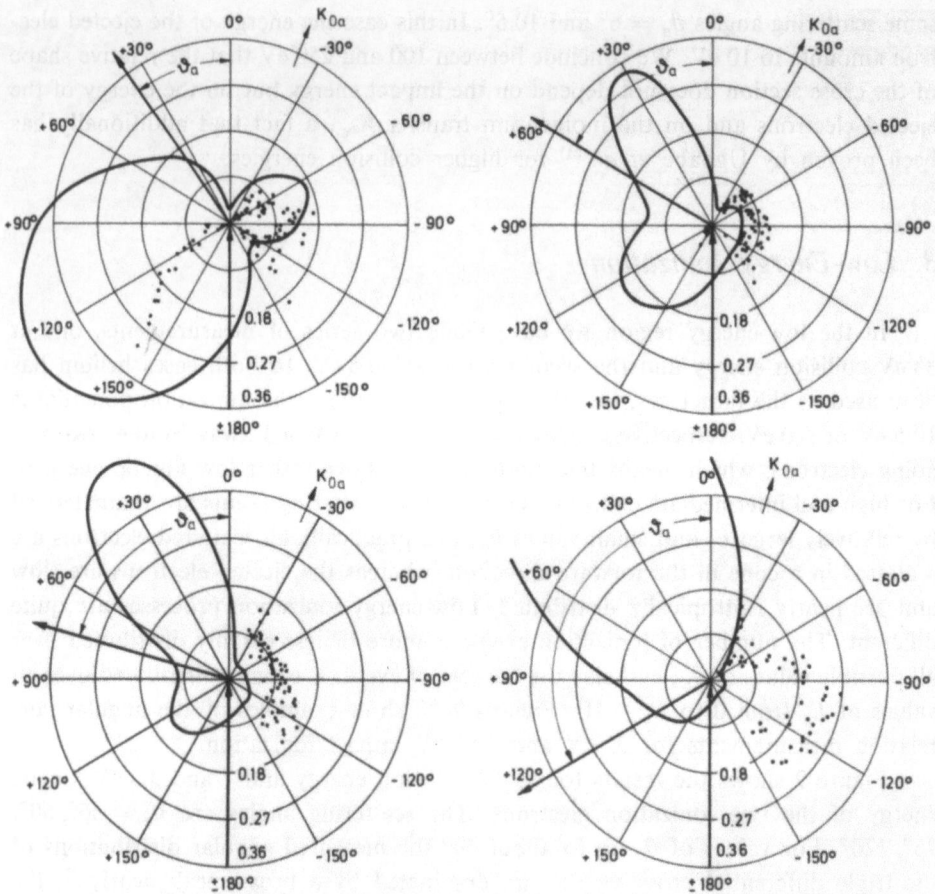

Figure 9. Angular dependences of the coincidence cross section for the single ionization of helium
at the collision energy $E_0 = 35$ eV. The data calculated by the Coulomb-projected Born approximation including exchange of Geltman are shown by the full line curves. $E_0 = 35$ eV,
$E_a = 7$ eV, $E_b = 3.5$ eV. Clockwise from top left: $\vartheta_a = 36°$, $60°$, $120°$, $75°$.

display two peaks, the greater being at about $30°$ and an order of magnitude larger
than the cross section on the negative ϑ_b side. The larger forward lobe cannot be
reached experimentally with our equipment.

In the lower right-hand side of Figure 9 ($\vartheta_a = 120°$) the shape of the measured
angular distribution shows some structure. There exists a well-pronounced minimum
at $\vartheta_b = -40°$. The cross section nearly reaches zero. Even the maximum of the
backward peak can be recognized at $\vartheta_b' = -100°$. Moreover, the strong rise of the
cross section in the forward direction is clearly visible for the first time. For large
angles ϑ_a the calculated results are governed by the strong peak around $\vartheta_b = 30°$.
For increasing ϑ_a the maximum of this peak increases too, the full width at half-
maximum gets broader, and the backward peak around $\vartheta_b = 130°$ decreases. In

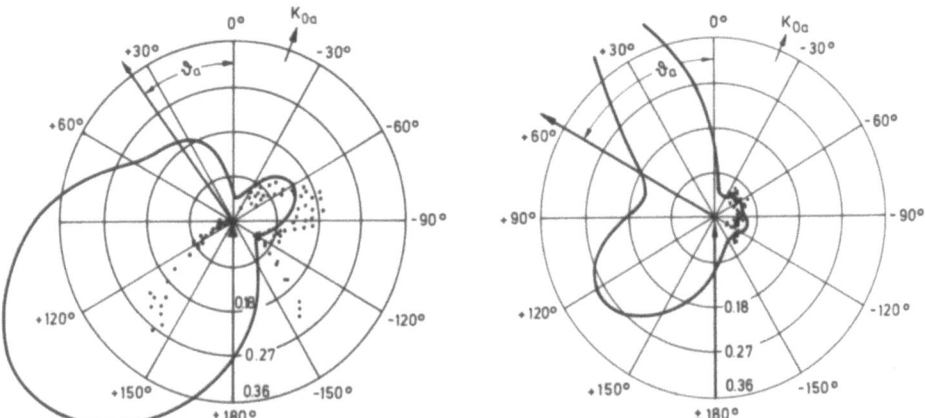

Figure 10. Same as for Figure 9, but in these measurements the energies of both outgoing electrons are equal. $E_0 = 35$ eV, $E_a = 5.3$ eV, $E_b = 5.3$ eV. Left: $\vartheta_a = 36°$. Right: $\vartheta_a = 60°$.

contrast to the predictions of the theory the experimental cross sections increase with increasing ϑ_a in the measured angular region.

Figure 10 shows two polar plots for $\vartheta_a = 36°$ and $60°$. In this case both outgoing electrons have the same energy $E_a = E_b$. The behavior of the measured as well as the calculated angular distributions is comparable to that in Figure 9. The dominant features are the broad backward peak and the pronounced peak with its maximum between $\vartheta_b = -80°$ and $\vartheta_b = -90°$ for $\vartheta_a = 36°$ and the nearly structureless and

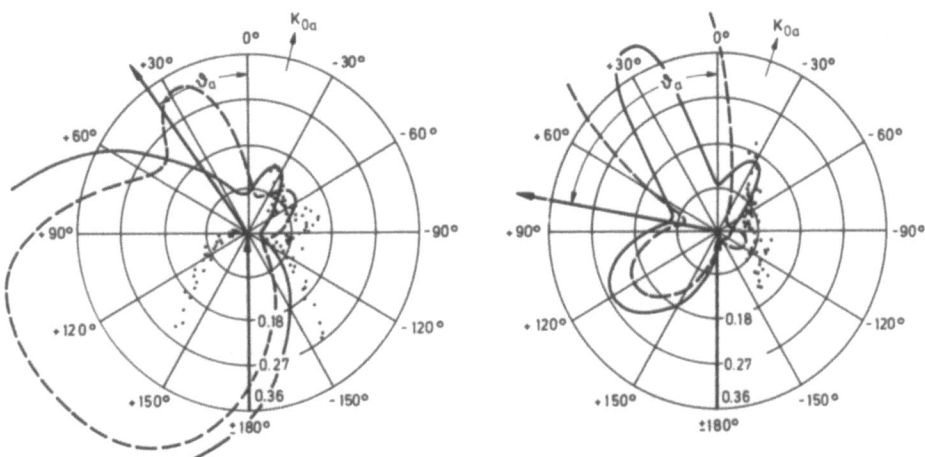

Figure 11. Coincidence cross section of helium as a function of the scattering angle ϑ_b. The ionization electron with lower energy has been detected at the fixed scattering angles $\vartheta_a = 36°$ (left) and $80°$ (right). The solid-line curve shows the calculated data of Geltman. In this approximation the outgoing electrons are described by Coulomb waves with the effective charges $Z_a = 1.6875$ and $Z_b = 2$. The dashed curves are obtained by exchanging these effective charges. $E_0 = 35$ eV, $E_a = 3.5$ eV, $E_b = 7$ eV.

constant angular distributions at $\vartheta_a = 60°$. The results of the Coulomb-projected Born approximation are almost independent of the energy partition of the excess energy $E_0 - $ IP between the two ionization electrons.

Until now we have discussed results for $E_a > E_b$ and $E_a = E_b$. In Figure 11 the slower of the outgoing electrons has been detected at the fixed scattering angle ϑ_a. The measured angular distribution at $\vartheta_a = 36°$ consists of three clearly visible peaks. The strong rise of the triple differential cross section in forward direction can already be recognized at this relatively small scattering angle ϑ_a whereas for $E_a > E_b$ this phenomenon can be found at $\vartheta_a \geq 90°$. At $\vartheta_a = 80°$ the coincidence cross section shows the expected behavior as for $E_a > E_b$ and for $E_a = E_b$ at comparable scattering angles ϑ_a.

In these two diagrams the full curve represents the calculated data with the effective charges $\lambda_a = 1.6875$ and $\lambda_b = 2$ seen by the electrons with the energies $E_a = 3.5$ eV and $E_b = 7$ eV, respectively. The data described by the dashed curve are obtained by exchanging the effective charges. The difference between the calculated data arising solely from the exchange of the values of the effective charges is quite remarkable. In both diagrams the number of peaks changes for different effective charges and the relative maxima of the calculated triple differential cross section are shifted by more than $30°$ except for the backward peak.

The measurements demonstrate that for fixed ϑ_a the angular dependences hardly vary with the partition of the excess energy $E_0 - $ IP between the emerging electrons. The threshold theories[16,27] predict such a behavior, namely, that the cross section does not depend on the ratio E_a/E_b. Our experimental test of this question is illustrated in Figure 12.

For the sake of clarity the dots have been replaced by curves representing the averaged data. The shape of each angular dependence at fixed $\vartheta_a = 36°$ is nearly the same and independent of the energies of the two emerging electrons. Also the angular distributions with different energies E_a and E_b do not differ in their magnitude.

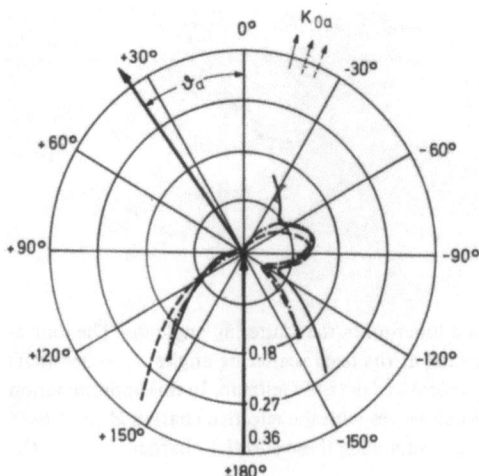

Figure 12. Measured angular distributions of the triple differential cross section of helium for three different ratios E_a/E_b. The electron a has always been detected at $\vartheta_a = 36°$. The measured dots have been replaced by curves representing the averaged data. The inserted error bar on the solid-line curve gives the estimated statistic uncertainties of the three curves. $E_0 = 35$ eV, $\theta_a = 36°$. Dashed line: $E_a = 7$ eV, $E_b = 3.5$ eV. Dash–dot line: $E_a = 5.3$ eV, $E_b = 5.3$ eV. Solid line: $E_a = 3.5$ eV, $E_b = 7$ eV.

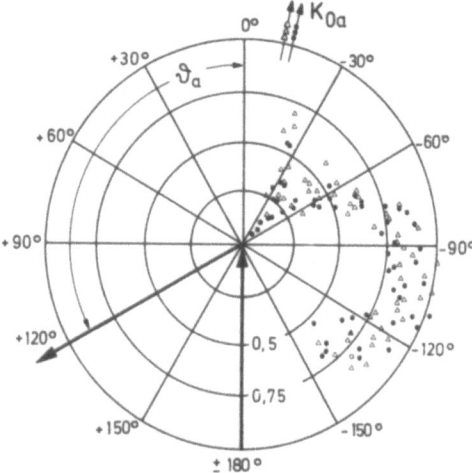

Figure 13. Measurements of the angular dependence of the triple differential cross section of helium at the collision energy $E_0 = 30.5$ eV. The solid circles and the open triangles represent the data for two different energy pairs E_a and E_b at the same scattering angle $\vartheta_a = 120°$. Solid circle: $E_a = 3$ eV, $E_b = 3$ eV. Open triangle: $E_a = 2$ eV, $E_b = 4$ eV.

For a primary energy of 30.5 eV the angular correlations are even more independent of the energy ratio E_a/E_b. This can be seen in Figure 13, which shows two measurements for the scattering angle $\vartheta_a = 120°$. In one case the energies of both outgoing electrons are equal (full circles), in the other case the electron b has twice the energy of electron a (open triangles). There is no essentially new behavior of the angular dependences in comparison to those at $E_0 = 35$ eV. But even for this rather low collision energy the other predictions of the threshold theories, namely, about the angular correlation, are not yet fulfilled.

Without question, the measurements at 35 eV collision energy show that the angular dependences of the triple differential cross section can no longer be understood in terms of binary encounter or recoil processes as for higher impact energies. Rapid variations of the angular distributions with changing scattering angle ϑ_a indicate interferences between the amplitudes of direct scattering and exchange scattering. (There are two different exchange amplitudes in the three-electron problem). The interference is possible, because in this energy range the different scattering amplitudes have similar magnitudes. Another low-energy phenomenon is that the momentum transfer $\mathbf{K}_{0a}$ loses its importance as a means of classification. This is due to the fact that neither of the exchange amplitudes possesses the momentum transfer vector as a symmetry axis. Therefore the angular distribution of the triple differential cross section for fixed ϑ_a, E_a, and E_b is no longer approximately symmetric with respect to the momentum transfer $\mathbf{K}_{0a}$.

4. Autoionization

After having investigated direct or nonresonant ionization processes over a wide energy range, it was interesting to investigate an autoionizing state in an electron–electron coincidence experiment. It is clear that the expected results will be very much connected with electron–photon coincidence measurements, since in both cases the decay of an aligned state is examined. The results of the electron–electron correlation measurements are somewhat more difficult to interpret because the resonant scattering amplitude interferes with that of the direct ionization.

Autoionizing states can be excited by different primary reactions, for example excitation by photons, protons, etc. If the excitation is done by colliding electrons in most experiments performed up to now[28–30] either the scattered or the emitted electrons are observed. In both cases all information concerning the second outgoing electron is lost. If only the scattered electron is registered, it is possible to determine the momentum transfer of the primary electron to the atom and therefore to know the quantization axis of the excited autoionizing state; but one does not know the angular distribution of the emitted electrons and cannot find out how the direct and the resonant scattering amplitudes interfere. If only the energy and angular dependences of the emitted electrons are observed, the measured cross section is an integration over all orientations of the excited atom and all magnetic sublevels m. One can obtain all possible information by the observation of the momenta of both electrons in the final state.

Because of limited intensity we could not work with excellent energy resolution. The overall energy resolution of our apparatus was approximately 300 meV. Therefore, we looked for a relatively isolated resonance with a fairly large natural linewidth. We decided to examine the autoionizing state $3s3p^64p$ of argon at quite high collision energy, namely, 250 eV, with the relatively small momentum transfer of $0.35a_0^{-1}$, which corresponds to a scattering angle $\vartheta_a = 3.5°$. For these scattering parameters the electrons behave almost like photons. Therefore the optical transition rules should essentially be valid. The excitation energy of this state is 26.6 eV, its natural linewidth approximately 80 meV.

This autoionizing state is formed by a one-electron excitation out of the $3s$ shell into the p orbitals. The excited state is split into a singlet and a triplet state. Madden, Ederer, and Codling[31] proved in 1969 that excitations of $3s$ electrons in argon, in contrast to those of $3p$ electrons, obey the LS coupling rules. Therefore, except for a small spin–flip contribution, essentially only singlet states can be excited. On the other hand, the ionic ground state is split into a $^2P_{3/2}$ and a $^2P_{1/2}$ state. In our experiment all electrons are analyzed with respect to their energies and we have chosen the energies in such a way that the ions after the collision are mostly in the lower $^2P_{3/2}$ state. Figure 14 shows the angular distribution of the pure direct ionization taken 200 meV away from the excitation energy of the resonance E_{res}.

The measured angular correlation shows all essential characteristics for high-energy electron impact ionization with strong binary and recoil peaks just opposite to each other and with deep minima in between. The symmetry axis is shifted from the direction of the momentum transfer axis to larger angles by about 10°.

At each of 11 angles (indicated in Figure 14 by small arrows) energy dependence of the coincidence cross section has been measured. Four of these energy dependences are given in Figure 15. To these experimental data we have fitted symmetric

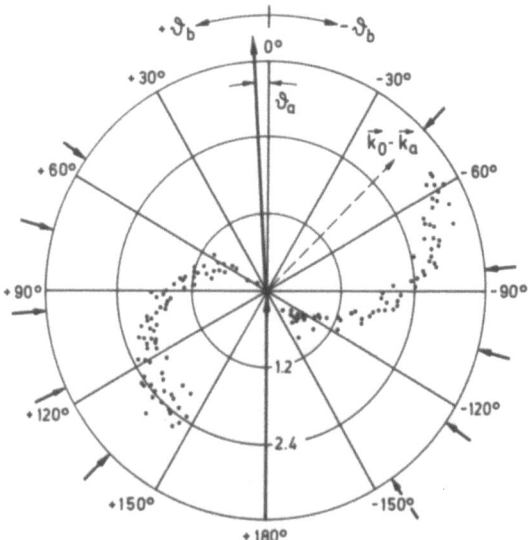

Figure 14. Angular dependence of the triple differential cross section for the ionization of argon. $E_0 = 250$ eV, $E_a = 223.2$ eV, $E_b = 11$ eV, $\vartheta_a = 3.5°$.

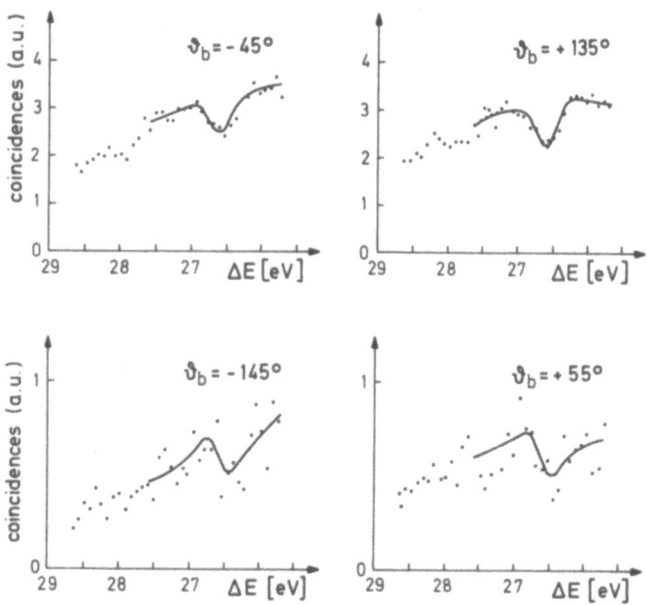

Figure 15. Energy dependences of the coincidence cross section measured at a collision energy of 250 eV as a function of the energy loss $\Delta E = E_0 - E_a$. The faster ionization electron has always been detected at the scattering angle $\vartheta_a = 3.5°$. The different scattering angles ϑ_b of the ejected electrons are given in the diagram.

and asymmetric Breit–Wigner profiles:

$$\sigma(\varepsilon) = \frac{b}{(1 + \varepsilon^2)} + \frac{a}{(1 + \varepsilon^2)} + \sigma_c, \qquad \varepsilon = \frac{E - E_{\text{res}}}{\Gamma/2}$$

with half-widths of 80 meV convoluted with a Gaussian shape of 300 meV half-width. The background of the direct ionization σ_c has been approximated by second-order polynomials. The fitted curves are shown by solid lines.

The energy dependence of the ionization cross section in the region of the autoionizing state can be described approximately by a coherent superposition of a nonresonant and a resonant amplitude with constant parameters, both on and off resonance. The phase of the resonating amplitude with respect to the direct scattering amplitude varies from η_{10} to $\eta_{10} + \pi$ with $\eta_1 = \eta_{10} + \arctan \varepsilon$. At the angles where the cross section of the direct ionization is large the autoionizing state causes a symmetric dip in the energy dependence (examples at $\vartheta_b = -45°$ and $\vartheta_b = 135°$ can be seen in Figure 15) whereas in the angular regions of the minima between binary and recoil peak the resonance structure is nearly totally asymmetric ($\vartheta_b = -125°$ and $\vartheta_b = 55°$). That means that off resonance the two amplitudes are nearly parallel to each other in the angular region of the momentum transfer direction ($\vartheta_b = -45°$ and $\vartheta_b = 135°$) and almost perpendicular to each other for $\vartheta_b \sim -135°$ and $\vartheta_b \sim 45°$.

The aligned $3s3p^64p$ state is a P state and therefore the angular dependence relative to the momentum transfer direction is $c_{1,-1}Y_{1,-1} + c_{1,0}Y_{1,0} + c_{1,1}Y_{1,1}$. The terms with $Y_{1,-1}$ and $Y_{1,1}$ are necessary to include transitions with $m = \pm 1$. The coefficients $c_{1,-1}$ and $c_{1,1}$ are equal because of the symmetry with respect to the scattering plane. The angular distribution of $|f_{\text{res}}|^2$ is shown in Figure 16 by tri-

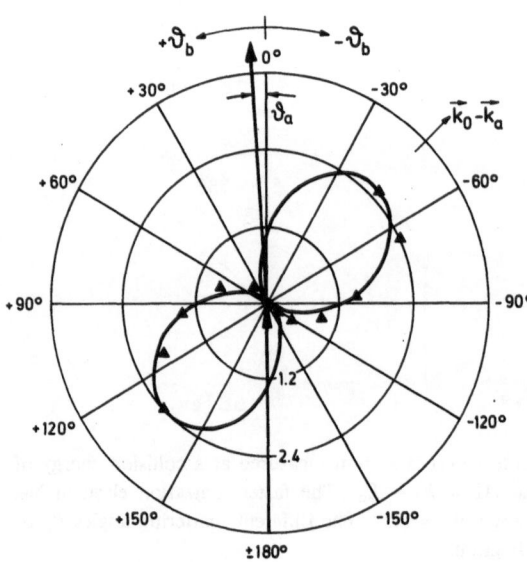

Figure 16. Angular dependence of the square of the resonant amplitude which describes the ionization via the aligned autoionization state $3s3p^64p$ of argon. The solid-line curve shows a pure $\cos^2 \vartheta$ distribution with a symmetry axis parallel to the direction of the momentum transfer $K_{oa} = k_o - k_a$. $E_0 = 250$ eV, $E_a = 223.2$ eV, $E_b = 10.8$ eV, $\vartheta_a = 3.5°$.

angular dots. The full curve represents a pure $\cos^2 \vartheta$ distribution with a symmetry axis parallel to the momentum transfer axis. The comparison of this distribution $[Y_{10}(\vartheta, \varphi)]$ with the measured data proves that one single partial wave describes the angular dependence within the experimental error. The cross section $|f_{res}|^2$ is very small perpendicular to the momentum transfer axis. Therefore we are quite sure that all contributions from the sublevels $m = +1$ and $m = -1$, which should be equally populated, are negligible for these scattering parameters. That means the excitation process for 250-eV electrons with a momentum transfer of $0.35a_0^{-1}$ can be handled by the Born approximation.

ACKNOWLEDGMENTS

The experimental work has been carried out with the financial support of the Deutsche Forschungsgemeinschaft. We would like to thank Dr. D.A.L. Paul for reading the manuscript.

References

1. H. S. W. Massey and E. H. S. Burhop, *Electronic and Ionic Impact Phenomena*, Vol. 1, Oxford University Press, Oxford (1969).
2. L. J. Kieffer and G. H. Dunn, *Rev. Mod. Phys.* **38**, 1–35 (1966).
3. H. Ehrhardt, M. Schulz, T. Tekaat, and K. Willmann, *Phys. Rev. Lett.* **22**, 89–92 (1969).
4. U. Amaldi, A. Egidi, R. Marconero, and G. Pizella, *Rev. Sci. Instrum.* **40**, 1001–1004 (1969).
5. K. Jung, E. Schubert, H. Ehrhardt, and D. A. L. Paul, *J. Phys. B* **9**, 75–87 (1976).
6. E. C. Beaty, K. H. Hesselbacher, S. P. Hong, and J. H. Moore, *Phys. Rev. A* **17**, 1592–1599 (1978).
7. E. C. Beaty, K. H. Hesselbacher, S. P. Hong, and J. H. Moore, in *Proceedings of the Xth ICPEAC Paris*, pp. 374–375, Commissariat a l'energie atomique, Paris (1977).
8. E. Weigold and I. E. McCarthy, *Adv. At. Mol. Phys.* **14**, 127–179 (1978).
9. S. T. Hood, A. Hamnett, and C. E. Brion, *J. Electron Spectrosc. Relat. Phenom.* **11**, 205–224 (1977).
10. R. Camilloni, A. Giardini-Guidoni, I. E. McCarthy, and G. Stefani, *Phys. Rev. A* **17**, 1634–1671 (1978).
11. M. A. Coplan, J. H. Moore, and J. A. Tossell, *J. Chem. Phys.* **68**, 329–330 (1978).
12. M. J. van der Wiel and C. E. Brion, *J. Electron Spectrosc. Relat. Phenom.* **1**, 439–441 (1974).
13. K. H. Tan and C. E. Brion, *J. Electron Spectrosc. Relat. Phenom.* **13**, 77–84 (1978).
14. S. Cvejanović and F. H. Read, *J. Phys. B* **7**, 1841–1852 (1974).
15. G. H. Wannier, *Phys. Rev.* **90**, 817–825 (1953).
16. A. R. P. Rau, *J. Phys. B* **9**, L283–L288 (1976).
17. R. Peterkop, *J. Phys. B* **4**, 513–521 (1971).
18. H. Ehrhardt, K. H. Hesselbacher, K. Jung, M. Schulz, and K. Willmann, *J. Phys. B* **5**, 2107–2116 (1972).
19. L. Vriens, in *Case Studies in Atomic Collision Physics*, E. W. McDaniel and M. R. C. McDowell, eds., pp. 337–398, North Holland, Amsterdam (1969).
20. A. E. Glassgold and G. Ialongo, *Phys. Rev.* **175**, 151–159 (1968).
21. V. Ya.Veldre and I. J. Vinkaln, in *Atomic Collisions*, V. Ya. Veldre, R. Ya. Damburg, and R. K. Peterkop, eds., pp. 114–118, Butterworth, London (1966).

22. E. C. Beaty, K. H. Hesselbacher, S. P. Hong, and J. H. Moore, *J. Phys. B* **10**, 611–620 (1977).
23. E. W. Knapp and M. Schulz, *J. Phys. B* **7**, 1875–1890 (1974).
24. A. Ugbabe, E. Weigold, and I. E. McCarthy, *Phys. Rev. A* **11**, 576–585 (1975).
25. S. Geltman, *J. Phys. B* **9**, 75–87 (1974).
26. E. Schubert, A. Schuck, K. Jung, and S. Geltman, *J. Phys. B* **12**, 967–978 (1979).
27. I. Vinkalns and M. Gailitis, in *Proceedings of the VIIth ICPEAC Leningrad*, pp. 648–650, Nauka, Leningrad (1967).
28. J. A. Simpson, G. E. Chamberlain, and S. R. Mielczarek, *Phys. Rev.* **139**, A1039–A1041 (1965).
29. L. Sanche and G. J. Schulz, *Phys. Rev. A* **5**, 1672–1683 (1972).
30. J. Fryar and J. W. McConkey, *J. Phys. B* **9**, 619–629 (1976).
31. R. P. Madden, D. L. Ederer, and K. Codling, *Phys. Rev.* **177**, 136–151 (1969).

4

Coincidence Investigations
of Electron–Hydrogen Collisions

ERICH WEIGOLD

Electron–photon and electron–electron coincidence experiments are reported on, respectively, electron impact excitation of H(2p) and electron impact ionization of atomic hydrogen. The electron–photon angular correlations (at 40, 54.4, and 100 eV and $\theta_e = 10°$, 15°, and 20°) are compared with calculations based on various approximations (such as hybrid close coupling, DWPO, or DWBA). None of the theories appears to be completely satisfactory over the entire range of energy and angles examined. The H(e, 2e)H$^+$ cross sections are similarly compared with various calculations. The results show the importance of correlations in the exit channel.

1. Introduction

Electron impact excitation and ionization of atomic hydrogen are interesting not only because of the fundamental quantum mechanical scattering processes involved, but also because an improved understanding of these fundamental processes should lead to better descriptions of other excitation and ionization processes. Since exact quantum mechanical solutions cannot be obtained for scattering problems involving more than two particles interacting via the long-range electromagnetic force, various approximation schemes have been adopted. These are best tested by applying them to the simplest possible system. Further, since the prime products of the calculations are scattering amplitudes, it is most desirable to make measurements that are as closely related to them as possible. In order to provide detailed

ERICH WEIGOLD ● Institute of Atomic Studies, Flinders University, Bedford Park, South Australia 5042.

tests it is obviously also necessary that the measured cross sections should involve as little integration as possible over observables and kinematical variables (e.g., spin, momenta, angles, etc.). For instance, measurement of differential cross sections avoids the loss of information due to integration over the scattering angle, and use of spin polarized targets and beams can give rise to the separation of direct and exchange amplitudes.

The electron–photon coincidence technique, first used in atomic physics by the Stirling group,[1] gives information on excitation amplitudes and in some cases on their relative quantum mechanical phases. The simplest example of an optically allowed transition is the case of the excitation of the H(2p) from the ground state, H(1s). In this process the scattered electron loses 10.2 eV in energy and a photon of wavelength 121.6 nm (Lyα) is subsequently emitted. In the absence of magnetic and electric fields, excitation of the metastable 2s state can be separated from 2p excitations through its much longer lifetime.

In electron impact ionization, on the other hand, the most detailed information is provided by the (e, 2e) reaction.[2-4] Since the collision kinematics are completely specified for single ionization in this reaction, it provides a fundamental testing ground for the different theoretical approximations to the ionization problem. The cross section is a measure of the probability that an incident electron of energy E_0 will produce on collision two final-state electrons of energies E_A and E_B emitted into the differential elements of solid angle Ω_A and Ω_B. If the polarizations of the particles are not measured, this cross section is in general a sixfold differential cross section. In the case of atomic hydrogen, where only one final ion state exists, the cross section is five dimensional. Whereas other ionization problems are often complicated by the use of different bound state target and ion wave functions and potentials, these are known for atomic hydrogen, therefore permitting direct comparison of the approximations.

2. Experiment

The experimental arrangements are discussed fully elsewhere[5-7] and need not be repeated in detail here. The atomic hydrogen beam is provided by a dc discharge (Wood's) tube, which operates with long-term stability and reliability.[7] Dissociation is typically 80% and the atomic density is of the order of 2×10^{18} m^{-3} in the collision region. The discharge tube is shown schematically in Figure 1.

Two cylindrical mirror analyzers are used to determine the energies and angles of the two emitted electrons in the H(e, 2e)H$^+$ measurements. In the (e, e'γ) experiments one of the analyzers is replaced by a photon detector consisting of a channeltron with a lithium fluoride window (see Figure 1). The detectors are mounted on two concentric coplanar turntables, and are rotated independently by stepping motors controlled by a PDP-11 computer. Standard fast timing electronics is employed[8] and the experiments are controlled by the computer.

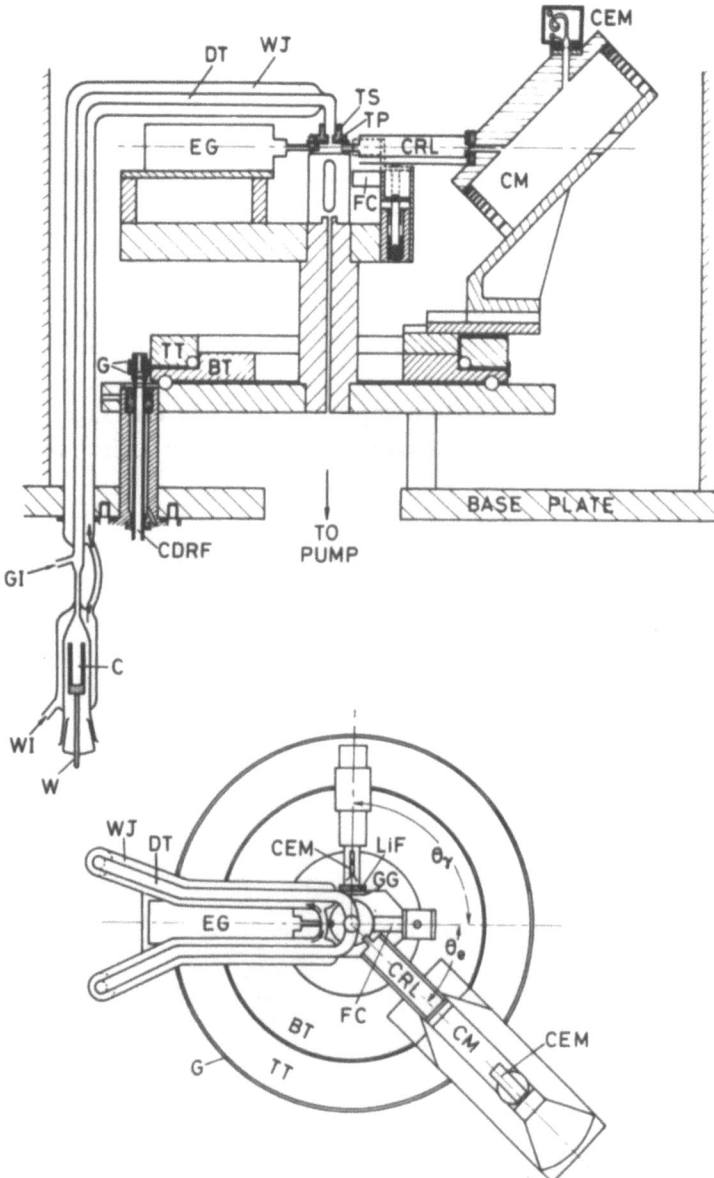

Figure 1. Side elevation and plan of the electron–photon scattering apparatus including the atomic hydrogen source. CEM, channel electron multiplier; CM, cylindrical mirror; CRL, collimator and retarding lens; FC, retractable Faraday cup; TP, tantalum aperture plate; TS, Teflon socket; EG, electron gun; WJ, water jacket; DT, discharge tube; WI, water inlet; GI, gas inlet; C, hollow aluminum cathode; W, tungsten feedthrough; CDRF, concentric double rotating feedthrough; TT, top turntable; BT, bottom turntable; G, driving gears; LiF, lithium fluoride window; GG, grounded grid.

3. Results and Discussion

3.1. H(e, 2e)H+

In the H(e, $2e$)H+ reaction an incident electron of momentum $\mathbf{k}_0$ ionizes a target hydrogen atom which is essentially stationary. The two final continuum electrons have energies and momenta E_A, $\mathbf{k}_A$ and E_B, $\mathbf{k}_B$, respectively. When the emitted electron energies are not equal we follow the convention of labeling the faster ("scattered") electron by subscript A.

Some of the angular correlations measured by Weigold et al.[5] are shown in Figures 2 and 3. The data are compared with differential cross sections calculated in a number of approximations.

The first is the plane-wave Born approximation with exchange (PWBE), which continues to provide a popular reference calculation for ionization calculations.

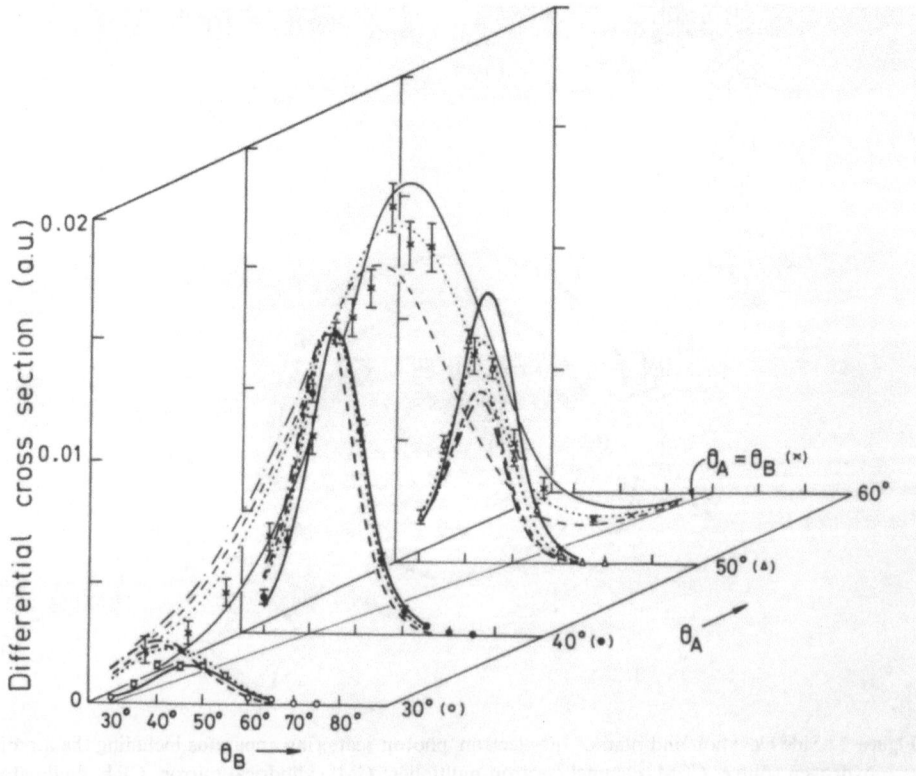

Figure 2. The H(e, $2e$)H+ differential cross section (in units of $a_0{}^2/2$Ry sr^2) at $E_0 = 413.6$ eV, $E_A = E_B = 200$ eV, $\phi_A - \phi_B = \pi$, plotted as a function of θ_A and θ_B. [5] The $\theta_A = 30°$, $40°$, $50°$, and θ_B experimental points are shown by ○, ●, △, and ×, respectively. The calculated curves,[5] which are discussed in the text, are ———, DWIA-1; ···, DWIA-2 ($\times 0.87$); ———, PWIA ($\times 0.66$); —·—·, PWBE ($\times 0.35$). The data and calculations are normalized to the peak height of the $\theta_A = 40°$ DWIA-1 differential cross section. When the two plane-wave curves coincide only the PWIA is shown.

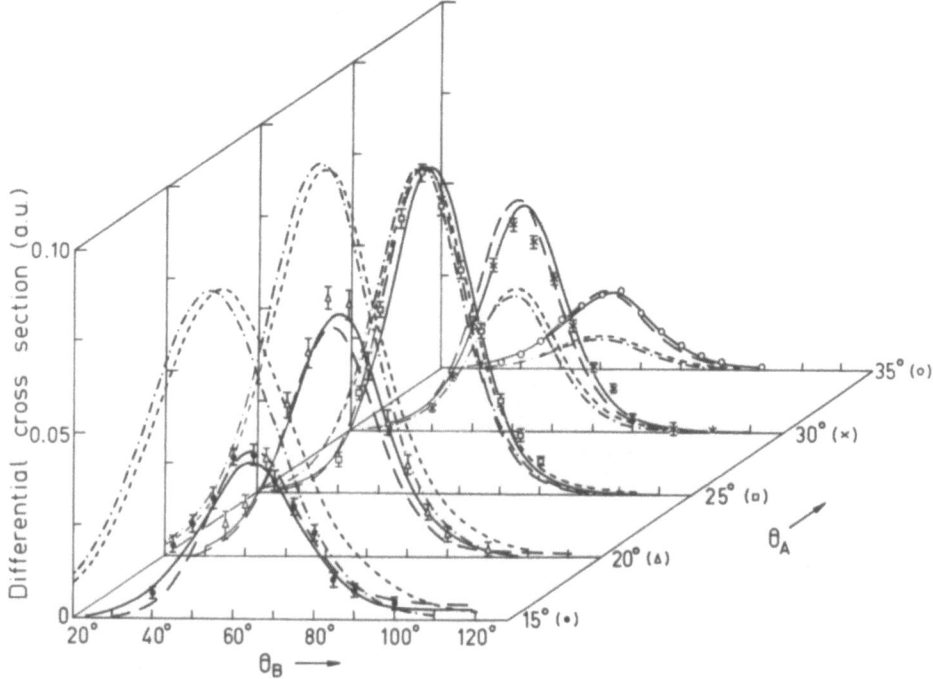

Figure 3. The H(e, $2e$)H$^+$ differential cross section (in units of $a_0^2/2\text{Ry sr}^2$) at $E_0 = 250$ eV, $E_A = 186.4$ eV, $E_B = 50$ eV, $\phi_A - \phi_B = \pi$, plotted as a function of θ_A and θ_B. The calculated cross sections are ——, DWIA-1; — —, DWIA-4 ($\times 1.38$); – – –, PWIA ($\times 0.54$); ·—·—, PWBE ($\times 0.25$). The data and calculations[5] are normalized to the peak height of the $\theta_A = 25°$ DWIA-1 differential cross section.

A much more difficult calculation is that using the half off-shell distorted-wave impulse approximation (DWIA) of McCarthy and Weigold.[3] On making the factorization approximation (which is exact in the eikonal or plane wave limits) the distorted-wave cross section is given by

$$\frac{d^5\sigma}{dE_A\,d\Omega_A\,d\Omega_B} = (2\pi)^4\,\frac{k_A k_B}{k_0}\,|T_M|^2\,\frac{1}{(2\pi)^3}\,|\langle\chi_A^{(-)}(\mathbf{k}_A)\chi_B^{(-)}(\mathbf{k}_B)\,|\,\psi_{1s}(\mathbf{r})\chi_0^{+}(\mathbf{k}_0)\rangle|^2$$

where $|T_M|^2$ is the half off-shell two electron (Mott) scattering cross section and $\chi^{(\pm)}$ are the optical model wave functions describing elastic scattering in the appropriate two-body subsystems. They are computed directly from optical model potentials by solving the distorted-wave equations for elastic scattering in the relevant channel in partial wave form and performing the radial integral and partial wave sums explicitly.

The optical model used to describe the incident electrons was that of Vanderpoorten.[8] This local energy-dependent central potential includes not only the static potential, but also polarization and exchange potentials, and an imaginary potential describing the loss of electrons from the elastic channel.

In the three-body final state the particles all interact via the long-range Coulomb force. In the spirit of DWIA the effective Coulomb charges Z_A and Z_B for the outgoing Coulomb waves should be simply unity. This is the DWIA-1 model in Figures 2 and 3. If, however, one wishes to remove the logarithmic singularity in asymptotic phase of the $(e, 2e)$ amplitude,[5,9,10] a constraint is imposed on the effective charges, and the effective charges then vary with angle. This is the DWIA-2 approximation in Figure 2.

Further, if $k_A \gg k_B$, the faster emitted electron feels the influence of the slower electron and proton diminish rapidly as it leaves the interaction region. The slower electron essentially screens the proton, and $Z_A \approx 0$ at large distances, the faster electron "seeing" effectively a neutral atom (in a continuum orbital). Therefore a further approximation was investigated (DWIA-4 in Figure 3) in which the fast out-going electron was treated in an analogous way to the incoming electron, and for the slow electron the distorted wave was taken to be a Coulomb wave ($Z_B = -1$).

Finally calculations were also carried out using plane waves (PWIA), an approximation which obviously still includes the effects of exchange.

Figure 2 shows the data at $E_0 = 413.6$ eV, $E_A = E_B = 200$ eV and the various calculated cross sections all normalized to the peak height of the DWIA-1 differential cross section at $\theta_A = 40°$. Clearly the DWIA-1 model gives the best overall description of the data, the plane-wave approximations being particularly poor even at this energy. It is interesting to note that although the DWIA-1 model is best at small angles (small momentum transfer $\mathbf{k} = \mathbf{k}_0 - \mathbf{k}_A$) the DWIA-2 model does significantly better at higher k. In other words the magnitude of the effective charges should decrease with increasing k, in contrast with the trend required by the asymptotic constraint (DWIA-2), which gives small values for Z_A and Z_B at forward angles.

Figure 3 shows the data and calculations for $E_0 = 250$ eV, $E_A = 186.4$ eV, $E_B = 50$ eV, $\phi_A - \phi_B = \pi$ normalized to the DWIA-1 peak height at $\theta_A = 25°$. The two plane-wave approximations are completely inadequate, whereas the two distorted-wave models give an excellent description of the data.

These results, as well as results at other energies,[5] show that the model with effective charges consistent with the asymptotic constraints (DWIA-2) does rather less well than models that may have the wrong asymptotic conditions but more nearly reflect the conditions in the region where all the particles are interacting strongly.

3.2. $H(e, e'\gamma)H$

Electron–photon angular correlations have been measured[6] at electron scattering angles of 10°, 15°, and 20° for incident energies of 40, 54.4, and 100 eV and at 10° for 200-eV electrons. These measurements are presently being extended to larger angles. Values of the parameters $\lambda = \sigma_0/\sigma$ and $R = \mathrm{Re}\langle a_0 a_1 \rangle/\sigma$ are extracted from the angular correlations and compared with various calculations.[6] $\sigma = \sigma_0 + 2\sigma_1$ $= \langle a_0 a_0 \rangle + 2\langle a_1 a_1 \rangle$, where the subscript 0 denotes the $m_l = 0$ substate and the subscript 1 the $m_l = \pm 1$ substate.

The results are summarized in Table 1, which gives the best fit values of λ and R

Table 1. Values of the Parameters λ and $\mathrm{Re}\langle a_0 a_1\rangle/\sigma$ Obtained from Various Calculations and the χ^2 Probabilities of the Resulting Fits to the Experimental Angular Correlations[6] at Various Incident Energies, Scattering Angles (θ_e), and Momentum Transfers $(k = |\mathbf{k}_0 - \mathbf{k}_A|)$

E_0 (eV) θ_e $k(a_0^{-1})$	Best fit	Born	DWBA[11]	DWPO[a(12)]	UGA[a(13)]	CPTM[14]	CPB[a(15)]	MCC[16]
40	0.450	0.500						
10°	0.352^b	0.354						
0.364	0.35	0.26						
40	0.289	0.357						
15°	0.320^b	0.339						
0.478	0.062	0.028						
40	0.275	0.291						
20°	0.309	0.321						
0.601	0.46	0.41						
54.4	0.242	0.340	0.325	0.38^c	0.29^c	0.664		0.288
10°	0.274	0.335	0.329	0.34	0.32	0.26		0.304
0.385	0.35	0.005	0.008	0.0001	0.043	d		0.12
54.4	0.280	0.235	0.216	0.30^c	0.18^c	0.39		0.182
15°	0.230	0.300	0.284	0.32	0.27	0.20		0.211
0.533	0.007	d	0.0001	d	0.0002	0.0004		0.0005
54.4	0.281	0.197	0.178	0.28^c	0.09^c	0.18		0.218
20°	0.249	0.282	0.253	0.30	0.19	0.07		0.123
0.688	0.16	0.045	0.04	0.03	d	d		d
100	0.194	0.141	0.134	0.20	0.11	0.19	0.01	
10°	0.175	0.246	0.239	0.26	0.22	0.15	0.05	
0.482	0.16	d	0.0002	d	0.0025	0.074	d	
100	0.183	0.106	0.099	0.17	0.06	0.09	0.01	
15°	0.148	0.218	0.202	0.25	0.17	0.02	0.05	
0.704	0.14	0.007	0.02	0.0001	0.03	d	d	
100	0.233	0.102	0.102	0.20	0.005	0.073	0.06	
20°	0.173	0.214	0.185	0.27	0.05	0.07	0.16	
0.928	0.66	0.52	0.55	0.37	0.013	0.09	0.40	
200	0.061	0.055	0.053	0.09	0.03	0.051		
10°	0.157	0.161	0.154	0.20	0.12	0.03		
0.667	0.47	0.46	0.46	0.25	0.27	0.0002		

[a] Estimated from published graphs.
[b] Constrained.
[c] Calculated at 50 eV incident energy.
[d] $P(\chi^2) < 10^{-5}$.

with corresponding probabilities together with the probabilities obtained using various theoretical values of λ and R. At 40 eV, where no other theories are available, the Born approximation gives a surprisingly adequate description of the data. At 54.4 and 100 eV the theories diverge from each other and from the data. None of the theories are in total accord with the experimental data. The Coulomb-projected Born[15] (CPB) and classical path T-matrix[14] (CPTM) approximations are in strong disagreement with the data.

4. Summary

These coincidence experiments show that our understanding of the simple three-body atomic system is far from complete. In the ionization process, the DWIA gives a very good description of the data, but final-state correlation effects between the three bodies are obviously important and must be taken into account more accurately. In the excitation process, none of the approximations proposed are entirely satisfactory, and further theoretical developments are required.

ACKNOWLEDGMENTS

This work was supported by the Australian Research Grants Committee. The author also wishes to acknowledge with thanks the contributions of his co-workers: A. J. Dixon, S. T. Hood, I. Fuss, and C. J. Noble.

References

1. M. Eminyan, K. B. MacAdam, J. Slevin, and H. Kleinpoppen, *Phys. Rev. Lett.* **31**, 576–578 (1973); and *J. Phys. B* **7**, 1519–1542 (1974).
2. H. Ehrhardt, K. H. Hesselbacher, K. Jung, and K. Willmann, *Case Stud. At. Phys.* **2**, 159–208 (1971).
3. I. E. McCarthy and E. Weigold, *Phys. Rep.* **27C**, 275–371 (1976).
4. E. Weigold, A. Ugbabe, and P. J. O. Teubner, *Phys. Rev. Lett.* **35**, 209–212 (1975).
5. E. Weigold, C. J. Noble, S. T. Hood, and I. Fuss, *J. Phys. B* **12**, 291–313 (1979); E. Weigold, S. T. Hood, I. Fuss, and A. J. Dixon, *J. Phys. B* **10**, L623–L627 (1977).
6. A. J. Dixon, S. T. Hood, and E. Weigold, *Phys. Rev. Lett.* **40**, 1262–1266 (1978); S. T. Hood, E. Weigold, and A. J. Dixon, *J. Phys. B.* **12**, 631–648 (1979).
7. S. T. Hood, A. J. Dixon, and E. Weigold, *J. Phys. E* **11**, 948–954 (1978).
8. R. Vanderpoorten, *J. Phys. B* **8**, 926–939 (1975).
9. M. R. H. Rudge, *Rev. Mod. Phys.* **40**, 564–590 (1968).
10. M. Schultz, *J. Phys. B* **6**, 2580–2599 (1973).
11. R. V. Calhoun, D. H. Madison, and W. N. Shelton, *Phys. Rev. A* **14**, 1380–1387 (1976).
12. L. A. Morgan and M. R. C. McDowell, *J. Phys. B* **8**, 1073–1081 (1975).
13. J. N. Gau and J. H. Macek, *Phys. Rev. A* **12**, 1760–1770 (1975).
14. M. J. Roberts, *J. Phys. B* **10**, 2219–2228 (1977).
15. L. A. Morgan and A. D. Stauffer, *J. Phys. B* **8**, 2342–2346.
16. J. Callaway, M. R. C. McDowell, and L. A. Morgan, *J. Phys. B* **9**, 2043–2051 (1976).

5

Correlation in Multiple Escape near Threshold

H. KLAR

The many-electron Coulomb interaction is shown to have stationary configurations that control the escape of strongly correlated electrons near threshold. The Wannier theory has been rederived both classically and quantum mechanically. Particular attention has been paid to the spin of the escaping electrons. In the case of two electrons the ionization probability near threshold into the triplet channel is shown to be much smaller than into the singlet channel. The analysis of three escaping electrons (e.g., double ionization by electron impact) shows that the quartet channel is depressed with respect to the doublets.

1. Introduction

In the last few years increasing attention has been paid to the ionization of atoms by low-energy electron impact.[1] Such experiments near threshold enable a detailed analysis of strongly correlated electrons.

In the framework of classical mechanics Wannier[2] has investigated the threshold ionization of H. He showed by careful inspection of the classical equations of motion that the only subspace relevant for double escape is that of the configuration space in which both electrons are at equal distances from the ion and in opposite directions. This comes from the fact that the potential surface shows a saddle in this configuration. This is easily seen by introducing hyperspherical coordinates

$$R = (r_1{}^2 + r_2{}^2)^{1/2}, \quad \alpha = \arctan(r_1/r_2), \quad \theta = \arccos(\mathbf{r}_1, \mathbf{r}_2) \tag{1}$$

H. KLAR • Fakultät für Physik der Universität Freiburg, Hermann-Herder-Strasse 3, 78 Freiburg im Breisgau, West Germany.

in which the saddle point has the coordinates $\alpha = \pi/4$ and $\theta = \pi$. The potential in the case of hydrogen near this point has the expansion

$$V = 2^{1/2}\left[-\frac{3}{2} - \frac{11}{4}\left(\alpha - \frac{\pi}{4}\right)^2 + \frac{1}{16}(\theta - \pi)^2 + \cdots\right]/R \qquad (2)$$

Inspection of equation (2) shows a motion stable in θ but unstable in α at constant R. Double escape results from an increase of the R coordinate accompanied by limited deviations of θ from π and of α from $\pi/4$. These deviations remain limited because V is stationary in α and θ at constant R. The instability of this configuration with respect to deviations in α has the limited effect of decreasing the cross section near threshold. Uncorrelated electrons would escape with a cross section linear in the energy E, but Wannier[2] found the cross section rose more slowly, as $\sigma \propto E^{1.127}$.

This result has also been derived quantum mechanically[3] for ^1S states and is in excellent agreement with the experiment of Cvejanović and Read.[1] Double escape in other states than $^1S^{even}$ as well as the escape of more than two electrons has been investigated elsewhere.[4]

In Section 2 we present a rederivation of Wannier's theory. We shall use, however, a slightly different set of coordinates than Wannier in order to make the mathematics as transparent as possible. In Section 3 we give a quantum mechanical treatment of the double escape taking spin, parity, and angular momentum of the pair of electrons into account. Finally, Section 4 generalizes the theory to more than two escaping electrons.

Throughout this paper we use atomic units.

2. Review of Wannier's Theory

First of all we assume that the threshold behavior of two electrons is controlled by vanishing total orbital angular momentum. The treatment of two electrons and the ion needs then only three coordinates. The mass of the ion we assume to be infinite. These three coordinates may be chosen as in equation (1). We prefer, however, a different set of hyperspherical angles originally introduced by Smith,[5] in which the electron positions in the body-fixed frame are given by

$$\mathbf{r}_1 = R\begin{pmatrix} \cos\psi\cos\frac{1}{2}[\phi + \frac{3}{2}\pi] \\ \sin\psi\sin\frac{1}{2}[\phi + \frac{3}{2}\pi] \\ 0 \end{pmatrix}, \qquad \mathbf{r}_2 = R\begin{pmatrix} \cos\psi\cos\frac{1}{2}[\phi - \frac{3}{2}\pi] \\ \sin\psi\sin\frac{1}{2}[\phi - \frac{3}{2}\pi] \\ 0 \end{pmatrix} \qquad (3)$$

with $0 \le \psi \le \pi/4$ and $0 \le \phi \le 2\pi$.

One easily verifies $R = (r_1^2 + r_2^2)^{1/2}$. The angular coordinates ψ and ϕ have the property of making the tensor of inertia diagonal. The diagonal elements read

$$\theta_1 = R^2\sin^2\psi, \qquad \theta_2 = R^2\cos^2\psi, \qquad \theta_3 = R^2$$

The electron–ion distances are

$$r_1 = R[(1 + \sin \phi \cos 2\psi)/2]^{1/2}$$
$$r_2 = R[(1 - \sin \phi \cos 2\psi)/2]^{1/2}$$

and the electron–electron distance becomes

$$r_{12} = R(1 - \cos \phi \cos 2\psi)^{1/2}$$

Note that the potential energy

$$V = -z\left(\frac{1}{r_1} + \frac{1}{r_2}\right) + \frac{1}{r_{12}}$$

can be written in the form $V = C(\psi, \phi)/R$, where $C(\psi, \phi)$ is independent of R. The kinetic energy of the nonrotating system is simply given by

$$T = \tfrac{1}{2}\dot{R}^2 + \tfrac{1}{2}R^2(\dot{\psi}^2 + \tfrac{1}{4}\dot{\phi}^2)$$

The exact Lagrange equations of the second kind read

$$\ddot{R} = R\left(\dot{\psi}^2 + \frac{1}{4}\dot{\phi}^2\right) + \frac{1}{R^2} C(\psi, \phi) \tag{4a}$$

$$\frac{d}{dt}(R^2\dot{\psi}) = -\frac{1}{R}\frac{\partial C}{\partial \psi} \tag{4b}$$

$$\frac{1}{4}\frac{d}{dt}(R^2\dot{\phi}) = -\frac{1}{R}\frac{\partial C}{\partial \phi} \tag{4c}$$

Following Wannier we solve now equations (4) near the saddle defined by

$$\frac{\partial C}{\partial \psi} = \frac{\partial C}{\partial \phi} = 0$$

In our coordinates the saddle is located at $\psi = 0$ and $\phi = \pi$ corresponding to the configuration $r_1 = r_2$ and $\theta = \pi$. The potential has there the expansion

$$C(\psi, \phi) = -C_0 + C_1\psi^2 - C_2(\phi - \pi)^2 + \cdots \tag{5}$$

with

$$C_0 = (4z - 1)/2^{1/2}$$
$$C_1 = 1/[2(2^{1/2})] \tag{6}$$
$$C_2 = (12z - 1)/[8(2^{1/2})]$$

We linearize the equations of motion with respect to ψ and ϕ, and replace equation (4a) by the expression of the total energy E. At threshold ($E = 0$) this gives the simple equation

$$(\dot{R})^2 = 2C_0/R$$

which can be solved exactly,

$$R(t) = (9C_0 t^2/2)^{1/3} \tag{7}$$

In equations (4b) and (4c) we replace the time variable by R, equation (7), and find

$$R^2\psi'' + \tfrac{2}{3}R\psi' + (C_1/C_0)\psi = 0$$
$$R^2\phi'' + \tfrac{2}{3}R\phi' - (C_2/C_0)(\phi - \pi) = 0$$

The primes indicate here differentiation with respect to R. These equations have the solutions

$$\phi = \pi + R^{-1/4}(D_1 R^{-\mu} + D_2 R^{\mu}) \tag{8a}$$
$$\psi = D_3 R^{-1/4} \cos(\varrho \ln R + D_4) \tag{8b}$$

with the characteristic exponents

$$\mu = \frac{1}{4}\left(\frac{100z - 9}{4z - 1}\right)^{1/2}$$
$$\varrho = \frac{1}{4}\left(\frac{9 - 4z}{4z - 1}\right)^{1/2} \tag{9}$$

Note that deviations in ψ from the equilibrium $\psi = 0$ are stable, see equation (5). The solution equation (8b) remains, therefore, finite for large values of R. The integration constants D_3 and D_4 are arbitrary. The motion in ϕ around $\phi = \pi$ is unstable; solution (8a) remains finite only if $D_2 = 0$.

In order to derive the threshold law we must now continue the zero-energy solution to small but finite values of the energy. It appears intuitively reasonable that our solution derived above actually holds also for finite values of the energy. Then, however, the constant D_2 need not to be zero. A bound for D_2 at finite energy E follows essentially from the observation[2] that the coordinate R can appear only in the combination RE or R/C_0. Inspection of equation (8a) then yields

$$(D_2)_{\text{max}} \propto E^{\mu - 1/4} \tag{10}$$

which is a relation for double escape between the range of D_2 and the energy.

Finally, it is not difficult to see that the volume of phase space responsible for double escape is limited in the same way. Therefore, the cross section is expected to rise as

$$\sigma \propto E^{\mu - 1/4} \tag{11}$$

One would expect that at increasing ion charge z correlation becomes less and less important. Note that for $z \to \infty$ we actually fall back to a linear law, see equation (9) for $\mu(z)$, which is the result for uncorrelated electrons. The numerical value of the exponent for $z = 1$ is $\mu - \frac{1}{4} = 1.12689$. Concluding this section we remark that the threshold law is completely determined by the vanishing of the diverging orbit. Therefore, only one of the two characteristic numbers, equation (10), is relevant for the cross section, namely, μ.

3. Quantum Mechanical Treatment of Two Escaping Electrons

In this section we use the same coordinates as above. The Hamiltonian then reads[4]

$$H = -\frac{1}{2}\left(\frac{\partial^2}{\partial R^2} + \frac{5}{R}\frac{\partial}{\partial R} - \frac{\Lambda^2}{R^2}\right) + \frac{C(\psi, \phi)}{R} \tag{12}$$

where Λ^2, the so-called grant angular momentum restricted to zero total orbital angular momentum $(L = 0)$, is given by

$$\Lambda^2 = -\frac{1}{\sin 4\psi}\frac{\partial}{\partial\psi}\left(\sin 4\psi\frac{\partial}{\partial\psi}\right) - \frac{4}{\cos^2 2\psi}\frac{\partial^2}{\partial\phi^2} \tag{13}$$

We now solve the Schrödinger equation for total energy $E = 0$ in the subspace relevant for double escape, i.e., near the saddle $(\psi = 0, \phi = \pi)$ and large R.

To this end we expand the expression for Λ^2, equation (13),

$$\Lambda^2 = -\frac{\partial^2}{\partial\psi^2} - \frac{1}{\psi}\frac{\partial}{\partial\psi} - 4\frac{\partial^2}{\partial\phi^2}$$

and use the expansion equation (5) for the potential. The saddle structure of the potential suggests a wave function[3,4]

$$\Psi = R^{n-5/2}\exp[i(8R)^{1/2}Q(\psi, \phi)] \tag{14}$$

with

$$Q(\psi, \phi) = q_0 + q_1\psi^2 + q_2(\phi - \pi)^2 \tag{15}$$

We require equation (14) to be a zero-energy solution, $H\Psi = 0$.

We take asymptotically all terms in $1/R$ and $1/R^{3/2}$ into account but neglect $1/R^2$ and higher-order terms. This determines the coefficients appearing in equation (15):

$$q_0 = C_0^{1/2}$$

$$q_1 = (q_0/16)(-1 + 4i\varrho) \tag{16}$$

$$q_2 = (q_0/64)(-1 \pm 4\mu)$$

In (16) we have discarded a solution with a negative imaginary part of Q because this would lead to an exponentially diverging wave function.

The exponent n in equation (14) is determined along the same lines. One finds two solutions,

$$n_{1,2} = -\tfrac{1}{8} \pm \tfrac{1}{2}\mu + i\varrho$$

corresponding to a converging and to a diverging orbit as in the classical analog.

Further considerations show (see, e.g., Rau[3]) that the threshold law reads

$$\sigma \propto E^{n_1 - n_2 - 1/4} = E^{\mu - 1/4} \tag{17}$$

in accordance with equation (11).

Note that the wave function equation (14) is symmetric with respect to electron exchange, and describes, therefore, the ^1S state.

The ^3S state needs a wave function antisymmetric in configuration space. It is easy to see that a ^3S function must have a node line in the Wannier configuration. Antisymmetrization requires $\psi(\mathbf{r}_1, \mathbf{r}_2) = -\psi(\mathbf{r}_2, \mathbf{r}_1)$. Even parity ($S$ state) yields $\psi(\mathbf{r}_1, \mathbf{r}_2) = +\psi(-\mathbf{r}_1, -\mathbf{r}_2)$. From these two relations if follows that in the position $\mathbf{r}_1 + \mathbf{r}_2 = 0$ the wave function vanishes.

Electron exchange reads in our coordinates $R' = R$, $\psi' = \psi$, and $\phi' = -\phi$. The wave function, equation (14), needs in the case of ^3S, therefore, an additional factor taking into account the nodal structure on the saddle. It has been shown elsewhere[4] that in the case under consideration this factor may be chosen as $\sin \phi$, which is odd and vanishes at $\phi = \pi$. We write now instead of equation (14)

$$\Psi = R^{n-5/2}(\sin \phi) \exp[i(8R)^{1/2}Q(\psi, \phi)] \tag{18}$$

and introduce again the expansion (15) for Q. Solving now the Schrödinger equation along the same lines as in the ^1S case, one obtains the same expression for Q but different values for the index n, and, therefore, another threshold law, namely,

$$\sigma(^3S) \propto E^{3\mu - 1/4} \tag{19}$$

The physical meaning of (19) is that the nodal subspace passing through the Wannier configuration depresses the ionization probability. Experimental investigations using polarized electrons and polarized target atoms are in progress.[6] The range of validity of these threshold laws is not well known. We know, however, that the experimental results of Cvejanović and Read[1] using He and unpolarized electrons are in excellent agreement with equation (11) up to about 1.5–2.0 eV excess energy.

From the explicit form of the wave function, equation (14), we get information about the charge distribution at large values of R. Using the expression for $Q(\psi, \phi)$ we obtain for fixed values of R

$$|\Psi|^2 \propto \exp[-2(8R)^{1/2} \operatorname{Im} Q] = \exp[-q_0\varrho\psi^2(2R)^{1/2}] \tag{20}$$

From the definition of coordinates, see Section 2, one derives the relation

$$\sin 2\psi = (2r_1r_2 \sin \theta)/R^2 \tag{21}$$

with

$$\theta = \arccos(\mathbf{r}_1, \mathbf{r}_2)$$

Near the saddle, i.e., $r_1 = r_2$ and $\theta = \pi$, we get the simple geometrical meaning of ψ, namely, $\psi = \pi - \theta$. Equation (20) should, therefore, describe the angular distribution of the electrons. Note that the wave function (14) solves the wave equation in the Coulomb zone, i.e., in the range of R where the particles are controlled by their mutual electrostatic interactions. This zone extends to infinite values of R at zero energy; at finite energy this zone stretches out to the Wannier radius[2]

$$R_W = C_0/E \tag{22}$$

Putting expressions (21) and (22) into equation (20) we derive

$$|\Psi|^2 \propto \exp[-C_0\varrho(\pi - \theta)^2(2/E)^{1/2}] \tag{23}$$

From this equation we expect the angular distribution of the two electrons to be a Gaussian near $\theta = \pi$. The width of the Gaussian is proportional to the fourth root of the energy, $\theta_{1/2} \propto E^{1/4}$. This theoretical prediction compares favorably with the experiment by Cvejanović and Read.[1] It has been pointed out by Rau[7] that the characteristic index ϱ appearing in (20), see equation (9), is real only for ion charges $z = 1$ and 2. For $z \geq 3$, however, ϱ becomes purely imaginary, and $Q(\psi, \phi)$, therefore, real. In this case equation (23) does not hold, the angular distribution becomes constant. Note that equation (23) does not depend on the energy distribution $\varepsilon = E_1/E_2$ of emitted electrons, but only on the total excess energy. Such a flat energy distribution has been observed by Ehrhardt and co-workers[1] already at higher energy where threshold laws do not hold.

Equation (23) derives from the singlet wave function. In the case of the triplet an additional factor of $(\varphi - \pi)^2$ appears in front of the exponential. From the definition of coordinates one finds analogous to (21) the expression

$$\tan \phi = (r_1^2 - r_2^2)/(2r_1r_2 \cos \theta)$$

which yields near the saddle

$$\phi - \pi \simeq (\varepsilon - 1)/(2\varepsilon^{1/2})$$

with $\varepsilon = (r_1/r_2)^2$.

In the framework of classical trajectories this ratio ε corresponds to the energy distribution E_1/E_2. The node of the wave function in configuration space generates a node line in the energy distribution of the differential cross section.

So far we have treated only[1,3] S states. Higher total orbital angular momenta do not change the threshold laws presented here. This comes from the fact that the threshold behavior is controlled asymptotically by Coulombic terms $(1/R)$ rather than by centrifugal terms $(1/R^2)$. Some other angular momentum/parity/spin states have been investigated,[4] leading to the conclusion that equation (17) holds if nodes are absent, and that equation (19) holds in the presence of nodes on the saddle.

4. Escape of More than Two Electrons

The developments presented above can easily be generalized to any number of escaping electrons.[4] The formal key to this extension is the fact that the many-particle Coulomb interaction

$$V = -z \sum_i \frac{1}{r_i} + \sum_{i<j} \frac{1}{r_{ij}} \tag{24}$$

has a stationary configuration, i.e., a multidimensional saddle. Equation (2) has shown in the case of two electrons that the Wannier configuration minimizes the mutual repulsion of the electrons and maximizes their attraction by the ion. The escape of N electrons at low velocity requires their radial distances r_i to remain nearly equal, as they do in the case for $N = 2$. We notice first that the attractive part of the potential alone is stationary at equal r_i and constant

$$R = \left(\sum_i r_i^2 \right)^{1/2} \tag{25}$$

This is seen by inspection of the function

$$\phi(r_1, \ldots, r_N, \lambda) = -z \sum_i \frac{1}{r_i} + \lambda(R^2 - r_1^2 - \cdots - r_N^2)$$

where λ is a Lagrange multiplier. This function becomes stationary for $r_i = R/N^{1/2}$ ($i = 1, \ldots, N$), and has there the series expansion

$$\phi = -(zN^{3/2}/R)\left[1 + (3/2R^2) \sum_i (r_i - R/N^{1/2})^2 + \cdots \right]$$

In this configuration, the electrons are then expected to arrange themselves so as to minimize their repulsion.

In order to take into account this "principle of stationariness" compactly, it is convenient to describe the electron positions by a set of $3N - 1$ angles ω supplemented by the radius R, equation (25),

$$\mathbf{r}_i = R\mathbf{f}_i(\omega), \qquad i = 1, \ldots, N \tag{26}$$

The potential (24) is then written in Coulomb form

$$V = (1/R)C(\omega) \tag{27}$$

and for a stationary system at constant R we must have

$$\nabla C(\omega) = 0 \tag{28}$$

Note that the function $C(\omega)$ actually depends only on $3N - 4$ angles because of rotational invariance. The general solution of equation (28) is difficult to obtain. Elementary investigations give, in special cases, the following results on equilibrium configurations. Three electrons form an equilateral triangle, four form a regular tetrahedron, and five form a symmetric bipyramid.

In Wannier's two-electron system we found one stable and one unstable mode of motion near the saddle. We must expect also in the N-electron system unstable modes. Instability of the equilibrium means that a highly excited system may decay so that not all electrons escape. The possibility of decay is already seen in the case of double excitation. The unstable motion leads in this case only to single escape. Thus it is clear that triple excitation will show two unstable modes because it may decay either into a doubly excited state with one electron falling back into the core or into a singly excited state with two electrons falling back. In the first alternative, the electrons that retain higher excitation may remain on a saddle of lower dimensionality.

It has been pointed out by Fano[8] that the Wannier configuration has an important role not only at small positive energy but also at small negative energy. An electron pair presumably attains states of high double excitation only by passing through the Wannier saddle region. Our generalization to more than two electrons extends the conjecture of Reference 8 that saddle structures of the function $C(\omega)$ control, quite generally, the approach of electrons to multiply excited states.

The Schrödinger equation for N electrons at zero total energy, in terms of the coordinates in equation (26), reads[9]

$$\left[R^{1-3N} \frac{d}{dR} R^{3N-1} \frac{d}{dR} - \frac{\Lambda^2}{R^2} - \frac{2C(\omega)}{R} \right] \Psi = 0 \tag{29}$$

where Λ^2 acts only in the ω variables. If node lines are absent on the saddle $\omega = \omega_0$, we solve equation (29) by

$$\Psi = R^{-(3N-1)/2+n} \exp[i(8R)^{1/2}Q(\omega)] \tag{30}$$

Repeating the procedure described in Section 3, we then find

$$Q^2 + 4(\nabla Q)^2 + C = 0 \tag{31}$$

and

$$n = \frac{1}{4} + \frac{\Lambda^2 Q}{Q} \bigg|_{\omega=\omega_0} \tag{32}$$

The WKB-like equation (31) is easy to solve for our purpose. Because we need only the value of $\Lambda^2 Q$ on the saddle, in equation (31) we expand Q and C up to quadratic terms in ω and compare the coefficients. Note that the operator Λ^2 as part of a Laplacian consists of second derivatives only.

Finally, we must incorporate spin variables and the Pauli principle for the total wave function.[9] The wave function classified by parity (π), orbital (L, M_L), and spin (S, M_S) angular momentum is given by

$$\Psi_{nLM_LSM_S}(R, \omega, \sigma) = \sum_{\varkappa} \Psi_{\varkappa}{}^{nLM_L}(R, \omega)\theta_{\varkappa}{}^{SM_S}(\sigma)$$

Here $\theta_{\varkappa}{}^{SM_S}(\sigma)$ is a spin eigenfunction of S^2 and S_z and $\varkappa$ labels alternative spin states degenerate in S^2. Each of the space wave functions $\Psi_{\varkappa}(R, \omega)$ is expected to have the form (30), but must, however, show the correct behavior with respect to electron permutation and parity. This requires some of the $\Psi_{\varkappa}$ to have a node on the Wannier saddle. Because the function Q depends on $|\omega - \omega_0|$ quadratically as the potential itself does, a factor has then to be added to the wave function (30) to represent the nodes, if any:

$$\Psi_{\varkappa}{}^{nLM_L}(R, \omega) = f_{\varkappa}{}^{nLM_L}(\omega)R^{-(3N-1)/2+n} \exp[i(8R)^{1/2}Q(\omega)] \tag{33}$$

The function $Q(\omega)$ is then still given by equation (31). The characteristic indices n, however, will depend on the nodal factor. Equation (32) still holds if there are no nodes on the saddle.

Triple escape as an example has been investigated by Schlecht.[4] We do not repeat here the complicated mathematics and give only the result for the threshold laws. In the doublets the cross section turns out to be

$$\sigma \propto E^{(\mu-1)/2} \tag{34}$$

with μ given by

$$\mu = \left[\frac{13z3^{1/2} - 11 + 2(1 + 108z^2)^{1/2}}{z3^{1/2} - 1} \right]^{1/2}$$

Double ionization of a negative ion, e.g., by electron impact, leads to an exponent of 2.83 ($z = 1$). For $z = 2$ the numerical value is 2.27 and should characterize the double ionization of a neutral atom by slow electron impact. Note that for $z = \infty$, equation (34) reduces to $\sigma \propto E^2$, which is the result to be expected in the absence of correlation.

The quartet wave function has a node on the saddle, analogous to the triplet in double escape, and leads to the cross section

$$\sigma \propto E^{\mu-1/2}$$

The numerical value of this exponent is large (about 5). The cross section in the quartet channel, therefore, is much smaller than in the doublet.

References

1. S. Cvejanović and F. H. Read, *J. Phys. B* **7**, 1841 (1974); H. Ehrhardt, K. Jung, and E. Schubert, Chapter 3 of this book, and references therein.
2. G. H. Wannier, *Phys. Rev.* **90**, 817 (1953).

3. R. Peterkop, *J. Phys. B* **4**, 513 (1971); A. R. Rau, *Phys. Rev. A* **4**, 207 (1971).
4. W. Schlecht, thesis, Universität Freiburg, Freiburg im Breisgau, West Germany, unpublished, (1976); H. Klar and W. Schlecht, *J. Phys. B* **9**, 1699 (1976).
5. F. T. Smith, *Phys. Rev.* **120**, 1058 (1960); *J. Math. Phys.* **3**, 735 (1962); see also A. J. Dragt, *J. Math. Phys.* **6**, 533 (1965).
6. M. J. Alguard, V. W. Hughes, M. S. Lubel, and P. F. Wainwright, *Phys. Rev. Lett.* **39**, 334 (1977); D. Hils and H. Kleinpoppen, *J. Phys. B* **11**, L283 (1978); D. Hils, K. Rubin, and H. Kleinpoppen, Chapter 54 of this book; M. S. Lubel, Chapter 53 of this book.
7. A. R. Rau, *J. Phys. B* **9**, 10 (1976).
8. U. Fano, *J. Phys. B* **7**, L401 (1974); U. Fano, Lecture delivered at the NATO Advanced Study Institute, Carry-le-Rouet, France, August 31–September 13, 1975; U. Fano and C. D. Lin, *Atomic Physics*, Vol. 4, pp. 47–70, Plenum, New York (1974).
9. D. L. Knirk, *J. Chem. Phys.* **60**, 66 (1974).

6

Interference Effects in the Multiphoton Ionization of Atoms

K. C. Mathur

Time-dependent perturbation theory is used to study the generalized cross sections for the two-photon ionization of a metastable helium atom. Multipole interference effects are considered. It is shown that for certain photon frequencies, significant enhancement of cross section occurs due to the quadrupole effects.

1. Introduction

The study of multiphoton processes is of prime importance in the understanding of laser-induced gas breakdown, nonlinear optical processes, and the study of excited states. With the rapid developments in the field of laser technology, it is becoming possible to produce tunable narrow-linewidth lasers which can be used to study the finer aspects of multiphoton excitation and ionization processes in atoms and molecules. A recent experiment on the multiphoton ionization in alkali atoms has been reported by Granneman and Van der Wiel.[1] On the theoretical side, the time-dependent perturbation theory has been widely used to study the multiphoton absorption process in laser–matter interaction.[2] Nonperturbative approaches, such as the momentum translation approximation and the space translation approximation have also been used. The various approximations are discussed in a recent review by Lambropoulos.[2]

In the present work we have used the time-dependent perturbation theory to study the two-photon ionization of helium from its first excited singlet metastable state. Such a study is important since photoionization from the excited states of

---

K. C. Mathur ● Physics Department, University of Roorkee, Roorkee 247672, India.

helium plays an important role in the transfer of radiation through laboratory plasmas and hot stellar atmospheres.

Most of the work in the study of the multiphoton ionization of atoms has been confined to the use of the electric dipole approximation. The early calculations making use of the dipole approximation are due to Zernik[3] and Bebb and Gold.[4] All such studies have ignored the effects of the higher multipoles. In some recent work[5-8] it has been shown that the higher multipoles play an important role in the study of multiphoton ionization of atoms. When one considers higher multipoles (such as quadrupoles) it becomes possible that the two-photon ionization of an atom takes place with one dipole and one quadrupole photon instead of the usual two-dipole photons. This could lead to interference between the dipole and the quadrupole contributions and result in significant changes in the generalized cross sections for a certain range of incident photon frequencies.

In our calculations we have included both the dipole and the quadrupole effects in the study of two-photon ionization from the 2^1S state of helium.[9] We find that by the inclusion of the quadrupole effects significant change in the cross sections occurs in the wavelength range 5042–5045 Å.

2. Theory

The total generalized cross section for the N-photon ionization of an atom from an initial state g to a final state f is given by[2,4]

$$\hat{\sigma} = \frac{(2\pi\alpha w)^N}{4\pi^2} \frac{mk}{\hbar} \int |K_{fg}^{(N)}|^2 \, d\Omega_{\hat{z}}.$$ (1)

where $\hbar w$ is the energy of an incident photon, k the wave number of the ejected electron, α the fine-structure constant, and m the electron mass. The transition matrix K, for the case of two-photon ionization, is given by

$$K_{fg}^{(2)} = \frac{\sum_{a_j} \langle a_f | R | a_j \rangle \langle a_j | R | a_g \rangle}{(w_{a_j a_g} - w + i\Gamma/2)}$$ (2)

where $w_{a_j a_g} = w_{a_j} - w_g$. w_{a_j} and w_g are the atomic frequencies corresponding to the intermediate state a_j and the initial state a_g. Γ is the level width of the intermediate state and R is an operator. For the helium target atom, and in the dipole approximation, one writes

$$R = \hat{\varepsilon} \cdot \mathbf{r_1} + \hat{\varepsilon} \cdot \mathbf{r_2}$$ (3)

where $\mathbf{r_1}$ and $\mathbf{r_2}$ denote the position vectors of the atomic electrons and $\hat{\varepsilon}$ is a unit vector in the direction of the polarization of the incident radiation. When one considers both the dipole and the quadrupole effects[7] in the interaction Hamiltonian, one can express

$$R = \sum_{l=1}^{2} (D_l + Q_l) \tag{4}$$

with $D_l = Z_l$ and $Q_l = 0.5\, ik_{\mathrm{ph}} Z_l X_l$, where K_{ph} is the incident photon wave vector. In arriving at equation (4) we have assumed the light to be linearly polarized along the z axis and propagating along the x axis. In the evaluation of the matrix elements in equation (2), the wave function for the final continuum state is written as

$$\psi_f = 2^{-1/2}[\psi_{1S}(\beta/\mathbf{r}_1)\psi_k(\mathbf{r}_2) + \psi_{1S}(\beta/\mathbf{r}_2)\psi_k(\mathbf{r}_1)] \tag{5}$$

where

$$\psi_{1S}(\beta/\mathbf{r}) = 2e^{-\beta r} Y_{00}(\hat{r}) \qquad \text{with } \beta = 2$$

and

$$\psi_k(\mathbf{r}) = 4\pi \sum_{lm} i^l \exp(i\eta_l) R_{kl}(r) Y_{lm}(\hat{r}) Y_{lm}^*(\hat{k})$$

and

$$R_{kl}(r) = |\,\Gamma(l+1-i/k)\,|\,[\Gamma(2l+2)]^{-1} \exp(\pi/2k - ikr)$$
$$\times (2kr)^l \,_1F_1(l+1+i/k, 2l+2, 2ikr)$$

with

$$\eta_l = \arg \Gamma(l+1-i/k)$$

The bound-state wave functions (a_g, a_j) used in the present calculations are taken from Chan and co-workers.[10,11] Further the infinite summation over the intermediate states in equation (2) is performed by including N' intermediate states exactly. The contribution of the remaining states is obtained using closure and assuming an average intermediate state energy equal to $\hbar w_{N'+1}$.

3. Results and Discussion

Figures 1 and 2 show our results of the generalized cross section versus wavelength in the wavelength range 5016–5045 Å, for the two-photon ionization of helium from the 2^1S state, in the electric dipole plus quadrupole approximation.

From Figure 1 it is observed that the cross section rises rapidly from 5016 Å onwards and acquires a peak value around 5017.1 Å and then drops rapidly. The increase in the generalized cross section occurs because the transition to the continuum from the initial 2^1S state proceeds through a number of real intermediate states. The maximum at 5017.1 Å occurs through resonance with the dipole-allowed 3^1P intermediate state. The resonance due to the higher dipole-allowed P states would occur towards lower wavelengths. From Figure 1 it is further noted that the cross section continues to decrease up to about 5040 Å, beyond which it shows signs of increasing.

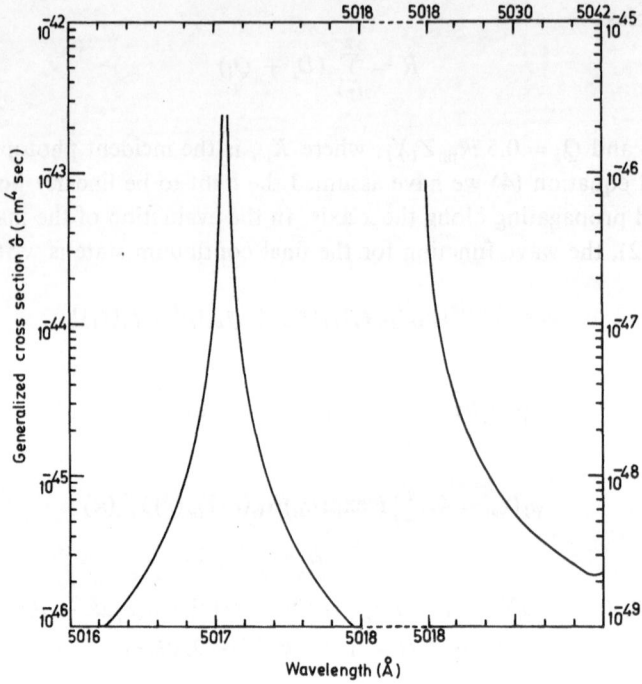

Figure 1. Two-photon ionization of He from the 2^1S state in the wavelength range 5016–5042 Å.
————, Present calculation in the electric dipole plus quadrupole approximation.

In order to show this increase explicitly we have shown in Figure 2 the generalized cross sections in the wavelength range 5042–5045 Å on a wider scale. We note that the cross section rises and acquires a maximum value at about 5043.5 Å, beyond which it again falls rapidly. In this figure we have also shown for comparison the generalized cross sections in the dipole approximation alone (indicated by the dashed curve). As seen from the Figure 2, the dipole contribution decreases very slowly and remains nearly constant. The rise in the cross section beyond 5040 Å (Figures 1 and 2) arises due to the interference between the dipole and the quadrupole contributions. The quadrupole resonance peak is obtained at 5043.5 Å. Beyond 5045 Å the quadrupole contribution becomes small and the total contribution to the cross section is again dominated by the dipole term alone. The reason for the large contribution due to quadrupole in the wavelength range 5042–5045 Å could be understood as follows. In the vicinity of the dipole-allowed 3^1P state lies the dipole-forbidden but quadrupole-allowed 3^1D state. Although the matrix element for this quadrupole-connected 3^1D state, i.e., $\langle 3^1D \mid Q \mid 2^1S \rangle$ is small compared to the dipole matrix element $\langle 3^1P \mid D \mid 2^1S \rangle$, the energy denominator in equation (2) in the wavelength range 5042–5045 Å for the 3^1D intermediate state becomes very small (smaller than that for the 3^1P state), thereby enhancing significantly the value of $K_{fg}^{(2)}$ and hence the cross sections. Thus the quadrupole contribution to the total cross section becomes significantly large in the above wavelength range where the dipole contribution is small. The quadrupole peak at 5043.5 Å is due to the denominator becoming resonant corresponding to the 3^1D intermediate state. The natural width of the 3^1D state is quite small and could be neglected.

Further, in the above wavelength range the interference with respect to the higher multipoles would be small because the energy denominators corresponding to, say, octupole-allowed transitions would be nonresonant and also the octupole matrix element would be much smaller.

In the above calculations, we have included explicitly five intermediate states, namely, 2^1P, 3^1P, 4^1P, 3^1D, and 4^1D and accounted for the remaining by using closure. It is found that in the vicinity of a resonance the contribution of the closure term is small, whereas away from resonance the closure contribution cannot be neglected.

From the above study we conclude that in a certain range of photon wavelengths the quadrupole plays a significant role in the estimation of the generalized cross sections. It may also be emphasized that the numerical values of the cross section with the inclusion of the quadrupole become substantial and hence the quadrupole peak could easily be detected with the present-day tunable lasers. Also the dipole and the quadrupole peaks are separated by 26.4 Å and could easily be resolved.

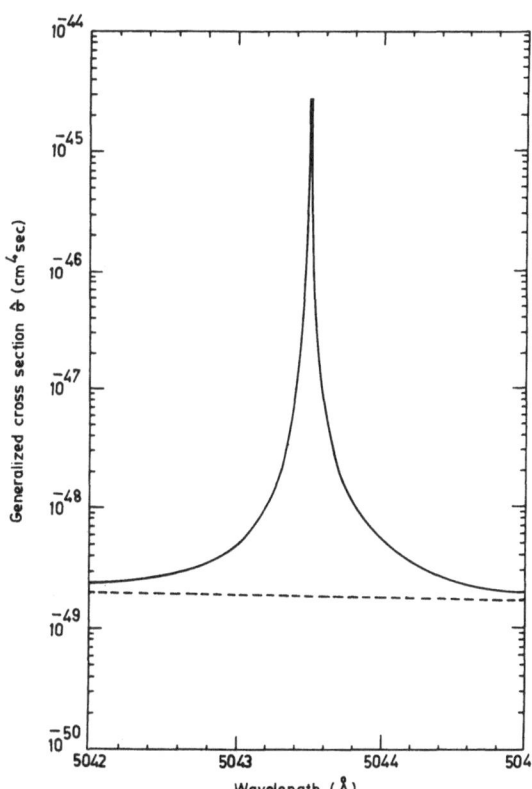

Figure 2. Two-photon ionization of He from the 2^1S state in the wavelength range 5042–5045 Å. ——, Present calculation in the dipole plus quadrupole approximation. – – –, Present calculation in the dipole approximation alone.

References

1. E. H. A. Granneman and M. J. van der Wiel, *J. Phys. B* **8**, 1617–1625 (1975).
2. P. Lambropoulos, *Adv. At. Mol. Phys.* **12**, 87–163 (1975).
3. W. Zernik, *Phys. Rev.* **133**, A117–A120 (1964).
4. H. B. Bebb and A. Gold, *Phys. Rev.* **143**, 1–24 (1966).
5. E. A. Power and T. Thirunamachandran, *J. Phys. B* **8**, L170–L172 (1975).
6. P. Lambropoulos, G. Doolen, and S. P. Rountree, *Phys. Rev. Lett.* **34**, 636–639 (1975).
7. A. Rachman, G. Laplanche, Y. Flank, and M. Jaouen, *Phys. Lett.* **58**, 155–157 (1976).
8. Y. Flank, G. Laplanche, M. Jaouen, and A. Rachman, *J. Phys. B* **9**, L409–L412 (1976).
9. K. C. Mathur, *Phys. Rev. A* **18**, 2170 (1978).
10. F. T. Chan and S. T. Chen, *Phys. Rev. A* **10**, 1151–1156 (1974); **9**, 2393–2397 (1974); **8**, 2191–2194 (1975).
11. F. T. Chan and C. H. Chang, *Phys. Rev. A* **12**, 1383–1392 (1975).

7

Correlation
in Electron–Atom Excitation

D. E. Golden and N. C. Steph

The use of delayed coincidences and photon polarization measurements to study correlation effects in electron–atom inelastic scattering is detailed. Sources of systematic error in experimental results with respect to determination of the correlation parameters λ and χ for helium are discussed. The results of several calculations are compared to the experimental data for the 2^1P_1 state of helium.

Correlation, which indicates a lack of internal independence, has not been discussed as such in the atomic physics literature until relatively recently. However, correlation has been discussed extensively in high-energy, nuclear, and solid-state physics. The underlying idea is that internal symmetries may be uncovered by fixing the external symmetries in the preparation of an experiment or calculation. This certainly must be true in any case where structure is attributed to an object. The trick is to figure out how to probe the structure. For example, by proper experimental design, one might be able to probe the excitation of fine and even hyperfine levels.

The subject of coherent excitation of different fine and hyperfine levels begins with the beam–foil measurements in the mid-1960's.[†] These experiments were aimed at the measurement of atomic lifetimes by looking at radiation from foils excited by ion impact. While the measurements were made under supposed zero external field conditions, oscillations in the light intensity were observed. These oscillations were attributed to Stark mixing due to a small electric field in the ion beam itself.[†]

[†] See, for example, Reference 1.

D. E. Golden and N. C. Steph ● Department of Physics and Astronomy, University of Oklahoma, Norman, Oklahoma 73019.

However, the correct explanation, given by Macek,[2] is that the oscillations were beats due to interferences between the various hyperfine levels. Thus Macek postulated the coherent excitation of fine and hyperfine levels.

Correlation effects in electron–atom excitation were first studied in a scattering experiment using the technique of delayed coincidence.[3] While this measurement technique has been used extensively in nuclear physics,[4] it is only relatively recently that it has become widely used in atomic physics. It was first used to study atomic lifetimes in 1955.[5] Since that time it has been used to study ionization,[6] excitation of metastables,[7] energy transfer from a metastable to radiating state,[8] and to separate excitation cross sections for levels that could not be separated in a scattered electron detector.[9] This last application has most recently been used to separate the 3^3D and 3^1D excitation cross sections in Kr by McGregor and Kleinpoppen at the University of Stirling. These levels are separated by only 0.0004 eV and so this experiment would not be possible within present electron energy analyzer technology without the use of the delayed coincidence technique.

The work of Macek[2] was extended by Macek and Jaecks[10] to point out that more insight regarding inelastic scattering could be obtained by studying angular correlations between inelastically scattered electrons and photons from the decay of an excited state than from the measurement of an inelastic cross section. The basis of the work of Macek and Jaecks[10] is the first consistent theoretical treatment of electron impact excitation due to Percival and Seaton[11] and the notion of Macek[2] that radiation from different fine and hyperfine levels introduces oscillatory terms into the radiative decay of atoms. Macek and Jaecks[10] took the magnetic substates to be excited coherently and developed a time-dependent theory. This theory has been reformulated by many others. Most recently the subject of electron–photon angular correlations in atomic physics has been reviewed by Blum and Kleinpoppen.[12] In this work the e^-–H correlation parameters are also developed for the first time.

The first electron–photon angular correlation measurements were reported by Eminyan et al.[3] for excitation of the 2^1P state of helium. We should point out that the kind of information to be obtained can also be obtained from experiments with laser-excited atoms such as have been performed at Kaiserslautern and New York University. However, the interpretation of the data is less clear-cut because the laser excites some distribution of excited states. Electron–photon angular correlations have since been studied in Ne and Ar by the Flinders group, in Kr and Hg by the Stirling group, in H_2 by the Kaiserlautern group, and in Ar and H by the Windsor group. However, the interpretation of the results of scattering from targets other than helium is incomplete, as was pointed by Slevin and Farago[13] for argon.

The standard way to treat the 2^1P state of helium is to describe it by a coherent superposition of the degenerate sublevels and neglect spin–orbit and spin–spin interactions in the collision. In addition, the 2^1P state will be excited from the 1^1S state in a field free region. With reference to Figure 1, the electrons are incident along the Z direction and both the scattered electron detector and the photon detector are free

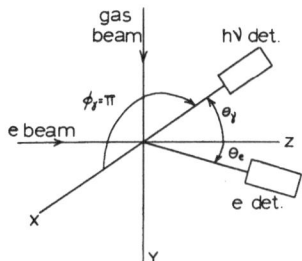

Figure 1. Schematic diagram of the electron–photon coincidence
experiment in the scattering plane.

to move in the X–Z plane, which is the scattering plane. In this case the excitation
amplitudes, a_m with $m = 0 \pm 1$, are only functions of the electron energy E and the
electron scattering angle θ_e. In addition, since there is a mirror symmetry through
the scattering plane, $a_{-1} = -a_1$ and the wave function can be normalized such that
the scattering amplitudes a_m are very simply related to the scattering cross sections
for excitation of the sublevels σ_m,

$$
\begin{aligned}
|a_0|^2 &= \sigma_0 \\
|a_1|^2 &= \sigma_1 \\
|a_0|^2 + 2|a_1|^2 &= \sigma
\end{aligned}
\tag{1}
$$

where σ is the differential cross section for exciting the 2^1P level. The relative phase
χ between a_0 and a_1 is simply given by $a_1 = |a_1| e^{i\chi}$ and $a_0 = |a_0|$. The wave function
at a given E and θ_e is completely described within an arbitrary phase factor by σ_0,
σ_1, and χ, and the scattering completely determined by a measurement of these
parameters. The parameters are determined with the exception of the sign of χ by
a measurement of the electron–photon coincidence rate $\dot{N}_c$, which was given by
Macek and Jaecks,[10]

$$
\frac{d\dot{N}_c}{d\Omega_e\, d\Omega_\gamma\, dz} = A\{\lambda \sin^2\theta_\gamma + (1 - \lambda)\cos^2\theta_\gamma - 2[\lambda(1 - \lambda)]^{1/2}\cos\chi \sin\theta_\gamma \cos\theta_\gamma\}
\tag{2}
$$

Here $\lambda = \sigma_0/\sigma$ and

$$
A = \frac{3}{8\pi}\, \frac{I_e}{e}\, \varrho(z)\, \frac{\gamma'}{\gamma}\, \varepsilon_e \varepsilon_\gamma \sigma
$$

where I_e is the incident electron beam current, $\varrho(z)$ is the density of the target atoms
in the interaction volume, γ'/γ is the branching ratio for the decay of the 2^1P state,
ε_e and ε_γ are the detector efficiencies, and e is the electron charge. Since the sign of
χ is not given by equation (2) this must be determined by a separate measurement
such as the polarization of the radiation. We can rewrite equation (2) as the sum of
two cosine functions as follows:

$$
\begin{aligned}
\frac{d\dot{N}_c}{d\Omega_e\, d\Omega_\alpha\, dz} &= \frac{A}{2}\, [(1 - \cos\chi)\cos^2(\theta_\gamma - \beta) + (1 + \cos\chi)\cos^2(\theta_\gamma + \beta)] \\
&= Af(\lambda, \chi, \theta_\gamma)
\end{aligned}
\tag{3}
$$

where $\sin \beta = \lambda$. When $\cos \chi \simeq 1$, the first term in equation (3) may be neglected so that

$$\frac{d\dot{N}_c}{d\Omega_e \, d\Omega_\gamma \, dz} \simeq \frac{A}{2} \, (1 + \cos \chi) \cos^2(\theta_\gamma + \beta) \qquad (4)$$

Thus, for small χ we have a simple periodic function whose amplitude depends only on χ and whose phase depends only on λ. This is instructive from the point of view of unfolding values of λ and χ from measurements of $\dot{N}_c$. In Figure 2, $f(\lambda, \chi, \theta_\gamma)$ is plotted as a function of θ_γ for $\lambda = 0.48$, $\chi = 0.20$ (solid line, 1); $\lambda = 0.48$, $\chi = 0.30$ (dotted line, 2); and for $\lambda = 0.58$, $\chi = 0.20$ (dash-dotted line, 3). The difference between curves 1 and 2 is due to about a 2% change in amplitude, which is caused by a 50% change in $|\chi|$ so that small errors in amplitude give very large errors in $|\chi|$. The situation in regard to λ is not quite so bad. The difference between curves 1 and 3 is about a 5.5° change in phase, which is due to a 20% change in λ.

When we perform an experiment, we will study coincidences with detectors that view finite solid angles for some time T so that equation (3) must be integrated over the solid angles of the detectors and the time. We perform the integration over the two solid angles and write

$$\frac{\dot{N}_c}{\dot{N}_e} = \frac{3}{8\pi} \, \varepsilon_\gamma \frac{J_c}{J_e} \, f(\lambda, \chi, \theta_\gamma) \qquad (5)$$

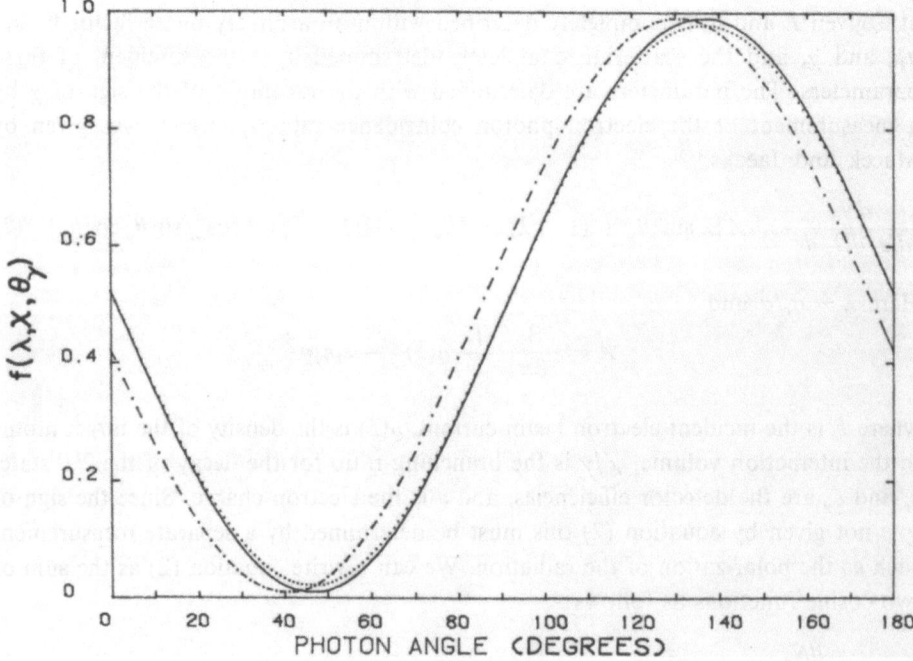

Figure 2. $f(\lambda, \chi, \theta_\gamma)$ vs. photon angle (θ_γ). (1) ——, $\lambda = 0.48$, $\chi = 0.20$; (2) $\cdots$, $\lambda = 0.48$, $\chi = 0.30$; (3) —·—, $\lambda = 0.58$, $\chi = 0.20$.

where

$$\frac{J_c}{J_e} = \frac{\int [\varrho(z)/\varrho_0] \, d\Omega_e \, d\Omega_\gamma \, dz}{\int [\varrho(z)/\varrho_0] \, d\Omega_e \, dz}$$

and $\dot{N}_e$ is the scattered electron rate. For the case where the atomic beam density is very large compared to the background gas density and uniform over the extent of the beam and for infinite angular resolution, equation (5) reduces to

$$\frac{\dot{N}_c}{\dot{N}_e} = \frac{3}{8\pi} \, \varepsilon_\gamma \, \Delta\Omega_\gamma f(\lambda, \chi, \theta_\gamma) \tag{6}$$

Then one counts for a time T at fixed θ_e for various values of θ_γ and fits to the equation

$$\frac{N_c}{N_e} = Bf(\lambda, \chi, \theta_\gamma) \tag{7}$$

One can then determine the parameters B, λ, and $\cos \chi$ as was done by Eminyan et al.[14] Alternatively, for $\theta_\gamma = \pi/2$, one may write

$$\frac{N_c}{N_e} = \frac{3}{8\pi} \, \varepsilon_\gamma \, \frac{J_c}{J_e} \left(\theta_e, \frac{\pi}{2}\right) \lambda \tag{8}$$

and coincidence measurements at this angle can be used to obtain λ as a function of E and θ_e, as was done by Sutcliffe et al.[15] In addition, the coincidence rate for photons observed perpendicular to the scattering plane is given by

$$\dot{N}_c = K\sigma[(1 - \lambda) \sin^2\phi_\gamma + \lambda] \tag{9}$$

Equation (9) was used by Tan et al.,[16] together with measurements of $\dot{N}_c$ at $\phi_\gamma = 0$ and $\pi/2$ to obtain values of λ. Also they used a linear polarization filter to obtain values of λ and $|\chi|$. In this later experiment they studied coincidences between electrons which had excited the 2^1P state of helium and photons perpendicular to the scattering plane whose linear polarization made an angle β with the incident electron beam. Then

$$\lambda = \frac{N_c(\beta=0)}{N_c(\beta=\pi/4) + N_c(\beta=3\pi/4)} \tag{10}$$

$$2\left(\frac{1 - \lambda}{\lambda}\right)^{1/2} \cos \chi = \frac{N_c(\beta=\pi/4) - N_c(\beta=3\pi/4)}{N_c(\beta=0)}. \tag{11}$$

Now let us consider the experimental questions: (1) What it the effect of the finite angular resolution of the detectors? (2) What is the effect of a nonnegligible background gas density? The effect of the finite resolution of the photon detector can be seen with the aid of Figure 3, where $f(\lambda, \chi, \theta_\gamma)$ is plotted for $\lambda = 0.48$ and $\chi = 0.20$ (solid curve). Suppose we assume that this is the "true" function and

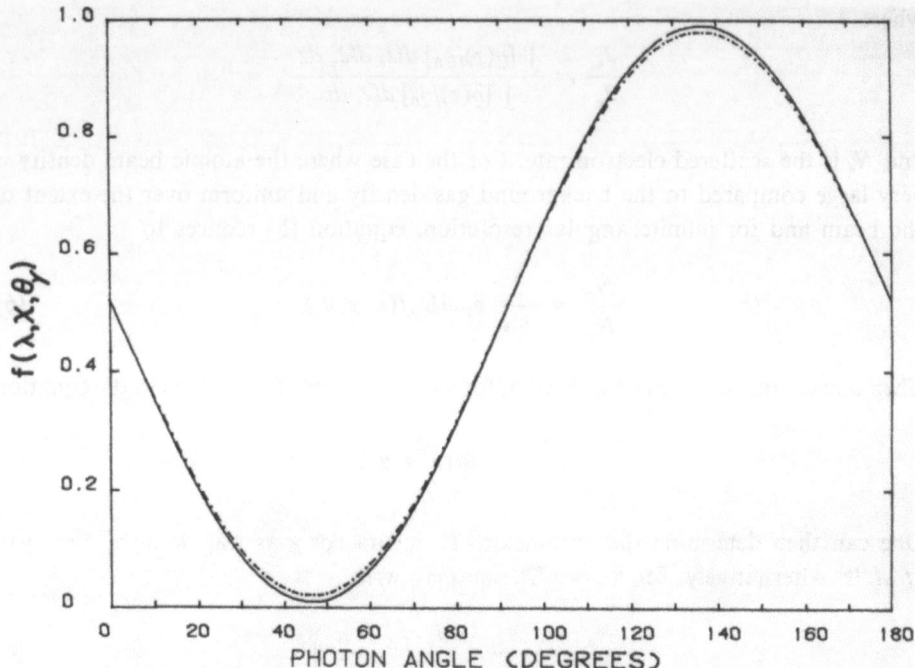

Figure 3. $f(\lambda, \chi, \theta_y)$ vs. photon angle (θ_y). ——, $\lambda = 0.48$, $\chi = 0.20$; —·—, obtained by averaging solid line over 20° intervals to simulate a flat detector response over 20°.

ask what is the effect of a flat detector response over, say, 20°. The effect would be to draw a new curve that is constructed from points obtained by averaging over 20° intervals from the solid curve of Figure 3. This curve is drawn as a dash-dotted line in Figure 3. The period of the curve remains the same, but the amplitude is decreased. As has been discussed above, for small χ the effect of decreasing the amplitude is to increase the value of χ. This change of about 2% in the amplitude looks like a 50% change in $|\chi|$. This effect, as well as a similar effect due to the detection of photons out of the scattering plane, was corrected for by Eminyan et al.[14] by using the following equation:

$$\frac{N_c}{N_e} = \frac{3}{8\pi} \, \varepsilon_y \, \varDelta\Omega_y \left[\varkappa f(\lambda, \chi, \theta_y) + \frac{2}{3} (1 - \varkappa) \right] \tag{12}$$

where $\varkappa$ is given by[†] $(1 - \varDelta\Omega_y/4\pi)(1 - \varDelta\Omega_y/2\pi)$. Of course it is possible to make $\varDelta\Omega_y$ sufficiently small so that the correction represented by equation (12) is unnecessary. The effect of a finite background gas density is to give too large a measured value of N_c/N_e at both small and large values of either θ_e or θ_y. This is a

---

[†] *Note added in proof.* It should be noted that this definition of $\varkappa$ is slightly different than that given in Reference 14. This typographical error was pointed out to us by Professor K. B. MacAdam.

correction that is symmetric about 90° and affects both λ and χ. This correction requires knowledge of the solid angles of the detectors, the variation of these solid angles with path length, and the atomic beam profile, and can be written as $(J_e/J_c)(\theta_e, \theta_\gamma)$. This expression has been evaluated by Sutcliffe et al.[15] for $\theta_\gamma = \pi/2$ from auxiliary measurements of elastic angular distributions with the atomic beam on and with the atomic beam off and the chamber flooded to the same background gas density as with the beam on. The correction to the data of Sutcliffe et al.[15] is plotted as $(J_e/J_c)(\theta_e, \pi/2)$ in Figure 4. One would expect a similar correction to be obtained if θ_e is kept fixed and θ_γ varied. It should be noted that in the range $40° < \theta_e < 140°$ the graph is relatively flat. Therefore, provided one makes the solid angle of the detector sufficiently small and avoids both small and large angles, this correction is unnecessary.

The problem of measuring the number of coincidences at a given θ_e, θ_γ, and E can be further complicated by accidental coincidences. These occur when the clock is started and stopped by electrons and photons from different scattering events and is given by the product of the two rates and the time window of the coincidence detector

$$\dot{N}_A = \dot{N}_e \dot{N}_\gamma \, \Delta t \tag{13}$$

An example from the work of Sutcliffe et al.[15] is shown in Figure 5. The background is due to the accidental coincidences spread out in time (channel number). Once a start pulse is obtained the a priori probability of a stop pulse is p. Then the probability of obtaining a stop in the ith channel is given by $P_i = (1 - p)^i p$. Then the background distribution due to accidental coincidences is given by

$$\dot{N}_{A_i} = \dot{N}_e \dot{N}_\gamma \, \Delta t p (1 - p)^i T \tag{14}$$

The number of true coincidences is obtained by fitting the background and sub-

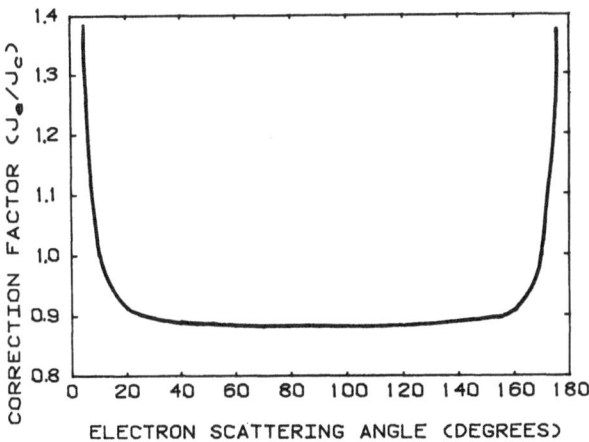

Figure 4. $(J_e/J_c)(\theta_e, \pi/2)$ vs. electron scattering angle (θ_e) from Reference 15.

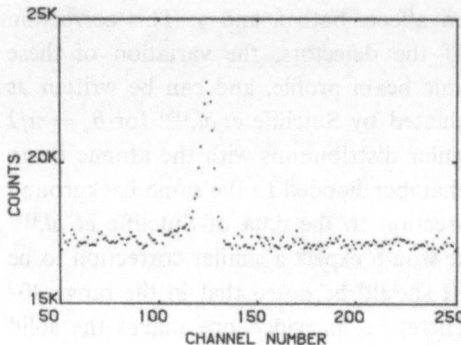

Figure 5. Coincidence peak for $\theta_\gamma = \pi/2$, $\phi_\gamma = \pi$, $\theta_e = 10°$, $\phi_e = 0$ from Reference 15.

tracting it from the signal. Since the coincidence rate divided by the accidental rate is inversely proportional to the gas beam density and the electron current, the background can be suppressed by lowering these quantities. However, the signal-to-noise ratio is increased as these quantities are increased, up to the point where resonance trapping takes place. Then, bearing in mind that not a great deal of attention has been paid as yet to some of these considerations, the data in helium for λ at 80 eV are presented in Figure 6. The data, for the most part, are in reasonably good agreement below 70°. That includes data from Oklahoma, Sutcliffe *et al.*,[15] Belfast,

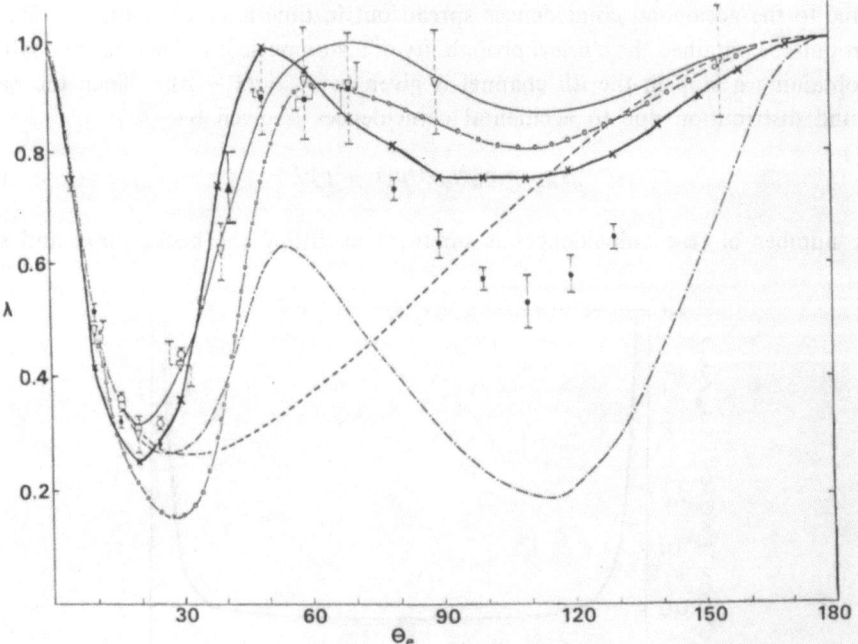

Figure 6. λ vs. θ_e for He($2^1P \rightarrow 1^1S$) at 80 eV. ▽, Sutcliffe *et al.*;[15] ●, Hollywood *et al.*;[17] ○, Eminyan *et al.*;[14] □, Ugbabe *et al.*;[18] ▲, Tan *et al.*;[16] ——, Madison and Calhoun;[19] —○—, Thomas *et al.*;[27] – – –, Born calculation; —·—, Baluja and McDowell;[20] —×—, Fon *et al.*[22]

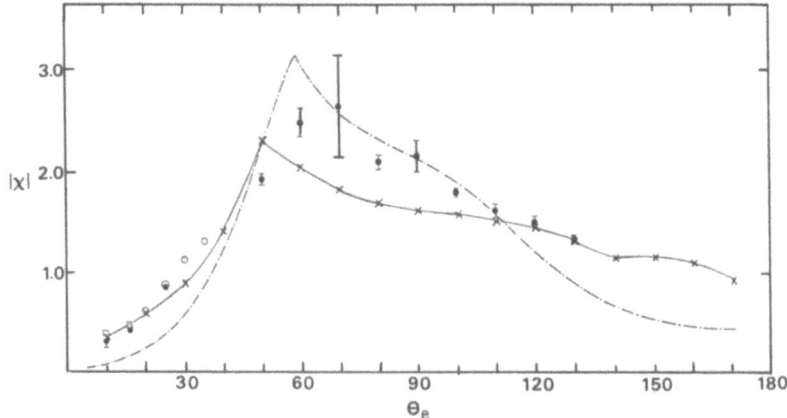

Figure 7. $|\chi|$ vs. θ_e for He($2^1P \rightarrow 1^1S$) at 80 eV. ●, Hollywood *et al.*;[17] ○, Eminyan *et al.*;[14] □, Ugbabe *et al.*;[18] —·—, Baluja and McDowell;[20] —×—, Fon *et al.*[22]

Hollywood and Williams,[17] Stirling, Eminyan *et al.*,[14] Flinders, Ugbade *et al.*,[18] and Windsor, Tan *et al.*[16] The large-angle data are another story, where on the face of it a considerable difference exists between the Belfast and Oklahoma data. However, the difference only involves the points at 80° and 90° thus far. (Both groups are repeating the measurements.) The distorted-wave calculations of Madison and Calhoun[19] agree with all of the small-angle data and the large-angle data of Sutcliffe *et al.*[15] The other calculations included on the plot are the distorted-wave calculation of Baluja and McDowell,[20] the many-body calculation of Thomas *et al.*,[21] a first Born approximation calculation, and the most recent *R*-matrix calculation of Fon *et al.*[22] The measurements of $|\chi|$ vs. θ_e at 80 eV in He are presented in Figure 7,

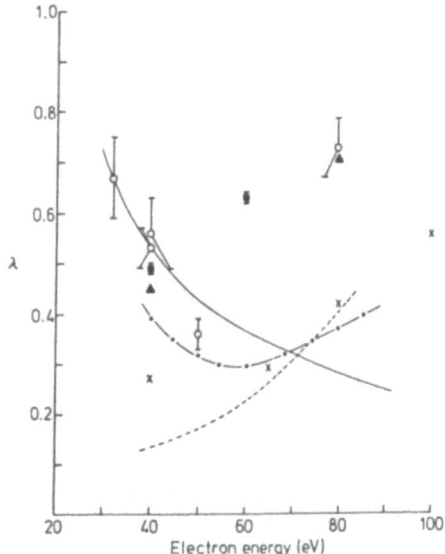

Figure 8. λ vs. electron energy (E) for He($2^1P \rightarrow 1^1S$) for $\theta_e = 42°$. ○, Tan *et al.*;[16] ■, Eminyan *et al.* (40°);[14] ▲, Madison and Shelton;[16] ×, Scott and McDowell;[23] – – –, Thomas *et al.*;[21] —·—, Flannery and McCann;[24] ——, first Born approximation.

together with the calculations of Baluja and McDowell,[20] and Fon *et al.*[22] Here the agreement is fair between the measurements and both of the calculations.

The variation of λ with electron energy for a fixed electron scattering angle of 42° is presented in Figure 8. As before, the picture is not completely clear as yet. Below 40 or 50 eV, the first Born approximation agrees with the data quite well, while at 80 eV the distorted-wave calculations of Madison and Calhoun[19] agree with the data. In between there is a minimum which is not very well predicted by the remaining calculations of Scott and McDowell,[23] Thomas *et al.*,[21] or Flannery and McCann.[24]

Linear and circular polarization measurements of the $3^1P \rightarrow 2^1S$ photons detected in delayed coincidence with electrons that have excited the 3^1P state of helium have been made by Standage and Kleinpoppen.[25] A schematic diagram of their apparatus is shown in Figure 9. The photons are detected perpendicular to the scattering place. The polarization vector is defined in terms of the intensity component at an angle β with respect to the electron direction

$$P_1 = N_c(\beta=0) - N_c(\beta=\pi/2)$$
$$P_2 = N_c(\beta=\pi/4) - N_c(\beta = 3\pi/4) \qquad (15)$$
$$P_3 = N_c(\text{RHC}) - N_c(\text{LHC})$$

where RHC and LHC denote left- and right-hand circular polarization. The measurements of the components of the polarization vector, the degree of polarization, and

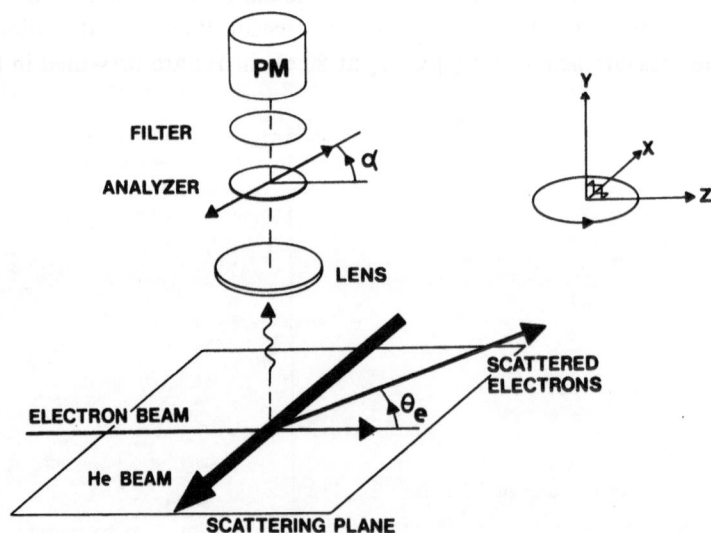

Figure 9. Schematic diagram of the polarization measurement of Reference 25. The X–Z plane is the scattering plane; the photons are detected by the photomultiplier (PM) along the Y axis. Scattering angle θ_e and linear polarizer angle α are measured in the X–Z plane. Positive scattering angle is shown.

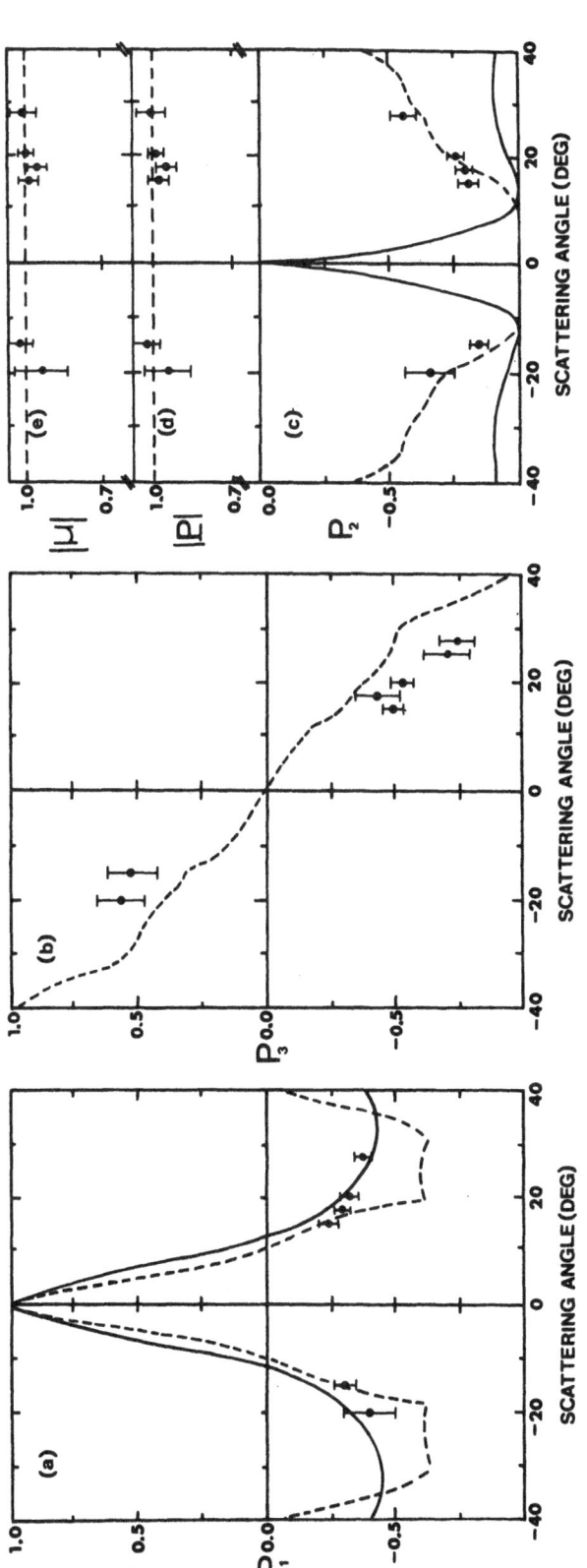

Figure 10. Experimental results of Reference 25. (a)–(c) Experimental data for the vector polarization components P_1, P_3, and P_2 respectively, of He ($3^1P \rightarrow 2^1S$) coincident photons at 80 eV incident electron energy vs. electron scattering angle. ——, first Born approximation; - - -, multichannel eikonal approximation. (d) Degree of polarization. (e) Degree of coherence.

the degree of coherence from the measurements of Standage and Kleinpoppen[25] are shown in Figure 10. These measurements show that the excitation is completely polarized and completely coherent.

ACKNOWLEDGMENT

This work was supported in part by the NSF and the AFOSR.

References

1. S. Bashkin and G. Beauchemin, *Can. J. Phys.* **44**, 1603 (1966).
2. J. Macek, *Phys. Rev. Lett.* **23**, 1 (1969); *Phys. Rev. A* **1**, 618 (1970).
3. M. Eminyan, K. B. MacAdam, J. Slevin, and H. Kleinpoppen, *Phys. Rev. Lett.* **31**, 576 (1973).
4. R. E. Bell, in *Alpha, Beta, and Gamma Ray Spectroscopy*, K. Siegbahn, ed., p. 905, North-Holland, Amsterdam (1966).
5. E. Brannen, F. R. Hunt, R. H. Adlington, and R. W. Nichols, *Nature (London)* **175**, 810 (1955).
6. H. Ehrhardt, M. Schulyz, T. Tekaat, and K. Willmann, *Phys. Rev. Lett.* **2**, 89 (1969).
7. D. E. Golden, D. J. Burns, and V. C. Sutcliffe, *Phys. Rev. A* **10**, 212 (1974).
8. D. J. Burns, D. E. Golden, and D. Galliard, *J. Chem. Phys.* **65**, 266 (1976).
9. A. Pochat, D. Rozuel, and J. Peresse, *J. Phys. (Paris)* **34**, 701 (1973).
10. J. Macek and D. H. Jaecks, *Phys. Rev. A* **4**, 2288 (1971).
11. I. Percival and M. Seaton, *Phil. Trans. R. Soc. London A* **251**, 113 (1958).
12. K. Blum and H. Kleinpoppen, *Phys. Rep.* **52**, 203 (1979).
13. J. Slevin and P. S. Farago, *J. Phys. B* **8**, L407 (1975).
14. M. Eminyan, K. B. MacAdam, J. Slevin, and H. Kleinpoppen, *J. Phys. B* **7**, 1519 (1974).
15. V. C. Sutcliffe, G. N. Haddad, N. C. Steph, and D. E. Golden, *Phys. Rev. A* **17**, 100 (1978).
16. K. H. Tan, J. Fryar, P. S. Farago, and J. W. McConkey, *J. Phys. B* **10**, 1073 (1977).
17. M. T. Hollywood, A. Crowe, and J. F. Williams, *J. Phys. B* **12**, 819 (1979).
18. A. Ugbabe, P. J. O. Teubner, E. Weingold, and H. Arriola, *J. Phys. B* **10**, 71 (1977).
19. D. H. Madison and R. V. Calhoun (unpublished).
20. K. L. Baluja and M. R. C. McDowell, *J. Phys. B* **12**, 835 (1979).
21. L. D. Thomas, G. Csanak, H. S. Taylor, and G, S. Yarlagadda, *J. Phys. B* **10**, 1073 (1977).
22. W. C. Fon, K. A. Berrington, and A. Kingston *J. Phys. B* **12**, L171 (1979).
23. T. Scott and M. R. C. McDowell, *J. Phys. B* **8**, 1851 (1975).
24. M. R. Flannery and K. J. McCann, *Phys. Rev. A* **12**, 846 (1975).
25. M. C. Standage and H. Kleinpoppen, *Phys. Rev. Lett.* **36**, 577 (1976).

8

Angular Correlation Parameters for Excitation of the 2^1P State of Helium and the 2p State of Atomic Hydrogen

M. T. Hollywood, A. Crowe, and J. F. Williams

Angular correlations between the scattered electrons resulting from the excitation of the 2^1P state of helium and the photon emitted in the deexcitation of the state have been measured using a coincidence technique for an incident electron energy of 81.2 eV. Similar correlations have been measured for the excitation of the 2p state of atomic hydrogen by 100-eV electrons. These measurements are the first for large-momentum-transfer scattering. The angular correlations are analyzed to yield the ratio of the differential cross sections for excitation of the degenerate magnetic sublevels of the state and, in helium, the phase difference between the corresponding excitation amplitudes. The Fano–Macek orientation and alignment parameters are also derived from the data. The measurements provide a more sensitive test of theoretical models than previous small-angle scattering data. None of the published theoretical values agrees with our data.

1. Introduction

Present "state-of-the-art" experimental techniques permit the study of electron impact excitation of atoms under single-collision conditions with specification of the momentum vectors of the incoming and outgoing particles. Then, for a fixed scattered electron momentum vector, the measurement of the angular correlation between that

M. T. Hollywood, A. Crowe, and J. F. Williams • Department of Pure and Applied Physics, Queen's University, Belfast BT7 1NN, Northern Ireland.

electron and the radiated photon permits the determination of the anisotropy of the excited atomic states.

The study of the excitation process can be traced via the work of Smit,[1] Percival and Seaton,[2] Series,[3] Kleinpoppen,[4] and Eminyan et al.,[5] for example, while the analysis of the angular correlations and the description of the excited state have been formulated by Macek and Jaecks,[6] Fano and Macek,[7] and Blum and Kleinpoppen.[8] The analysis closely resembles the theory of perturbed angular correlations in nuclear physics, for example, Frauenfelder and Steffen.[9] The most recent survey of this field in relation to atomic processes has been given by Blum and Kleinpoppen.[10]

In the formalism of Fano and Macek,[7] the anisotropy of the excited state can be described by the "source" parameters consisting of an orientation vector $\mathbf{O}$ proportional to the average angular momentum $\langle J \rangle$ of the given state and of an alignment tensor $\mathbf{A}$, whose components are proportional to the mean values of expressions quadratic in the components of the total angular momentum J. These parameters are related to the observables of the subsequent decay, namely, the photon angular distribution and polarization. The intensity of the emitted radiation described in the detector frame is

$$I = \frac{C}{3} S \left\{ 1 - \frac{1}{2} h^{(2)} A_0^{\text{det}} + \frac{3}{2} h^{(2)} A_{2+}^{\text{det}} \cos 2\beta + \frac{3}{2} h^{(1)} O_0^{\text{det}} \sin 2\beta \right\} \qquad (1)$$

where A_0^{det} and A_{2+}^{det} are elements of $\mathbf{A}$ and O_0^{det} is a component of $\mathbf{O}$. The constants $h^{(k)}$ ($k = 1, 2$) depend on the given transition, C depends on the photon frequency and distance from the source to the detector, S represents the averaged emission over all polarizations and directions, and the angle β is defined by the unit polarization vector $\varepsilon = (\cos \beta, i \sin \beta, 0)$.

The determination of $\mathbf{A}$ and $\mathbf{O}$ follows from equation (1) by determining I as a function of β with appropriate choice of the detector frame. However, this paper studies the $2^1P \rightarrow 1^1S$ (58.4 nm) transition in helium and the $2p \rightarrow 1s$ (121.6 nm) transition in atomic hydrogen for which there are no efficient measures of circular polarization. Rather than measure linear polarizations to determine the components of $\mathbf{A}$ through equation (1), it was more convenient for our existing apparatus to measure, in the scattering plane, the photon angular distributions.

If the spin–orbit and spin–spin interactions are ignored the angular correlation— i.e., the joint probability density for scattering of the electron in a direction (θ_e, ϕ_e) with subsequent emission of the photon in a direction $(\theta_\gamma, \phi_\gamma)$, between the scattered electron and photon for any s–p transition after summing over the photon polarizations—is given by

$$\frac{d^2 P_c}{d\Omega_e \, d\Omega_\gamma} = \frac{\sigma}{\Sigma} \cdot \frac{dP_c}{d\Omega_\gamma}$$

Here Σ and σ are the total and differential cross sections, respectively, for excitation of a given state at a particular energy E. Also $dP_c/d\Omega_\gamma$ is the probability for photon emission after electron scattering and, in this collision reference frame, for coplanar geometry with $\phi_e = 0$, $\phi_\gamma = \pi$, is given by

$$\frac{dP_c}{d\Omega_\gamma} \propto N$$

where

$$N = \tfrac{2}{3} + \tfrac{1}{3}A_0^{\text{col}}(3\cos^2\theta_\gamma - 1) - A_{1+}^{\text{col}}\sin 2\theta_\gamma + A_{2+}^{\text{col}}\sin^2\theta_\gamma \tag{2}$$

Then

$$A_0^{\text{col}} = \langle 3L_z^2 - L^2\rangle/[L(L+1)\sigma] = \tfrac{1}{2}(1 - 3\lambda)$$

$$A_{1+}^{\text{col}} = \langle L_xL_z + L_zL_x\rangle/[L(L+1)\sigma] = 2^{1/2}\,\text{Re}\langle a_0a_1\rangle/\sigma \tag{3}$$

$$A_{2+}^{\text{col}} = \langle L_x^2 + L_z^2\rangle/[L(L+1)\sigma] = \tfrac{1}{2}(\lambda - 1)$$

where $\lambda = \sigma_0/\sigma$ and $\sigma_\mu = \langle a_\mu a_\mu\rangle$ is the differential cross section for exciting the $m_L = \mu$ magnetic sublevel and σ is the total differential cross section. If more than one channel contributes then the $\langle\ \rangle$ denotes a sum over final states and an average over initial states.

The above two $p \to s$ transitions are considered separately since the 2^1P decay in helium involves only one allowed spin channel with a consequently simpler analysis than the $2p$ decay in atomic hydrogen in which the total system can exist in either a singlet or triplet state. In helium, where only one channel contributes, we follow Eminyan et al.[5] and define a_0 to be real and $a_1 = |a_1|e^{i\chi}$ so that $\langle a_0a_1\rangle = [\lambda(1 - \lambda)]^{1/2}e^{i\chi}$. Then $2^{1/2}\,\text{Re}\langle a_0a_1\rangle = \sigma A_{1+}^{\text{col}}$ and $2^{1/2}\,\text{Im}\langle a_0a_1\rangle = -\sigma O_{1-}^{\text{col}}$, where O_{1-}^{col} is the orientation vector and the angular correlation function N is given by

$$N = \lambda\sin^2\theta_\gamma + (1 - \lambda)\cos^2\theta_\gamma - [\lambda(1 - \lambda)]^{1/2}\cos\chi\sin^2\theta_\gamma \tag{4}$$

In atomic hydrogen the matrix elements $\langle a_0a_1\rangle$ do not simplify, so that by defining $R = \text{Re}\langle a_0a_1\rangle/\sigma$ (after Morgan and McDowell[11]) the function $dP_c/d\Omega_\gamma$ is given by $dP_c/d\Omega_\gamma \propto N$, where, for the present geometry,

$$N = 3\lambda + 4 + 3(1 - 2\lambda)\cos^2\theta_\gamma - 3(2^{1/2})R\sin 2\theta_\gamma \tag{5}$$

No information can be deduced about $\text{Im}\langle a_0a_1\rangle$ in the manner adopted for helium.

Previous studies of the $2^1P - 1^1S$ transition in helium have been made by Eminyan et al.,[5,12] Tan et al.,[13] and Ugbabe et al.[14] within the electron scattering angles from $10°$ to $40°$, at the energies of 40–200 eV. Their experimental values of λ were in agreement with the first Born approximation (FBA) only for small θ_e (at 80 eV, $\theta_e \lesssim 10°$) and the experimental χ values were nonzero. Their work also indicated that the discrepancy between theory and experiment increased with electron scattering angle. The recent work of Sutcliffe et al.[15] determined relative values of λ, not from angular correlation measurements over a range of photon angles, but from coincidence count rates at a fixed photon, and variable electron, scattering angles.

This method gives less information than the present experiment as it also does not determine χ.

In atomic hydrogen, angular correlations for 100-eV incident energy electrons were reported several years ago by Williams[16] for $\theta_e = 10°$, $15°$ and recently by Dixon et al.[17] for $\theta_e = 5°$, $10°$, and $15°$. In both studies the angular range of θ_e was too small and limited to small values of θ_e where all theoretical approximations give similar values of λ. The aim of the present studies was to extend the in-plane angular correlation measurements for both He and H up to as large an angle as possible.

2. Apparatus and Experimental Technique

Descriptions of the apparatus for the hydrogen measurements[16] and the helium measurements[18] have been given previously. Both apparatuses are of the crossed electron–atom beam type and use energy analysis of the scattered electron and the time coincidence technique to specify the single collision from which the inelastically scattered electron and the radiated photon originated.

For the hydrogen measurements, the scattered electrons were energy analyzed by a 127° cylindrical electrostatic analyzer and detected by a Mullard B318 BL channel electron multiplier. The 121.6-nm photons were detected by a similar Mullard B419 BL multiplier preceded by magnesium fluoride windows. Grids with suitable potentials prevented charged particles from entering the photon detector. The helium apparatus was basically similar except that the scattered electron energy analyzer was a 180° hemispherical analyzer and no filter could be used to isolate the 58.4-nm radiation.

The electron and photon detectors both produced negative pulses of base width about 15 nsec which were subsequently amplified and their time difference measured by a time-to-amplitude converter and a pulse height analyzer. The time correlation peak was generally of the order of 5 nsec FWHM. For large electron scattering angles, an accumulation time of up to two weeks was required to obtain good statistical accuracy of the number of true coincidence events. The number of true coincidence events as a function of the photon scattering angle was normalized to the number of scattered electrons detected. In this way variations in the incident electron and atom beam intensities and in the electron detector efficiency, for example, were taken into account. Studies have been made[18] of the effect of resonance trapping and of other possible sources of error on the data.

Because the differential cross section decreases by about three orders of magnitude as θ_e increases from $10°$ to $130°$,[19] the coincidence count rate becomes very small at large angles. As an indication of the optimization of the present apparatus for helium (and similarly for the atomic hydrogen apparatus) Figure 1 shows a time coincidence spectrum at an incident electron energy of 81.2 eV, $\theta_e = 10°$, $\theta_y = 140°$, and a channel width of 0.39×10^{-9} sec.

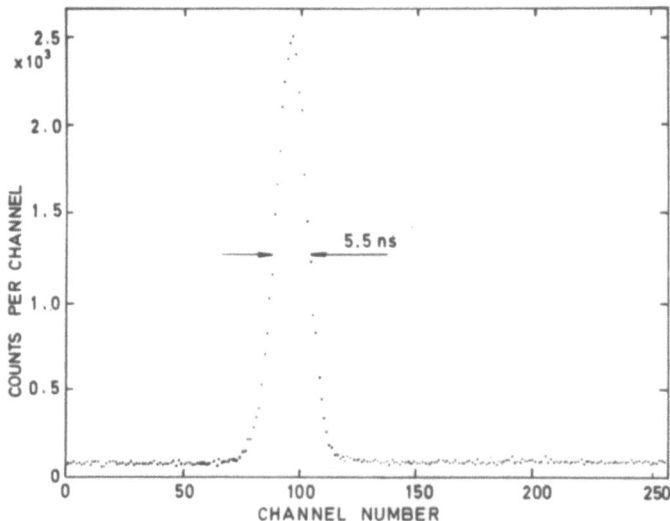

Figure 1. A time coincidence spectrum for an incident electron energy of 81.2 eV, electron scattering angle, $\theta_e = 10°$, photon detector angle, $\theta_\gamma = 140°$, and a channel width of 0.39×10^{-9} sec.

3. Results and Discussion

3.1. Helium

Table 1 lists values of λ, $|\chi|$, the components of the alignment tensor, A^{col}, the orientation vector $|O^{col}_{1-}|$, and the minimum photon angle of the angular correlation function θ_{min} for each electron scattering angle from 10° to 130°. Figure 2 compares typical angular correlations obtained at 16° and 100° with those calculated using the first Born approximation. At all the studied angles there are differences between the experimental data and the calculated FBA values in the angular position of the minimum and in the amplitude of the angular correlation. Even at 16°, the FBA curve lies outside the 70% confidence limits of the experimental data. A fuller discussion of the failure of the FBA to describe the angular correlations has been given elsewhere.[18]

Figure 3 shows the dependence of λ on electron scattering angle θ_e. For angles less than 40° the present values are in agreement with all previous experimental data when allowance is made for the one standard deviation size of error bars quoted for the data. The only other experimental data above 40° are those of Sutcliffe *et al.*[15] Unfortunately they did not make a systematic study of λ beyond 70°, and only their 90° point overlaps the present large angle range. At 90° they find $\lambda = 0.9 \pm 0.1$, which is considerably larger than the present value of 0.613 ± 0.021.

The present data indicate a significant minimum in λ around 110°. The large-angle behavior of λ is an important region to study since the various theoretical predictions vary most in that region. None of the predicted values are in good agree-

Table 1. Values of the Parameters Derived from the Measured Angular Correlations as a Function of the Electron Scattering Angle θ_e for an Incident Electron Energy of 81.2 eV

θ_e (deg)	λ	$|\chi|$ (rad)	θ_{min} (deg)	$|O_{1-}^{col}|$	A_0^{col}	A_{1+}^{col}	A_{2+}^{col}
10	0.516 ± 0.018	0.306 ± 0.068	44.1 ± 0.6	0.150 ± 0.033	−0.274 ± 0.018	0.477 ± 0.027	−0.242 ± 0.005
16	0.323 ± 0.008	0.415 ± 0.030	56.2 ± 0.4	0.188 ± 0.013	0.016 ± 0.008	0.428 ± 0.013	−0.339 ± 0.003
25	0.283 ± 0.007	0.844 ± 0.017	63.0 ± 0.4	0.336 ± 0.007	0.076 ± 0.007	0.299 ± 0.010	−0.359 ± 0.003
50	0.895 ± 0.013	1.924 ± 0.040	−7.5 ± 0.8	0.288 ± 0.019	−0.842 ± 0.013	−0.106 ± 0.020	−0.053 ± 0.006
60	0.892 ± 0.020	2.484 ± 0.141	−16.0 ± 1.6	0.190 ± 0.049	−0.838 ± 0.020	−0.246 ± 0.030	−0.054 ± 0.013
70	0.907 ± 0.045	2.637 ± 0.492	−16.0 ± 3.9	0.140 ± 0.154	−0.861 ± 0.045	−0.254 ± 0.068	−0.047 ± 0.033
80	0.729 ± 0.019	2.091 ± 0.042	−22.0 ± 1.4	0.386 ± 0.016	−0.593 ± 0.019	−0.221 ± 0.028	−0.136 ± 0.008
90	0.613 ± 0.021	1.973 ± 0.038	−29.7 ± 2.2	0.448 ± 0.010	−0.419 ± 0.021	−0.191 ± 0.031	−0.194 ± 0.009
100	0.570 ± 0.019	1.797 ± 0.028	−28.8 ± 2.9	0.482 ± 0.004	−0.355 ± 0.019	−0.111 ± 0.028	−0.215 ± 0.007
110	0.530 ± 0.046	1.619 ± 0.064	−19.5 ± 23.8	0.499 ± 0.003	−0.295 ± 0.046	−0.024 ± 0.069	−0.235 ± 0.016
120	0.578 ± 0.032	1.511 ± 0.044	10.4 ± 7.5	0.493 ± 0.005	−0.368 ± 0.032	0.030 ± 0.048	−0.211 ± 0.011
130	0.647 ± 0.028	1.287 ± 0.039	21.1 ± 2.6	0.495 ± 0.011	−0.471 ± 0.028	0.134 ± 0.042	−0.177 ± 0.009

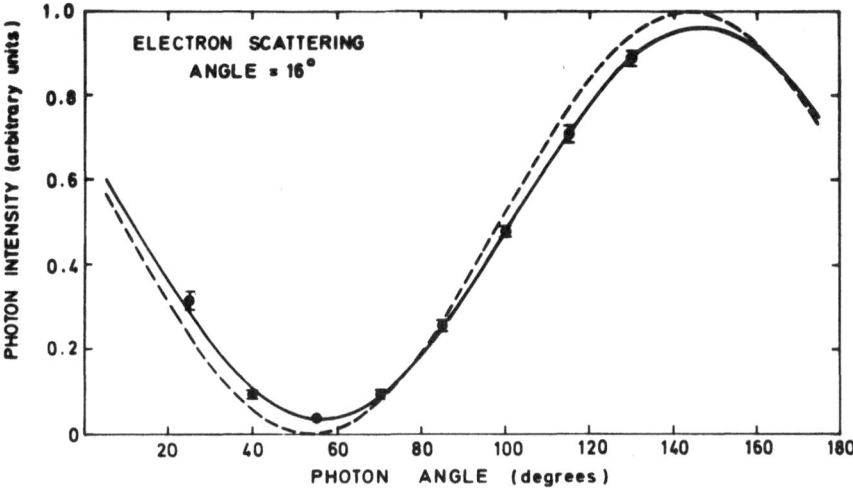

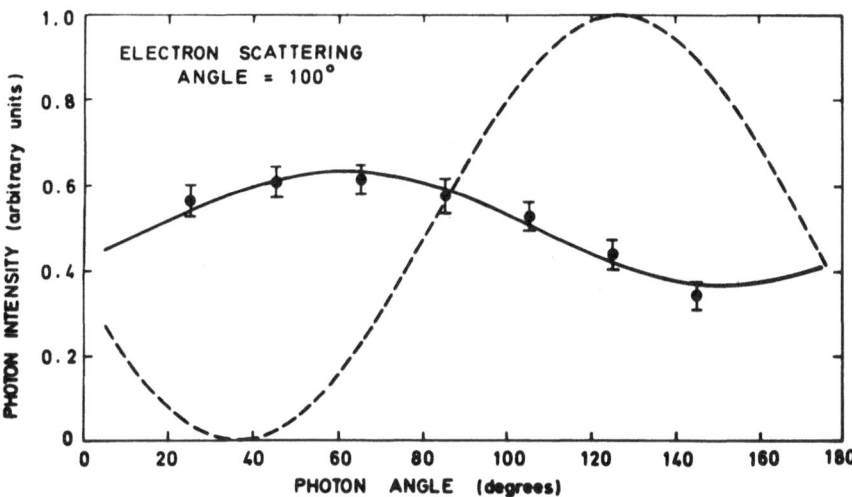

Figure 2. Electron–photon angular correlations in helium showing the normalized coincidence count rate as a function of photon scattering angle for electron scattering angles of 16° and 110° and an incident electron energy of 81.2 eV. The solid-line curve is a chi-squared optimization of equation (4) to the experimental data. The dashed curve is the prediction of the first Born approximation.

ment with our experimental data. The FBA predicts a slowly rising value of λ above 70° with no minimum near 110°. Figure 3 shows theoretical predictions from the distorted-wave approximation of Madison et al. (from Sutcliffe et al.[15]), the distorted-wave polarized orbital approximation of Scott and McDowell,[20] the first-order many-body theory of Thomas et al.,[21] and the five-state R-matrix calculation of Fon et al.[22] For the sake of clarity, the results of Flannery and McCann[23]

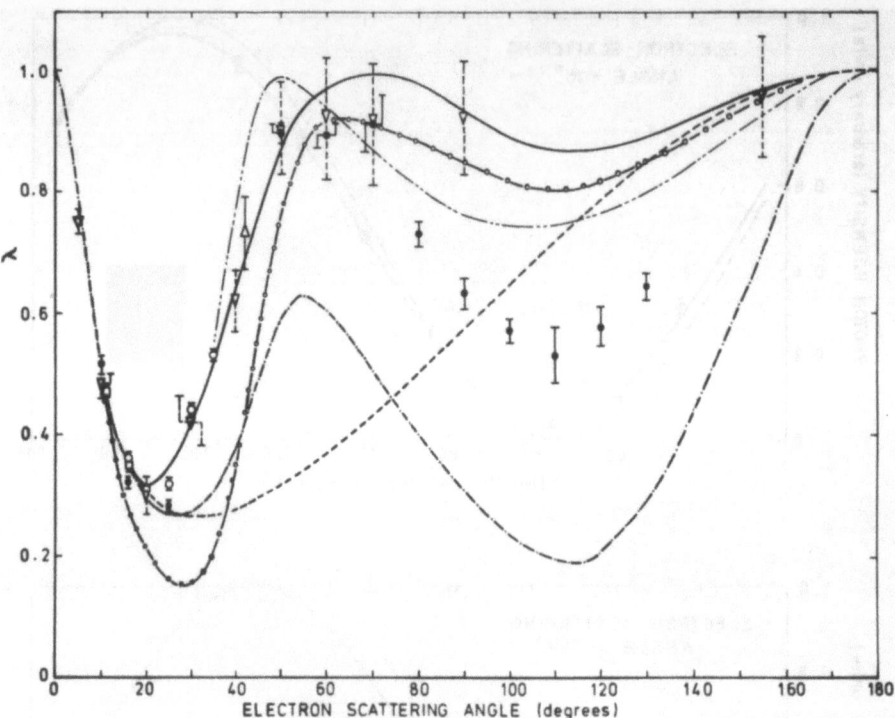

Figure 3. The dependence of λ on electron scattering angle for an incident electron energy of 81.2 eV in helium. The experimental data are the present data ($\bullet$); Eminyan et al.[5] ($\bigcirc$); Ugbabe et al.[14] ($\square$); Tan et al.[13] ($\triangle$); Sutcliffe et al.[15] ($\triangledown$). The theoretical predictions are the fiirst Born approximation (— —); Madison et al. (from reference 15) (——); Thomas et al.[21] (—$\bigcirc$—); Scott and McDowell[20] (—·—); and Fon et al.[22] (—---—).

and Bransden and Winters,[24] which were reported over a limited range of small scattering angles, are not shown. The recent data of Meneses et al.[25] are also not shown since they are similar to those of Thomas et al.[21] All these theories predict the maxima and minima at about the same angles as the observed values. However, there is little quantitative agreement with the observed values. The five-state R-matrix values at large angles are in closest agreement with experiment.

The dependence of $|\chi|$ on electron scattering angle, as shown in Figure 4, is in excellent agreement with the earlier data of Eminyan et al.[5] and Ugbabe et al.[14] at small angles. There is also fair agreement between the general shape predicted by the theories of Scott and McDowell[20] and by Fon et al.[22] and by the present experimental data.

An alternative view of the excitation process can be obtained from the variation of the alignment tensor and orientation vector with electron scattering angle as shown in Figure 5. Since our in-plane angular correlation technique determines only $|\chi|$ we can deduce only $|O_{1-}^{col}|$. From the relationship $2O_{1-}^{col} = P^{circ} = \langle L_y \rangle$ it is deduced that the radiation emitted normal to the scattering plane ($\theta_\gamma = \phi_\gamma = \pi/2$ and $\phi_e = 0$)

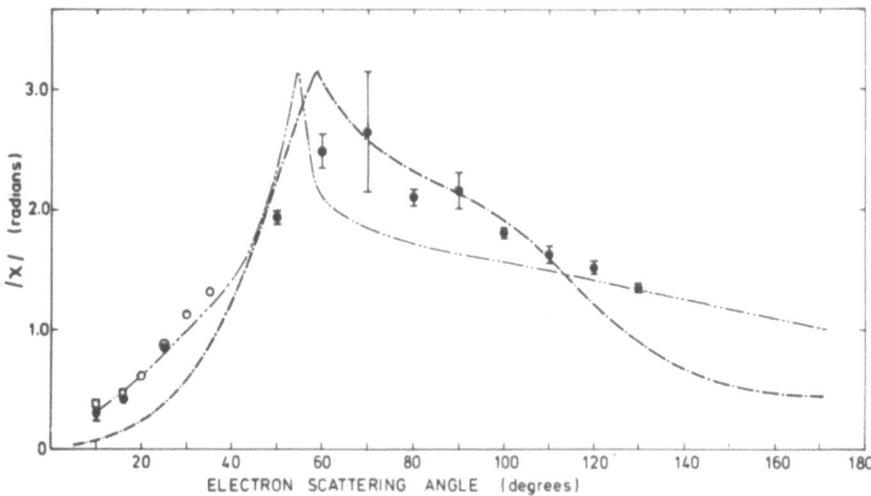

Figure 4. The dependence of $|\chi|$ on electron scattering angle is shown for an incident electron energy of 81.2 eV for the excitation of the 2^1P state of helium. The present experimental data are indicated ($\bullet$). Previous data are Eminyan *et al.*[5] ($\bigcirc$) and Ugbabe *et al.*[14] ($\square$). The theoretical predictions are by Scott and McDowell[20] (—·—) and Fon *et al.*[22] (— — — —).

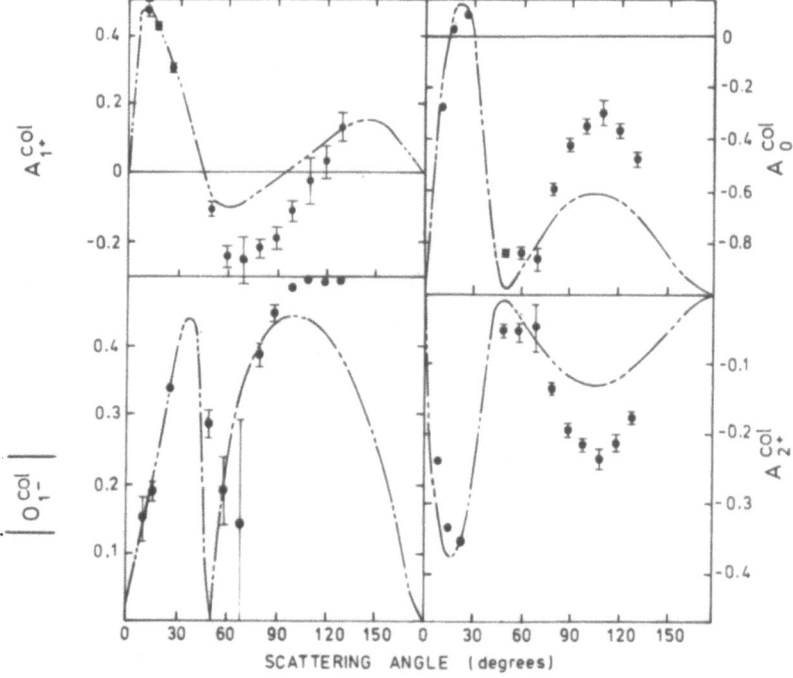

Figure 5. The dependence of the components of the alignment tensor and the orientation vector on the electron scattering angle for 81.2-eV electrons incident on helium atoms. For clarity, only the present data ($\bullet$) and the recent calculation of Fon *et al.*[22] (— — — —) are shown.

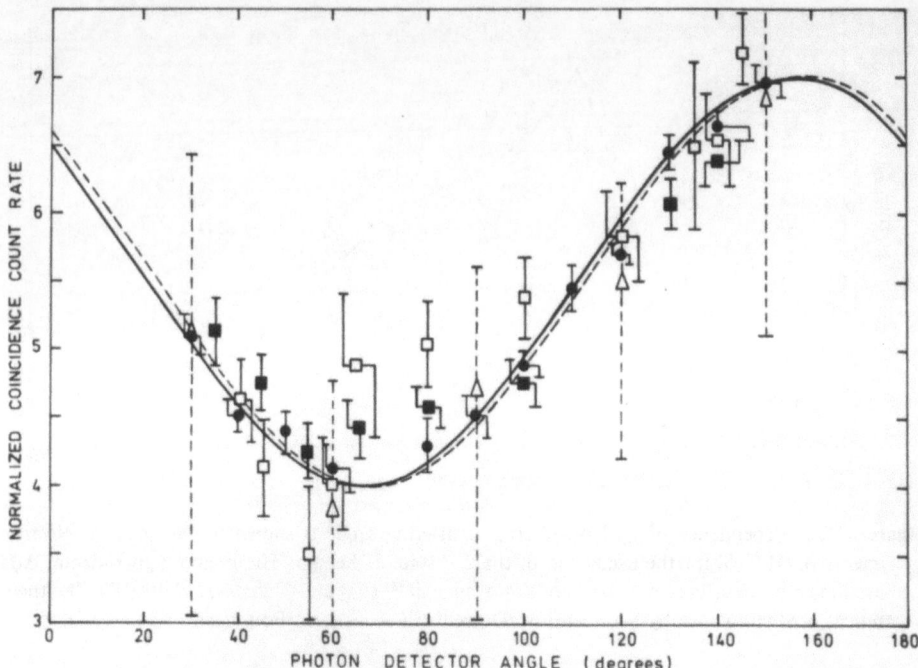

Figure 6. Angular correlations between 10.2-eV energy-loss electrons and 121.6-nm photons for 100-eV electrons incident on atomic hydrogen. The coincidence count rate (normalized) is shown as a function of photon detector angle for an electron scattering angle of 10°. The solid-line curve is a chi-squared optimization of equation (5) to the present experimental data (●). Other experimental data are from Williams[16] (△), Dixon et al.[17] (□), and Hood et al.[26] (■). The dashed curve is the prediction of the first Born approximation.

would be expected to be 100% circularly polarized at angles in the region of 35° and 110°. If we assume the sign of χ from the theory of Scott and McDowell,[20] this radiation is right-hand circularly polarized at $\theta_e \sim 35°$ and left-hand circularly polarized at $\theta_e \sim 110°$. Also from Figure 3, the radiation has zero circular polarization at $\theta_e \sim 70°$. It is also seen at $\theta_e \sim 35°$ and 110° that χ passes through $\pi/2$, A_{1+}^{col} passes through zero, and there is minimum photon intensity in the forward direction.

A_0^{col} is a measure of the anisotropy of the population of the magnetic substates referred to the z axis as quantization axis. For $\theta_e = 0°$ or 180° the conservation of angular momentum requires only the $m_L = 0$ substates to be excited and $A_0^{col} = -1$. Figure 5 shows that $A_0^{col} = 0$ when $\theta_e = 16°$ and 27°. At these scattering angles, then, the three substates are equally populated. This behavior of A_0^{col} alone would imply an isotropic photon angular distribution for an incoherent excitation process, but for the present case both A_{1+}^{col} and A_{2+}^{col} are nonzero at these angles and the observed anisotropic distributions result from interference between the coherent excitation amplitudes.

3.2. The 2p–1s Transition in Atomic Hydrogen

Figure 6 shows a typical angular correlation curve between the 10.2-eV energy loss electrons and the 121.6-nm photons from the $2p$ state of atomic hydrogen for 100-eV incident electrons. The electron scattering angle is 10°. The error bars denote plus and minus one standard deviation. The decrease in the size of the present error bars compared with those of the data of Williams[16] reflects the improvements in the apparatus during the past three years. Also shown for comparison are the recent data of Hood et al.[26] and Dixon et al.[17] The solid curve is a chi-squared optimization of equation (2) to the present data. The dashed curve is the prediction of the FBA, which agrees closely with the best fit to the present data.

Figure 7 shows the dependence of the λ parameter on electron scattering angle for 100-eV electrons incident on atomic hydrogen. The values of Hood et al.[26] at $\theta_e = 10°$, 15°, and 20° are also shown with error bars estimated to represent the chi-squared probability shown in their paper. With measurements at only four values of θ_e it is possible to say that the FBA predicts values of λ at small θ_e ($\lesssim 10°$) and large θ_e ($\gtrsim 150°$) that are in agreement with the experimental values.

All the theoretical results shown in Figure 7 are due to Morgan and Stauffer[27] and Morgan and McDowell.[11] They have used the distorted-wave polarized orbital approximation (McDowell et al.[28]), the Born approximation, and the Coulomb-projected Born approximation (Geltman,[29] and references therein) both with and without the inclusion of exchange. In each of the latter two approximations an electron–nuclear screening parameter ζ was varied between 0 and 1. The effects of exchange on λ were small unless ζ was near zero. However, it is seen that for the

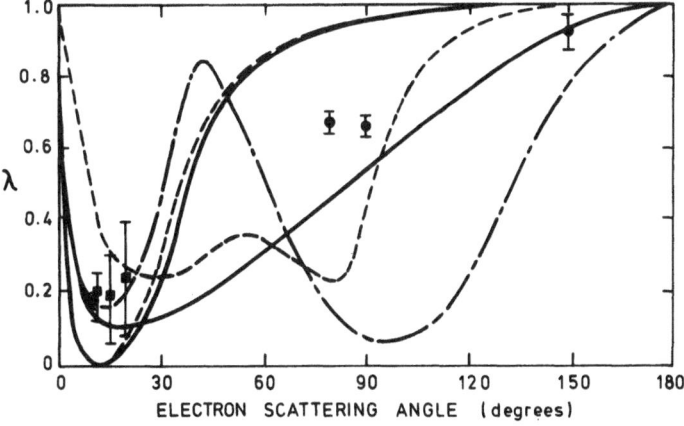

Figure 7. The dependence of the λ parameter on electron scattering angle for 100-eV electrons incident on atomic hydrogen ($1s$–$2p$). The experimental data are the present values (●) and Hood et al.[26] (■). The theoretical predictions are: the first Born (lower solid-line curve) and Coulomb-projected Born approximations (upper solid-line curve); the Born–Oppenheimer (lower dashed curve) and Coulomb-projected Born exchange approximations (upper dashed curve), all from Morgan and Stauffer;[27] and the distorted-wave polarized orbital approximation of McDowell et al.[28] (— – —).

80° and 90° data no theory adequately describes either the shape or the values of the experimental data.

In conclusion it may be seen that although a number of theoretical approximations have previously been shown to fairly well describe the total and differential cross sections for the 2^1P excitation of helium and the $(2p + 2s)$ excitation of atomic hydrogen, no theory gives an adequate description of the coincidence measurements.

Future work will look at the $n = 3$ state excitation of atomic hydrogen, for which a more general approach using the density-matrix formalism, parametrized either with electric and magnetic multipoles and their time derivatives[30] or state multipoles[10] has been used to accurately describe the possible interference effects.

References

1. J. A. Smit, *Physica* **2**, 104 (1935).
2. I. C. Percival and M. J. Seaton, *Phil. Trans. R. Soc. London* **251**, 113 (1958).
3. G. W. Series, in *Physics of One and Two Electron Atoms*, F. Bopp and H. Kleinpoppen, eds., pp. 268–295, North-Holland, Amsterdam (1969).
4. H. Kleinpoppen, in *Physics of One and Two Electron Atoms*, F. Bopp and H. Kleinpoppen, eds., pp. 612–631, North-Holland, Amsterdam (1969).
5. M. Eminyan, K. B. McAdam, J. Slevin, and H. Kleinpoppen, *J. Phys. B* **7**, 1519 (1974).
6. J. Macek and D. H. Jaecks, *Phys. Rev. A* **4**, 2288 (1971).
7. U. Fano and J. H. Macek, *Rev. Mod. Phys.* **45**, 553 (1973).
8. K. Blum and H. Kleinpoppen, *J. Phys. B* **8**, 922 (1975).
9. H. Frauenfelder and R. M. Steffen, in *Alpha, Beta and Gamma Ray Spectroscopy*, K. Siegbahn, ed., Chap. 19, North-Holland, Amsterdam (1965).
10. K. Blum and H. Kleinpoppen, *Phys. Rep.* **52**, 203–261 (1979).
11. L. A. Morgan and M. R. C. McDowell, *J. Phys. B* **8**, 1073 (1975).
12. M. Eminyan, K. B. McAdam, J. Slevin, and H. Kleinpoppen, *Phys. Rev. Lett.* **31**, 576 (1973).
13. K. H. Tan, J. Fryar, P. S. Farago, and J. W. McConkey, *J. Phys. B* **10**, 1073 (1977).
14. A. Ugbabe, P. J. O. Teubner, E. Weigold, and H. Arriola, *J. Phys. B* **10**, 71 (1977).
15. V. C. Sutcliffe, G. N. Haddad, N. C. Steph, and D. E. Golden, *Phys. Rev. A* **17**, 100 (1978).
16. J. F. Williams, in *The Physics of Electronic and Atomic Collisions: Invited Lectures of IXth ICPEAC*, J. S. Risley and R. Geballe, eds., pp. 139–150, Washington University Press, Seattle, Washington (1975).
17. A. J. Dixon, S. T. Hood, and E. Weigold, *Phys. Rev. Lett.* **40**, 1262 (1978).
18. M. T. Hollywood, A. Crowe, and J. F. Williams, *J. Phys. B* **12**, 819 (1979).
19. A. Chutjian and S. K. Srivastava, *J. Phys. B* **8**, 2360 (1975).
20. T. Scott and M. R. C. McDowell, *J. Phys. B* **9**, 2235 (1976).
21. L. D. Thomas, G. Csanak, H. S. Taylor, and B. S. Yarlagadda, *J. Phys. B* **7**, 1719 (1974).
22. W. C. Fon, K. A. Berrington, and A. E. Kingston, *J. Phys. B* **12**, L171 (1979).
23. M. R. Flannery and K. J. McCann, *J. Phys. B* **8**, 1716 (1975).
24. B. H. Bransden and K. H. Winters, *J. Phys. B* **9**, 1115 (1976).
25. G. D. Meneses, N. T. Padial, and G. Csanak, *J. Phys. B* **11**, L237 (1978).
26. S. T. Hood, E. Weigold, and A. J. Dixon, *J. Phys. B* **12**, 631 (1979).
27. L. A. Morgan and A. D. Stauffer, *J. Phys. B* **8**, 2342 (1975).
28. M. R. C. McDowell, L. A. Morgan, and V. P. Myerscough, *J. Phys. B* **8**, 1053 (1975).
29. S. Geltmann, in *Electron and Photon Interactions with Atoms*, M. R. C. McDowell and H. Kleinpoppen, eds., pp. 387–396, Plenum, New York (1975).
30. G. Gabrielse and Y. B. Band, *Phys. Rev. Lett.* **39**, 697 (1977).

9

New Aspects in Electron–Photon Coincidence Experiments

H. Kleinpoppen and I. McGregor

Results are given of some recent studies that extend the range of topics that are now investigated using the electron–photon coincidence method. The topics discussed are the use of this method to yield submillivolt electron energy resolution in electron differential cross-section studies for the excitation of the helium 3^3D state, the extension of angular correlation data to krypton, an atom with which spin–orbit coupling plays a role, and a theoretical approach to coherent excitation of degenerate excited states.

1. Introduction

In the short period since the development of electron–photon coincidence techniques for the study of electron impact excitation of atoms, several breakthroughs have been achieved which can be considered as characteristic features of this novel method. When the first ideas on electron–photon angular correlations came up,[1] an analogy to impact threshold polarization and to magnetic sublevel excitation was the guiding concept. In the course of the development of the theory of electron–photon coincidences[2] an amplitude representation for exciting substates was introduced which led to angular correlation parameters λ and χ, where λ is the ratio of magnetic substate cross section σ for exciting the $m_l = 0$ substate to the total differential cross sections σ and where χ is the phase between the excitation amplitudes a_0 and a_1 for exciting the magnetic sublevels $m_l = 0$ and $m_l = \pm 1$ of ^1P states.[3,4] The parameters λ and χ proved to be crucial test parameters for theoretical approx-

H. Kleinpoppen and I. McGregor • Institute of Atomic Physics, University of Stirling, Stirling, Scotland. Dr. Kleinpoppen's temporary address is Fakultät für Physik and Zentrum für interdisziplinäre Forschung of Universität Bielefeld, 48 Bielefeld, West Germany.

imations (particularly the phase χ, a very sensitive test quantity for theory). Calibration of $\lambda = \sigma_0/\sigma$ to σ provided sets of data for the magnetic sublevel cross sections σ_0 and σ_1[5]. In addition to this relation between angular correlation data and scattering amplitudes and phases, the description of such data in terms of "target" parameters like the orientation vector, alignment tensors,[6] and also the state multipoles[7] opened another wide range of new aspects. A complete analysis of the target parameters for orientation and alignment of helium ^1P states was possible by applying data sets of λ and χ (including the sign[8]).

A thorough study of the Stokes parameters of the coincident photons with regard to the consistency and the validity of the model of coherent excitation for the magnetic substates of the ^1P state of helium led to an unambiguous conclusion that the coherent sublevel excitation is not only complete (i.e., the degree of coherency for the excitation of the ^1P magnetic sublevels was measured to 100%) but that the coincident photons display the phase χ between the amplitudes a_0 and a_1 as a macroscopic measurable phase difference between two orthogonal light vectors.[8]

Several experimental groups have recently taken up the angular correlation method for impact studies of atoms. These applications confirmed and extended the measurements of the Stirling group. The range of angular variation for electron–photon angular correlations of helium was extended;[9–11] applications to other atomic systems were also undertaken[12–15] (see Section 3 in this chapter). Considerations with regard to spin-dependent processes in electron–photon angular correlations from hydrogen[16] and mercury[17] were undertaken. Such experiments will allow separation of singlet–triplet excitation and will reveal further details on spin–orbit interactions.

In this chapter we report on two further extensions of experimental studies of electron–photon angular correlations.

In Section 2 we discuss the use of this method to give enhanced resolution in differential electron cross-section data. Finally, in Section 4 we would like to mention the predictions of interesting "even–odd" parity coherence effects[18] for coherent excitation of hydrogenic states with opposite parity. Such effects should be measurable in electron–Lyman-α coincidence experiments. Furthermore, coherence effects from the excitation of $n = 3$ states of atomic hydrogen[19] should be observable through cascading effects leading to electron–Lyman-α coincidences.

2. A Method for Measuring the Electron Differential Cross Section (on an Absolute Scale) for Excited States of Helium Separated by 0.5 meV from Adjacent States

In general, if one wishes to measure a differential cross section for inelastic electron scattering, one would measure the scattered electron intensity for a particular energy loss. This method is only meaningful if the apparatus has an energy resolution adequate to pick out only those electrons corresponding to the state of interest and to reject electrons from all other possible excitation states. Such a measurement can be interpreted by the following equation:

$$I(\theta) = K'V\frac{d\sigma}{d\theta} \tag{1}$$

where $d\sigma/d\theta$ is the unknown differential cross section, V relates to the scattering volume, K corresponds to all other relevant parameters (e.g., system detection efficiency, atom density, electron current, etc.) and $I(\theta)$ is the measured scattered electron intensity.

If the resolution of the apparatus is not good enough to resolve the electron energy loss spectrum into all of its states, then the method outlined above would be of little use. However, if a coincidence measurement is performed between the scattered electron and the photon emitted from the decay of the state being studied, use can then be made of the additional resolution given by the photon.[20] A narrow-line interference filter would be necessary to ensure that the only overlap between the photons that are detected and the electrons that are detected corresponds to the state being investigated. This process can be understood by

$$N_c(\theta) = K''V\frac{d\sigma}{d\theta} \tag{2}$$

which is similar to equation (1), differing only in the terms included in the constant K'' and with the measured quantity now the coincident count. If the photons are detected normal to the scattering plane, the photon detection is insensitive to the electron scattering angle and consequently equation (2) gives a measure of the electron differential cross section.

In order to establish the absolute scale for such a measurement and to linearize any instrumental effect, the parameters K'' and V and any angular dependence must be known in full. An alternative method is to combine a coincidence measurement [represented by equation (2)] for the unresolved state with an intensity measurement [represented by equation (1)] for a well-resolved state under identical experimental conditions. By forming the ratio for the experimentally observed quantities, all terms in K' and K'' relating to the experimental conditions cancel and the remaining factor can be determined by normalizing to any suitable cross-section data for the state in question which is known either theoretically or experimentally.

This latter method has been applied to resolve the 3^3D state of helium from its adjacent 3^1D and $3^{1,3}P$ states with a total resolution from the scattered electrons of 120 meV (see Figure 1). The experimental arrangement (Figure 2) comprises an electron gun, a 127° monochromator, and a lens system that produces a focused beam of electrons with an energy resolution of about 90 meV onto a gas beam of diameter 1 mm produced by a multicapillary array. The electrons scattered through a known angle are collected and energy analyzed by a 127° analyzer whose resolution is similar to that of the monochromator, before being detected by a channeltron

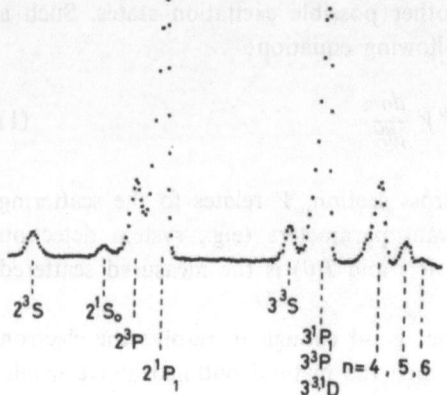

Figure 1. Electron energy loss spectrum for scattering from helium from a system with overall electron energy resolution of 120 meV.

multiplier. The light emitted from the decay of the 3^3D state is collected by a lens positioned normal to and above the scattering plane and detected by an XP1017 photomultiplier. The coincidence circuit used is a typical arrangement[4] comprising a fast amplifier and a constant fraction discriminator for both channels, whose outputs are fed into a time-to-amplitude converter and then to a pulse-height analyzer.

The electron analyzer was tuned for the energy loss corresponding to the 3^1P state. The experimental conditions were thus maintained constant for both the coincidence measurement for the excitation of the helium 3^3D state and for the in-

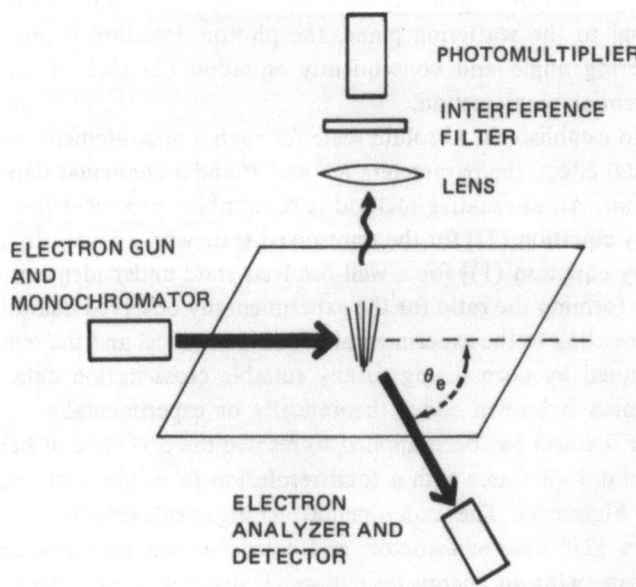

Figure 2. Schematic diagram of the experimental arrangement used to measure the electron differential cross section for excitation of the helium 3^3D state using an electron–photon coincidence method.

Table 1. Absolute Differential Cross Section for Electron Scattering at an Energy of 39.7 eV
for Excitation to the 3^3D State of Helium (in Units of 10^{-20} cm^2/sr)

θ (deg)	Measured values	Theoretical values[a]
30	4.3	4.2
35	3.8	3.6
40	2.9	2.8
50	2.4	2.4
60	1.7	1.7
75	0.7	0.7
82	0.3	0.6
90	0.3	0.5

[a] Reference 21.

tensity measurement for the excitation of the combined 3^1P, 3^1D, and 3^1D states. Using combined states such as the 3^1P, $3^{1,3}D$ states rather than a single well-resolved state such as the 2^1P state should not result in any additional error provided the relevant cross-section data are available, and if, as here, the combined state includes the state under study, it ensures that identical experimental conditions prevail for both measurements. The cross-section data used here are all taken from Chutjian and Thomas,[21] who measured the differential cross section for electron scattering and excitation of the combined 3^1P, $3^{1,3}D$ state of helium, and who calculated, using a first-order many-body theory, the differential cross section for electron scattering and excitation of the 3^3D state.

Preliminary results are shown in Table 1 and may be compared with the theoretical values of Chutjian and Thomas. The error representing the statistical error (one standard deviation) in our measurements is less than 10% for all angles except 20° and 90°, for which it is 15%. The error in normalizing to the known data and in the known data is estimated to be about 40% for the angular range covered. Although the angular range covered is limited and the error fairly large, it does illustrate another aspect of the versatility of the electron–photon coincidence method in filtering out particular aspects of interest in atomic physics.

3. Angular Correlations in Electron Excitation of Krypton

This study of krypton is an extension of previous studies in helium,[22,11] neon,[14] and argon.[13,15] Krypton is an atom in which spin–orbit coupling during the collision process and the deexcitation process can no longer be ignored if a full understanding is to be made. The krypton "singlet," 5^1P_1, and "triplet," 5^3P_1, states have been studied. These states are well defined in energy, being separated by 600 meV, and may be thought of as mixtures of pure states described by the LS coupling scheme. To date, only the limiting cases (pure LS and pure JJ coupling schemes) have been considered.[2] Although it may be more meaningful to analyze the data in terms of a multipole expansion, which would avoid the necessity of making theoretical assumptions in the initial state, the analysis in this paper will be confined to using terms of the λ and χ formulation based on the LS coupling scheme. Since even isotopes account for almost 90% of naturally occurring krypton, hyperfine effects are neglected in the analysis. The preliminary data presented here are for incident electron energies of 36 and 60 eV and for electron scattering angles of 20°, 30°, and 40°.

The experimental arrangement is similar to that described in the initial helium studies carried out at Stirling.[22] The only significant difference is that the photons are detected using a Bendix uv detector (BX 762) which has a magnesium fluoride window with a cutoff at 114.0 nm. The wavelengths of the two lines that have been studied are just above the cutoff at 123.6 nm and 116.5 nm for the triplet $(4p^55s\ ^3P_1)$ and the singlet $(4p^55s\ ^1P_1)$ states, respectively. The solid angle for detection of the photons is 0.03 sr and the photon detector could be rotated in the scattering plane through the angular range of 35°–120°.

To avoid resonance trapping of the radiation, the background pressure during the runs was maintained at 2×10^{-7} mbar, compared to the base pressure of 5×10^{-8} mbar with the beam off. The absence of resonance trapping was confirmed by performing a series of measurements for both the singlet and the triplet states at an electron scattering angle of 40° for an incident electron energy of 36.6 eV. The photon angle was set to correspond to the maximum in the angular correlation curve, which is the point most sensitive to resonance trapping. The effect of resonance trapping would be to decrease the amplitude of the correlation curve and/or to alter the θ_{min} value. The onset of trapping then corresponds to a decrease in the normalized coincidence count at this photon angle. To ensure that for these and other scattering parameters resonance trapping can safely be neglected, the background pressure (and beam density) was reduced by a factor of 2 from the onset of trapping as measured above.

An overall energy resolution of 175 meV as measured by the electron analyzer was employed since the gain in primary electron current from the monochromator and in throughput of the analyzer is more important than high resolution. The presence of the metastable triplet $^3P_{0,2}$ states does not present difficulties since the timing circuit prevents them giving rise to any real coincidences. A check is at present being carried out to ensure that the magnesium fluoride window does not exhibit any birefringence effects, which could result in errors being introduced in the measured angular correlation due to polarization of the photons.

For coplanar geometry the angular correlation function is

$$N_{1,3} = \lambda_{1,3}\sin^2\theta_\gamma + (1-\lambda)\cos^2\theta_\gamma - 2[\lambda_{1,3}(1-\lambda_{1,3})]^{1/2}\cos\chi_{1,3}\sin\theta_\gamma\cos\theta_\gamma \quad (3)$$

Table 2. Values of λ, χ and θ_{min} Obtained for Krypton at the Given Electron Energies and Scattering Angles for the Excitation of the Resonant 5^1P_1 and 5^3P_1 States, (denoted by S and T, respectively)

Energy (eV)	Electron scattering angle (deg)	State	λ	χ (rad)	θ_{min} (deg)
36	20	S	0.39 ± 0.03	1.93 ± 0.05	119.0 ± 4.9
		T	0.36 ± 0.02	1.93 ± 0.03	114.9 ± 3.6
	30	S	0.79 ± 0.02	1.70 ± 0.03	174.9 ± 0.8
		T	0.72 ± 0.03	1.65 ± 0.04	175.4 ± 1.6
	40	S	0.73 ± 0.03	1.06 ± 0.02	21.2 ± 2.4
		T	0.75 ± 0.03	1.12 ± 0.03	18.3 ± 2.4
60	20	S	0.29 ± 0.02	1.77 ± 0.03	11.7 ± 2.0
		T	0.29 ± 0.02	1.74 ± 0.02	10.1 ± 1.9
	30	S	0.96 ± 0.11	1.40 ± 0.18	1.5 ± 2.9
		T	0.75 ± 0.09	1.35 ± 0.06	10.5 ± 5.3

where the subscripts 1 and 3 refer to the singlet and triplet transitions, respectively. λ is the ratio of the differential cross section for excitation of the $M = 0$ sublevel to the total differential cross section, i.e., $\lambda = \sigma_0/\sigma$, and χ is the relative phase between the scattering amplitudes a_0 and a_1 and can be defined by $a_1 = |a_1| e^{i\chi}$ if a_0 is real and positive.

The data were fitted to the angular correlation function N multiplied by a normalizing factor by means of a least-squares routine and the results for λ, χ, and θ_{min} are shown in Table 2. The data could also have been interpreted in terms of the alignment tensor and orientation vector[6] whose terms are related to λ and χ. The major characteristic of these results is that for any given excitation parameters, the values obtained for λ, χ, and θ_{min} for the singlet and triplet are equal, or almost so, over the angular and energy range studied. This contrasts with those results of Malcolm and McConkey[15] for argon whose values are for smaller electron scattering angles than studied here. This contrast may be explained by the fact that krypton, rather than argon, is more likely to be affected by the breakdown in the LS coupling scheme, and that for large impact parameters (i.e., small-angle scattering), spin–orbit coupling of the incident electron to the atom will be small. These results would seem to confirm the breakdown of the LS coupling scheme for krypton and imply by their equalities that the two states studied can be considered as mixtures of coherently excited singlet and triplet states.

4. Coherent Excitation of Degenerate Atomic States

Coherent impact excitation has been reported or predicted for various types of experiments: in particle–photon angular correlations as summarized in several chapters of this book,[15,11,23,24] in total excitation cross sections of hydrogen line radiation under the condition of Stark mixing,[25] and in impact excitation with polarized electrons and polarized atoms.[26] While excitation amplitudes or the Fano–Macek alignment and orientation parameters are adequate quantities for describing coherent excitation of states with sharp angular momentum, a generalization into applications of state multipoles (which may be coherently excited) is necessary for excitation processes of degenerate states with different angular momentum.

Gabrielse and Band[27] parametrized the density matrix of coherently excited states in terms of the various electric and magnetic multipoles and their time derivatives. They particularly showed that the "coherent" multipoles, characterizing the interference terms, provide new information about the scattering process, and they stressed the importance of a determination of these "interference parameters."

Blum and Kleinpoppen[28] described an equivalent method in terms of state multipoles following earlier work by Fano[29] in nuclear physics. We consider as an example of this method coherent electron impact excitation of hydrogenic states which have the same principal quantum number n but different orbital angular momentum L and component M. We denote the spin-averaged scattering amplitudes by

$$\langle f_{L'M'} f_{LM}^{*} \rangle = \sum_{S} (2S + 1) f_{L'M'}^{(S)} f_{LM}^{(S)*} \tag{4}$$

where S is the total spin of projectile and atomic target. We normalize the scattering amplitudes to the partial differential cross sections for exciting a level with quantum numbers n, L, M averaged over all spins:

$$\tfrac{1}{4} \sum_{S} (2S + 1) \, | f_{LM}^{(S)} |^{2} = \sigma_{LM} \tag{5}$$

The quantities of equation (4) are elements $\langle L'M' | \varrho | LM \rangle$ of the density matrix ϱ describing the atomic ensemble. Quantities with $L \neq L'$, $M' \neq M$ in equation (4) are in interference terms due to coherent excitation of these states. It can be shown that there is, for each pair of L, L', a relationship between "state multipoles" (Fano[29]) and the quantities of equation (4):

$$\langle T(L'L)_{KQ}^{+} \rangle = \sum_{MM'} \langle f_{L'M'} f_{LM}^{*} \rangle (-1)^{LM} (LM, L' - M' | K - Q) \tag{6}$$

with $(LM, L' - M' | K - Q)$ as a standard Clebsch–Gordan coefficient (Blum and Kleinpoppen[28]). From the properties of the Clebsch–Gordan coefficients follow the limitations $K \leq L + L'$, $-K \leq Q \leq K$. Note the relations

$$\langle T(L'L)^{+}_{KQ} \rangle = \langle T(L'L)_{KQ} \rangle^{*} \tag{7a}$$

where the asterisk stands for complex conjugation, and

$$\langle T(L'L)^{+}_{KQ} \rangle = (-1)^{L'+L+K+Q} \langle T(L'L)^{+}_{K-Q} \rangle \tag{7b}$$

which follows from reflection invariance in the scattering plane, and

$$\langle T(L'L)^{+}_{KQ} \rangle = (-1)^{K} \langle T(LL')^{+}_{KQ} \rangle \tag{7c}$$

which follows from equations (7a) and (7b).

The state multipoles of equation (6) can be related to the Fano–Macek "orientation vector" (O_{1-}) and "alignment tensors" (A_0, A_{1+}, A_{2+}) of atomic states with sharp angular momentum L. For example, the quantities $\langle T(LL)^{+}_{1Q} \rangle$ and $\langle T(LL)^{+}_{2Q} \rangle$ are proportional to the spherical components of the "orientation vector" $\langle L_Q \rangle$ and of the "alignment tensor."

As an example we list here the state multipoles and their connections to the scattering amplitude (f_{LM}), the orientation vector (O_{1-}), and the alignment tensor (A_{2+}, A_{1+}, A_0), which contribute to the transition from the coherently excited hydrogenic $n = 3$ levels to the $n = 2$ level in the field free case:

$$\langle T(2, 2)^{+}_{22} \rangle = (8/7)^{1/2} \operatorname{Re}\langle f_{20} f^{*}_{22} \rangle - (3/7)^{1/2} \langle | f_{21} |^2 \rangle$$

$$\langle T(22)^{+}_{21} \rangle = -(12/7)^{1/2} \operatorname{Re}\langle f_{21} f^{*}_{22} \rangle - (2/7)^{1/2} \operatorname{Re}\langle f_{20} f^{*}_{21} \rangle$$

$$\langle T(22)^{+}_{22} \rangle = (8/7)^{1/2} \langle | f_{22} |^2 \rangle - (2/7)^{1/2} \langle | f_{21} |^2 \rangle - (2/7)^{1/2} \langle | f_{20} |^2 \rangle$$

$$\langle T(22)^{+}_{11} \rangle = i(4/5)^{1/2} \operatorname{Im}\langle f_{21} f^{*}_{22} \rangle + i(5/6)^{1/2} \operatorname{Im}\langle f_{20} f^{*}_{21} \rangle$$

$$\langle T(22)^{+}_{00} \rangle = (1/5)^{1/2} \sigma(3d)$$

$$\langle T(20)^{+}_{22} \rangle = \langle f_{22} f^{*}_{00} \rangle$$

$$\langle T(20)^{+}_{21} \rangle = \langle f_{21} f^{*}_{00} \rangle$$

$$\langle T(20)^{+}_{20} \rangle = \langle f_{20} f^{*}_{00} \rangle$$

$$\langle T(00)^{+}_{00} \rangle = \sigma(3s)$$

$$\langle T(11)^{+}_{11} \rangle = \sigma(3p) O_{1-}$$

$$\langle T(11)^{+}_{21} \rangle = -\sigma(3p) A_{1+}$$

$$\langle T(11)^{+}_{20} \rangle = (2/3)^{-1/2} \sigma(3p) A_0$$

The observation of the light, emitted in the decay of the atomic states, in coincidence with the scattered electrons allows one to extract more detailed information (in particular about the interference terms) than the measurement of the differential

Table 3. Numerical Values of Multipole Parameters of $n = 3$ States of Atomic Hydrogen for Various Electron Energies (E) and Electron Scattering Angles (θ)

E (eV)	θ	$\langle T(0,0)_{00}^{\pm}\rangle$	$\langle T(2,0)_{20}^{\pm}\rangle$	$\langle T(2,0)_{21}^{\pm}\rangle$	$\langle T(2,0)_{22}^{\pm}\rangle$	$\langle T(2,2)_{00}^{\pm}\rangle$	$\langle T(2,2)_{20}^{\pm}\rangle$	$\langle T(2,2)_{21}^{\pm}\rangle$	$\langle T(2,2)_{22}^{\pm}\rangle$
200	2	1.31-1	8.97-2	-3.84-3	-1.10-1	1.10-1	1.97-1	5.27-3	-8.06-2
	4	1.22-1	1.69-2	-6.59-3	-9.44-2	8.70-2	1.55-1	8.31-3	-6.36-2
	6	1.08-1	5.95-2	-7.68-3	-7.33-2	5.97-2	1.06-1	8.54-3	-4.35-2
	8	8.89-2	4.18-2	-7.23-3	-5.17-2	3.62-2	6.39-2	6.89-3	-2.64-2
	10	6.83-2	2.69-2	-5.86-3	-3.35-2	1.99-2	3.48-2	4.72-3	-1.44-2
	12	4.91-2	1.61-2	-4.24-3	-2.02-2	1.01-2	1.75-2	2.88-3	-7.32-3
	14	3.34-2	9.12-3	-2.83-3	-1.15-2	-4.88-3	8.36-3	1.62-3	-3.52-3
	16	2.17-2	4.95-3	-1.77-3	-6.31-3	2.27-3	3.84-3	8.58-4	-1.63-3
	18	1.37-2	2.61-3	-1.07-3	-3.36-3	1.04-3	1.73-3	4.40-4	-7.42-4
	20	8.40-3	1.35-3	-6.23-4	-1.77-3	4.71-4	7.68-4	2.21-4	-3.34-4
400	2	1.28-1	8.52-2	-3.65-3	-1.04-1	-1.02-1	1.82-1	4.86-3	-7.45-2
	4	1.11-1	6.28-2	-5.39-3	-7.71-2	6.42-2	1.15-1	6.13-3	-4.70-2
	6	8.39-2	3.81-2	-4.92-3	-4.70-2	3.15-2	5.60-2	4.51-3	-2.30-2
	8	5.53-2	1.96-2	-3.39-3	-2.43-2	1.28-2	2.26-2	2.44-3	-9.33-3
	10	3.23-2	8.90-3	-1.94-3	-1.11-2	4.60-3	8.05-3	1.09-3	-3.34-3
	12	1.73-2	3.73-3	-9.81-4	-4.67-3	1.54-3	2.67-3	4.38-4	-1.11-3
	14	8.72-3	1.49-3	-4.63-4	-1.88-3	5.00-4	8.57-4	1.66-4	-3.61-4
	16	4.27-3	5.87-4	-2.10-4	-7.48-4	1.63-4	2.75-4	6.14-5	-1.17-4
	18	2.07-3	2.31-4	-9.45-5	-2.98-4	5.39-5	8.96-5	2.28-5	-3.85-5
	20	1.01-3	0.27-5	-4.27-5	-1.21-4	1.84-5	3.01-5	8.64-6	-1.31-5
500	2	1.27-1	8.31-2	-3.56-3	-1.02-1	9.28-2	1.75-1	4.68-3	-7.16-2
	4	1.05-1	5.68-2	-4.87-3	-6.98-2	5.54-2	9.90-2	5.29-3	-4.05-2
	6	7.35-2	3.07-2	-3.96-3	-3.78-2	2.33-2	4.15-2	3.34-3	-1.70-2
	8	4.35-2	1.37-2	-2.37-3	-1.70-2	7.98-3	1.41-2	1.52-3	-5.81-3
	10	2.25-2	5.39-3	-1.17-3	-6.70-3	2.42-3	4.24-3	5.75-4	-1.76-3
	12	1.07-2	1.97-3	-5.17-4	-2.46-3	6.93-4	1.20-3	1.97-4	-5.01-4
	14	4.82-3	6.95-4	-2.16-4	-8.78-4	1.96-4	3.37-4	6.51-5	-1.42-4
	16	2.14-3	2.46-4	-8.81-5	-3.13-4	5.69-5	9.60-5	2.15-5	-4.08-5
	18	9.55-4	8.85-5	-3.61-5	-1.14-4	1.71-5	2.84-5	7.23-6	-1.22-5
	20	4.33-4	3.28-5	-1.51-5	-4.28-5	5.38-6	8.77-6	2.52-6	-3.82-6

cross section.[28] States with different parity cannot decay to the same final state. Thus, no information about interference terms with $L' + L$ odd can be obtained as long as no external field is present that mixes states with different parity. If levels with $n = 3$ have been excited and the light emitted in the $n = 3 \rightarrow n = 2$ decay is observed, the intensity and polarization of the radiation depends on all the terms $\langle T(LL)^+_{KQ} \rangle$ with $L = 0, 1, 2$ and $K \leq 2$ and on the multipoles describing s–d-interference, that is, $\langle T(0, 2)^+_{KQ} \rangle$ with $K = 2$. That is, in order to describe the properties of the emitted light one has to calculate the relevant terms $\langle T(LL)^+_{KQ} \rangle$ related to orientation vector and alignment tensor and in addition the interference terms $\langle T(0, 2)^+_{KQ} \rangle$.

Numerical results for the excitation of the $n = 3$ levels have been given recently in Reference 30. However, these authors did not calculate terms corresponding to $\langle T(0, 2)^+_{KQ} \rangle$. We present in Table 3 results for the relevant multipoles obtained in first-order Born approximation. Note that in Born approximation the terms $\langle T(LL)^+_{1Q} \rangle$ and $\langle T(0, 2)^+_{1Q} \rangle$ are zero.

ACKNOWLEDGMENTS

We would like to gratefully acknowledge the assistance of Dr. A. Chutjian during the cross-section work and of Drs. S. Al-Shamma, R. Hippler, and J. F. Williams in the preparation for and in the collection of data in the krypton studies. This work has been supported in part by the Science Research Council.

References

1. H. Kleinpoppen, unpublished report, Columbia University, New York (1967).
2. J. Macek and D. H. Jaecks, *Phys. Rev. A* **4**, 2288 (1971).
3. M. Eminyan, K. B. MacAdam, J. Slevin, and H. Kleinpoppen, *Phys. Rev. Lett.* **31**, 576 (1973).
4. M. Eminyan, K. B. MacAdam, J. Slevin, and H. Kleinpoppen, *J. Phys. B* **7**, 1519 (1974).
5. A. Chutjian, *Bull. Am. Phys. Soc.* (1975); *J. Phys. B* **9**, 1749 (1976).
6. U. Fano and J. Macek, *Rev. Mod. Phys.* **45**, 553 (1973).
7. K. Blum and H. Kleinpoppen, in *Electron and Photon Interactions with Atoms*, H. Kleinpoppen and M. R. C. McDowell, eds., p. 501, Plenum Press, New York (1976).
8. M. C. Standage and H. Kleinpoppen, *Phys. Rev. Lett.* **36**, 577 (1975).
9. K. H. Tan, J. Fryar, P. S. Farago, and J. W. McConkey, *J. Phys. B* **10**, 1073 (1977).
10. V. C. Sutcliffe, G. N. Haddad, N. C. Steph, and D. E. Golden, *Phys. Rev. A* **17**, 100 (1978).
11. M. T. Hollywood, A. Crowe, and J. F. Williams, *J. Phys. B.*, to be published (1979); Chapter 8 of this book.
12. J. F. Williams, in *Proceedings of the IXth ICPEAC Conference*, Seattle, Washington, J. S. Risley and R. Geballe, eds., p. 138, University of Washington Press, Seattle (1976); A. J. Dixon, S. T. Hood, and E. Weigold, *Phys. Rev. Lett.* **19**, 1262 (1978).
13. A. Ugbabe, P. J. O. Teubner, E. Weigold, and H. Arriola, *J. Phys. B* **10**, 71 (1977).
14. H. Arriola, P. J. O. Teubner, A. Ugbabe, and E. Weigold, *J. Phys. B* **8**, 1275 (1975).
15. I. C. Malcolm and J. W. McConkey, *J. Phys. B*, to be published (1979); see also J. W. McConkey and I. C. Malcolm, Chapter 12 of this book.

16. M. R. C. McDowell, L. A. Morgan, and V. P. Myerscough, *J. Phys. B* **8**, 1053 (1975).
17. G. F. Hanne, Chapter 47 of this book.
18. K. Blum and H. Kleinpoppen, *J. Phys. B* **10**, 3282 (1977).
19. K. Blum, E. E. Fitchard, and H. Kleinpoppen, *Z. Phys.* **A287**, 137 (1978).
20. J. Peresse, A. Pochat, and F. Gelebart, *Phys. Lett.* **41A**, 135 (1972); also, A. Pochat, D. Royuel, and J. Peresse, *J. Phys. Paris* **34**, 701 (1973).
21. A. Chutjian and L. D. Thomas, *Phys. Rev. A* **11**, 1583–1595 (1975).
22. M. Eminyan, H. Kleinpoppen, J. Slevin, and M. C. Standage, in *Electron and Photon Interactions with Atoms*, H. Kleinpoppen and M. R. C. McDowell, eds., pp. 455–483, Plenum Press, New York (1976).
23. D. E. Golden and N. C. Steph, Chapter 7 of this book.
24. K. Blum, Chapter 11 of this book.
25. A. H. Mahan and S. J. Smith, *Phys. Rev. A* **16**, 1789 (1977).
26. H. Kleinpoppen, Chapter 55 of this book.
27. G. Gabrielse and Y. B. Band, *Phys. Rev. Lett.* **39**, 697 (1977).
28. K. Blum and H. Kleinpoppen, *J. Phys. B* **10**, 3283 (1977).
29. U. Fano, *Phys. Rev.* **90**, 577 (1953).
30. K. Blum, E. E. Fitchard, and H. Kleinpoppen, *Z. Phys.* **A287**, 137 (1978).

10

The Possibility of a Complete Determination of Scattering Amplitudes for S → D Excitation by Electron–Photon Coincidence Measurements

GERARD NIENHUIS

It is well known that the density matrix of an atomic ¹P state can be fully determined by investigating the polarization and anisotropy of its spontaneous emission. By means of electron–photon coincidence techniques a complete determination of ¹S → ¹P excitation amplitudes is possible, as demonstrated by several groups. We point out that a complete measurement of a 3×3 polarization matrix in electron–photon coincidence experiments permits one to determine at most five real parameters of the excited-state density matrix for every scattering angle. In the case of an S → D excitation this is sufficient information to determine completely the complex scattering amplitudes. The effect of unresolved hyperfine structure or fine structure in the Russell–Saunders coupling approximation is easily incorporated into the description by inserting two reduction coefficients, one for the orientation and one for the alignment. These coefficients depend exclusively on geometrical factors.

1. Introduction

Electron–photon coincidence studies of electron–atom collisions can give much more detailed information on scattering amplitudes than is provided in more tradi-

GERARD NIENHUIS • Fysisch Laboratorium, Rijksuniversiteit te Utrecht, Utrecht, The Netherlands.

tional electron–atom scattering experiments.[1,2] The polarization and the angular
distribution of the detected photon yields information on the magnetic substates of
the atom excited by the coincidently detected electron. It has been demonstrated
that in the special case of a $^1S \rightarrow {}^1P$ excitation the information contained in the
angular distribution of the photons emitted in the scattering plane, without regard
to polarization, is sufficient to determine the amplitudes for exciting the magnetic
substates of the ^1P state, including their relative phases.[3–6] In this special case the
same information is provided by a measurement of the 2×2 polarization matrix
of the photons emitted orthogonal to the scattering plane.[7,8]

In the present contribution we point out that for each scattering angle the coincidence rate of photons with an arbitrary polarization emitted in an arbitrary direction
is given by a generalized 3×3 polarization matrix, regardless of the angular momenta
of the excited and the final atomic state. As a result of basic symmetry properties
of the collision system, a complete measurement of this polarization matrix for a
certain scattering angle provides at most five real parameters characterizing the
excited state of the atom. The experimental determination of both the angular
distribution of photons emitted in the scattering plane and the 2×2 polarization
matrix of photons emitted orthogonal to this plane would constitute such a complete
measurement. We illustrate the result in some special cases. In particular we show
that a complete measurement permits one to determine completely the scattering
amplitudes of the magnetic substates in the case of a $^1S \rightarrow {}^1D$ excitation. Additional complexities due to unresolved hyperfine structure or fine structure are briefly
discussed.

2. The Polarization Matrix

We consider the case of a beam of electrons with velocity $\mathbf{v}_0$ and density n_e
incident on a gas of atoms with density n_A in the initial ground state $|i\rangle$. Atoms
that are excited by electron impact to the excited state $|e\rangle$ can subsequently decay
by photon emission to the final state $|f\rangle$. The atomic levels have angular mo-
menta J_i, J_e, and J_f. The incident electrons and the atoms in the initial state are
unpolarized, and the spin of the scattered electron is unobserved. The detectors for
the scattered electrons and for the emitted photon cover solid-angle elements $d\Omega_e$
and $d\Omega_p$. A polarizer in front of the photodetector selects photons with polarization $\boldsymbol{\epsilon}$.
Then the coincidence rate per unit scattering volume is[1]

$$N(\boldsymbol{\epsilon}) = \frac{\omega^3}{2\pi\hbar c^3} \, d\Omega_e \, d\Omega_p \, \mathrm{Tr}_f \boldsymbol{\epsilon}^* \cdot \boldsymbol{\mu}_{fe} \varrho_e \boldsymbol{\mu}_{ef} \cdot \boldsymbol{\epsilon} \tag{1}$$

where ω is the frequency of the emitted photons, $\boldsymbol{\mu}_{ef}$ is the atomic electric dipole
operator between the excited state $|e\rangle$ and the final state $|f\rangle$. The effective den-

sity matrix ϱ_e of the excited atoms is defined to have matrix elements

$$\langle J_e M_e | \varrho_e | J_e M_e' \rangle = n_e n_A v \int_0^T d\tau e^{-\Gamma_e \tau} \frac{1}{2(2J_e + 1)} \sum_{mm_0} \sum_{M_i}$$
$$\times a(J_e M_e; m, \mathbf{v} \leftarrow J_i M_i; m_0, \mathbf{v}_0) a^*(J_e M_e'; m, \mathbf{v} \leftarrow J_i M_i; m_0, \mathbf{v}_0) \tag{2}$$

where m_0 and m are the z components of the electron spin before and after the collision, $\mathbf{v}_0$ and $\mathbf{v}$ are the electron velocity before and after the collision, and a denotes the scattering amplitudes. The time domain T indicates the resolution time of the coincidence apparatus, and Γ_e is the spontaneous decay rate of the excited state. The density matrix ϱ_e is normalized to the density of excited atoms that have lived less than the resolution time T.

We introduce the 3×3 Cartesian matrix

$$\mathbf{C} = \frac{\omega^3}{2\pi\hbar c^3} d\Omega_e \, d\Omega_p \, \mathrm{Tr}_f \mu_{fe} \varrho_e \mu_{ef} \tag{3}$$

so that the coincidence rate (1) for photons with polarization $\boldsymbol{\epsilon}$ is $\boldsymbol{\epsilon}^* \cdot \mathbf{C} \cdot \boldsymbol{\epsilon}$. The matrix $\mathbf{C}$ determines the photon emission in every direction with any polarization, and it may be looked upon as a generalized polarization density matrix of the emitted photons. The common 2×2 polarization matrix or, equivalently, the Stokes parameters[9] of the radiation in a certain direction are found from the matrix elements of $\mathbf{C}$ between polarization vectors orthogonal to this direction. For example, the xy submatrix of $\mathbf{C}$ gives the 2×2 polarization matrix of the radiation in the z direction. One notices that the photon emission rate is determined by the polarization $\boldsymbol{\epsilon}$, independent of the direction of emission. This is typical for dipole radiation.

It is the matrix $\mathbf{C}$ that contains precisely the information on the excited atoms that may be extracted from the polarization dependence and the angular distribution of the emitted radiation. Therefore the introduction of $\mathbf{C}$ is particularly suited to the analysis of the information contained in the emitted radiation. We turn to this analysis in the next sections.

3. Expansion in Irreducible Tensors

It is clear from equation (3) that any anisotropy of the polarization matrix $\mathbf{C}$ must be the result of a similar anisotropy of the excited-state density matrix ϱ_e. This similarity is best borne out if we expand ϱ_e and $\mathbf{C}$ in irreducible spherical tensors, constructed by recoupling angular-momentum states or spherical unit vectors. The complete orthonormal set of spherical tensor operators between states with angular momenta J and J' is defined by[10,11]

$$T_q^k(JJ') = \sum_{MM'} |JM\rangle \left(\frac{2k+1}{2J+1} \right)^{1/2} \langle JM | J'M'; kq \rangle \langle J'M' | \tag{4}$$

for $|J - J'| \leq k \leq J + J'$, $-k \leq q \leq k$. Likewise we introduce a complete ortho-
normal set of nine irreducible matrices

$$\mathbf{S}_q^k = \sum_{\sigma,\sigma'} \mathbf{u}_\sigma \left(\frac{2k + 1}{3} \right)^{1/2} \langle 1\sigma \,|\, 1\sigma'; kq \rangle \mathbf{u}_{\sigma'}^* \tag{5}$$

for $k = 0, 1, 2$, $-k \leq q \leq k$. The spherical unit vectors $\mathbf{u}_\sigma$ are given by

$$\mathbf{u}_{\pm 1} = 2^{-1/2}(\mp 1, -i, 0), \qquad \mathbf{u}_0 = (0, 0, 1) \tag{6}$$

Explicit expressions for the 3×3 matrices $\mathbf{S}_q^k$ are listed by Carrington.[12] As a
result of the completeness and the orthogonality of the sets we may expand

$$\varrho_e = \sum_{kq} \varrho_{kq} T_q^{k\dagger}(J_e J_e)$$

where

$$\varrho_{kq} = \mathrm{Tr}_e \, \varrho_e T_q^k(J_e J_e) \tag{7}$$

and

$$\mathbf{C} = \sum_{kq} c_{kq} \mathbf{S}_q^{k\dagger}$$

where

$$c_{kq} = \mathrm{Tr} \, \mathbf{C} \cdot \mathbf{S}_q^k \tag{8}$$

By applying the Wigner–Eckart theorem and some standard recoupling techniques
one may show from (3) that[13]

$$c_{kq} = \frac{\omega^3}{2\pi \hbar c^3} \, d\Omega_e \, d\Omega_p b_k(J_e J_f) \varrho_{kq} \tag{9}$$

where

$$b_k(J_e J_f) = (-1)^{J_e + J_f + k + 1} |\langle J_e || \mu || J_f \rangle|^2 \begin{Bmatrix} 1 & k & 1 \\ J_e & J_f & J_e \end{Bmatrix} \tag{10}$$

So the expansion coefficients c_{kq} of the polarization matrix $\mathbf{C}$ are proportional to
the expansion coefficients ϱ_{kq} of the density matrix of the excited atoms, and the proportionality constant is independent of the magnetic quantum numbers q. Since $\mathbf{C}$
is fully determined by the nine coefficients c_{kq} for $k = 0, 1, 2$ we conclude that a
complete measurement of the polarization matrix determines the expansion coefficients ϱ_{kq} of the excited-state density matrix for $k = 0, 1, 2$, which is equivalent to
the population, the orientation, and the alignment of the excited state. This determines
ϱ_e completely only if $J_e \leq 1$. This conclusion is in line with the results of Fano and
Macek.[2] Expansion coefficients ϱ_{kq} with a rank k which is larger than 2 do not show
up in the polarization properties of the radiation, and cannot be determined by
investigating the properties of the emitted radiation. However, as we shall see later
on, if we have additional information on ϱ_e such as its symmetry or coherence
properties, a full measurement of $\mathbf{C}$ may be sufficient to reconstruct ϱ_e even if J_e
is larger than 1.

4. Symmetry Considerations

The 3×3 Hermitian polarization matrix $\mathbf{C}$ is determined by nine real parameters. However, as a result of reflection symmetry with respect to the scattering plane these parameters are not independent. The symmetry properties of $\mathbf{C}$ arise from the symmetry behavior of the scattering amplitudes.

The parity invariance of the electron–atom interaction yields the relation

$$a(J_e M_e; m, \mathbf{v} \leftarrow J_i M_i; m_0, \mathbf{v}_0) = \eta_e \eta_i a(J_e M_e; m, -\mathbf{v} \leftarrow J_i M_i; m_0, -\mathbf{v}_0) \qquad (11)$$

where η_e and η_i are the parity of the atomic levels $|e\rangle$ and $|i\rangle$. Further the electron–atom interaction is rotationally invariant. If the scattering plane is chosen to be the x–z plane, we find from a rotation over π about the y axis

$$a(J_e M_e; m, \mathbf{v} \leftarrow J_i M_i; m_0, \mathbf{v}_0) = (-1)^{J_e - M_e + 1/2 - m}(-1)^{J_i - M_i + 1/2 - m_0}$$
$$\times a(J_e - M_e; -m, -\mathbf{v} \leftarrow J_i - M_i; -m_0, -\mathbf{v}_0) \quad (12)$$

After combining (11) and (12) we obtain the result of reflection symmetry with respect to the scattering plane

$$a(J_e M_e; m, \mathbf{v} \leftarrow J_i M_i; m_0, \mathbf{v}_0) = (-1)^{J_e - M_e + 1/2 - m}(-1)^{J_i - M_i + 1/2 - m_0}$$
$$\times \eta_e \eta_i a(J_e - M_e; -m, \mathbf{v} \leftarrow J_i - M_i; -m_0, \mathbf{v}_0) \quad (13)$$

for scattering in the x–z plane.

After substituting (13) in (2) we obtain the reflection symmetry of ϱ_e

$$\langle J_e M_e | \varrho_e | J_e M_e' \rangle = (-1)^{M_e' - M_e} \langle J_e - M_e | \varrho_e | J_e - M_e' \rangle \qquad (14)$$

for scattering in the xz plane. The hermiticity of ϱ_e is expressed by

$$\langle J_e M_e | \varrho_e | J_e M_e' \rangle = \langle J_e M_e' | \varrho_e | J_e M_e \rangle^* \qquad (15)$$

The relations (14) and (15) may be translated into relations between the expansion coefficients ϱ_{kq}. After using symmetry properties of the Clebsch–Gordan coefficients we obtain from (14), (4), and (7)

$$\varrho_{kq} = (-1)^{k-q} \varrho_{k-q} \qquad (16)$$

for scattering in the x–z plane. From the hermiticity relation (15) we obtain

$$\varrho_{kq}^* = (-1)^q \varrho_{k-q} \qquad (17)$$

Then the expansion coefficients c_{kq} of the polarization matrix obey the same relations

$$c_{kq} = (-1)^{k-q} c_{k-q} \qquad (18)$$

for scattering in the x–z plane, and

$$c_{kq}^{*} = (-1)^{q} c_{k-q} \tag{19}$$

For convenience we express the coefficients c_{kq} in terms of five real parameters, which determine $\mathbf{C}$ completely, and which are rather directly related to coincidence-rate measurements. We write

$$
\begin{aligned}
c_{00} &= 3^{1/2} N_0, & c_{11} &= c_{1-1} = -\tfrac{1}{2} i P \\
c_{20} &= -\tfrac{3}{2}^{1/2} D_0, & c_{21} &= -c_{2-1} = \tfrac{1}{2} D_1, & c_{22} &= c_{2-2} = \tfrac{1}{2} D_2
\end{aligned}
\tag{20}
$$

From the explicit expressions for $\mathbf{S}_q^{k(12)}$ one finds that $\mathbf{C}$ is given by

$$
\mathbf{C} = \begin{pmatrix}
N_0 - \tfrac{1}{2}(D_0 + D_2) & 0 & \tfrac{1}{2}(D_1 + iP) \\
0 & N_0 - \tfrac{1}{2}(D_0 - D_2) & 0 \\
\tfrac{1}{2}(D_1 - iP) & 0 & N_0 + D_0
\end{pmatrix}
\tag{21}
$$

which is a Cartesian Hermitian matrix obeying reflection symmetry with respect to the x–z plane. The relation between the parameters occurring in (21) and coincidence-rate measurements is directly found. If we denote as $\hat{x}$, $\hat{y}$, and $\hat{z}$ the unit vectors in the x, y, and z direction, we find from (1), (3), and (21)

$$
\begin{aligned}
N_0 &= \tfrac{1}{3}[N(\hat{x}) + N(\hat{y}) + N(\hat{z})] \\
P &= N[2^{-1/2}(i, 0, 1)] - N[2^{-1/2}(-i, 0, 1)] \\
D_0 &= \tfrac{2}{3} N(\hat{z}) - \tfrac{1}{3}[N(\hat{x}) + N(\hat{y})] \\
D_1 &= N[2^{-1/2}(1, 0, 1)] - N[2^{-1/2}(1, 0, -1)] \\
D_2 &= N(\hat{y}) - N(\hat{x})
\end{aligned}
\tag{22}
$$

Hence N_0 is the average coincidence rate for three orthogonal polarizations of the photons, P is the difference in coincidence rate for right-hand and left-hand circularly polarized photons in the y direction, etc.

The measurement of the five quantities (22) constitutes a complete determination of the information contained in the coincidence experiment. In general neither the four Stokes parameters of the photon emission orthogonal to the scattering plane nor the angular distribution of photons in the scattering plane is in itself sufficient to determine these five parameters. The four Stokes parameters of the emission in the y direction determine the 2×2 polarization matrix

$$
\begin{pmatrix}
N_0 - \tfrac{1}{2}(D_0 + D_2) & \tfrac{1}{2}(D_1 + iP) \\
\tfrac{1}{2}(D_1 - iP) & N_0 + D_0
\end{pmatrix}
\tag{23}
$$

which is the x–z submatrix of $\mathbf{C}$. On the other hand, the angular distribution of the photons, without regard to their polarization in the scattering plane as a function of the polar angle θ_p is given by

$$I(\theta_p) = N(\hat{y}) + N[(\cos \theta_p, 0, -\sin \theta_p)] \tag{24}$$

which is the coincidence rate for photons summed over two orthogonal polarizations perpendicular to the direction of emission. From (21) and (24) we find for this angular distribution

$$I(\theta_p) = 2N_0 - \tfrac{1}{4}(D_0 - D_2) - \tfrac{1}{4}(3D_0 + D_2) \cos 2\theta_p - \tfrac{1}{2}D_1 \sin 2\theta_p \tag{25}$$

Hence from a measurement of this distribution only three parameters can be extracted.

Equations (22), (20), and (9) relate the expansion coefficients ϱ_{kq}, $k = 0, 1, 2$ to measurable quantities. The relation between ϱ_{kq} and the scattering amplitudes can be evaluated from (7) and (2) in specific cases. This we shall do in Section 5.

5. The Case of $^1S \to {}^1P$ and $^1S \to {}^1D$ Excitation

In the case that the initial state is a ^1S and the excited state has zero spin, we may assume to a very good approximation that the electron spin component m does not change at the collision. If we further assume that the scattering amplitude is independent of m, for a certain scattering angle the only remaining variable is M_e; we denote the amplitudes as a_{M_e}, and the magnetic substates of the excited state as $|M_e\rangle$. The excited-state density matrix ϱ_e as given by (2) is then

$$\varrho_e = \frac{n_e n_A v}{\Gamma_e} (1 - e^{-\Gamma_e T}) \sum_{M_e M_{e'}} |M_e\rangle a_{M_e} a_{M_{e'}}^* \langle M_{e'}| \tag{26}$$

which is clearly a pure state.

5.1. $^1S \to {}^1P$

The case of $^1S \to {}^1P$ excitation with successive radiative decay to a ^1S state has been studied experimentally by several authors.[3-6] For illustrative purposes we give for this case the relation between the coefficients ϱ_{kq} and the scattering amplitudes. Then (20) and (9) fix the relation between coincidence rates and scattering amplitudes for arbitrary photon polarization. For scattering in the x–z plane the symmetry relation (13) reads

$$a_1 = -a_{-1} \tag{27}$$

in the common case that the ^1S and the ^1P state have opposite parity. We introduce the usual parametrization of the amplitudes

$$|a_0|^2 + 2|a_1|^2 = \sigma \tag{28}$$

$$|a_0|^2/\sigma = \lambda \tag{29}$$

$$\arg a_1/a_0 = \chi \tag{30}$$

These three parameters determine the three amplitudes apart from an unobservable overall phase. One should notice that, in contrast to what has been sometimes suggested, the relation (27) is invalid if the scattering does not occur in the x–z plane. The amplitudes a_1 and a_{-1} depend not only on the scattering angle θ between the scattered velocity and the incoming velocity, but also, be it in a trivial way, on the azimuthal angle φ. Likewise the value of the relative phase χ is different for a different scattering plane.

From (26) and (7) we evaluate the coefficients ϱ_{kq}. For brevity we introduce

$$B = \frac{n_e n_A v}{\Gamma_e} (1 - e^{-\Gamma_e T}) \tag{31}$$

We find

$$\varrho_{00} = 3^{-1/2} B(|a_1|^2 + |a_0|^2 + |a_{-1}|^2) = 3^{-1/2} B\sigma \tag{32}$$

$$\varrho_{11} = -2^{-1/2} B(a_0 a_1{}^* + a_{-1} a_0{}^*) = iB\sigma[\lambda(1 - \lambda)]^{1/2} \sin \chi = \varrho_{1-1} \tag{33}$$

$$\varrho_{20} = 6^{-1/2} B(|a_1|^2 - 2|a_0|^2 + |a_{-1}|^2) = 6^{-1/2} B\sigma(1 - 3\lambda) \tag{34}$$

$$\varrho_{21} = -2^{-1/2} B(a_0 a_1{}^* - a_{-1} a_0{}^*) = -B\sigma[\lambda(1 - \lambda)]^{1/2} \cos \chi = -\varrho_{2-1} \tag{35}$$

$$\varrho_{22} = Ba_{-1}a_1{}^* = -\tfrac{1}{2} B\sigma(1 - \lambda) = \varrho_{2-2} \tag{36}$$

The relation between ϱ_{kq} and c_{kq} is particularly simple in the case that $J_e = 1$ and $J_f = 0$, since the factors $b_k(J_e J_f)$ defined by (10) are independent of k:

$$b_k(1\ 0) = \tfrac{1}{3} |\langle 1||\mu||0\rangle|^2 \tag{37}$$

The polarization matrix $\mathbf{C}$ represents now a pure state, in the sense that $\mathbf{C} \cdot \mathbf{C}$ is proportional to $\mathbf{C}$. From (9), (20), and (21) this matrix can be directly evaluated if we apply (32)–(37). The result is

$$\mathbf{C} = F \frac{\sigma}{3} |\langle 1||\mu||0\rangle|^2 \begin{pmatrix} 1 - \lambda & 0 & -[\lambda(1 - \lambda)]^{1/2} e^{i\chi} \\ 0 & 0 & 0 \\ -[\lambda(1 - \lambda)]^{1/2} e^{-i\chi} & 0 & \lambda \end{pmatrix} \tag{38}$$

where

$$F = \frac{\omega^3 n_e n_A v}{2\pi\hbar c^3 \Gamma_e} d\Omega_e\, d\Omega_p (1 - e^{-\Gamma_e T}) \tag{39}$$

One sees directly from (38) that the radiation emitted in the y direction is completely (elliptically) polarized. This results from the fact that the excited-state density matrix ϱ_e (26) is a pure state, and the final state $|f\rangle$ has no undetermined magnetic substates in this special case. The entire polarization matrix is now determined by the three parameters σ, χ, and λ. Their values are fixed by a measurement of the angular distribution in the scattering plane, or by a measurement of the Stokes parameters of the radiation in the y direction.

5.2. $^1S \rightarrow {}^1D$

In this case the symmetry relation (13) reduces to

$$a_2 = a_{-2}, \qquad a_1 = -a_{-1} \tag{40}$$

for scattering in the x–z plane, provided that the ^1S and the ^1D state have the same parity. The scattering amplitudes a_{M_e} are now fixed by five real parameters, which we choose in analogy to (28)–(30)

$$|a_0|^2 + 2|a_1|^2 + 2|a_2|^2 = \sigma \tag{41}$$

$$|a_0|^2/\sigma = \lambda \tag{42}$$

$$2|a_1|^2/\sigma = \mu \tag{43}$$

$$\arg a_1/a_0 = \chi \tag{44}$$

$$\arg a_2/a_1 = \psi \tag{45}$$

The coefficients ϱ_{kq} are found from (26) and (7)

$$\varrho_{00} = 5^{-1/2}B(|a_2|^2 + |a_1|^2 + |a_0|^2 + |a_{-1}|^2 + |a_{-2}|^2)$$
$$= 5^{-1/2}B\sigma \tag{46}$$

$$\varrho_{11} = 5^{-1/2}B(a_1a_2^* + \tfrac{3}{2}^{1/2}a_0a_1^* + \tfrac{3}{2}^{1/2}a_{-1}a_0^* + a_{-2}a_{-1}^*)$$
$$= -i5^{-1/2}B\sigma\{[\mu(1-\lambda-\mu)]^{1/2}\sin\psi + [3\lambda\mu]^{1/2}\sin\chi\} = \varrho_{1-1} \tag{47}$$

$$\varrho_{20} = 14^{-1/2}B(2|a_2|^2 - |a_1|^2 - 2|a_0|^2 - |a_{-1}|^2 + 2|a_{-2}|^2)$$
$$= 14^{-1/2}B\sigma(2 - 4\lambda - 3\mu) \tag{48}$$

$$\varrho_{21} = 7^{-1/2}B(3^{1/2}a_1a_2^* + 2^{1/2}a_0a_1^* - 2^{1/2}a_{-1}a_0^* - 3^{1/2}a_{-2}a_{-1}^*)$$
$$= 7^{-1/2}B\sigma\{[3\mu(1-\lambda-\mu)]^{1/2}\cos\psi + 2[\lambda\mu]^{1/2}\cos\chi\} = -\varrho_{2-1} \tag{49}$$

$$\varrho_{22} = 7^{-1/2}B(2^{1/2}a_0a_2^* + 3^{1/2}a_{-1}a_1^* + 2^{1/2}a_{-2}a_0^*)$$
$$= 7^{-1/2}B\sigma\{2[\lambda(1-\lambda-\mu)]^{1/2}\cos(\chi+\psi) - \tfrac{1}{2}(3^{1/2})\mu\} = \varrho_{2-2} \tag{50}$$

From the preceding section we conclude that these five quantities (46)–(50) may be measured in an electron–photon coincidence experiment. Since these relations (46)–(50) may be looked upon as five independent equations of the five parameters σ, λ, μ, χ, and ψ, we arrive at the conclusion that a complete measurement of the photon emission at a fixed scattering angle is sufficient to determine the scattering amplitudes a_{M_e} for $^1S \rightarrow {}^1D$ excitation, including their relative phases.

For the sake of completeness we express the polarization matrix $\mathbf{C}$ in terms of the five parameters. If the photons are emitted by radiative decay from the ^1D to a ^1P state, the relation between c_{kq} and ϱ_{kq} is fixed by the factors (10)

$$b_k(21) = 15^{-1/2}|\langle 2||\mu||1\rangle|^2 \qquad \text{if } k = 0$$
$$= \tfrac{1}{2}(5^{1/2})|\langle 2||\mu||1\rangle|^2 \qquad \text{if } k = 1$$
$$= \tfrac{1}{10}(\tfrac{7}{3})^{1/2}|\langle 2||\mu||1\rangle|^2 \qquad \text{if } k = 2 \tag{51}$$

From (46)–(50), (9), (51), and (20) we evaluate explicitly the quantities N_0, P, D_0, D_1, and D_2.

The result is

$$N_0 = G/3 \tag{52}$$

$$P = G\{[\mu(1 - \lambda - \mu)]^{1/2} \sin \psi + [3\lambda\mu]^{1/2} \sin \chi\} \tag{53}$$

$$D_0 = -G(2 - 4\lambda - 3\mu)/6 \tag{54}$$

$$D_1 = G\{[\mu(1 - \lambda - \mu)]^{1/2} \cos \psi + 2[\tfrac{1}{3}\lambda\mu]^{1/2} \cos \chi\} \tag{55}$$

$$D_2 = G\{2(3^{-1/2})[\lambda(1 - \lambda - \mu)]^{1/2} \cos(\chi + \psi) - \tfrac{1}{2}\mu\} \tag{56}$$

where

$$G = \frac{\omega^3 n_e n_A v}{2\pi\hbar c^3 \Gamma_e} \, d\Omega_e \, d\Omega_p (1 - e^{-\Gamma_e T}) \frac{\sigma}{5} |\langle 2 || \mu || 1 \rangle|^2 \tag{57}$$

Substituting (52)–(56) into (21) gives an explicit expression for **C** in terms of the amplitude parameters. So we have related these parameters directly to measurable quantities. It is easy to verify that the emitted radiation is not completely polarized in this case, even though the emitting ^1D state is a pure state.

6. Effects of Unobserved (Hyper)fine-Structure Coupling

The expansion in irreducible spherical tensors is particularly appropriate to include hyperfine or fine-structure coupling. This we shall illustrate by incorporating the effect of a nuclear spin I.

We consider the usual case that the nuclei are initially unpolarized, and that the duration of the collisions is short compared to the hyperfine precession time. Then immediately after the collision the nuclei are still unpolarized, since the electron–atom interaction affects only the electronic part of the atom, and the coupling between the electronic and the nuclear angular momenta has had no time to become effective. The density matrix corresponding to the electronic state of the atom at this instant is therefore given by the integrand of equation (2) at time zero, whereas the isotropic normalized density matrix of the nuclei is given by $T_0^{0\dagger}(II)/(2I + 1)^{1/2}$. During the lifetime of the excited state there is ample time for the hyperfine precession to occur. The total density matrix has a time-dependent behavior described by the Hamiltonian H_e of the excited atom, with eigenstates $|FM_F\rangle$. These may be written as linear combinations of eigenstates of $\mathbf{J}^2$, J_z, $\mathbf{I}^2$, and I_z in terms of Clebsch–Gordan coefficients

$$|FM_F\rangle = \sum_{M_e} \sum_{M_I} |J_e M_e; IM_I\rangle\langle J_e M_e; IM_I | FM_F\rangle \tag{58}$$

which merely indicates that the total angular momentum F of the atom is the sum of the electronic angular momentum J and the nuclear spin I.

The effective density matrix that replaces (2) is equal to

$$\varrho_e = n_e n_A v \int_0^T d\tau e^{-\Gamma_e \tau} \frac{1}{2(2J_e + 1)(2I + 1)^{1/2}} \sum_{M_e M_e'} \sum_{m m_0} \sum_{M_i} \text{Tr}_I \, e^{-iH_e \tau / \hbar} \, | J_e M_e \rangle$$

$$\times a(J_e M_e; m, \mathbf{v} \leftarrow J_i M_i; m_0, \mathbf{v}_0) a^*(J_e M_e'; m, \mathbf{v} \leftarrow J_i M_i; m_0, \mathbf{v}_0)$$

$$\times \langle J_e M_e' \, | \, T_0^\dagger (II) e^{iH_e \tau / \hbar} \tag{59}$$

where Tr_I denotes the trace over the nuclear spin. It is this density matrix that gives the correct expression for the coincidence rate when substituted in (1), since the nuclear-spin operator commutes with the electric dipole moment. The integrand in (59) contains a product of an operator on the electronic angular momentum and a unit operator on the nuclear spin. The electronic part may be expanded in irreducible spherical tensors $T_q^k(J_e J_e)$ as in Section 3. Furthermore we apply the identity[11]

$$T_q^{k\dagger}(J_e J_e) T_0^\dagger (II)(2I + 1)^{1/2} = \sum_{FF'} [(2F + 1)(2F' + 1)]^{1/2} (-1)^{J_e + F' + k + I}$$

$$\times \begin{Bmatrix} J_e & k & J_e \\ F & I & F' \end{Bmatrix} T_q^{k\dagger}(FF') \tag{60}$$

Since the states $| FM_F \rangle$ are eigenstates of the Hamiltonian H_e, the operation of the evolution operators on the right-hand side of (60) is simply given by

$$e^{-iH_e \tau / \hbar} T_q^{k\dagger}(FF') e^{iH_e \tau / \hbar} = e^{i\omega_{FF'} \tau} T_q^{k\dagger}(FF') \tag{61}$$

where $\omega_{FF'} = (E_F - E_{F'})/\hbar$ is the frequency separation between the hyperfine levels with angular momentum F and F'. Finally we apply the equality[11]

$$\text{Tr}_I \, T_q^{k\dagger}(FF') = [(2F + 1)(2F' + 1)]^{1/2} (-1)^{J_e + F' + k + I} \begin{Bmatrix} J_e & k & J_e \\ F & I & F' \end{Bmatrix} T_q^{k\dagger}(J_e J_e) \tag{62}$$

If we substitute successively (60), (61), and (62) in (59) and perform the time integration, we find that the resulting expansion of ϱ_e in the tensors $T_q^{k\dagger}(J_e J_e)$ is directly related to the corresponding expansion in the absence of hyperfine coupling, according to

$$\varrho_e = \sum_{kq} r_k \varrho_{kq} T_q^{k\dagger}(J_e J_e) \tag{63}$$

where ϱ_{kq} are the expansion coefficients of the density matrix (2) for zero nuclear spin, and r_k are reduction coefficients given by

$$r_k = \sum_{FF'} \frac{(2F + 1)(2F' + 1)}{(2I + 1)} \begin{Bmatrix} J_e & k & J_e \\ F & I & F' \end{Bmatrix}^2 \frac{\Gamma_e}{\Gamma_e - i\omega_{FF'}} \frac{1 - e^{-\Gamma_e T + i\omega_{FF'} T}}{1 - e^{-\Gamma_e T}} \tag{64}$$

This reduction coefficient, which is independent of the magnetic quantum number q, is a real number between zero and one. It attains the value 1 for $k = 0$, which indicates that the excited-state population is not affected by the hyperfine coupling. By using the sum rule for $6j$ symbols

$$\sum_{FF'} \frac{(2F+1)(2F'+1)}{(2I+1)} \begin{Bmatrix} J_e & k & J_e \\ F & I & F' \end{Bmatrix}^2 = 1 \qquad (65)$$

one verifies that r_k is unity also for $k \geq 1$ in the special case that the hyperfine split-tings $\omega_{FF'}$ are much smaller than the decay rate Γ_e, or than the inverse resolution time T^{-1}. In these cases the hyperfine precession has no time to occur. In the more common case that the splittings $\omega_{FF'}$ are large compared to Γ_e and to T^{-1}, the pre-cession has ample time to occur, and the off-diagonal terms with $F = F'$ in (64) become negligible. Then the reduction factor r_k attains the value

$$r_k = \sum_F \frac{(2F+1)^2}{2I+1} \begin{Bmatrix} J_e & k & J_e \\ F & I & F \end{Bmatrix}^2$$

The resulting effect of the hyperfine coupling on the polarization matrix $\mathbf{C}$ is most easy to assess. From (9) it is obvious that a reduction of ϱ_{kq} with a factor r_k gives rise to a corresponding reduction of c_{kq} with the same factor. According to (20) the sole effect of the hyperfine structure is the reduction of the parameter P with the factor r_1, and of the parameters D_0, D_1, and D_2 with the factor r_2. The information content of a complete measurement of the polarization-dependent coincidence rate is unaffected by hyperfine coupling, even though the sensitivity may be reduced as a result of the depolarizing effect of the hyperfine structure.

Exactly the same conclusion holds for the case of fine-structure splitting in the Russell–Saunders approximation. If the scattering amplitudes do not depend on the electron spin, we may repeat the derivations of this section while replacing J, I, and F by L, S, and J. One may conclude that also the scattering amplitudes for $S \to D$ excitation may be completely determined by an electron–photon coincidence experiment, even if the D state is a doublet or a triplet.

References

1. J. Macek and D. H. Jaecks, *Phys. Rev. A* **4**, 2288–2300 (1971).
2. U. Fano and J. H. Macek, *Rev. Mod. Phys.* **45**, 553–573 (1973).
3. M. Eminyan, K. B. MacAdam, J. Slevin, and H. Kleinpoppen, *Phys. Rev. Lett.* **31**, 576–579 (1973).
4. M. Eminyan, K. B. MacAdam, J. Slevin, and H. Kleinpoppen, *J. Phys. B* **7**, 1519–1542 (1974).
5. H. Arriola, P. J. O. Teubner, A. Ugbabe, and E. Weigold, *J. Phys. B.* **8**, 1275–1279 (1975).
6. A. Ugbabe, P. J. O. Teubner, E. Weigold, and H. Arriola, *J. Phys. B* **10**, 71–79 (1977).
7. K. Blum and H. Kleinpoppen, *J. Phys. B.* **8**, 922–925 (1975).
8. M. C. Standage and H. Kleinpoppen, *Phys. Rev. Lett.* **36**, 577–580 (1976).
9. M. Born and E. Wolf, *Principles of Optics*, Pergamon Press, New York (1975).
10. U. Fano, *Phys. Rev.* **90**, 577–579 (1953).
11. U. Fano and G. Racah, *Irreducible Tensorial Sets*, Academic Press, New York (1959).
12. C. G. Carrington, *J. Phys. B* **4**, 1222–1229 (1971).
13. G. Nienhuis, *Physica* **81C**, 381–391 (1976).

Threshold and Pseudothreshold Excitation of Molecules by Electron Impact

K. Blum

Threshold and pseudothreshold excitation of diatomic molecules are considered. In particular it is shown that certain geometrical factors, which suggest a preferential population of particular magnetic substates, can be separated from the excitation amplitudes. The case of excitation is discussed in detail.

1. Introduction

In this chapter we give a discussion of threshold and pseudothreshold excitation of diatomic molecules by electron impact. For simplicity only singlet–singlet transitions are considered and all explicit relativistic effects are neglected.

The nature of the problem to be discussed is outlined in Section 2. The main problem that we consider is whether particular rotational substates will be predominantly excited at threshold. Previous theoretical treatments of this question[1,2] are briefly discussed in Sections 3.1 and 3.2. In Section 3.3 we develop a theory for homonuclear molecules, based on a suggestion by McConkey and co-workers.[3] We show that certain geometrical factors (depending of the quantum numbers of initial and final states) can be separated from the excitation amplitudes. These factors show a preferential population of magnetic substates. The case of $\Sigma_g^+ \rightarrow \Sigma_u^+$ excitation is discussed in detail. In Section 4 we briefly consider pseudothreshold excitation of molecules following a treatment given in a recent publication.[4]

K. BLUM ● Institut für Theoretische Physik I, Universität Münster, 44 Münster, West Germany.

This paper is closely connected with the experimental work reported in Reference 3 (see also Chapter 12 in this book, by J. W. McConkey and I. C. Malcolm).

2. Threshold Excitation of Molecules: General Discussion

We denote the molecular states by the symbol $| \Lambda V N M_N \rangle$, where V is the vibrational quantum number, N is the total orbital angular momentum of the molecule, M_N its z component, and Λ its component with respect to the molecular axis n. In diatomic molecules Λ is equal to the projection of the electronic angular momentum e on the molecular axis. In the united atom approximation the precession of the electronic angular momentum around the molecular axis is relatively slow and e can be considered as a good quantum number.

We will discuss the process

$$e + (\Lambda_0 V_0 N_0) \rightarrow e' + (\Lambda_1 V_1 N_1)$$
$$ \textsf{L} \rightarrow (\Lambda_f V_f N_f) + \gamma$$

where an initial state with sharp quantum numbers $\Lambda_0 V_0 N_0$ is excited by electron impact to a state with sharp quantum numbers $\Lambda_1 V_1 N_1$, which then decays to a state $\Lambda_f V_f N_f$ by photon emission. The emitted photons will be observed and their polarization measured. We will assume at first that the rotational levels with quantum numbers N_0 and N_1 can be resolved in the experiment, which is at present possible only for the lightest molecules. An average over all magnetic substates will be made.

Conservation of total angular momentum of electron and molecule and of its z component gives the selection rule

$$M_{N_0} + m_0 = M_{N_1} + m_1 \tag{1a}$$

where m_0 and m_1 denote the z component of the angular momentum of initial and final electrons, respectively.

The condition (1a) follows from the invariance of the total transition operator against rotations around the z axis. We will choose the incoming beam axis as z axis. The orbital angular momentum of the incident electron is perpendicular to its axis of motion, which gives $m_0 = 0$. Close to threshold the scattered electron has practically zero momentum $\mathbf{p}_1 \approx 0$ and thus zero angular momentum $m_1 = 0$. Consequently we have the threshold selection rule

$$M_{N_0} = M_{N_1} \tag{1}$$

Condition (1) restricts the number of possible magnetic substates that can be excited from an initial molecular state with sharp angular momentum N_0. However, equation (1) is not as restrictive as in the case of threshold polarization of atoms. Here one usually excites from the atomic ground state with orbital angular momentum zero.

Therefore, only a single magnetic substate with quantum number $M_N = 0$ can be populated at threshold.

In molecular excitation processes the situation is complicated because at room temperature the initial state is a statistical mixture of many rotational states excited by thermal motion with a distribution governed by Boltzmann and nuclear statistics. If the rotational levels can be resolved in the experiment we pick out one of these levels with angular momentum N_0 which has $(2N_0 + 1)$ different magnetic substates. All these substates will be excited in general to final states with the same magnetic quantum number in accordance with the selection rule (1). Consequently, $(2N_0 + 1)$ magnetic substates will be populated at threshold instead of a single one as in atomic excitation.

The radiation emitted by these substates will then overlap incoherently and this will in general result in a reduced threshold polarization.

To see this more explicitly we define the polarization as usual by the expression

$$P = \frac{I_{\parallel} - I_{\perp}}{I_{\parallel} + I_{\perp}} \tag{2}$$

where it is assumed that the emitted light is observed perpendicular to the incident beam axis and where $I_{\parallel}$ and $I_{\perp}$ denote the intensity of light with linear polarization parallel and perpendicular to the beam axis. We obtain an expression for the threshold polarization by specializing equation (41) of Reference 4 to singlet excitation. Applying condition (1) we obtain

$$P_{\text{thr}} = \frac{\sum_{M_{N_0}} \left[3 \begin{pmatrix} N_1 & 1 & N_f \\ M_{N_0} & 0 & -M_{N_0} \end{pmatrix}^2 - \frac{1}{2N_1 + 1} \right] Q_{M_{N_0}}}{\sum_{M_{N_0}} \left[\begin{pmatrix} N_1 & 1 & N_f \\ M_{N_0} & 0 & -M_{N_0} \end{pmatrix}^2 + \frac{1}{2N_1 + 1} \right] Q_{M_{N_0}}} \tag{3}$$

Here $(\cdots)$ denotes a $3j$ symbol and $Q_{M_{N_0}}$ is the total cross section for the transition

$$\Lambda_0 V_0 N_0 M_{N_0} \to \Lambda_1 V_1 N_1 M_{N_1} = M_{N_0}$$

at threshold.

This expression may be compared with the corresponding one for atomic threshold excitation where only a single magnetic substate $M_N = 0$ is excited. In this case we obtain from equation (3) the expression

$$P_{\text{thr}} = 3 \begin{pmatrix} N_1 & 1 & N_f \\ 0 & 0 & 0 \end{pmatrix}^2 - \frac{1}{2N_1 + 1} \Bigg/ \left[\begin{pmatrix} N_1 & 1 & N_f \\ 0 & 0 & 0 \end{pmatrix}^2 + \frac{1}{2N_1 + 1} \right] \tag{4}$$

Here the threshold polarization is independent of dynamical factors and depends only on the angular momentum coupling scheme as emphasized in the classic paper by Percival and Seaton.[5]

In the molecular case we have to observe the radiation emitted from $(2N_0 + 1)$ magnetic substates. Consequently we have to sum incoherently over these states, and because each one is multiplied by the corresponding inelastic cross section Q_{MN_0} the threshold polarization (3) depends on the dynamics of the excitation process and can in general not be determined from symmetry reasons alone. Therefore the question arises whether it is possible to apply simplifying assumptions to enable one to calculate the threshold polarization without detailed knowledge of the scattering dynamics. We will discuss this question in the following sections.

3. Threshold Excitation: Special Models

3.1. The United Atom Approximation

The dependence of the total cross section on the angular momentum quantum numbers can be given explicitly by transforming from the laboratory system to the system fixed in the molecule (i.e., it rotates with the molecule), with the molecular axis taken as the quantization axis. Such transformations are well known in molecular spectroscopy. More recently they have been applied to scattering processes, particularly in connection with Fano's frame transformation theory.[6] We find that the cross section for threshold excitation is given up to a numerical factor by the expression

$$Q_{MN_0} \sim \sum_{L_0} (2N_1 + 1) \begin{pmatrix} N_1 & L_0 & N_0 \\ -M_{N_0} & 0 & M_{N_0} \end{pmatrix}^2 \begin{pmatrix} N_1 & L_0 & N_0 \\ -\Lambda_1 & \Lambda_1 - \Lambda_0 & \Lambda_0 \end{pmatrix}^2$$
$$\times |\langle \Lambda_1 V_1 | T | \Lambda_0 V_{01} L_{01} \Lambda_1 - \Lambda_0 \rangle|^2 \tag{5}$$

Here L_0 denotes the angular momentum of the incident electron; the sum is taken over all partial waves. T denotes the transition operator. The T-matrix element is independent of the molecular angular momentum and contains the details of the dynamics. The first $3j$ symbol expresses angular momentum conservation in the laboratory frame; the second one characterizes the coupling of the internal quantum numbers. This equation is exact apart from using the Born–Oppenheimer approximation. We will discuss such transformations in more detail in Section 3.3.

In the united atom approximation[1] l_0 and l_1 are good quantum numbers and we have an additional coupling rule

$$l_0 + L_0 = l_1$$

at threshold. Here l_0 and l_1 denote the electronic orbital angular momentum of the initial and final molecule, respectively. If the molecule is excited from the ground state we have $l_0 = 0$, and therefore only a single partial wave

$$l_1 = L_0$$

contributes to the cross section (5).

In this case the T-matrix element cancels in the expression for the threshold polarization and we obtain from equations (3) and (5)

$$P_{\text{thr}} = \frac{\sum_{M_{N_0}} \left[3 \begin{pmatrix} N_1 & 1 & N_f \\ M_{N_0} & 0 & -M_{N_0} \end{pmatrix}^2 - \frac{1}{2N_1 + 1} \right] \begin{pmatrix} N_1 & L_0 & N_0 \\ -M_{N_0} & 0 & M_{N_0} \end{pmatrix}^2}{\sum_{M_{N_0}} \left[\begin{pmatrix} N_1 & 1 & N_f \\ M_{N_0} & 0 & -M_{N_0} \end{pmatrix}^2 + \frac{1}{2N_1 + 1} \right] \begin{pmatrix} N_1 & L_0 & N_0 \\ -M_{N_0} & 0 & M_{N_0} \end{pmatrix}^2} \tag{5a}$$

with $L_0 = l_1$. This expression can be further simplified by performing the sum over M_{N_0}. Equation (5a) is more complicated than in the atomic case but it allows the calculation of the threshold polarization without detailed knowledge of the inelastic cross sections.

Baltayan and Nedelec[1] developed this model and applied it to excitation of molecular hydrogen for singlet and nonsinglet transitions. They found reasonable agreement with their experimental results.

3.2. The Model of Jette and Cahill

Jette and Cahill[2] have applied a different procedure. They used the following condition as selection rule at threshold:

$$\langle \Lambda_0 N_0 M_{N_0} V_0 \,|\, L_z \,|\, \Lambda_0 V_0 N_0 M_{N_0} \rangle = \langle \Lambda_1 V_1 N_1 M_{N_1} \,|\, L_z \,|\, \Lambda_1 V_1 N_1 M_{N_1} \rangle$$

where L_z is the z component of the angular momentum operator of the molecular electrons. The expectation value can easily be calculated, and Jette and Cahill obtained

$$\frac{M_{N_0} \Lambda_0^2}{N_0(N_0 + 1)} = \frac{M_{N_1} \Lambda_1^2}{N_1(N_1 + 1)}$$

Thus, if $\Lambda_0 = 0$ the right-hand side must be zero as well and if $\Lambda_1 \neq 0$ it follows $M_{N_1} = 0$. Therefore only a single magnetic substate will be populated at threshold and the expression for the threshold polarization is similar to the atomic case.

However, it should be noted that the conditions of Reference 2 are not in agreement with the exact threshold selection rule (1).

3.3. The Model of McConkey and Co-workers

Recently Malcolm, Dassen, and McConkey[3] suggested another procedure for an approximate treatment of threshold excitation. Following Dunn[7] they concluded that in $\Sigma \rightarrow \Sigma$ transitions at threshold, substates with $M_{N_0} = M_{N_1} = 0$ will be predominantly excited, and in $\Sigma \rightarrow \Pi$ transitions, substates $M_{N_0} = M_{N_1} = \pm N_0$ (see also Chapter 12 in this book, by J. W. McConkey and I. C. Malcolm). In this section we give a derivation and development of these ideas.

We will consider states of *homonuclear* molecules with definite parity $P = \pm 1$ and definite values of π, where $\pi = 0$ corresponds to "gerade," $\pi = 1$ to *ungerade* electronic states, which we will denote by the symbol

$$|\Gamma\rangle = |\Lambda V N M_N P \pi\rangle$$

States of definite parity have been constructed, for example, by Rubin[8] (see also the appendix of Reference 4). Threshold excitation of states $|\Gamma_0\rangle \rightarrow |\Gamma_1\rangle$ by electron impact is characterized by the corresponding element of the T operator, which we will denote by $\langle\Gamma_1\mathbf{p}_1 \approx 0 | T | \Gamma_0\mathbf{p}_0\rangle$ where $\mathbf{p}_0$ and $\mathbf{p}_1$ denote the momentum of initial and scattered electrons, respectively. Partial wave expansion gives

$$\langle\Gamma_1\mathbf{p}_1 \approx 0 | T | \Gamma_0\mathbf{p}_0\rangle = \frac{1}{4\pi} \sum_{L_0} (2L_0 + 1)^{1/2}\langle\Gamma_1 L_1 = 0 | T | \Gamma_0 L_0 m_0 = 0\rangle P(\theta)_{L_0} \quad (6)$$

where L_0 and m_0 denote the orbital angular momentum of the initial electron and its z component respectively and $P(\theta)_{L_0}$ is a Legendre polynomial. Parity invariance gives for each partial wave the restriction (at threshold)

$$P_1 = P_0(-1)^{L_0} \quad (7a)$$

For Σ states we have $P = (-1)^{N+s}$ with $s = 0$ for Σ^+ and $s = 1$ for Σ^-, and from equation (7a) we obtain for a Σ-Σ transition

$$(-1)^{N_1+s_1+N_0+s_0+L_0} = 1 \quad (7b)$$

Because the nuclei are identical and with the assumption that the nuclear spins are unaffected by the collision the symmetric ("s") or antisymmetric ("a") character of initial and final nuclear states will remain unchanged during the collision. This results in the condition

$$P_1(-1)^{\pi_1} = P_0(-1)^{\pi_0} \quad (7c)$$

Equations (7a) and (7c) give the condition

$$(-1)^{\pi_1+\pi_0+L_0} = +1 \quad (7)$$

The restriction (7) shows that to g–g and u–u transitions only even partial waves, to g–u transitions only odd partial waves can contribute. This will be essential for the following discussions.

We will now derive an expression for the contribution $A(\mathbf{n})$ of molecules with a given direction $\mathbf{n} = (\beta, \alpha)$ of the molecular axis to the total amplitude (6). Here β denotes the angle between the molecular axis $\mathbf{n}$ and the z axis and α denotes the azimuth angle of $\mathbf{n}$. By transforming to the molecule-fixed system we obtain[6]

$$A(\mathbf{n}) = \frac{1}{(4\pi)^2} [(2N_0 + 1)(2N_1 + 1)]^{1/2} D(\mathbf{n})^{(N_1)^*}_{\Lambda_1 M N_1}$$

$$\times \left[\sum_{L_0} (2L_0 + 1)^{1/2}\langle\Lambda_1 V_1\pi_1 | T(\mathbf{n}) | \Lambda_0 V_0\pi_0, L_0 m_0 = 0\rangle \cdot P(\theta)_{L_0}\right] D(\mathbf{n})^{(N_0)}_{\Lambda_0 M N_0}$$

$$(8a)$$

Here $D(\mathbf{n})_{\Lambda M_N}^{(N)}$ denotes an element of the rotation matrix for the transformation from the laboratory system to the molecule-fixed system. These elements are also the probability amplitudes for finding the molecule with axis pointing in the $\mathbf{n}$ direction if the molecule has angular momentum quantum numbers $\Lambda N M_N$. Thus, we can interpret equation (8a) in the following way. The amplitude $A(\mathbf{n})$ is the product of the amplitude $D(\mathbf{n})_{\Lambda_0 M_{N_0}}^{(N_0)}$ for finding in the initial state $\Lambda_0 N_0 M_{N_0}$ a molecule with axis $\mathbf{n}$ times the amplitude for a transition to a final state $\Lambda_1 V_1 \pi_1$ in the molecule-fixed system times the amplitude $D(\mathbf{n})_{\Lambda_1 M_{N_1}}^{(N_1)*}$ for finding a final state with rotational quantum numbers $N_1 M_{N_1}$. By integrating equation (8a) over all directions of n we obtain the total amplitude (6).

Finally, by transforming $|L_0 m_0 = 0\rangle$ to the molecule-fixed system we obtain the expression (M_0 is the component of L_0 in the direction n)

$$A(\mathbf{n}) = \frac{1}{4\pi^2} [(2N_0 + 1)(2N_1 + 1)]^{1/2} D(\mathbf{n})_{\Lambda_1 M_{N_1}}^{(N_1)*}$$

$$\times \left[\sum_{L_0 M_0} (2L_0 + 1)^{1/2} \langle \Lambda_1 V_1 \pi_1 | T | \Lambda_0 V_0 \pi_0 L_0 M_0 \rangle D(\mathbf{n})_{M_0 0}^{(L_0)} P(\theta)_{L_0} \right] \cdot D(\mathbf{n})_{\Lambda_0 M_{N_0}}^{(N_0)}$$

$$(8b)$$

where

$$D(\mathbf{n})_{M_0 0}^{(L_0)} = (-1)^{M_0} \left(\frac{4\pi}{2L_0 + 1} \right)^{1/2} Y(\beta \alpha)_{L_0 M_0}$$

$$(8c)$$

is the corresponding rotation matrix element and $Y(\beta, \alpha)_{L_0 M_0}$ is a spherical harmonic.

We note that the T-matrix element in equation (8b) is independent of $\mathbf{n}$ because all quantum numbers occurring in the element are now defined in the molecule-fixed system with $\mathbf{n}$ as quantization axis and the transition matrix element is then independent of the direction of $\mathbf{n}$ in the laboratory system. The $\mathbf{n}$ dependence is given explicitly by the rotation matrix elements in equation (8b).[6]

From the axial symmetry of the T-matrix element around the molecular axis we obtain the condition

$$M_0 = \Lambda_1 - \Lambda_0 \qquad (9)$$

and the sum over M_0 in equation (8b) is superfluous. We obtain

$$A(\mathbf{n}) = \left(\frac{1}{4\pi} \right)^{3/2} [(2N_0 + 1)(2N_1 + 1)]^{1/2} D(\mathbf{n})_{\Lambda_1 M_{N_1}}^{(N_1)*}$$

$$\times \left[\sum_{L_0} (-1)^{\Lambda_1 - \Lambda_0} Y(\beta, \alpha)_{L_0 \Lambda_1 - \Lambda_0} \langle \Lambda_1 V_1 \pi_1 | T | \Lambda_0 V_0 \pi_0 L_0 \Lambda_1 - \Lambda_0 \rangle P(\theta)_{L_0} \right]$$

$$\times D(\mathbf{n})_{\Lambda_0 M_{N_0}}^{(N_0)} \qquad (10)$$

First we will inspect the $\mathbf{n}$ dependence of the term in square brackets in equation (10), which is the amplitude for the indicated transition in the molecule-fixed system. For dissociative processes the corresponding amplitude has been investigated by Dunn[7] for the special cases of **n** parallel to z and **n** perpendicular to z. Dunn's results can be generalized to molecular excitation for any direction of **n** in the following way.

From the explicit formulas of the spherical harmonics (see for example Reference 9) follows (up to a numerical factor)

$$Y(\beta, \alpha)_{lm} \sim P(x)_{lm} e^{im\alpha}$$

where $P(x)_{lm}$ is an associated Legendre polynomial

$$P(x)_{lm} \sim (1 - x^2)^{m/2} \frac{d^{l+m}}{dx^{l+m}} (x^2 - 1)^l$$

with $x = \cos \beta$. Because $(x^2 - 1)^l$ is a polynomial containing only terms $\sim x^n$ with n even, any odd derivative contains only terms $\sim x^m$ with m odd. Therefore, a common factor x can be separated if $l + m$ is odd. Consequently, we can write the spherical harmonics in the following form:

$$Y(\beta, \alpha)_{lm} \sim \begin{cases} \cos \beta (\sin \beta)^m \widetilde{Y(\beta, \alpha)}_{lm} & \text{if } l + m \text{ is odd} & \text{(11a)} \\ (\sin \beta)^m \widetilde{Y(\beta, \alpha)}_{lm} & \text{if } l + m \text{ is even} & \text{(11b)} \end{cases}$$

where we defined

$$\widetilde{Y(\beta, \alpha)}_{lm} \sim \begin{cases} \dfrac{1}{\cos \beta} \dfrac{d^{l+m}}{dx^{l+m}} (x^2 - 1)^l e^{im\alpha} & \text{if } l + m \text{ is odd} & \text{(11c)} \\ \dfrac{d^{l+m}}{dx^{l+m}} (x^2 - 1)^l e^{im\alpha} & \text{if } l + m \text{ is even} & \text{(11d)} \end{cases}$$

$\tilde{Y}_{lm}$ can be written as a sum of terms $\sim (\cos \beta)^n$. *The important point to note is that in both cases, (11c) and (11d), $\tilde{Y}_{lm}$ contains one term independent of β.* By including the numerical factors left out in the equations given above we can write equality signs instead of the proportionality sign in equations (11) by taking the numerical factor into $\tilde{Y}_{lm}$. With the help of equation (11) we write the term in square brackets in equation (10) in the form

$$\sum_{0L} (-1)^{A_1 - A_0} Y(\beta, \alpha)_{L_0 A_1 - A_0} \langle A_1 V_1 \pi_1 | T | A_0 V_0 \pi_0 L_0 A_1 - A_0 \rangle P(\theta)_{L_0}$$

$$= \begin{cases} \cos \beta (\sin \beta)^{A_1 - A_0} \sum_{L_0} (-1)^{A_1 - A_0} \widetilde{Y(\beta, \alpha)}_{L_0 A_1 - A_0} \\ \times \langle A_1 V_1 \pi_1 | T | A_0 V_0 \pi_0 L_0 A_1 - A_0 \rangle P(\theta)_{L_0} \\ \qquad\qquad\qquad\qquad \text{if } L_0 + A_1 - A_0 \text{ is odd} & \text{(12a)} \\ \\ (\sin \beta)^{A_1 - A_0} \sum_{L_0} (-1)^{A_1 - A_0} \widetilde{Y(\beta, \alpha)}_{L_0 A_1 - A_0} \\ \times \langle A_1 V_1 \pi_1 | T | A_0 V_0 \pi_0 L_0 A_1 - A_0 \rangle P(\theta)_{L_0} \\ \qquad\qquad\qquad\qquad \text{if } L_0 + A_1 - A_0 \text{ is even} & \text{(12b)} \end{cases}$$

Equations (12) generalize the results obtained by Dunn. For the special cases $\beta = 0$ (**n** parallel to z) and $\beta = \pi/2$ (**n** perpendicular to z) we can rederive Dunns results for homonuclear molecules from equations (12) (see Reference 7, Table 1). Whether molecules with axes parallel to z or molecules with axes perpendicular to z will be predominantly excited depends on (i) the value of $\Lambda_1 - \Lambda_0$, and (ii) whether the transition under consideration is g–g or u–u (L_0 even) or g–u (L_0 odd). Condition (7) is therefore essential for the whole treatment.

As an example let us consider a transition $\Sigma_g^+ \to \Sigma_u^+$ for homonuclear molecules. Because of condition (7) only odd values of L_0 contribute to the sum in (12), and with $\Lambda_1 = \Lambda_0 = 0$ we have from equation (12) with $\pi_1 = 1$, $\pi_0 = 0$

$$\sum_{L_0 \text{ odd}} Y(\beta, \alpha)_{L_0 0} \langle V_1 \pi_1 \mid T \mid V_0 \pi_0 L_0 0 \rangle P(\theta)_{L_0}$$

$$= \cos \beta \sum_{L_0 \text{ odd}} \overline{Y(\beta, \alpha)}_{L_0 0} \langle V_1 \pi_1 \mid T \mid V_0 \pi_0 L_0 0 \rangle P(\theta)_{L_0} \tag{13}$$

That is, molecules with **n** parallel to z will predominantly contribute to the excitation whereas molecules with **n** perpendicular to z cannot be excited.

We will now return to the amplitude (10) for a transition between states with definite rotational quantum numbers in order to discuss the contribution of particular magnetic substates. We will concentrate on the case $\Lambda_0 = \Lambda_1 = 0$, for which the rotation matrix elements reduce to the corresponding spherical harmonics. Applying equations (12) we have

$$Y(\beta, \alpha)_{NM_N} = \begin{cases} \cos \beta (\sin \beta)^{M_N} \overline{Y(\beta, \alpha)}_{NM_N} & \text{if } N + M_N \text{ is odd} \tag{14a} \\[2mm] (\sin \beta)^{M_N} \overline{Y(\beta, \alpha)}_{NM_N} & \text{if } N + M_N \text{ is even} \tag{14b} \end{cases}$$

where the quantities $\tilde{Y}_{NM_N}$ are given by equation (11c) and equation (11d), respectively. Condition (7b) shows $N_1 + N_0$ is odd, and by combining equations (14a) and (14b) we can write

$$Y(\beta, \alpha)^*_{N_1 M_{N_0}} Y(\beta, \alpha)_{N_0 M_{N_0}} = \cos \beta (\sin \beta)^{2M_{N_0}} \overline{Y(\beta, \alpha)}^*_{N_1 M_{N_0}} \overline{Y(\beta, \alpha)}_{N_0 M_{N_0}} \tag{15}$$

Inserting equations (13) and (15) into equation (10) we obtain

$$A(\Sigma_g^+ \to \Sigma_n^+, \mathbf{n}) = \left(\frac{1}{4\pi}\right)^{3/2} [(2N_0 + 1)(2N_1 + 1)]^{1/2}$$

$$\times (\cos \beta)^2 (\sin \beta)^{2M_{N_0}} \overline{Y(\beta, \alpha)}^*_{N_1 M_{N_0}} \overline{Y(\beta, \alpha)}_{N_0 M_{N_0}} \sum_{L_0 \text{ odd}} \overline{Y(\beta, \alpha)}_{L_0 0}$$

$$\times \langle V_1 \pi_1 \mid T \mid V_0 \pi_0 L_0 0 \rangle P(\theta)_{L_0} \tag{16}$$

where $\overline{Y(\beta, \alpha)}_{NM_N}$ is a sum of terms $\sim(\cos \beta)^n$ (with one nonvanishing term independent of β) multiplied by the α-dependent exponential function.

Equation (16) shows that (i) the amplitude vanishes for **n** perpendicular to z

and (ii) if $M_{N_0} \neq 0$ the amplitude vanishes also for **n** parallel to z. Consequently, substates with $M_{N_1} = M_{N_0} = 0$ will be preferentially excited at threshold.

By integrating the amplitude (16) over all directions of **n**, squaring the obtained expression, and integrating over all electronic angles θ, we obtain the total cross section (5), which determines the threshold polarization (3) for states with definite quantum numbers N_0 and N_1. Because of the preferential population of the substates with small magnetic quantum numbers the emitted light will be polarized. Similar conclusions can be drawn for non-Σ transitions.

4. Pseudothreshold Excitation

In practise it is very difficult to make a reliable measurement of polarization close to threshold because of intensity limitations and because of disturbing effects of resonances and cascades. These disturbing effects can be avoided by applying the "pseudothreshold" technique of King et al.[10] in which high incident energies well away from threshold are used. In this method electrons scattered in the forward direction with a given energy are observed in coincidence with the photons emitted in the subsequent decay. Because of this coincidence method the observation is restricted to radiation emitted by those molecules only that "scattered" the electrons in the forward direction. Therefore, a certain subensemble of molecules is selected in the experiment, so to speak, and only the radiation from this subensemble is observed.

Because both, incident and scattered electrons, move along the z axis, both have angular momentum component zero with respect to this axis. Consequently, we have the same selection rule (1) as in the case of threshold excitation and the same substates will be excited.

This coincidence method has been applied in Reference 3 to excitation of the Werner bands in molecular hydrogen without resolving the rotational states. We will briefly discuss what kind of information on the excited states can be extracted from such measurements. We will consider the case that the rotational levels have not been resolved and discuss the excitation of both, hetero- and homonuclear molecules.

A detailed discussion of electron–photon coincidence experiments for the molecular case has been given in a recent paper.[4] By specializing equation (45) given in that paper to forward scattering and singlet transitions we obtain for the angular distribution $I(\theta)$ for the emitted photons the expression

$$I(\theta) = \overline{C(\omega)} \, | \langle A_f V_f | r_{A_1 - A_f} | A_1 V_1 \rangle |^2 \, [\tfrac{2}{3} \langle T_{00} \rangle + 6^{-1/2} \langle T_{20}^+ \rangle (3 \cos^2 \theta - 1)] \quad (17)$$

where θ is the polar angle measured from the z axis and A_f, V_f denote electronic and vibrational quantum number of the final molecular states.

The factor in front of the square brackets is essentially the line strength for a radiative transition $A_1 V_1 \rightarrow A_f V_f$. The angular distribution is expressed in terms of two parameters $\langle T_{00} \rangle$ and $\langle T_{20}^+ \rangle$ which contain the information of the excited states which can be extracted from a measurement of the angular distribution. $\langle T_{00} \rangle$ is essentially the differential cross section for forward scattering averaged over all rotational states. $\langle T_{20}^+ \rangle$ is an alignment parameter. It measures the extent to which different magnetic substates have been preferentially excited. If initial and final states have sharp angular momenta N_0 and N_1, this parameter is the usual alignment parameter as defined in nuclear and atomic physics.[11,12] In our case of interest we have to sum incoherently over all rotational quantum numbers N_0 and N_1 in order to obtain the total alignment $\langle T_{20}^+ \rangle$.

$\langle T_{20}^+ \rangle$ can be obtained experimentally by measuring the angular distribution $I(\theta)$ for several angles θ, which allows then the separation of the contributions of the two terms in equation (17).

We stress the point that the parametrization of the angular distribution (and of the Stokes parameters characterizing the polarization) in terms of the "multipole" parameters $\langle T_{00} \rangle$ and $\langle T_{20}^+ \rangle$ is the most convenient one from the experimental standpoint because this allows the separation and determination of the multipoles immediately from the measurement of $I(\theta)$ (or the Stokes parameters) for several angles. Expressions for the measured quantities in terms of the excitation amplitudes mix from the beginning on the experimental results with additional theoretical assumptions (for example on the angular momentum coupling scheme).

Therefore we suggest the following procedure. Firstly, the various multipole parameters should be extracted from the measurements and listed for various energies, angles, etc. Secondly, the multipoles should be related to the amplitudes characterizing the excitation process and the underlying theoretical assumptions carefully discussed.

In conclusion, by measuring the angular distribution of the emitted photons in coincidence with forward scattered electrons the total alignment parameter averaged over all rotational quantum numbers N_0 and N_1 can be determined from the measurements. If electrons scattered in other than the forward direction are observed in coincidence with the emitted photons, two additional parameters can be determined corresponding to two further components of the "alignment tensor." By measuring the circular polarization of the radiation a fifth parameter, characterizing the "orientation," can be obtained.[4]

ACKNOWLEDGMENTS

I wish to thank J. W. McConkey and H. Jakubowicz for a helpful discussion.

References

1. P. Baltayan and O. Nedelec, *J. Phys. B* **4**, 1332 (1971); P. *Phys. (Paris)* **36**, 125 (1975).
2. A. N. Jette and P. Cahill, *Phys. Rev.* **176**, 186 (1968).
3. I. C. Malcolm, J. W. McConkey, *J. Phys. B* **12**, L67 (1979); I. C. Malcolm, H. W. Dassen, and J. W. McConkey, *J. Phys. B* **12**, 1003 (1979).

4. K. Blum and H. Jakubowicz, *J. Phys. B* **11**, 909 (1978).
5. I. C. Percival and M. J. Seaton, *Phil. Trans. R. Soc. London Ser. A* **251**, 113 (1958).
6. E. S. Chang and U. Fano, *Phys. Rev. A* **6**, 173 (1972).
7. G. H. Dunn, *Phys. Rev. Lett.* **8**, 62 (1962).
8. P. L. Rubin, *Opt. Spectr.* **20**, 325 (1966).
9. A. R. Edmonds, *Angular Momentum in Quantum Mechanics*, Princeton University Press, Princeton, New Jersey (1960).
10. G. C. M. King, A. Adams, and F. H. Read, *J. Phys. B* **5**, L254 (1972).
11. U. Fano, *Phys. Rev.* **90**, 577 (1953).
12. K. Blum and H. Kleinpoppen, *Phys. Rep.* **52**, 203 (1979).

Polarization Correlation Measurements in Ar and H₂

J. W. McConkey and I. C. Malcolm

Electron-polarized photon coincidence techniques have been used to study the excitation of the resonance lines of Ar and Werner band radiation of H₂ at an incident energy of 50 eV. Measurement of the Stokes parameters of the radiation enabled the appropriate polarization ellipses to be obtained. Different λ and χ parameters were demonstrated for the two Ar lines. Total coherence of the 106.7-nm line was demonstrated at an electron scattering angle of 5°. "Threshold" polarizations measured using coincidence techniques were compared with values obtained using conventional techniques and with estimates based on simple angular momentum conservation and symmetry arguments. Lyman and Werner band polarizations were measured in the energy range from threshold to 300 eV. The Stokes parameters for the Werner emission were related to the relevant multipole moments.

1. Introduction

In the past few years, since the pioneering paper of Macek and Jaecks,[1] an increasing amount of effort both theoretically and experimentally has gone into the subject of electron–photon angular correlations[2–14] and related topics.[15,16]

Since such studies yield very basic information about either the magnitudes and relative phases of the scattering amplitudes, which define the scattering process, or about the alignment and orientation parameters, which define the excited target particle, they have proved to be perhaps the most sensitive available test of existing theoretical approximations.[17–22]

To date, most of the experimental work has been confined to He[5–11] with some

J. W. McConkey and I. C. Malcolm ● Physics Department, University of Windsor, Windsor, Ontario, Canada N9B 3P4.

data on H[13,14] and some on Ne[10] and Ar.[12] In He where LS coupling exists and there are no complications due to fine or hyperfine structure, the situation is relatively straightforward, but with other systems this is by no means the case. For example, in the heavier rare gases where, due to the existence of fine structure, more than one resonance line occurs for a given value of principal quantum number, various assumptions have been made to try to analyze the available data. Arriola *et al.*[12] assumed that the so-called λ and χ parameters were the same for the two resonance lines that they were unable to resolve, and in later work by this group[10] experimental parameters were chosen to make the contribution due to the excitation of the predominantly triplet level negligible. This reduced the problem to a He-like situation again, but the question as to the relative values of λ and χ for the singlet and triplet states remained. Part of the present work was aimed at answering this question by making separate measurements on the singlet and triplet levels. The first part of this report deals with this work and further details are given in a separate publication.[24]

As this field has developed it was only a question of time before an extension to molecules became a reality. Very recently, Blum and Jakubowicz[23] have applied the density-matrix formulation of electron–photon coincidence experiments to the case of electron scattering from molecules obeying Hund's case (b) coupling scheme. On the experimental side, no previous coincidence measurements have been reported, and, in fact, very few straightforward polarization measurements of molecular lines have been made. The second half of this paper presents the first attempt to redress this situation, however crudely. Again, additional details of part of this work may be found in a separate publication.[25] Further theoretical insights into these problems are given in Chapter 11 in this volume, by K. Blum.

2. Excitation of Argon

In the work reported here the excitation of the $3p^5\,4s$ and $3p^5\,4s'$ levels are studied. These levels are based on the $3p^5\,{}^2P_{3/2}$ and $3p^5\,{}^2P_{1/2}$ levels of Ar^+, respectively. The decay of these levels is accompanied by emission of photons at 106.7 and 104.8 nm and these are detected in a direction perpendicular to the scattering plane in coincidence with the inelastically scattered electron which was responsible for the initial excitation. The polarization state of the photons is also determined and, in fact, the experiment is a polarization correlation experiment as distinct from the more usual angular correlation one, where the photons are detected in the scattering plane. A comparison of the angular and polarization correlation methods is set out by Tan *et al.*[9] Equivalent information is obtained from the two types of experiment.

2.1. Basic Theory

In this work we measure the electron–photon coincidence rates $N_c(\beta)$ for photons emitted in a direction perpendicular to the scattering plane. β is related to the detected photon polarization e by

$$e = e_{\theta_\gamma} \cos \beta + e_{\phi_\gamma} \sin \beta \tag{1}$$

The angles β, θ_γ, and ϕ_γ are defined in Figure 1. Inserting the values appropriate to our geometry ($\theta_\gamma = \pi/2$, $\phi_\gamma = \pi/2$) in the general result of Macek and Jaecks,[1] we find

$$N_c(\beta) \propto \langle a_0^* a_0 \rangle \cos^2 \beta + 2\langle a_1^* a_1 \rangle \sin^2 \beta - 2^{1/2} \, \mathrm{Re}\langle a_0^* a_1 \rangle \sin 2\beta \tag{2}$$

where a_{M_J} is a scattering amplitude for the excitation of a substate with total angular momentum projection M_J and $\langle \ \rangle$ denotes an average over the spin states of the incident electrons and a sum over the unobserved spin states of the scattered electrons. $\langle a_{M_J}^* a_{M_J} \rangle$ is thus the differential cross section σ_{M_J} for excitation of the appropriate sublevel. The parameter λ is defined as

$$\lambda = \frac{\sigma_0}{\sigma_0 + 2\sigma_1} \tag{3}$$

and the parameter χ by

$$(\sigma_0\sigma_1)^{1/2} \cos \chi = \mathrm{Re}\langle a_0^* a_1 \rangle \tag{4}$$

This reduces to the usual definition of χ when impulse excitation of an LS-coupled ^1P state is considered.[6] Expression (2) for $N_c(\beta)$ can now be written in terms of

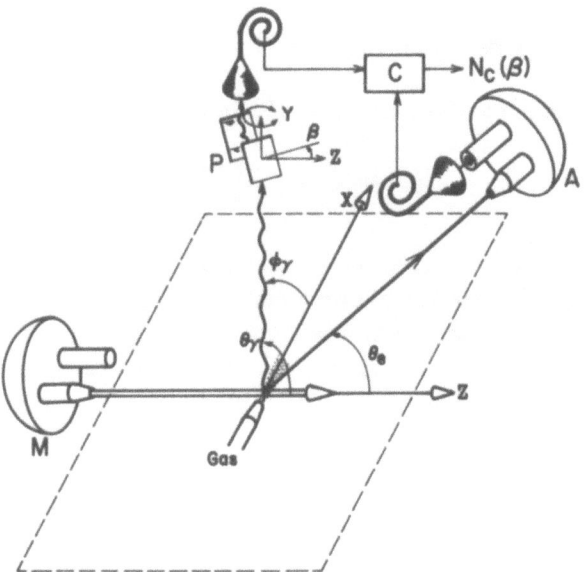

Figure 1. Schematic diagram of coincidence apparatus; M and A, hemispherical electron monochromator and analyzer; P, double reflection polarizer; C, coincidence electronics. The azimuthal angle ϕ_γ is measured with respect to the electron scattering plane. The direction of photon emission is perpendicular to this plane ($\theta_\gamma = \phi_\gamma = \pi/2$).

λ and χ as follows:

$$N_c(\beta) \propto (1 - \lambda) \sin^2 \beta + \lambda \cos^2 \beta - [\lambda(1 - \lambda)]^{1/2} \sin 2\beta \cos \chi \qquad (5)$$

The polarization correlation $P_c(\beta)$ is defined by

$$P_c(\beta) = \frac{N_c(\beta) - N_c(\beta + \pi/2)}{N_c(\beta) + N_c(\beta + \pi/2)} \qquad (6)$$

and the parameters λ and χ are given in terms of $P_c(\beta)$ by

$$\lambda = \tfrac{1}{2}[1 + P_c(0)] \qquad (7)$$

$$\cos \chi = -[1 - P_c^2(0)]^{-1/2} P_c(\pi/4) \qquad (8)$$

Thus a determination of $P(0)$ and $P(\pi/4)$ provides equivalent information to that obtained from an angular correlation scan in coplanar observation.

Traditionally, the values obtained for λ and χ are discussed in terms of an appropriate set of classical dipoles lying in the scattering plane.[1,6] Using the result of Macek and Jaecks,[1] the radiation pattern in the scattering plane of this set of dipoles is given by putting $\phi_y = \pi$ and summing over both polarization states. The coincidence rate $N_c(\theta_y)$ for photons without regard to polarization, as a function of θ_y (the photon emission angle in the plane of scattering relative to the incident Z direction), can be written

$$N_c(\theta_y) \propto [\lambda^{1/2} \sin \theta_y - (1 - \lambda)^{1/2} \cos \theta_y e^{i\chi}]^2 \qquad (9)$$

The direction of minimum intensity $\theta_y^{\min}$ (Eminyan et al.)[6] is given by

$$\tan 2\theta_y^{\min} = \frac{-P_c(\pi/4)}{P_c(0)} \qquad (10)$$

Fano and Macek[3] define a set of orientation and alignment parameters describing the state of the excited source atoms. The values of some of them can be obtained from our measurements. These parameters are defined in terms of the expectation values of various combinations of the total angular momentum (J) and its components. These, in turn, can be related to λ and χ or, in our case, $P_c(\beta)$, as below:

$$\langle J_X^2 \rangle = \tfrac{1}{2}[1 + P_c(0)] \qquad (11)$$

$$\langle J_Z^2 \rangle = \tfrac{1}{2}[1 - P_c(0)] \qquad (12)$$

$$A_0^{\mathrm{col}} \equiv \langle 3J_Z^2 - J^2 \rangle [J(J + 1)]^{-1} = -\tfrac{1}{4}[1 + 3P_c(0)] \qquad (13)$$

$$A_{1+}^{\mathrm{col}} \equiv \langle J_X J_Z + J_Z J_X \rangle [J(J + 1)]^{-1} = -\tfrac{1}{2} P_c(\pi/4) \qquad (14)$$

$$A_{2+}^{\mathrm{col}} \equiv \langle J_X^2 - J_Y^2 \rangle [J(J + 1)]^{-1} = \tfrac{1}{4}[P_c(0) - 1] \qquad (15)$$

The components $\langle J_X \rangle$, $\langle J_Z \rangle$ and the orientation parameter O_0^{col} are identically zero; $\langle J_Y^2 \rangle$ is identically 1; and $\langle J^2 \rangle$ is identically 2. Except for the special case given later, $\langle J_Y \rangle$ and the orientation parameter O_{-1}^{col} cannot be determined from our data. A complete determination of these requires also the measurement of the circular polarization correlation such as was done by Standage and Kleinpoppen[8] in their study of He(3^1P) excitation.

If we can measure the coincidence rates $N_c(r)$ and $N_c(l)$ for right and left circularly polarized light, respectively, then we may define a polarization correlation for circularly polarized light as

$$P(\text{circ}) = \frac{N_c(r) - N_c(l)}{N_c(r) + N_c(l)} \tag{16}$$

In practice, we were not able to measure $P(\text{circ})$ because of the experimental problems that one encounters in the vacuum ultraviolet spectral region. We note in passing that $P(0)$, $P(\pi/4)$, and $P(\text{circ})$ may be identified with the three Stokes' parameters[8,26,27] which characterize the vector polarization of the emitted light. The degree of polarization is given by

$$P = [P^2(0) + P^2(\pi/4) + P^2(\text{circ})]^{1/2} \leq 1 \tag{17}$$

If the light is perfectly coherent, $P = 1$; if partial polarization, $P < 1$, is observed, then incomplete coherence may be deduced. A completely incoherent beam of light is unpolarized, $P = 0$.

In the λ and χ parametrization, inability to measure the circular polarization is equivalent to being unable to determine the sign of χ.

A final parameter of interest is the polarization correlation for forward-scattered electrons. It has been pointed out by King et al.[28] that a measurement of this parameter in certain instances reproduces threshold conditions as far as polarization is concerned. This is valuable because it enables one to measure "threshold" polarization at an energy well removed from threshold where problems, due to low light intensity and the existence of resonance and cascade effects, are absent.

Putting $\beta = 0$ in equation (6) and using threshold values (denoted by the subscript T) for the quantities appearing in equation (2) we obtain the following expression for the threshold polarization P_T:

$$P_T = \frac{\langle a_0^* a_0 \rangle_T - 2\langle a_1^* a_1 \rangle_T}{\langle a_0^* a_0 \rangle_T + 2\langle a_1^* a_1 \rangle_T} \tag{18}$$

Assuming LS coupling and taking the quantization direction to lie along the incident (Z) axis, then only amplitudes for which $M_L = 0$ will contribute to the excitation at threshold or for forward scattering.[28] In these cases, $M_J = M_S$ and if singlet excitation only is considered $M_J = M_S = 0$. This implies that only the term $\langle a_0^* a_0 \rangle_T$ in equation (18) is nonzero and hence $P_T = +1$.

2.2. Experimental

Figure 1 is a schematic diagram of the apparatus. The details have been given in other publications[24,29] and so will only be briefly summarized here. Electrons from a hemispherical monochromator are incident on Ar atoms emanating from a single capillary needle. The inelastically scattered electrons are monitored by a hemispherical analyzer, which can be positioned to select the electron scattering angle θ_e. For this work θ_e was normally 5°. The system was operated at an overall resolution of about 90 meV, which was adequate to resolve the two states under investigation.

Photons emitted from the interaction region in a direction perpendicular to the electron scattering plane undergo polarization analysis by double reflection from gold surfaces.[9]

Electrons and photons are detected in Bendix 4039c channeltrons and pulses from there are routed via standard coincidence electronics to a time–amplitude converter (TAC). The TAC output is sampled by a pulse-height analyzer (PHA). The polarization analyzer is rotated between two orthogonal positions by means of a motor. A timer determines the period for which the analyzer remains in each position. This system is linked to the memory subgroup selection of the PHA so that the electron–photon time delays are recorded in the memory half appropriate to the position of the analyzer. Up to 200 cycles of the analyzer position are performed during a run to average out the effects of any drifts in the detector efficiencies, head pressure, etc. Systematic errors due to these effects are thus rendered negligible. Runs of many hours (130) are often necessary because of the low coincidence rates. Thus the convenience of deriving both λ and $|\chi|$ from measurements of polarization correlation is offset by the reduction in counting efficiency produced by the polarization analyzer. This device cuts down the photon flux by a factor of 50, so although the electron count rate is in the range 2–5 kHz that of the photons is only of the order of 200–300 Hz.

It sometimes happens that, because of electron beam focusing conditions, the center of the interaction region does not lie exactly on the axis of the rotation of the polarization analyzer. This is reflected in an asymmetry in the photon signal at angles β and $\beta + \pi$. The difference is usually less than 1%, but where necessary a first-order correction can be obtained by taking the true signal at angle β to be the mean of the measured ones at β and $\beta + \pi$. As the automatic data-logging system only samples at two angles (β and $\beta + \pi/2$) a correction factor F was obtained before and after each run by measuring the singles rates I in the photon channel at angles of the analyzer β, $\beta + \pi/2$, $\beta + \pi$, $\beta + 3\pi/2$. F is given by

$$F = \frac{I(\beta + \pi/2)[I(\beta) + I(\beta + \pi)]}{I(\beta)[I(\beta + \pi/2) + I(\beta + 3\pi/2)]} \tag{19}$$

and was always within 4% of unity. Because F is determined from the photon singles rates it can be measured in a very short time compared to that necessary for measuring coincidence rates. If the measured coincidence rate at angle β of the polarization analyzer is $N_c'(\beta)$ then the polarization correlation $P_c(\beta)$ defined in equation (6) is given by

$$P_c(\beta) = \frac{1}{q} \cdot \frac{F \cdot N_c'(\beta) - N_c'(\beta + \pi/2)}{F \cdot N_c'(\beta) + N_c'(\beta + \pi/2)} \tag{20}$$

where q is the polarization efficiency of the analyzer.[9,24] A value for q of 0.80 was calculated using published data for the optical constants and this was checked by direct measurement using a method due to Sampson.[30]

The possible error in this figure is approximately 5% due to the spread in the published optical data. This is reflected in a systematic error of the same magnitude in our values of λ and $|\chi|$. Additional consistency checks on our method of evaluating q are obtained from polarization measurements, using a similar polarizer, on the He resonance lines[31] and on Lyman α from H_2.[25]

The polarization obtained from the photon singles rates was found to be independent of head pressure up to the value used to obtain data, suggesting that our measurements are free from pressure-dependent effects. The background pressure of 5×10^{-8} Torr was increased to 8×10^{-6} Torr when the gas jet was on.

To prevent buildup of insulating layers, which might cause instability, the interior of the apparatus was maintained at 150°C by means of a tungsten–halogen lamp.

2.3. Results and Discussion

2.3.1. Polarization Correlations

Values of λ, cos χ, and $|\chi|$ derived from our measurements at 50 eV incident electron energy and scattering angle of 5° are given in Table 1. Statistical uncertainties are given corresponding to one standard deviation. For the 106.7-nm line no uncertainty is quoted on $|\chi|$ as the value of zero given seems to be the only one compatible with our measurement of cos χ since our systematic errors are only 5%. The parameters of alignment (and, where possible, orientation) and the expectation values of components of total angular momentum, are also given in Table 1.

The data in Table 1 show clearly the different λ and, especially, $|\chi|$ values for the two lines. It is interesting to note that the values of λ obtained are quite close to 0.73, which would be predicted by the Born approximation assuming the dominance of singlet excitation. This prediction is contained in the relationship $\lambda = \cos^2 \theta_K$ (θ_K defining the direction of the momentum transfer vector K) given by Eminyan et al.[6] Very recently, Csanak[32] has carried out a first-order many-body calculation for the electron impact excitation of these levels, and has obtained a value for λ of 0.72 for both states at 50 eV impact energy and 5° scattering angle. In this calculation spin–orbit interaction was incorporated into the wave functions, but not into the scattering orbitals.

The radiation patterns in the scattering plane, obtained by substituting our values of λ and $|\chi|$ into equation (9), are shown in Figures 2 and 3 for the 104.8-

Table 1. Measured and Evaluated Parameters Obtained for an Incident Electron Energy of 50 eV[a]

Parameter	104.8 nm	106.7 nm
$P_c(0)$	0.43 ± 0.10	0.20 ± 0.10
$P_c(\pi/4)$	0.49 ± 0.14	-1.16 ± 0.15
λ	0.72 ± 0.06	0.60 ± 0.05
$\cos \chi$	0.55 ± 0.16	1.18 ± 0.16
$|\chi|$	0.99 rad ± 0.19	0.0^b
$\langle J_y \rangle$	—	0.0^b
$\langle J_x^2 \rangle$	0.72 ± 0.06	0.60 ± 0.05
$\langle J_z^2 \rangle$	0.28 ± 0.06	0.40 ± 0.05
A_{2+}^{col}	-0.14 ± 0.03	-0.20 ± 0.02
A_{1+}^{col}	0.25 ± 0.07	0.58 ± 0.08
A_0^{col}	-0.58 ± 0.09	-0.40 ± 0.08
O_{1-}^{col}	—	0.0^b
θ_y^{min}	$24° \pm 6$	$40° \pm 4$
P_T	0.41 ± 0.13	0.94 ± 0.14

[a] Total angular momentum expectation values and alignment parameters are in units of $\hbar$. Statistical uncertainties are given, corresponding to one standard deviation. Additional possible systematic errors are not included (see text).
[b] See text.

and 106.7-nm lines, respectively. The orientations of the patterns relative to θ_y^{min} and θ_K are also displayed. It is apparent from equation (9) and Figure 2 that at the energy and scattering angle studied, the radiation pattern of the 104.8-nm line is identical to that produced by two linear dipoles oscillating coherently in the x and z directions, with a phase difference of ± 0.99 rad. The ratio of the amplitude along z to that along x, $[\lambda/(1 - \lambda)]^{1/2}$, is 1.6, and θ_y^{min} is 24°. For the 106.7-nm line, $\chi = 0$ (as is assumed in the Born approximation). This emission, therefore, is such as would be produced by a single linear dipole oriented in the direction $\theta_y^{min} = 40°$, as shown in Figure 3. Such a simple situation has not been observed previously. In this case, it is clear that there is total coherence, and a full analysis such as that carried out by Standage and Kleinpoppen[8] is not necessary to establish this. The total coherence here indicates that the scattering amplitudes relating to different unobserved spin states and occurring in the sums of equation (2) are identical if nonzero. Such is the case, for example, in the impulse excitation of ^1P states in the earlier work on He. As a result of this total coherence of the 106.7-nm line we are able to deduce the values of $\langle J_y \rangle$ and O_{-1}^{col} shown in Table 1. Again the values of zero given are the only ones compatible with our measurement of $\cos \chi$.

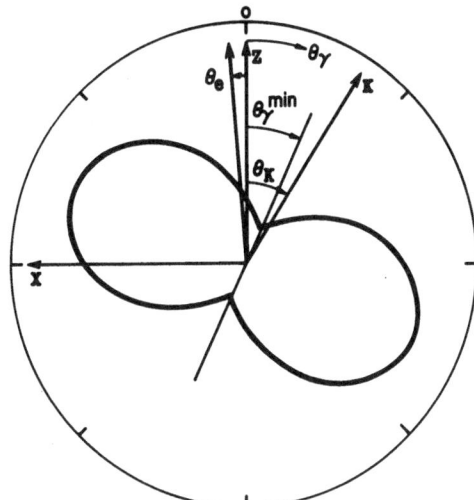

Figure 2. Radiation pattern in the scattering plane for Ar 104.8-nm line, derived from the polarization correlation. Electron scattering angle $\theta_e = 5°$ relative to incident (Z) direction. Incident energy, 50 eV. θ_y^{min} is the direction of minimum intensity, θ_K the direction of the momentum transfer vector ($31°$).

The values of θ_y^{min} are also shown in Table 1 and they may be compared to θ_K ($= 31°$). In the earlier studies of He, Ne, and Ar, at electron scattering angles less than 20°, θ_y^{min} was always somewhat larger than θ_K. In the Born approximation, θ_y^{min} and θ_K should be identical for the simplified situation of $^1S \rightarrow {}^1P$ excitation.[6] Very recent unpublished work of Hollywood, Crowe, and Williams on He at 80 eV incident energy demonstrates that θ_y^{min} and θ_K tend to diverge rapidly as one progresses to larger scattering angles.

An alternative way of discussing the results is in terms of the polarization of the light, equation (17). Total coherence of the excitation, which, in the absence of fine or hyperfine interactions, leads to total coherence of the radiation, is reflected by total polarization of the emitted radiation, $P = 1$. Assuming this to be so, then a

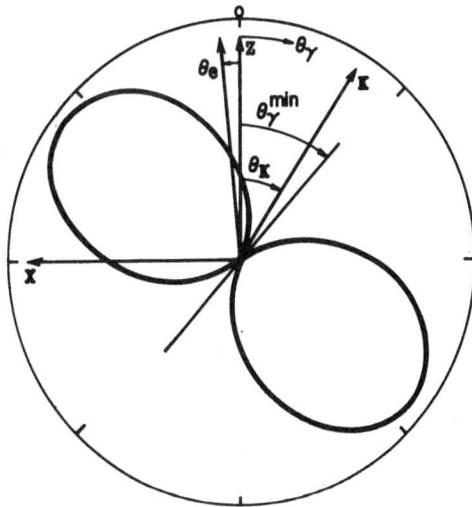

Figure 3. Radiation pattern for Ar 106.7-nm line, derived from the polarization correlation. Other conditions and parameters as defined in Figure 2.

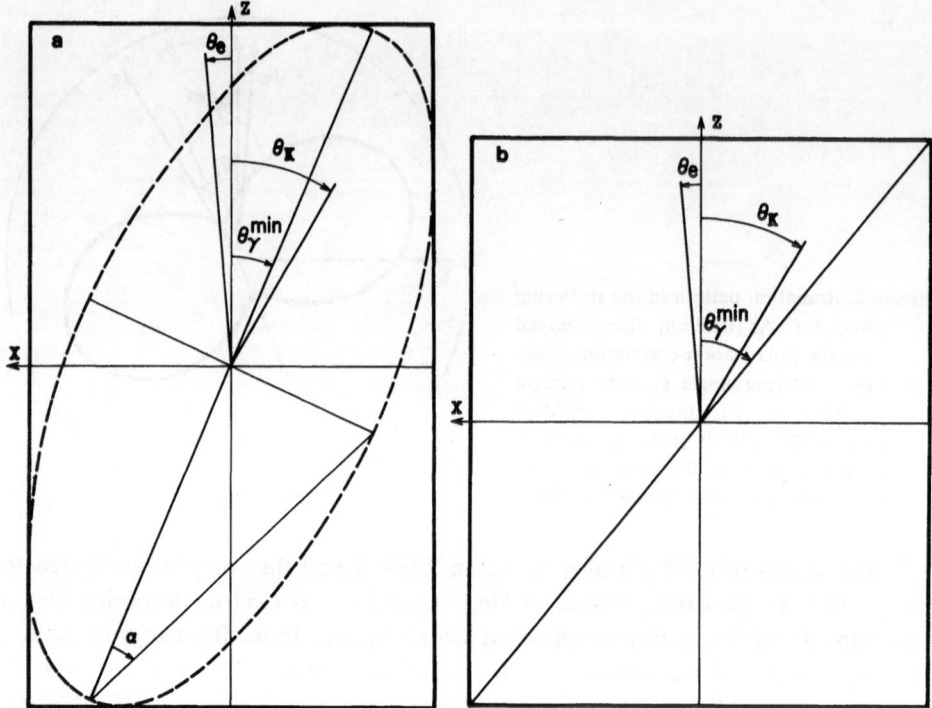

Figure 4. (a) Polarization ellipse appropriate to 104.8-nm line. Conditions and parameters as defined in Figure 2. α is defined in text, equation (21). (b) Polarization ellipse appropriate to the 106.7-nm line. Other conditions and parameters as defined in Figure 2. Note that because $\chi = 0$ in this instance the ellipse has degenerated into a straight line, the diagonal in the figure.

polarization ellipse can be constructed for each line, Figures 4a and 4b, following Born and Wolf.[26] The orientation of the ellipse can be identified with θ_γ^{min}, which is related to the phase shift χ between the two dipoles that are responsible for the radiation, and the ellipticity is related to the relative amplitudes of the two oscillators. Specifically, in terms of our correlation functions $P_c(\beta)$, we have

$$P_c(0) = \cos 2\alpha \cos 2\theta_\gamma^{min} \tag{21a}$$

$$P_c\left(\frac{\pi}{4}\right) = \cos 2\alpha \sin 2\theta_\gamma^{min} \tag{21b}$$

$$P(\text{circ}) = \sin 2\alpha \tag{21c}$$

where α is an auxiliary angle $(-\pi/4 < \alpha < \pi/4)$ shown on Figure 4a which specifies the ellipticity. (For the 104.8-nm line $|\alpha|$ was found to be 24°.) Just as the sign of χ is indeterminate from an angular correlation experiment, so the sign of α is indeterminate unless the measurement of circular polarization is made. The sign of α determines whether the radiation is right or left circularly polarized. If only $P(0)$

and $P(\pi/4)$ are measured then $|\alpha|$ can still be evaluated from

$$\cos 2\alpha = [P_c^2(0) + P_c^2(\pi/4)]^{1/2} \qquad (22)$$

The total coherence of the 106.7-nm line for the particular experimental parameters chosen is evident from Figure 4b. The fact that radiation is observed that is totally linearly polarized means that the assumption of total polarization or total coherence is justified in this case.

2.3.2. Threshold Polarization

Dassen et al.[31] in their investigation of the impact polarization of the unresolved resonance lines of Ar, obtained a value of 0.006 ± 0.005 at 12.5 eV incident electron energy—the lowest energy at which they were able to work. The polarization correlations obtained in the present work for scattering at $\theta_e = 0° \pm 1.5$ (the FWHM angular resolution of the spectrometer) are shown in Table 1. They are consistent with high positive polarizations at threshold, particularly for the 106.7-nm line. This apparently disagrees with the result of Dassen et al., which indicates a threshold polarization close to zero. It must be remembered, however, that the energy spread in that work was close to 1 eV and substantial changes in the measured polarization might be introduced as a result of resonance or cascade effects. In addition, there is some doubt about the justification of the comparison since LS coupling is very questionable and impulse excitation conditions may not exist, i.e., it may not be correct to neglect the precessions of the individual spin and orbital angular momenta during the collision so that separate conservation of these quantities, crucial to the argument of King et al.,[28] may not apply. It is likely that impulse excitation conditions do apply in our "pseudothreshold" measurements because at an incident electron energy of 50 eV the excitation time should be short compared to the various precession times in the atom.

Our measurement of "threshold" polarization gives a value close to $+1$ for the 106.7-nm line, implying that $\langle a_1^* a_1 \rangle$, equation (18), is close to zero due either to a fortuitous choice of energy and angle or to a dominance of singlet character for the line. The former explanation is more likely since significant triplet characteristic of this line has been demonstrated by Hippler and Schartner[33] from optical oscillator strength measurements and by Lee and Lu[34] from a detailed analysis of the Ar absorption spectrum.

3. Excitation of H_2

Although until recently most of the electron impact excitation cross sections for molecular emissions were measured without regard for possible polarization of the radiation, there is considerable evidence in the literature[35-39] to indicate that sometimes molecular radiation from such sources may be quite strongly polarized. However, there has been no attempt to systematically investigate the polarization of molecular radiation either experimentally or theoretically, presumably because of the greatly increased difficulty of the problem over the atomic case.

The work presented here involves the excitation of the $B\,^1\Sigma_u^+$ and $C\,^1\Pi_u$ states of H_2 from the ground $^1\Sigma_g^+$ state and the subsequent relaxation of these states back to the ground state with the emission of the Lyman and Werner bands, respectively. There are a number of advantages in choosing H_2 as a starting point in molecular polarization studies. On the experimental side, it has the advantage that only a few rotational levels of the ground state are significantly populated at room temperature. This in turn leads to minimal rotational development in the excited states. In addition, the rotational structure of the emission bands is relatively well resolved so that, in principle at least, there is the possibility of observing single rotational lines. The Lyman and Werner bands have the additional advantage of lying in the vacuum *uv* spectral region where low noise detectors such as the channeltron can be conveniently used. Theoretically, H_2, being the simplest molecule, should be the most amenable to study. In addition, the particular states being investigated have symmetry properties that impose simplifying restrictions on the excitation process under certain conditions as discussed later.

It might be noted in passing that atomic radiation following dissociative excitation of molecules can be polarized significantly,[40–43] as was predicted by Van Brunt and Zare.[44]

The present work encompasses two separate experiments, one in which the polarization of the radiation was measured as a function of incident electron energy for particular vibrational–rotational transitions and the other in which the integrated radiation from a single vibrationally excited level was detected in coincidence with the inelastically scattered electrons that had excited that level. In the following section, we summarize some basic theoretical ideas that are relevant to the above experiments. First we consider the steady state case in which the polarization of the emitted light is monitored without regard for the scattered electron. In particular, we consider excitation both very close to threshold and at high incident energy. We then present the basic equations governing the coincidence experiment following Reference 23.

3.1. Basic Theory

3.1.1. Steady State Polarization

Calculations of the polarization of molecular radiation are complicated by a number of factors that are not present for atoms. The additional degrees of freedom available in a molecule can mean very often that states of quite large values of total angular momentum J, with a consequent large number of possible magnetic substates, must be considered. In addition the initial excitation process is more complicated because of the larger number of possible initial states, possible variations in electronic transition probability within the rotational envelope, and the fact that the population distribution among the initial states is governed by a number of factors in addition to the statistical weighting. Such factors as the temperature of the source, and for homonuclear diatomics the nuclear spin statistics and molecular symmetries involved, must be taken into account.[45]

Previous calculations of diatomic molecular polarization by Jette and Cahill[46] and Baltayan and Nedelec[37] have essentially been an extension of the Percival and Seaton[47] treatment of atomic polarization. The former authors concluded that threshold polarization could be determined when a change of electronic angular momentum along the internuclear axis occurred, simply by considering the symmetry of the total molecular wave functions involved, and without having to consider the excitation cross sections of the different magnetic sublevels. In other instances, for example, $\Sigma \rightarrow \Sigma$ excitation, the polarization of the emitted light was determined by the excitation cross sections of the individual magnetic sublevels. These authors calculated the threshold polarization of selected rotational lines of the Fulchur bands of H_2. The basic assumption inherent in their work is that the excitation time is much shorter than the rotational period and hence the orbital angular momentum of the electron being excited can be considered as essentially decoupled from the other movement of the molecule.

Baltayan and Nedelec[37] have used a density-matrix approach and applied the united atom approximation to calculate some threshold polarizations and the Born approximation to obtain some polarizations at high incident energies. They make the assumption that the total orbital angular momentum of the molecule is conserved in the excitation process. It has been suggested[23] that discrepancies between these two sets of calculations[37,46] are due to the different assumptions made. Threshold polarization of molecular radiation thus depends on the dynamics of the scattering process, and hence assumptions about excitation and relaxation times and about applicable coupling schemes need to be made with much care.

Recently, Blum and Jakubowicz[23] obtained expressions for the steady state polarization of the molecular radiation using a density-matrix formulation in which they first obtained the photon density matrix for the time correlated case in which the exciting electron and the resulting decay photon were detected in coincidence and then carried out an integration over all observation times and scattered electron directions. The polarization of a particular rotational line was related to the total cross sections for excitation of the individual magnetic sublevels, and the square of the dipole matrix element for the particular transition involved. They point out that the information that is obtained from a straightforward polarization measurement corresponds to a determination of certain of the multipole moments of the excited state averaged over all times and directions of electron emission. These multipole moments are related to the alignment and orientation parameters discussed in Section 2.

It is instructive to carry this discussion of the evaluation of threshold polarization one step further because of the possibility of measuring the "threshold" polarization using coincidence techniques[28] (see earlier). This can be done in an energy region well away from threshold where the dynamics of the excitation process should be more clearly defined.

One factor that does not seem to have been considered previously is the fact that symmetry considerations may in certain cases place severe restrictions on the possible magnetic sublevels that may be excited. Obviously this can dominate the polarization of the emitted light.

As far back as 1962 Dunn[48] demonstrated that symmetry arguments when applied to excitation of molecular states led to considerable restrictions regarding the possible orientations of the target molecule relative to the exciting electron beam, i.e., to the quantization (Z) axis. He applied these ideas to the dissociation of molecules and predicted strong anisotropies at threshold in the dissociation products relative to the Z direction for certain transitions. Numerous papers by various authors[49,50] have developed and extended these ideas. Applying these concepts to the present problem we find that for a $\Sigma_g^+ \to \Sigma_u^+$ transition (Lyman band excitation) the target molecules with axes parallel to the exciting beam direction will be preferentially excited whereas for a $\Sigma_g^+ \to \Pi_u$ transition (Werner band excitation) target molecules with axes oriented perpendicular to the exciting beam will be preferred. This, of course, means that certain values of M_{J_0} will be preferred in the excitation process, $M_{J_0} = 0$ for Σ_u^+ excitation and $M_{J_0} = \pm J_0$ for Π_u excitation.

Assuming that total orbital angular momentum is conserved and that, at threshold, the exciting electron carries no Z component of angular momentum into or away from the collision, then the above discussion predicts that at threshold preferential population of the magnetic sublevels of the upper state will also occur with consequent polarization of the emitted light when these states decay. For example, we now expect $M_{J_1} = 0$ substate excitation for the Σ_u^+ state and $M_{J_1} = \pm J_0$ for Π_u excitation. We note that the above conclusions are entirely different to those suggested by Jette and Cahill.[46]

Knowing the population distribution among the upper M_{J_1} levels, the relative intensities of the $\Delta M_J = 0, \pm 1$ components may be evaluated readily using appropriate Clebsch–Gordon coefficients and hence the threshold polarization of the different rotational lines may be obtained.

If more than one rotational line is contained in the acceptance aperture of the spectrometer (and this is often the case for reasons of both intensity and available resolution) then the resultant threshold polarization may be calculated taking account of the relative excitation of the different values of J_1 from a distribution of ground-state rotational levels.

It is clear also that the above restrictions on the relative orientation of the molecule at the threshold for excitation are oversimplified. In practice a $\sin^2 \theta$ or $\cos^2 \theta$ distribution will result relative to the Z direction. This will result in a relaxation of the allowable values of M_{J_1} and hence a change (reduction) in the observed polarization. A reduction will also be expected due to the effects of hyperfine structure, which are neglected in the above discussion. Values predicted will thus represent upper limits. For a more quantitative discussion of these ideas see Chapter 11 in this volume, by K. Blum.

At high incident energies one may make use of the Born approximation, which predicts that the excitation probability depends on $| \boldsymbol{\mu} \cdot \mathbf{K} |^2$, where $\boldsymbol{\mu}$ is the electric dipole moment and $\mathbf{K}$ is the momentum transfer vector. At high incident electron energies $\mathbf{K}$ is predominantly directed perpendicular to the incident, Z, direction and so preferential excitation will occur for molecules and transitions where $\boldsymbol{\mu}$ is also in this direction. In the case of $X^1\Sigma_g^+ \rightarrow B^1\Sigma_u^+$ excitation $\boldsymbol{\mu}$ is aligned along the internuclear axis, whereas for $X^1\Sigma_g^+ \rightarrow C^1\Pi_u$ excitation it is perpendicular to this axis. This means that for $B(C)$ state excitation at high energy, molecules oriented perpendicular (parallel) to the Z direction will be preferred. Thus under these conditions we should expect preferential population of $M_{J_1} = \pm J_0$, (0) for B, (C) state excitation, respectively. The expected polarization of the different rotational lines involved may now be calculated readily using the appropriate Clebsch–Gordan coefficients as before.

However, we must modify these intuitive ideas for the reasons outlined by Zare.[52] He pointed out that when the integration over all scattered electron directions (momentum transfers) is carried out for the initial excitation process, a strong weighting occurs for momentum transfer directions that lie at significantly smaller angles than 90° to the incident z direction. This, and the effects of hyperfine structure, will produce a significant reduction in the anisotropies and resultant polarizations obtained. Integration over a number of rotational lines will reduce any effects even more and so it is likely, certainly at the highest energies (few hundred eV) used in this experiment, that rather small values of polarization will be observed. This is in fact the case, as shown later.

3.1.2. Polarization Correlations

The equations relevant to the present experiment may be obtained directly from Blum and Jakubowicz.[23] They obtain expressions for the different Stokes parameters of the molecular radiation emitted in a specified direction following excitation by an electron which is subsequently scattered also in a specific direction. These expressions relate the Stokes parameters describing the polarization states of the photons to so-called state multipoles (which, in turn, are related to the appropriate scattering amplitudes), dipole matrix elements, geometrical factors, and perturbation coefficients which allow for the effect of fine or hyperfine structure. In our particular case, where we do not resolve the individual rotational states, we must define a new set of multipoles $\langle T_{KQ}^{\dagger} \rangle$ in which a formal summation over the relevant rotational quantum numbers has been carried out and which include the perturbation coefficients mentioned previously.

Recognizing that our $P(\pi/4)$, $P(\text{circ})$, and $P(0)$ [equations (6) and (16)] may be identified with η_1, η_2, and η_3, respectively, of Blum and Jakubowicz, we obtain from equation (32) of their paper the following equations:

$$I = c \mid M \mid^2 \left[-\frac{2}{3^{1/2}} \langle T_{00}^\dagger \rangle - \langle T_{22}^\dagger \rangle \sin^2 \theta \cos 2\phi \right.$$

$$\left. + 2 \langle T_{21}^\dagger \rangle \sin \theta \cos \theta \cos \phi - \langle T_{20}^\dagger \rangle \frac{(3 \cos^2 \theta - 1)}{6^{1/2}} \right] \tag{23a}$$

$$IP_c(\pi/4) = -2c \mid M \mid^2 [\langle T_{22}^\dagger \rangle \cos \theta \sin 2\phi + \langle T_{21}^\dagger \rangle \sin \theta \sin \phi] \tag{23b}$$

$$IP(\text{circ}) = -2ic \mid M \mid^2 \langle T_{11}^\dagger \rangle \sin \theta \sin \phi \tag{23c}$$

$$IP_c(0) = c \mid M \mid^2 \left[\langle T_{22}^\dagger \rangle (1 + \cos^2 \theta) \cos 2\phi + \langle T_{21}^\dagger \rangle \sin 2\theta \cos \phi \right.$$

$$\left. + \left(\frac{3}{2} \right)^{1/2} \langle T_{20}^\dagger \rangle \sin^2 \theta \right] \tag{23d}$$

I is the total intensity obtained without regard for polarization, c is a constant which contains the acceptance solid angle of the photon detector, M is the dipole matrix element relevant to the decay of the excited state, and θ and ϕ define the direction of photon emission (equivalent to θ_y and ϕ_y of Figure 1). $\langle T_{22}^\dagger \rangle$, $\langle T_{21}^\dagger \rangle$, and $\langle T_{20}^\dagger \rangle$ correspond to the alignment of the system, whereas $\langle T_{11}^\dagger \rangle$ is the orientation parameter. In our case, where only $P_c(0)$ and $P_c(\pi/4)$ are determined and where $\theta = \phi = \pi/2$, the relevant equations become

$$P_c \left(\frac{\pi}{4} \right) = -2 \langle T_{21}^\dagger \rangle \left[-\frac{2}{3^{1/2}} \langle T_{00}^\dagger \rangle + \langle T_{22}^\dagger \rangle + \frac{1}{6^{1/2}} \langle T_{20}^\dagger \rangle \right]^{-1} \tag{24a}$$

$$P_c(0) = \left[-\langle T_{22}^\dagger \rangle + \left(\frac{3}{2} \right)^{1/2} \langle T_{20}^\dagger \rangle \right] \left[-\frac{2}{3^{1/2}} \langle T_{00}^\dagger \rangle + \langle T_{22}^\dagger \rangle + \frac{1}{6^{1/2}} \langle T_{20}^\dagger \rangle \right]^{-1} \tag{24b}$$

Clearly for a complete determination of relative values of all five multipole moments involved a minimum of four independent measurements have to be carried out, always including a measurement of the circular polarization (to get information about $\langle T_{11}^\dagger \rangle$). These could be achieved by means of an additional polarization correlation measurement in a different direction, for example, $\theta = 54°44'$, $\phi = 90°$ or by carrying out a series of angular correlation measurements in more than one plane. For more detailed discussion of the physical significance of the state multipoles the reader is referred to Chapter 11 in this volume, by K. Blum.

3.2. Experimental

Two experimental arrangements were used. The coincidence data were taken using the apparatus described previously and precautions similar to those in the Ar work were taken. The straightforward polarization data were obtained using a crossed electron–gas beam system similar to that described previously.[51] A vacuum monochromator sampled photons emitted orthogonal to both beams. The polarization analyzer, which was mounted behind the exit slit of the monochromator, was a triple reflection device.[30] The reflecting surfaces were gold-coated, $\lambda/4$ optically flat, Pyrex laser mirrors. The detector rotated with the polarizer, thus eliminating any possible anisotropic effects due to variation of detector sensitivity as a function of the orientation of the electric vector of the incident photon relative to the detector surface.

Pulses from the channeltron were fed into a standard pulse-counting system. Four different orientations of the polarizer, 90° apart, were sampled sequentially. By repetitive sampling at these four positions systematic variations due to fluctuation in pressure and beam current were minimized. By using four positions of the polarizer rather than two, and averaging the data obtained, small mechanical misalignments of the system were allowed for.

A number of effects had to be carefully considered in order to ensure reliability of the measurements. These and other experimental details have been considered fully by Malcolm *et al.*[25]

3.3. Results and Discussion

3.3.1. Polarization of Lyman and Werner Bands

Figures 5 and 6 show examples of the data obtained for the polarization of selected Lyman and Werner emissions and its variation with incident electron energy. The (6, 1) Lyman band at 107 nm was investigated with a spectrometer passband of 0.35 nm (FWHM). Under these conditions the $R(0)$, $R(2)$, and $P(1)$ lines all contributed to the measured intensity. An attempt to isolate individual rotational lines was unsuccessful owing to intensity limitations. It is apparent from Figure 5 first that the data appear to be approaching a threshold value of approximately 0.2 and secondly that some broad secondary structure occurs around 175 eV. It is apparent that P will not become negative until incident energies in excess of 300 eV are reached. This is consistent with the arguments expressed in Section 3.1.1. Using the simplified ideas discussed in Section 3.1.1, a threshold polarization of just over 0.6 was calculated for the rotational lines under consideration here. An attempt to measure the threshold polarization of this Lyman band using the "pseudothreshold" coincidence technique[28] was unsuccessful because of intensity problems.

Figure 6 shows the polarization function of the (0, 1) Werner band at 105.5 nm. It was investigated with a 1-nm spectrometer passband but the data taken using narrower slits were very similar. The polarization function was also independent of the vibrational transition studied as might be expected. A pronounced maximum of approximately 0.23 is apparent at an incident energy of 20 eV. By 50 eV the polarization has fallen to 0.15 and by 300 eV it is quite small, but still positive.

The measurements suggests a threshold polarization value of 0.15 or less though

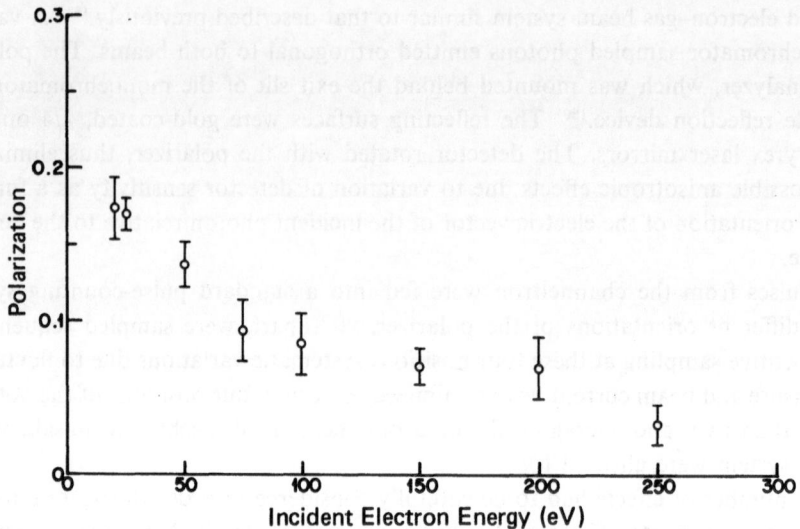

Figure 5. Polarization of (6, 1) Lyman band as a function of incident electron energy. Instrumental resolution 0.35 nm (FWHM). The error bars represent one standard deviation.

the rather rapid falloff in the polarization as the threshold is approached may be caused by resonance effects. Using the electron–photon coincidence apparatus a "pseudothreshold" polarization value of 0.17 ± 0.03 was obtained. Electrons that had excited $C\,^1\Pi_u$ ($v' = 0$) were detected in coincidence with decay photons. No wavelength selection was used in the photon channel apart from that which occurred

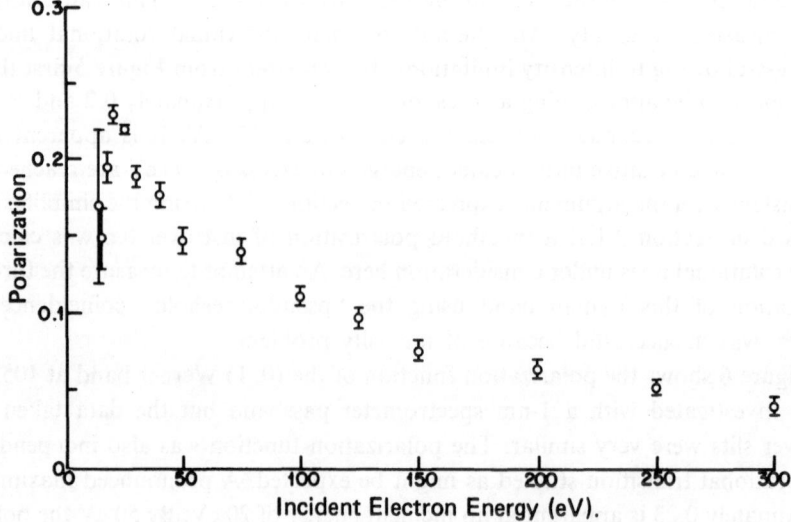

Figure 6. Polarization of (0, 1) Werner band as a function of incident electron energy. Instrumental resolution 1 nm (FWHM). The error bars represent one standard deviation.

because of the rapid falloff in sensitivity of the channeltron detector towards longer wavelengths. For this reason a number of vibrational bands were almost certainly contributing to the observed coincidence signal. This should not be a major problem, however, as no dependence of P on vibrational transition is expected.

Using the simplified theory of Section 3.1.1 a negative threshold polarization is predicted for the integrated vibrational band. However, the calculation is complicated owing to the perturbations which are known to occur in the P and R branches of the Werner system because of mixing with the $B'\,{}^1\Sigma_u{}^+$ state.[53] The relative contributions of the P, Q, and R branches to the total intensity are crucial when the polarization has to be estimated.

3.3.2. Stokes Parameters for the Werner $(0, v'')$ Bands

$P_c(0)$ and $P_c(\pi/4)$ [see equation (6)] were measured at an incident electron energy of 50 eV and at an electron scattering angle of 5°. $P_c(0)$ was also measured at 0° and 10°. In the electron channel only those that were inelastically scattered following excitation of $C\,{}^1\Pi_u$ $(v' = 0)$ were detected. No wavelength selection was possible in the photon channel so, as in the pseudothreshold measurement already discussed, a number of vibrational transitions contributed to the observed signal. The data were obtained at a pressure where the polarization of the integrated radiation, monitored without regard for coincidences, was constant as a function of pressure. This suggests that the data should be free from resonance trapping or collisional depolarizing effects.

The values obtained for the Stokes parameters are listed in Table 2 as are other relevant parameters. The 5° data are displayed in Figure 7, which shows the polarization ellipse appropriate to this electron scattering angle. Here we have made the same assumptions as were made in the case of Ar in order to treat the data in this way. We note that $\theta_\gamma{}^{\min}$ is again rather close to the momentum transfer direction θ_K.

The pattern obtained is equivalent to what would be obtained from a source consisting of two classical dipoles oscillating in the z and x directions with amplitudes a and b, respectively, and relative phase χ where, from equation (23) a, b,

Table 2. Measured and Evaluated Parameters for $H_2(C\,{}^1\Pi_u, v' = 0)$ Excitation at 50 eV

Electron scattering angle, θ_e (deg)	$P_c(0)$	$P_c(\pi/4)$	$\theta_\gamma{}^{\min}$ (deg)	θ_K (deg)	$\cos\chi$	α (deg)
0	0.17 ± 0.03	—	—	—	—	—
5	0.29 ± 0.06	-0.24 ± 0.13	34	29	0.25	37
10	-0.06 ± 0.10	—	—	33	—	—

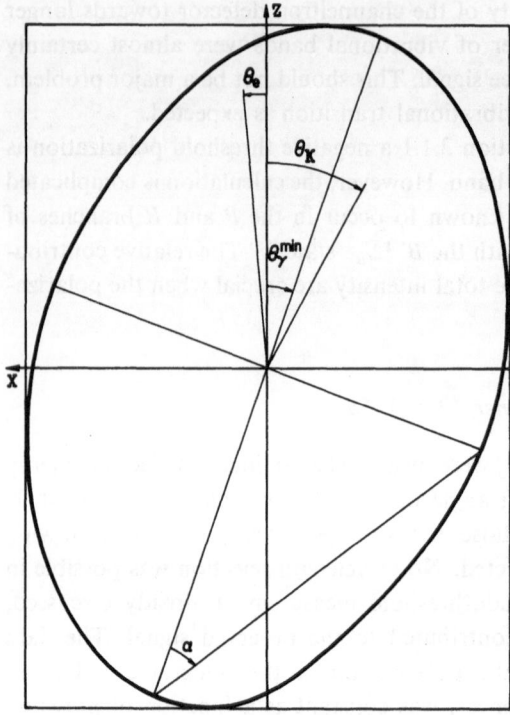

Figure 7. Polarization ellipse appropriate to Werner $(C\,^1\Pi_u,\ v'=0 \to X\,^1\Sigma_g{}^+, v'')$ radiation. Electron scattering angle 5°; incident energy 50 eV. See text and Table 2 for significance and actual values of $\theta_\gamma{}^{\min}$, θ_K, and α.

and $|\chi|$ are given by

$$a \sim \left(-\frac{1}{3^{1/2}}\langle T_{00}^\dagger\rangle + \frac{2}{6^{1/2}}\langle T_{20}^\dagger\rangle\right)^{1/2} \tag{25a}$$

$$b \sim \left(-\frac{1}{3^{1/2}}\langle T_{00}^\dagger\rangle + \langle T_{22}^\dagger\rangle - \frac{1}{6^{1/2}}\langle T_{20}^\dagger\rangle\right)^{1/2} \tag{25b}$$

$$\cos\chi = \langle T_{21}^\dagger\rangle\left[\frac{1}{3}\left(-\frac{1}{2^{1/2}}\langle T_{00}^\dagger\rangle + \langle T_{20}^\dagger\rangle\right)\left(-2^{1/2}\langle T_{00}^\dagger\rangle + 6^{1/2}\langle T_{22}^\dagger\rangle - \langle T_{20}^\dagger\rangle\right)\right]^{-1/2} \tag{25c}$$

The factor $C\,|\,M\,|^2$ has been normalized to unity for convenience. With only three measured quantities and four unknowns in equation (25), it is clear that the individual multipoles cannot be evaluated. As discussed earlier additional measurements will be necessary before this can be achieved.

The sharp variation in $P(0)$ from 0.17 at zero degrees to 0.29 at 5° and to -0.06 at 10° should be noted.

4. Summary and Conclusions

This paper reports on polarization measurements in Ar and H_2. In the Ar work we have managed to study the two resonance lines separately and demonstrate that the λ and χ parameters were different for the two lines at an electron scattering angle of 5°. Orientation and alignment parameters describing the excited states have been evaluated. The measured Stokes parameters have been used to obtain radiation patterns in the scattering plane and also the relevant polarization ellipses. The radiation pattern of the 104.8-nm line is identical to that produced by two linear dipoles oscillating coherently, but that of the 106.7-nm line is such as would be produced by a single linear dipole. Total coherence of this radiation is thus demonstrated and the implications of this have been discussed. "Threshold" polarizations have been measured for the two lines and the values obtained are quite different from those obtained by conventional techniques. The significance of this has been discussed.

The H_2 work involves the excitation of the $B\,^1\Sigma_u^+$ and $C\,^1\Pi_u$ states and the subsequent emission of the Lyman and Werner bands, respectively. The molecular radiation has been shown to be strongly polarized under certain conditions and the polarizations of selected bands have been measured. To our knowledge this is the first detailed study of the polarization of molecular radiation excited by electron impact that has appeared in the literature. A brief summary of existing theoretical approaches to molecular polarization has been given and some additional symmetry arguments are presented suggesting that threshold polarizations should be capable of evaluation under certain conditions. Comparisons have been made between experiment and theory, but few conclusions could be drawn. Some Stokes' parameters have been measured for Werner radiation from $C\,^1\Pi_u$, $v' = 0$ and these are presented in Table 2. This is the first time that such data have been obtained for molecules. The limitations of the present measurements are highlighted.

ACKNOWLEDGMENTS

We are pleased to acknowledge financial assistance from the National Research Council of Canada and NATO Division of Scientific Affairs. I. C. Malcolm also wishes to acknowledge financial aid from the Foyle Trust. We wish to thank Dr. G. W. F. Drake, Dr. W. E. Baylis, Professor P. S. Farago, Dr. G. C. M. King, Mr. H. W. Dassen, and other members of the electron collisions group for helpful discussions. We also acknowledge the expert technical assistance of our Workshop staff.

References

1. J. Macek and D. H. Jaecks, *Phys. Rev. A* **4**, 2288–2300 (1971).
2. J. Wykes, *J. Phys. B* **5**, 1126–1137 (1972).
3. U. Fano and J. Macek, *Rev. Mod. Phys.* **45**, 553–573 (1973).
4. K. Blum and H. Kleinpoppen, *J. Phys. B* **8**, 922–925 (1975).
5. M. Eminyan, K. B. MacAdam, J. Slevin, and H. Kleinpoppen, *Phys. Rev. Lett.* **31**, 576–579 (1973).
6. M. Eminyan, K. B. MacAdam, J. Slevin, and H. Kleinpoppen, *J. Phys. B* **7**, 1519–1542 (1974).
7. M. Eminyan, K. B. MacAdam, J. Slevin, M. C. Standage, and H. Kleinpoppen, *J. Phys. B* **8**, 2058–2066 (1975).

8. M. C. Standage and H. Kleinpoppen, *Phys. Rev. Lett.* **36**, 577–580 (1976).

9. K. H. Tan, J. Fryar, P. S. Farago, and J. W. McConkey, *J. Phys. B* **10**, 1073–1082 (1977).

10. A. Ugbabe, P. J. O. Teubner, E. Weigold, and H. Arriola, *J. Phys. B* **10**, 71–79 (1977).

11. V. C. Sutcliffe, G. N. Haddad, N. C. Steph, and D. E. Golden, *Phys. Rev. A* **17**, 100–107 (1978).

12. H. Arriola, P. J. O. Teubner, A. Ugbabe, and E. Weigold, *J. Phys. B* **8**, 1275–1279 (1975).

13. J. F. Williams, in *The Physics of Electronic and Atomic Collisions—Invited Lectures*, J. R. Risley and R. Geballe, eds., pp. 139–150, University of Washington Press, Seattle, Washington (1975).

14. A. J. Dixon, S. T. Hood, and E. Weigold, *Phys. Rev. Lett.* **40**, 1262 (1978).

15. I. V. Hertel and W. Stoll, *J. Phys. B* **7**, 570–592 (1974).

16. J. Macek and I. V. Hertel, *J. Phys. B* **7**, 2173–2188 (1974).

17. L. D. Thomas, G. Csanak, H. A. Taylor, and B. S. Yarlagadda, *J. Phys. B* **7**, 1719–1733 (1974).

18. M. R. Flannery and K. J. McCann, *J. Phys. B* **8**, 1716–1733 (1975).

19. B. H. Bransden and K. H. Winters, *J. Phys. B* **9**, 1115–1120 (1976).

20. T. Scott and M. R. C. McDowell, *J. Phys. B* **9**, 2235–2254 (1976).

21. G. D. Meneses, N. T. Padial, and G. Csanak, *J. Phys. B* **11**, L237–L242 (1978).

22. D. H. Madison and W. N. Shelton, *Phys. Rev. A* **7**, 499–513 (1973); D. H. Madison and R. V. Calhoun, quoted in Reference 11.

23. K. Blum and H. Jakubowicz, *J. Phys. B* **11**, 909–925 (1978).

24. I. C. Malcolm and J. W. McConkey, *J. Phys. B* **12**, 511–519 (1979).

25. I. C. Malcolm, H. W. Dassen, and J. W. McConkey, *J. Phys. B* **12**, 1003–1018 (1979).

26. M. Born and E. Wolf, *Principles of Optics*, Pergamon Press, Oxford (1959).

27. W. H. McMaster, *Rev. Mod. Phys.* **33**, 8–28 (1961).

28. G. C. M. King, A. Adams, and F. H. Read, *J. Phys. B* **5**, L254–L257 (1972).

29. J. A. Preston, M. A. Hender, and J. W. McConkey, *J. Phys. E* **6**, 661–666 (1973).

30. J. A. R. Samson, *Techniques of Vacuum Ultraviolet Spectroscopy*, Wiley, New York (1967).

31. H. W. Dassen, I. C. Malcolm, and J. W. McConkey, *J. Phys. B* **10**, L493–L495 (1977).

32. G. Csanak, private communication (1978).

33. R. Hippler and K.-H. Schartner, *Z. Phys.* **270**, 225–228 (1974).

34. C.-M. Lee and K. T. Lu, *Phys. Rev. A* **8**, 1241–1257 (1973).

35. T. A. R. Irwin and F. W. Dalby, *Can. J. Phys.* **43**, 1766–1775 (1965).

36. P. Cahill, R. Schwartz, and A. N. Jette, *Phys. Rev. Lett.* **19**, 283–286 (1967).

37. P. Baltayan and O. Nedelec, *J. Phys. B* **4**, 1332–1342 (1971).

38. P. Baltayan and O. Nedelec, *J. Phys. (Paris)* **36**, 125–128 (1975).

39. J. Watson, Jr. and R. J. Anderson, *J. Chem. Phys.* **66**, 4025–4030 (1977).

40. W. R. Ott, W. E. Kauppila, and W. L. Fite, *Phys. Rev. A* **1**, 1089–1098 (1970).

41. G. R. Möhlmann, S. Tsurubuchi, and F. J. de Heer, *Chem. Phys.* **18**, 145–152 (1976).

42. G. J. Fisanick-Englot, D. E. Donohue, and R. S. Freund, *Bull. Am. Phys. Soc.* **22**, 1330 (1977).

43. C. Karolis and E. Harting, *J. Phys. B* **11**, 357–370 (1978).

44. R. J. Van Brunt and R. N. Zare, *J. Chem. Phys.* **48**, 4304–4308 (1968).

45. G. Herzberg, *Molecular Spectra and Molecular Structure I: Spectra of Diatomic Molecules* 2nd ed., Van Nostrand, Princeton, New Jersey (1950).

46. A. N. Jette and P. Cahill, *Phys. Rev.* **176**, 186–193 (1968).

47. I. C. Percival and M. J. Seaton, *Phil. Trans. R. Soc. London* **A251**, 113–138 (1958).

48. G. H. Dunn, *Phys. Rev. Lett.* **8**, 62–64 (1962).

49. R. N. Zare and D. R. Herschbach, *Proc. I.E.E.E.* **51**, 173–182 (1963).

50. R. J. Van Brunt, *J. Chem. Phys.* **60**, 3064–3070 (1973).

51. F. G. Donaldson, M. A. Hender, and J. W. McConkey, *J. Phys. B* **5**, 1192–1210 (1972).

52. R. N. Zare, *J. Chem. Phys.* **47**, 204–215 (1967).

53. E. J. Stone and E. C. Zipf, *J. Chem. Phys.* **56**, 4646–4652 (1972).

13

Effect of Optical Potentials on Angular Correlation Parameters

D. H. MADISON

Angular correlation parameters and differential cross sections are calculated in the distorted-wave approximation for 40–200-eV electron impact excitation of the $2'P$ state of helium. Initial- and final-channel distorted waves are calculated as eigenfunctions of a complex optical potential. The optical potential consists of a static Hartree-Fock atomic potential, a polarization potential term, an exchange distortion term, and an imaginary term representing electron absorption into other open channels. The appropriate inelastic scattering T matrix is obtained for non-Hermitian operators. Effects of polarization, exchange, and absorption are examined individually and collectively and the results are compared with experimental data.

1. Introduction

The area of electron impact excitation of atoms has recently been receiving considerable attention in the literature. The quantities of fundamental importance to this process are differential cross sections for exciting magnetic sublevels of atoms, which have recently become the subject of intense experimental investigation.[1–7] The results of these and other experimental works have stimulated many elaborate theoretical calculations using widely differing approaches—distorted-wave, many-body, close-coupling, second-order potential, pseudostate expansions, multichannel eikonal. A review of this work has recently been completed by Bransden and McDowell.[8] None of these calculations, however, have given good quantitative agreement with experimental data over a broad energy and angular range for the various measured quantities.

D. H. MADISON ● Drake University, Des Moines, Iowa 50311.

If experimental electron–atom cross sections for elastic scattering are compared with inelastic cross sections for excitation of allowed transitions, strong similarities in the shape of the two cross sections are readily noticed. This observation indicates that elastic scattering may have a dominant effect on inelastic scattering. The distorted-wave (DW) model for inelastic scattering represents the inelastic event as a transition between two elastic scattering states. Early elementary DW calculations[9] produced medium-energy differential cross sections and ratios (λ) of magnetic sublevel cross sections that were in reasonable agreement with experimental data for electron excitation of the ^1P state of He. The observed differences between the complex phases (χ) for magnetic sublevel amplitudes were in less satisfactory agreement with the DW calculation. The elastic scattering distorted waves used in that calculation were obtained from a static atomic potential calculated from Hartree–Fock wave functions for an isolated atom. The encouraging results obtained from this elementary DW calculation would suggest that an improved DW calculation might yield even better agreement with the experimental data. An obvious improvement in the previous DW calculation could be achieved by using more accurate elastic scattering distorted waves in calculating the DW amplitude. Previous work on the elastic scattering problem provides suggestions for obtaining better initial- and final-state distorted waves for the inelastic problem.

Over the last several years, there has been a considerable effort in elastic scattering research directed at obtaining a localized central (optical) potential to represent nonlocal and second-order effects.[10–17] This complex optical potential represents such physical effects as atomic polarization, electron exchange, and electron absorption into other channels. These works demonstrated that it is possible to construct optical potentials that produce elastic scattering cross sections that are in good agreement with experimental data. In the present work, we have used these optical potentials to calculate initial- and final-state distorted waves for the inelastic scattering amplitude. Baluja et al.[18] have recently reported a DW calculation for electron excitation of H and He⁺ which included a real polarization and exchange potential in the calculation of the distorted waves. Bransden and Winters[19] have reported calculations in which the initial-channel distorted wave contained effects of polarization, exchange, and absorption. The present work represents a generalization of the work of Baluja et al. to complex potentials and a generalization of the work of Bransden and Winters to include final-state effects of the optical potential.

The use of a complex optical potential for the calculation of distorted waves introduces non-Hermitian Hamiltonians. The effects of non-Hermitian operators on the DW T matrix are examined in Section 2. In Section 3, the results of the present calculation are compared with experimental differential cross sections, λ and χ parameters for excitation of the 2^1P state of helium for electron impact energies in the range 40–200 eV. The effects of polarization, exchange, and absorption are examined both individually and collectively. The conclusions are contained in Section 4.

2. Theory

In a distorted-wave treatment of the scattering process, the inelastic process is viewed as a transition between elastic scattering states. From a theoretical viewpoint, one is free to include any portion of the interaction potential that may seem desirable in the calculation of the elastic scattering wave functions. The remaining portion of the interaction potential which is not used in the calculation of the elastic states is viewed as the perturbation potential which causes the atomic transition. Since the distorted-wave method represents the leading term of an expansion in powers of the interaction potential, it can be argued that reducing the magnitude of the interaction potential should increase the accuracy of the resulting cross sections. Reducing the magnitude of the interaction potential implies including more of the total interaction in the calculation of the elastic scattering states (distorted waves). The use of optical potentials in the calculation of distorted waves represents an attempt to minimize the perturbation potential by including effects such as polarization, exchange, and absorption into other channels in the calculation of the distorted waves.

The theoretical development of the distorted-wave T matrix and cross sections is well known for Hermitian operators.[9] Use of optical potentials with complex parts introduces non-Hermitian operators. In this section, we shall obtain the distorted-wave T matrix for non-Hermitian distorting potentials. The total (Hermitian) Hamiltonian for the system may be expressed

$$H = H_0 + V_a = H_0' + V_b \tag{1}$$

H_0 would typically represent the sum of the Hamiltonian for the isolated target and the kinetic energy operator for the projectile such that the eigenfunctions of H_0 could be expressed as product wave functions:

$$(H_0 - E)\beta_a(0)\psi_a(1, \ldots, n) = 0 \tag{2a}$$

$$(H - E)\Psi(0, \ldots, n) = 0 \tag{2b}$$

Here $\beta_a\psi_a$ is the solution for the asymptotic noninteracting system (plane wave times target wave function) and Ψ is a solution for the full Hamiltonian. The post form of the "exact" T matrix for Hermitian operators is given by

$$T_{ba} = (n + 1)\langle \beta_b(0)\psi_b(1, \ldots, n) \,|\, V_b A(0, \ldots, n) \,|\, \Psi_a^{(+)}(0, \ldots, n)\rangle \tag{3a}$$

while the prior form of the same amplitude is

$$T_{ba} = (n + 1)\langle \Psi_b^{(-)}(0, \ldots, n) \,|\, A(0, \ldots, n)V_a \,|\, \beta_a(0)\psi_a(1, \ldots, n)\rangle \tag{3b}$$

Here $A(0, \ldots, n)$ is the operator that antisymmetrizes particles 0 through n. Obviously, the antisymmetrizing operator is not needed in these expressions since $\Psi_{a(b)}$ will be fully antisymmetric but it is useful to explicitly include it since Ψ is typically approximated as a nonantisymmetric product of projectile and target wave functions in a practical calculation. The formal solution for Ψ may be written

$$\Psi_a^{(+)} = (1 + G^{(+)}V_a)\beta_a\psi_a \tag{4}$$

where

$$G^{(\pm)} = (E - H \pm i\varepsilon)^{-1}$$

The T matrix can be cast into a form appropriate for a distorted-wave calculation if the Hermitian operator V_b is split into two non-Hermitian parts $U_b + W_b$, where U_b would represent the optical potential. The final-state distorted waves would then be solutions of

$$(H_0' + U_b - E)\phi_b\psi_b = 0 \tag{5}$$

A formal solution to this equation may be written

$$\phi_b^{(\pm)}\psi_b = \beta_b\psi_b + (E - H_0' - U_b \pm i\varepsilon)^{-1}U_b\beta_b\psi_b \tag{6}$$

providing the Green's function can be formed for the non-Hermitian operator. It can be shown by standard techniques employing homogeneous boundary conditions that the Green's function for a non-Hermitian operator of the form

$$\hat{O} = -\frac{\hbar^2}{2m}\nabla^2 + U$$

is given by

$$G(\mathbf{r}, \mathbf{r}') = \sum_l \frac{\tilde{\chi}_l(\mathbf{r})\chi_l(\mathbf{r}')^*}{E_l - E} \tag{7}$$

where χ_l is an eigenfunction of $\hat{O}$ and $\tilde{\chi}_l$ is an eigenfunction of the adjoint $\hat{O}^+$. If the post and prior forms of the T matrix, (3a) and (3b), are equated it may be shown that

$$\langle \beta_b\psi_b \mid AV_a - V_bA \mid \beta_a\psi_a \rangle = 0 \tag{8}$$

Using (4) along with

$$\phi_b^{(-)}\psi_b = \beta_b\psi_b + G_0'^{(-)}U_b\phi_b^{(-)}\psi_b \tag{9}$$

and

$$G^{(+)} = G_0'^{(+)}[1 + V_bG^{(+)}] \tag{10}$$

where

$$G_0'^{(\pm)} = (E - H_0' \pm i\varepsilon)^{-1}$$

and the post–prior relation (8), it may be shown that the post form of the T matrix may be expressed as

$$T_{ba} = (n + 1)\langle \phi_b^{(-)}\psi_b \mid (V_b - U_b)A \mid \Psi_a^{(+)} \rangle$$
$$+ (n + 1)\langle \phi_b^{(-)}\psi_b \mid (U_b - V_b)A + AV_a \mid \beta_a\psi_a \rangle \tag{11}$$

where $\bar{\phi}_b$ is the solution of Hermitian conjugate distorting potential. In a similar manner, the prior form of the T matrix may be expressed

$$T_{ba} = (n+1)\langle \Psi_b^{(-)} | A(V_a - U_a) | \phi_a^{(+)}\psi_a\rangle$$
$$+ (n+1)\langle \beta_b\psi_b | A(U_a - V_a) + V_b A | \phi_a^{(+)}\psi_a\rangle \tag{12}$$

The second term of (10) will vanish providing the Wronskian relation

$$\lim_{r \to \infty} W(\bar{\phi}_b^{(-)*}, \eta_n) = 0 \tag{13}$$

where

$$W(U, V) = \frac{\partial U}{\partial r} V - U \frac{\partial V}{\partial r}$$

and where η_n is the single-particle wave function for the atomic electron that is changed by the collision. Likewise, the second term of (12) will vanish if

$$\lim_{r \to \infty} W(\eta_n^*, \phi_a^{(+)}) = 0 \tag{14}$$

Conditions (13) and (14) will be satisfied for excitation processes in which η_n is a bound state. The vanishing of these second terms is not automatic for ionization of continuum states. These terms will vanish even for ionization if it is assumed that the wave packets of the two continuum electrons do not overlap. It is readily seen that the results for non-Hermitian operators are identical to the Hermitian results if the replacement $\bar{\phi}_b \to \phi_b$ is made. For Hermitian U_b, $\bar{\phi}_b$ reduces to ϕ_b and the standard two-potential formula is obtained. The distorted-wave approximation is obtained by setting

$$\Psi_{a(b)} = \phi_{a(b)}\psi_{a(b)} \tag{15}$$

3. Results

3.1. Numerical Procedure

For the present calculation, we have taken

$$U_b = U_a = V_s + V_{\text{pol}} + V_{\text{ex}} + V_{\text{abs}} \tag{16}$$

where V_s is the spherical average of the static atomic potential for an isolated atom, V_{pol} is the polarization potential, V_{ex} is the exchange potential, and V_{abs} is the imaginary absorption potential. The atomic wave functions used in obtaining these potentials were calculated from Fisher's[20] Hartree–Fock program in the frozen-core approximation. Naturally, these same atomic wave functions were used in evaluating the DW T-matrix elements. Setting $U = V_s$ corresponds to the FG calculation of Reference 9. For the other terms in the optical potential, we have used the adiabatic polarization potential of Temkin and Lamkin,[21] the exchange potential of Furness and McCarthy[10,13] and the imaginary absorption potential of McCarthy et al.[16] Vanderpoorten and Winters[22] have recently examined various approximations for the optical potential for the $1S \rightarrow 2S$ transition in hydrogen. We have chosen these simplified potentials since (a) they are easily calculable, and (b) our primary interest was in the general effect of the various terms.

As indicated by equation (16), both the initial- and final-state distorted waves were calculated using ground-state atomic wave functions[9,23] in V_{ex} and V_{abs}. It has recently been argued by Winters[24] that the historic approach of Mott and Massey[25] should be followed in which ground-state potentials are used to calculate incident-channel distorted waves and excited state potentials are used to calculate exit-channel distorted waves. We have found that using the excited-state charge density in V_{ex} for the exit channel gave almost the same results as those obtained using the ground-state charge density in V_{ex} for the exit channel. McCarthy et al.[16] noted a similar weak dependence on atomic charge density for the absorption potential for elastic scattering.

Both the direct and exchange scattering T matrices have been evaluated using the LS scheme partial wave expansions of Calhoun et al.[26] instead of the $(LS)J$ scheme of Reference 9. The former scheme is appropriate for low-Z atoms and represents a significant reduction in the amount of computer time required for the calculation. There was not any significant difference between the results of the two schemes (a small difference resulted from the spin–orbit term included in the J-scheme calculation).

3.2. λ Parameter

In Figures 1 and 2, theoretical and experimental values for the λ parameter are given for exciting the 2^1P state in the energy range 40–200 eV. Each figure presents the distorted-wave results using the static atomic potential and the full optical potential. Use of the full optical potential instead of the static potential in calculating distorted waves did not produce a large change for the λ parameter. All of the calculations were essentially identical for angles less than 10°, reflecting the fact that the $m = \pm 1$ cross section is becoming small.

In examining the relative effects of the individual terms in the optical potential, we found the following:

(a) The polarization potential was responsible for most of the change near the small-angle minimum but was relatively unimportant at other angles.

(b) At angles away from the small-angle minimum, most of the change produced by the optical potential resulted from the exchange potential term. We found that the exchange potential made the width of the small-angle minimum more narrow

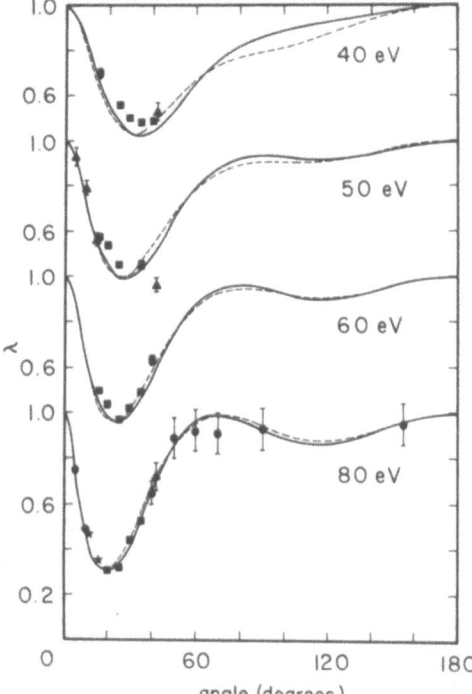

Figure 1. Angular correlation parameter λ for electron impact excitation of the 2^1P state of helium. The theoretical curves are distorted-wave calculations using ———, static atomic potential, and – – –, full optical potential. The experimental values are ●, Sutcliffe et al.,[7] ■, Eminyan et al.,[3] ★, Ugbabe et al.,[5] and ▲, Tan et al.[6]

without significantly changing the absolute value of the minimum. Baluja et al.[18] observed a similar behavior for the exchange potential for excitation of hydrogen except that they found a more significant increase in the absolute value of the small-angle minimum. We find that with increasing energy the effect of exchange becomes

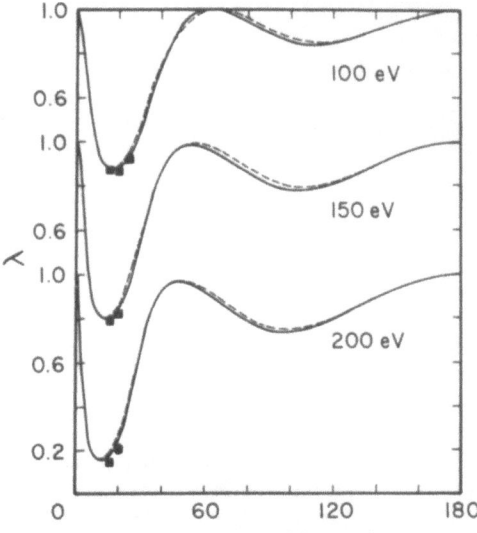

Figure 2. Same as Figure 1 except for higher incident electron energies.

less significant, while Baluja *et al.* found the exchange potential to have a larger effect at 100 eV than at 54 eV. In that calculation, it was found that for 100-eV excitation of hydrogen the exchange term increased the λ parameter near the large-angle minimum by about 10%. In the present calculation, we find the corresponding increase for helium to be only 2–3%.

(c) The absorption potential produced relatively small changes at all angles.

In terms of agreement with experimental data, the significance of the small change produced by the full optical potential is difficult to evaluate. Optimistically, it can perhaps be argued that the agreement with the low-energy experimental data is somewhat better. At the higher energies, the small change is towards worse agreement with the experimental data.

3.3. χ Parameter

Present experimental data for the χ parameter are limited to angles less than about 45°. Results of the present calculation are compared with experiment in Figures 3 and 4. Each figure contains the results of the distorted-wave calculation using the static atomic potential and the full optical potential. The results using the static potential alone are consistently larger than the experimental data in this energy region. For angles less than about 40°, use of the full optical potential produced

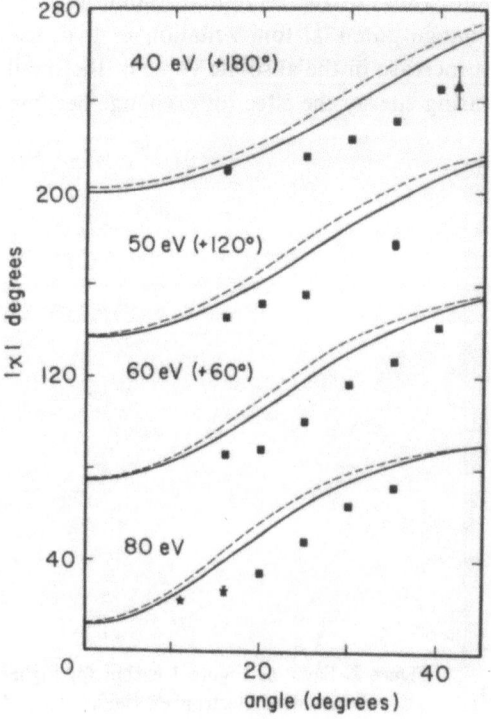

Figure 3. Angular correlation parameter χ for electron impact excitation of the 2^1P state of helium. The legend is the same as Figure 1.

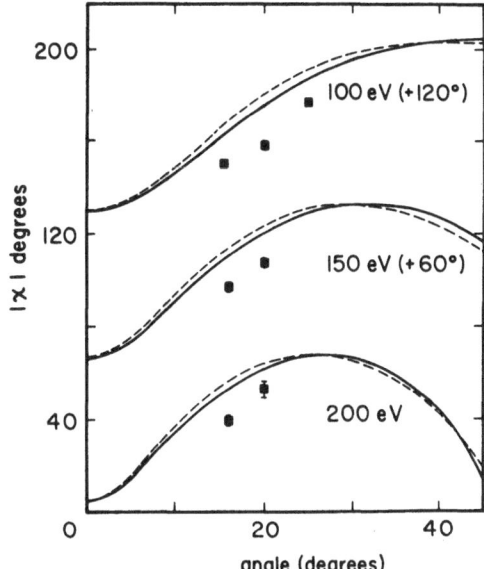

Figure 4. Same as Figure 3 except for higher incident electron energies.

even larger χ parameters, making agreement with experimental data even worse. The relative effect of the full optical potential decreased with increasing incident electron energy. The full optical potential produced a somewhat larger change in the χ parameter than was seen for the λ parameter. Unfortunately, this change made agreement with experimental data worse instead of better.

When the contributions from the individual terms of the optical potential were examined, it was again found that the absorption term was relatively unimportant. For the angles displayed in the figures, exchange and polarization produced comparable changes, while for the larger angles (not shown), the exchange potential produced most of the change.

3.4. Differential Cross Sections

The results of the calculation of differential cross sections for the various forms of the optical potential are shown in Figure 5 for 40- and 80-eV incident electrons and in Figure 6 for 100- and 200-eV electrons. These four energies were selected for comparison since the largest amount of experimental data was available for these energies. Including the full optical potential gave small changes except at the lowest energies.

In examining the effects of the individual terms in the optical potential, we found that exchange and polarization generally increased the large-angle cross sections without significantly changing the small-angle cross sections. Baluja et al.[18] saw a similar effect for excitation of hydrogen. However, in that case, the increase in the

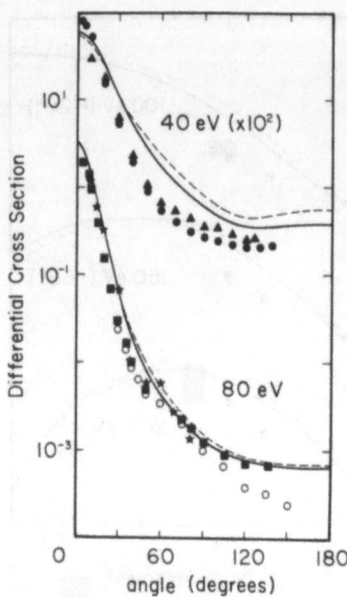

Figure 5. Differential cross sections for electron impact excitation of the 2^1P state of helium in units of a_0^2/sr. The theoretical curves are distorted-wave calculations using ——, static atomic potential, and ---, full optical potential. The experimental values are ●, Truhlar *et al.*,[27] ▲, Hall *et al.*,[28] ■, Chutjian and Srivastava,[29] ○, Opal and Beaty,[30] and ★, Truhlar *et al.*[31]

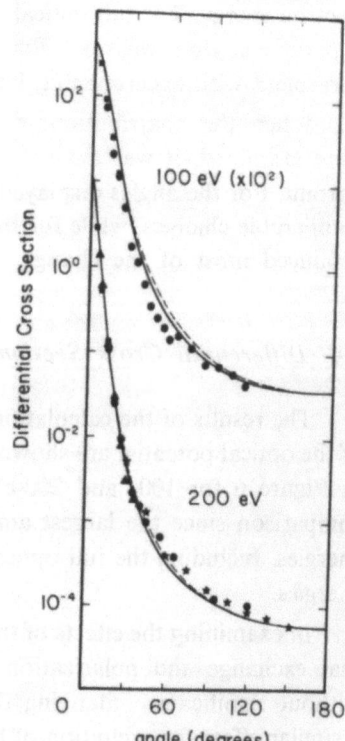

Figure 6. Same as Figure 5 except for higher energies. The experimental values are ●, Suzuki and Taka-yanagi,[32] ▲, Dillon and Lassettre,[33] ■, Vriens *et al.*,[34] and ★, Opal and Beaty.[30]

cross section improved agreement with experimental data since the original DW calculations lie below the data while for helium the increase worsens agreement with experimental data since the original DW calculation was generally either through the data or above it. Adding the absorption potential further increased the large-angle cross section. Similar to the angular correlation results, the absorption potential generally represented a smaller effect than exchange. The biggest effect is observed for the larger angles reflecting the sensitivity of the large-angle cross section to the atomic potential. As the energy of the incident electron increases, all three of the additional optical terms become less important. By 200 eV, the additional potential terms did not make a significant contribution. Including the full optical potential worsens agreement with experimental differential cross sections for all energies in this range except perhaps at 200 eV, where the original calculation was below the experimental data.

4. Conclusion

We have calculated differential cross sections and angular correlation parameters for electron excitation of the 2^1P state of helium in the DW approximation and have examined the effect of adding polarization, exchange, and absorption potential terms to the static atomic potential used in the calculation of the distorted waves. The effects of polarization, exchange, and absorption were examined individually and collectively. In general, it was found that the optical potential produced a relatively small change in the calculated results. Of the three potential terms added to the static potential, it was found that overall the exchange term generally produced the largest change. The effects of all the additional terms decreased with increasing incident electron energy.

Relative to agreement with experimental data, including the optical potential made agreement for the χ parameter worse, agreement for the differential cross section slightly worse except for the highest energy, where there was a slight improvement, and agreement for the λ parameter slightly worse except at the lowest energies, where there was perhaps a slight improvement. When this work was started, it was anticipated that the additional potential terms would have a small effect on the differential cross section and λ parameter, where the original static potential gave results in reasonable agreement with experimental data, while having a more significant effect on the χ parameter, which did not agree as well with the experimental data. This general behavior has been observed. Unfortunately, the change in the χ parameter was away from the experimental data.

ACKNOWLEDGMENT

This work was supported by the Research Corporation.

References

1. M. Eminyan, K. B. MacAdam, J. Slevin, and H. Kleinpoppen, *Phys. Rev. Lett.* **31**, 576–579 (1973); *J. Phys. B* **7**, 1519–1542 (1974).
2. M. Eminyan, H. Kleinpoppen, J. Slevin, and M. C. Standage, in *Electron and Photon Interactions with Atoms*, H. Kleinpoppen and M. R. C. McDowell, eds., pp. 455–483, Plenum, New York (1976).
3. M. Eminyan, K. B. MacAdam, J. Slevin, M. C. Standage, and H. Kleinpoppen, *J. Phys. B* **8**, 2058–2066 (1975).
4. M. C. Standage and H. Kleinpoppen, *Phys. Rev. Lett.* **36**, 577–580 (1976).
5. A. Ugbabe, P. J. O. Teubner, E. Weigold, and H. Arriola, *J. Phys. B* **10**, 71–79 (1977).
6. K. H. Tan, J. Fryar, P. S. Farago, and J. W. McConkey, *J. Phys. B* **10**, 1073–1082 (1977).
7. V. C. Sutcliffe, G. N. Haddad, N. C. Steph, and D. E. Golden, *Phys. Rev. A* **10**, 100–107 (1978).
8. B. H. Bransden and M. R. C. McDowell, *Phys. Rep.* **30C**, 207–303 (1977).
9. D. H. Madison and W. N. Shelton, *Phys. Rev. A* **7**, 499–513 (1973).
10. J. B. Furness and I. E. McCarthy, *J. Phys. B* **6**, 2280–2291 (1973).
11. F. W. Byron and C. J. Joachain, *Phys. Lett. A* **49**, 306–308 (1974); *Phys. Rev. A* **15**, 128–146 (1977).
12. R. Vanderpoorten, *J. Phys. B* **8**, 926–939 (1975).
13. M. E. Riley and D. G. Truhlar, *J. Chem. Phys.* **63**, 2182–2191 (1975).
14. B. H. Bransden, M. R. C. McDowell, C. J. Noble, and T. Scott, *J. Phys. B* **9**, 1301–1317 (1976).
15. C. J. Joachain, R. Vanderpoorten, K. H. Winters, and F. W. Byron, *J. Phys. B* **10**, 227–238 (1977).
16. I. E. McCarthy, C. J. Noble, B. A. Phillips, and A. D. Turnbull, *Phys. Rev. A* **15**, 2173–2185 (1977).
17. M. R. H. Rudge, *J. Phys. B* **10**, 2451–2457 (1977).
18. K. L. Baluja, M. R. C. McDowell, L. A. Morgan, and V. R. Myerscough, *J. Phys. B* **11**, 715–726 (1978).
19. B. H. Bransden and K. H. Winters, *J. Phys. B* **8**, 1236–1244 (1975).
20. C. Froese Fischer, *Comput. Phys. Commun.* **4**, 107–116 (1972).
21. A. Temkin and J. C. Lamkin, *Phys. Rev.* **121**, 788–794 (1961).
22. R. Vanderpoorten and K. Winters, *J. Phys. B* **12**, 473–488 (1979).
23. L. D. Thomas, G. Csanak, H. S. Taylor, and B. S. Yarlagadda, *J. Phys. B* **7**, 1719–1733 (1974).
24. K. H. Winters, *J. Phys. B* **11**, 149–165 (1978).
25. N. F. Mott and H. S. W. Massey, *The Theory of Atomic Collisions*, 3rd ed., Clarendon Press, Oxford (1965).
26. R. V. Calhoun, D. H. Madison, and W. N. Shelton, *Phys. Rev. A* **14**, 1380–1387 (1976); R. V. Calhoun, D. H. Madison, and W. N. Shelton, *J. Phys. B* **10**, 3523–3533 (1977).
27. D. G. Truhlar, S. Trajmar, W. Williams, S. Ormonde, and B. Torres, *Phys. Rev. A* **8**, 2475–2482 (1973).
28. R. I. Hall, G. Joyez, J. Mazeau, J. Reinhardt, and C. Schermann, *J. Phys. (Paris)* **34**, 827–843 (1973).
29. A. Chutjian and S. K. Srivastava, *J. Phys. B* **8**, 2360–2368 (1975).
30. C. B. Opal and E. C. Beaty, *J. Phys. B* **5**, 627–635 (1972).
31. D. G. Truhlar, J. K. Rice, A. Kupperman, S. Trajmar, and D. C. Cartwright, *Phys. Rev. A* **1**, 778–802 (1970).
32. H. Suzuki and T. Takayanagi, Abstracts VIII ICPEAC (Beograd), pp. 286–287 (1973).
33. M. A. Dillon and E. N. Lassettre, *J. Chem. Phys.* **62**, 2373–2390 (1975).
34. L. Vriens, J. A. Simpson, and S. R. Mielczarek, *Phys. Rev.* **165**, 7–15 (1968).

14

The Calculation of Orientation and Alignment Parameters for the Electron Impact Excitation of the 3^1P State of He in the First-Order Many-Body Theory

G. D. Meneses and Gy. Csanak

The first-order many-body theory (FOMBT) has been applied to the calculation of orientation and alignment parameters (λ and χ) in the case of electron impact excitation of the 3^1P state of He at 50 and 80 eV impact energies. The results obtained are compared with experimental results and with those of other theoretical models.

In previous works the first-order many-body theory (FOMBT) for electron–atom inelastic scattering[1] has been applied to the calculation of the differential and total inelastic scattering cross sections of the $n = 2$ and $n = 3$ states of helium,[2-4] as well as to the calculation of orientation and alignment parameters, λ and χ, of the 2^1P state.[3,4] Since experimental results are available at various energies for these parameters in the case of the 3^1P state also[5,6] and distorted wave polarized orbital (DWPO)[7] and multichannel eikonal theory (MET)[8] results have also been reported, it appeared to be worthwhile to calculate the λ and χ parameters in the FOMBT for the 3^1P state.

G. D. Meneses and Gy. Csanak ● Instituto de Física "Gleb Wataghin," Universidade Estadual de Campinas, 13.100 Campinas, S.P., Brazil.

Table 1. Orientation and Alignment Parameters for 3^1P Excitation of He at 50 eV Electron Impact Energy

θ (deg)	λ (exp)[a]	λ (calc)		$|\chi|$ (exp)[a]	$|\chi|$ (calc)	
		Present	DWPO[b]		Present	DWPO[b]
0	—	1.000	1.000	—	—	—
5	—	0.935	0.939	—	8.50, -2^c	7.71, -2^c
10	—	0.784	0.800	—	1.06, -1	9.57, -2
15	0.57 ± 0.03	0.622	—	0.62 ± 0.03	1.43, -1	—
20	0.50 ± 0.02	0.486	0.532	0.68 ± 0.02	2.02, -1	1.78, -1
25	0.57 ± 0.01	0.378	—	0.66 ± 0.02	2.92, -1	—
30	0.58 ± 0.03	0.291	0.368	0.90 ± 0.04	4.29, -1	3.56, -1
35	—	0.225	—	—	6.46, -1	—
40	—	0.175	0.263	—	9.96, -1	7.44, -1
45	—	0.163	—	—	1.51	—
50	—	0.223	0.229	—	2.08	1.58
55	—	0.361	—	—	2.52	—
60	—	0.533	0.376	—	2.84	2.66
65	—	0.683	—	—	3.11	—
70	—	0.782	0.532	—	2.84	3.41
75	—	0.835	—	—	2.70	—
80	—	0.854	0.563	—	2.52	3.82
85	—	0.852	—	—	2.39	—
90	—	0.843	0.528	—	2.30	4.04
95	—	0.835	—	—	2.26	—
100	—	0.839	0.458	—	2.15	2.10
105	—	0.854	—	—	2.06	—
110	—	0.873	0.370	—	1.95	1.94
120	—	0.902	0.296	—	1.74	1.67
130	—	0.904	0.284	—	1.62	1.29
140	—	0.911	0.381	—	1.61	8.98, -1
150	—	0.951	0.573	—	1.59	6.42, -1
160	—	0.990	0.784	—	1.34	5.02, -1
170	—	0.999	0.941	—	4.11, -1	4.32, -1
180	—	1.000	1.000	—	—	—

[a] Eminyan et al.[5]
[b] T. Scott and M. R. C. McDowell.[7]
[c] Number n following comma after entry means multiplication by 10^n.

From recent measurements of Standage and Kleinpoppen[6] the sign of the χ parameter can also be extracted from the polarization parameters P_1, P_2, P_3 that are measured directly and, according to Blum and Kleinpoppen,[9] are related to λ and χ as follows:

$$P_1 = 2\lambda - 1, \qquad P_2 = -[\lambda(1 - \lambda)]^{1/2} \cos \chi, \qquad P_3 = 2[\lambda(1 - \lambda)]^{1/2} \sin \chi$$

Table 2. Orientation and Alignment Parameters for 3^1P Excitation of He at 80 eV Impact Energy

θ (deg)	λ (exp)[a,b]	λ (calc) MET[c]	λ (calc) Present	λ (calc) DWPO[d]	$\|\chi\|$ (exp)[a,b]	$\|\chi\|$ (calc) Present	$\|\chi\|$ (calc) DWPO[d]
0	—	1.000	1.000	1.000	—	—	—
5	—	—	0.816	0.822	—	6.42, −2[e]	5.22, −2[e]
10	0.51 ± 0.01	0.463	0.540	0.564	0.54 ± 0.02	1.07, −1	8.50, −2
15	0.38 ± 0.01	—	0.363	—	0.55 ± 0.02	1.85, −1	—
20	0.36 ± 0.02	0.173	0.263	0.328	0.64 ± 0.05	3.17, −1	2.26, −1
25	0.31 ± 0.02	0.167	0.207	—	0.75 ± 0.05	5.29, −1	—
30	0.44 ± 0.02	0.163	0.187	0.283	1.11 ± 0.02	8.57, −1	5.24, −1
35	—	0.241	0.220	—	—	1.29	—
40	—	0.398	0.339	0.349		1.73	1.12
45	—	0.579	0.541	—	—	2.11	—
50	—	0.712	0.737	0.542	—	2.44	2.13
60	—	0.841	0.912	0.599	—	3.14	3.18
70	—	0.896	0.928	0.520	—	2.50	3.74
80	—	0.936	0.908	0.430	—	2.16	4.01
90	—	0.958	0.881	0.347	—	1.96	4.20
100	—	0.965	0.852	0.278	—	1.83	1.87
110	—	0.972	0.850	0.239	—	1.65	1.58
120	—	0.982	0.860	0.248	—	1.50	1.26
130	—	0.989	0.885	0.320	—	1.38	9.64, −1
140	—	0.993	0.916	0.457	—	1.28	7.46, −1
150	—	0.996	0.949	0.624	—	1.21	6.09, −1
160	—	0.998	0.976	0.807	—	1.15	5.31, −1
170	—	0.999	0.994	0.953	—	1.12	4.98, −1
180	—	1.000	1.000	1.000	—	—	—

[a] Eminyan et al.[5]
[b] Experimental polarization data[6] not shown here agree well with angular correlation data.[5]
[c] MET calculation of Flannery and McCann.[8]
[d] T. Scott and M. R. C. McDowell.[7]
[e] ,n means multiplication by 10^n.

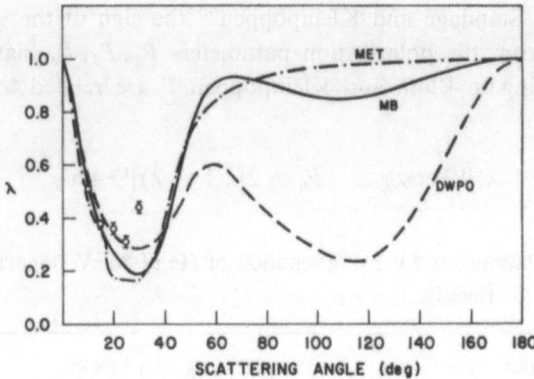

Figure 1. Variation of λ with scattering angle for He 3^1P excitation at 80 eV. $\bigcirc$, experiment;[5] ——, present calculations; – – –, distorted-wave polarized orbitals method of Scott and McDowell;[7] —·—, multichannel eikonal treatment (Flannery and McCann[8]).

Calculations have been carried out in the FOMBT for λ and χ at 50 and 80 eV impact energies for the 3^1P state of He. The calculational procedure is the same as been described previously.[4] The Hartree–Fock scattering orbitals have been generated by a program of Bates.[10] The direct part of the T matrix was calculated by adding and subtracting the Born T matrix. In calculating each partial wave contribution of the direct T matrix, numerical integration was carried out until $R = 50$ a.u., and the "tail correction," i.e., the integral between $R = 50$ a.u. and $R = \infty$, was calculated approximately analytically. The Hartree–Fock orbitals for $l = 0$–7 were included where convergence has been achieved to at least two figures accuracy for the energies considered.

Tables 1 and 2 give the present results for λ and $|\chi|$ along with the experimental results of Eminyan et al.,[5] the DWPO theory results of Scott and McDowell,[7] and the MET results of Flannery and McCann[8] at 50 and 80 eV incident energies, respectively. These results are also shown in Figures 1–4. Figures 5–7 show the present results for P_1, P_2, and P_3 along with the experimental results of Standage and Kleinpoppen[6] and the results of MET of Flannery and McCann.[8] Figures 8 and 9 show the present results for the total differential cross section (DCS) along with the experimental results of Chutjian[11] as well as the DWPO[7] and MET[8] results.

Figure 2. Variation of λ with scattering angle for He 3^1P excitation at 50 eV. $\bigcirc$, experiment;[5] ——, present calculations; – – –, calculations of Scott and McDowell; —·—, calculations of Scott and McDowell.[7]

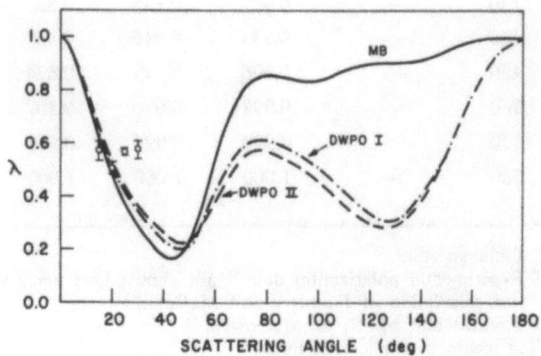

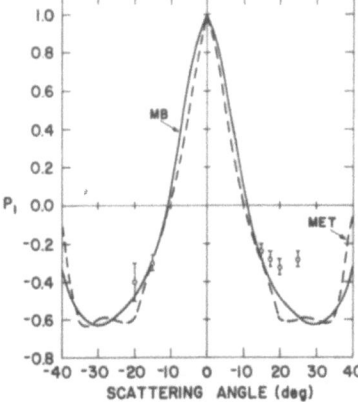

Figure 3. Variation of $|\chi|$ with scattering angle for He 3^1P excitation at 80 eV. $\bigcirc$, experiment;[5] $\square$, experiment;[5] $\triangle$, experiment;[5] ——, present calculations; – – –, DWPO I calculations of Scott and McDowell;[7] —·—, DWPO II calculations of Scott and McDowell;[7] —··—, MET calculations of Flannery and McCann.[8]

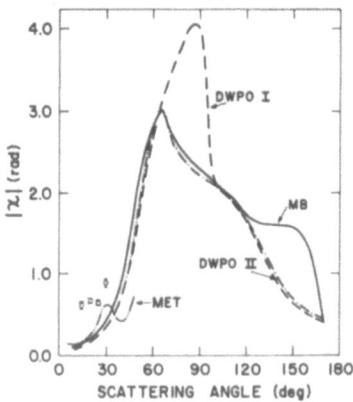

Figure 4. Variation of $|\chi|$ with scattering angle for the 3^1P excitation at 50 eV, otherwise same as in Figure 3. [See note added in proof on page 186.]

Figure 5. Variation of P_1 with scattering angle for He 3^1P excitation at 80 eV. $\bigcirc$, experiment;[5] ——, present calculation; – – –, calculation of Flannery and McCann.[5]

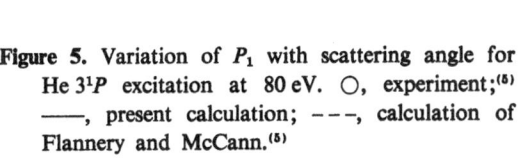

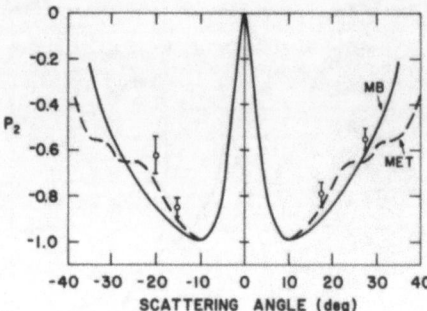

Figure 6. Same as Figure 5 except for P_2 instead of P_1.

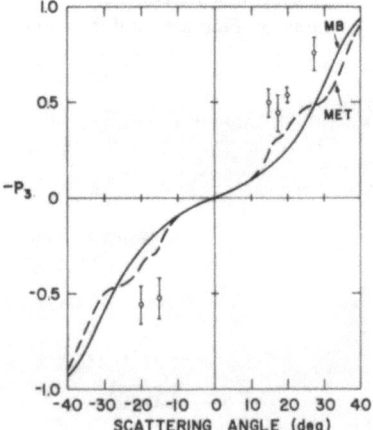

Figure 7. Same as Figure 5 except for P_3 instead of P_1.

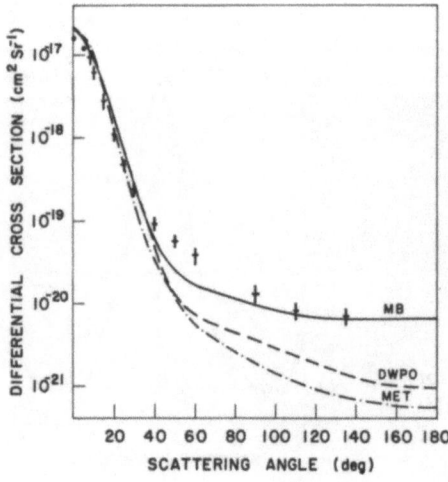

Figure 8. Absolute differential cross section for electron impact excitation of the 3^1P state of He at an incident energy of 80 eV. †, experiment;[3] ———, present calculations; – – –, DWPO calculations;[7] —·—, MET calculations.[8] The two heavy dots in the upper left-hand corner are extrapolated experimental results.

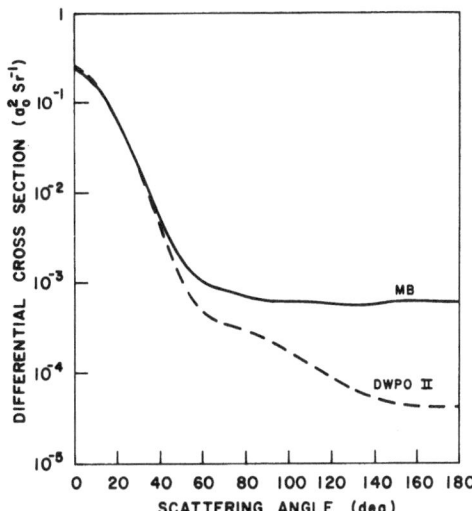

Figure 9. Same as in Figure 8, except that the incident energy is 50 eV.

The tables and figures show that the FOMBT gives slightly a better result for $|\chi|$ than the DWPO model does and a slightly worse result for λ than the same model for the energies considered. The results of the MET are practically equal in quality to the FOMBT results for λ and χ. Both DWPO and MET give poor agreement with experiment for the DCS for angles $\theta > 40°$, whereas FOMBT gives good agreement with the experimental DCS for practically all angles.

ACKNOWLEDGMENTS

The authors gratefully acknowledge the financial support of Conselho Nacional de Desenvolvimento Científico e Tecnológico, Brasil. They would like to thank Professor Sérgio Porto for his interest in and support of this work. They are very much indebted to Drs. Nick Winter (Lawrence Livermore Laboratory), David Cartwright (Los Alamos Scientific Laboratory), and N. T. Padial (UNICAMP) for generous help in various stages of this work, and Drs. T. Scott and M. R. Flannery for sending their numerical results.

References

1. Gy. Csanak, H. S. Taylor, and R. Yaris, *Phys. Rev. A* **3**, 1322–1326 (1971).
2. L. D. Thomas, Gy. Csanak, H. S. Taylor, and B. S. Yarlagadda, *J. Phys. B* **7**, 1719–1733 (1974).
3. A. Chutjian and L. D. Thomas, *Phys. Rev. A* **11**, 1583–1595 (1975).
4. G. D. Meneses, N. T. Padial, and Gy. Csanak, *J. Phys. B* **11**, L237–L242 (1978).
5. M. Eminyan, K. B. McAdam, J. Slevin, M. C. Standage, and H. Kleinpoppen, *J. Phys. B* **8**, 2058–2066 (1975).
6. M. C. Standage and H. Kleinpoppen, *Phys. Rev. Lett.* **36**, 577–580 (1976).

7. T. Scott and M. R. C. McDowell, *J. Phys. B* **9**, 2235–2254 (1976); private communication of DWPO II data.
8. M. R. Flannery and K. J. McCann, *J. Phys. B* **8**, 1716–1733 (1975); private communication.
9. K. Blum and H. Kleinpoppen, *J. Phys. B* **8**, 922–925 (1975).
10. G. N. Bates, *Comput. Phys. Commun.* **8**, 220–229 (1974).
11. A. Chutjian, *J. Phys. B* **9**, 1749–1756 (1976).

Note Added in Proof

After the proof of this work had been prepared, a small numerical error was found in the 50 eV calculation. The corrected values of λ and χ are similar to the ones presented above. The correction for λ in the 0–35° and 70–180° angular range is less than 3% and in the 40–65° interval is less than 25%. The correction for χ is less than 8% at every angle except at 170°, where the correct value of 1.39 eliminates the sudden drop at this angle on Figure 4.

15

Electron–Photon Correlations in Bremsstrahlung Processes

Werner Nakel

This survey is concerned with measurements of electron–photon correlations in the atomic-field (electron–nucleus) bremsstrahlung process and in the electron–electron bremsstrahlung process. The atomic-field bremsstrahlung angular correlation measurements that have been reported involve measurements of photon angular distributions for several fixed outgoing electron directions and one distribution of decelerated outgoing electrons for a fixed photon direction. Detailed measurements of spectral distributions have been reported for fixed electron and photon directions. In the linear polarization correlation measurements in the atomic-field bremsstrahlung process which involve measuring the angular dependence of the photon linear polarization for a fixed direction of outgoing electrons, the radiation has been found to be almost completely polarized. A strong decrease of the polarization caused by electron spin–flip radiation has been observed for a photon emission angle near the minimum of the bremsstrahlung cross section. The processes of electron–electron bremsstrahlung and atomic-field bremsstrahlung have been isolated by means of a kinematically overdetermined measurement. The photon angular distribution for a fixed direction of outgoing electrons has been measured.

1. Introduction

The emission of bremsstrahlung in the scattering of electrons by atoms or nuclei has already been investigated in detail as a function of many parameters. In these experiments, an electron beam passes through a target foil, which should be so thin that bremsstrahlung is produced only by one single collision between an incoming

Werner Nakel • Physikalisches Institut der Universität Tübingen, 7400 Tübingen, West Germany.

electron and an atom or a nucleus. However, since these measurements were made without regard to the outgoing electrons, the results are necessarily averaged over all electron scattering angles and important details are lost in this averaging process.

When, on the other hand, the bremsstrahlung photons are detected in coincidence with decelerated electrons scattered in a particular direction, new information on the elementary collision process can be obtained and a check of the theoretical work in its barest form becomes possible.

Although such a coincidence measurement of the bremsstrahlung process had been suggested as early as 1932 by Scherzer,[1] the first electron–photon correlation measurement was reported only in 1966 by Nakel.[2–4] The measurements show strong angular correlations, with the maximum emission probability of the photon occurring on the same side of the initial beam in which the outgoing electron is also scattered.

A measurement of the angular distribution of decelerated outgoing electrons for a fixed photon direction has been reported by Hub and Nakel in 1967.[5]

Measurements of the absolute triply differential cross section for various collision parameters have been made by Nakel and Sailer in 1970,[6] by Kreuzer and Nakel in 1971,[7] by Aehlig and Scheer in 1972,[8] by Faulk and Quarles in 1973,[9] and by Aehlig et al. in 1977.[10]

In 1978 Behncke and Nakel[11] reported a linear polarization correlation measurement in atomic-field bremsstrahlung which involves measuring the angular dependence of the photon linear polarization for fixed direction of outgoing electrons. Since, according to a simple classical picture, the orbital plane of the radiating electron is determined by the coincidence measurement, one would expect the radiation to be always completely linearly polarized in this experiment. However, at relativistic energies the effect of electron spin becomes important and the emission of bremsstrahlung can take place not only as a result of a change of momentum, but as a result of a change of spin orientation.[12] In measuring the polarization correlation it was possible to discern the appearance of the spin–flip radiation by a deviation from complete linear polarization.

An electron passing through matter can produce bremsstrahlung in the Coulomb field of a nucleus and in a collision with an atomic electron. However, the electron–electron bremsstrahlung component is in general very small compared to the electron–nucleus (atomic-field) bremsstrahlung. Experimental investigations show[13,14] that it is difficult to isolate the electron–electron bremsstrahlung from the electron–nucleus bremsstrahlung in measurements in which the final electrons are not observed. In 1972 Nakel and Pankau[15] used the electron–photon coincidence technique to differentiate exactly between the two bremsstrahlung components by means of a kinematically overdetermined measurement and measured the triply differential cross section[16,17] and the angular correlation.[18]

All correlation measurements reviewed here have been performed in the intermediate-energy region (a few hundred keV). Two high-energy absolute cross-section points were reported in 1969 by Bernardini et al.[19] and by Sieman et al.[20]

In the following the term electron–nucleus bremsstrahlung is used besides the term atomic-field bremsstrahlung, especially in distinction from electron–electron bremsstrahlung.

2. Angular Correlations in the Atomic-field Bremsstrahlung Process

In the radiative collision, the initial momentum of the incident electron becomes shared between the momenta of three particles: the outgoing electron, the atomic nucleus, and the emitted photon. Therefore, the photon may be radiated in any direction, regardless of the direction of the outgoing electron.

Let us consider the electron–photon angular correlations for this radiation process. Experimentally one records the number of coincidences between the photons and the outgoing electrons as a function of the scattering angles or of the photon energy.

The experiments reported here consist of coincidence measurements of the relative or absolute cross sections for the atomic-field bremsstrahlung process, including its dependence upon the electron and photon emission angles and the photon energy. This triply differential cross section is usually called the fully differential cross section, although it might be noted that the measurements and formulas are summed or averaged over the directions of the electron spins and the photon polarization.

For theoretical predictions of the bremsstrahlung process, one has to calculate the probability that the incident electron will make a transition to a different electron state with a photon emitted while in the Coulomb field of a nucleus. The perturbation causing this transition is the interaction of the electron with the Coulomb field and the radiation field. Whereas presently the interaction of electrons with the radiation field can be treated only by perturbation theory, their interaction with the Coulomb field of the atomic nucleus can, in principle, be handled exactly. In the latter case one has to use exact wave functions, which describe an electron in a screened, nuclear Coulomb field. This is particularly important when the atomic field is strong, i.e., for high atomic numbers.

It is not possible to solve the Dirac wave equation in closed form for an electron in a Coulomb field in the continuum state. An exact calculation of the bremsstrahlung process was performed by Tseng and Pratt.[21] They solved the Dirac equation numerically while using partial-wave expansion, but their results did not yet include the triply differential cross section, only the integrated cross sections. Therefore, until now, any measured angular correlations can only be compared to approximate calculations. These calculations differ by various approximate electron wave functions (Born approximation, Sommerfeld–Maue functions, etc.), the relativistic or non-relativistic procedures, the kind and degree of approximations and the degree of screening used, etc. Consequently, the results often cover different energy ranges and different ranges of atomic numbers for the target nucleus.

The status of bremsstrahlung calculations as of 1959 was summarized in a review paper by Koch and Motz.[22] Since this time several new calculations for the triply differential cross section have been published.[23,24]

The classical Bethe–Heitler formula[25] describes the production of bremsstrahlung in the first Born approximation. The region of validity is roughly $\alpha Z/\beta \ll 1$, where α is the fine-structure constant, β the velocity of the outgoing electron in units of the light velocity, and Z is the atomic number of the target atom. The region of validity of the calculations of Elwert and Haug[23] (Sommerfeld–Maue wave functions) can be expressed by $\alpha Z \ll 1$. This does not depend on the energy of the incident electron or of the photon energy. For high Z, neither the Bethe–Heitler formula nor the Elwert–Haug calculation can be expected to be accurate.

2.1. Photon Angular Distributions for a Fixed Direction of Outgoing Electrons

Measurements of photon angular distributions for fixed directions of the outgoing electrons have been reported beginning in 1966 by Nakel.[2–4] The incident electron energy was 300 keV. With a magnetic spectrometer outgoing electrons with an energy of 170 keV and scattering angles of 0°, 5°, and 10° were selected[4] (Figure 1, a–c). A gold foil ($Z = 79$) served as a target. The solid curves represent the theory of Elwert and Haug.[23] As the first experimental values were only relative, they have been normalized to the theoretical curves at the maximum of the angular distributions.

The measurements show strong angular correlations, with the maximum emission probability of the photon occurring on the same side of the initial beam in which the outgoing electron is also scattered. This behavior can be understood classically by considering the radiation of the electron along the hyperbolic orbit around the nucleus. At the smallest distance from the nucleus the radiation is strongest. Here, the velocity and the acceleration are perpendicular to each other. Because of relativistic effects the emission takes place predominantly into a lobe along the direction of motion, as in the case of synchrotron radiation. Thus one obtains most of the radiation towards the deflection of the outgoing electron.

For outgoing radiating electrons that are not deflected (0° direction) the corresponding photon angular distribution has always to be symmetric about the beam (Figure 1a).

Although the condition $\alpha Z \ll 1$ is not fulfilled for $Z = 79$, the measured angular correlations are in qualitative agreement with the Elwert–Haug theory. This is not true for the value of the absolute cross section. In 1971 Kreuzer and Nakel[7] measured the triply differential cross section for $Z = 79$ for forward scattered electrons and for photons emitted at 13°. Here the calculation underestimates the cross section by a factor of 1.7.

For aluminum ($Z = 13$) experiment and theory have been found in agreement in a measurement by Nakel and Sailer[6] in 1970.

In 1972 a detailed investigation of the absolute cross section for silver ($Z = 47$)

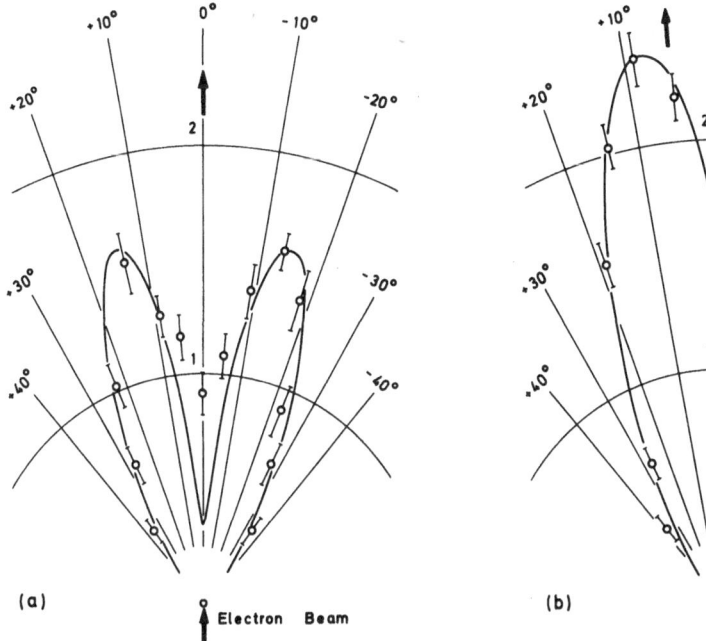

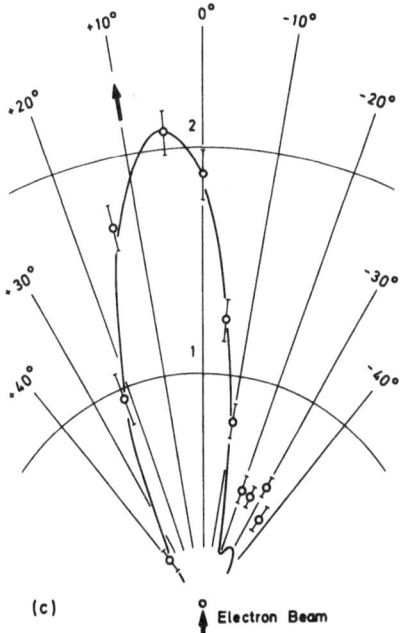

Figure 1. (a)–(c) Measured (circles) and calculated (full line) photon angular distributions for fixed directions of outgoing electrons (Nakel,[4] p. 176). Incident electron energy 300 keV, outgoing electron directions of (a) 0°, (b) 5°, and (c) 10°, and energy of 170 keV. Atomic number $Z = 79$. The curves are predicted by the theory of Elwert and Haug.[23]

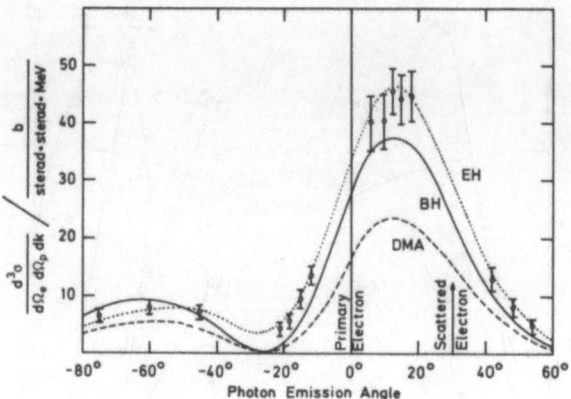

Figure 2. Photon angular distribution for fixed direction of outgoing electrons on silver (Aehlig and Scheer,[8] p. 244). $T_0 = 180$ keV; photon energy $= 80$ keV; EH: Elwert–Haug calculation; BH: Bethe–Heitler calculation; DMA: Deck–Moroi–Alling calculation.

was reported by Aehlig and Scheer.[8] They measured a photon angular distribution for outgoing electrons with an energy of 100 keV and a scattering angle of 30°. The incident electron energy was 180 keV (Figure 2). Their results were in agreement with the calculations of Elwert and Haug[23] within the experimental uncertainties.

2.2. Electron Angular Distribution for a Fixed Photon Direction

An angular distribution of decelerated outgoing electrons for a fixed photon direction of 2° had been reported in 1967 by Hub and Nakel.[5] The incident electron energy was 300 keV, and atomic number $Z = 13$. A magnetic spectrometer selected outgoing electrons of 220 keV, i.e., the energy of the coincident photons was relatively small.

The distribution (Figure 3) is strongly peaked in the forward direction. Photons with an energy that is far from the high-energy limit are radiated preferably by electrons that are deflected through relatively small angles in the nuclear Coulomb field.

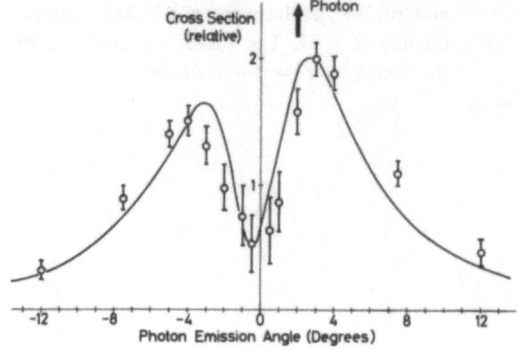

Figure 3. Angular distribution of decelerated outgoing electrons for a fixed photon direction of 2° (Hub and Nakel,[5] p. 601). The curve is the theory of Elwert and Haug.[23] Incident electron energy 300 keV, outgoing electron energy 220 keV, and atomic number $Z = 13$.

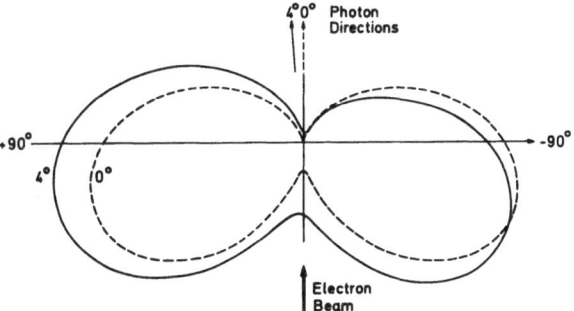

Figure 4. Calculated angular distribution of decelerated outgoing electrons for fixed photon directions at the short-wavelength limit (Haug,[26] p. 62). Incident electron energy 300 keV and atomic number $Z = 13$.

An example for the correlations at the short-wavelength limit in which all the incident electron energy is radiated is shown in Figure 4. The distributions were calculated by the Elwert–Haug theory.[26] Although the outgoing electrons have velocity zero, the correlation is dependent on the directions of the final electrons.

2.3. Spectral Distributions for Fixed Electron and Photon Directions

First measurements of the spectral distribution for fixed electron and photon directions were reported in 1973 by Faulk and Quarles.[9] The incident electron energy was 140 keV and the scattering materials were thin films of aluminum and gold. The data are compared with the Elwert–Haug theory and with the Bethe–Heitler theory. Both theories generally give satisfactory agreement for aluminum.

In 1977 Aehlig, Metzger, and Scheer[10] reported an extensive investigation of the spectral distribution of the absolute cross section for gold. The incident electron

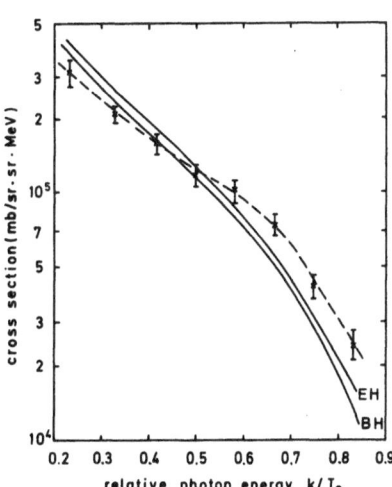

Figure 5. Spectral distribution for electron scattering angle 20° and photon emission angle 10° (Aehlig et al.,[10] p. 208). Atomic number $Z = 79$. BH: Bethe–Heitler calculation; EH: Elwert–Haug calculation; $\times$: experiment.

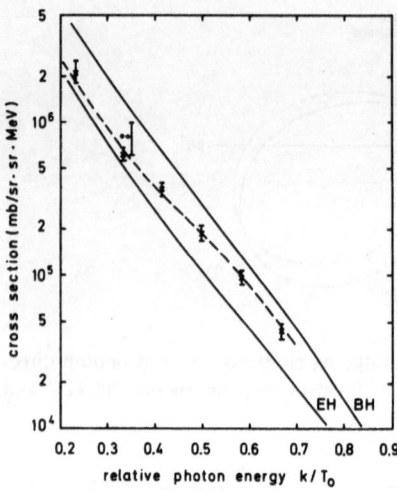

Figure 6. Spectral distribution for electron scattering angle 0° and photon emission angle 10° (Aehlig et al.,[10] p. 208). Atomic number $Z = 79$. BH: Bethe–Heitler calculation; EH: Elwert–Haug calculation; ×: experiment; ● : measurement of Kreuzer and Nakel (converted).

energy was 300 keV. Measurements were made for four different combinations of electron scattering angle and photon emission angle over a wide range of the relative photon energy. The measured cross sections deviate from the calculated ones by a factor of up to 2. On the average, the deviations from the Elwert–Haug theory calculations are somewhat less than those from the Bethe–Heitler formula (Figures 5 and 6).

Recently, Pratt and Lee[27] have performed a full relativistic partial-wave expansion calculation of the triply differential cross section of bremsstrahlung from 300-keV electrons impinging on gold atoms.

3. Angular Correlations in the Electron–Electron Bremsstrahlung Process

In contrast to the electron–nucleus system the electron–electron system has no electric dipole moment. Therefore, the (nonrelativistic) electron–electron bremsstrahlung consists predominantly of electric quadrupole radiation.

Only very few experimental results are available on electron–electron (e–e) bremsstrahlung.

In colliding-beam experiments[28] high-energy electron beams are crossed to generate e–e bremsstrahlung. Another possible way to study e–e bremsstrahlung is to let an electron beam pass through matter and to use the collisions with the orbital electrons of the atoms. If the energy transfer to the electron is much greater than the binding energy of the electron, one can neglect the binding energy and treat the problem approximately as a collision between free electrons. However, there is a problem, in that an electron passing through matter can produce bremsstrahlung both in the Coulomb field of a nucleus and in collision with an atomic electron. Also, the e–e bremsstrahlung in general gives such a small contribution to the total bremsstrahlung emission that it is not taken into account in most measurements of atomic-field bremsstrahlung.

Especially in the case of high-Z targets, the experiments give almost pure nuclear bremsstrahlung, since the electron–nucleus bremsstrahlung is closely proportional to Z^2, whereas the e–e bremsstrahlung is proportional to the number of electrons. Therefore, to investigate e–e bremsstrahlung, one has to use a low-Z target.

An important guide to discern the e–e contribution in the total bremsstrahlung spectrum is the maximum photon energy. Whereas the energy the nucleus receives can be neglected, the energy of the recoil electron may be considerable. Therefore, the e–e bremsstrahlung spectrum is distributed up to a maximum value less than the end point of the electron–nucleus bremsstrahlung spectrum; moreover, it is angle dependent. The different upper bounds of the spectra had been used to isolate the e–e contribution from the total spectrum in noncoincident experiments (see for example Reference 13). However, the experimental results are still inconclusive[14] so that other explanations of the enhanced spectra cannot be excluded. It is interesting to note that Hackl[29] did not find any evidence for a contribution of e–e bremsstrahlung in his experiment on lithium ($Z = 3$).

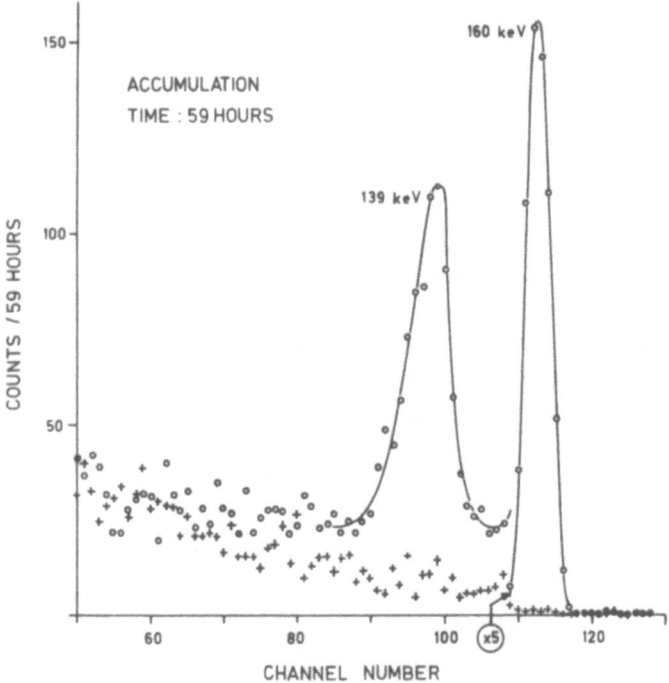

Figure 7. Bremsstrahlung spectrum (pulse-height distribution) from carbon at photon emission angle $-35°$ observed in coincidence with outgoing electrons of 140 keV at scattering angle 20° showing the coincidence peaks of e–e bremsstrahlung (139 keV) and e–nucleus bremsstrahlung (160 keV). The crosses give the random coincidence measured simultaneously (Nakel and Pankau,[18] p. 322).

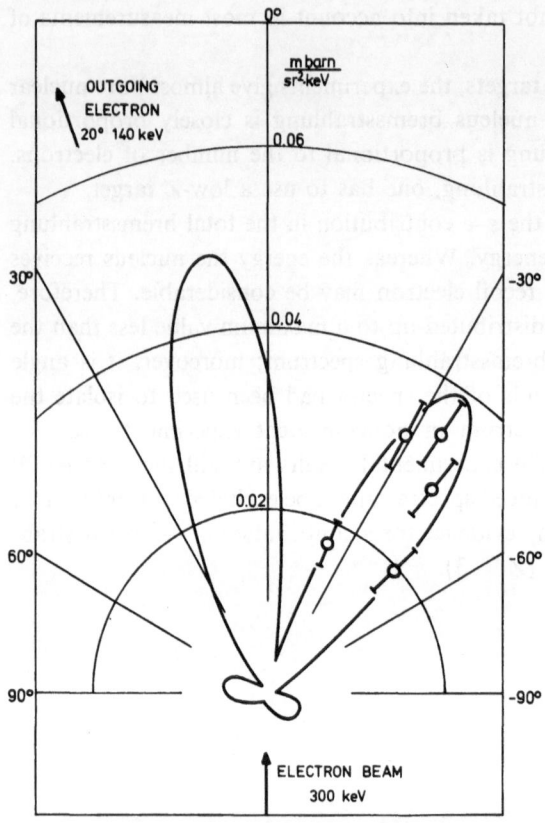

Figure 8. Photon angular distribution of e–e bremsstrahlung for fixed direction of outgoing electrons (Nakel and Pankau,[18] p. 323). The solid curve is the theoretical prediction.[30,31] Atomic number $Z = 6$. The cross section is given per atom, e.g., for six target electrons.

These experimental investigations show that it is difficult to isolate the e–e bremsstrahlung component from the total spectrum when only the photon spectrum is observed. In 1972 Nakel and Pankau[15] used the electron–photon coincidence technique to differentiate exactly between the two bremsstrahlung components by means of a kinematically overdetermined measurement and measured the triple differential cross section[16,17] and the angular correlation.[18]

Photon Angular Distribution for a Fixed Direction of the Outgoing Electrons

In the final state of the e–e bremsstrahlung process there are two outgoing electrons and one photon. The three particles have $3 \times 3 = 9$ degrees of freedom. In a kinematically complete experiment, all nine momentum components are determined. Four out of nine momentum components of the final state are determined by the initial state via momentum and energy conservation. There remain $9 - 4 = 5$ momentum components of the outgoing particles to be measured.

In an electron–photon coincidence experiment Nakel and Pankau[15–18] measured the momenta of one of the outgoing electrons and of the photon (six quantities). The three-body final state is thus once overdetermined. So it is possible, in principle, to differentiate between the processes of *e–e* bremsstrahlung and electron–nucleus bremsstrahlung.

The measuring method is based on a coincidence observation between the photon spectrum emitted in a definite direction and outgoing electrons of definite energy and direction. The *e–e* bremsstrahlung photons are well separated from the electron–nucleus bremsstrahlung photons since the former have an energy that is reduced by the recoil energy of the second electron.

Figure 7 gives coincident photon spectra measured with a Ge(Li) detector for a total of 59 h of observing time. The true coincident photon spectrum (open circles) shows the peak of *e–e* bremsstrahlung at 139 keV and the peak of *e*–nucleus bremsstrahlung at 160 keV. The random coincidences (crosses) are measured simultaneously. 300-keV electrons are used incident on a carbon target ($Z = 6$). A magnetic spectrometer selected outgoing electrons of 140 keV and scattering angle 20°.

In Figure 8 a photon angular distribution of *e–e* bremsstrahlung for a fixed direction of outgoing electrons at 20° is shown. In contrast to the *e*–nucleus bremsstrahlung the photon energy of *e–e* bremsstrahlung is dependent on the photon emission angle for kinematical reasons (e.g., from 136 keV at −23° to 134 keV at −47°). The solid line is the theoretical prediction.[30,31]

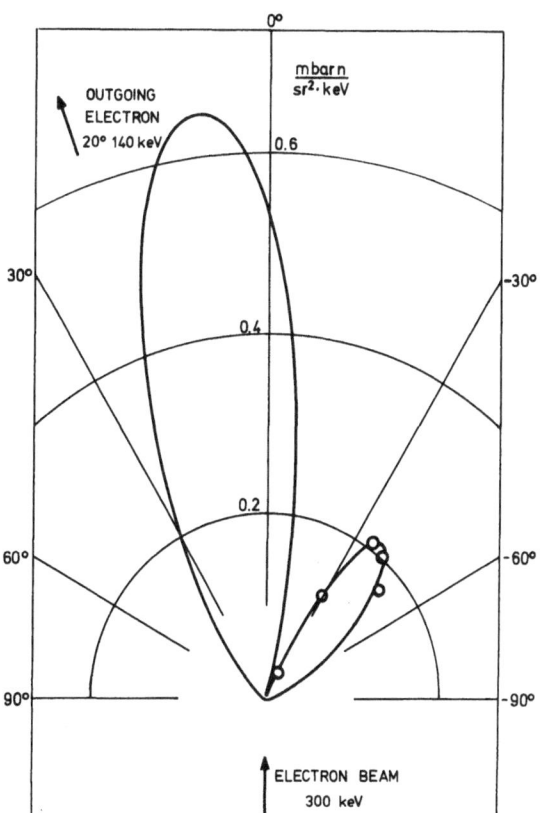

Figure 9. Photon angular distribution of atomic-field bremsstrahlung for fixed direction of outgoing electrons (Nakel and Pankau,[18] p. 323). The solid curve is the theoretical prediction.[23] Atomic number $Z = 6$.

The calculation of e–e bremsstrahlung cross sections is a straightforward application of quantum electrodynamics. However, compared to the Bethe–Heitler formula for the corresponding process in the Coulomb field of a nucleus, these expressions are extremely complicated owing to recoil and exchange effects.[32] Therefore the common practice was to use approximations. In 1973 an exact calculation in lowest-order quantum electrodynamics was carried out by two independent groups. Mack and Mitter[30] solved the problem by using computer programs for formula manipulations. Haug[31] used the traces calculated by Anders.[33]

In Figure 9 the photon angular distribution of e–nuclear bremsstrahlung is shown, measured simultaneously with the e–e bremsstrahlung.

Measurements in the region of the larger lobe of the e–e distribution have not been done until now because of background problems due to the emission direction of electron and photon being too close. To reduce background, a triple coincidence measurement with the second outgoing electron is planned.

4. Polarization Correlations in the Atomic-Field Bremsstrahlung Process

Bremsstrahlung produced by a beam of unpolarized electrons in the Coulomb field of an atomic nucleus exhibits linear polarization properties. Many measurements have already been made to investigate the dependence of the polarization on a variety of parameters (as, for instance, the photon energy, the photon emission angle, the initial electron energy, and the atomic number of the target[34–36]). However, in all these measurements the decelerated outgoing electrons are not observed. Consequently, this polarization is produced by averaging over many single bremsstrahlung processes with different directions of the outgoing electrons.

In 1978 Behncke and Nakel[11] reported a linear polarization correlation measurement which involves measuring the angular dependence of the photon linear polarization for fixed direction of outgoing electrons.

According to a simple classical picture,[34] the bremsstrahlung is caused by a change of momentum in the orbital motion of the electron and consists of electric dipole radiation with the electric vector in the plane containing the vector of the electron acceleration and the direction of emission. The direction of the electron acceleration is given by the difference of the momenta of the incoming and outgoing electron. Since the direction of the outgoing electron is determined by the coincidence measurement, one would expect the radiation to be always completely linearly polarized in the reaction plane in these experiments.

However, at relativistic energies the effect of electron spin becomes important and the emission of bremsstrahlung can take place not only as a result of a change of momentum, but as a result of a change of spin orientation.[12] In measuring the polarization correlation it was possible to discern the appearance of the spin–flip radiation by a deviation from the complete linear polarization.

4.1. Angular Dependence of the Photon Linear Polarization for a Fixed Direction of Outgoing Electrons

In order to measure the polarization correlation, the photons are analyzed in a Compton polarimeter and detected in coincidence with outgoing electrons scattered in a particular direction. A schematic view of the experimental arrangement of Behncke and Nakel[11] is shown in Figure 10.

The degree of linear polarization is defined by the expression

$$P = (I_\perp - I_\parallel)/(I_\perp + I_\parallel) \tag{1}$$

where $I_\perp$ and $I_\parallel$ are the bremsstrahlung intensity components with electric vectors perpendicular and parallel, respectively, to the reaction plane. The quantity that was measured directly is the ratio of the counting rates of the true coincidences $N_\perp/N_\parallel$. In this case $N_\perp$ and $N_\parallel$ denote the true coincidence counting rates of scattered photons in the direction perpendicular and parallel, respectively, to the reaction plane. To obtain the linear polarization P from the measured data the asymmetry ratio R of the Compton polarimeter must be known. The ratio of the coincidence rates $N_\perp/N_\parallel$, asymmetry ratio R, and the polarization P are connected by the relation

$$P = \frac{R+1}{R-1} \frac{1 - N_\perp/N_\parallel}{1 + N_\perp/N_\parallel} \tag{2}$$

R was calculated by a Monte Carlo program.

The result of the measurement is shown in Figure 11. According to definition (1) the negative values of the polarization indicate that the radiation is polarized predominantly parallel to the emission plane.

The theoretical curve (solid line) has been calculated for the present experimental situation and may be compared directly with the experimental results. Starting point

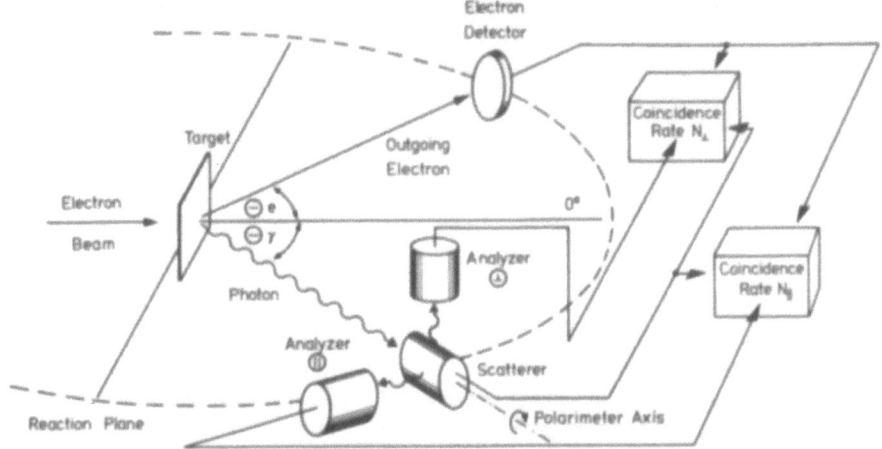

Figure 10. Schematic view of the experimental arrangement (Behncke and Nakel,[11] p. 1680).

for the calculation of the curve was the theory of Elwert and Haug.[23,37] Atomic–electron screening was neglected.

The polarization, calculated according to the Elwert–Haug theory and presented first without taking corrections into account, is shown in Figure 12a (solid line). From this curve the solid line of Figure 11 is obtained by making the following corrections:

(a) Corrections were made for the finite solid angles of the photon and electron detector and for the finite energy width of the magnetic spectrometer used in the experiment.

(b) Depending on the finite thickness of the target, there is a certain probability that the radiating electron is scattered elastically before and/or after the bremsstrahlung process. The influence of this plural scattering on the polarization was calculated with a Monte Carlo program.

(c) Correction was made for e–e bremsstrahlung, since this contribution could not be separated from the electron–nucleus bremsstrahlung with the NaI(Tl) detectors used in the experiment. Therefore, the polarization correlation of e–e bremsstrahlung was calculated,[30,38] yielding the result shown in Figure 12a (dashed line). The corresponding triply differential cross section (for six target electrons) is shown in Figure 12b (dashed line). The polarization resulting from these two elementary processes by taking into account the corresponding cross sections is represented by the dashed curve of Figure 11. From this curve, with corrections (a) and (b), one obtains the solid line curve of Figure 11, which must be compared with the measured values.

For comparison with the polarization in the elementary process, Fig. 12a (dotted line) shows the polarization integrated over all directions of motion of the outgoing electrons. This curve was calculated in the Born approximation[39] and corresponds to a measurement in which the outgoing electrons are not observed. Here the radiation in the forward direction (0°) shows a complete absence of linear

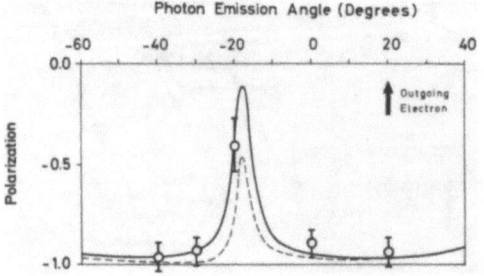

Figure 11. Linear photon polarization in the elementary bremsstrahlung process as a function of the photon emission angle for outgoing electrons of +20° and 140 keV. Primary electrons of 300 keV are used, incident on a carbon target. The experimental values are shown by open circles. The theoretical curve (solid line) has been calculated for the present experimental situation with the Elwert–Haug theory. The dashed curve shows the theoretical polarization without corrections for the finite solid angles, the finite energy width of the magnetic electron spectrometer, and the electron plural scattering in the target. The correction, however, for a contribution of electron–electron bremsstrahlung has been made (Behncke and Nakel,[11] p. 1682).

Figure 12. (a) Theoretical photon linear polarization (uncorrected) in the elementary process of electron–nucleus bremsstrahlung (———) and electron–electron bremsstrahlung (– – –) as a function of the photon emission angle, calculated for the experimental parameters. For comparison with the elementary process, the theoretical polarization of electron–nucleus bremsstrahlung is shown, integrated over all directions of motion of the outgoing electron (···). (b) Theoretical triply differential cross section of electron–nucleus bremsstrahlung (———) and electron–electron bremsstrahlung (– – –) as a function of the photon emission angle, calculated for the experimental parameters and with corrections for the finite solid angles and the finite energy width of the magnetic electron spectrometer. (The cross section of electron–electron bremsstrahlung is multiplied by a factor of 6 according to the number of electrons in the carbon atom) (Behncke and Nakel,[11] p. 1683.)

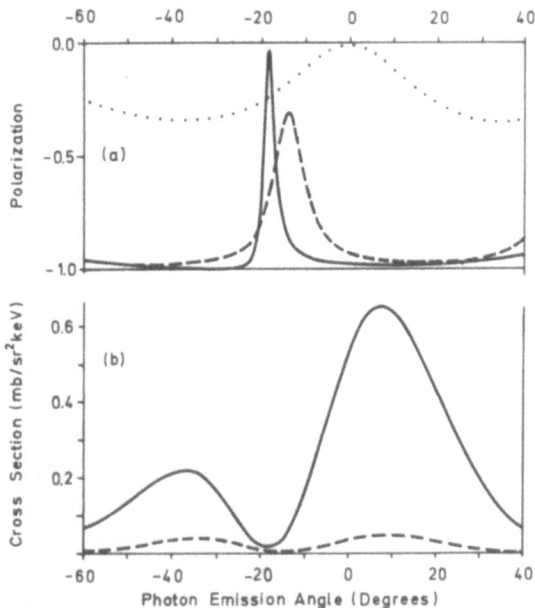

polarization because of symmetry requirements in the superposition of many single processes, whereas the polarization of the single process is nearly complete in this direction. On the other hand, the sharp change of the polarization disappears in the noncoincident case. Thus, the integration over the outgoing electron direction covers up the fundamental behavior of the bremsstrahlung process. The general features of the integral polarization are discussed in terms of orbital motion and spin effects by Motz and Placious.[34]

4.2. Discussion

Looking at the angular dependence of the polarization of the electron–nucleus bremsstrahlung (Figure 12a, solid line) and at the corresponding cross section (Figure 12b, solid line), one sees that the greatest part of the radiation is almost completely polarized with the electric vector in the reaction plane, containing the momenta of the incoming and outgoing electron and the photon. This is in accordance with the conception of the simple classical model that the radiation emitted in the plane of the orbit of the electron is linearly polarized with its electric vector in this plane. However, such a simple model cannot explain the deviation from the complete polarization, especially the strong decrease of the polarization in the minimum of the cross section.

In a theoretical analysis Fano, McVoy, and Albers[12] investigated the contributions to the bremsstrahlung coming from a change of momentum in the orbital motion of the electron and a change of spin orientation. They point out that momentum and spin changes are not independent and that the contributions of the change of momentum and spin orientation yield, respectively, linearly and circularly polarized bremsstrahlung. Interference effects of both contributions yield observable features even when the incident electron beam is unpolarized, e.g., that the dominant component of the electric vector is perpendicular to the reaction plane. However, when the momenta of the incoming and outgoing electron and the photon all lie in the same plane, the entire contribution of the orbital motion is polarized in that plane. Therefore, deviations from the complete polarization in the reaction plane must be caused by the influence of the spin–flip radiation. The strong decrease of the polarization in the minimum of the cross section shows that here the influence of the spin is particularly strong. Behncke and Nakel could confirm the decrease of the polarization by a measured value at $-20°$ (measuring time, 300 h).

Also, in the other parts of the polarization curve, as shown in Figure 12a, the degree of polarization generally does not reach -1. Only when the momentum transfer to the nucleus is parallel to the photon emission direction is the photon completely linearly polarized. This was shown by Olsen[40] in a Born-approximation calculation. With the parameters of the experiment, this angle appears at $-29°$.

The deviations from complete polarization in the reaction plane should not occur in the radiation process of a spinless particle. To verify this expectation, Behncke and Nakel have calculated the bremsstrahlung polarization of a hypothetical spinless electron under the same external conditions with a formula[41] which follows from the Duffin–Kemmer equation. For all photon emission angles the calculation gives a complete polarization with the electric vector in the reaction plane.

For higher target atomic numbers the calculations of Elwert and Haug show that the decrease of the polarization in the minimum of the cross section diminishes strongly with increasing atomic number. For gold this decrease nearly vanishes and the polarization is nearly complete. However, because the condition $\alpha Z \ll 1$ is required, the calculation cannot be expected to be accurate. Measurements for high atomic numbers are in progress and could show the influence of Coulomb effects in their barest form. For comparison, calculations of the polarization by means of partial-wave expansions[21] would be highly desirable.

Up to now bremsstrahlung electron–photon correlation measurements were still summed or averaged over the directions of the spins of the incoming and outgoing electrons. One might conceive of a future ideal experiment employing a polarized electron beam and detectors sensitive not only to the polarization of the photons but to the polarization of the outgoing electrons.

ACKNOWLEDGMENT

The author is indebted to Dr. E. Haug for his critical reading of the manuscript.

References

1. O. Scherzer, *Ann. Phys.* (*Leipzig*) **13**, 137 (1932).
2. W. Nakel, *Phys. Lett.* **22**, 614 (1966).
3. W. Nakel, *Phys. Lett.* **25A**, 569 (1967).
4. W. Nakel, *Z. Phys.* **214**, 168 (1968).
5. R. Hub and W. Nakel, *Phys. Lett.* **24A**, 601 (1967).
6. W. Nakel and U. Sailer, *Phys. Lett.* **34A**, 181 (1970).
7. K. Kreuzer and W. Nakel, *Phys. Lett.* **34A**, 407 (1971).
8. A. Aehlig and M. Scheer, *Z. Phys.* **250**, 235 (1972).
9. J. D. Faulk and C. A. Quarles, *Phys. Lett.* **44A**, 317 (1973); *Phys. Rev. A* **9**, 732 (1974).
10. A. Aehlig, L. Metzger, and M. Scheer, *Z. Phys.* **A281**, 205 (1977).
11. H.-H. Behncke and W. Nakel, *Phys. Rev. A* **17**, 1679 (1978).
12. U. Fano, K. W. McVoy, and J. R. Albers, *Phys. Rev.* **116**, 1159 (1959).
13. D. H. Rester, *Nucl. Phys.* **A118**, 129 (1968).
14. E. Haug, *Phys. Lett.* **54A**, 339 (1975).
15. W. Nakel and E. Pankau, *Phys. Lett.* **38A**, 307 (1972).
16. W. Nakel and E. Pankau, *Phys. Lett.* **44A**, 65 (1973).
17. W. Nakel and E. Pankau, *Z. Phys.* **264**, 139 (1973).
18. W. Nakel and E. Pankau, *Z. Phys.* **A274**, 319 (1975).
19. C. Bernardini, F. Felicetti, R. Querzoli, V. Silverstrini, and G. Vignola, *Lett. Nuovo Cimento* **1**, 15 (1969).
20. R. H. Siemann, W. W. Ash, K. Berkelman, D. L. Hartill, C. A. Lichtenstein, and R. M. Littauer, *Phys. Rev. Lett.* **22**, 421 (1969).
21. H. H. Tseng and R. H. Pratt, *Phys. Rev. A* **3**, 100 (1971).
22. H. W. Koch and J. W. Motz, *Rev. Mod. Phys.* **31**, 920 (1959).
23. G. Elwert and E. Haug, *Phys. Rev.* **183**, 90 (1969).
24. R. T. Deck, D. S. Moroi, and W. R. Alling, *Nucl. Phys.* **A133**, 321 (1969).
25. H. Bethe and W. Heitler, *Proc. R. Soc.* (*London*) *Ser. A* **146**, 83 (1934).
26. E. Haug, Dissertation, Universität Tübingen, 1966 (unpublished).
27. R. H. Pratt and C. M. Lee, *Bull. Am. Phys. Soc.* **22**, 83 (1977).
28. P. I. Golubnichii, E. A. Kushnirenko, A. P. Onuchin, and V. A. Sidorov, *Yad. Fiz.* **7**, 1240 (1968).
29. P. Hackl, Dissertation, University of Vienna, 1970 (unpublished); H. Aiginger and E. Unfried, *Acta Phys. Austriaca* **35**, 331 (1972).
30. D. Mack and H. Mitter, *Phys. Lett.* **44A**, 71 (1973).
31. E. Haug (unpublished); cf. *Z. Naturforsch* **30a**, 1099 (1975).
32. I. Hodes, Ph.D. Thesis, University of Chicago, 1953 (unpublished).
33. T. Anders, Dissertation, Universität Freiburg (1961); *Nucl. Phys.* **59**, 127 (1964).
34. J. W. Motz and R. C. Placious, *Nuovo Cimento* **15**, 571 (1960).
35. R. W. Kuckuck and P. J. Ebert, *Phys. Rev. A* **7**, 456 (1973).
36. W. Lichtenberg, A. Przybylski, and M. Scheer, *Phys. Rev. A* **11**, 480 (1975).
37. E. Haug, private communication; cf. *Phys. Rev.* **188**, 63 (1969).
38. D. Mack, private communication.
39. R. L. Gluckstern and M. H. Hull, *Phys. Rev.* **90**, 1030 (1953).
40. H. A. Olsen, *Springer Tracts Mod. Phys.* **44**, 83 (1968).
41. A. I. Aichieser and W. B. Berestezki, *Quantenelektrodynamik*, B. G. Teubner, Leipzig (1962)

References

16

Angular Distribution and Polarization of x-Ray Radiation by Electron Impact on Free Atoms

M. Aydinol, R. Hippler, I. McGregor, and H. Kleinpoppen

Angular distributions of characteristic x-ray line radiation and bremsstrahlung photons, produced in collisions between electrons and krypton and xenon atoms, have been measured in the energy range from 5 to 15 keV. The x-rays have been detected by a Si(Li) detector with an energy resolution of 200 eV. While the bremsstrahlung photons display a highly anisotropic angular distribution, the combined $L\alpha_{1,2}$ characteristic line radiation of xenon only shows a small anisotropy. The polarization of the $L\alpha_{1,2}$ radiation which results from this anisotropy approaches the theoretically predicted threshold polarization of the characteristic $L\alpha_{1,2}$ x-ray radiation of xenon. Bremsstrahlung anisotropies of Xe and Kr are compared with theoretical predictions.

1. Introduction

Although the discovery of x-ray radiation by Röntgen[1] in 1895 was a milestone for the development of modern physics, it took more than a dozen years before the first x-ray spectrum had been reported.[2] It turned out that x-ray spectra induced by electron bombardment of sufficiently thin targets in general consist of a continuum part with some characteristic lines superimposed on it. (In fact, characteristic *K* and *L* lines had been discovered a few years prior to this by the use of an absorption technique.[3]) It was soon recognized that the continuous part of the spectrum,

M. Aydinol, R. Hippler, I. McGregor, and H. Kleinpoppen • Institute of Atomic Physics, University of Stirling, Stirling, Scotland.   R. Hippler's permanent address: Fakultät für Physik and Zentrum für interdisziplinäre Forschung of Universität Bielefeld, Postfach 8640, D-4800 Bielefeld 1, West Germany.

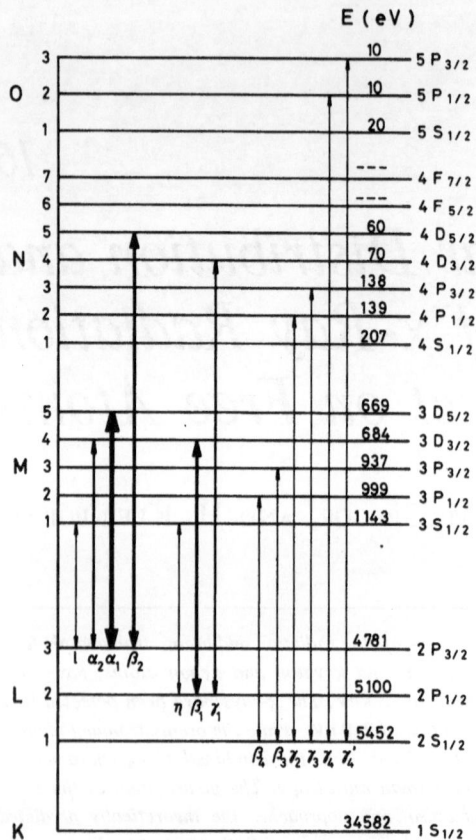

Figure 1. Schematic diagram for the *L* x-ray transitions in xenon.

which has a high-energy limit at a photon energy equal to the incident electron energy, is due to the slowing down of the incident electrons in the fields of nuclei or electrons (bremsstrahlung).

From the first measurements of the angular distribution of the bremsstrahlung spectrum it turned out that the radiation is emitted anisotropically. Following Sommerfeld,[4] with the modifications given by Kulenkampff *et al.*,[5] an analytical expression for the angular distribution can be given:

$$I(\theta)/I(90°) = [(1 - P \cdot \beta^2)(1 - \beta \cos \theta)^4]^{-1}[(1 - \beta \cos \theta)^2 - P(\cos \theta - \beta)^2] \qquad (1)$$

where $I(\theta)$ is the intensity observed under a photon emission angle θ measured with respect to the incident electron beam direction, $\beta = \bar{v}/c$ (the ratio of the average electron velocity $\bar{v}$ and the speed of light c), and P is the degree of polarization defined as

$$P = \frac{I_{\parallel} - I_{\perp}}{I_{\parallel} + I_{\perp}} \qquad (2)$$

with $I_{\parallel}$ and $I_{\perp}$ being the photon intensities with the electric vector parallel and perpendicular to the electron beam direction, respectively.

For characteristic radiation, the emission may be anisotropic if it results from transitions in which the total angular momentum quantum number of the initial state is $j > \frac{1}{2}$.[6] The angular distribution can then be expressed as

$$I(\theta)/I(90°) = 1 - P \cos^2 \theta \qquad (3)$$

In this paper we report on an angular distribution study of characteristic and bremsstrahlung radiation from free atoms (Kr, Xe) in the gaseous phase induced by 5–15 keV electron impact. Until now, all of the bremsstrahlung measurements have been performed with solid targets, where the influence of the solid on either the incident electron direction (straggling) or the inelastic energy loss is at best only vaguely known. The characteristic lines under study are the L transitions in Xe. An energy diagram showing the Xe x-ray transitions of interest is given in Figure 1. Total cross sections for some of these transitions are given elsewhere.[7]

2. Apparatus

The experiment consists of crossed electron/atom beams inside a vacuum chamber and an x-ray detector outside the chamber (Figure 2). The electron gun is a triode gun system with a tungsten filament, providing electron beam currents of about 100 μA measured with a Faraday cup. To monitor the focusing of the gun, the Faraday cup consists of three coaxial cylinders. Both the electron gun and the Faraday cup are mounted on a turntable rotatable inside the vacuum chamber, allowing the x-ray observation angle to be varied from $-135°$ to $+135°$. The x-ray detector, which is fixed in position, is a lithium-drifted silicon [Si(Li)] detector with an energy resolution of about 200 eV. The atom beam density is about 10^{13} atoms cm^{-3}.

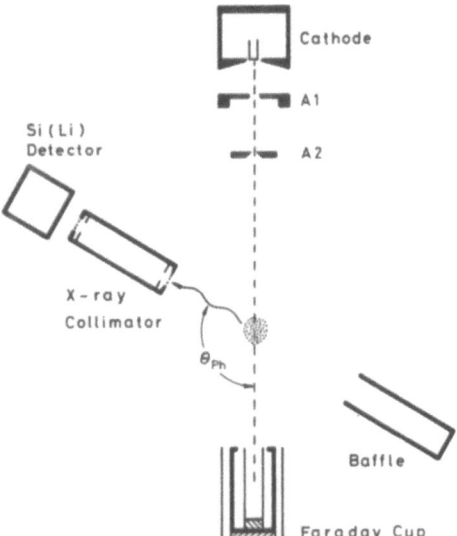

Figure 2. Experimental setup (schematic).

In order to reduce background signals resulting from collisions of electrons with solid materials, an x-ray collimator is placed in front of the Si(Li) detector inside the vacuum chamber. It consists of a steel pipe of length 240 mm with two knife-edge-shaped diaphragms of 8 mm diameter at each side. Both diaphragms are covered by a 25-μm beryllium window; the first prevents electrons from entering the collimator, the second serves as vacuum-tight x-ray window.

3. Results

A typical x-ray spectrum obtained from 8-keV electron impact on Xe is given in Figure 3. It shows the Xe-L diagram lines superimposed on a continuous bremsstrahlung spectrum with a high-energy cutoff at the incident electron energy. The observed lines are identified as the $Ll(2P_{3/2} \to 3S_{1/2})$, $L\alpha_{1,2}(2P_{3/2} \to 3D_{5/2;3/2})$, $L\beta_1(2P_{1/2} \to 3D_{3/2})$, and $L\beta_{3,4}(2S_{1/2} \to 3P_{3/2;1/2})$, $L\beta_{2,15}(2P_{3/2} \to 4D_{5/2;3/2})$ and $L\gamma_1 \times (2P_{1/2} \to 4D_{3/2})$ transitions. Owing to the limited energy resolution of the Si(Li) detector, the α_1 and α_2 lines having an energy difference of 15 eV and the β_1, β_3, and β_4 lines having energy differences of about 80 eV are not resolved. Fortunately, the three $\beta_{1,3,4}$ lines result from either the $2P_{1/2}$ or the $2S_{1/2}$ initial states and hence can be used for normalization since no anisotropy is expected from them.

In order to extract the information about the relative intensities $I(\theta)$ the x-ray spectra have been deconvoluted by a computer program, fitting the peaks by Gaussians and with the bremsstrahlung part being represented in the neighborhood of the peaks by a second-order polynomial. All peak intensities have been normalized to the $L\beta_{1,3,4}$ peak, which is expected to be isotropic as all the lines result from initial levels with $j = \frac{1}{2}$. In Figure 4 the normalized intensity ratio $I(\theta)/I(90°)$ of the combined $L\alpha_{1,2}$ line of Xe is plotted over $\cos^2 \theta$. The incident electron energy was 6 keV. A least-squares fit through the data points, accounting for the statistical error bars,

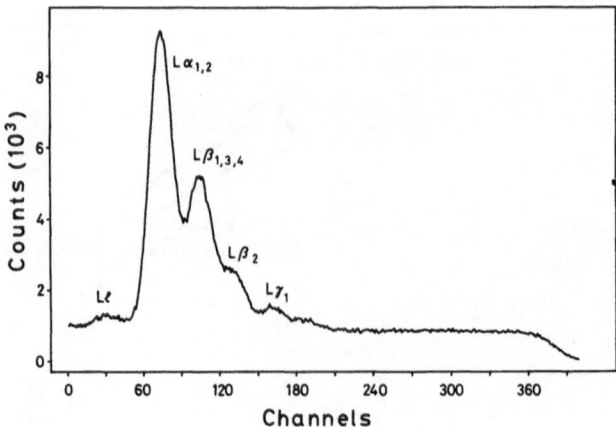

Figure 3. Photon energy spectrum of xenon following 8-keV electron bombardment.

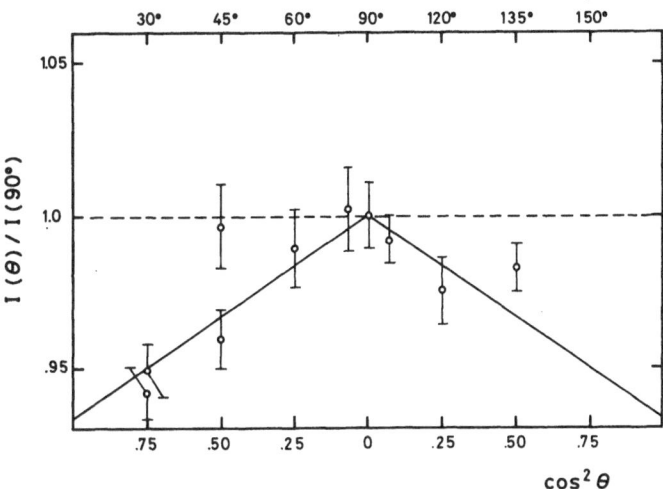

Figure 4. Angular distribution of the combined xenon $L\alpha_{1,2}$ line. The bombarding electron energy was 6 keV.

gives the straight line in Figure 4. The slope of this line is directly related to the polarization degree P.

4. Discussion

In Figure 5 the degree of polarization P of the combined $L\alpha_{1,2}$ line is given as a function of the incident electron energy. It turns out that the P values are quite small ($|P| < 2\%$), except near threshold, where values of $+6.6\%$ and $+4.8\%$ have been observed at incident electron energies of 6 and 6.5 keV, respectively. This observation is in agreement with expectations for the threshold polarization. Expressing the polarization degree P for the Ll, $L\alpha_1$, and $L\alpha_2$ as[8]

$$
\begin{aligned}
P(l) &= 3 \frac{\sigma_0 - \sigma_1}{5\sigma_0 + 7\sigma_1} \\
P(\alpha_1) &= (\sigma_0 - \sigma_1)/(7\sigma_0 + 13\sigma_1) \\
P(\alpha_2) &= 3(\sigma_1 - \sigma_0)/(4\sigma_0 + 11\sigma_1)
\end{aligned}
\tag{4}
$$

with σ_0, σ_1 being the cross sections for exciting the $m_l = 0$, $m_l = \pm 1$ magnetic sublevels, respectively, the expected threshold values for P, assuming $\sigma_1 = 0$ at threshold, are

$$
\begin{aligned}
P^{\text{thr}}(l) &= +60\% \\
P^{\text{thr}}(\alpha_1) &= +14.3\% \\
P^{\text{thr}}(\alpha_2) &= -75\%
\end{aligned}
\tag{4a}
$$

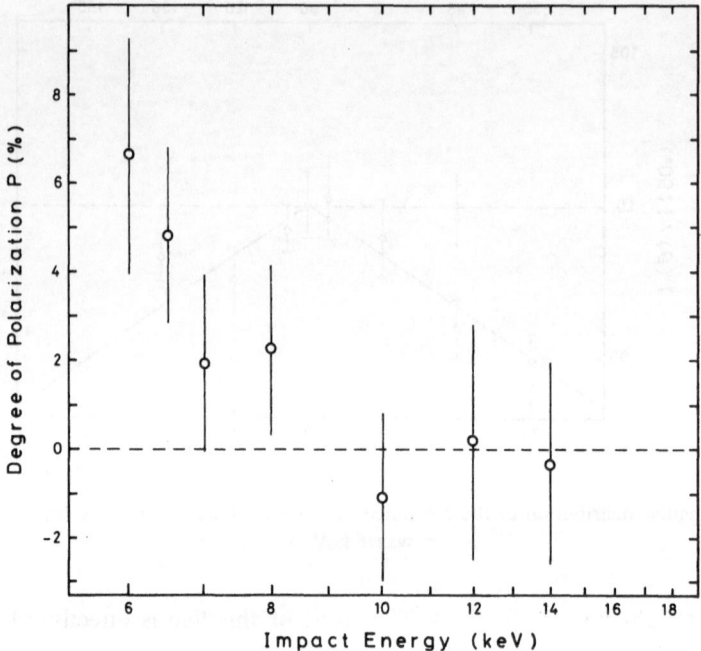

Figure 5. Polarization degree P of the combined $L\alpha_{1,2}$ line of xenon as a function of the electron impact energy.

Taking into account the transition probabilities Γ for the α_1 and α_2 transitions being[9]

$$\Gamma(\alpha_1) = 2.95 \times 10^{14}/S$$

$$\Gamma(\alpha_2) = 0.33 \times 10^{14}/S$$

the average threshold polarization for the combined $L\alpha_{1,2}$ line can be calculated as

$$P^{thr}(\alpha_1, \alpha_2) = +7.2\%$$

This value is very close to our experimental finding of $+6.6\%$ at 6 keV.

To our knowledge, there are no electron impact x-ray data to compare our data with. The only known experimental investigation of the α_1 transition has been performed by Hrdý *et al.*[10] with mercury, giving -14% at an incident electron energy of about 30 keV. There are theoretical calculations of McFarlane[8] and Berezhko and Kabachnik[11] to compare our data with. Both calculations have been performed within the frame of the first Born approximation and hence are not expected to give the correct energy dependence at low incident electron velocities. At the higher energies the accuracy of our polarization data are insufficient to test the calculations quantitatively.

Our data are supported by polarization measurements of gaseous and solid targets excited by proton impact.[12,13] The measurements of Schöler and Bell[12] for the combined Ll, $L\alpha_{1,2}$ transitions of (solid) Cu and Ge excited by 100-keV proton impact give values of

$$P_{\text{Cu}}(l, \alpha_{1,2}) = +3.7\%$$

$$P_{\text{Ge}}(l, \alpha_{1,2}) = +3.2\%$$

The measurements of Lutz et al.[13] for the Ll and the $L\alpha_{1,2}$ lines of (gaseous) Xe excited by 300-keV proton impact gave values of

$$P_{\text{Xe}}(l) = +22.6\%$$

$$P_{\text{Xe}}(\alpha_{1,2}) = +4.5\%$$

From the measurements of the angular distribution of Auger electrons following inner-shell ionization by electron impact the alignment parameter A_2 defined as

$$A_2 = \frac{\sigma_0 - \sigma_1}{\sigma_0 + 2\sigma_1} \tag{5}$$

can be obtained directly. As far as the same initial level is concerned, the alignment parameter and the polarization degree are directly related to each other. We therefore can compare our x-ray data with Auger electron measurements. If we do so with measurements of Sandner and Schmitt[14] for LMM Auger transitions in Ar, we find disagreement both in magnitude and in sizes. Whereas our data give a value of

$$A_2^x = 0.90 \pm 0.35$$

the A_2^A values of Sandner and Schmitt[14] become negative near threshold and are of the order of a few percent. It should be noted here that Sandner and Schmitt[14] give some qualitative arguments to state that for an ionization process a threshold law should not be valid. However, our own measurements and also the x-ray measurements by proton impact support the threshold law. We therefore feel that other reasons may be responsible for the different behavior of the x-ray and the Auger data. One reason may be the interaction between the scattered electron and the emitted Auger electron. Such an interaction is neglected in the calculations and may be of importance at incident electron energies not much above threshold. In fact, van der Wiel et al.[15] have measured the ratio of Ar^{2+} to Ar^{3+} ions produced by the Auger decay of Ar L vacancies following 8-keV electron bombardment. This ratio rapidly varies between values of 9 to 11 for an incident electron energy loss of about 8 eV above the L threshold, which is attributed to postcollisional interaction (PCI). The observed effect is of the order of 10%; the influence on the angular distribution of the Auger electrons may be even larger.

The angular distribution of the bremsstrahlung spectrum shows a marked difference when compared with the diagram lines. A shift of the angle of maximum intensity towards angles smaller than 90° is observed. The shift is predicted by

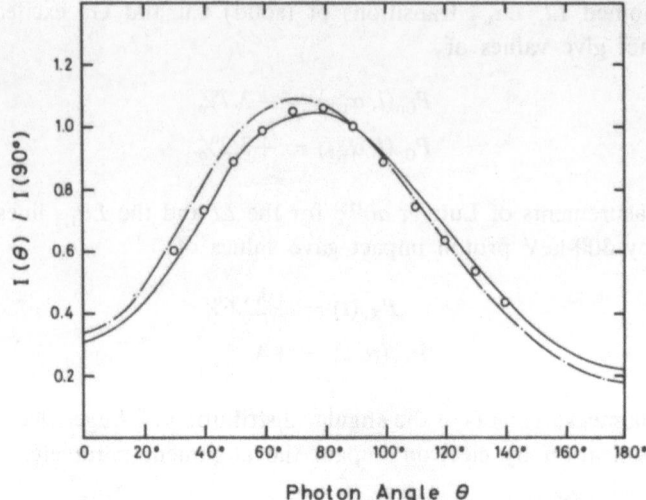

Figure 6. Angular distribution of bremsstrahlung radiation at a photon energy of 9.5 keV, induced by 10-keV electron impact on xenon; ○, present data; —·—, Lee *et al.*;[16] ——, modified Sommerfeld formula [equation (1)] for $P = 75\%$.

theory and due to the retardation of the potentials of the electromagnetic field. Figure 6 shows the results for 10-keV electron impact on Xe and the emission of 9.5-keV photons near the high-energy limit. A comparison is made with a theoretical calculation of Lee *et al.*[16] for Ag ($Z = 47$). The difference in Z between our experiment ($Z = 54$) and theory ($Z = 47$) should have no large influence on the angular distribution.[17] Then the theory overestimates the shift of the angle of maximum intensity, which is experimentally found to be 80°, whereas from theory 70° are expected. From

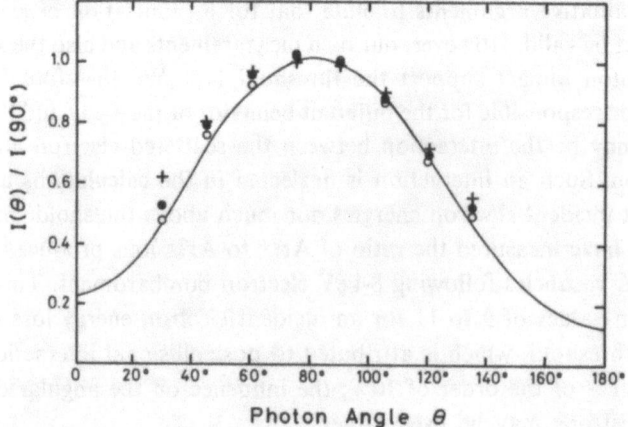

Figure 7. Angular distribution of bremsstrahlung radiation following 6-keV electron impact on krypton at photon energies of: +, 2.4 keV; ●, 4.0 keV; and ○, 5.4 keV. ——, modified Sommerfeld formula [equation (1)] for $P = 85\%$.

the comparison with the modified Sommerfeld formula (1), a value for the polarization degree P of $P = 75\%$ can be derived, although the experimentally found angular distribution is not fully represented by that formula. This may be due to the fact that in equation (1) a mean value for the relative electron velocity β has been used instead of performing an integration over the period of slowing down. In Figure 7 a comparison is made for some parts of the bremsstrahlung spectrum induced by 6-keV electron impact on Kr. The three curves obtained for photon energies of 2.4, 4.0, and 5.4 keV can be represented best by equation (1) with values for P of 65%, 75%, and 85%, respectively. The decrease of P with decreasing photon energy is in qualitative agreement with calculations of Kirkpatrick and Wiedmann.[18]

5. Conclusions

To our knowledge the report given in this paper shows for the first time that the observed anisotropy or polarization of characteristic x-ray radiation approaches the theoretically predicted threshold polarization ($L\alpha_{1,2}$ transition in xenon). Furthermore the measured anisotropies of bremsstrahlung radiation from free Kr and Xe atoms represent the most sensitive tests (which are independent of solid-state effects) for theoretical predictions.

ACKNOWLEDGMENTS

Part of this work was supported by the Science Research Council. One of us (R.H.) gratefully acknowledges the hospitality of the Institute of Atomic Physics of the University of Stirling. During this study one of us (M. A.) was in receipt of a scholarship from the Ministry of Education, Turkey.

References

1. W. C. Röntgen (trans. A. Stanton), *Nature (London)* **53**, 276 (1896).
2. W. H. Bragg and W. L. Bragg, *Proc. R. Soc. (London)* **88A**, 428 (1913).
3. C. G. Barkla, *Proc. Cambridge Philos. Soc.* **15**, 257 (1909).
4. A. Sommerfeld, *Ann. Phys. (Leipzig)* **11**, 257 (1931).
5. H. Kulenkampff, M. Scheer, and E. Zeitler, *Z. Phys.* **157**, 275 (1959).
6. W. Mehlhorn, *Phys. Lett.* **26A**, 166 (1968).
7. M. Aydinol, M. Sc. thesis, University of Stirling, Scotland (1976); M. Aydinol, I. McGregor, and H. Kleinpoppen, Proceedings of the Second International Conference on Inner Shell Ionization Phenomena, Freiburg, p. 214 (1976).
8. S. C. McFarlane, Ph.D. thesis, University of Stirling, Scotland (1972).
9. J. H. Scofield, *At. Data Nucl. Data Tables* **14**, 121 (1974).
10. J. Hrdý, A. Henins, and J. A. Bearden, *Phys. Rev. A* **2**, 1708 (1970).
11. E. G. Berezhko and N. M. Kabachnik, *J. Phys. B* **10**, 2467 (1977).
12. A. Schöler and F. Bell, *Z. Phys.* **A286**, 163 (1978).

13. H. O. Lutz, N. Luz, S. Sackmann, W. Jitschin, and R. Hippler, Chapter 26 of this book.
14. W. Sandner and W. Schmitt, *J. Phys. B* **11**, 1833 (1978).
15. M. J. van der Wiel, G. R. Wight, and R. R. Tol, Proceedings of the Second International Conference on Inner Shell Ionization Phenomena Freiburg, p. 123 (1976).
16. C. M. Lee, L. Kissel, R. H. Pratt, and H. K. Tseng, *Phys. Rev. A* **13**, 1714 (1976).
17. R. H. Pratt, private communication (1976); R. H. Pratt, H. K. Tseng, C. M. Lee, L. Kissel, C. MacCallum, and M. Riley, *At. Data Nucl. Data Tables* **20**, 175 (1977).
18. P. Kirkpatrick and L. Wiedmann, *Phys. Rev.* **67**, 321 (1945).

17

Energy Dependence of the Alignment in Inner-Shell Ionization by Electron Impact

W. Sandner, M. Weber, and W. Mehlhorn

We have measured the alignment in inner-shell ionization of the M_4 and M_5 shell of krypton via the nonisotropic angular distribution of $M_{4,5}$-$N_{2,3}N_{2,3}(^1S_0)$ Auger electrons for electron impact in the energy range $T = 1$–50 keV. The experimental A_2 coefficients agree well with theoretical values calculated by Berezhko and Kabachnik and Omidvar using Hartree–Slater wave functions, but disagree with those values calculated with hydrogenic wave functions. The general energy dependence of the A_2 coefficient for electron impact ionization of all subshells studied so far (L_3 of Ar and Mg and M_4, M_5 of Kr) will be discussed.

1. Introduction

In contrast to the alignment of atoms after impact excitation,[1] it was only recently realized that inner-shell ionization by particle[2,3] or photon[4] impact in general also leads to an alignment of the ions. The alignment manifests itself through polarization or nonisotropic angular distribution of characteristic x-radiation[2,5-8] and nonisotropic angular distribution of Auger electrons.[2,3,9,10] After the first theoretical and experimental investigation of the angular distribution of Auger electrons[3,9,10] this method turned out to be a powerful tool for studies of the alignment after inner-shell ionization by electron[11-14] or ion impact.[13] Experimentally, systematic investigations over an electron impact energy range of about two orders

W. Sandner, M. Weber, and W. Mehlhorn • Fakultät für Physik, Albert-Ludwigs-Universität, D 7800 Freiburg im Breisgau, West Germany.

of magnitude provided accurate data for a comprehensive picture of the energy dependence of the alignment;[12,14] theoretically, the application of angular correlation formalisms elucidated which basic quantities can be determined by these experiments.[15,16] In particular it was shown that the high symmetry of the experiment (no detection of electron polarization, and no detection of scattered particles) reduces the set of independent parameters to $j + 1$ and $j + \frac{1}{2}$, when j, the total angular momentum of the intermediate ionic state, is integer and half integer, respectively. These parameters are most conveniently described by alignment tensors $\mathscr{A}_{k0}$ (k even), which can be expressed (with the normalization $\mathscr{A}_{00} = 1$) as linear combinations of magnetic substate cross sections $\sigma(nlj \mid m_j \mid)$, divided by the total shell cross section $\sigma(nlj)$. They can be measured via the angular distribution of selected Auger transitions, which is given by[3,16]

$$I(\theta) \propto 1 + \sum_{k=2,4,\ldots} A_k P_k(\cos \theta) \tag{1}$$

where P_k are Legendre polynomials of even order and A_k are the anisotropy coefficients. In the special case of the final ionic state A^{2+} having vanishing orbital angular momentum $L_f = 0$ (only one Auger partial wave is ejected), the A_k are proportional only to the alignment tensors $\mathscr{A}_{k0}$.

In this chapter we present our latest experiment of this type, the measurement of the A_2 coefficients of the M_4 and M_5 shell of krypton for electron impact energies from 1 to 50 keV. In addition, the high-energy behavior of the alignment, which is found to be similar for all systems that have been investigated so far, is discussed by connecting both the high-energy approximations for the cross sections and the well-known space transformation properties of the tensors $\mathscr{A}_{k0}$.

2. Experimental Procedure

In the experiment, which will be described elsewhere in detail,[14] we measured the angular distribution of the M_4- and M_5-$N_{2,3}N_{2,3}(^1S_0)$ Auger electrons of krypton. The coefficients A_2 and A_4, calculated from magnetic substate cross sections in the (nlm_l) scheme,[3] may be taken from Table 1. We note, first, that in the case of M_4-shell ionization with $j = \frac{3}{2}$ only one parameter, the coefficient A_2, is left [apart from the total shell cross section $\sigma(3d) \propto \mathscr{A}_{00}$, which is normalized to unity in equation (1)]; second, that both A_2 coefficients differ only by a constant factor $\frac{4}{5}$. Recent calculations predicted the A_4 coefficient of the M_5 shell to be less than 10^{-2} in this energy region,[16,17] which is below our experimental uncertainty. To eliminate the influence of this unobserved parameter, the Auger electron detector was positioned at those angles θ where $P_4(\cos \theta)$ vanishes ($\theta = 30.5°, 70.1°, 109.9°,$ and $149.5°$). The detector was a 30° parallel-plate electrostatic analyzer with 0.14% resolution, which could be rotated inside the μ-metal shielded vacuum chamber from $\theta = 15°$ to 165°.

Table 1. Anisotropy Coefficients A_2 and A_4 for the M_4 and M_5 Shell of Krypton, Calculated from Magnetic Substate Cross Sections $\sigma(nl \mid m_l \mid)$

$M_4 \ (3d, j = \frac{3}{2})$	$M_5 \ (3d, j = \frac{5}{2})$
$A_2 = \dfrac{\sigma(3d, 0) + \sigma(3d, 1) - 2\sigma(3d, 2)}{\sigma(3d, 0) + 2\sigma(3d, 1) + 2\sigma(3d, 2)}$	$A_2 = \dfrac{8}{7} \dfrac{\sigma(3d, 0) + \sigma(3d, 1) - 2\sigma(3d, 2)}{\sigma(3d, 0) + 2\sigma(3d, 1) + 2\sigma(3d, 2)}$
$A_4 = 0$	$A_4 = \dfrac{2}{7} \dfrac{3\sigma(3d, 0) - 4\sigma(3d, 1) + \sigma(3d, 2)}{\sigma(3d, 0) + 2\sigma(3d, 1) + 2\sigma(3d, 2)}$

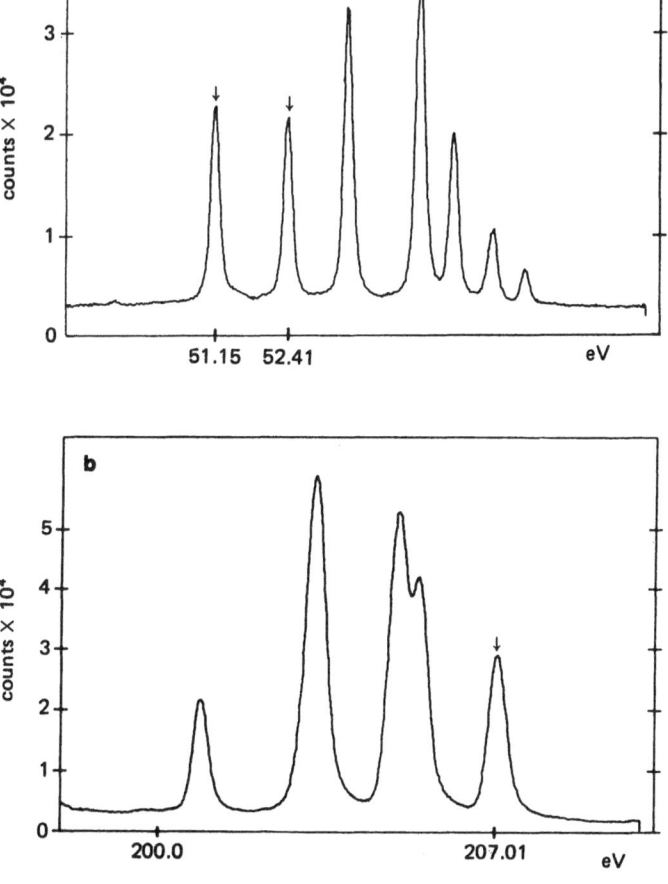

Figure 1. (a) $M_{4,5}$-$N_{2,3}N_{2,3}$ Auger spectrum of krypton and (b) $L_{2,3}$-$M_{2,3}M_{2,3}$ Auger spectrum of argon, obtained at $\theta = 150°$, $T = 2$ keV. The M_5- and M_4-$N_{2,3}N_{2,3}(^1S_0)$ lines of Kr (at 51.15 and 52.41 eV) and the isotropic L_2-$M_{2,3}M_{2,3}$ ($^3P_{0,1,2}$) line of argon (at 207.01 eV) are indicated.

The target gas consisted of a mixture of krypton and argon; the pressure ratio was about 80:20. This allowed the normalization of the measured krypton Auger intensities to the intensity of the isotropic argon L_2-$M_{2,3}M_{2,3}(^3P_{012})$ line to control the angular-dependent target volume variations. Figure 1 shows both the obtained $M_{4,5}$-$N_{2,3}N_{2,3}$ Auger spectrum of krypton and the $L_{2,3}$-$M_{2,3}M_{2,3}$ Auger spectrum of argon, with the Auger lines under consideration being specified. Great care was taken to determine the actual density ratio in the target region, which was found to depend on the vertical distance from the gas inlet capillary: A second parallel-plate analyzer, mounted at a fixed angle ($-90°$), acted as a monitor for this purpose. The target gas pressure was about 2×10^{-3} Torr; by careful centering of the gas inlet it could be achieved that the influence of the different total absorption of krypton Auger electrons and argon Auger electrons (with energies of about 50 and 200 eV, respectively) became negligible for the angular distribution measurement. The alignment has been measured at seven different impact energies between 1 and 50 keV.

3. Results and Discussion

3.1. Alignment of M_4 and M_5 Shell of Krypton

In figure 2, the evaluated A_2 values are plotted as a function of the incident electron energy, which is given in keV and in units of the $M_{4,5}$-shell binding energy

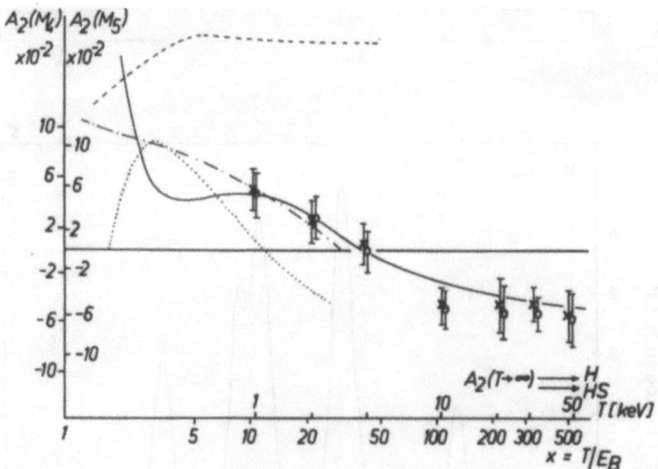

Figure 2. Experimental A_2 coefficient for the M_4 shell ($\bigcirc$) and M_5 shell ($\times$) of krypton.[14] The incident energy T is given in keV and in units of $M_{4,5}$-shell binding energy $E_B \approx 94$ eV, respectively. The A_2 scales for the M_5 and M_4 shells differ by a factor of 8/7; thus the theoretical A_2 curves, calculated in BA with hydrogenic (H) and Hartree–Slater (HS) wave functions, coincide. H ($---$) from Omidvar et al.;[18] HS ($-\cdot-$) from Omidvar;[17] and H ($\cdots$) and HS ($\longrightarrow$) from Berezhko and Kabachnik.[16] $A_2(T \to \infty)$ is the nonrelativistic high-energy limit, calculated with H and HS dipole matrix elements.

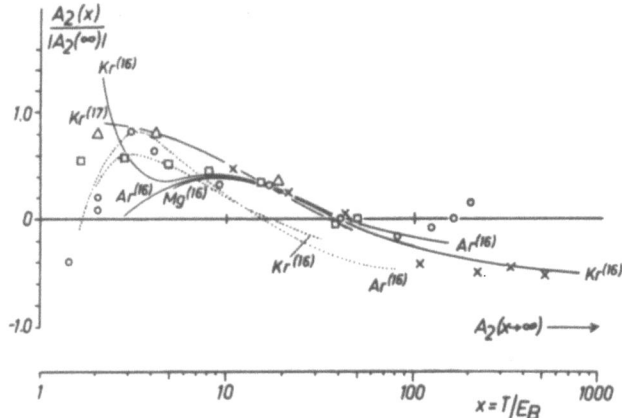

Figure 3. Experimental A_2 coefficients for argon L_3 shell[10,12] ($\triangle$, $\bigcirc$), krypton $M_{4,5}$ shell ($\times$), and magnesium L_3 shell[13] ($\square$), plotted as a function of incident energy T in units of the binding energy E_B. The A_2 values are normalized to the theoretical high-energy limit $|A_2(x \to \infty)|$. Calculated curves are obtained in BA with hydrogenic ($\cdots$) and Hartree–Slater (——) wave functions.

$E_B \approx 94$ eV, respectively. Two different scales are used for the A_2 coefficients for the M_4 shell and the M_5 shell in such a way (the M_4 scale is $\frac{7}{8}$ of the M_5 scale) that the plotted values of both coefficients should coincide. This is obviously reproduced by the experiment. Several calculated curves for the energy dependence of the alignment can be compared with our experimental results. Because they all have been obtained in first Born approximation (BA), they essentially reflect the influence of the different wave functions employed. In general, hydrogenlike (H) wave function calculations deviate from the experimental result; in particular they appear to be sensitive to the choice of the effective charge Z_{eff} of the continuum electron: the dashed line has been obtained with $Z_{eff} = 1$[18] and the dotted line with Z_{eff} being equal to the bound state Z_{eff},[16] which is about 19. Excellent agreement with the experimental results in the investigated energy region could be obtained with Hartree–Slater (HS) wave-function calculations[16,17] represented by the solid line and dash–dot line, respectively. Also it can be seen from Figure 2 that the nonrelativistic high-energy limit $A_2(T \to \infty)$, which is calculated using dipole matrix elements (H and HS) and assuming the momentum transfer $\mathbf{K}$ is perpendicular to the incident beam,[14] yields an A_2 value that is still considerably lower than the values at $T = 500E_B$ (50 keV).

3.2. High-Energy Behavior of Alignment

For a more comprehensive discussion of the high-energy behavior of the alignment following electron impact ionization, we present a picture where all experimental results that have been obtained so far are summarized together with the corresponding calculated curves (Figure 3). The A_2 coefficients of argon L_3 shell, krypton

$M_{4,5}$ shell, and magnesium L_3 shell are plotted as a function of incident electron energy x, given in units of the corresponding binding energy E_B. Both the theoretical and experimental A_2 coefficients are plotted as reduced values $A_2(x)/|A_2(x \to \infty)|$. For reduction of the experimental results we used the $A_2(x \to \infty)$ obtained with HS dipole matrix elements, yielding -0.06 (Ar L_3 shell), -0.075 (Mg L_3 shell), and -0.11 and -0.13 (Kr M_4 and M_5 shell), respectively. In the following, we want to focus our attention only on the energy region above $10E_B$, where the Born approximation is expected to be valid.

First we note that in fact in this energy region the BA calculations, employing HS wave functions, are in all cases in excellent agreement with the experiment (we disregard for the moment the deviation of the experimental A_2 values of Ar above $100E_B$, which will be discussed separately). The use of hydrogenlike wave functions, with equal Z_{eff} for bound and continuum states, reproduces the shape of the alignment curve, but it appears to be shifted by a factor of ~ 0.5 toward lower impact energies. Moreover, we find that in the reduced scale $A_2(x)/|A_2(x \to \infty)|$ both the experimental values and the HS wave-function calculations nearly coincide in one universal alignment curve, which is valid at least from $T = 10E_B$ to $100E_B$.

For a discussion of this feature we have to remember that in the Born approximation the basic quantity being calculated is the double differential cross section $d^2\sigma(nl\mu)/(dK\, dE)$, where K is the amount of momentum transfer, E the energy transfer, and μ the magnetic quantum number in a frame with $\mathbf{K}$ as quantization axis. Since in our noncoincidence experiments the quantization axis is the incident beam direction $\mathbf{Z}$ rather than the unobserved $\mathbf{K}$ axis, the double differential cross sections have to be transformed to the $\mathbf{Z}$ axis. If we construct a tensor from double differential cross sections, e.g., an alignment tensor $\mathscr{A}_{20}^K(K, E)$, its transformation becomes particularly simple and consists in multiplication by $P_2(\cos \lambda)$, where λ is the angle between the $\mathbf{K}$ and $\mathbf{Z}$ axes. The experimentally observed A_2 coefficient is then obtained by integration over all K and E[12,16]:

$$A_2 \propto \int dK\, dE\, \mathscr{A}_{20}^K(K, E)P_2(\cos \lambda) \tag{2}$$

From this we can write the reduced values $A_2(x)/|A_2(x \to \infty)|$ as

$$\frac{A_2(x)}{|A_2(x \to \infty)|} = \frac{A_2^K}{|A_2(x \to \infty)|} \frac{1}{2}(3\langle \cos^2 \lambda \rangle - 1) \tag{3}$$

Here $A_2^K \propto \int dK\, dE\, \mathscr{A}_{20}^K(K, E)$ and $\langle \cos^2 \lambda \rangle$ is the average of $\cos^2 \lambda$, weighted by $\mathscr{A}_{20}^K(K, E)$. Equation (3) predicts for nonrelativistic kinematics one zero crossing of $3\langle \cos^2 \lambda \rangle - 1$, because at threshold $\lambda = 0$ and for infinitely high energies $\lambda \to \pi/2$. This has already been pointed out by Fano and Macek[19] in connection with the alignment following electron impact excitation. In our case this kinematical zero crossing occurs at $T \approx 40E_B$. Thus additional zero crossings for $x < 40$ are due not to kinematical effects but to zeros of A_2^K. The observation that the reduced A_2

curves for the different shells (L_3 and $M_{4,5}$ shell) nearly coincide in the region of Born approximation is mainly caused by the fact that in equation (3) the weights $\mathscr{A}_{20}^K(K, E)$ of $\langle \cos^2 \lambda \rangle$ and thus the double differential cross sections $d^2\sigma(nl\mu)/(dK\, dE)$ must have a similar dependence of K and E (in units of E_B). This, in turn, may be traced back to the similar shape of the $2p$ and $3d$ radial wave functions of the atoms under consideration.

The relation between the energy dependence of the alignment and collision kinematics is most clearly seen in the region of the Bethe approximation, which is expected to be valid above several $10E_B$. To demonstrate this, we rewrite equation (3) using certain approximations,[12] which have been proved to be not very restrictive in the case of argon L_3-shell alignment, as follows:

$$\frac{A_2(x)}{|A_2(x \rightarrow \infty)|} \approx \frac{A_2^K}{|A_2(x \rightarrow \infty)|} \frac{1}{2} (3 \overline{\cos^2 \lambda} - 1) \tag{4}$$

where $\overline{\cos^2 \lambda}$ is now the average of $\cos^2 \lambda$ weighted by $d^2\sigma(nl)/(dK\, dE)$ and thus can be identified with the average direction of the momentum transfer occurring in an (nl)-shell ionization process. Then, in a final step we apply the Bethe approximation to the quantity A_2^K, which contains only integrated cross sections $\sigma_\mu(nl) = \int_{K,E} d^2\sigma(nl\mu)$ without transformation coefficients. Here the energy dependence of the cross sections $\sigma_\mu(nl)$ is mainly governed by the dipole contributions,[20] which allows, again under certain approximations, the factor A_2^K to be related to the high-energy limit $A_2(x \rightarrow \infty)$[12]:

$$A_2^K \approx -2A_2(x \rightarrow \infty) \tag{5}$$

Combining equations (4) and (5), we can immediately write the reduced A_2 coefficient as

$$\frac{A_2(x)}{|A_2(x \rightarrow \infty)|} \approx 3 \overline{\cos^2 \lambda} - 1 \tag{6}$$

With the result of equation (6) an interpretation of the high-energy part of the universal curve of Figure 3 is now possible. The reduced value $A_2(x)/|A_2(x \rightarrow \infty)|$ reflects for each impact energy x in the energy region of the Bethe approximation the averaged direction of the momentum transfer $\mathbf{K}$. As a consequence we find that even at such high impact energies as $500E_B$ the averaged momentum transfer is not perpendicular to the incident beam. In the case of argon L_3-shell alignment above $100E_B$, relativistic effects seem to shift the averaged direction of $\mathbf{K}$ back towards the incident beam direction.[12]

ACKNOWLEDGMENT

The authors wish to thank Drs. M. Rødbro, R. DuBois, and V. Schmidt for sending their experimental data prior to publication.

References

1. I. C. Percival and M. J. Seaton, *Phil. Trans. R. Soc. London Ser. A* **251**, 113–138 (1958). [For a summary of recent developments in this field see, for instance, H. Kleinpoppen, *Adv. Quantum Chem.* **10**, 77 (1977) and references therein.]
2. W. Mehlhorn, *Phys. Lett.* **26A**, 166–167 (1968).
3. B. Cleff and W. Mehlhorn, *J. Phys. B* **7**, 593–604 (1974).
4. S. Flügge, W. Mehlhorn, and V. Schmidt, *Phys. Rev. Lett.* **29**, 7–9 (1972).
5. S. C. McFarlane, *J. Phys. B* **5**, 1906–1915 (1972).
6. J. Hrdý, A. Henins, and J. A. Bearden, *Phys. Rev. A* **2**, 1708–1711 (1970).
7. K. A. Jamison, P. Richard, F. Hopkins, and D. L. Matthews, *Phys. Rev. A* **17**, 1642 (1978).
8. A. Schöler and F. Bell, *Z. Phys.* **A286**, 163–168 (1978).
9. B. Cleff and W. Mehlhorn, *Phys. Lett.* **37A**, 3–4 (1971).
10. B. Cleff and W. Mehlhorn, *J. Phys. B* **7**, 605–611 (1974).
11. E. Döbelin, W. Sandner, and W. Mehlhorn, *Phys. Lett.* **49A**, 7–8 (1974).
12. W. Sandner and W. Schmitt, *J. Phys. B* **11**, 1833–1848 (1978).
13. M. Rødbro, R. DuBois, and V. Schmidt *J. Phys. B* **11**, L551 (1978).
14. M. Weber, W. Sandner, and W. Mehlhorn (to be published).
15. J. Eichler and W. Fritsch, *J. Phys. B* **9**, 1477–1489 (1976).
16. E. G. Berezhko and N. M. Kabachnik, *J. Phys. B* **10**, 2467–2477 (1977); and private communication.
17. K. Omidvar, *J. Phys. B* **10**, L55–L61 (1977).
18. K. Omidvar, H. L. Kyle, and E. C. Sullivan, *Phys. Rev. A* **5**, 1174–1187 (1972); and private communication.
19. U. Fano and J. H. Macek, *Rev. Mod. Phys.* **45**, 533–573 (1973).
20. M. Inokuti, *Rev. Mod. Phys.* **43**, 297 (1971).

Identification of States Excited by Collision: Current State of Theory

U. Fano

Circumstances that have hindered the theoretical analysis of collision experiments are illustrated, with reference to the sorting out of geometrical and dynamical aspects of each process. It is pointed out how stationary states of a target are superposed coherently immediately following a collision but become progressively incoherent as the partition of total energy between target and scattered particle becomes increasingly precise. Recent progress in the geometrical characterization of target states is outlined with reference to still unpublished work of G. Gabrielse and T. Baer.

1. Introduction

The task of theory in atomic collision physics has been expanded greatly by the steadily increasing wealth of detailed experimental data. Experiments now provide joint distributions—in time, angle, and polarization—of pairs of radiations or particles ejected by collision products. Such sophisticated collection of data has great importance to the extent that particular details may be linked specifically and sensitively to particular dynamical effects; examples of such close links are emerging.

The full characterization of the intermediate states of collision products has a central role in this correlation of observations with the underlying dynamics, as indicated by the title of this chapter. More specifically, however, this chapter reduces to two groups of comments concerning, respectively, general aspects of the theory and particular developments that have occurred recently within my range of vision.

U. FANO ● University of Chicago, Chicago, Illinois 60637, U.S.A.

The general comments are meant to illustrate circumstances that have made the progress of theory slow and cumbersome; the particular examples might then show how advances have resulted by formulating each step of the theory in terms of observable parameters.

2. General Aspects of Theory

Theory proceeds along two opposite but converging approaches, the *ab initio* calculation of observables and the extraction of significant parameters from experimental results. In either approach collision theory sorts out truly dynamical parameters from purely geometrical elements (such as directions of momenta, of spins, etc.). The ingredients of this analysis have been well known, in essence, since about 1950; nevertheless their intermixing in a bewildering variety of ways still chokes the progress of seemingly straightforward developments.

The *geometrical aspects* of collisions are treated by Wigner–Racah algebra, whose elements are widely known. They are often analyzed in terms of *multipole moments* of atoms, molecules, or radiations. However, two major circumstances have complicated matters, particularly by preventing the standardization of analytical procedures:

(a) The number of geometrical elements to be combined is *large*, the more so as each of these elements appears twice in quantum calculations, namely, in each probability amplitude and again in its complex conjugate.

(b) Numerous noncommuting symmetries (e.g., under rotations and reflections) prove relevant concurrently. A continuing effort has thus been required to readapt general procedures in order to deal properly with newer combinations of geometrical elements. The bothersome occurrence of alternative phase conventions may well be an artifact of failure to consider all relevant symmetries explicitly, e.g., in the classification of a base set of states. Indeed the complete description of a physical phenomenon should be made *unique*, by specifying an adequate set of boundary and subsidiary conditions, as contrasted to fragmentary mathematical statements that may leave their substrate largely unspecified.

Once the geometrical elements have been separated out, the *dynamical parameters* of theory are invariant under space rotations and inversions, in keeping with the invariance of physical laws. Typically these parameters are expressed as the "reduced matrix elements" of operators, which emerge from application of the Wigner–Eckart theorem. They can also be classified according to parity under time reversal, a symmetry that has not been exploited fully even though knowledge of its importance is ancient. Note that the very structure of the time evolution operator $\exp(i \int V \, dt)$ implies that sign reversal of an interaction V amounts to reversal of the time variable; accordingly analysis of the time dependence of experiments should ascertain the sign of interactions. The evaluation of dynamical parameters appears as the appropriate meeting point of phenomenologic analysis of measurements and of *ab initio* theory. This evaluation has been progressing only slowly, which is not surprising in view of the increasing complexity of collision processes.

Note finally how the historical development of quantum theory may have obscured important aspects of collision physics. The state of an atomic target immediately after a collision is actually represented by a coherent superposition of all of its stationary states that are consistent with conservation laws. Incoherence between these component states—intuitively expected as the normal outcome—results only eventually, from the partitioning of the total energy between the target and the scattered particle. This partitioning, in fact, sharpens itself only gradually in the course of time; it displays an inherent residual uncertainty ΔE, which decreases inversely to the time interval Δt after the collision. States of different configurations thus generally become incoherent after a short time; incoherence then affects different terms of the same configuration, and finally individual levels of the fine and hyperfine structures.

3. Examples of Geometrical Analysis

In a process of spontaneous dipole emission the observable angular and polarization distributions of the radiation depend on expressions that are quadratic in the transition dipole moment of the emitter. These expressions characterize, in turn, the electric quadrupole moment ("alignment") and the magnetic dipole moment ("orientation") of the emitter's initial excited state of orbital motion. The explicit relation between these parameters of the emitter and the distribution of radiation has been described by Fano and Macek in 1973,[1] but only for initial states that are eigenstates of the squared angular momentum. (The subsequent removal of this restriction is mentioned below.) Observations of the radiation thus determine two tensorial parameters of the emitter's state at the time of emission. Variations of these parameters as functions of the time elapsed between an exciting collision and the emission process constitute the "quantum beats." The multipole moments of a free-standing emitter vary independently, but they become interrelated in the presence of external fields. Fano and Macek stressed how in this case the experimental study of the time dependence of emitted radiation may serve to measure the values of many additional parameters of the emitter at earlier times. Interest in this analysis is just beginning to develop.

The Fano–Macek treatment of dipole emission may serve as a prototype for the treatment of more general collision processes. A complication arises from the numerous alternative values of the orbital momentum of colliding particles as contrasted to the single small value ($j_r = 1$) that occurs for dipole emission. This circumstance has prevented rapid fruition of a recent start by Kohmoto and Fano,[2] but it is mitigated by restrictions on the magnitude of the angular momentum *transferred* to the target ($\mathbf{j}_t$), which are inherent to the internal mechanics of the target and

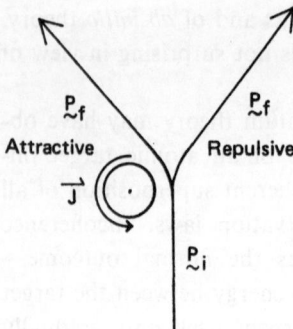

Figure 1. Diagram showing how attractive or repulsive forces lead to collisions with opposite values of $(\mathbf{p}_i \times \mathbf{p}_f) \cdot \mathbf{J}$.

are often quite sharp $[j_t = 0(1)]$. A formalism designed to take advantage of these restrictions has now been developed[3] but remains to be applied. The initial aim of Kohmoto and Fano was to identify the sign relationship between the orientation of the magnetic dipole moment induced in a target by collision, the deflection of a scattered particle, and the effective interaction between particle and target (Figure 1). They succeeded only in showing that the orientation is reversed by a sign reversal of the interaction.[2]

This sign relationship—mediated by the sign of phase shifts—belongs to a class of symmetry effects that have been noted repeatedly in collision physics. As a prototype example one may recall the early calculations of the joint contributions to nuclear γ-ray emission by magnetic dipole and electric quadrupole processes; their respective matrix elements might well be complex but their relative phases are necessarily ± 1.

An analysis that appears to sift out such effects systematically has been developed quite recently by Gabrielse[4] in a broader context. Gabrielse's initial aim was to extend the Fano–Macek treatment to dipole emissions excited by e-H collisions and hence to remove the restriction to emission by excited states with specified angular momentum. To this end he had to identify the emitter's excited state by an extended set of multipole moments, e.g., by the expectation values of dipole matrix elements $(lm \mid \mathbf{r} \mid l'm')$ off-diagonal as well as diagonal in the orbital quantum numbers (l, l'). Following Lombardi's lead,[5] he focused on *real* components of multipole moments, defined as the mean values of *Hermitian* tensorial operators and hence directly observable, with a definite parity under the time reversal operator K. These components are joint eigenvectors of the four symmetry operators J^2, J_z^2, $R_y(180°)$, and K. Further, he adapted the Wigner–Racah methods so as to deal throughout with expressions that are eigenvectors of this set of operators.

Thereby Gabrielse's approach extends the familiar use of invariance under space rotations—which leads, e.g., to the combination rules of angular momenta and of multipole moments—to invariance under a maximal set of rotation and *reflection* operations *in space and time*. The study of symmetries under reflections of the time variable as well as of space variables and of their several *combinations* can thus be conducted more systematically than in the past. Collisions, whose dynamics is invariant under time reversal, can nevertheless yield parameters odd under this operation because their observational frame is itself nonsymmetric, being designed to detect differences between the initial and final states of the reactants. These odd parameters depend on sine, rather than on cosine, functions either of the time elapsed since a collision or of differences between phase shifts. Observation of a sine dependence on the time between collision and light emission is the earmark of a time-odd parameter. Conversely the measurement of a time-odd parameter serves a sensitive indicator of phase differences.

Gabrielse has stressed that the *product* of the parities under time reversal, K, and under space inversion, π, is most relevant to the analysis of collisions because both the initial and final momentum eigenstates of the colliding particle are even under $K\pi$. Collision parameters that are odd under time reversal and even under space inversion of the *whole process* thus reveal a sign reversal of the *target's $K\pi$*. This argument explains a connection noted in Reference 2 between the orientation of a target, which is odd under $K\pi$, and the differences of phase shifts. Similarly, in the analysis of the quantum beats of light emission following a collision, the coefficient of a sine component measures a process in which the target's parity under $K\pi$ is reversed, such are the production of an electric dipole moment of orbital motion or of a spin orientation in a previously unoriented target.

The Fano–Macek treatment has now been extended by Tom Baer in a new direction, namely, to deal with the *stimulated* emission of dipole radiation in the context of photon echo experiments.[6] In the case of stimulated emission, the observed distribution of light relates directly to the transition dipole moment of the emitter rather than to an expression quadratic in this moment. The more direct character of this relation stems from the constructive interference of the radiation emitted by different atoms of a sample; it also reflects the circumstance that both the initial and the final state of each elementary emission process are represented by a coherent superposition of the upper and lower levels of the transition. Observation of the angular and polarization distributions of the emitted light thus determines the electric dipole moment of the gas sample in its state of coherent oscillation.

Here, as in the spontaneous emission by atoms excited by collision in an external field, the value of the dipole moment—or of any other multipole moment—of the emitter at the time of observation depends generally on the values of all multipole moments that had been generated at earlier times in the course of excitation. In a photon echo experiment, the excitation involves exposure to two intense laser pulses. The repeated processes of dipole absorption and emission during each pulse establish nonzero values of 2^k-pole moments with all values of $k \leq j_1 + j_2$, where j_1 and j_2 are the angular momenta of the upper and lower levels of the transition. The magnitude of each of the 2^k-pole moments depends on the "area" of the exciting pulse in a novel manner which has also been studied by Baer. Both the 2^k-pole moments generated by the first pulse and those generated by the second contribute to the eventual dipole moment which manifests itself through the echo emission.

ACKNOWLEDGMENT

This work was supported by the U.S. Department of Energy, Division of Basic Energy Sciences under contract No. COO-1674-147.

References

1. U. Fano and J. H. Macek, *Rev. Mod. Phys.* **45**, 553–573 (1973).
2. M. Kohmoto and U. Fano, *Tenth International Conference on the Physics of Electronic and Atomic Collisions, Abstract of Papers*, p. 516, Commissariat a L'Energie Atomique, Paris (1977).
3. M. Kohmoto and U. Fano, unpublished.
4. G. Gabrielse, submitted to *Phys. Rev. A* (1979).
5. M. Lombardi, *J. Phys.* (*Paris*) **30**, 631–642 (1969); M. Giroud, M. Lombardi, and J. C. Pebay-Peyroula, *J. Phys.* (*Paris*) **30**, 789–794 (1969); M. Lombardi, thesis, Grenoble (1969); A. Omont, *Prog. Quantum Electron.* **5**, 69–138 (1977).
6. T. Baer, *Phys. Rev. A* **18**, 2570 (1978).

19

Angular Correlation between Autoionization Electron and Scattered Ion in Slow He⁺–He Collisions

R. Morgenstern and A. Niehaus

We have investigated the excitation of autoionization states in He⁺/He collisions at impact energies between 1 and 6 keV by energy analysis of electrons, ejected at $\vartheta = 180°$, and by measurement of coincidences between electrons, ejected at $\vartheta = 135°$ and different azimuthal angles ϕ, and ions, scattered into $\theta = 6°$. This yields complex population amplitudes for magnetic sublevels of the excited states. It turns out that mainly the He(2p²)¹D state is excited, and that the measured electron distribution is nearly rotationally symmetric around a direction that coincides with the internuclear axis of the collision partners at the distance of closest approach. If this axis is taken as the quantization axis, the m = 0 sublevel is nearly exclusively populated. This implies the following interpretation: near the distance of closest approach the electron cloud is "blown up" by the electron promotion via the $2p\sigma_u$ orbital. Since the internuclear distance is small at this instant the electron cloud does not follow the rotation of the internuclear axis but stays fixed in space. In a rotating frame this corresponds to a $2p\sigma$–$2p\pi$ rotational coupling.

1. Introduction

The analysis of electrons ejected from collisionally excited autoionizing atoms can yield detailed information about the excitation mechanism. Whereas a pure

R. Morgenstern and A. Niehaus ● Rijksuniversiteit Utrecht, Fysisch Laboratorium, Princetonplein 5, Postbus 80000, NL 3508 TA Utrecht, The Netherlands.

energy analysis can only answer the question of which autoionizing states were excited, the determination of angular correlations between electrons and scattered projectiles may lead to a description of the excited atoms in terms of a density matrix or of complex population amplitudes for magnetic sublevels. This allows a more detailed comparison with theoretical excitation models.

At impact energies of only a few electron kilovolts the ion–atom collisions may be called slow in the sense that a description of the system in terms of potential curves is justified. The projectile velocity is too small for a direct excitation or ionization of the electrons by the Coulomb interaction. Therefore the excitation of higher states is due to transitions at potential curve crossings. The Fano–Lichten promotion model,[1] which is based on electron correlation diagrams, is expected to provide at least a qualitative understanding of the occurring excitations. The experimental results will therefore be compared with this model. We have chosen the He+/He collision system for the investigation of angular correlations. This has several advantages: (i) The autoionizing states of He have been determined by various methods and can be regarded as well known.[2] (ii) The superexcited He** atoms autoionize into He+$(1s)^2S$, i.e., into an isotropic state. This facilitates the interpretation of the measured angular correlations. (iii) Only a few He** states are excited in He+/He collisions at impact energies below 10 keV. As will be shown below, mainly the states He**$(2p^2)^1D$ and He**$(2s2p)^1P$ contribute to the measured spectra. (iv) The correlation diagram for the He+/He system looks relatively simple. Excitation of autoionizing states is due to promotion of two electrons via the $2p\sigma_u$ orbital. It is generally assumed that the $2p\sigma_u$–$2p\pi_u$ rotational coupling is the main mechanism leading to excited electrons in this system. Gerber et al.[3] discussed in more detail the processes leading to excitation of the two states ^1D and ^1P mentioned above. For all these reasons one expects the He+/He collision system to be the most suitable one for the experiments and the comparison with theory.

Unfortunately there are some facts that complicate the evaluation of the measurements. Figure 1 shows a noncoincident energy spectrum of electrons from 6 keV He+/He collisions, ejected at $\vartheta = 135°$ with respect to the beam direction. One observes mainly two groups of electrons. The one around 35 eV is due to excitation and decay of the target atom, and the one near 28 eV is due to emission from the fast projectile excited in a charge exchange process. The same states are excited in target

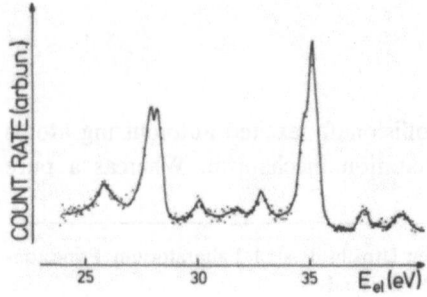

Figure 1. Energy spectrum of electrons, ejected at $\vartheta = 135°$ in 6-keV He+/He collisions. Two groups of electrons are observed, due to emission from the target (around 35 eV) and from the charge exchanged projectile (around 28 eV). The two prominent peaks contain unresolved contributions of He$(2p^2)^1D$ and He$(2s2p)^1P$.

and projectile, but electrons from the projectile suffer a large energy shift due to the Doppler effect. The large peaks contain contributions from both He**$(2p^2)^1D$ and He**$(2s2p)^1P$, with nominal energies of 35.30 eV and 35.54 eV, respectively. Although the spectrometer resolution was sufficient to resolve two peaks that are 240 meV apart, the contributions of these states are not resolved. This is mainly due to the so-called "postcollision interaction (PCI)": since the lifetimes of the autoionization states are very short ($\sim 10^{-15}$ sec), the electron emission takes place in the Coulomb field of the slowly receding collision partner. Therefore instead of a well-localized narrow peak a broad and shifted electron peak results from autoionization of one state. In addition, the Doppler effect causes—due to the small He mass—considerable shifts and broadenings of the electron peaks, even at the low recoil energies of the target atoms. For these reasons it is not sufficient to measure angular correlations between ions and electrons at a certain electron energy, but one has to measure coincident electron energy spectra from which by a careful analysis the contributions of the various autoionizing states can be deduced.

We have performed two types of measurements:

(a) Noncoincident energy analysis of electrons, emitted in the backward direction, i.e., at $\vartheta = 180°$ with respect to the He⁺ beam. At this angle the Doppler broadening of electron peaks can nearly be neglected. The resulting electron spectra look very complicated owing to interferences of contributions from different states caused by the PCI effect. The successful analysis of these spectra, which yields population amplitudes of the contributing states and their relative phase, proved the validity of our semiclassical description of the PCI effect, which also complicates the coincidence electron spectra and their analysis in terms of angular correlations.

(b) Measurement of electron energy spectra at different ejection angles in coincidence with scattered ions. Using the knowledge obtained from the 180° spectra we are able to analyze these coincident energy spectra and to determine the contributions of ^1D and ^1P state to the spectra at the different angles. From this we obtain the angular correlations between the scattered ions and electrons ejected from one single autoionizing state.

2. Noncoincident Electron Spectra at $\vartheta = 180°$

A He⁺ beam of variable energy collides on a thermal He gas target at single collision conditions. The main parts of the experimental setup were described earlier.[4] The electrons ejected at $\vartheta = 180°$ are analyzed by a double hemispherical analyzer the first sphere of which is passed by the He⁺ beam. Figure 2 shows electron energy spectra at various collision energies.[5] At high impact energies one can clearly distinguish electrons ejected from the target atom around 35 eV and electrons ejected from the projectile, shifted to lower energies by the Doppler effect, depending on the projectile velocity. At low collision energies, however, contributions from target and

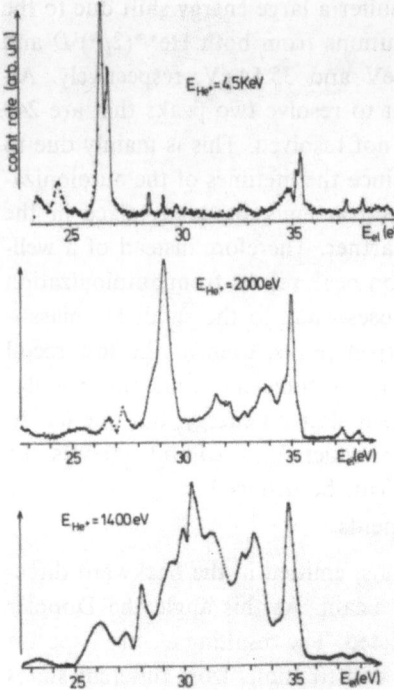

Figure 2. Energy spectra of electrons emitted in the backward direction ($\vartheta = 180°$) in He⁺/He collisions. At high collision energies one observes well-separated peaks which can be ascribed to various autoionization states. At low collision energies peaks are broadened and new structures occur which are due to interferences of different contributions.

projectile seem to overlap and there are more peaks in the spectra than possible autoionizing states in He. Morgenstern *et al.*[6] have explained these structures quantitatively by a semiclassical model for the description of PCI effects. A potential curve diagram as used in this model is shown in Figure 3. Excitation is assumed to occur at time $t = 0$ at crossings of the potential curves for the initial state and the excited states with energies E_n and $E_{n'}$. The potential energy for the excited system is assumed to be independent of the internuclear distance, and for the final state

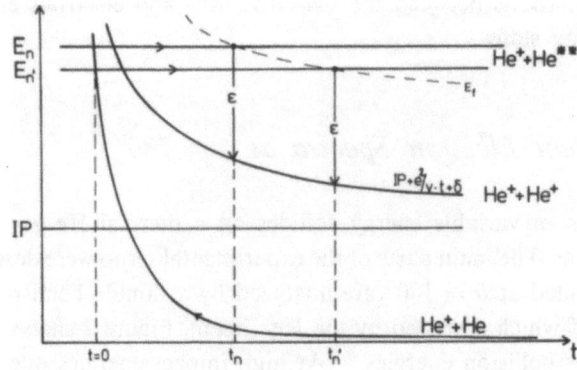

Figure 3. Potential curve diagram used to describe excitation and decay of autoionizing states in a semiclassical model. Transitions from state E_n at time t_n and from E_n' at t_n' lead to the same electron energies and may interfere, if the autoionization states are populated coherently.

He$^+$ + He$^+$ a pure Coulomb curve is assumed. If electrons of energy ε are detected, they may be due to autoionizing transitions from state E_n at time t_n or from state $E_{n'}$ at time $t_{n'}$. Classically one would add the corresponding intensities, but in a semiclassical description one has to add the corresponding transition amplitudes coherently, if the excitation of the autoionization states occurred coherently. Such a coherence is assumed in the model and therefore the excited-atom wave function is represented as a coherent mixture of autoionization state wave functions:

$$\psi_i(t) = \sum_n a_n(t)\psi_n \exp(-iE_n t) \tag{1}$$

Assuming an exponential decay of the autoionizing states with lifetime τ_n it is shown in Reference 6 that the spectral intensity $P(\varepsilon)$ of the electrons is given by

$$P(\varepsilon) = \left| \sum_n \frac{a_n(0) \exp[-\varepsilon/2v\tau_n \varepsilon_0{}^n(\varepsilon_0{}^n - \varepsilon)]}{(v\tau_n)^{1/2}(\varepsilon_0{}^n - \varepsilon)} \right.$$
$$\left. \times \exp\{-iv^{-1}[1 - (\varepsilon_0{}^n - \varepsilon)/E_a + \ln(\varepsilon_0{}^n - \varepsilon)/E_a]\} \right|^2$$
$$= \left| \sum_n C_n(\varepsilon_0{}^n, v, \varepsilon) \exp[-i\alpha_n(\varepsilon_0{}^n, v, \varepsilon)] \right|^2 \tag{2}$$

Here $a_n(0)$ are the initial population amplitudes of the autoionizing states, $\varepsilon_0{}^n = E_n - \text{IP}$ (IP is the ionization potential of the atom) are the "nominal" energies the electrons would have in case of ejection at infinite separation of the collision partners, v is the relative velocity of the collision partners, and E_a is an (unknown) energy, related to the internuclear distance δ at which the excitation takes place by $E_a = \delta^{-1}$.

Formula (2) is valid for electron emission from an atom at rest in the laboratory frame. It has to be modified for a quantitative comparison with the experimental spectra, since in our experiment the electrons are emitted from moving atoms—either from the fast projectile or from the slowly moving target atom. If the Doppler shifts of electron energies are designated by $D_s = \varepsilon_s - \varepsilon_{\text{lab}}$ and $D_f = \varepsilon_f - \varepsilon_{\text{lab}}$ for the cases of electron emission from the slow and the fast atoms, respectively, the Doppler shift is correctly taken into account when the electron energy ε in formula (2) is replaced by $\varepsilon_{\text{lab}} + D_s$ and $\varepsilon_{\text{lab}} + D_f$, respectively. D_s and D_f are known functions of the electron laboratory energy so that the spectral electron intensity is given by

$$P(\varepsilon_{\text{lab}}) = \left| \sum_n \{C_n{}^s(\varepsilon_0{}^n, v, \varepsilon_{\text{lab}} + D_s) \exp[-i\alpha_n{}^s(\varepsilon_0{}^n, v, \varepsilon_{\text{lab}} + D_s)] \right.$$
$$\left. + C_n{}^f(\varepsilon_0{}^n, v, \varepsilon_{\text{lab}} + D_f) \exp[-i\alpha_n{}^f(\varepsilon_0{}^n, v, \varepsilon_{\text{lab}} + D_f)]\} \right|^2 \tag{3}$$

In the case of the spectrum arising from He$^+$/He collisions at 1.4 keV we tried to reproduce the experimental spectrum with formula (2) by taking moduli and phases of the initial population amplitudes $a_n(0)$ as fit parameters. Three autoionizing states were taken into account: $(2s^2)^1S$, $(2p^2)^1D$, and $(2s2p)^1P$ with "nominal" electron energies of 33.23, 35.30, and 35.54 eV, respectively. The result is shown in Figure 4,

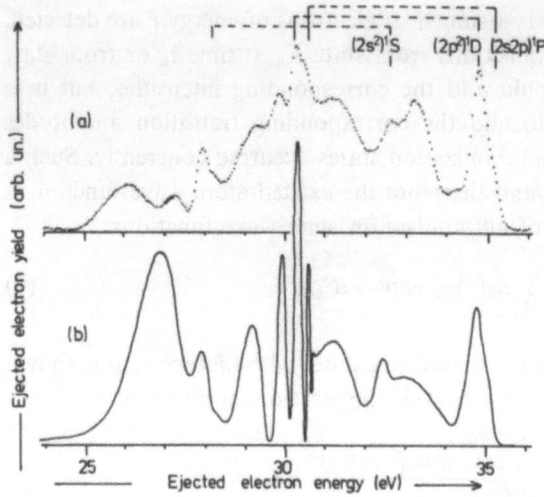

Figure 4. Energy spectrum of electrons from He$^+$/He collisions at 1400 eV, ejected at $\vartheta = 180°$. (a) Experiment; (b) calculated spectrum using formula (3) with fit parameters given in Table 1.

the solid line being the fit curve to the experimental points. The parameters used are given in Table 1.

There are some discrepancies between the experimental and the calculated spectrum. These are discussed in Reference 6. However, it is important to realize that there are two types of interferences in the theoretical spectrum, which are in fact observed also in the measurement: (i) Interferences between contributions from different states. The broad peaks at 31.2 and 33.2 eV, for example, are due to such interferences. (ii) Interferences between contributions from the slow and from the fast particle, respectively. The fast oscillations near 30 eV, for example, are due to this type of interference.

The agreement between experimental and theoretical spectrum in this respect justifies the assumption of coherence of the various excitations which was made in the model. Such a coherence can be understood by assuming that all excitations are caused by closely spaced curve crossings of only one molecular potential curve with

Table 1. Population Amplitudes for Different States, Used as Fit Parameters to Calculate the Spectrum of Figure 4b[a]

State	$\mid a_n(0) \mid$	χ_n^s/π	χ_n^f/π	τ_n (a.u.)
^1S	1	0.5	0.75	60
^1D	4	0.5	0.75	80
^1P	2	0	0.25	200

[a] $\mid a_n(0) \mid$ are the moduli and χ_n^s and χ_n^f are the phases—different for slow and fast particles—of the amplitudes for the different states. τ_n are the autoionization lifetimes used in the calculation.

the curves corresponding to the excited atomic states at infinite internuclear separation. In the diabatic correlation diagram for He⁺ + He there is in fact only one curve leading to doubly excited states, namely, the $(1s\sigma_g)(2p\sigma_u{}^2)^2\Sigma_g$ curve.

In the case of the coincidence electron spectra the PCI analysis need only be carried out in the region of electron energies around the nominal energies for the ^1P and ^1D state. Since these parts of the spectra correspond to ionization at large separation $(R \sim \Delta\varepsilon^{-1})$, where the assumptions made in our description of PCI should be valid, one may conclude that the PCI analysis does not introduce a severe error into the analysis of the coincidence spectra to be discussed below.

3. Electron Energy Spectra in Coincidence with Scattered Ions

For a determination of angular correlations we measured energy spectra of electrons, emitted at different angles, in coincidence with ions scattered through $\theta = 6°$. We chose to keep the electron ejection angle $\vartheta = 135°$ with respect to the beam axis constant, and to vary the azimuthal angle ϕ of electron ejection with respect to the scattering plane. This has two advantages: (i) At $\vartheta = 135°$ we can observe the energy spectra of electrons from the target undisturbed by electrons from the projectile, since the latter have much lower energy. (ii) Because of the rotational symmetry there is no change in the fraction of the scattering volume "seen" by the detectors when the observation angle is changed.

The electrons are energy analyzed in an electrostatic hemispherical condensor at $\vartheta = 135°$. The projectiles scattered through $\theta = 6°$ are energy analyzed in an electrostatic analyzer described by Gerber et al.,[7] which consists mainly of two unipotential lenses and has a rotational symmetry around the beam axis. An aperture with a sector opening of $\Delta\phi = 15°$, which is rotatable around the beam axis, allows only those ions to be analyzed that are scattered into a certain azimuthal angle $\phi \pm 7.5°$. The plane, containing beam and electron detector, defines $\phi = 0°$. Using an electronic setup similar to that described by van der Wiel and Wiebes,[8] the electron energy spectra are measured in coincidence with scattered ions, which suffered an inelastic energy loss of $Q = 60$ eV, corresponding to an excitation of the autoionization states. First results of these measurements were reported by Kessel et al.[9]

Figure 5 shows energy spectra of electrons resulting from 2-keV He⁺/He collisions ejected at $\vartheta = 135°$. The one on top is a noncoincident spectrum which shows again two groups of electrons due to emission from the target and—at lower electron energies—from the projectile. In the coincident spectra at $\phi = 0°$, $90°$, and $180°$ only electrons from the target are analyzed in the energy range where the ^1D and ^1P state contribute to the spectra. The coincident spectra differ in three respects from each other: (i) they appear at slightly different energies; (ii) the intensities, which are normalized to the number of scattered ions, are much different; and (iii) the shape of the spectra varies with the azimuthal angle. The different energies are caused by different Doppler shifts for different directions of the recoil velocities at

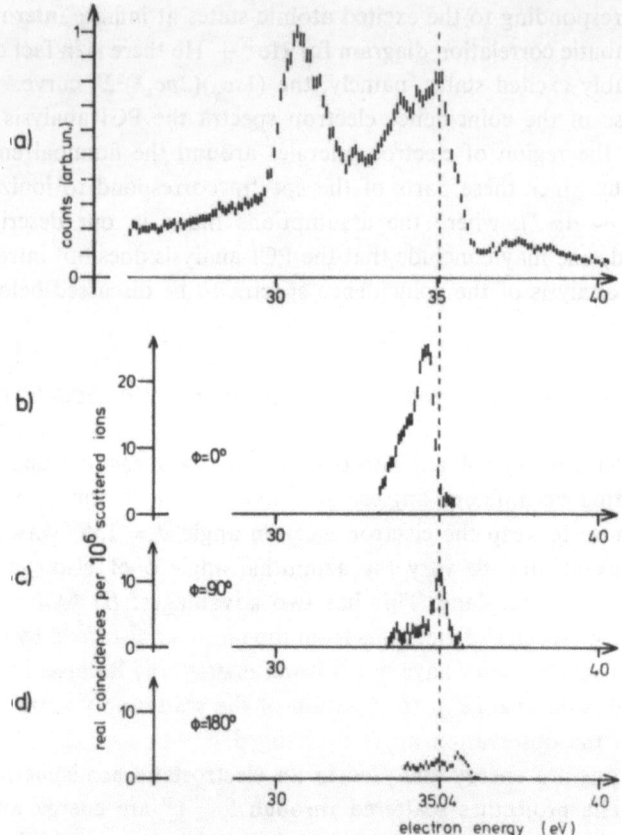

Figure 5. Measured energy spectra of electrons from He$^+$/He collisions at 2000 eV, ejected at
$\vartheta = 135°$. (a) Noncoincident spectrum; (b)–(d) energy spectra measured in coincidence with
ions scattered into $\theta = 6°$ and azimuthal angles of $\phi = 0°$, 90°, and 180°, respectively.

different orientations of the scattering plane. They can easily be calculated. The
variation of intensities is due to a deviation of the angular distribution of electrons
from cylindrical symmetry with respect to the beam axis. As demonstrated in the 180°
measurements, the shapes of the energy spectra are due to interferences between
autoionizing transitions from different states. Since the various contributions are
dependent on the emission angle, one has to expect different shapes of the energy
spectra.

For a quantitative description of these spectra one has to know the angular
dependence of electron emission on the parameters that describe the excited atom.
If one assumes a completely coherent excitation of pure states during the collision,
the angular dependence of electron emission can be described in terms of complex
population amplitudes for magnetic sublevels. Such a description was developed in
our group.[10] For a comparison with the experiments we represent the excited-atom
wave function as a coherent superposition of ^1D and ^1P magnetic substates with

population amplitudes d_m and p_m, respectively:

$$\psi_i = \sum_{m=-2}^{2} d_m \psi_{D,m} + \sum_{m=-1}^{1} p_m \psi_{P,m} \tag{4}$$

The final state is represented by the product of a wave function for He+(1s) and a plane wave for the ejected electron with wave vector **k**:

$$\psi_f = \psi_{\text{ion}} e^{i\mathbf{k} \cdot \mathbf{r}} \tag{5}$$

The transition amplitude for autoionization is given by the matrix element

$$M = \langle \psi_f \, | \, V_c \, | \, \psi_i \rangle \tag{6}$$

with V_c being the transition operator for autoionization. With the usual approximation of V_c as a scalar operator and the relation

$$a_m = (-1)^m a_{-m} \tag{7}$$

for the population amplitudes, which follows from reflection symmetry of the interaction with respect to the scattering plane, the matrix element becomes

$$M = M_D + M_P$$

with

$$M_D = 5^{-1/2}[\tfrac{1}{2} \, | \, d_0 \, | \, \exp(i\varrho_0)(3\cos^2\vartheta - 1) - (\tfrac{3}{2})^{1/2} \, | \, d_1 \, | \, \exp(i\varrho_1)\sin 2\vartheta \cos\phi$$
$$+ (\tfrac{3}{2})^{1/2} \, | \, d_2 \, | \, \exp(i\varrho_2)\sin^2\vartheta \cos 2\phi] \tag{8}$$

$$M_P = 3^{-1/2}[| \, p_0 \, | \, \exp(i\chi_0)\cos\vartheta - 2^{1/2} \, | \, p_1 \, | \, \exp(i\chi_1)\sin\vartheta \cos\phi]$$

Here $| \, d_m \, |$, ϱ_m and $| \, p_m \, |$, χ_m are moduli and phases of the complex population amplitudes. $| \, M_D \, |^2$ and $| \, M_P \, |^2$ would represent the angular distribution of ejected electrons if only the D state or the P state were excited. It is worth noting that the angular intensity distribution of the outgoing electron is described by the spherical harmonics corresponding to the angular momentum of the autoionizing state. This is so because both the operator causing the autoionization and the final ion state have no angular dependence.

Eichler and Fritsch provided a more general theory of angular correlations[11] which includes spin effects. They do not assume a coherent excitation of pure states but describe the excited system by a density matrix. In our experiment, however, it may be justified to neglect spin forces in the excitation processes between the He target and the He+ projectile. Then the formulas of the general theory reduce to those of our description as given in Equation (8).

In our derivation of a formula for the angular-dependent electron spectra we neglect any possible alterations of the angular distribution due to PCI effects, because in the electron energy range where we determine the angular distributions the changes

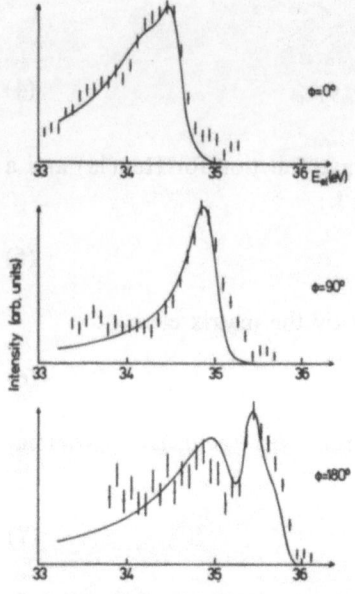

Figure 6. Experimental energy spectra of electrons measured in coincidence with scattered ions together with the fit curves (solid lines) which were calculated with formula (2) modified as described in Section 3. Fit parameters are given in Table 2.

of electron energies due to PCI are small ($<5\%$). We then arrive at a proper formula for the observed angular-dependent electron spectra if in equation (2) the initial population amplitudes $a_n(0)$ are replaced by the angular-dependent quantities M_D and M_P from equation (8).

With formula (2) modified in the way described one should be able to describe the measured spectra quantitatively. We have carried out fit calculations by taking the moduli and phases of population amplitudes for the ^1D and ^1P state as fit parameters. Figure 6 shows three experimental energy spectra at various azimuthal angles together with the corresponding fit curves. We have determined a total of 13 such spectra with their fit curves at different ϕ angles. The spectra are normalized to the same height. Therefore only the energy position and the shape of the spectra should be compared here. The fit curve for the angular intensity variation is shown in Figure 7. As a measure of the intensity we have integrated the experimental and the calculated spectra over an energy range of 0.4 eV around the maximum. These integral values are compared in Figure 7. The solid line is from the calculations, the crosses

Figure 7. Azimuthal angular distribution of coincident electrons. ×, Measurements between 0° and 180°; ·, measurements between 180° and 360°; ——, theoretical curve based on the energy spectra which were calculated using the parameters of Table 2.

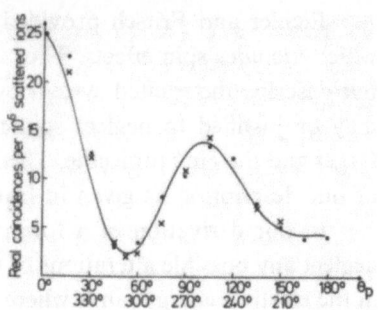

Table 2. Initial Complex Population Amplitudes for Magnetic Sublevels of ^1D and ^1P State, Used as Fit Parameters to Calculate the Energy Spectra of Figure 6 and the Angular Distribution of Figure 7[a]

State	m	$\lvert p_m(0)\rvert$	χ_m/π	$\lvert d_m(0)\rvert$	ϱ_m/π
^1P	0	5.4	0.6		
	1	7.6	1		
^1D	0			23	0
	1			25	1.7
	2			39	0.9

[a] Parameters are given in a frame with the beam direction as z axis. Although formula (8) for the angular distributions contains only the cosine of the relative phases, we are able to determine the phases including their sign. This is due to the interference of ^1D and ^1P contributions, which depends on the sign of the phases.

represent measurements between $\phi = 0°$ and $180°$, and the points spectra between $\phi = 180°$ and $360°$. The agreement of these two sets is theoretically expected because of the reflection symmetry with respect to the scattering plane[(1)] and provides a good test for the accuracy of the measurements.

The set of parameters used for all the fit curves shown in Figures 6 and 7 are given in Table 2. It is remarkable that the excitation probability—proportional to the squares of the amplitudes—of the ^1D state is $\sim 97\%$ while only $\sim 3\%$ is due the ^1P state.

It is quite interesting to see that another type of measurements yielded similar results. Bordenave-Montesquieu et al.[(12)] have investigated He+/He collisions at 15 keV and measured the angular distributions of autoionization electrons with

Table 3. Comparison of Relative Excitation Probabilities for the Magnetic Sublevels as Derived from Table 2 with Values from Bordenave-Montesquieu et al.,[(12)] which Were Found in the Case of He+/He Collisions at 15 keV

State	$\lvert m\rvert$	Parameter	Relative population probabilites, %	
			This work	Results from Reference 12
^1D	0	$\lvert d_0\rvert^2$	11	12
	1	$2\lvert d_1\rvert^2$	25	23
	2	$2\lvert d_2\rvert^2$	61	65
^1P	0	$\lvert p_0\rvert^2$	0.6	Neglected
	1	$\lvert 2P_1\rvert^2$	2.4	Neglected

respect to the beam axis (i.e., the ϑ dependence) in noncoincidence measurements. They neglected contributions from the ^1P state and determined population probabilities for the ^1D substates. The comparison of their results with the probabilities deduced from the amplitudes of Table 2 is given in Table 3 and shows a remarkable similarity of both data sets.

4. Interpretation of the Experimental Results

In the following we take only the ^1D state into account since it is excited with a relative probability of 97%. We calculated the complete electron angular distribution from this state using the ^1D amplitudes of Table 2, and Figure 8 may give an impression of the result. It shows a near rotational symmetry around an axis that lies in the scattering plane and is tilted with respect to the beam axis by an angle of 73°. This axis coincides nearly with the momentum transfer axis, which has an angle of 75.6° to the beam axis. As discussed before, the angular dependence of the ejected electron is described by the spherical harmonic corresponding to the angular momentum of the autoionizing state. Therefore the electron density in the excited atom has the same symmetries as the ejected electron intensity and the excited atom is oriented in space in the same way as the measured distribution shown in Figure 8. Looking at Figure 8 one can quantitatively understand the measured angular distribution shown in Figure 7. The electron intensity is measured at $\vartheta = 135°$ and various ϕ angles. The high intensity at $\phi = 0°$ is due to the big lobe. At $\phi = 105°$ the "ring" of the distribution causes a maximum, and at $\phi = 180°$ and 60° we observe low intensities due to the minima between the lobe and the ring.

The symmetry axis of the electron distribution seems to be a quite natural quantization axis and therefore we have calculated the complex population amplitudes of the ^1D state with respect to this axis. The new amplitudes are given in Table 4. Now the $m = 0$ sublevel is mainly populated and there are only small contributions

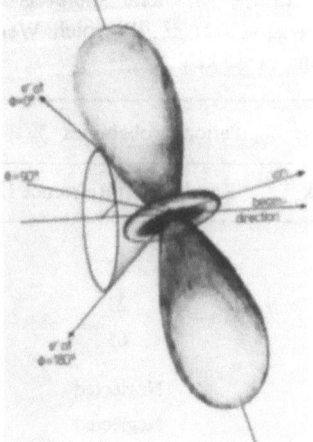

Figure 8. Image of the intensity distribution of ejected electrons, which in the case of He** autoionization has the same angular dependence as the electron density of the excited electron cloud. The position of the electron detector at $\vartheta = 135°$ and $\phi = 0°$ and 180° is indicated. The measured azimuthal angular distribution of electrons as shown in Figure 7 may be qualitatively understood from this picture.

Table 4. Complex Population Amplitudes of Table 2 for State ^1D Calculated in a New Coordinate Frame, Whose z Axis is Rotated in the Scattering Plane by $-73°$ with Respect to the Beam Direction

m	0	1	2
$|d_m|$	66	14.7	3.3
ϱ_m/π	0	-0.5	0.05

of $|m| = 1$ and $|m| = 2$—in fact Figure 8 looks like an electron cloud corresponding to a D_0 state. Also there is a remarkable relation between the relative phases of the $m = 1$ and the $m = 2$ amplitudes, which are $-\pi/2$ and nearly 0, respectively. Because of this phase relation one can describe the excited state as a coherent superposition of two pure $m = 0$ substates along two axes oriented symmetrically to the axis used for the values of Table 4.

This implies the following interpretation of the excitation process: (i) The excitation takes place at a certain internuclear distance R_c, e.g., at a potential curve crossing on the "way in" and on the "way out." In a trajectory picture as shown in Figure 9 this corresponds to an excitation at two different orientations of the internuclear axis. (ii) The excitation occurs into pure $m = 0$ substates (internuclear axis as quantization axis). (iii) The cloud of excited electrons does not follow the rotation of the internuclear axis. Taking $\bar{z}$ (see Figure 9) as quantization axis, the combined state, resulting from a coherent superposition of the two contributions, contains admixtures of $|m| = 1$ and $|m| = 2$. The admixtures are dependent on the spreading angle 2β, but they have always relative phases of $\pm\pi/2$ and 0, respectively. The same is true if the transitions are not well localized, but occur over certain ranges symmetrically to the $\bar{z}$ axis.

In this picture the amount of $m \neq 0$ admixtures yields a valuable piece of information, namely, the spreading angle 2β between the two orientations of the internuclear axis at which the excitations occur. In our case we obtain a value of $2\beta = 46°$. Since the He⁺/He interaction potential is known,[13] we can calculate the trajectory leading to a scattering angle of $\theta = 6°$. In connection with the knowledge of the spreading angle 2β we can therefore localize the internuclear distance R_c at which the excitation takes place. We obtain a value of $R_c = 0.5a_0$.

Figure 9. Trajectory diagram for the He⁺/He collision. The measured spectra can be explained by a coherent superposition of two pure $m = 0$ substates of the ^1D state, which were excited at the internuclear distance R_c at two different orientations of the internuclear axis on the way in and out. With $\bar{z}$ as quantization axis the resulting state has admixtures of $|m| = 1$ and $|m| = 2$ substates, which depend on the spreading angle 2β.

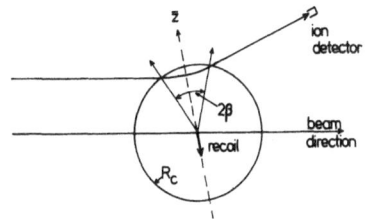

As mentioned in the Introduction, the excitation in He$^+$/He collisions is usually regarded as due to the $2p\sigma$–$2p\pi$ coupling. Our results allow a more detailed description of such a rotational coupling: In the incoming channel the system behaves adiabatically, i.e., the electron clouds follow the motions of the nuclei. At small internuclear distances the electrons are promoted via the $2p\sigma$ orbital. Because of this the electron density is high at large distances from the nuclei. The "blown-up" electron cloud is no longer able to follow the rotation of the internuclear axis, i.e., the electrons behave diabatically with respect to rotation. Since the excited electron cloud is resting it has to be described by a mixture of sublevels whose amplitudes vary, if the reference frame is rotated with the internuclear axis.

There is another axis of interest, namely, the y axis, which is perpendicular to the scattering plane. Any angular momentum transferred from the motion of the heavy particles to the electron cloud produces a nonvanishing expectation value of angular momentum $\langle L_y \rangle$, whereas $\langle L_z \rangle = \langle L_x \rangle = 0$ due to the reflection symmetry with respect to the scattering plane. From the complex amplitudes of Table 2 or Table 4 one obtains a value of $\langle L_y \rangle = -0.95\hbar$. This means that in fact angular momentum is transferred to the electron cloud, and that its amount as well as its sign is in qualitative agreement with the expectation from a simple classical picture.

ACKNOWLEDGMENTS

We would like to acknowledge the contributions of Ulrich Thielmann, who performed the 180° measurements, and Bernhard Müller and Quentin C. Kessel, who participated in the coincidence studies.

References

1. W. Lichten, *Phys. Rev.* **164**, 131–141 (1967).
2. P. J. Hicks and J. Comer, *J. Phys. B* **11**, 1866–1879 (1975).
3. G. Gerber, R. Morgenstern, and A. Niehaus, *J. Phys. B* **6**, 493–510 (1973).
4. G. Gerber, R. Morgenstern, and A. Niehaus, *J. Phys. B* **5**, 1396–1411 (1972).
5. U. Thielmann, Ph.D. thesis, Freiburg (1977).
6. R. Morgenstern, A. Niehaus, and U. Thielmann, *J. Phys. B* **10**, 1039–1058 (1977).
7. G. Gerber, A. Niehaus, and B. Steffan, *J. Phys. B* **6**, 1836–1848 (1973).
8. M. J. van der Wiel and G. Wiebes, *Physica* **53**, 225–255 (1971).
9. Q. C. Kessel, R. Morgenstern, B. Müller, A. Niehaus, and U. Thielmann, *Phys. Rev. Lett.*, 645–648 (1978).
10. R. Morgenstern, *Proceedings of the Ninth International Conference on Physics of Electronic and Atomic Collisions*, Seattle, Washington 1975, J. S. Risley and R. Geballe, eds., Invited lectures, pp. 345–358, University of Washington Press, Seattle (1975).
11. J. Eichler and W. Fritsch, *J. Phys. B* **9**, 1477–1489 (1976).
12. A. Bordenave-Montesquieu, P. Benoit-Cattin, A. Gleizes, and H. Merchez, *J. Phys. B* **8**, L350–L354 (1975).
13. R. P. Marchi and F. T. Smith, *Phys. Rev. A* **139**, 1025–1038 (1965).

20

Coherence Effects
in Postcollision Interactions
in Electron Impact Experiments

FRANK H. READ AND JOHN COMER

The degree of interference between overlapping PCI structures is discussed. The theoretical models used in analyzing these structures are briefly described and examples of observed PCI structures are given. It is argued that experimental results should in principle be fitted by formulas corresponding to partial interference between different autoionizing states and their backgrounds, although it is acknowledged that in practice this would usually require too many adjustable parameters.

1. Introduction

Most studies of coherence effects are concerned with simple two-stage processes in which a projectile (photon, electron, or ion) excites a target atom or molecule, which subsequently decays by emitting a photon or electron. The temporal separation of the excitation and decay stages is usually sufficiently large that the only coupling between the stages is that caused by the relative populations and coherence of the intermediate states of the atom or molecule. It is also usually the case that the experimental parameters limit the observed excitation to a single excited state (which may of course have more than one magnetic sublevel). Investigations of this type are therefore concerned with the populations and coherence of the magnetic sublevels, or more precisely, with the density matrix or statistical tensor that describes the

FRANK H. READ AND JOHN COMER ● Department of Physics, Schuster Laboratory, University of Manchester, Manchester M13 9PL, U.K.

formation of the excited state. The theoretical basis of such two-stage studies is now reasonably well established (see, for example, Eichler and Fritsh[1]). A familiar example of this type of study is that in which electron–photon correlations in the two-stage reaction

$$e + A \rightarrow A^* + e$$

$$A^* \rightarrow A + h\nu$$

are used to deduce the density matrix of the intermediate excited state A^*.

A fundamentally different situation arises when additional coupling effects exist between the excitation and decay states (these stages can also be referred to as the entrance and exit channels of the reaction). This happens for example when the entrance and exit channels both contain charged particles and when the lifetime of the excited state is short enough to allow an appreciable Coulomb interaction between the channels. This interaction has been described[2] as a postcollision interaction (PCI). Its effect is to destroy the two-stage nature of the collision process, causing particle energies and other parameters of the exit channel to be influenced by the entrance channel. The concept of a well-defined intermediate excited atom then loses its validity, as does the concept of a density matrix for the excited atom or the concept of coherence between magnetic sublevels or between different excited states. The existence of a direct background also destroys the two-stage nature of the collision. It is nevertheless often useful to retain the two-stage model as a first approximation to such reactions, with the postcollision interaction being added as a perturbation, in which case it is still possible to enquire about the coherence introduced by the excitation stage. This is the approach used in the present paper. Up to the present time a full theoretical treatment of postcollision interactions has proved rather intractable, but two simplified models have been successfully used in interpreting experimental observations. We start by outlining these models, and then go on to discuss what can be learned about the degree of coherence between different intermediate excited states.

2. Theoretical Models of PCI

The earlier work on postcollision interactions has been reviewed by Read[3] and Spence;[4] more recent work and further references can be found in the latest publications.[5–8] Briefly, postcollision interactions have been observed in reactions in which an atom A is excited to an autoionizing state A^{**} by a charged projectile p (an electron or atomic ion)

$$p + A \rightarrow A^{**} + p \tag{1a}$$

after which it decays by ejecting an Auger electron

$$A^{**} \rightarrow A^+ + e \tag{1b}$$

The associated direct process is the direct ionization reaction

$$p + A \rightarrow A^+ + p + e \tag{1c}$$

The postcollision interaction occurs between the scattered projectile in (1a) and the Auger electron in (1b) causing changes of energy in both of them. Alternatively an autoionizing state of an ion can be excited by collision with a particle or by the absorption of a real or virtual photon, for example,[9,10]

$$h\nu + A \rightarrow A^{+**} + e \tag{2a}$$

and this is again followed by the ejection of an Auger electron

$$A^{+**} \rightarrow A^{++} + e \tag{2b}$$

A recent example of the PCI effects observed in the first type of reaction is shown in Figure 1. Here the $3s3p^64s(^3S)$ state of argon is excited by electron impact, after which it decays to the $^2P_{1/2}$ and $^2P_{3/2}$ states of Ar$^+$. The lower spectrum, corresponding to near-threshold excitation, shows considerable structure extending to energies much higher than $E_0 + \Gamma$, where E_0 ($= 25.03$ eV) is the nominal energy of the $3s3p^64s(^3S)$ state and Γ ($= 0.08$ eV) is its natural decay width.

The two theoretical models which have been used to analyze experimental data of this sort are the quantal "shake-down" model[3] and the semiclassical model of Morgenstern et al.[11-13] In the shake-down model the probability with which a scattered electron of final momentum **k** (in atomic units) is produced in autoionization

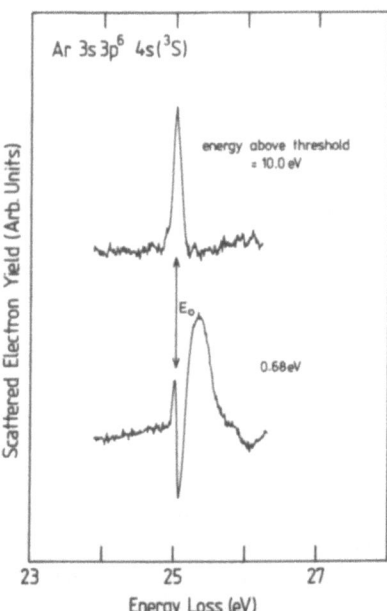

Figure 1. Measured electron energy loss spectra in argon in the region of the $3s3p^64s$ (^3S) state. In each spectrum the energy at which the scattered electron is detected is kept constant (at 10.0 or 0.68 eV) while the incident energy is varied. The energy loss (the difference between the incident and scattered energy) is then the apparent energy of the autoionizing atom, or more correctly, the sum of the energies of the ejected electron and the residual ion. The energy E_0 ($= 25.03$ eV) is the nominal energy of the $3s3p^64s$ (^3S) state. Adapted from the data of Wilden et al.[8]

processes of the type (1) is proportional to $|q(\mathbf{k})|^2$, where the overlap integral q is given by

$$q(\mathbf{k}) \sim \int \psi_f{}^*(\mathbf{k}, \mathbf{r})\psi_i(\mathbf{k}_0, \mathbf{r}) \exp(-r/2\tau k_0) \, d\mathbf{r} \tag{3}$$

and where τ is the lifetime of the autoionizing state, $\mathbf{k}_0$ and ψ_i are the momentum and wave functions that the electron would have in the absence of autoionization of the residual atom, and ψ_f is the wave function for the scattered electron of final momentum $\mathbf{k}$ in the field of the atomic ion alone. The radial part of ψ_i has been taken[14,15] to be proportional to that of a free outgoing wave, namely, $r^{-1} \exp(ik_0 r)$ and the radial part of ψ_f to be proportional to the asymptotic form of a Coulomb wave, namely, $r^{-1} \exp[i(kr + k^{-1} \ln 2kr - \frac{1}{2}l\pi + \alpha_l)]$, where l is the angular momentum of the scattered electron and α_l is the argument of $\Gamma(1 + l - i/k)$. This form of ψ_f is suitable if the scattered electron has a final energy $E_s \gtrsim 1$ eV, since the important part of the integrand in equation (3) is then at large radii, but it would not be suitable when E_s is negative or nearly zero. In the sudden approximation the angular momentum of the scattered electron is not changed by the autoionization process, in which case evaluation of (3) gives the amplitude q for the momentum change $k_0 \rightarrow k$,

$$q(l, k_0 \rightarrow k) = \exp[i(\alpha_0 - \alpha_l)]\left[k^2 k_0 \tau \sinh\left(\frac{\pi}{k}\right)\right]^{-1/2}$$

$$\times (2k)^{-i/k}\left[\frac{1}{2\tau k_0} + i(k - k_0)\right]^{-1+i/k} \tag{4}$$

In the limit in which the energy shifts caused by postcollision interactions are small, namely, the limit of large $\frac{1}{2}k_0{}^2$ and large τ, the value of $|q|^2$ can be shown to be

$$|q|^2 = (\tau k_0 \Delta^2)^{-1} \exp[-(\tau k_0 \Delta)^{-1}] \tag{5}$$

to the first order in $\Delta/(\frac{1}{2}k_0{}^2)$ and to the second order in $(\tau\Delta)^{-1}$, where Δ is the energy shift $\frac{1}{2}(k_0{}^2 - k^2)$. The right-hand side of (5) is that given by the purely classical model of Barker and Berry[16] (who were the first to observe PCI effects, in He^+ scattering experiments).

Figure 2 shows examples of the calculated real and imaginary parts of the overlap integral q as a function of the ejected electron energy, for two autoionizing states of the helium atom excited at a constant incident energy. The structure is clearly similar to that shown in Figure 1; in both cases the yield changes sharply near the nominal onset energy, and then oscillates with decreasing frequencies towards higher energies. The oscillation is caused by the changing degree of overlap of the oscillating wave functions ψ_i and ψ_f representing the scattered electron before and after the shake-down event. One feature of note, in both Figures 1 and 2, is that the initial peak is narrower than the natural width of the autoionizing state. This is caused by the fact that small PCI energy shifts occur only for events in which the autoionizing atom survives for more than one mean lifetime.[8] This mechanism for effectively selecting the longer-lived atoms leads to a corresponding decrease in the observed energy uncertainty. The structures shown in Figure 2 are also similar to those observed in many other types of PCI experiments.

The fact that the observed structures resemble the real or imaginary parts of q (or a linear combination of these), rather than $|q|^2$, indicates that the PCI amplitude interferes with an amplitude for the direct transition from the entrance to the exit channel [see equation (1c)]. In general we can suppose that the direct amplitude consists of two parts, $b + B$, the first of which interferes with the PCI amplitude and the second of which does not. The observed yield is then proportional to

$$P(E) = |aq(E) + b(E)|^2 + |B|^2$$
$$= |aq|^2 + |b|^2 + 2[\text{Re}(aq)\,\text{Re}(b) + \text{Im}(aq)\,\text{Im}(b)] + |B|^2 \qquad (6)$$

The other model that has been used to analyze experimental data is the semiclassical model of Morgenstern et al.[11-13] In the application of this model to electron impact studies[12] the initial state of the system (after collision) is taken to consist of an autoionizing atom and a scattered electron, separated by a distance δ. The electron recedes from the atom with a velocity v (which corresponds to the energy that it would have in the absence of PCI), and at some later moment the system undergoes a transition to a continuum state in which the atom has ejected an electron, and in which this ejected electron has taken up the Coulomb energy between itself and the scattered electron. The transition amplitude between the two states of the

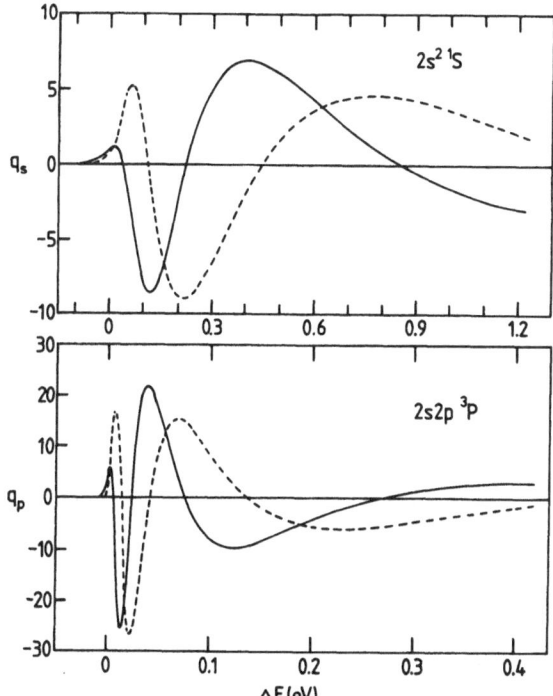

Figure 2. Values of the real (full curve) and imaginary (broken curve) parts of the overlap integral q [equations (3) and (4) with $l = 0$] for the states $(2s^2)\,{}^1S$ and $(2s2p)\,{}^3P$ of helium as a function of the difference ΔE between the energy E_j of the ejected electrons and the nominal mean energy that they would have in the absence of postcollision interactions. The incident electron energy is 60.2 eV and the states have been assumed to have decay widths of 138 and 14 meV, respectively, and mean energies of 57.82 and 58.30 eV, respectively. From Read.[8]

system is calculated by a first-order perturbation treatment, with the stationary phase approximation being used in an integration over the scattered electron position. This gives an ejected electron yield proportional to

$$P(E) = |b(E) + \sum_{n} C_n(E_0{}^n, E) \exp[-i\alpha_n(E_0{}^n, E)]|^2 + |B|^2 \tag{7}$$

where b and B are again the parts of the direct amplitude that are respectively coherent and incoherent with the PCI amplitude, and the summation is over all the auto-ionizing states that are excited by the collision. E and $E_0{}^n$ are, respectively, the observed energy of the ejected electron and the energy that it would have in the absence of PCI. The amplitude C_n is given (in atomic units) by

$$C_n(E_0{}^n, E) = \frac{a_n(0) \exp[-(\delta^{-1} + E_0{}^n - E)\delta/2\nu_n\tau_n(E - E_0{}^n)]}{(\nu_n\tau_n)^{1/2}(E - E_0{}^n)} \tag{8}$$

where $a_n(0)$ is the amplitude with which the autoionizing state n is formed, and τ_n is its decay lifetime. The phase α_n is given by

$$\alpha_n(E_0{}^n, E) = \nu_n{}^{-1}\{1 - \delta(E - E_0{}^n) + ln[\delta(E - E_0{}^n)]\} \tag{9}$$

A more accurate form of α_n has also been used by Morgenstern et al.[13] These forms of the amplitude and phase correspond to experiments in which the incident electron energy is fixed and the ejected electron energy is scanned, but similar expressions apply to the other forms of PCI experiments.[7,11–13,17] This semiclassical model also gives structures similar in shape to those shown in Figure 2 and has been found to fit the available data well.[7,11–13,17]

3. The Degree of Coherence in Overlapping PCI Structures

As can be seen from Figures 1 and 2, postcollision interactions can cause shifts in the energy of observed ejected or scattered particles. These energy shifts are especially large near the excitation thresholds of the autoionizing states. When the shifts are large enough to cause an overlap in the PCI structures of the two neigh-boring states, as happens in the example shown in Figure 3, the question arises of whether or not the observed PCI structures add coherently or incoherently. As discussed in the Introduction, this is not a straightforward question because of the presence of the direct contribution and the lack of separation of the entrance and exit channels, and because of the consequently imprecise definition of the intermediate excited states. We will continue, however, to treat reactions in which PCI is observed as though they are approximately two-stage reactions.

The first point to note is that the presence of oscillatory structure is not itself an indication of the existence of coherence between different autoionizing states. This is clear from the spectrum shown in Figure 1, in which the initial oscillation is

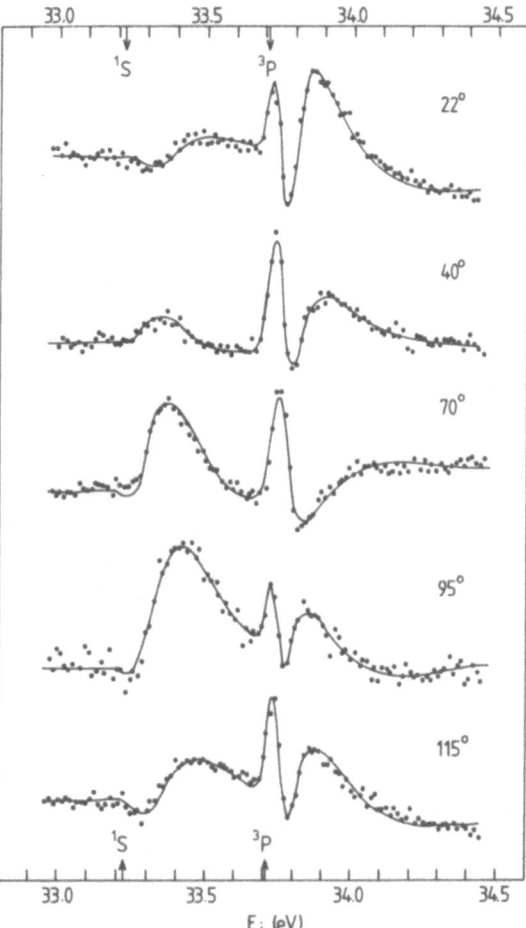

Figure 3. The points are the ejected-electron yields obtained by Hicks et al.[2] with electron impact on helium at an incident energy of 60.2 eV and at five different angles of observation. The full curves show the best fits obtained using the shake-down model and assuming that the amplitudes corresponding to autoionization of the $(2s^2)\,{}^1S$ and $(2s2p)\,{}^3P$ states are incoherent with each other while being coherent with their respective parts of the direct ionization amplitude [equation (12)]. The arrows show the nominal mean ejection energies of the two autoionizing states. The vertical scales are arbitrary and different. From Read.[13]

caused[8] by a single isolated state, as well as from the overlap integrals shown in Figure 2.

Kessel et al.,[7] Morgenstern et al.,[11–13] and Niehaus[17] have assumed, in applying the semiclassical model to electron impact experiments (and also to ion impact experiments), that the system immediately after the collision consists of a coherent superposition of excited states,

$$\Psi_i(t) = \sum_n a_n(t)\psi_n \exp(-iE_n t) \tag{10}$$

where the ψ_n are the wave functions of the excited states and the E_n are their energies. This leads naturally to the presence of coherent interference between the PCI structures of neighboring overlapping states, as given by equation (7). This assumption will be discussed below, together with the effect of the detection process on the degree of coherence in the observed spectra.

In the case of the shake-down model the same assumption about coherence between different states would lead to the generalization

$$P(E) = | b + \sum_n a_n q_n |^2 + | B |^2 \tag{11}$$

of equation (6). If on the other hand it is assumed that the autoionizing states are excited or observed incoherently, and that each interferes with a separate part of the direct background, the yield takes the form

$$P(E) = \sum_n | b_n + a_n q_n |^2 + | B |^2 \tag{12}$$

Equations (11) and (12) represent the extremes of complete coherence and complete incoherence in the observed spectra. The point at issue is to decide whether either of these extremes is correct, or whether an intermediate degree of coherence should be assumed when interpreting experimental data.

Before going into the details of this question it is important to distinguish between the degree of coherence induced by the excitation process and the degree of coherence actually observed when the decay particles are detected. The observed cross section for the two-stage process 1 (a, b) can be written in general[1,18] as

$$P(\mathbf{p}, \mathbf{k}) = \mathrm{Tr}\{\varrho^{\mathrm{eff}}(\mathbf{k})\varrho^{\mathrm{st}}(\mathbf{p})\} \tag{13}$$

where $\mathbf{p}$ and $\mathbf{k}$ are the momenta of the projectile and ejected electron, respectively, and the ϱ's are density matrices. The statistical matrix ϱ^{st} defines the degree of coherence induced by the excitation of the autoionizing states, while the efficiency matrix ϱ^{eff} is a function of the detection system. We see from this that the degree of interference actually observed between structures belonging to different states depends on ϱ^{eff} as well as on ϱ^{st}.

Starting with a consideration of the excitation process, we follow Fano[19] in supposing that the state of an atomic target immediately after a collision is represented by a coherent superposition of all its stationary states that are consistent with conservation laws, and that incoherence between the states develops in time, after the collision, as a result of the increasing sharpness in the partitioning of the total energy between the target and the scattered particle. The uncertainty ΔE in the partitioning is inversely proportional to the time interval Δt after the collision, so that the most closely spaced states, such as those separated only by hyperfine interactions, are the last to become incoherent.

We may go further and assume that the initial relative phase of the coherently excited states depends on the angle through which the projectile p [equation (1a)] is scattered (as happens for inelastic electron–photon coincidence experiments[20]). In the case of the ion impact experiments of Morgenstern et al.[11–13] the range of scattering angles is small and the range of relative phases is therefore presumably also small. In the electron-impact experiments, on the other hand, the scattered electrons have a very low energy and cover the whole range of angles. To estimate the resultant range $\Delta\phi$ in the initial relative phase of two states n and m differing in energy by E_{nm} it seems reasonable to associate with the range of scattering angles a range Δt in the effective time of the collision. Since each stationary state has the time dependence $\exp(iE_n t/\hbar)$, the range of initial times leads to a range

$$\Delta\phi = E_{nm} \cdot \Delta t/\hbar \tag{14}$$

in the initial phase difference. In the absence of any other criterion, we take as a crude estimate of the difference Δt between forward and backward scattering events, the value

$$\Delta t = 2r/v \tag{15}$$

where r is an effective collision radius (~ 0.2 nm) and v is the velocity of the scattered electron ($\sim 6 \times 10^5$ ms^{-1} for an electron of energy ~ 1 eV). In the case of the levels $2s^2(^1S)$ and $2s2p(^3P)$ of helium for example, for which $E_{nm} = 0.48$ eV (see Figure 3), we find that

$$\Delta\phi \sim 0.5 \text{ rad}$$

For those experiments in which the scattered electron is not observed (as in the experiments relating to Figure 3), we interpret $\Delta\phi$ as the uncertainty in the relative phase, which thus gives rise to a lack of complete coherence between the excited states. On the other hand for those experiments in which the scattered electron (but not the ejected electron) is observed, we would expect the degree of coherence to be higher, but a systematic study of the differences in the results from these two types of experiments has not yet been carried out.

The nonobservation of other experimental parameters, such as polarizations, can also decrease the observed degree of interference between different PCI structures belonging to different states. In the case of an ideal experiment in which the incident particle p and the target atom A [equation (1)] are prepared in definite magnetic sublevels M_{pi} and M_i, respectively, and the detectors are sensitive only to the definite sublevels M_{pf}, M_f, and m_s of the scattered projectile, the residual ion, and the ejected electron, respectively, the differential cross section for the two-state process represented by equation (1) is given by[1]

$$\frac{d^4\sigma}{d\Omega_{pf}\,d\Omega_k} \sim P(M_{pi}, M_i \rightarrow M_{pf}, M_f, m_s)$$
$$= |\sum_{M_a} \langle \mathbf{p}_i, M_{pi}M_i \rightarrow \mathbf{p}_f, M_{pf}, M_a\rangle\langle \mathbf{p}_f, M_{pf}, M_a \rightarrow \mathbf{p}_f, \mathbf{k}, M_{pf}, M_f, m_s\rangle|^2 \tag{16}$$

Here the matrix elements represent the excitation and decay stages, and M_a is the unobserved magnetic quantum number of the intermediate excited state. For example if we consider the $2s^2(^1S)$ and $2s2p(^3P)$ states of He as being degenerate (as in the energy regions in Figure 3 at which the PCI structures overlap), and treat their excitation and decay as a two-stage process, and if $M_{pi} = \frac{1}{2}$ (for an incident electron or helium ion) and $M_{pf} = -\frac{1}{2}$ (which implies that an exchange reaction has taken place), we find that $M_a = 1$ and that therefore only the ^3P state has been excited. If on the other hand the value of M_p is not changed by the collision, then $M_a = 0$ and both the singlet or triplet states can be excited, possibly coherently. In practice the values of M_{pi} and M_{pf} have not been defined in the experiments performed so far, and so the observed yields are presumably the sum of a part representing pure ^3P excitation and a part representing a possibly coherent mixture of ^1S and ^3P excitations.

Another point to consider is the role of the direct background. As mentioned before, its presence means that strictly we are unable to use the arguments that apply to simple two-stage processes. This is particularly important in the case of the electron impact experiments, since the height of the background is typically of the same order as the height of the PCI structures. The effect of this on the degree of interference between overlapping PCI structures is unknown.

In discussing PCI structures and their interference we have so far ignored the possible role of intermediate negative ion resonances in the electron collision process, on the grounds (i) that the nonresonant models give a satisfactory explanation, (ii) that the PCI phenomenon has been seen in all the autoionizing states in all the atoms for which it has been sought so far, and (iii) that energy shifts of individual auto-ionizing states are seen over wide ranges of incident electron energy (about a few electron volts). For a resonance model to give a valid explanation of the PCI phenomena would require less specificity and greater resonance widths than have been found so far in the case of resonances in other energy ranges. It can also be argued that if the underlying resonances do indeed have widths corresponding to the range of incident energies over which PCI effects are seen, then the lifetimes of the resonances are not only much smaller than the autoionization lifetimes but become comparable to or less than the collision time, thus considerably weakening the usefulness of the resonance model. If, however, it is eventually established that the creation of intermediate resonances is important in determining the strength, and perhaps also the shape, of PCI structures, as has been suggested,[21,22] then we should again be unable to use the simple arguments given above concerning coherence and interference, and would again not have a satisfactory alternative with which to replace them.

Having discussed the various difficulties in establishing the degree of interference between overlapping PCI structures, we are left with the view that almost complete coherence between different autoionizing states exists in ion impact excitation, but not in electron impact excitation. Since it does not seem possible to establish the degree of incoherence in the latter case by theoretical reasoning we must try to establish it from the experimental results.

An essential difficulty in doing this is that formulas (11) and (12), representing complete coherence and complete incoherence, respectively, both contain a large number of adjustable parameters (5 and 6, respectively, disregarding the incoherent background $|B|^2$), so that good fits can be obtained to data even when an incorrect

Table 1. The Parameters of Equation (11) (Coherent) and Equation (12) (Incoherent) Derived by Least-Squares Fits to the Experimental Data of Figure 3[a]

θ (deg)	$\lvert a_s \rvert$	$\lvert a_p \rvert$	$\chi_s - \phi$	$\chi_p - \phi$	$\chi_p - \chi_s$	$\lvert b \rvert$	$\lvert B \rvert$
			(a) Coherent				
22	0.9	4.7	0.30	0.80	0.50	158	423
40	1.0	3.0	1.00	0.60	−0.40	48	132
70	3.9	1.7	1.10	1.70	0.60	13	83
95	3.3	1.1	0.65	1.75	1.10	4.2	67
115	2.9	1.6	0.05	0.85	0.80	6.8	61

Normalized $\chi^2 = 2.38$

			(b) Incoherent				
22	4.0	4.6	0.20	0.80	29	158	422
40	2.3	2.9	1.10	0.60	21	45	131
70	3.2	1.6	0.95	0.10	15	27	78
95	2.7	1.2	0.60	0.85	9	21	63
115	2.0	1.7	0.35	0.70	11	28	53

Normalized $\chi^2 = 1.68$

[a] The absolute normalization is arbitrary, but is the same for the five scattering angles. The phases $\chi_{s,p}$ and $\phi_{s,p}$ are the relative phases of the amplitudes $a_{s,p}$ and $b_{s,p}$, respectively, and are measured in units of π. The uncertainties in these phases are of the order of ± 0.1, and the uncertainties in the relative magnitude $\lvert a_{s,p} \rvert$, $\lvert b_{s,p} \rvert$ and $\lvert b \rvert$ are of the order of $\pm 20\%$. The unknown phases $(\alpha_o - \alpha_i)$ (see equation (4)) are assumed to be zero. From Read.[13]

formula is used. Formulas representing an intermediate degree of coherence would contain even more adjustable parameters. The approach by Read[15] has therefore been to fit formulas (11) and (12) by a least-squares procedure to a set of spectra in which only one experimental parameter is varied, namely, the spectra shown in Figure 3. The results obtained are summarized in Table 1. Both equations fit the data reasonably well, the average values of the normalized chi-squared being 2.38 and 1.68, respectively. Both these values are higher than that expected for the 107 data points (approximately 1.0 ± 0.2), but that for incoherence can be regarded as significantly better than that for coherence, even when allowance is made for the extra parameter in the incoherent case. The fits obtained assuming incoherence are shown as the full curves in Figure 3. The variable parameters in the least-squares fits are the relative magnitudes $\lvert a_s, p \rvert$ of the autoionizing amplitudes $a_{s,p}$, the relative magnitudes $\lvert b_{s,p} \rvert$ or $\lvert b \rvert$ of the coherent direct ionization background, the relative height $\lvert B \rvert^2$ of the incoherent background, and the phases of the $a_{s,p}$ relative to b, namely, $\chi_s - \phi$ and $\chi_p - \phi$.

Rather than considering the values of the normalized chi-squared, a better understanding is obtained by considering the dependences of the derived parameters on the angle of observation of the ejected electron. For example, the value of $|a_s|$ is expected to be constant because of the isotropy associated with an S state, and it varies over a smaller range (2.0–4.0) in the incoherent case than in the coherent case (0.9–3.7). Another test is provided by the phase differences $\chi_s - \phi_s$ and $\chi_p - \phi_p$, which vary slightly more smoothly and over smaller ranges in the incoherent case than the values of $\chi_s - \phi$ and $\chi_p - \phi$ in the coherent case. The parameter $|a_p|$ has a similar dependence on the scattering angle in both cases (varying approximately as $|\cos\theta|$, as found by Morgenstern et al.[13]). The direct ionization amplitude $|b_s|$ is more isotropic than $|b_p|$, as might be expected. A final test is provided by the value of $\chi_p - \chi_s$, which is expected to be constant in the coherent case (and was restricted by Morgenstern et al. to have the same value for four of the five scattering angles), but it can be seen that this behavior has not been found. This evidence is therefore insufficient to definitely distinguish between the two assumptions although it tends to favor the assumption of incoherence.

4. Summary

To summarize, some coherence between different autoionizing states is certainly present in the electron impact experiments, but the actually observed degree of interference between overlapping PCI structures is reduced by the process of detection. In general therefore experimental results should be fitted by formulas corresponding to partial interference between different autoionizing states and their background. In practice this would involve too many adjustable parameters, and would lead to possibly arbitrary and nonunique values of these parameters. A smaller number of parameters is required for the extreme assumptions of complete coherence and complete incoherence, and of these the latter has been found to give the better fit to one set of data. Clearly more work is needed before more definite guidance can be given.

References

1. J. Eichler and W. Fritsch, *J. Phys. B* **9**, 1477–1489 (1976).
2. P. J. Hicks, S. Cvejanović, J. Comer, F. H. Read, and J. M. Sharp, *Vacuum* **24**, 573–580 (1974).
3. F. H. Read, *Radiat. Res.* **64**, 23–36 (1975).
4. D. Spence, *Comments Atom. Mol. Phys.* **5**, 159–172 (1976).
5. D. Roy, A. Delâge, and J.-D. Carette, *J. Phys. B* **11**, 895–908 (1978).
6. D. Spence, *J. Chem. Phys.* **68**, 2980–2981 (1978).
7. Q. C. Kessel, R. Morgenstern, B. Müller, A. Niehaus, and U. Thielmann, *Phys. Rev. Lett.* **40**, 645–648 (1978).
8. D. G. Wilden, J. Comer, and P. J. Hicks, *Nature* **273**, 651–653 (1978).
9. M. J. Van der Wiel, G. R. Wright, and R. R. Tol, *J. Phys. B* **9**, L5–L9 (1976).

10. V. Schmidt, N. Sandner, W. Melhorn, M. Y. Adam, and F. Wuilleumier, *Phys. Rev. Lett.* **38**, 63–66 (1977).
11. R. Morgenstern, A. Niehaus, and U. Thielmann, *Phys. Rev. Lett.* **37**, 199–202 (1976).
12. R. Morgenstern, A. Niehaus, and U. Thielmann, *J. Phys. B* **9**, L363–L367 (1976).
13. R. Morgenstern, A. Niehaus, and U. Thielmann, *J. Phys. B* **10**, 1039–1058 (1977).
14. G. C. King, F. H. Read, and R. C. Bradford, *J. Phys. B* **8**, 2210–2224 (1975).
15. F. H. Read, *J. Phys. B* **10**, L207–L212 (1977).
16. R. B. Barker and H. W. Berry, *Phys. Rev.* **151**, 14–19 (1966).
17. A. Niehaus, *J. Phys. B* **10**, 1845–1857 (1977).
18. S. Devons and L. B. J. Goldfarb, *Encyclopedia of Physics*, Vol. 42, S. Fluegge, ed., pp. 362–554, Springer-Verlag, New York (1957).
19. U. Fano, Chapter 18 in this volume.
20. H. Kleinpoppen, Chapter 55 in this volume.
21. H. S. Taylor and R. Yaris, *J. Phys. B* **8**, L109–L113 (1975).
22. R. K. Nesbet, *Phys. Rev. A* **14**, 1326–1332 (1976).

Collision-Stimulated Autoionization

V. I. Matveev and E. S. Parilis

The field of a slow charged particle receding from or passing by an atom that is in a recently excited autoionizing state causes the stimulated decay of that state, which can be described by wave-function deformation in terms of time-dependent mixing of different autoionizing states. Stimulated decay by a fast charged particle is small and is described within perturbation theory. The cross section for stimulated autoionization is estimated. Some typical examples of field-stimulated autoionization are discussed; namely, (1) the decay of an autoionizing heliumlike ion stimulated by its image charge in a metal; (2) the postcollision interaction of heavy atoms; (3) the "carambole" atom–molecule collisions.

1. Introduction

In some collision events the atom produced in an autoionizing state is influenced by the field of a receding charged particle. In "carambole"-type atom–molecule collisions or beam–foil experiments the moving autoionizing atom passes by a charged center or crosses a metal surface receding from the image charge. The interaction with the field provokes a stimulated decay of the autoionizing state.

In contrast with the large increase of the Auger rate in a quasimolecule during a close heavy-ion collision caused by the molecular orbital effects[1,2] or the Doppler broadening and the broadening due to Coulomb interaction at different internuclear distances,[3] the field-stimulated autoionization is caused by polarization of the autoionizing state and can be described by time-dependent mixing of different auto-ionizing states. It occurs at distant collisions with charged particles and is a part of the postcollision interaction.[4-6]

V. I. Matveev and E. S. Parilis • Arifov Institute for Electronics, Tashkent, USSR.

2. General Theory

2.1. Deformation of Autoionizing State by a Slow Charged Particle

Let a charged particle (with velocity v and charge Z_1) pass by an autoionizing two-electron atom or ion with nuclear charge Z_2. For a heavy particle moving along a certain trajectory such that the distance $R(t)$ between the particle and the atom is larger than atomic dimensions the atomic electron–particle interaction is given by (in atomic units)

$$V \approx -\frac{Z_1}{R(t)} - \frac{Z_1}{R^3(t)} \mathbf{R}(t) \cdot \mathbf{r} \tag{1}$$

where $\mathbf{r}$ are electron coordinates whose origin coincides with the atomic nucleus. Electronic states in the atom will be described by single-electron wave functions φ.

A slow charged particle may cause the deformation of the wave functions of the autoionizing states; for convenience the deformation is treated as for the Stark effect. We confine our case to autoionizing states with main quantum number $n = 2$.

Let us consider the influence of the perturbation (1) on single-electron wave functions φ_k, $k = 1, 2, 3, 4$ for states $2S, 2P_0, 2P_{\pm 1}$ respectively (the subscript denotes the orbital moment projection). If the separation of other levels with $n \neq 2$ is large, we can expand the field-dependent single-electron wave function Ψ in the functions φ_k:

$$\Psi = \sum_{k=1}^{4} a_k(t)\varphi_k e^{-i\varepsilon_k t}$$

where ε_k are single-electron energies. Then

$$i\frac{d}{dt} a_j = \sum_k V_{jk}a_k \exp[-i(\varepsilon_k - \varepsilon_j)t], \qquad j, k = 1, 2, 3, 4 \tag{2}$$

where

$$V_{jk} = \langle \varphi_j | V | \varphi_k \rangle$$

Let the electron states be adiabatically moved to the charged-particle position; then in (2) the coefficients V_{jk} will be time independent and will depend on the distance R between the particle and nucleus. Then if we direct the z axis along $\mathbf{R}$,

$$V_{jk} = -\frac{Z_1}{R} \cdot \delta_{jk} - \frac{Z_1}{R^2} \langle k | z | j \rangle \tag{3}$$

where only $\langle 1 | z | 2 \rangle = \langle 2 | z | 1 \rangle^* \approx 3/Z_2$ are not zero and δ_{jk} is the Kronecker symbol.

Thus the perturbation (1) shifts all states at $-Z_1/R$ and mixes the states $2s$ and $2p_0$. Consequently the stationary solution of the system (2), which adiabatically turns into the state $2s$ at $R \to \infty$, is

$$\Psi_1(t) = e^{-i\omega_1 t}[A(R)\varphi_{2s} + B(R)\varphi_{2p_0}] \equiv e^{-i\omega_1 t}\Phi_1 \tag{4}$$

where

$$\omega_1 = \frac{(\tilde{\varepsilon}_1 + \tilde{\varepsilon}_2)}{2} - \frac{1}{2}[(\varepsilon_1 - \varepsilon_2)^2 + |V_{12}|^2 \cdot 4]^{1/2} \tag{5}$$

$$\tilde{\varepsilon}_i = \varepsilon_i + V_{ii}$$

and since $|\varepsilon_1| > |\varepsilon_2|$, then

$$\omega_1 \xrightarrow[R\to\infty]{} \varepsilon_1$$

and

$$A(R) = B(R)\frac{(\tilde{\varepsilon}_2 - \omega_1)}{V_{12}}, \qquad B(R) = -V_{12}[|V_{12}|^2 + (\omega_1 - \tilde{\varepsilon}_2)^2]^{-1/2} \tag{6}$$

The functions

$$\Phi_1 \xrightarrow[R\to\infty]{} \varphi_{2s}, \qquad \Phi_1 \xrightarrow[R\to0]{} 2^{-1/2}(\varphi_{2s} - \varphi_{2p_0})$$

are the same as in the linear Stark effect.

The solution, which turns into $2p_0$ at $R \to \infty$ is

$$\Psi_2(t) = e^{-i\omega_2 t}[A(R)\varphi_{2p_0} - B(R)\varphi_{2s}] \equiv e^{-i\omega_2 t}\Phi_2 \tag{7}$$

$$\omega_2 = \frac{(\tilde{\varepsilon}_1 + \tilde{\varepsilon}_2)}{2} + \frac{1}{2}[(\tilde{\varepsilon}_1 - \tilde{\varepsilon}_2)^2 + 4|V_{12}|^2]^{1/2} \tag{8}$$

Thus instead of states $2s$ and $2p_0$, in the presence of the field the stationary states Φ_1 and Φ_2 arise with energies ω_1 and ω_2, respectively. Therefore if in the absence of the field there was, for example, the autoionizing state 2^2S, then in the presence of the field it is necessary to change its wave function $\varphi_{2S}(\mathbf{r}_1)\varphi_{2S}(\mathbf{r}_2)$ to $\Phi_1(\mathbf{r}_1)\Phi_1(\mathbf{r}_2)$. By analogy for other configurations

$$\varphi_{2s}\varphi_{2p_0} \to \Phi_1\Phi_2, \qquad \varphi_{2s}\varphi_{2p_{\pm1}} \to \Phi_1\varphi_{2p_{\pm1}}, \qquad \varphi_{2p_0}\varphi_{2p_0} \to \Phi_2\Phi_2$$

2.2. Change of Autoionization Decay Rate

Let us consider now the R dependence of the autoionization rate. The $1s$ state wave function deformation caused by the field may be neglected and then the presence of the field results only in an energy shift of $-Z_1/R$. The Auger electron wave function is chosen as follows[7]:

$$\psi_p^{(-)} = \frac{1}{(2\pi)^{3/2}} e^{(\pi/2)(Z^*/p)}\Gamma\left(1 + i\frac{Z^*}{p}\right){}_1F_1\left[-\frac{i}{p}, 1, -i(pr + \mathbf{p}\cdot\mathbf{r})\right]e^{i\mathbf{p}\cdot\mathbf{r}}$$

where $\mathbf{p}$ is the Auger electron momentum and Z^* is the effective charge of the ion continuum. Consider a transition from the autoionizing state $2s2p_{\pm1}$. The initial wave function is

$$\Psi_i^{a,s}(\mathbf{r}_1, \mathbf{r}_2) = 2^{-1/2}[\Phi_1(\mathbf{r}_1)\varphi_{2p_{\pm1}}(\mathbf{r}_2) \mp \Phi_1(\mathbf{r}_2)\varphi_{2p_{\pm1}}(\mathbf{r}_1)]$$

and the initial state energy is

$$E_i = \omega_1 + \varepsilon_{2p_{\pm1}} - Z_1/R$$

The plus sign corresponds to the state Ψ^s and the minus sign corresponds to the state Ψ^a. The final energy is $E_f = \varepsilon_{1s} - Z_1/R + E_p$, where $E_p = p^2/2 - Z_1/R$ is the Auger electron energy. The Auger transition amplitude at fixed R is

$$A_{if} = \left(\Psi_f^{a,s}, \frac{1}{|\mathbf{r}_1 - \mathbf{r}_2|} \Psi_i^{a,s} \right)$$

Therefore the decay rate is given by

$$w(R) = A^2(R)w^{a,s}(2s2p_{\pm1}) + B^2(R)w^{a,s}(2p_02p_{\pm1}) \tag{9}$$

where according to (6)

$$B(R) \xrightarrow[R\to\infty]{} 0, \qquad A^2 = 1 - B^2, \qquad B^2(R) \xrightarrow[R\to0]{} \tfrac{1}{2}$$

and $w^{a,s}(2s2p_{\pm1})$ has the same meaning as for the decay of the state $2s2p_{\pm1}$ in the absence of a field, the only difference being that the electron energies are R dependent. Thus as $R \to \infty$, $w(2s2p_{\pm1})$ tends to the decay rate in the absence of the field and $w(2p_02p_{\pm1})$ has the same meaning. From (9) it follows that, if $w(2p_02p_{\pm1}) > w(2s2p_{\pm1})$, then as the field increases (R decreasing) the decay rate will increase and vice versa.

If one chooses the initial state $2p_02p_{\pm1}$, then the probability of its decay in the presence of the field will be

$$w(R) = A^2(R)w(2p_02p_{\pm1}) + B^2(R)w(2s2p_{\pm1}) \tag{10}$$

It should be noted that the autoionizing decay of the state $2p^2\ ^3P$ is forbidden by selection rules. In the presence of the field this restriction is removed and the auto-ionization rate is

$$w(R) = B^2(R)w(2s2p_{\pm1}\ ^3P) \tag{11}$$

By an analogous method one can get the decay rate for states $2s$, $2s2p_0$, etc.

A more consistent calculation of wave functions of autoionizing states in the outer field region may be carried out for helium or heliumlike ions. For this purpose we expand the field-deformed wave function over the two-electron wave functions of autoionizing states $2s^2\ ^1S$, $2s2p\ ^{1,3}P$, $2p^2\ ^1S$, ^1D, ^3P of a heliumlike ion. Taking into account the degeneracy over the projections of total orbital angular momentum (M_L) and spin, we have 16 such functions. Consequently we obtain the system of equations similar to (2) and consisting of 16 equations. This system can be solved analytically. As a result, for the decay rate of the states $2s2p\ ^1P(M_L = \pm1)$ we have

$$w(R) = A^2(R)w(2s2p\ ^1P) + B^2(R)w(2p^2\ ^1D) \tag{9a}$$

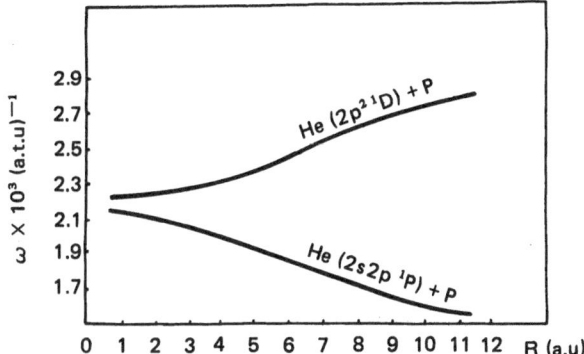

Figure 1. Autoionization decay rate $W(R)$ versus interatomic distance R in the encounter of a proton with a helium atom for two autoionizing states: (a) $He(2p^2\,{}^1D) + p$; (b) $He(2s2p\,{}^1P) + p$.

and for $2p^2\,{}^1D(M_L = \pm 1)$

$$w(R) = A^2(R)w(2p^2\,{}^1D) + B^2(R)w(2s2p\,{}^1P) \qquad (10a)$$

Analogous formulas can be derived for the decay rates for other autoionizing helium states.

Using equations (9a) and (10a) for a He atom in the field of a proton the rate $w(R)$ was calculated and is shown in Figure 1 as a function of the distance R.

The coefficient

$$B^2(R) = \frac{w(R) - w(2p^2\,{}^1D)}{w(2s2p\,{}^1P) - w(2p^2\,{}^1D)}$$

which is a measure of the mixing for the two autoionizing states is plotted versus R in Figure 2. It is calculated from a formula of the type of equation (6).

Therefore a slow field causes a mixing of the configurations, and when decay rates for different configurations differ a large change of the decay rate might be expected, either increasing or decreasing. The mixing of configurations may lead also to a change of the Auger electron angular distribution. At a fixed R the Auger electron spectrum is given by

$$\frac{dn}{dE_p} = \frac{1}{2\pi}\frac{w(R)}{[\varDelta E(R) - E_p]^2 + w^2(R)/4} \qquad (12)$$

Figure 2. $B^2(R)$ versus R for the two autoionizing states of Figure 1.

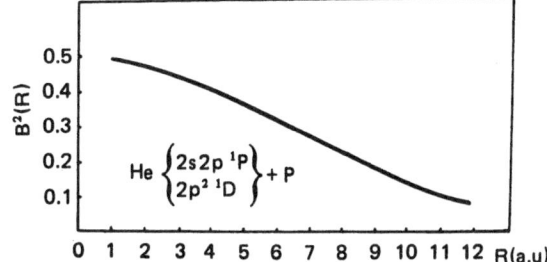

where $\Delta E(R) = E_i - (\varepsilon_{1s} - Z_1/R)$. From (12) it follows that in the presence of the field the spectrum line is shifted and changes its width.

The statement of autoionization stimulation in the field of a point charge will be valid also for a homogeneous outer field. Actually, the interaction with a homogeneous outer field is expressed as $V = V_0 - \mathbf{E} \cdot \mathbf{r}$, which formally coincides with equation (1).

2.3. The Cross Section of Stimulated Autoionization

If the autoionizing state is formed at time t_0, then the probability of decay at time t is given by the integration along the trajectory $R(t)$:

$$W(t) = \int_{t_0}^{t} w(t) \exp\left[-\int_{t_0}^{t} w_{tot}(t)\,dt\right] dt \tag{13}$$

where $w(t) \equiv w[R(t)]$ and

$$\exp\left[-\int_{t_0}^{t} w_{tot}(t)\,dt\right]$$

is the probability that the autoionizing state survives till time t, and $w_{tot}(t) = w(t) + w_x(t)$ where $w_x(t)$ is the radiative decay rate. The spectrum of Auger electrons ejected at time t may be determined by

$$\frac{dN}{dE_p} = \int_{t_0}^{t} \frac{dn}{dE_p} w(t) \exp\left[-\int_{t_0}^{t} w_{tot}(t)\,dt\right] dt \tag{14}$$

It is seen that spectrum broadening (in comparison with the case when the field is absent) is due not only to the change with R of the autoionizing term, but also to the function $w(t)$. To describe the influence of the field on the decay rate, we introduce its change $\tilde{w}(t) = w(t) - w_0$, where w_0 is the decay rate in the absence of the field. If $\tilde{w} > 0$, then the field accelerates the decay, while if $\tilde{w} < 0$ it decelerates the decay. By analogy one can introduce the total probability of stimulated transition

$$\tilde{W}(t) = \int_{t_0}^{t} \tilde{w}(t) \exp\left[-\int_{t_0}^{t} w_{tot}(t)\,dt\right] dt \tag{15}$$

By integrating over all impact parameters for $t = \infty$ a cross section for stimulation of decay may be obtained:

$$\tilde{\sigma} = 2\pi \int_{0}^{\infty} \varrho\,d\varrho \int_{t_0}^{\infty} \tilde{w}(t)\,dt \tag{16}$$

The cross section for stimulation of autoionization decay has the following meaning. The flux of atoms in an autoionizing state decays according to the rule $dI(t) = -w_0 I(t)\,dt$, where w_0 is the rate of spontaneous decay. When passing through matter the flux decay is

$$dI = -(w_0 + n\tilde{\sigma}v)I\, dt$$

where n is the atomic density, v is the flux velocity, and $\tilde{\sigma}$ is the cross section of stimulated autoionization. For acceleration $\tilde{\sigma} > 0$, while for deceleration $\tilde{\sigma} < 0$.

2.4. Stimulation of Autoionization by a Fast Charged Particle

Deformation of an autoionizing state as the result of a passing fast charged particle is small and may be described by perturbation theory. For simplicity we confine ourselves to the autoionizing states of $2s2p$ type. In this case the interaction with a projectile particle may be described by potential (1). The stimulated decay is a second-order effect of perturbation theory and its cross section is

$$\tilde{\sigma} = \frac{8\pi Z_1^2}{v^2} \sum_f \left(\ln \frac{v}{|\omega_{fi}|r_0} \right) \frac{|\mathbf{T}_{if}|^2}{3} \tag{17}$$

where $\omega_{fi} = \varepsilon_f - \varepsilon_i$ is the energy difference between the initial and final states, v is the velocity and Z_1 is the charge of the projectile, r_0 is a parameter of atomic dimensions for which the expansion (1) will be valid [a similar parameter is introduced in the theory of inelastic scattering of fast particles by atoms (see, for example, Reference 7], and

$$|\mathbf{T}|^2 = \left| \sum_n \frac{(\psi_f, |\mathbf{r}_1 - \mathbf{r}_2|^{-1}\psi_n)(\psi_n, (\mathbf{r}_1 + \mathbf{r}_2)\psi_i)}{\varepsilon_i - \varepsilon_n + i\delta} \right.$$
$$\left. + \sum_n \frac{(\psi_f, (\mathbf{r}_1 + \mathbf{r}_2)\psi_n)(\psi_n, |\mathbf{r}_1 - \mathbf{r}_2|^{-1}\psi_i)}{\varepsilon_f - \varepsilon_n + i\delta} \right|^2 \tag{18}$$

Here the $\psi_{i,n,f}$ are the initial, intermediate, and final atomic states, and the summation is over intermediate states; $|\mathbf{T}|^2$ is a purely atomic characteristic that can be calculated using equation (18) or in the shake-off approximation. For example, if $Z_1 = 8$, $Z_2 = 10$, $v = 20$, then $\tilde{\sigma} = 2.7 \times 10^{-5}$ a.u. As an interesting comparison, the cross section of stimulated autoionization under a slow charged particle [from equation (16) with $v = 0.1$, $Z_1 = 8$, $Z_2 = 10$] varies $\tilde{\sigma} = 0.1$–4.8 a.u. for different autoionizing states. There is a possibility of complete quenching of an autoionizing state by a projectile. The cross section of such a transition is given by equations (17) and (18) with only one difference—that ψ_f is now a wave function of the state $1s^2$. For $Z_2 = 10$, $Z_1 = 8$, and $v = 20$, $\tilde{\sigma} = -0.7 \times 10^{-7}$ a.u.

3. Some Typical Examples

Let us consider some typical experimental situations in which the above-mentioned effects are displayed. ·

3.1. Beam–Foil Event

Let an atom become excited into an autoionizing state during its passage through a thin film or by scattering from a metal surface. Then the decay of the autoionizing state will occur in the field of the retreating image charge (Figure 3). In Table 1 the dependence of autoionization rates on the distance R from the metal surface, calculated by equations (9) and (10), is given. Using these data and equation (13) one can estimate the total decay probability and the Auger electron spectrum using equation (14).

The calculations were made for heliumlike ions with atomic numbers $Z_2 = 5$ and 10; the image charge is then $Z_1 = -3$ and -8, respectively. From the table it is seen that the spectrum shape for Auger electrons ejected by ions after their passage through the film changes as they retreat from it. In addition some peaks broaden and some narrow. The experimental investigation of this effect is of great interest.

3.2. Postcollision Interaction in Atomic Collisions

In heavy-atom collisions the electron vacancies are formed in inner shells at close approach and then the atoms separate in excited states.

The excitation cross section is obtained by integrating the excitation probability $W_{ex}(\varrho, v)$ over the impact parameter ϱ:

$$\sigma_{ex} = 2\pi \int_0^\infty \varrho \, d\varrho \, W_{ex}(\varrho, v) \tag{19}$$

Let the interaction occur at time $t = 0$; then the autoionization probability may be obtained by integration along the trajectories $R(t)$ [in analogy with (13)]:

$$W_a(\varrho, v) = W_{ex}(\varrho, v) \int_0^\infty w_a(t) \exp\left[-\int_0^t w_{tot}(t) \, dt\right] dt \tag{20}$$

Here $w_a(t) \equiv w_a[R(t)]$ is the autoionization rate and the total decay rate is $w_{tot}(t) = w_a(t) + w_x(t)$, where $w_x(t)$ is the radiative decay rate.

In analogy the probability for decay via radiation is given by

$$W_x(\varrho, v) = W_{ex}(\varrho, v) \int_0^\infty w_x(t) \exp\left[-\int_0^t w_{tot}(t) \, dt\right] dt \tag{21}$$

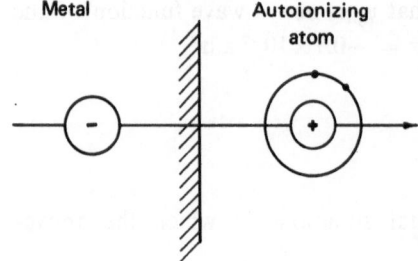

Figure 3. Autoionizing atom and its image charge.

Table 1. Autoionization Rate $w(R) \times 10^8$ per a.u. for a Heliumlike Ion in the Field of Its Image Charge

R (a.u.)	$2s^2$	$2s2p_0(s)$	$2s2p_{\pm1}(a)$	$2s2p_{\pm1}(s)$	$2p_0^2$	$2p_02p_{\pm1}(a)$	$2p_02p_{\pm1}(s)$
			$Z_1 = -8$, $Z_2 = +10$				
5	4.53	4.69	0.199	4.67	5.65	0.0008	7.08
4	4.50	4.75	0.198	4.68	5.61	0.0022	7.07
3	4.42	4.92	0.193	4.74	5.47	0.0066	7.01
2	4.14	5.54	0.175	4.96	5.00	0.0244	6.80
1	3.79	6.56	0.128	5.54	4.10	0.0722	6.21
			$Z_1 = -3$, $Z_2 = +5$				
6	4.36	5.16	0.189	4.79	5.36	0.011	6.97
5	4.22	5.52	0.181	4.88	5.14	0.019	6.86
4	4.03	6.07	0.166	5.07	4.78	0.034	6.69
3	3.85	6.68	0.145	5.33	4.35	0.055	6.42
2	3.79	7.07	0.122	5.61	4.03	0.078	6.14
1	3.82	7.18	0.105	5.81	3.88	0.094	5.94

The cross sections are, respectively,

$$\sigma_a = 2\pi \int_0^\infty \varrho \, d\varrho \, W_a(\varrho, v) \qquad \text{and} \qquad \sigma_x = 2\pi \int_0^\infty \varrho \, d\varrho \, W_x(\varrho, v)$$

The fluorescence yield is

$$\omega_x = \frac{\sigma_x}{\sigma_x + \sigma_a} = \frac{1}{\sigma_{\text{ex}}} 2\pi \int_0^\infty \varrho \, d\varrho \, W_x(\varrho, v) \tag{22}$$

We use for the excitation probability the following approximation:

$$W_{\text{ex}}(\varrho, v) = \exp(-\varrho^2/\varrho_0^2), \qquad \varrho_0 = \varrho_0(v) \tag{23}$$

For simplicity we assume that the dependence $w_a(R)$ is

$$w_a(R) = \begin{cases} w_a^{(1)} & R < R_0 \\ w_a^{(0)} & R > R_0 \end{cases} \tag{24}$$

where $w_a^{(0)}$ is the autoionization rate in isolated atoms and $w_a^{(1)}$ is the rate at close approach; then

$$\omega_x = \omega_x^{(1)} + (\omega_x^{(0)} - \omega_x^{(1)})[e^{-R_0^2/\varrho_0^2} + e^{-w_a^{(1)}R_0/v}(1 - e^{-R_0^2/\varrho_0^2})] \tag{25}$$

where $\omega_x^{(0)} = w_x/(w_a^{(0)} + w_x)$ is the fluorescence yield in an isolated atom and $\omega_x^{(1)} = w_x/(w_a^{(1)} + w_x)$ is the fluorescence yield at close approach.

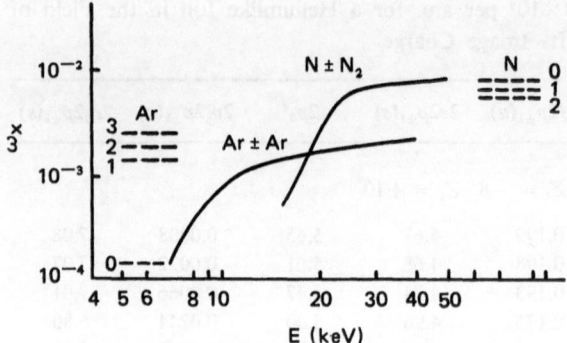

Figure 4. Fluorescence yield near the threshold, after Afrosimov et al.[9]

It should be noted that the smoothing of the step [equation (24)] slightly changes the formula for the fluorescence yield.

As mentioned in the Introduction, the effects of molecular orbitals and of the field-stimulated polarization cause a large increase in the Auger rate in the quasi-molecule at close approach; then $w_a^{(1)} \gg w_a^{(0)}$ and therefore $\omega_x^{(1)} \ll \omega_x^{(0)}$ (if w_x changes slowly with R). Then the formula (25) describes an increase in the fluorescence yield with increasing velocity of the receding collision particles. The increase is very large near the threshold for formation of an electron vacancy v_t, where $\varrho_0(v) \sim (v - v_t)^n$ is near zero. Another source of the increase is the increase of electron population in the outer shells of the autoionizing atom due to its postcollision charge exchange with the collision partner. It gives a nonadiabatic decrease of $w_a^{(1)}$ with v. The decay of the autoionizing state continues even after the collision via the interatomic Auger effect along three channels.[2] Evidently a great part of the large increase of ω_x near the threshold of K and L vacancy formation in heavy-ion collisions observed by Afrosimov et al.[9] (Figure 4) is caused by the postcollision increase of the Auger rate.

3.3. Stimulated Autoionization in a Carambole Collision

The carambole[8] collision of a fast atom with a diatomic molecule occurs at certain molecular orientations, when two successive collisions with atoms I and II of a molecule are possible (Figure 5).

Figure 5. Carambole collision. ϱ_1, ϱ_2 are impact parameters; ψ is the angle between the molecular axis and the atomic trajectory; R_1, R_2 are the distances from the projectile atom to atoms I and II.

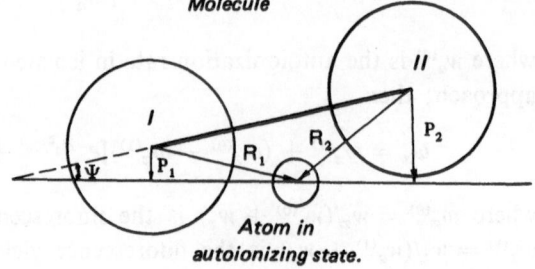

If the collision with atom I is a violent one, an excitation of the projectile atom into an autoionizing state occurs with probability $W_{ex}^I(\varrho_1, v)$. Then the autoionization probability [as in equation (13)] is

$$W_a(\varrho_1, v, \psi) = W_{ex}^I(\varrho_1, v) \int_0^\infty w_a(t) \exp\left[-\int_0^t w_{tot}(t)\, dt\right] dt \qquad (26)$$

Considering the interaction with atoms I and II as independent we may take $w_a(t) \equiv w_a(R_1(t), R_2(t))$ as

$$w_a(t) = w_1(R_1(t)) + \tilde{w}(R_2(t)) \qquad (27)$$

where $w_1(R_1(t))$ is a postcollision autoionization rate after the collision with atom I (as discussed in Section 2) and $\tilde{w}(R_2(t))$ is the rate of stimulation of autoionization on passing by atom II [as in equation (15)].

Since the excitation of the autoionizing state occurs at small ϱ_1 we can integrate (26) over ϱ_1 and average it over the molecular orientations (assuming ψ to be small).

As a result, the total cross section for formation and decay of an autoionizing state in carambole atom–molecule collisions is

$$\sigma^I = \sigma_a^I + \sigma_x^I \frac{\tilde{\sigma}_{st}^{II}}{4\pi D^2} \qquad (28)$$

where D is the distance between atoms I and II, and

$$\sigma_a^I + \sigma_x^I = \sigma_{ex}^I = 2\pi \int_0^\infty \varrho_1\, d\varrho_1\, W_{ex}^I(\varrho_1, v)$$

is the cross section for excitation of the autoionizing state in the collision with atom I, and

$$\sigma_a^I = 2\pi \int_0^\infty \varrho_1\, d\varrho_1\, W_{ex}^I(\varrho_1, v) \int_0^\infty w_I(t) \exp\left[-\int_0^t w_{tot}(t)\, dt\right] dt$$

is the cross section for autoionization in the collision only with atom I. Also

$$\tilde{\sigma}^{II} = 2\pi \int_0^\infty dt \int_0^\infty \tilde{w}(t)\varrho_2\, d\varrho_2$$

is the cross section for autoionization stimulation by the atom II. If the collision with atom II is a violent one and precedes the passage by atom I, then the carambole cross section will be

$$\sigma^{II} = \sigma_a^{II} + \sigma_x^{II} \frac{\tilde{\sigma}_{st}^I}{4\pi D^2} \qquad (29)$$

The atoms I and II are assumed in general to be different. The total cross section is

$$\sigma = \sigma^I + \sigma^{II} = \sigma_a^I + \sigma_a^{II} + \frac{\sigma_x^I \tilde{\sigma}_{st}^{II} + \sigma_x^{II} \tilde{\sigma}_{st}^I}{4\pi D^2} \qquad (30)$$

The effect described above is of great interest for experimental investigation. It provides a unique possibility for studying the rapidly decaying autoionizing state at a time $t = D/v \approx 10^{-14}$–10^{-15} after its formation.

It is possible that a part of the large change of fluorescence yield observed by Afrosimov et al.[9] in heavy-ion collisions results from the fact that in their experiments molecular nitrogen was used as a target.

In a more detailed investigation one could calculate the stimulated decay for different trajectories in double atom–molecule scattering that differ by elastic energy losses.

4. Conclusion

The aim of this chapter was to discuss some effects connected with collision-stimulated autoionization. The experiments studying such effects are expected to give rise to a more detailed theoretical investigation.

References

1. V. V. Afrosimov, Yu. S. Gordeev, A. N. Zinoviev, D. H. Rasulov, and A. P. Shergin, *JETP Lett.* **24**, 33 (1976).
2. L. M. Kishinevsky and E. S. Parilis, *Fifth International Conference on Physics of Electronic and Atomic Collisions*, Abstracts of papers, p. 100, Nauka, Leningrad (1967).
3. R. B. Barker and H. W. Berry, *Phys. Rev.* **151**, 14 (1966).
4. A. Niehaus, *Tenth International Conference on Physics of Electronic and Atomic Collisions*, Invited papers and progress reports, G. Watel, ed., p. 185, North-Holland Publishing Company, Amsterdam (1978).
5. F. H. Read, *Ninth International Conference on Physics of Electronic and Atomic Collisions*, Invited papers, p. 176 (1975).
6. V. N. Ostrovsky, *JETP* **72**, 2079 (1977).
7. L. D. Landau and E. M. Lifshitz, *Quantum Mechanics*, Addison-Wesley, Reading, Mass. (1963).
8. C. Foster and F. W. Saris, *Eighth International Conference on Physics of Electronic and Atomic Collisions*, Abstracts of papers, B. C. Čobič and M. V. Kurepa, eds., p. 716, Institute of Physics, Belgrade (1973).
9. V. V. Afrosimov, Yu. S. Gordeev, A. N. Zinoviev, G. G. Meskhi, and A. P. Shergin. *Tenth International Conference on Physics of Electronic and Atomic Collisions*, Abstracts of papers, p. 202, Commissariat a L'Energie Atomique, Paris (1977).

22

Electronic Correlation in Electron–Atom Scattering

R. K. NESBET

The quantitative theory of electronic correlation effects in low-energy electron scattering by complex target atoms is reviewed.

1. Introduction

In electron–atom scattering, electronic correlation produces not only directly observable qualitative effects (such as the postcollision interaction when an autoionizing state is excited), but also quantitative changes in cross sections. Correlation between the incident electron and electrons of the target atom results in the long-range polarization potential that dominates low-energy scattering by neutral atoms. A quantitative theory of structural features such as resonances and threshold effects must also include electronic correlation within the target atom. Accurate absolute cross-section calculations, needed to calibrate experimental data, also require a systematic quantitative treatment of electronic correlation effects.

This article reviews the present status of the quantitative theory, with particular reference to special problems relevant to low-energy electron scattering by open-shell target atoms. The theory of polarization potentials is considered in Section 2, and of internal target atom correlation in Section 3. A critique of quantitative theoretical methods for complex atoms is given in Section 4.

R. K. NESBET • IBM Research Laboratory, San Jose, California 95193, U.S.A.

2. Polarization Potentials

Electron scattering by an N-electron atom can be described by a stationary-state wave function[1,2]

$$\Psi = \sum_p \mathscr{A}\theta_p\psi_p + \sum_\mu \Phi_\mu c_\mu \tag{1}$$

Here θ_p is a normalized N-electron target state wave function, ψ_p is a one-electron *channel orbital*, antisymmetrized into θ_p by the operator $\mathscr{A}$, and Φ_μ is an $(N+1)$-electron function constructed from quadratically integrable one-electron orbital basis functions. The functions ψ_p are not quadratically integrable if $E_p \leq E$, where E is total energy and E_p is an eigenvalue of $(\theta_p \mid H \mid \theta_q)$, assumed to be diagonalized. The open-channel orbitals ψ_p are oscillatory for large r. Their asymptotic phases determine scattering matrices and cross sections. The part of Ψ containing open-channel orbitals remains distinct from the quadratically integrable part for any calculation using a finite set of functions $\{\Phi_\mu\}$.

In close-coupling theory,[3] part of the quadratically integrable or *bound component* of Ψ is represented in the form

$$\sum_\gamma \mathscr{A}\theta_\gamma\psi_\gamma \tag{2}$$

analogous to the open-channel part of equation (1), except that the functions θ_γ represent target states with $E_\gamma > E$, corresponding to closed channels. The closed-channel orbital functions ψ_γ are quadratically integrable.

The target-atom functions θ_γ need not correspond to specific stationary states, although they should be orthogonal to the open-channel states $\{\theta_p\}$. In close-coupling theory polarization effects are represented by using *pseudostates*[4-6] $\theta_{\gamma(p)}$ that correspond to the first-order perturbation of θ_p by a polarizing field. In the polarized orbital method,[7] the closed-channel orbital ψ_γ to be combined with a pseudostate $\theta_{\gamma(p)}$ is replaced by a function $\chi_{(i)}\psi_p$, where $\chi_{(i)}$, which depends on the coordinates of two electrons, describes the modulation of ψ_p due to induced polarization of the target atom. The antisymmetrized function

$$\mathscr{A}\sum_i \theta_{\gamma(p)}\chi_{(i)}\psi_p \tag{3}$$

is quadratically integrable.

In the matrix variational method,[1,2] target-atom polarization effects are represented by including suitable basis functions $\{\Phi_\mu\}$ in the bound component of Ψ. Variational theory gives a system of inhomogeneous linear equations for the coefficients c_μ. Formal solution of these equations is equivalent to a partitioning technique used in resonance theory.[8] The bound component of Ψ defines a projection operator Q such that

$$\Psi_Q = Q\Psi = \sum_\mu \Phi_\mu(\Phi_\mu \mid \Psi) = \sum_\mu \Phi_\mu c_\mu \tag{4}$$

If operator P is the orthogonal complement of Q, then

$$\Psi_P = P\Psi \cong \sum_p \mathscr{A}\theta_p \psi_p \tag{5}$$

The modified Schrödinger equation is

$$M'_{PP}\Psi_P = [M_{PP} - M_{PQ}(M_{QQ})^{-1}M_{QP}]\Psi_P = 0 \tag{6}$$

where M denotes $H - E$. The operator M_{QQ}^{-1} is a linear integral operator with kernel

$$\sum_{\mu\nu} \Phi_\mu (H - E)^{-1}_{\mu\nu} \Phi_\nu^* \tag{7}$$

Equation (6) provides a common basis for various practical methods of computing correlation effects between the external electron and the target electrons. The physical nature of these effects can be examined by deriving effective one-electron equations. The matrix operator acting on channel orbitals is

$$m^{pq} = (\theta_p \mid M'_{PP} \mid \mathscr{A}\theta_q) \tag{8}$$

The terms in M'_{PP} arising from $(M_{QQ})^{-1}$ define a matrix optical potential. In m^{pq} this operator acts on channel orbitals and describes correlation and polarization effects. In its full form the optical potential is an energy-dependent integral operator. For the terms arising from virtual electric dipole excitations of the target atom, detailed analysis shows that in the limit of large radial coordinate r this operator is equivalent to a polarization potential of the form $-\alpha/2r^4$. The parameter α is the electric dipole polarizability of the target atom. In general, virtual target atom excitations of multipole index l produce a multipole polarization potential of the asymptotic form $-\alpha_l/2r^{2l+2}$, where α_l is a generalized multipole polarizability.

In practice, only a few functions θ_y are included in close-coupling calculations. Pseudostate polarization functions $\theta_{y(p)}$ can give accurate values of the static electric dipole polarizability of a given target state θ_p. These functions are known in closed form for hydrogen,[4] and have been computed variationally for complex atoms.[6]

Close-coupling equations for the external open- and closed-channel orbitals ψ_p and $\psi_{y(p)}$ are obtained from the matrix Hamiltonian operator m^{pq}, equation (8). Matrix elements between $\mathscr{A}\theta_p \psi_p$ and $\mathscr{A}\theta_{y(p)}\psi_{y(p)}$ arise from the electronic Coulomb interaction. When expanded in spherical polar coordinates of two electrons, the Coulomb potential depends on the radial coordinates through a factor

$$r_<^\lambda / r_>^{\lambda+1} \tag{9}$$

multiplying spherical harmonics of degree λ in the angular variables. Here $r_<$ is the lesser of r_1, r_2 and $r_>$ is the greater. Off-diagonal matrix elements of spherical harmonics λ connect a target-atom open-channel state of given L to pseudostates with

$$L' = |L - \lambda|, \quad |L - \lambda| + 2, \ldots, L + \lambda \tag{10}$$

In the close-coupling equations such matrix elements produce an effective off-diagonal potential proportional to $1/r^{\lambda+1}$, connecting the external closed-channel and open-channel orbitals. The partitioning transformation indicated in equation (6) causes this off-diagonal element to contribute quadratically to an effective polarization potential acting in the open channel. This potential is proportional to $1/r^{2\lambda+2}$ for large r. When $\lambda = 1$, this is the electric dipole polarization potential.

3. Target–Atom Electronic Correlation

It is customary to refer to corrections to the Hartree–Fock approximation as electronic correlation effects. Correlation between the external electron and electrons of the target atom directly affects electron scattering through the long-range polarization potential. This is qualitatively and quantitatively more significant than the expected effects of electronic correlation within the target atom. Nevertheless, for accurate quantitative predictions it is necessary to consider target-atom electronic correlation. Although most electron–atom scattering calculations have not included such correlation, work of this kind is becoming feasible. Recent calculations by O'Malley et al.[9] on e^-–He scattering below 19 eV represent one of the first serious studies of such effects.

Since the Coulomb interaction is a two-electron operator, the dominant effect of electronic correlation is the modification of electron-pair components of the wave function. Bound-state calculations of atomic energy levels that include electronic pair correlation quantitatively appear to be adequate (energy differences accurate to a few percent) for excitation energies and ionization potentials.[10,11] However, it has been necessary to include three-electron and higher-order correlation terms in order to compute electron affinities of comparable accuracy.[11]

In elastic scattering, the absolute value of the target state energy is not significant if the effective potential acting on the scattered electron is computed accurately. The principal requirement is accurate computation of the static polarizability. Correlation corrections to Hartree–Fock polarizabilities are relatively small. For a closed-shell atom, the physical effect of electronic correlation, which represents a weakening of the electronic Coulomb repulsion, is contraction of the wave function. This should systematically reduce the polarizability (by a dimensional argument) and weaken the polarization potential. The effect is to reduce low-energy scattering phase shifts (modulo π), as found in recent calculations.[9]

Electronic correlation is described by admixture into a Hartree–Fock wave function of other electronic configurations, described by virtual electronic excitations. This will tend to reduce matrix elements contributing to the decay of resonant states, so electronic correlation should reduce computed resonance widths. In the case of e^-–He resonances, correlation reduces the coefficient of the configuration $1s^2$ in the He ground-state wave function. The dominant configurations $1s2s^2$ and $1s2p^2$ of the ^2S resonance state at 19.36 eV interact through Coulomb matrix elements with the configuration $1s^2ks$ of the ground-state continuum, but not with perturbing configurations $(nl)^2 ks$ unless $n = 2$. New calculations are needed to explore this expected effect of electronic correlation on the resonance width.

In the case of inelastic scattering, the relative spacing of target energy levels must be determined relatively accurately. In general this will require quantitative inclusion of target-atom electron-pair correlation effects, not yet possible in a scattering calculation for atoms beyond helium. Calculations on helium, in the $n = 2$ and $n = 3$ excitation regions, have taken advantage of the fact that the dominant electronic configurations are those of two electrons outside a tightly bound He^+ core. Except for the ground-state polarization potentials (dipole and quadrupole), the only important correlation is between the excited target electron and the external electron. Recent variational calculations[12] show that the relative spacing of the target-atom excited energy levels can be obtained adequately by considering the He^+ core as frozen.

The situation is more complicated for open-shell atoms.[2,13] In the case of C, N, and O atoms, the open-shell structure $2s^2 2p^k$ produces residual target correlation energy effects due to one-electron virtual excitations $2s \rightarrow ns$, nd or $2p \rightarrow np$. In contrast to closed-shell configurations, these effects cannot be made to vanish by a Hartree–Fock calculation. In computing the ground-state polarizability or polarization potential, the level of approximation represented by single virtual excitations from the ground-state configuration tends to underestimate the polarizability. The ground-state energy H_{00} is stabilized by one-electron "correlation" effects, not included in the polarization function Ψ_1. Hence the energy denominator $H_{11} - H_{00}$ in the perturbation formula for the polarizability is too large, and the polarizability is systematically underestimated.[6]

The second problem encountered with open-shell target-atom states is due to the ambiguous nature of an open-shell orbital such as $2p$ in configurations $2s^2 2p^k$. When $2p$ occurs as a component of the orbital wave function for an external p wave, it introduces configurations into the scattering wave function Ψ, corresponding to virtual excitations of $2s^2 2p^{k+1}$, that are of the same form as virtual excitations of $2s^2 2p^k$ that describe correlation effects in the target wave function Ψ_0. A balanced representation of correlation effects in Ψ_0 and Ψ requires either that Ψ be restricted in form to avoid such correlation effects, if absent from Ψ_0, or that Ψ_0 be augmented to include all virtual excitation effects implied by the structure of Ψ. However, if Ψ_0 is augmented in this way new structure is introduced into Ψ, and the imbalance reappears at a higher level of virtual excitations. This imbalance tends to overestimate the binding energy of negative ion states or scattering resonances.[2]

4. Critique of Methods for Complex Atoms

The use of polarization pseudostates $\theta_{\gamma(p)}$ simplifies the structure of the bound component of the scattering wave function. An alternative approach, using a hierarchy of continuum Bethe–Goldstone equations,[14] is to restrict the pattern of virtual excitations considered in the set of quadratically integrable functions $\{\Phi_\mu\}$. In practice, calculations have been limited to configurations in which single virtual excitations of a reference target state configuration are coupled to an external orbital function. Calculations with this structure have been carried out with the matrix variational method.[1,2]

The essential difference between these two approaches is that in the pseudostate method relative coefficients of the component functions used to construct $\theta_{\gamma(p)}$ are computed in advance of the scattering calculation. In the matrix variational method, these relative coefficients are determined variationally as part of the scattering calculation. By limiting the number of coupled equations to reasonable size, the pseudostate method greatly simplifies scattering calculations by close-coupling or R-matrix[15] methods.

For open-shell target atoms, the pseudostate method enforces compatibility between the levels of electronic correlation in the target and scattering wave functions. In this "polarized frozen-core" approximation[6,16] virtual excitations of the reference state occur only in the frozen functions θ_p for the target-atom state and $\theta_{\gamma(p)}$ for its corresponding polarization pseudostate. Target correlation effects other than those explicitly included in θ_p are excluded from the scattering wave function. This enforced compatibility shows clearly in the very satisfactory results of calculations of electron affinities, where the "scattering" wave function refers to a negative ion state.[16]

However, the rigidity of structure of the wave function used in the pseudostate method may make it unsuitable for calculations intended to meet strict quantitative criteria of accuracy. The pseudostate function $\theta_{\gamma(p)}$ is computed variationally to represent the static polarizability of a specific target state. In inelastic scattering, the electric dipole scattering effect is analogous to photoexcitation with respect to energy transfer, and should be described by a frequency-dependent polarizability. The static electric dipole polarizability is one energy moment of the oscillator strength distribution function. Representation of the frequency-dependent polarizability requires accurate values of at least several low-order moments, including the dipole oscillator strength sum. In calculations on open-shell atoms it has been shown that this sum is not given accurately by variational calculations satisfactory for the polarization pseudostate.[17] This implies that some allowance should be made for modification of $\theta_{\gamma(p)}$ with excitation energy transfer in calculations of inelastic scattering beyond the threshold region.

Another difficulty with the pseudostate method is that $\theta_{\gamma(p)}$ is computed for a physical situation in which the external electron is sufficiently remote from the target atom that it can be represented by uniform electric field. Short-range correlation effects, requiring modification of $\theta_{\gamma(p)}$ for small r, are not taken into account.

The alternative approach,[14] using the matrix variational method and a hierarchy of virtual excitations to construct the bound-state basis $\{\Phi_\mu\}$, removes these restrictions due to the structural rigidity of the pseudostate method. However, for open-shell target states, a satisfactory way to maintain consistency between computed correlation energies of target and scattering states has not yet been found.[2,13]

In practice, and *ad hoc* net correlation energy correction has been included in scattering calculations.[2] A more systematic method is needed for converged quantitative calculations on open-shell atoms.

References

1. R. K. Nesbet, *Adv. Quantum Chem.* **9**, 215 (1975).
2. R. K. Nesbet, *Adv. At. Mol. Phys.* **13**, 315 (1978).
3. P. G. Burke, *Adv. At. Mol. Phys.* **6**, 288 (1973); P. G. Burke and M. J. Seaton, *Methods Comput. Phys.* **10**, 1 (1971).
4. R. J. Damburg and E. Karule, *Proc. Phys. Soc.* (*London*) **90**, 637 (1967); R. J. Damburg and S. Geltman, *Phys. Rev. Lett.* **20**, 485 (1968).
5. P. G. Burke and J. F. B. Mitchell, *J. Phys. B* **7**, 665 (1974).
6. Vo Ky Lan, M. LeDourneuf, and P. G. Burke, *J. Phys. B* **9**, 1065 (1976).
7. A. Temkin, *Phys. Rev.* **107**, 1004 (1957); R. J. Drachman and A. Temkin, in *Case Studies in Atomic Collision Physics*, E. W. McDaniel and M. R. C. McDowell, eds., Vol. 2, p. 399, American Elsevier, New York (1972).
8. H. Feshbach, *Ann. Phys.* (*N.Y.*) **5**, 357 (1958); **19**, 287 (1962).
9. T. F. O'Malley, P. G. Burke, and K. A. Berrington, *J. Phys. B* **11** (1978).
10. A. Hibbert, *Rep. Prog. Phys.* **38**, 1217 (1975).
11. C. M. Moser and R. K. Nesbet, *Phys. Rev. A* **4**, 1336 (1971).
12. R. K. Nesbet, *J. Phys. B* **11**, L21 (1978).
13. R. K. Nesbet, *Int. J. Quantum Chem. Symp.* **11**, 263 (1977).
14. R. K. Nesbet, *Phys. Rev.* **156**, 99 (1967).
15. P. G. Burke and W. D. Robb, *Adv. At. Mol. Phys.* **11**, 143 (1975)
16. M. LeDourneuf and Vo Ky Lan, *J. Phys. B* **10**, L97 (1977).
17. R. K. Nesbet, *Phys. Rev. A* **16**, 1 (1977).

23

Observation of the Postcollision Interaction in the Scattered-Electron Spectra of Helium

D. Roy

This paper reports measurements of differential electron energy-loss spectra in helium, involving residual scattered-electron energy from 1 eV to about 40 eV. Strong influence of the postcollision interaction is observed at low residual energy, mainly for the feature around 58 eV. Its line shape is quite similar to those obtained in ejected-electron spectra. Comparison is made with such spectra measured in the same conditions.

1. Introduction

Among the scattered-electron ("energy-loss") measurements carried out in helium, only the integral spectra reported by Spence[1] clearly evidenced the effects of the postcollision interaction (PCI) between the ejected and the scattered electrons, following the autoionization process. The few differential spectra available for the 60-eV autoionization region are those of Silverman and Lassettre[2] at 500-eV incident electron energy, Simpson et al.[3,4] at energies between 90 and 400 eV, Boersch et al.[5] at 25 keV, and Dillon and Lassettre[6] at 400 eV. All those measurements were thus carried out at energies where the PCI is imperceptible. However, as recently shown by Wilden et al.[7] in argon, it is interesting to correlate differential ejected-electron spectra with scattered-electron spectra for a better understanding of the PCI process; but so far this has not been possible in helium.

D. Roy • Département de Physique, et Centre de Recherches sur les Atomes et les Molécules (CRAM), Université Laval, Québec, P. Qué. G1K 7P4, Canada.

This paper reports the measurements of a series of differential electron energy-loss spectra involving residual scattered-electron energy (E_s) ranging from 1 eV to about 40 eV. All the measurements were carried out in the forward direction ($\theta = 2°$), with a high-resolution electron spectrometer, involving a static gas target.[8,9] Two modes of measurements were used: the "constant incident energy" mode, and the "constant residual energy" mode. In the former, only the energy loss is swept (this is a standard energy-loss spectrum[9]), while in the latter, which is preferred for the near-threshold measurements, both the energy loss and the incident-electron energy are swept.

2. Results and Discussion

Seven of these spectra are reproduced in Figure 1. The two upper spectra were measured at a constant incident-electron energy of 100 and 80 eV, respectively ($E_s \simeq 40$ and 20 eV), while the others were at constant residual energy. In the upper spectrum, the features evidently correspond to the four autoionizing states lying in this region: the $(2s^2)\,^1S$, $(2s2p)\,^3P$, $(2p^2)\,^1D$, and $(2s2p)\,^1P$ states. In the lower spectra, however, the shapes, relative magnitudes, and energy positions of the features are drastically modified under the influence of the PCI. While the line shapes of the features in the spectrum measured at 100 eV are quite standard for energy losses following autoionizing state excitation, they become very complex when the PCI is involved.

It is interesting to compare these scattered-electron spectra with ejected-electron spectra measured in the corresponding conditions. Figure 2 presents such spectra for a residual electron energy (E_s) ranging from 0 to 10 eV. Three among these seven spectra were presented in a preceding paper dealing with the PCI.[9] All were measured in the forward direction, and in the mode "constant energy loss," for which only the incident-electron energy is swept; the residual electron energy (E_s) is the difference between the value of the energy loss and the ionization energy.[9] In order to make clearer which autoionizing state is responsible for the features, the ejected-electron energy scale is shifted by an amount corresponding to the ionization energy (24.59 eV).

Since the two sets of spectra were obtained in the same direction, it was expected not to obtain the same line shapes nor the same magnitudes of the features because of the correlation effects between the two outgoing electrons. In both cases, it is clear that a complex interaction occurs when the residual electron energy is lowered. Particularly the feature around 58 eV exhibits oscillations and multiple extrema, still more pronounced in the scattered-electron spectra; this line shape is even more complex than in the integral spectra measured by Spence.[1] In fact the profiles of this feature are quite similar to those obtained in the differential measurements of ejected-electron spectra of Hicks et al.,[10] taken at various ejection angles, and later

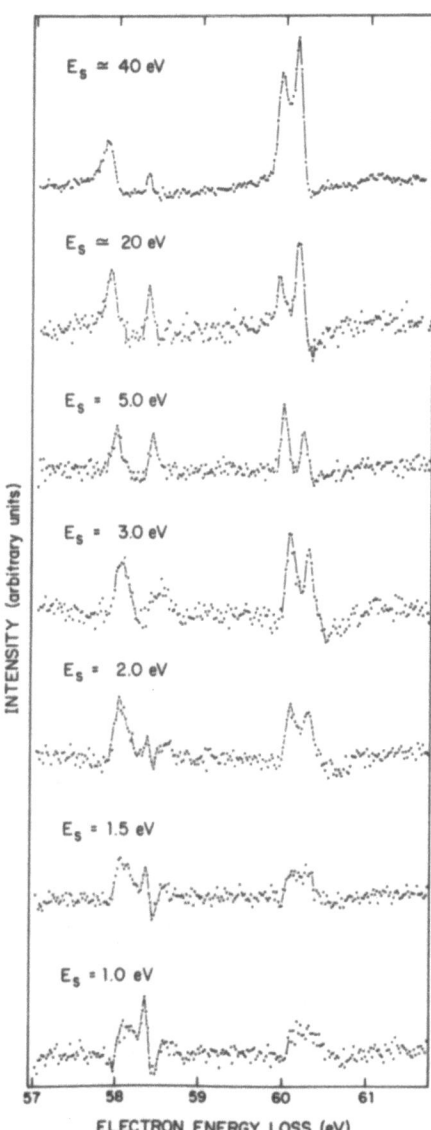

Figure 1. Differential energy-loss spectra in helium, measured in the forward direction, for various values of the residual scattered-electron energy.

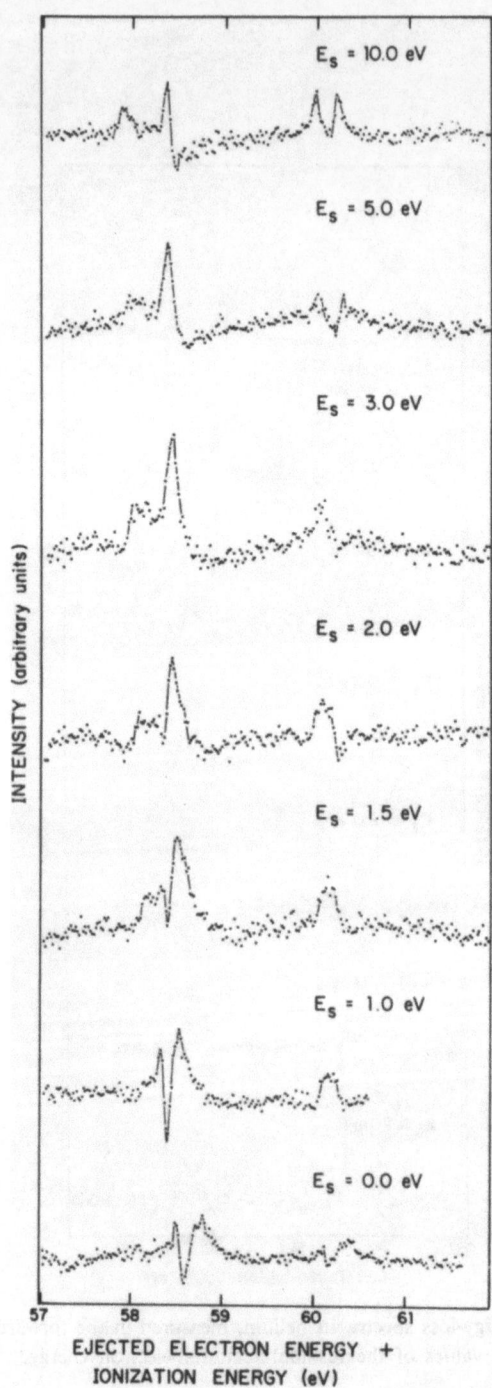

Figure 2. Differential ejected-electron spectra in helium, measured in the forward direction, for various values of the residual scattered-electron energy.

reproduced by means of the models proposed by Morgenstern *et al.*[11] and Read.[12] Systematic coincidence measurements would be quite helpful for a complete description of the phenomenon.

References

1. D. Spence, *Phys. Rev. A* **12**, 2353–2360 (1975).
2. S. M. Silverman and E. N. Lassettre, *J. Chem. Phys.* **40**, 1265–1271 (1964).
3. J. A. Simpson, S. R. Mielczarek, and J. Cooper, *J. Opt. Soc. Am.* **54**, 269–270 (1964).
4. J. A. Simpson, G. E. Chamberlain, and S. R. Mielczarek, *Phys. Rev.* **139**, A1039–A1041 (1965).
5. H. Boersch, J. Geiger, and B. Schröder, *Physics of the One- and Two-Electron Atoms*, pp. 637–641, North-Holland, Amsterdam (1969).
6. M. A. Dillon and E. N. Lassettre, *J. Chem. Phys.* **62**, 4240–4241 (1975).
7. D. G. Wilden, J. Comer, and P. J. Hicks, *Nature* **273**, 651–653 (1978).
8. D. Roy, A. Delâge, and J.-D. Carette, *J. Phys. E* **8**, 109–114 (1975).
9. D. Roy, A. Delâge, and J.-D. Carette, *J. Phys. B* **11**, 895–908 (1978).
10. P. J. Hicks, S. Cvejanović, J. Comer, F. H. Read, and J. M. Sharp, *Vacuum* **24**, 573–580 (1974).
11. R. Morgenstern, A. Niehaus, and U. Thielmann, *J. Phys. B* **10**, 1039–1058 (1977).
12. F. H. Read, *J. Phys. B* **10**, L207–L212 (1977).

24

Excitation of Autoionizing States in Noble Gases by Fast Electrons

V. V. Balashov, A. N. Grum-Grzhimailo, N. M. Kabachnik,
A. I. Magunov, and S. I. Strakhova

The basic results of the theoretical elaboration of the coincidence (e, 2e) method are summarized and the most interesting outstanding problems of the theory of this method are discussed. The characteristics of $(ns)^{-1}[(n+1)p]^1P$ autoionizing states in Ne and Ar are presented.

1. Introduction

A theory has been suggested[1,2] to unify the description of effects involving the excitation and decay of autoionizing states in the interaction of atoms with fast charged particles. The theory was formulated specifically for the study of the properties of such states by the electron–electron angular correlation method [the (e, 2e) coincidence technique].[2]

Of all the known methods of spectroscopy of autoionizing states, the coincidence method (e, 2e) is, in principle, the most versatile and sensitive one. However, many questions relating to both its potentialities and the optimal conditions for its application have still been scantily explored.

Some of the questions that arise in the theoretical description of the (e, 2e) study of autoionizing states are similar to those that are characteristic of the theory of electron–photon correlations in the process of deexcitation of the discrete atomic states excited by inelastic electron scattering.[3] The spectroscopy of autoionizing

V. V. Balashov, A. N. Grum-Grzhimailo, N. M. Kabachnik, A. I. Magunov, and S. I. Strakhova ● Institute of Nuclear Physics, Moscow State University, Moscow 117234, U.S.S.R.

states is more complicated than the spectroscopy of discrete excited states. The theory requires a reliable description not only of the resonance transitions but also of the direct ionization of atoms in the adjacent continuum. The kinematic possibilities of the (e, 2e) method permit one to vary the relation between the contribution of the direct and resonance processes and the character of interference between them.

The first applications of that theory involved experiments with He atoms.[4-6] Subsequently, resonances in the ejected electron spectra from the electron impact ionization of alkali atoms were analyzed.[7] From the theoretical point of view in both cases we deal with resonances of the simplest kind, that is autoionizing states in helium such as $(2s2p)$ ^1P, $(2s)^2$ ^1S, etc., which are between the first and the second ionization thresholds, and autoionizing states in the alkali atoms with $(np)^{-1}[(n+1)s]^2$ ^1P configuration which interact with only one open channel.

In References 8 and 9 the theory proposed in References 1 and 2 was generalized to the case of the decay of an autoionizing state through several (in particular, two) channels, and this generalized variant of the theory was applied to the description of the excitation and decay of the $(ns)^{-1}[(n + 1)p]$ ^1P states in noble gas (neon and argon) atoms.

Until now most of our calculations have been made within the framework of the plane-wave Born approximation for the incoming and scattered electron. This fact certainly limits the scope of application of many of the results and it does not permit their quantitative comparison with the experimental data using electron beams of medium energy. In Reference 10 an attempt was made to go beyond the limits of the Born approximation by using a variant of Feshbach's many-channel diffraction theory,[11] but this problem still requires further comprehensive study.

In the present work we summarize the basic results of the theoretical description of the coincidence (e, 2e) method for the study of autoionizing states and discuss the most interesting outstanding problems of the theory of this method. We note that all our investigations in the field of collision spectroscopy of autoionizing states of atoms, including those that involve the coincidence method, are inseparable from the study of similar problems of nuclear physics. The problem of a correct unified description of the direct and resonance processes in nuclear disintegration by high-energy particles is especially urgent.[12,13]

2. Direct Ionization of Atoms by Electron Impact at Small Momentum and Energy Transfers

Progress in the study of the spectroscopy of atomic autoionizing states excited by fast electrons requires a special investigation of the mechanism of direct ionization of atoms by electron impact at small energy and momentum transfers.[14]

The well-known (e, 2e) experiments on the quasifree knock-out of an electron[15] are used to determine the properties of the ground-state target atom. This method of the angular correlation of the scattered and ejected electrons can also give valuable information on the properties of the wave function for a highly excited atom in the continuum.

At present, such measurements of triple differential cross sections (TDCS) in the (e, 2e) experiments are available for helium[16] and for the outer shells of neon and argon.[17,18] The experimental data, obtained for Ne at fixed values of momentum

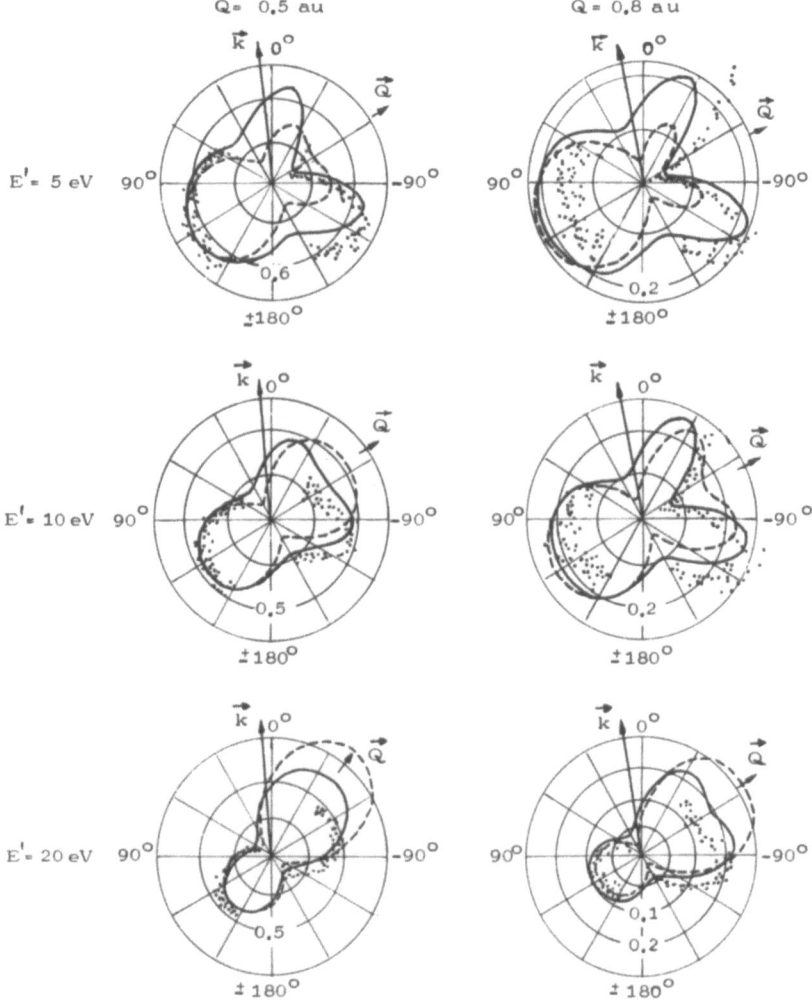

Figure 1. The triple differential cross section for the ionization of the $2p$ subshell of Ne for several values of ejected electron energy E' and momentum transfer Q. The ejected electron emission angle is measured from the direction of the incident beam $\mathbf{k_0}$. The direction of the scattered electron is $\mathbf{k}$ and $\mathbf{Q}$ is the direction of the momentum transfer given by $\mathbf{Q} = \mathbf{k_0} - \mathbf{k}$. The solid curve shows the results of the calculation in the Born approximation using the Herman–Skillman wave function for the ejected electron.[20] The dashed curve is the calculation in the VIA approximation by Knapp and Schulz.[19] Experimental points are from the work by Jung *et al.*[17] The values of E' (eV) and Q (a.u.) are indicated. The data are normalized in the recoil direction at a minimal value of $Q = 0.5$.

and energy transfers, show that the shape of the TDCS as a function of the emission angle of the ejected electron varies slightly with the incident electron energy.[17] Therefore, in theoretical calculations one can use the first Born approximation for the incident electron.

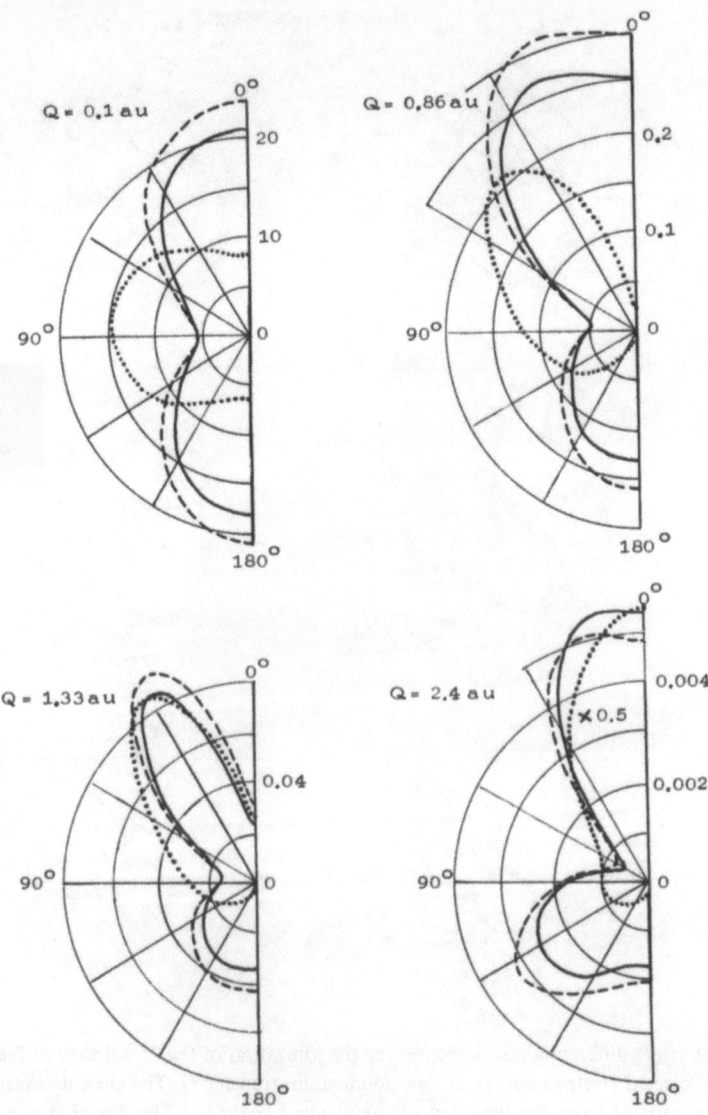

Figure 2. The triple differential cross section $[(d^3\sigma/d\Omega\, d\Omega'\, dE)_{\text{direct}}\, (a_0^2/\text{sr}^2\, \text{Ry})]$ for the direct ionization of the $2p$ subshell of Ne at the ejected electron energy $E' = 1.764$ Ry, corresponding to the $(2s)^{-1}(3p)^1P$ resonance, for different momentum transfers Q. ——, HF calculation; ---, HS calculation; $\cdots$, the plane-wave calculation including the orthogonalization of the wave functions of the initial and final states. The angles are measured from the direction of the momentum transfer.

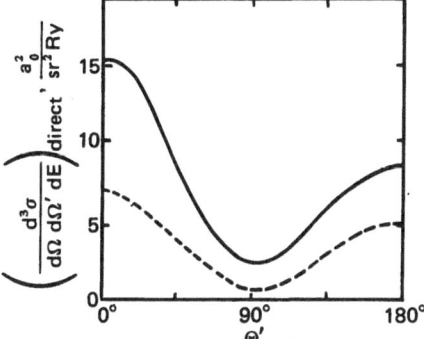

Figure 3. The triple differential cross section for the direct ionization of the $3p$ subshell of Ar at the ejected electron energy $E' = 0.798$ Ry for $Q = 0.346$ a.u. The solid curve is the HF calculation and the dashed curve is the HS calculation.

For such calculations[16,19] where the Coulomb interaction of the ejected electron with the ion and the exchange with Ne^+ electrons were taken into account, the results show a qualitative discrepancy with experiment at some of energy and momentum transfer values. It is therefore important to take into account the distortion of the ejected-electron wave function more accurately. Figure 1 shows the results of our calculation[20] of the TDCS for the ionization of the $2p$ subshell of Ne using the Born approximation and the functions calculated in the Herman–Skillman (HS) potential model for the distorted wave functions of the ejected electron. Similar results were obtained using more elaborate Hartree–Fock (HF) wave functions. Comparison of the calculated TDCS with the experimental data shows that better account of the ejected-electron–residual-ion interaction allows one to obtain better qualitative agreement with experiment and, in particular, of the binary-to-recoil intensity ratio.

We note that the experimental shape of the TDCS has characteristic features, such as the shift of the symmetric pattern with respect to the direction of momentum transfer to the region of larger angles which cannot be obtained in the Born approximation. This effect could be accounted for by the distorted-wave approximation.

Figure 2 shows the results of calculations[8,9] of the angular correlation for the direct $(e, 2e)$ process at an ejected-electron energy corresponding exactly to the decay of the autoionizing state $(2s)^{-1}(3p)$ ^1P in neon. Figure 3 shows the same for the $(3s)^{-1}(4p)$ ^1P in argon.

3. Comparative Analysis of the Feature of the Resonance Ionization for Helium and Neon Atoms

Theoretical investigations[2] indicated, and experiments later confirmed,[6] that in an $(e, 2e)$ experiment the profile of the autoionizing resonance $(2s2p)$ ^1P of helium varied greatly with the electron ejection angle. In particular it was shown that in the directional change from the angles close to the direction of the momentum transfer vector $\mathbf{Q} = \mathbf{k}_0 - \mathbf{k}$ to the hemisphere which is "backward" relative to this direction ("antiquasielastic kinematics"), the resonance becomes much more symmetric because the contribution of direct transitions decreases with such variation of the kinematics.

In the case of the $(2s)^{-1}(3p)$ ^1P resonance in neon no such behavior is observed[8] (see Figure 4). The shape of the resonance in the $(e, 2e)$ process varies slightly with the ejection angle and reproduces the general features of a resonance shape in the scattered electron spectra at the corresponding values of Q (see Figure 5). What is the cause of such a strong difference between helium and neon?

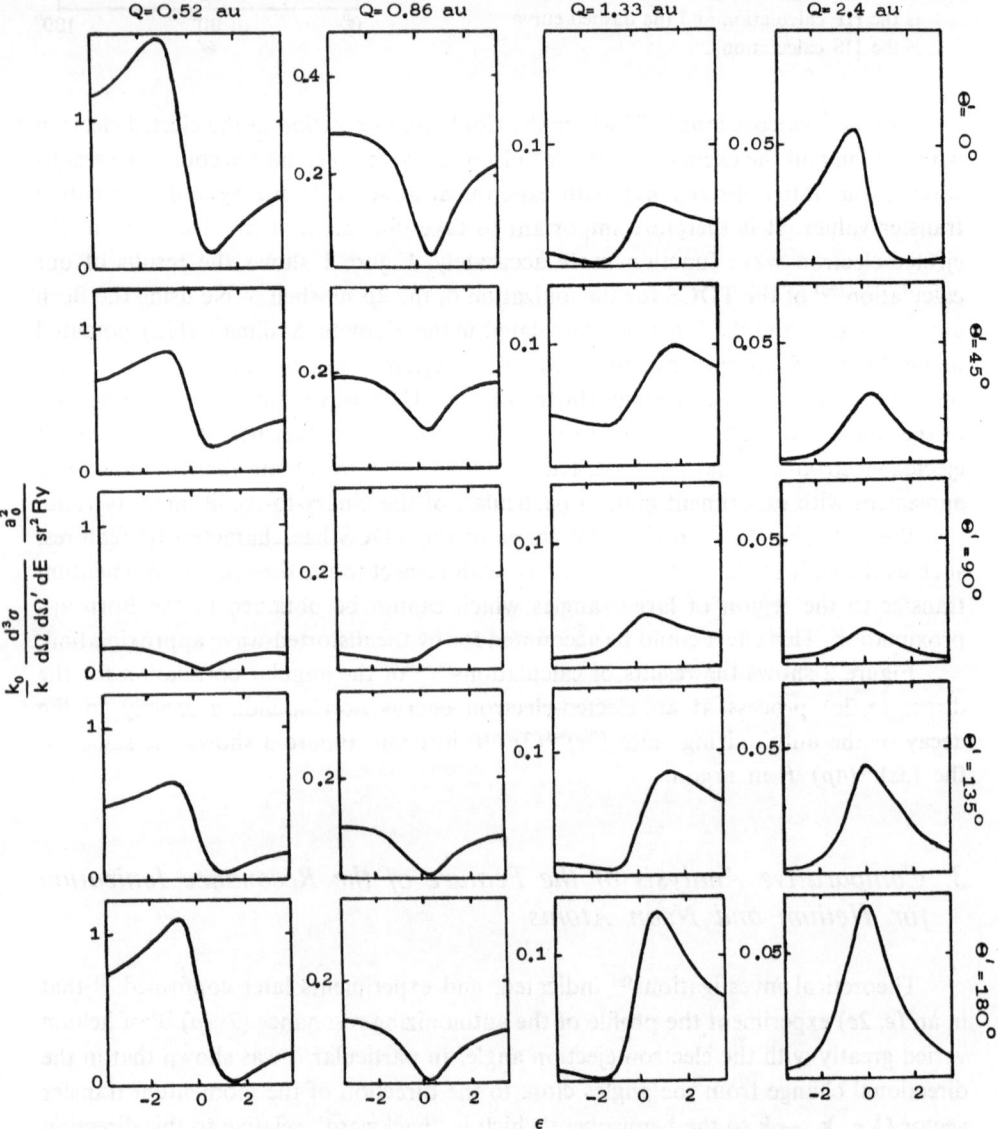

Figure 4. The profiles for the $(2s)^{-1}(3p)^1P$ resonance of Ne in the spectrum of the $(e, 2e)$ process.

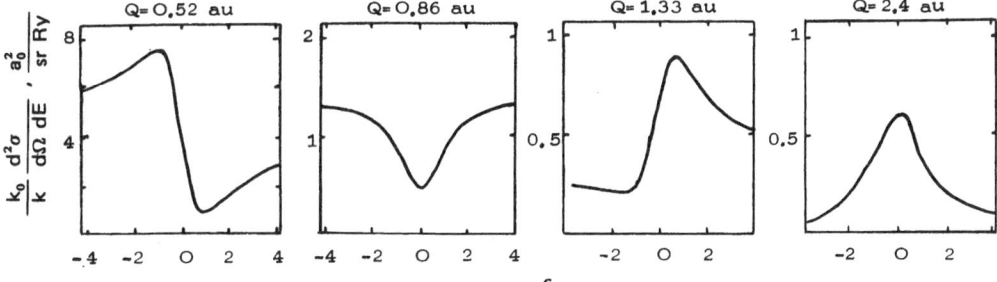

Figure 5. The profiles for the $(2s)^{-1}(3p)^1P$ resonance of Ne in the spectrum of scattered electrons for different values of momentum transfer.[8]

The main cause is that in neon, at the ejected-electron energy corresponding to the autoionizing state $(2s)^{-1}(3p)$ ^1P, the direct ionization does not exhibit a strong forward–backward asymmetry relative to the direction of the **Q** vector. This was demonstrated by the calculations, which were made using different variants of the final-state wave function for the system Ne$^+$ + e, namely, Herman–Skillman calculations and frozen-core Hartree–Fock calculations (see Figure 2). However, the consideration of the interaction in the final state between the ejected electron and the ion is very important for the description of the angular correlation in the direct ionization. The plane-wave approximation for all of the three electrons, namely, the incoming, scattered, and ejected electrons, gives quite a different shape for the

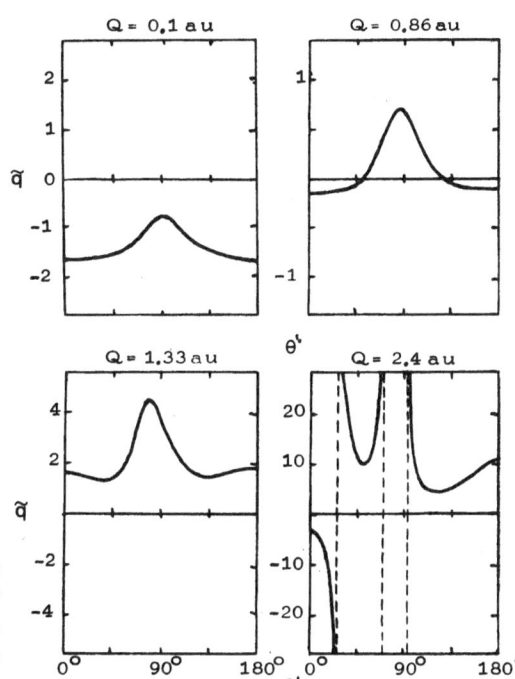

Figure 6. The dependence of the profile index $\bar{q}(\theta', Q)$ for the $(2s)^{-1}(3p)^1P$ resonance of Ne on the emission angle of the ejected electron at various values of the momentum transfer Q.

angular correlation. It is worth noting another interesting feature of the decay of the $(2s)^{-1}(3p)$ ^1P autoionizing state in neon. As one can see from Figure 4, the character of asymmetry of the resonance profile for the $(e, 2e)$ experiment should be reversed in the transition from small to large values of the momentum transfer somewhere in the region of $Q = 0.86$ a.u. This fact can easily be illustrated by showing, as in Figure 6, a change with momentum transfer in the resonance profile index that occurs in the Fano formula:

$$\frac{d^3\sigma}{d\Omega\, d\Omega'\, dE} = \sigma_b(\theta', Q) + \sigma_a(\theta', Q) \frac{[\bar{q}(\theta', Q) + \varepsilon]^2}{1 + \varepsilon^2} \qquad (1)$$

Here θ' is the angle between the momentum $\mathbf{k}'$ of the ejected electron and the momentum transfer $\mathbf{Q}$. Note that the general character of the dependence of the resonance profile index on the momentum transfer is conserved after the integration of the triple differential cross section over all emission directions of the ejected electron, which corresponds to observation of the resonance in the energy-loss spectra of fast electrons in the single-arm experiment (e, e') (see Figure 7). At very small momentum transfers the profile index for the $(2s)^{-1}(3p)$ ^1P state is negative; in the region of $Q = 0.8$ a.u. the value $q(Q)$ changes sign and then increases sharply with Q. A

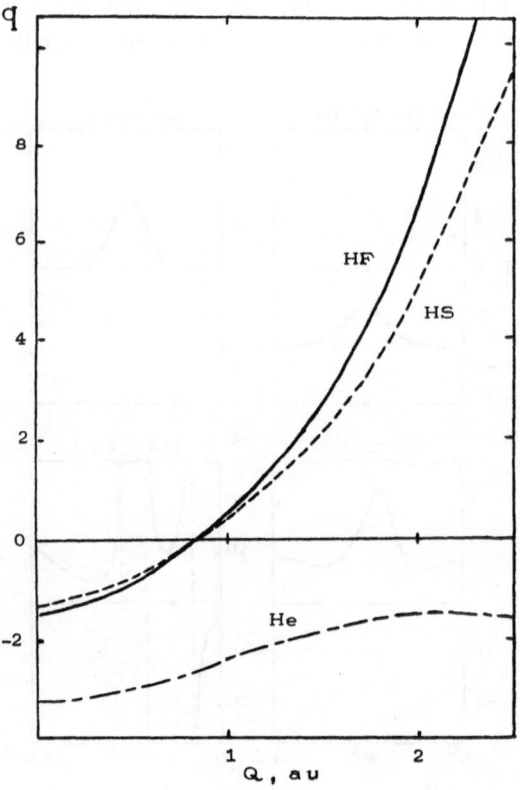

Figure 7. The Q dependence of the profile index for the $(2s)^{-1}(3p)^1P$ resonance of Ne considered in the energy-loss spectrum of the fast electrons.[8] For comparison the Q dependence of the profile index for the $(2s2p)^1P$ resonance of He is also shown.[22]

similar curve was also obtained in the RPAE (random phase approximation with exchange) in Reference 21. Thus, in contrast to helium, the character of the asymmetry for the $(2s)^{-1}(3p)$ ^1P resonance of neon is reversed in the transition from the "optical limit" to the momentum transfer of the order of 1.5 a.u. In the middle part of this interval $|q(Q)| < 1$ the resonance appears as a "window resonance."

4. Sensitivity of the Results to the Choice of the Wave Function for the Ejected-Electron–Residual-Ion System

Turning now from helium and neon to heavier inert gas atoms, we are faced with a new aspect of the problem. Even in the case of argon the main characteristics of the resonance corresponding to the $(3s)^{-1}(4p)$ ^1P autoionizing state turn out to be very sensitive to the method of description of the interaction of the ejected electron with the residual ion Ar^+. This problem will be dealt with in detail elsewhere; here we show only some examples.

Figure 8 shows the dependence of $(3s)^{-1}(4p)$ ^1P resonance parameters in argon on the angle between the directions of emission of the ejected electron and the momentum transfer. The parameters were calculated for the $Ar^+ + e$ system using two different wave functions, obtained with the Herman–Skillman potential and with the frozen-core Hartree–Fock method. For parametrization of the resonance we use, in addition to the Fano formula, the equivalent formula suggested by Shore:

$$\frac{d^3\sigma}{d\Omega\, d\Omega'\, dE} = F(\theta', Q) + \frac{A(\theta', Q)\varepsilon + B(\theta', Q)}{1 + \varepsilon^2} \tag{2}$$

Figure 3 shows the dependence of the parameter

$$F(\theta', Q) \equiv \left(\frac{d^3\sigma}{d\Omega\, d\Omega'\, dE}\right)_{direct}$$

corresponding to the direct $(e, 2e)$ process in the region of the autoionizing state. It should be noted that the parameter $B(\theta', Q)$, which is usually interpreted as the "yield" of the resonance, comprises the term that is due to interference of the resonance amplitude and the amplitude for the direct transition to the continuum. Therefore the angular θ' dependence of the $B(\theta', Q)$ parameter generally is not symmetric about a plane perpendicular to the vector Q.

The parameters $A(\theta', Q)$ and $\tilde{q}(\theta', Q)$ defining the character (sign) of the resonance asymmetry vary especially strongly from one variant of the calculation to another. In going from argon to the heavier inert gas Kr and Xe atoms one can expect even a higher sensitivity of the calculated results to the choice of the final-state wave function, particularly because of the important role of the wide single configurational resonances.[23] This, however, has not yet been treated properly.

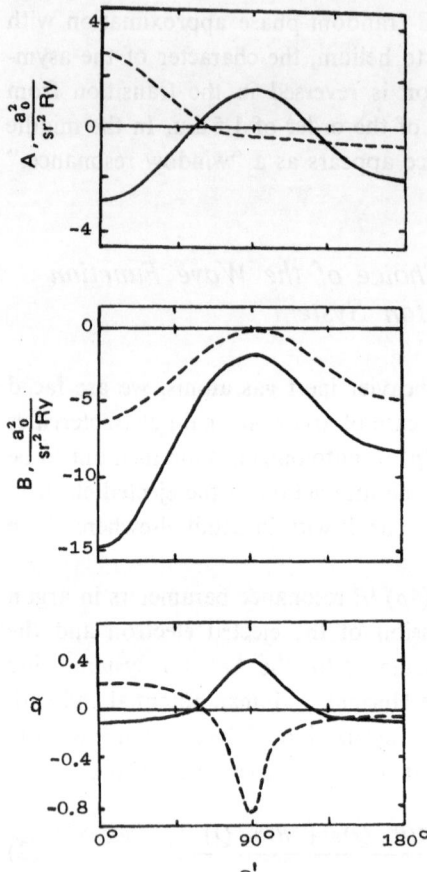

Figure 8. The parameters of the $(3s)^{-1}(4p)^1P$ resonance in the $(e, 2e)$ process[9] for $Q = 0.346$ a.u. Solid curves are the HF calculations and the dashed curves are the HS calculations.

5. Comparison of Calculations with Experiment

The above theory can be, and has already been, used for a unified description of the resonances in the three kinds of experiments: (a) in measurements of the energy-loss spectra of fast monochromatic electrons, (b) in measurements of the spectra of electrons ejected from atoms in collisions with fast charged particles, and (c) in the $(e, 2e)$ experiments. A comparison of theory and experiment for (a) and (b) was made earlier. Here we shall only discuss the interpretation of the coincidence experiments of the $(e, 2e)$ type.

Apart from the well-known and already discussed experimental data obtained by Weigold *et al.*[6] for helium, only preliminary results of Ehrhard's group[24] relating to the study of the $(3s)^{-1}(4p)$ resonance in argon by the $(e, 2e)$ method are available. A special theoretical analysis of these results was made elsewhere[9] by the present authors. Among the general conclusions of this analysis that are of interest for future experiments, the following is especially important. The calculations show that the resonance parameters F, B, and A depend differently on the ejected-electron emission angle. In this connection the angular correlation between the scattered and the ejected electron in the region of an autoionizing state has a different experimental shape for various parts of the resonance profile and therefore it depends strongly on the accuracy of determination of the atomic excitation energy in the $(e, 2e)$ experiment (see Figure 9).

A direct comparison of our results with the data of Ehrhard *et al.* is given in Figure 10. The upper part of the figure shows the energy-averaged profiles for the $(3s)^{-1}(4p)\ ^1P$ resonance for various directions at which the ejected electron was detected. The averaging interval was taken to be $\Delta E = 0.2$ eV, which is close to the experimental resolution typical of the modern experiments of interest. It is important to stress that in the calculation the consideration of the dependence of the direct part of the ionization amplitude on the ejected-electron energy was essential. At the same time we ignored the fact that in addition to the singlet $(3s)^{-1}(4p)\ ^1P$ state of interest, the triplet $(3s)^{-1}(4p)\ ^3P$ state falls within the given averaging interval of the argon excitation spectrum.

The data shown in Figure 10 indicate the obvious general resemblance between theory and experiment. However, there is no close agreement between them. Evidently, the electron energy 250 eV, used in the experiment by Ehrhardt *et al.*, is not

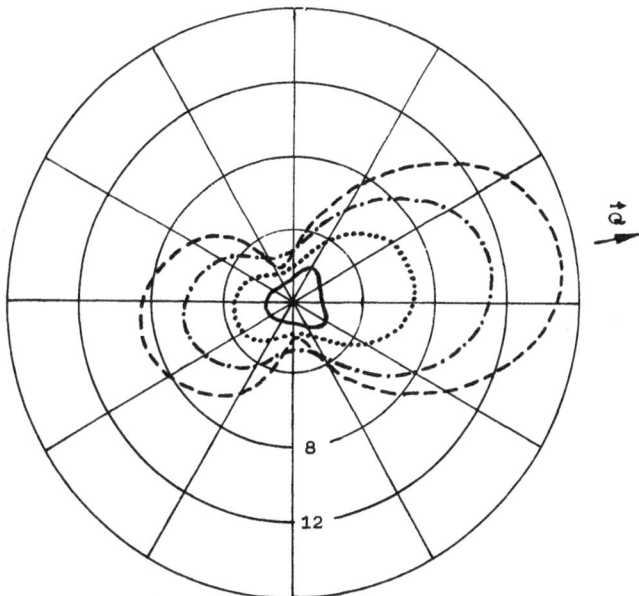

Figure 9. The ejected electrons angular distributions for different parts of the resonance profile for the $(3s)^{-1}(4p)^1P$ resonance in Ar.[9] ——, $\varepsilon = 0$ (multiplied by the factor 2); $\cdots$, $\varepsilon = 1$; —·—, $\varepsilon = 2$; – – –, $\varepsilon = 10$. The arrow shows the direction of the momentum transfer ($Q = 0.346$ a.u.).

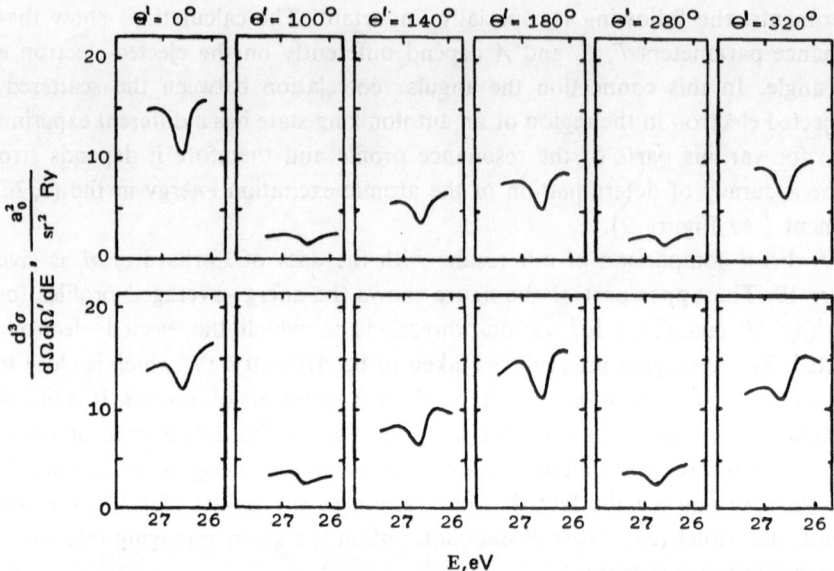

Figure 10. The profiles of the $(3s)^{-1}(4p)^1P$ resonance for various directions of the emission of the ejected electron (the angle between $\mathbf{k}'$ and $\mathbf{Q}$ is indicated); $Q = 0.346$ a.u. The upper part shows the energy-averaged calculated results.[9] The lower part shows the experimental results[24] normalized to the theoretical ones at 0°.

high enough for analyzing the corresponding results within the framework of the plane-wave Born approximation. This conclusion directly correlates with that which was drawn in the analysis of the (*e*, 2*e*) experiments on the direct ionization of neon atoms by 250-eV electrons (see Section 1).

6. *Conclusions*

The study of the properties of the autoionizing states in atoms by the electron–electron (*e*, 2*e*) angular correlation method has just started. For this method to become a reliable tool for the spectroscopy of autoionizing states it is important to try to make correlation experiments at significantly higher energies of electron beams than in the current experiments.[6,24] Then one can rely on the calculations performed within the plane-wave Born approximation for fast electrons, since these calculations are much more reliable and simpler in technique than all other methods. At the same time it is necessary to generalize the above theory to the case of the ionization of atoms by medium-energy electrons by adding the effects of distortion of the wave functions of the incoming and outgoing electrons and the exchange effects. We are going to give special consideration to this problem in the near future.

References

1. V. V. Balashov, S. S. Lipovetski, and V. S. Senashenko, *Sov. Phys. JETP* **63**, 1622 (1972).
2. V. V. Balashov, S. S. Lipovetski, and V. S. Senashenko, *Phys. Lett.* **39A**, 103 (1972).
3. H. Kleinpoppen, K. Blum, and M. C. Standage, *Proceedings of the Ninth International Conference on Physics of Electronic and Atomic Collisions*, Seattle, p. 641 (1975).
4. N. Oda, F. Nishimura, and S. Tahira, *Phys. Rev. Lett.* **24**, 42 (1970).
5. H. Suzuki, A. Konishi, M. Yamamoto, and K. Wakiya, *J. Phys. Soc. Jpn.* **28**, 534 (1970).
6. E. Weigold, A. Ugbabe, and P. J. O. Teubner, *Phys. Rev. Lett.* **35**, 209 (1975).
7. C. Theodosiou, *Phys. Rev. A* **16**, 2232 (1977).
8. V. V. Balashov, A. N. Grum-Grzhimailo, N. M. Kabachnik, A. I. Magunov, and S. I. Strakhova, *J. Phys. B* **12**, 2233 (1979).
9. V. V. Balashov, A. N. Grum-Grzhimailo, N. M. Kabachnik, A. I. Magunov, and S. I. Strakhova, *Phys. Lett.*, **67A**, 266 (1978).
10. V. V. Balashov and V. N. Mileev, *Tenth International Conference on Physics of Electronic and Atomic Collisions*, Abstracts, Commissariat à l'Energie Atomique, Paris, p. 668 (1977).
11. H. Feshbach and J. Hüfner, *Ann. Phys.* (*N.Y.*) **56**, 208 (1970).
12. V. V. Balashov, S. I. Grishanova, N. M. Kabachnik, V. M. Kulikov, and N. N. Titarenko, *Nucl. Phys.* **A216**, 574 (1973).
13. V. V. Balashov, E. F. Kislyakov, V. L. Korotkikh, and R. Vjunsh, preprint of the JINR, P4-11565, Dubna (1978).
14. V. V. Balashov, A. N. Grum-Grzhimailo, N. M. Kabachnik, A. I. Magunov, and S. I. Strakhova, *Tenth International Conference on Physics of Electronic and Atomic Collisions*, Abstracts, Commissariat à l'Energie Atomique, Paris, p. 669 (1977).
15. Yu. F. Smirnov and V. G. Neudatchin, *Sov. Phys. JETP Lett.* **3**, 192 (1966); I. McCarthy and E. Weigold, *Phys. Rep.* **27**, 275 (1976).
16. H. Ehrhardt, K. H. Hesselbacher, K. Jung, and K. Willmann, *J. Phys. B* **5**, 1559 (1972).
17. K. Jung, E. Schubert, H. Ehrhardt, and D. A. L. Paul, *J. Phys. B* **9**, 75 (1976).
18. H. Ehrhardt, K. H. Hesselbacher, K. Jung, F. Schubert, and K. Willmann, *J. Phys. B* **7**, 69 (1974).
19. E. W. Knapp and M. Schulz, *J. Phys. B* **7**, 1875 (1974).
20. V. V. Balashov, A. N. Grum-Grzhimailo, N. M. Kabachnik, A. I. Magunov, and S. I. Strakhova, *Phys. Lett. B* **12**, L27 (1979).
21. M. Ja. Amusia, V. K. Ivanov, and M. Yu. Kutchiev, in *Autoionization Phenomena in Atoms*, Moscow University Publishing House, Moscow, p. 68 (1976).
22. V. V. Balashov, S. S. Lipovetski, A. V. Pavlichenkov, A. N. Poljudov, and V. S. Senashenko, *Opt. Spectrosk.* **XXXII**, 10 (1972).
23. V. V. Balashov, N. M. Kabachnik, I. P. Sazhina, and S. I. Strakhova, in *Autoionization Phenomena in Atoms*, Moscow University Publishing House, Moscow, p. 92 (1976).
24. K. Jung, E. Schubert, and E. Ehrhardt, *Tenth International Conference on Physics of Electronic and Atomic Collisions*, Abstracts, Commissariat à l'Energie Atomique, Paris, p. 670 (1977).

25

The Postcollision Interaction as a Manifestation of Many-Electron Correlations

M. Ya. Amusia, M. Yu. Kuchiev, and S. A. Sheinerman

A discussion is given of special phenomena that occur in inelastic scattering of electrons by atoms or in inner-shell photoionization when the incoming electron energy is near the threshold energy of a nonstationary excited state. It is demonstrated that the interaction of the slow electron with the vacancy in the atom (the so-called "postcollision interaction") significantly affects the characteristics of the process. The inclusion of PCI in calculations of specific processes has achieved acceptable agreement with experimental data.

It is a great pleasure for us to present a paper in a volume dedicated to the seventieth birthday anniversary of Sir Harrie Massey, whose pioneer researches are so significant for the physics of atomic collisions.

1. Introduction

In studies of the inelastic scattering of electrons by helium atoms a rather interesting phenomenon has been observed[1] when the energy of the incoming electron just exceeds the autoionization energy. It appears that the energy of the electron that is ejected from the atom due to the decay of the autoionizing state depends on the

M. Ya. Amusia, M. Yu. Kuchiev, and S. A. Sheinerman • Ioffe Physical-Technical Institute of the Academy of Sciences, Leningrad, U.S.S.R.

energy of the scattered electron. The shape of the spectral line of the ejected electron that arises from this autoionizing state also depends upon the incoming electron energy. This effect has been called the "postcollision interaction" (PCI) in the literature.

Qualitatively the effect is explained by the fact that the incoming electron loses almost all its energy in the excitation of the autoionizing state and leaves the atom slowly. If the velocity of this slow electron is sufficiently low, its field affects all the characteristics of the decay of the autoionizing state and its influence is strongest when the slow electron is closest to the atom. The PCI shows itself most clearly when the lifetime of the excited autoionizing state is short, that is, when the width of the level is large. If it is comparable with the energy of the slow electron, it is impossible to assign a definite energy to the state. In this case the excitation and decay of the autoionizing state is a combined process and cannot be considered as two successive steps.

The PCI manifests itself not only in inelastic scattering of comparatively slow electrons but in other ionization processes. For example, if the photon energy just exceeds the inner-shell photoionization threshold, a deep hole and a slow electron form after the absorption of the photon, and both of them take part in the subsequent Auger or radiative decay of the inner vacancy. Therefore the Auger electron energy near the ionization threshold depends upon the frequency of the incident photon.[2]

As demonstrated in this paper, the interaction of charged particles in the final state is significant for the ionization by slow electrons and for inner-shell ionization. Then there are three particles: the slow electron, the ion, and the fast electron, which is ejected by the decay of the excited state. The interaction of the fast electron with the two other particles is small and may be neglected compared with the slow-electron–ion interaction. In this way this three-body problem is reduced to the consideration of the influence of the field of the ion upon the slow electron. This interaction, as will be shown below, is considerable and leads to a redistribution of the energy between the two outgoing electrons, the slow electron being decelerated, and the fast electron accelerated. This result is the opposite of that of the well-known Wannier regime,[3] where there are also two electrons and an ion in the final state, but where the energies of the outgoing electrons tend to become equal due to the electron interaction. The difference is determined by the fact that in the Wannier regime both electrons are slow and the interaction of all three particles should be taken into account, whereas in the case under consideration only the slow-electron–ion interaction is essential.

In this paper the excitation of the autoionizing state or vacancy and their decay are considered as a unified process. In this way we are able to describe the energy shift which comes from the PCI, and to calculate the cross section for the process.

Section 2 discusses the most important diagrams of the many-body theory, which describe the interaction of the slow electron with the atomic particles.

In Section 3 it is shown qualitatively how the use of these diagrams describes the shift, broadening, and asymmetry of the peak which corresponds to the autoionizing state in the ejected-electron spectrum. The asymmetry increases as the energy of the incoming electron approaches the excitation threshold of the autoionizing level. Also it is found that the wings of the profile oscillate.

In Section 4 the PCI is considered in the example of the production of a singly charged ion Ar^+ when the energy transferred to the atom is close to the $2p^6$ subshell ionization threshold. It is demonstrated that the PCI effect leads to the appearance of a maximum in the cross section for singly charged ion production at the threshold of $2p^6$ ionization. The result of calculating the absolute cross section and the shape of the peak are in satisfactory agreement with experimental data.[4] In Section 5 another example of PCI is presented, that of electron scattering from an Ar atom near the threshold of the autoionization level $3s4p(^3p)$. It is shown that the PCI effect manifests itself in the variation of the slow-electron energy as well as in the angular distribution.

2. Many-Body Diagrams of the PCI Effect

The simplest diagram, which describes the excitation of a single-particle auto-ionization level by electron impact and its subsequent decay is given in Figure 1a.

The thin line with an arrow to the right (left) describes the electron (hole) propagation, the double arrow line describes an electron in a discrete excited level, and the wavy line denotes the Coulomb interactions. The ordinary rules for comparing analytical expressions to the diagrams[5] are used.

The diagram in Figure 1a is a resonance, because its energy denominator, which corresponds to the virtual excitation of an autoionizing level, tends to zero if the energy transferred to the atom by the incoming electron is equal to the excitation energy of the discrete level. Therefore it is necessary to sum the contributions of all resonance diagrams in higher orders of the perturbation theory. One should also take into account the possibility of the decay of the autoionizing state which leads to an imaginary addition $i\Gamma/2$ to the energy of the autoionizing level whose width is Γ. Also the excitation and decay of this discrete state may proceed via virtual excitation of some other atomic configurations. Let us take this into account by substituting the effective interaction W instead of the Coulomb interaction.

Summing up all the resonance diagrams, we came to the diagram of Figure 1b, in which the dashed line denotes the interaction W and the solid line of the discrete state denotes that the width of this state is taken into account. The analytical expression $A_0(k_s)$ for the amplitude of the process presented by Figure 1b is as follows:

$$A_0(k_s) = \frac{\langle k_s, n \mid W \mid p, j\rangle\langle j, k_f \mid W \mid n, i\rangle}{\varepsilon_p - \varepsilon_s - E_{ex} + i(\Gamma/2)} \tag{1}$$

Here ε_p is the incoming electron energy, k_s and ε_s are the momentum and energy of the slow electron, k_f, ε_f are the momenta and the energy for the fast electron, E_{ex}

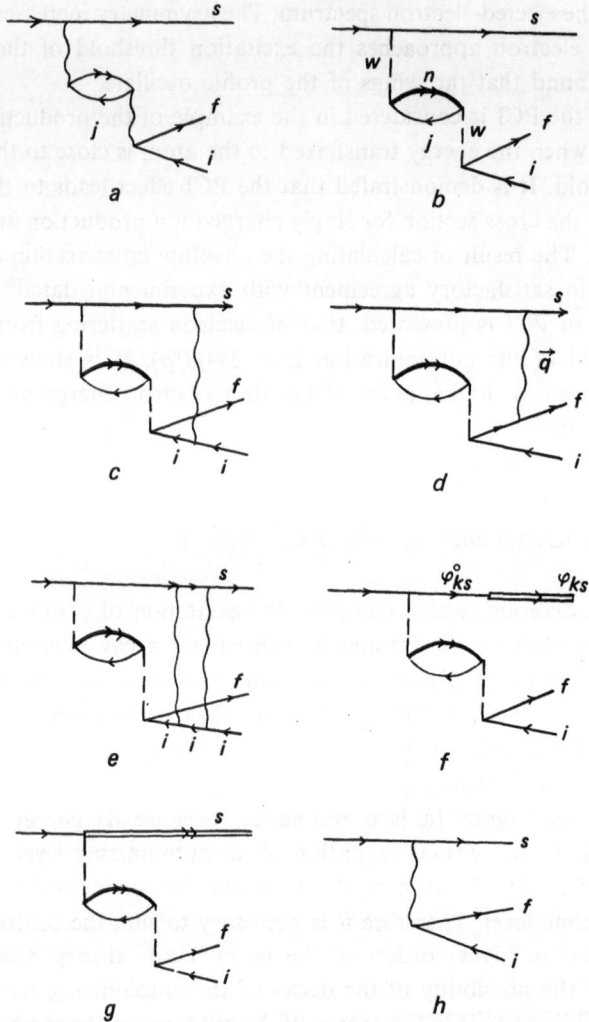

Figure 1. The diagrams (a)–(g) describe in different approximations the resonance process of electron inelastic scattering on an atom with the excitation and following decay of the auto-ionization state j. The diagram (f) takes into account the PCI effect. One of the background process diagrams is shown in (h).

and Γ are the excitation energy and width of the autoionization level, E_i denotes the energy of the hole i, and the symbols n and j denote the electron and hole states which form the electron–hole autoionization level. The energies of the incoming and outgoing electrons satisfy the energy conservation law

$$\varepsilon_p = \varepsilon_s + \varepsilon_f + E_i \tag{2}$$

If in the resonance region the dependence on energy of the matrix elements on the right-hand side of (1) may be neglected, then the amplitude $A_0(k_s)$ describes the Lorentz profile of the ejected-electron peak with its maximum at the energy

$$\varepsilon_f = E_{\text{ex}} - E_i \tag{3}$$

and with a width Γ.

Let us consider now the interaction of the show electron with the fast one and with the hole, both of which are produced by the decay of the autoionizing state. Of course, this is a three-body problem. But since one of the electrons is fast, its interaction with the other electron is negligible. Indeed, the direct estimation of corrections that come from diagrams presented in Figures 1c and 1d shows that if the condition[†]

$$k_s \ll 1 \ll k_f \tag{4}$$

is fulfilled, the contribution of Figure 1d is negligible compared with the contribution of Figure 1c.

The meaning of this result is quite obvious: the fast electron, which leaves the atom due to the decay of the autoionizing state has insufficient time to interact with the slow one.

In what follows we assume that the condition (4) is fulfilled. Therefore we limit ourselves to consideration of only the interaction of the slow electron with hole i, and this interaction should be taken into account nonperturbatively. It is also necessary to take into consideration the interaction of the slow electron with the field of the virtual autoionizing state.

The influence of the field of either the excited level or the i vacancy on the slow electron can be taken into account by the choice of the Hartree–Fock electron wave function.[6] Then the electron interaction with another field will be a specific correlation process. Note that this class of correlations has many common features with intershell correlations in the RPAE approximation where the electron interaction with the fields of the different vacancies has to be taken into consideration.[6]

Turning to our problem, let us choose the wave functions $\phi^{\circ}_{k_s}$ in the field of the autoionizing state. The thin line, which corresponds to the slow electrons in Figure 1b, denotes just the motion of this electron in such a field.

Let us take into account the interaction of the final-state slow electron with the ion i by summing up the "ladder" of diagrams analogous to that given in Figure 1e. Denoting by $A(k_s)$ the contribution of the sum of these diagrams we find

$$A(k_s) = A_0(k_s) + \int \frac{\langle \phi^{\circ}_{k_s} \mid V \mid \phi^{\circ}_{k_s'} \rangle}{\varepsilon_{k_s} - \varepsilon_{k_s'}} A_0(k_s')\, dk_s' + \cdots$$

$$= \int dk_s' \left[\langle \phi^{\circ}_{k_s} \mid + \frac{\langle \phi^{\circ}_{k_s} \mid V \mid}{\varepsilon_{k_s} - \varepsilon_{k_s'}} + \cdots \right] \mid \phi^{\circ}_{k_s'} \rangle A_0(k_s') \tag{5}$$

[†] In this chapter we use the atomic units $\hbar = m = e = 1$, the energy being in Ry.

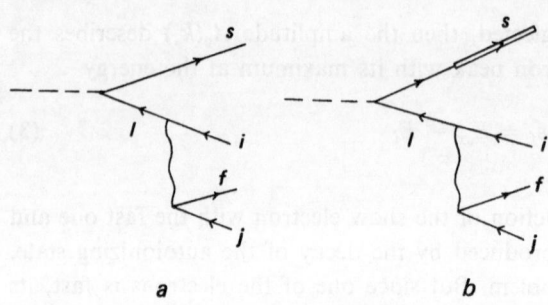

Figure 2. The diagrams (a) and (b) describe the photoionization of the inner shell with its following Auger decay. The graph (b) takes into account the PCI effect.

where $A_0(k_s)$ is determined by (1), and V is the potential of the hole i (more exactly, this is the difference between the field of the vacancy i and the excited atom field). The sum in square brackets in the last equation is the slow-electron wave function ϕ_{k_s} in the hole i field. Therefore, for $A(k_s)$ we obtain

$$A(k_s) = \int A_0(k_s')\langle \phi_{k_s} | \phi_{k_{s'}}^0 \rangle \, dk_s' \tag{5'}$$

Corresponding to this equality is the diagram of Figure 1f, in which the double line denotes an electron that moves in the field with the hole i, and the conjunction of ordinary and double lines denotes the overlap integral $\langle \phi_{k_s} | \phi_{k_{s'}}^0 \rangle$. The domain of validity of (5') will be considered below.

The amplitude (5'), which is obtained subject to the condition (4), describes the process of instantaneous decay of the autoionizing state in which the slow electron feels the sudden variation of the field.

An analogous method of taking account of the sudden field variation is applicable to some other processes in which a slow electron and a nonstationary state are created. For example, in the photoionization of an inner shell a vacancy is created which may then decay via an Auger process. The diagram of photoionization with subsequent vacancy decay due to an Auger process is presented in Figure 2a. Near the photoionization threshold the main correction to this diagram just as in the previous example comes from the interaction of the slow electron with the field of the ion in the final state, which is taken into account by diagram in Figure 2b. The thin line for the slow electron describes its motion in the field of the inner vacancy 1, whereas the double line denotes its motion in the total field of i and j holes.

3. The Observable Manifestations of PCI

Let us demonstrate that Figure 1f leads to a shift of the maximum and variation of the shape of the autoionization peak in the Auger electron spectrum. For simplicity we limit ourselves to the case $\Gamma \ll 1$. If this condition is fulfilled, the slow electron is localized far from the atom, as demonstrated below, and the problem becomes much simpler.

Interchanging in (5') the integration over coordinate (the overlap integral) and energy, we obtain

$$A(k_s) = \int \phi_{k_s}^*(r)\phi(r)\, dr \tag{6}$$

where $\phi_{k_s}(r)$ is the final state wave function, and $\phi(r)$, according to (1), is determined by the equation

$$\phi(r) = \int \frac{\phi_{k_s}^\circ(r)\langle k_s, n \mid W \mid p, j\rangle}{\varepsilon_p - E_{\text{ex}} - \varepsilon_s + i(\Gamma/2)}\, dk_s \langle j, k_f \mid W \mid n, i\rangle \tag{7}$$

The sign k_s in the matrix element $\langle k_s, n \mid W \mid p, j\rangle$ denotes the slow-electron wave function in the intermediate state $\phi_{k_s}^\circ$. The matrix element $\langle j, k_f \mid W \mid n, i\rangle$ is almost independent of k_s [see formula (2)] in the resonance region which we are interested in and therefore is taken out of the integral. This is why the integration over k_s in (7) may be performed using the "retarded" Green's function, which describes the motion of the slow electron in the field of the excited atom. Using the asymptotic properties of Green's function,[7] we obtain from (7)

$$\phi(r) \to A_0(k_0)\, \frac{e^{ik_0 r}}{r} + O\!\left(\frac{1}{r^2}\right), \qquad r \to \infty \tag{8}$$

where the "momentum" k_0 is complex:

$$k_0 = k_1 + ik_2 = [2(\varepsilon_p - E_{\text{ex}}) + i\Gamma]^{1/2} \tag{9}$$

and A_0 is the amplitude (1) of inelastic scattering without PCI.

The first term in (8) is the "divergent" wave. Taking account of the finite width of the level leads to its exponential damping, due to which the divergent wave become localized in the region $r \lesssim 1/k_2$. If the condition $\Gamma \ll 1$ is fulfilled, this region exceeds the atomic dimensions $1/k_2 \gg 1$. Besides that, the terms $O(1/r^2)$ at such distances proved to be smaller than the divergent wave and may be neglected.

Therefore the main contribution to the integral (6) comes from distances $1 < r < 1/k_2$ at which the wave function $\phi(r)$ is well approximated by the first term on the right-hand side of (8). At such distances the slow-electron motion in the Coulomb field is quasiclassical. Therefore, the overlap integral (6) may be evaluated by the stationary phase method.[8]

Let us consider the process in which the slow electron in the final state is part of the continuous spectrum. The wave function $\phi_{k_s}(r)$ describing the motion of the electrons in the field of the ion oscillates. The overlap integral (6) reaches its maximum value if with variation of r (in the region $r \lesssim 1/k_2$ which is of interest for us) the phases of both oscillating functions $\phi_{k_s}(r)$ and $\phi(r)$ vary almost identically:

$$\frac{d}{dr}\left[\alpha_{k_s}(r) - k_1 r\right] = 0, \qquad r = 1/ck_2$$

c is a constant of the order of 1, $k_1 r$ is the phase of $\phi(r)$, and $\alpha_{k_s}(r)$ is the phase of $\phi_{k_s}(r)$. At large distances the field of the ion is pure Coulombic. Using the phase of the Coulomb wave function, we obtain from the last equation

$$\varepsilon_s = \frac{k_1^2}{2} - ck_2 \tag{10}$$

We observe that taking account of PCI results in the shift of the maximum of the amplitude modulus as a function of energy ε_s by the value $\delta\varepsilon_s = -ck_2$.[8] Because of the fulfillment of the energy conservation law (2) the maximum of the electron spectral line, which arises from the autoionizing level, shifts in the opposite direction $\delta\varepsilon_f = ck_2$. Therefore the PCI leads to a redistribution of energy between the outgoing electrons—the "fast" electron accelerates and the slow electron decelerates. This result is not surprising, because the energy exchange between the two outgoing electrons is not a result of their direct interaction (the diagram of Figure 1d is small) but is a consequence of the attraction of the slow electron to the ion.

If k_s deviates from the value determined by (10) the arguments of $\phi_{k_s}(r)$ and $\phi(r)$ are changed, and the modulus of the integral (6) decreases as the phase difference increases. Let us note that $\partial^2 \alpha_{k_s}(r)/\partial k_s^2 > 0$ for $r \sim 1/k_2$ and $\varepsilon_p - E_{ex} > \Gamma$. Therefore the phase of $\phi_{k_s}(r)$ varies faster with an increase, rather than a decrease, of k_s. That is why the overlap integral slowly decreases on the lower energy side. Consequently, due to PCI the contour of the autoionization profile in the Auger spectrum becomes nonsymmetric: the profile wing decreases more slowly than the Lorentz profile without PCI on the higher ε_f (lower ε_s) energy side, and the profile decreases more quickly than the Lorentz one on the lower energies of ε_f. As the quantity $\partial^2 \alpha_{k_s}(r)/\partial k_s^2$ is positive we conclude that the broadening of one wing is more prominent than the narrowing of the other. Therefore, overall, the profile becomes broader. The effects considered above, that is the shift of the autoionizing profile, its broadening, and its asymmetry, were connected with the addition to the phase of $\phi_{k_s}(r)$ of a quantity, which is determined by the Coulomb field of the ion. This addition increases with decrease of the slow electron energy. That is why these effects are most prominent near the threshold for excitation of an autoionizing level or for innershell ionization.

Up to now we have taken into account only the amplitude of the resonance process. Let us consider the result of interference between resonance and nonresonance amplitudes. The simplest example of a nonresonance amplitude is given in Figure 1h. Without taking account of PCI the phase of the resonance amplitude varies by an extra π in the region of resonance, and the cross section acquires the form of the so-called Fano profile.[10]

Let us now take into account the influence of PCI upon the phase of $A(k_s)$. As a function $\phi(r)$ we use the asymptotic form (8), whereas as a final state wave function $\phi_{k_s}(r)$ we use a Coulomb wave function. Both approximations are valid at least for an estimation, because the region of electron localization is of the order of $r \sim 1/k_2$ and for small $\Gamma(k_2 \ll 1)$ far exceeds the atomic dimension. Then for

$A(k_s)$ we obtain from (6)

$$A(k_s) = \frac{e^{\Psi_1 + i\Psi_1}}{\varepsilon_p - E_{ex} - \varepsilon_{k_s} + i\Gamma/2} \tag{11}$$

where

$$\Psi_1(k_s) = -\frac{1}{k_s} \arctan \frac{2k_s k_2}{k_1^2 + k_2^2 - k_s^2}$$

$$\Psi_2(k_s) = \frac{1}{2k_s} \ln \frac{(k_s - k_1)^2 + k_2^2}{(k_s + k_1)^2 + k_2^2}$$

k_1 and k_2 are determined by (9). Earlier[11] the expression (11) was obtained using the quasimolecular adiabatic consideration of PCI.

We observe that PCI adds to the phase of $A(k_s)$ the magnitude $\psi_2(k_s)$, which may strongly vary in the resonant region.

When interference of the resonance amplitude (11) with a background one is taken into account the fast change of $\Psi_2(k_s)$ leads to some features in the cross section.[9] Let us suppose for simplicity that the background amplitude is a constant b in the vicinity of the resonance. The full amplitude is equal to $b + A(k_s)$, and for the cross section we get the formula

$$\sigma(k_s) = \sigma_0 \left| 1 + \alpha \frac{\Gamma/2}{\varepsilon_p - E_{ex} - \varepsilon_s + i(\Gamma/2)} e^{\Psi_1 + i\Psi_2} \right|^2 \tag{12}$$

where σ_0 is the nonresonant cross section and $\alpha = 2/B\Gamma$. Figure 3 illustrates the σ dependence on k_s according to (12) when $\alpha = \pm 1$, $\Gamma = 0.1$ eV, $\varepsilon_p - E_{ex} = 1$,

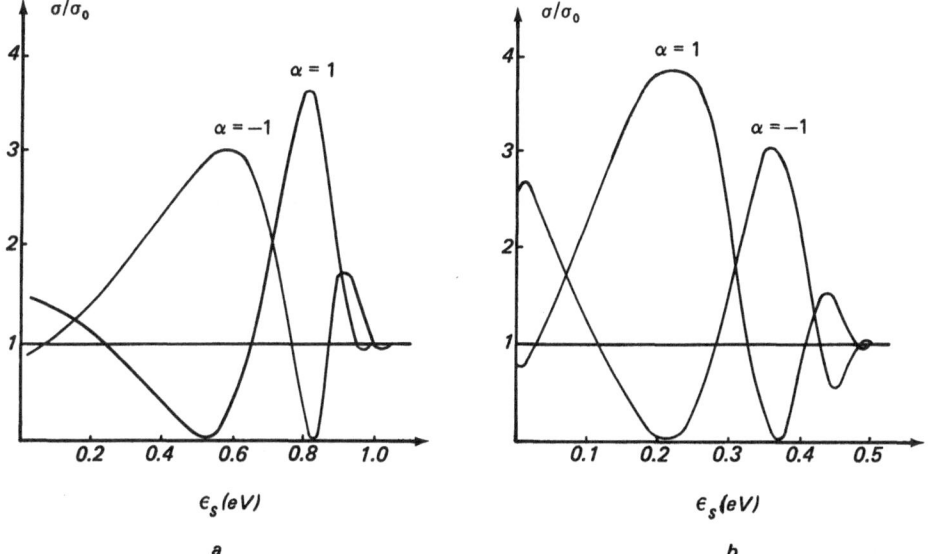

Figure 3. The differential cross section versus the energy of the slow electron taking into account the PCI effect. $\Gamma = 0.1$ eV, the energy excess over the threshold is 1 eV in (a), 0.5 eV in (b).

and 0.5 eV. It can be seen that PCI leads to the oscillatory structure in the auto-ionization line profile. This structure becomes larger as the excess energy $\varepsilon_p - E_{ex}$ above the threshold decreases.

In order to be able to observe these oscillations, it is necessary to have an incoming electron beam with an energy spread much smaller than the width of the autoionization level.

Up to now, we have considered the slow electron in the final state to be in a continuous spectrum. The decrease of the slow-electron energy may be so large that the electron may be captured into a discrete bound level, and in the final state an excited atom may be formed. The investigation analogous to that performed above demonstrates that in this case the characteristic manifestations of PCI in the spectrum of ejected electrons also exist, that is the autoionization profile shifts, becomes asymmetric and broader, and, in the wings of the profile, oscillations appear. For the transitions into discrete states the calculations using the formulas (1) and (6) are simplified. If the discrete state wave function decreases faster than the function $\phi(r)$ (this takes place only for $k_2 < \varkappa_n$, where $\varkappa_n = (-2E_n)^{1/2}$ and E_n is the discrete level energy) we are able according to (6) and (8) to neglect the level width.

It is rather interesting to study how PCI influences the cross section for excited atom formation. Resonant scattering with formation of an excited atom may take place without PCI, as is demonstrated by Figure 1g. The cross section for the process determined by this diagram has a maximum at resonances $\varepsilon_p = E_{ex} + E_s$ ($E_s < 0$ being the discrete state energy) with width which is equal to that of the autoionizing level and all these resonances are below the ionization threshold. Taking into account PCI changes the situation completely. Due to PCI the slow-electron energy decreases, and it may be captured to the discrete level, even if the incoming electron energy is above the reaction threshold. Therefore the increase of the yield of neutral atoms should take place above the ionization threshold. The calculations of chapter four support this statement.

We have considered the consequences of taking into account the diagram of Figure 1f for the case of inelastic scattering. The same consideration is correct for Figure 2b, which describes the PCI in photoionization near the inner-shell threshold. All the main conclusions about the ejected electron spectrum are valid for the electron line which corresponds to Auger decay. The discrete states of the outgoing slow electron correspond in this case to excited levels of the single charged ion and the PCI effects must manifest themselves strongly in the yield of singly charged ions.

All the results of this chapter were obtained for $\Gamma \ll 1$. However, the formula (5'), which is in fact the basis of our consideration, is valid even if this condition is not fulfilled. The derivation of (5'), apart from the inequality (4) is based on the assumption that the contribution to the cross section for the neglected nonresonance diagram, such as for example given in Figure 1h, may be well separated from the contribution of the resonance diagrams, which are taken into account. This assumption is correct if the width Γ is less than the energy interval in which the nonresonance amplitude varies, i.e., if $\Gamma \ll \varepsilon_f$. For $\Gamma \sim 1$, it follows from (8) and (9) that the slow electron is localized at distances $r \sim 1$ and the simple asymptotical form of wave functions is here insufficient. Therefore in this case it is impossible to obtain simple analytical formulas that describe the variation of the line shape as a function of the incoming electron energy. However, it may be rigorously proved (see the Appendix) that in this case PCI also leads to the line shift.

4. PCI Effect: Singly Charged Ion Production at the Threshold of $2p^6$ Ionization

Let us consider the single charged ion yield in inelastic electron scattering from Ar, when the energy ω transfered to the atom is close to the $2p^6$ subshell ionization threshold. According to Section 2 the PCI in this case is taken into account by the diagrams of Figure 4 in which the thin line of the slow electron corresponds to its motion in the $2p$-hole field, whereas the double line describes the electron motion in the field of two holes.

The differential cross section for single-charge ionization of an atom by fast electrons is connected with the density of the generalized oscillator strength (GOS) $F(q, \omega)$.[12] Integrating over angles and summing over spin projection we obtain for $F(q, \omega)$, corresponding to the diagram in Figure 3, the following expression:

$$F = \sum_p 4\omega \frac{3(2l_5 + 1)}{2p + 1} \begin{pmatrix} l_5 & l_1 & 1 \\ 0 & 0 & 0 \end{pmatrix}^2$$
$$\times \left| \langle n_1 l_1, \varepsilon_3 l_3 \| V_p \| n_2 l_2, n_4 l_4 \rangle \sum_{\varepsilon_6} \frac{B_{n_1 l_1 \to \varepsilon_6 l_5}(q) \langle \widetilde{n_5 l_5} \| \varepsilon_6 l_5 \rangle}{\omega - \varepsilon_6 + \varepsilon_{n_1 l_1}} \right|^2 \qquad (13)$$

Here p is the orbital angular momentum, transferred in the Auger decay, $\langle \widetilde{n_5 l_5} \| \varepsilon_6 l_5 \rangle$ is the overlap integral of the radial HF wave functions of the "slow" electron in two different self-consistent fields, $\langle \| V_p \| \rangle$ is the reduced matrix element of the Auger decay, $B_{n_1 l_1 \to \varepsilon_6 l_5}(q)$ is the reduced matrix element of $n_1 l_1 \to \varepsilon_6 l_5$ transition under the action of fast scattered electron; the summation over ε_6 includes also the integration over the states of the continuous spectrum. In (13) only dipole transitions are taken into account because we consider below only small momentum q transferred to the atom by the fast incoming electron. That is why the slow electron may be in either s or d states. In the derivation of (13) we neglect the width of the $2p$ level in accor-

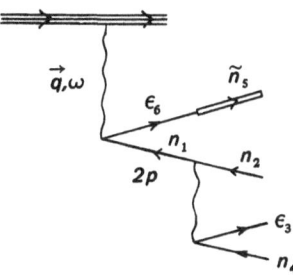

Figure 4. The diagram describing the Ar $2p$ subshell ionization by fast electrons with PCI effect taken into account.

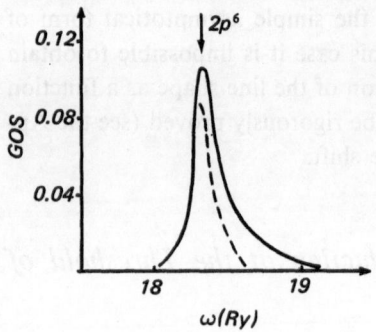

Figure 5. The GOS density for the process given in Figure 4. - - -, Experiment (Reference 4); ——, theory.

dance with the discussion given in Section 3. For the wave functions of the fast Auger electron in the $\varepsilon_3 l_3$ state we used the HF $\phi_i^{N(LS)}$ functions calculated in the "frozen" core with the hole i.[6] (The other hole is screened by the electron that is on the discrete level.) The HF wave functions of the electron in the intermediate state $\varepsilon_6 l_5$ were calculated in the field of the completely rearranged ion with the hole in the $2p$ state. The discrete electron ns and nd wave functions were determined by self-consistent calculations of configurations, which take into account the existence of two holes in the k_2 and k_4 states.

As demonstrated by the calculations, the probability of the processes with transitions to discrete ns levels is, on average, two orders smaller than that of the corresponding processes with transitions to discrete nd levels. The probability of electron transitions into the first eight discrete d levels was calculated. The results of calculations of the GOS density dipole component of the processes with inclusion of PCI are given in Figure 5. In accordance with the qualitative considerations given in this paper, the inclusion of the particular part of electron interactions due to the PCI leads to a sharp increase in the yield of single charged ions at the $2p^6$ subshell threshold. The width of the peak obtained is $\sim$2.5–3 eV. In the same figure we draw the experimental curve for the yield of singly charged ions which is extracted from the measurements of Van der Wiel.[4] The experimental energy loss curve presented[4] in relative units, was normalized by us to that its maximum coincides with the maximum of the experimental curve for the $2p^6$-subshell ionization quoted by Hudson and Kieffer.[13]

The figure demonstrates a satisfactory agreement between theoretical and experimental data. Since this agreement is achieved only by inclusion of PCI, the essential role of PCI in the formation of the Ar$^+$ ion near the L-shell threshold may be considered as proved.

The possible important role of PCI in the above process was recognized by Van der Wiel et al.[4] Other mechanisms of Ar$^+$ ion production near the L-shell threshold, in particular the direct ionization of the M shell and the ionization of the M shell under the action of virtual excitations of the L shell, were discussed by Amusia et al.[14] The contribution of these mechanisms in Ar$^+$ ion formation near L-shell threshold proved to be negligible compared with the considered processes of PCI.

5. PCI Effect: Electron Scattering near the Threshold of the Autoionization Level

Let us consider now a concrete example of PCI effects in the case when the slow electron in the final state is in a continuum spectrum. We investigate the process of inelastic-electron–Ar-atom scattering with excitation of the autoionizing state $3s4p(^3P)$ when the incoming electron energy is close to the excitation threshold (26.46 eV) of this level. As shown in Sections 2 and 3, the PCI effect is then described by the diagram in Figure 1f. In this case (a triplet autoionizing state) only the exchange diagram that is given in Figure 6a should be taken into account. The contribution of this diagram was calculated using formulas (1) and (5').

The HF wave functions calculated for a rearranged core with a $3p$ hole were used as the final-state slow-electron wave functions. The slow-electron wave functions in the intermediate state were calculated in the excited-state field.

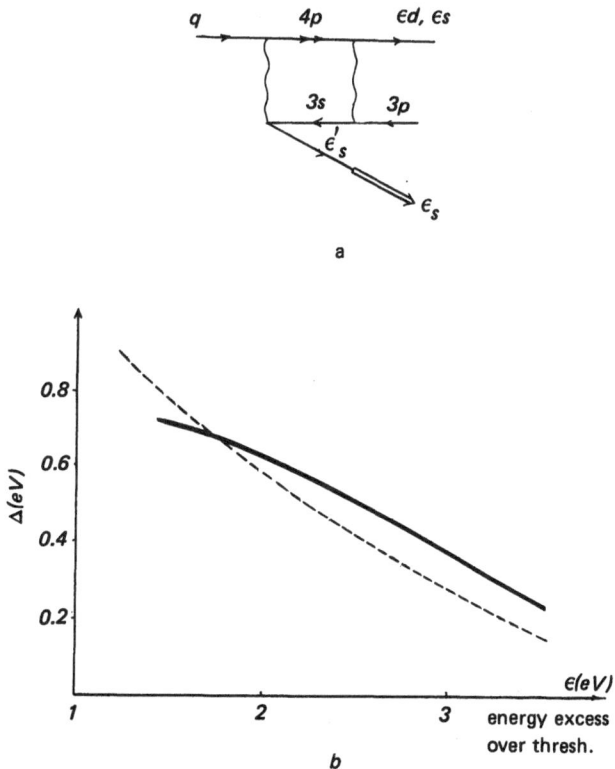

Figure 6. Ionization of Ar by slow electrons near the excitation threshold of the $3s4p(^3P)$ autoionization triplet state. (a) The diagram of this process. (b) The scattered slow-electron energy shift Δ due to the PCI effect versus the incident electron energy ε excess over the $3s4p(^3P)$ level threshold. $---$, Experiment (Reference 15); ——, calculation.

The incoming electron functions were determined in the field of the "frozen" atom, whereas the functions of the electron ejected due to autoionization decay were taken in the rearranged field of the core with a $3p$ hole.

The main difficulty in obtaining the Figure 6a diagram contribution lies in the evaluation of the overlap integral between slow-electron wave functions, which belong to the spectra of different continua [see (5′)]. For these functions we used the finite cutoff radius $R = 100$. For fixed energy ε_{k_s} the overlap integral as a function of the intermediate electron energy $\varepsilon_{k'}$ rapidly varies in the vicinity of $\varepsilon_{k'} = \varepsilon_{k_s}$. Therefore in the integration over k' [see (5′)] it was necessary to take rather many wave functions of the intermediate slow electron.

The calculations have demonstrated that the matrix element $\langle 3s, \varepsilon_f l_f \| V \| 3p, 4p \rangle$ for Auger decay with the transition to the state $l_f = 2$ exceeds essentially the matrix element of the transition to the s state and depends rather weakly upon the energy of the autoionization electron in the energy region of interest. In the lowest order of perturbation theory this matrix element determines the width of the autoionization level:

$$\Gamma = \frac{4\pi}{3} \, | \langle 3s, \varepsilon_f d \| V \| 3p, 4p \rangle \, |^2$$

which proved to be equal to $\Gamma = 0.2$ eV. In the following calculations this magnitude determines the imaginary part of the energy denominator in (1). The diagram in Figure 6a was calculated for the slow-electron angular momenta $l_s = 0, 1, 2$. The role of higher momenta is of minor importance because the process is considered near the threshold. The calculation demonstrates that the matrix elements of inelastic scattering $\langle \varepsilon_{k'} l_s, 4p \| V \| 3s, q l_q \rangle$ depend on the energy $\varepsilon_{k'}$ in the usual manner. At the threshold $\varepsilon_{k'} = 0$ only the monopole transition is possible and with increasing $\varepsilon_{k'}$ the dipole and quadrupole transitions contributions become important. Thus, for example, at $\varepsilon_{k'} = 5$ eV the reduced matrix element of the quadrupole transition already reaches the value of the monopole transition. Therefore without taking account of PCI the slow-electron angular distribution is isotropic at the threshold, and the anisotropy coefficient β in the differential cross section

$$\frac{d\sigma}{d\Omega} = \sigma_0[1 + \beta p_2(\cos \theta)] \tag{14}$$

would be equal to zero.

Our calculation demonstrates that the overwhelming contribution to the diagram in Figure 6a in all the region close to the ionization threshold is given by the quadrupole transition. This corresponds to $\beta = +1$. Therefore, in this process PCI effects essentially alter the slow-electron angular distribution.

Using the amplitude that corresponds to Figure 6a the differential cross section for inelastic resonance scattering with the formation of a single charged ion was calculated. For a fixed energy of the incoming electron the maximum of the cross section as a function of the Auger electron energy was determined. In Figure 6b the Auger electron energy shift Δ due to PCI is presented as a function of the excess of the incoming electron energy ε over the excitation threshold of the autoionization level.

In the same figure the experimental dependence, obtained by Fryar and McConkey[15] is also presented. In this experiment the total cross section for the production of the singly charged ion Ar^+ as well as of the excited atom Ar^* was measured. From the figure it is seen that for $\varepsilon > 1$ eV the experimental and theoretical curves are in satisfactory agreement while for lower energies they are different. This deviation is a result of the fact that for such small energies the energy shift due to PCI becomes comparable to or even larger than the energy of the slow electron. Therefore the slow electron is mainly captured to the discrete levels in the final state (see Section 4). Such transitions were taken into account in the experimental cross section and gave the main contribution to it for $\varepsilon \to 0$. But in our calculations of the scattering cross section for Ar^+ ion production such transitions were neglected.

6. PCI Theory: Future Development

In this paper the quantitative quantum mechanical theory is developed for the phenomenon that is called in the literature the postcollision interaction. This theory is based on the assumption that the variation of the field, in which the slow electron moves, proceeds for such a small interval of time that the interaction of fast and slow electrons is negligible.

The necessary numerical computations have been made and the theoretical predictions have been compared with experimental data for specific examples. Almost all the possible manifestations of PCI are demonstrated: the shift of the maximum in the energy distribution of fast (or slow) electrons, the asymmetry of the contour and its broadening, and the variation of the slow-electron angular distribution. Unfortunately it proved to be rather difficult to make the detailed numerical calculations of the interference between the resonant and nonresonant amplitudes because of the fast variation of the resonant amplitude phase and the difficulties in obtaining the nonresonant part.

The extension of the theory presented in this paper is connected with the consideration of the cases in which there are not one but two show electrons interacting with the changing ion field. Such a process has recently been observed.[16]

Of course, without any simplifications this is a real three-body problem with all its difficulties. However, in order to be able to satisfy the demand of experiments it is necessary to develop a method of approximation for the calculation of this phenomenon.

The next direction for extension of the theory presented in this paper is to consider the autoionizing states with $\Gamma \gtrsim 1$ (in atomic units). This problem is interesting not only from a pure theoretical point of view but also from the experimental point of view, since the states with large widths do exist in atoms. For large Γ all the PCI effects are much more impressive. For example, the energy shift would become extremely large. However, according to (8) much shorter distances (of the order of an atomic radius) are essential in this case and therefore the asymptotic form of the slow-electron wave function is insufficiently accurate for this case. Of course, the contribution of the diagrams that are responsible for PCI can be calculated numerically as was done in this paper. But for $\Gamma \gtrsim 1$ it is difficult to prove that it is possible to neglect the nonresonance diagrams. Therefore the problem becomes much more difficult due to the necessity of taking into account the nonresonance processes neglected in this paper.

The third interesting point is to consider the fact that the final state field (for example with two holes) is nonspherically symmetric. Thus a rather effective exchange of angular momentum between the slow electron and the final state of the ion is possible mainly when the width $\Gamma \gtrsim 1$ and therefore this electron is localized near the atom.

The question of the influence of PCI on the angular distribution of the slow electron is rather interesting. First of all the PCI can differently change the different partial amplitudes: it can intensify one part of them and decrease the other. This effect directly changes the angular distribution. The example of such a phenomenon is given in Section 5.

But there exists another possibility for PCI to manifest itself in the angular distribution. It is connected with an exchange of momentum between the electron and the vacancy. Such an exchange may be intensive when the electron feels the difference of the ion field from the spherical one. This is possible when the electron is localized at the distance $r \lesssim 1$, which according to (8) corresponds to the width $\Gamma \gtrsim 1$.

ACKNOWLEDGMENT

Finally, we would like to thank Dr. Chernysheva L. V. for permitting us to use her computational programs in our numerical calculations.

Appendix

Let us demonstrate for a general case that PCI leads to a shift of the line that corresponds to the autoionization decay state in the spectrum of ejected electrons. The differential cross section for inelastic scattering is proportional to $|A(k_s)|^2$, where $A(k_s)$ is given by the expression (5). Let us define the mean energy of the slow electron as a value averaged over the cross section:

$$\bar{\varepsilon}_s = \frac{\int \varepsilon_s \, d\sigma(\varepsilon_s)}{\int d\sigma(\varepsilon_s)} = \frac{\int \varepsilon_s \, |A(k_s)|^2 \, dk_s}{\int |A(k_s)|^2 \, dk_s} \tag{A.1}$$

The mean energy is determined by the energy conservation law. Using (6) we transform (A.1) into

$$\bar{\varepsilon}_s = \frac{\langle \phi \mid \hat{H} \mid \phi \rangle}{\langle \phi \mid \phi \rangle} \tag{A.2}$$

where $\phi(r)$ is determined by (7) and $\hat{H}$ is the Hamiltonian that describes the slow-electron motion in the final state. In derivation of (A.2) it is taken into account that ϕ_{k_s} [see (5')] is the eigenfunction of the operator $\hat{H}$. Without taking into account PCI we obtain analogously

$$\overline{\varepsilon_s^\circ} = \frac{\langle \phi \mid \hat{H}_0 \mid \phi \rangle}{\langle \phi \mid \phi \rangle} \tag{A.3}$$

where $\hat{H}_0$ is the Hamiltonian that describes the slow-electron motion in the autoionization state field. Therefore the energy shift due to PCI is determined by

$$\delta\varepsilon_s = -\delta\varepsilon_f = \bar{\varepsilon}_s - \bar{\varepsilon}_s^\circ = \frac{\langle \phi \mid \hat{H} - \hat{H}_0 \mid \phi \rangle}{\langle \phi \mid \phi \rangle} = \frac{\langle \phi \mid V \mid \phi \rangle}{\langle \phi \mid \phi \rangle} \tag{A.4}$$

where $V = \hat{H} - \hat{H}_0$ is the variation of Hamiltonian due to the excited state decay. The potential V corresponds to attraction, and therefore its mean value is negative. Hence, the slow-electron energy decreases whereas the fast-electron energy increases.

References

1. P. J. Hicks, S. Cvejanovic, J. Comer, F. H. Read, and J. M. Sharp, *Vacuum* **24**, 573 (1974).
2. F. H. Read, *Radiat. Res.* **64**, 23 (1975).
3. G. H. Wannier, *Phys. Rev.* **90**, 817 (1953).
4. M. J. Wan der Wiel, G. R. Wight, and R. R. Tol, *J. Phys. B* **9**, L5 (1976).
5. A. A. Abrikosov, L. P. Gorkov, and I. E. Dzaloshinsky, *Methods of Quantum Field Theory in Statistical Physics* (in Russian), Fizmat, Moscow (1963).
6. M. Ya. Amusia, N. A. Cherepkov, and L. V. Chernysheva, *Sov. Phys. JETP* **60**, 160 (1971).
7. A. I. Baz, Ya. B. Zeldovich, and A. M. Perelomov, *Scattering, Reaction and Decays in Non-relativistic Quantum Mechanics* (in Russian), Nauka, Moscow (1971).
8. R. B. Barker and H. W. Berry, *Phys. Rev.* **151**, 14 (1966).
9. R. Morgenstern, A. Niehaus, and U. Thielmann, *J. Phys. B* **10**, 1039 (1977).
10. U. Fano, *Phys. Rev.* **124**, 1866 (1961).
11. V. N. Ostrovsky, *Sov. Phys. JETP* **72**, 2079 (1977).
12. H. Bethe, *Ann. Phys. (Leipzig)* **5**, 325 (1930).
13. R. D. Hudson and L. J. Kieffer, *At. Data* **2**, 205 (1971).
14. M. Ya. Amusia, M. Yu. Kuchiev, S. A. Sheinerman, and S. I. Sheftel, *J. Phys. B* **10**, L535 (1977).
15. J. Fryar and J. W. McConkey, *J. Phys. B* **9**, 619 (1976).
16. S. Ohtani, H. Nishimura, H. Suzuki, and K. Wakyia, *Phys. Rev. Lett.* **36**, 863 (1976); V. M. Mikushkin, I. P. Flax, and G. N. Ogurcov, *Seventh All-Union Conference on Physics of Electron and Atomic Collisions (USSR)*, Petrozavodsk, p. 62 (1978).

26

Coincidence Studies
of Inner-Shell Excitation
in Ion–Atom Collisions

H. O. Lutz, N. Luz, S. Sackmann, W. Jitschin, and R. Hippler

The theoretical description of inner-shell excitation in ion–atom collisions has focused on two regions: "Coulomb ionization" by structureless ions, and "molecular excitation" via a quasimolecule formed during the collision. The mechanisms involved can be investigated in detail by studying correlation effects. For this purpose techniques such as detection of coincidences between scattered ions and the emitted reaction product, or observation of anisotropies of line radiation can be used. By measuring the impact parameter dependence in molecular K-shell ionization, the basic role of the 2pσ–2pπ rotational coupling was established to be in agreement with theory. At higher collision velocities characteristic deviations between the experimental results and the theoretical predictions were found. In the Coulomb regime the polarization of proton-induced line radiation was studied using Xe and Au targets. The observed effect was smaller than predicted by PWBA calculations.

1. Survey

Characteristic x-ray emission lines have been seen at a rather early stage in atomic collision studies, thus arousing the natural curiosity of the researchers. As a result, inner-shell excitation in atomic collisions has been investigated in great detail experimentally as well as theoretically. Many reasons stimulated this interest. Inner-

H. O. Lutz, N. Luz, S. Sackmann, W. Jitschin, and R. Hippler ● Fakultät für Physik, Universität Bielefeld, West Germany.   N. Luz' present address: Uranit GmbH., 5170 Jülich, West Germany.

shell excitation and ionization studies have been a testing ground for theoretical techniques since the early years of quantum mechanics.[1] Quantitative knowledge of inner-shell processes have become important in several different fields, for example, astrophysics,[2] trace element analysis,[3] search for superheavy elements,[4] positron creation in supercritical fields,[5] and x-ray laser development.[6] Two main regions of interest have emerged, each encompassing the development of successful theoretical descriptions. Since they represent approximations to potentially rather complex interactions, their distinction remains more a practical matter than a matter of principle.

"Coulomb ionization" of atoms (Z_2) by structureless charged particles (Z_1) is characterized by the use of one-center (separated atom) electron wave functions. The atomic electrons (classical orbital velocity u) are assumed to be only slightly perturbed by the projectile (velocity v). For inner shells, this condition is fulfilled if $Z_1 \ll Z_2$ or $v \gg u$. Widely used models are[7] the semiclassical approximation (SCA), the plane-wave Born approximation (PWBA), and the binary encounter approximation (BEA). Detailed studies have elucidated the interaction process to great depth. This is true in particular for K-shell excitation and ionization. Because of the rather reliably known wave functions and the applicability of the independent electron model, total cross sections are now well understood[8,9] and, even for the impact-parameter dependence of ionization, good agreement between theory and experiment has been reached.[9]

The situation in L-shell ionization is not quite as satisfactory. The main reasons are the uncertainty in the wave functions used in the calculation, and the experimental problems associated with the existence of different L subshells. In the latter case, aside from the fluorescence yield problems, which are more severe than for K shells, the measured characteristic x-ray or Auger intensities are not simply proportional to the subshell ionization yields, but are connected through the Coster–Kronig yields. Total cross sections are fairly well accounted for by theory.[8,9] The more sensitive impact-parameter-dependent ionization probabilities, however, yield in general only qualitative agreement between experiment and theory.[9] The situation in higher shells is even worse; only few theoretical as well as experimental data are available in this regime.

At low ion velocities $(v \lesssim u)$ and comparable nuclear charges $(Z_1/Z_2 \sim 1)$, the internal structure of the projectile and its influence on the target-atom electrons cannot be neglected. During the collision a quasimolecule is formed and excitation may occur by way of coupling between promoted levels. The treatment of this "molecular excitation" is based on two-center (molecular) wave functions. The Fano–Lichten model[10] makes use of general principles well known from molecular physics and was of basic significance for the development of the theory.[11] As in Coulomb ionization, the largest progress in quantitative understanding of the phenomena has been made for K-shell excitation. The basic mechanism is thought to be coupling of the atomic $1s$ to the $2p$ state via the molecular $2p\sigma$–$2p\pi$ rotational coupling at small internuclear separations. Within the framework of the independent electron model, the emerging scaling properties allow a unified description of molecular K-shell excitation spanning the range from slow H$^+$–H to very heavy ion–atom collisions.[12] Total K-shell excitation cross sections generally appear to be well under control[11] and the behavior of the impact-parameter-dependent excitation corroborates the model.[13] Recent studies, however, revealed significant differences between the results of total[14] and differential excitation[15–17] cross section measurements and the corresponding theoretical predictions.[12] The now generally used quantitative picture, which rests on the basic $2p\sigma$–$2p\pi$ coupling mechanism, should therefore be put to more rigorous tests.

L-shell excitation, as in Coulomb ionization, is in a less satisfactory state, even though it originally provided the starting point for developing the concept of inner-shell molecular excitation.[10] Reasons are found in the electron wave functions, and particularly in the large number of level crossings involved. For example, excitation via the steeply promoted $4f\sigma$ level is generally assumed to follow a step function providing unit probability for internuclear separations smaller than a critical value R_c, because the many level crossings involved cannot be handled individually. Other excitation channels, e.g., via $3d\sigma$, $3d\pi$, and $3d\delta$, have been included explicitly, though quantitative treatment is yet limited to the Ar–Ar system.[18] The situation is complicated by the fact that multiple L excitation is the rule rather than the exception; e.g., in Ar–Ar for internuclear distances $R < R_c$, two vacancies are created, and if R becomes sufficiently small to activate the $3d\pi$–$3d\delta$ coupling, the number of vacancies may even become considerably larger.

One complication must be mentioned which frequently plays an important role in the interpretation of molecular excitation data. In Coulomb ionization well above threshold, the density of final states is greatest for the continuum, and the ionization channel has indeed highest probability. In inner-shell molecular excitation, electron promotion generally occurs between bound states. Thus, excitation can only occur if the upper level contains at least one vacancy. This "exit channel effect" may effectively block the excitation process if the upper level is filled prior to collision. However, even if the upper level is filled in the projectile, a vacancy may be created on the incoming part of the ion trajectory by a long-range interaction.[19]

Because of this "exit channel effect," total excitation cross-section data must be interpreted with caution. In general, the differential excitation cross section (impact-parameter-dependent excitation probability) allows a less ambiguous test of the theory. This is greatly facilitated in molecular K-shell excitation since here the impact-parameter dependence shows a very characteristic doubly peaked structure: The broad "adiabatic maximum" at large impact parameters is (at least at higher impact velocities) well separated from the sharp "kinematic maximum" at small impact parameters which arises from a rapid rotation of the internuclear axis through 90°.

Some effort has been made to introduce concepts that may help to bridge the gap between the Coulomb ionization regime and the molecular model. For example, in Coulomb ionization the change in binding of the target electrons due to the presence of the projectile has been taken into account (cf. Reference 20).

In the region of molecular excitation, transitions involving a large number of states have been treated on a statistical basis,[18,21] and coupling of the molecular $1s$ and $2p$ states over finite energy gaps to higher states is being considered with particular attention to transitions into the continuum.[22–24] It must be stated, however, that no comprehensive quantitative picture has yet evolved in this very difficult regime.

Investigations have so far concentrated on establishing models of the excitation or ionization process. The next step would be to refine our quantitative knowledge of the mechanisms involved and to use this knowledge to obtain information on atomic parameters as, e.g., wave functions. The information required may be found in correlation effects associated with the excitation process. The study of impact-parameter-dependent excitation probabilities already belongs to this group as it establishes a correlation between a particle scattered through a certain angle and the reaction product. Aiming at a description of the excitation process that should be as complete as possible, one may introduce a further sophistication by differentiating the reaction product according to its direction of emission. In heavy-atom K-shell excitation only states of total angular momentum $j = \frac{1}{2}$ are involved in emission of reaction products, meaning they are emitted isotropically, except in the case of satellites caused by additional higher-shell excitation; corresponding anisotropies have been found, for example, by Jamison and Richard[25] in proton-induced Al K x-ray satellites originating from the decay of KL vacancies. This turns our attention to L-shell excitation. It is known that excitation of states with $j > \frac{1}{2}$ may indeed result in a nonisotropic emission of Auger electrons.[26,27] Similarly, nonisotropic emission of L x-rays has been reported.[28] These experiments are difficult to interpret since the observed effects are rather small—of the order of a few percent. Unfortunately, only few experiments have been reported so far. In particular, only Auger emission following Ar L_3 electron-impact ionization has been studied systematically,[27] giving fair agreement between Born approximation calculations and the experimentally observed anisotropy.

Anisotropies and polarization effects are much greater if the scattered projectile and the reaction product are detected in coincidence.[29] Moreover, this technique gives much greater sensitivity as to the approximations and wave functions used. As it is, no such experiments on inner shells in heavy atoms have been reported yet, though for outer shells and in light atoms large effects have been found; see, for example, Kessel *et al.*,[30] and Rødbro *et al.*[31]

In the following we will choose a few selected topics. In particular, we will discuss the impact-parameter dependence of K-shell excitation in the molecular regime in more detail, where deviations from the standard $2p\sigma$–$2p\pi$ rotational coupling theory are found. Furthermore, we will also report on a few measurements of the alignment of the ion-excited heavy atom after collision in Coulomb ionization as well as in molecular excitation.

2. Experimental

2.1. Impact-Parameter Dependence

Only the basic setup will be described here. Details are given elsewhere.[13] A schematic of the experiment is shown in Figure 1. Monoenergetic ion beams were injected into dilute gaseous targets after collimation to better than 0.1°. Scattered ions could be detected in coincidence with x-rays or Auger electrons excited in the collision. The gas pressure in the target region was sufficiently low to ensure single-collision conditions.

Ions scattered through laboratory angles ϑ were detected in an arrangement of silicon surface barrier detectors placed behind an annular aperture. This detector arrangement was aligned coaxially with the primary beam. By moving it along the beam axis, the scattering angle ϑ was varied from 45° down to less than 1°.

The x-rays were detected at 90° to the incident beam direction in a proportional counter which viewed the target through a cone-shaped collimator. This restricted the effectively observed collision region to 3.5 mm (FWHM) in length, thus limiting the mean angular width of the detected coincident ions.

Auger electrons were detected at 90° to the incident beam direction in an electro-static electron spectrometer placed opposite to the proportional counter. The energy acceptance width of the spectrometer ($\Delta E/E \sim 3\%$) was well below the peak width of the K Auger electrons emitted in slow heavy-ion–atom collisions (of the order 50 eV). To ensure detection of all K electrons, the spectrometer was therefore scanned rapidly and repeatedly over a region extending 100 eV to either side of the peak maximum.

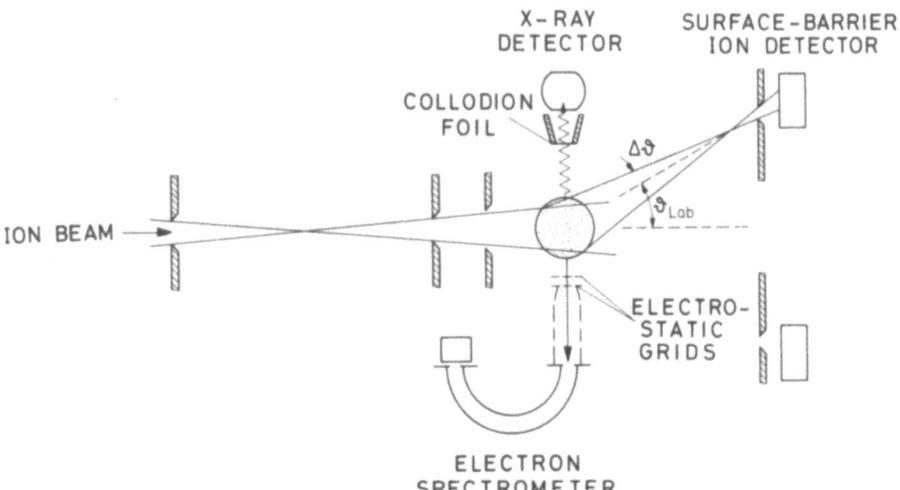

Figure 1. Schematic experimental setup for measuring the impact-parameter-dependent excitation probability in ion–atom encounters.

The ion detectors could not discriminate between primary and recoil atoms in such nearly symmetric collisions. The excitation probability $\bar{P}$ represents therefore the average for impact parameters corresponding to scattering angles ϑ and $\pi/2 - \vartheta$, weighted according to their respective laboratory system (LS) scattering cross section σ_1 and σ_2. The excitation probability $\bar{P}(\vartheta)$ is given by[16]

$$\bar{P}(\vartheta) = \frac{\sigma \cdot N}{\Omega(\sigma_1 + \sigma_2 \omega_2/\omega_1)}$$

with σ the total excitation cross section, N the number of coincidences relative to the noncoincident number of counts (x-ray or Auger electrons, respectively), Ω the LS ion detector solid angle, ω_1 and ω_2 the CM solid angle elements for scattering of a primary and a recoil atom into the ion detector and for symmetric systems $\omega_1 = \omega_2$.

The conversion of scattering angles ϑ into impact parameters b, as well as the calculation of the scattering cross sections, were performed using a Bohr (exponentially screened Coulomb) potential in a numerical integration of the scattering equation.

2.2. Anisotropic Emission of Reaction Products

Anisotropic emission due to nonstatistical population of magnetic subshells was studied employing coincidence and noncoincidence techniques.

The noncoincidence approach was used in cases in which the cross section involved did not yet allow coincidence experiments. An electron spectrometer and a Si(Li) x-ray detector could be moved about the target in a plane containing the incident beam direction. In the experiments to be reported here, only x-ray detection was employed. The degree of polarization P of the emitted radiation is then given by[32]

$$I(\theta)/I(90°) = 1 - P \cdot \cos^2 \theta$$

with $I(\theta)$ the x-ray intensity of the transition studied, and θ the x-ray emission angle relative to the beam direction.

In the coincidence experiments, the setup was very much as described in Section 2.1, with the exception that the coincidence signals from each ion detector were processed separately. Four solid-state surface barrier detectors were mounted behind the annular aperture, displaced 90° from each other about the circumference.

3. Results and Discussion

3.1. Impact-Parameter Dependence of K-Shell Excitation

The dominating decay of K vacancies in light atoms occurs via electron emission. The competitive x-ray decay contributes only a few percent or less. Therefore, the K

vacancy production will be discussed mainly on the basis of Auger electron data. Collision velocities are given in "reduced units" (r.u.) as introduced by Taulbjerg et al.,[12] essentially normalizing the ion velocity to the orbital velocity of the inner-shell electron.

To avoid the difficulties associated with an unknown $2p\pi$ vacancy occupation probability N_π before the $2p\sigma–2p\pi$ transition ("exit channel effect"), we have overlaid in the following figures the experimental $\bar{P}(b)$ data and the theoretical curve in the region of the adiabatic maximum. Thus, the validity of the theory will be judged by the difference between the shapes of the theoretical and the experimental $\bar{P}(b)$ curves. As a consequence, it is primarily at small impact parameters where deviations become apparent.

The first unambiguous experimental evidence for the validity of the two-state $2p\sigma–2p\pi$ coupling theory, in particular the appearance of the steep kinematic peak at small impact parameters, was obtained in a study of the Ne⁺–Ne systems (Figure 2). In this early work, x-ray emission was used as a measure of the collisional excitation. The Ne⁺–Ne system has now been investigated again. The results are also displayed in Figure 2, in addition to old and new x-ray data. The agreement between the different experiments and the theory is quite good. It follows from these and the other results to be presented here that the existence of the steep "kinematic" peak in the excitation probability at small impact parameters may be regarded as established.

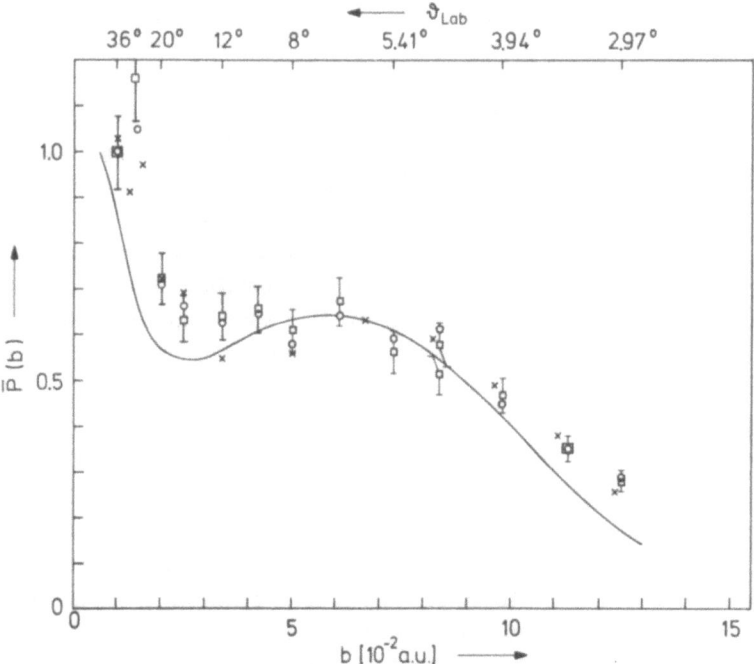

Figure 2. $\bar{P}(b)$ function for 363-keV Ne⁺–Ne ($v = 0.091$ r.u.). Solid curve: theory.[12]

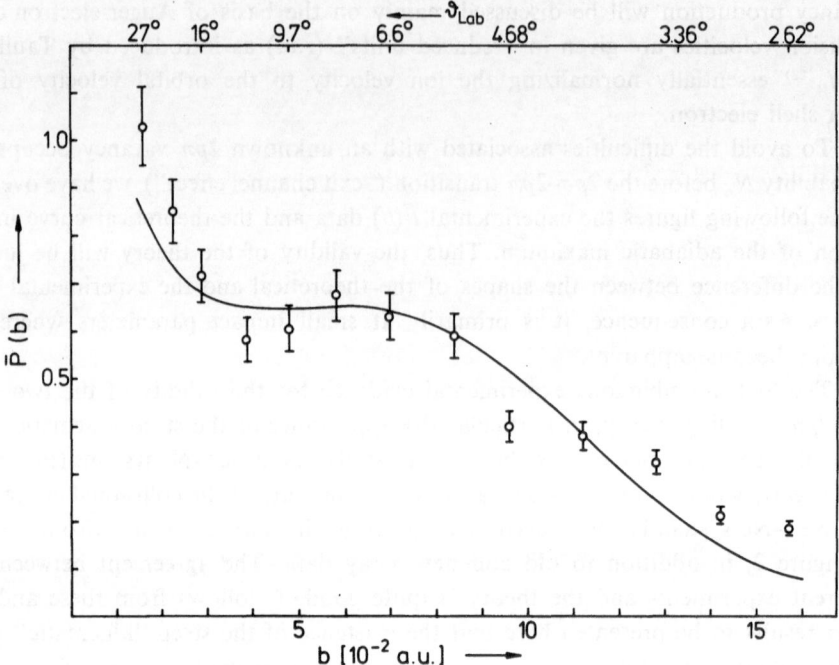

Figure 3. $\bar{P}(b)$ function for 250-keV Ne$^+$–O$_2$ ($v = 0.084$ r.u.). Solid curve: theory.[12]

For asymmetric collisions it has been proposed[12] that K excitation (in the lighter atoms) should also proceed predominantly by $2p\pi$–$2p\sigma$ rotational coupling as in symmetric collisions.

Figure 3 shows $\bar{P}(b)$ for oxygen K-shell excitation in 250-keV Ne$^+$–O$_2$ collisions ($v = 0.084$ r.u.). As for the symmetric system Ne$^+$–Ne, good agreement with the calculations is found. At higher collision velocity, the theory yields a pronounced valley between the adiabatic and the kinematic peak. Experimental data of 385-keV Ne$^+$–O$_2$ ($v = 0.105$ r.u.) show, however, a tendency to fill this valley (Figure 4). The separation between the adiabatic and the kinematic peak becomes even more pronounced for increasing collision velocity. We found that also in 400-keV O$^+$–O$_2$ collisions ($v = 0.135$ r.u.), the experimental data hardly show any tendency to reproduce the valley. This effect becomes quite evident at still higher collision velocities. Figure 5 shows results for 400-keV N$^+$–N$_2$ ($v = 0.167$ r.u.).

Apparently, the impact-parameter-dependent excitation probability is reproduced by the simple two-state $2p\sigma$–$2p\pi$ coupling theory only at low relative ion velocities ($v \lesssim 0.1$ r.u.). At $v > 0.1$ r.u., the steep rise from the adiabatic peak to the kinematic peak still remains quite evident in the experimental data, even though the valley in between is not followed. The general behavior confirms the basic role of the $2p\sigma$–$2p\pi$ rotational coupling in such systems. However, it also shows that the theory[12] is not yet sufficiently developed to describe the finer details of the excitation process.

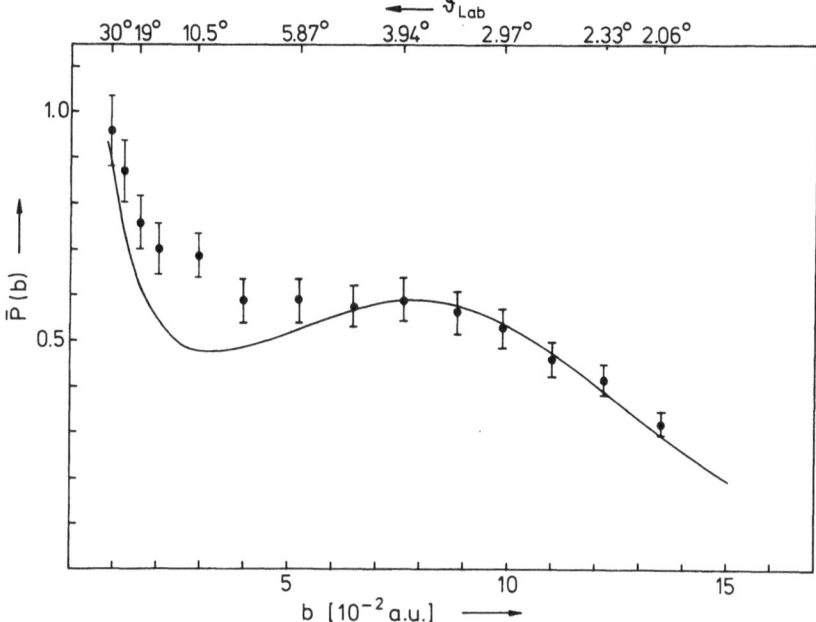

Figure 4. $\bar{P}(b)$ function for 385-keV Ne$^+$–O$_2$ ($v = 0.105$ r.u.). Solid curve: theory.[12]

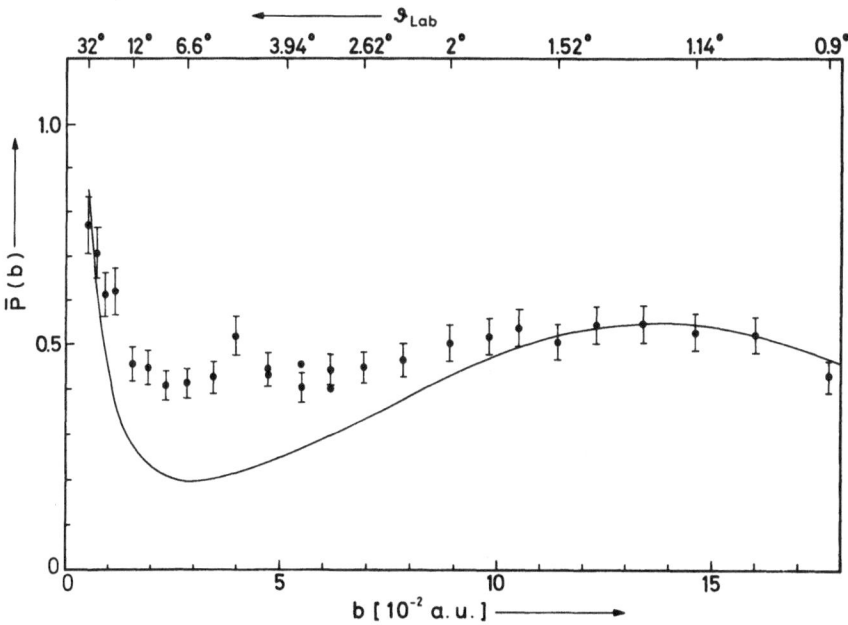

Figure 5. $\bar{P}(b)$ function for 400-keV N$^+$–N$_2$ ($v = 0.167$ r.u.). Solid curve: theory.[12]

An explanation may be found in the existence of additional level couplings, though at present it is difficult to judge which levels might be involved and which of the amplitudes involved add coherently or incoherently. For example, for the long-range coupling mechanism responsible for $2p\pi$ vacancy creation in the early stage of the collision it is generally assumed that any phase relationship with the $2p\sigma-2p\pi$ coupling is destroyed, and both processes are treated independently. However, situations may exist in which coupling of the molecular $2p$ states to higher states occurs at sufficiently small internuclear distances, rendering a separation of the $2p\sigma-2p\pi$ transition and the creation of $2p\pi$ vacancies difficult or impossible. The same argument holds for coupling of $2p\sigma$ to higher (continuum) states. The $2p\sigma$ binding energy is smallest at small distances; additionally, rotational effects may be of importance, occurring preferentially at large scattering angles, i.e., small impact parameters. For example, Taulbjerg[33] has shown that extending the $2p\sigma-2p\pi$ coupling calculation by including incoherently the $2p\sigma-3p\sigma$ radial coupling in the case of 385-keV Ne^+–O_2 exactly reproduces the experimental $\bar{P}(b)$ function even in the region between the kinematic and adiabatic peak. However, in a coherent super-position of $2p\sigma-2p\pi$, $2p\pi-3p\sigma$, and $2p\sigma-3p\sigma$ this agreement is lost again. At present, it seems to be very difficult to control such contributions from couplings to higher states. It is, however, evident from these theoretical estimates that a detailed under-standing of the excitation process cannot be achieved without consideration of such couplings.

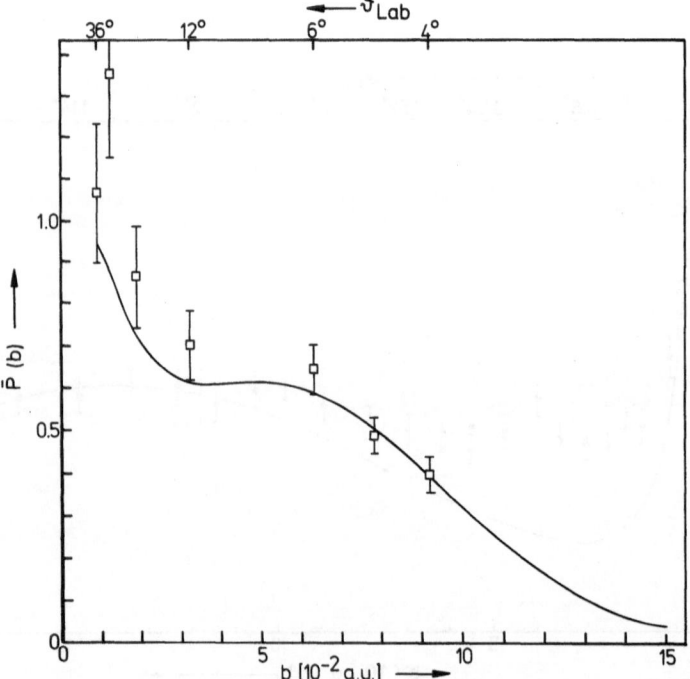

Figure 6. $\bar{P}(b)$ function for 420-keV Na^+–Ne ($v = 0.087$ r.u.). Solid curve: theory.[12]

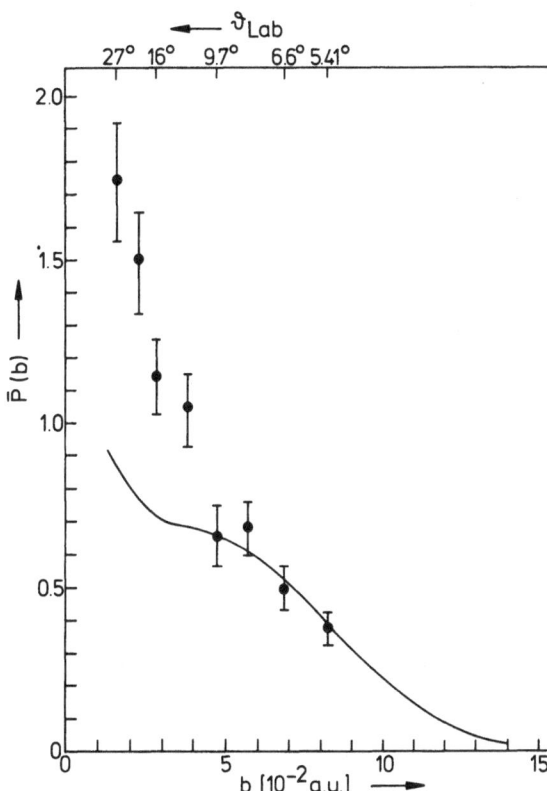

Figure 7. $\bar{P}(b)$ function for 380-keV Mg$^+$–Ne ($v = 0.077$ r.u.). Solid curve: theory.[12]

From the experimental point of view, processes other than the $2p\sigma$–$2p\pi$ coupling are expected to be particularly apparent in systems in which the *a priori* presence of $2p\pi$ vacancies is suppressed. Figures 6 and 7 show the experimental impact-parameter-dependent excitation for 420-keV Na$^+$–Ne ($v = 0.087$ r.u.) and 380-keV Mg$^+$–Ne ($v = 0.077$ r.u.). For the Na$^+$–Ne system, only x-ray data were measured. The $\bar{P}_x(b)$ data in Na$^+$–Ne were used previously[34] to show that K excitation in this system proceeds predominantly by $2p\pi$ vacancy creation at large internuclear separations, followed by a well-separated $2p\sigma$–$2p\pi$ rotational transition. In reference to slight deviations at small impact parameters between the Na$^+$–Ne and a Ne$^+$–Ne experiment, the possible influence of couplings between the molecular $2p$ levels at small internuclear separations and higher unoccupied states has been pointed out. In Figure 6 the experimental data are now compared to the proper theoretical $\bar{P}(b)$ function.[12] The deviations noted earlier become smaller, but are still noticeable. The Mg$^+$–Ne system (Figure 7) shows such differences between theory and experiment quite clearly. The experimental data at small impact parameters rise more steeply than the theoretical curve. This increase of the experimental excitation probability over the prediction of the rotational coupling theory at small impact parameters has also been established for the Ar–Ar collision system at 2.5, 4.5, and 8 MeV ion energy.[16] At all three energies, $\bar{P}(b)$ was found to rise sharply at small

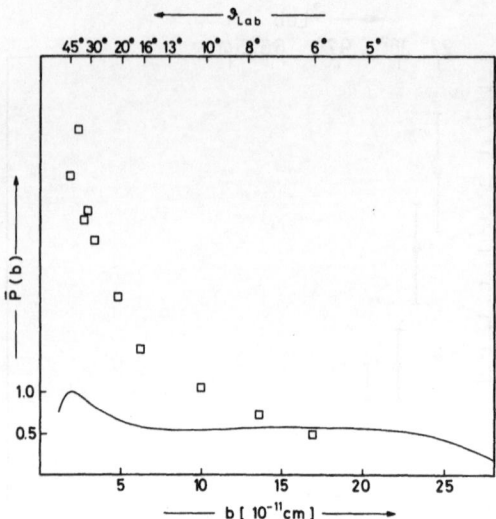

Figure 8. $\bar{P}(b)$ function for 2.5-MeV Ar–Ar ($v = 0.095$ r.u.). Solid curve: theory.[12]

values of b, the relative increase being most pronounced at the lowest ion energy, 2.5 MeV (Figure 8). This has been interpreted by a significant contribution of an excitation process occurring at small impact parameters which has not yet been considered quantitatively. This process has an ion energy dependence different from that responsible for the opening of the exit channel by creation of $2p\pi$ vacancies and is dominant at small energies. At increasing relative ion velocities, or low atomic $2p$ electron binding energies, the creation of $2p\pi$ vacancies at large internuclear distances becomes of rapidly increasing importance. This results in $\bar{P}(b)$ functions that tend to resemble more and more the predictions of the rotational coupling theory.[12] However, even in cases where N_π is of order unity, deviations of the experiment from the rotational coupling theory remain, the most apparent being the filling of the valley between the kinematic and the adiabatic peak in the $\bar{P}(b)$ function.

3.2. Anisotropy of Reaction Products

We have investigated the angular distribution of the noncoincident Ll, $L\alpha_{1,2}$, $L\beta_1$, $L\beta_2$, and $L\gamma_1$ characteristic x-rays emitted from thin Cs and Au foils, as well as Xe gas ($\sim$1 mbar). The targets were bombarded by 350-keV protons. Figure 9 shows a typical Xe x-ray spectrum obtained with a Si(Li) detector. The lines of interest are clearly resolved. The Ll line, which was expected to show the highest polarization, is rather weak, sitting on a strong background induced in the Si(Li) detector mainly by the strong $L\alpha$ and $L\beta$ lines. To overcome this difficulty, a 15-μm Ca absorber was placed in front of the x-ray detector. Its K absorption edge lies at 4.04 keV, thus effectively enhancing the Ll line relative to the other lines at higher x-ray energies as well as to the background. The K fluorescence radiation induced in the Ca absorber has an energy close to the Ll line of Xe. Its contribution is only a few percent and can be corrected for. The $L\beta_1$ and $L\gamma_1$ transitions are assumed to be isotropic as they originate from the decay of L_2 vacancies ($j = \frac{1}{2}$). They can therefore be used as intensity references. This is a less ambiguous situation than encountered in previous experiments.[28,35] Schöler and Bell[28] used a proportional counter to detect the proton-induced integral Ll, $L\alpha_{1,2}$ x-rays from Cu and Ge. As a result, x-rays emitted in transitions from different upper levels into the L_3 subshell could not be distinguished; the integral polarization (a few percent) was rather difficult to extract from data. Hrdý et al.[35] measured the polarization of $L\alpha_1$ x-rays from electron-impact ionization of Hg employing Bragg reflection from a quartz crystal. This necessitated a somewhat complicated analysis of the experimental data; the resulting polarization of about 14% appears considerably too high if compared to corresponding theories.[36,37]

The degree of polarization found in our experiments (Figure 10) is smaller than predicted by PWBA calculations.[28] For example, the measured Ll polarization for 350-keV proton impact on Xe is about 23%, far below the theoretical prediction of 41%. From the experimentally determined polarization, the atomic alignment parameter A_2[37] can be obtained. The A_2 values derived from different transitions were equal within the experimental errors as expected, giving support to the method used. The experimental results have to be corrected for the depolarizing influence of Coster–Kronig transitions between the different L subshells. Using known L_1, L_2, and L_3 subshell ionization cross sections (for Au,[38,39] for Xe[40]) and Coster–Kronig transition rates,[41] this can be calculated. For example, in case of 350-keV protons on Au, the experimental alignment parameter A_2 is too low by a factor of 0.66. The alignment parameters corrected accordingly are $A_2 = -(36 \pm 7)\%$ for

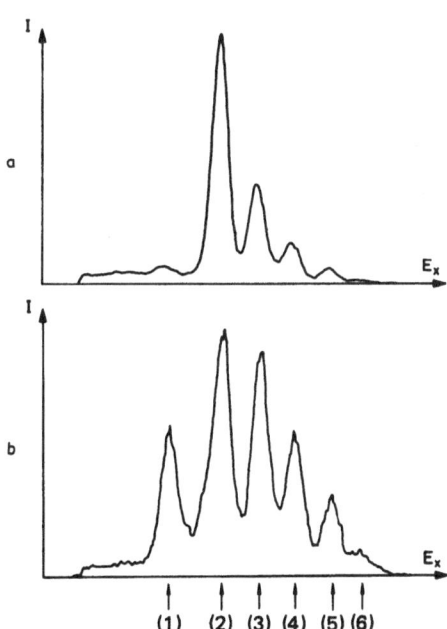

Figure 9. L x-ray spectra from a Xe gas target bombarded by 350-keV protons. (a) No absorber; (b) 15-μm Ca absorber. The peaks are: (1) Ll, 3638 eV; (2) $L\alpha_{1,2}$, 4110 eV; (3) $L\beta_1$, 4422 eV; (4) $L\beta_2$, 4720 eV; (5) $L\gamma_1$, 5036 eV; (6) $L\gamma_{2,3}$, 5318 eV.

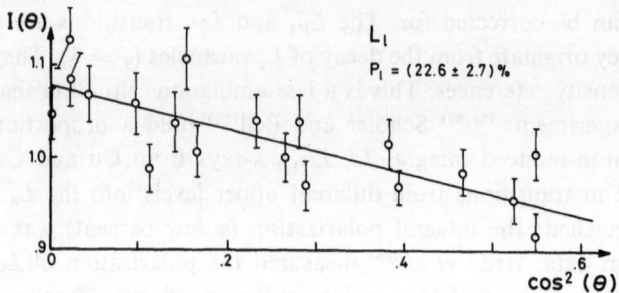

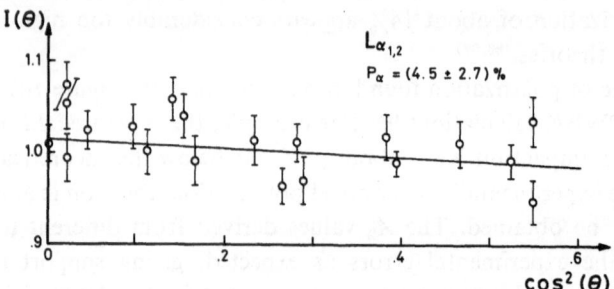

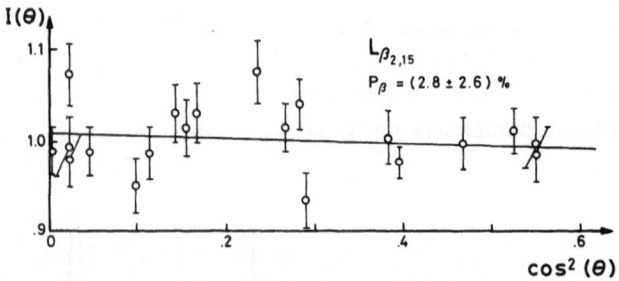

Figure 10. Angular distribution of characteristic x-ray lines excited by 350-keV protons in a Xe gas target. The straight lines are least-squares fits to experimental data.

Xe and $A_2 = -(47 \pm 8)\%$ for Au, whereas PWBA calculations yield $A_2 = -62\%$ and -74%, respectively.[28]

In the above experiments, the anisotropy of x-rays averaged over all ion trajectories was investigated. We have also performed some preliminary measurements of the azimuthal angular correlation between Ar LMM Auger electrons and scattered ions in 0.3–1 MeV Ar–Ar collisions. The electrons were detected under 90° with respect to the incident beam direction. The Ar ion scattering angles were 3° and 10°. So far, within an experimental uncertainty of about 10%, no azimuthal angular correlation could be observed. This result may not be so surprising if we recall the excitation mechanism. Many quasimolecular states are involved, providing a large number of reaction channels; the LMM Auger spectrum appears only as a very broad peak without any structure. As a result, strong deviations from a statistical population of magnetic subshells are not to be expected at impact energies well above excitation threshold.

4. Summary

In conclusion, our data on the impact-parameter-dependent K excitation probability in slow heavy-particle collisions emphasize the basic role of the $2p\sigma–2p\pi$ rotational coupling mechanism. We found, however, that the theory is not yet sufficiently developed to describe the finer details of the interaction. In particular, coupling to higher states does not seem to be under control. In systems in which the $2p\sigma–2p\pi$ channel is closed, $2p\pi$ vacancies may nevertheless be created by a long-range interaction in the first half of the collision. If this process is further suppressed by going to still heavier atoms, a new excitation mechanism shows up at small impact parameters.

Our data on the angular distribution of proton-induced characteristic Ll, $L\alpha_{1,2}$, and $L\beta_2$ lines show polarization, though less than predicted by PWBA calculations. The discrepancy is outside the experimental uncertainty and points to the need for more detailed calculations in the Coulomb ionization regime.

In the region of molecular inner-shell excitation in heavy atoms, no theory exists to describe angular correlation effects. Our preliminary experimental data on the azimuthal angular correlation of molecular L excitation in Ar–Ar collisions well above excitation threshold indicate no such effects. The reason may be found in the rather complicated nature of the excitation process favoring a statistical population of magnetic subshells.

ACKNOWLEDGMENTS

We are indebted to Professor H. Kleinpoppen and Dr. K. Taulbjerg for very stimulating discussions. Financial support by the Deutsche Forschungsgemeinschaft is gratefully acknowledged.

References

1. E. Merzbacher, in *Proceedings of the Second International Conference on Inner Shell Ionization Phenomena*, Book of Invited Papers, W. Mehlhorn and R. Brenn, eds., p. 1, Freiburg (1976).
2. T. E. Bunch, L. J. Caroff, and H. Mark, in *Atomic Inner Shell Processes*, Vol. II, p. 187, B. Crasemann, ed., Academic Press, New York (1975).
3. F. Folkmann, C. Gaarde, T. Huus, and K. Kemp, *Nucl. Instrum. Methods* **116**, 487 (1974).
4. P. H. Mokler, H. J. Stein, and P. Armbruster, *Phys. Rev. Lett.* **29**, 827 (1972).
5. B. Müller, V. Oberacker, J. Reinhardt, G. Soff, W. Greiner, and J. Rafelski, in *Proceedings of the International Conference on Nuclear Structure*, T. Marumori, ed., p. 838, Tokyo (1977).
6. P. Jaeglé, in *Proceedings of the Second International Conference on Inner Shell Ionization Phenomena*, Books of Invited Papers, W. Mehlhorn and R. Brenn, eds., p. 379, Freiburg (1976).

7. D. H. Madison and E. Merzbacher, in *Atomic Inner Shell Processes*, B. Crasemann, ed., Vol. I, p. 1, Academic, New York (1975).

8. P. Richard, in *Atomic Inner Shell Processes*, B. Crasemann, ed., Vol. I, p. 73, Academic, New York (1975).

9. H. O. Lutz, *Proceedings of the Second International Conference on Inner Shell Ionization Phenomena*, Book of Invited Papers, W. Mehlhorn and R. Brenn, eds., p. 104, Freiburg (1976).

10. U. Fano and W. Lichten, *Phys. Rev. Lett.* **14**, 627 (1965); W. Lichten, *Phys. Rev.* **164**, 131 (1967).

11. J. S. Briggs, *Rep. Prog. Phys.* **39**, 217 (1976).

12. K. Taulbjerg, J. S. Briggs, and J. Vaaben, *J. Phys. B* **9**, 1351 (1976).

13. S. Sackmann, H. O. Lutz, and J. S. Briggs, *Phys. Rev. Lett.* **32**, 805 (1974); N. Luz, S. Sackmann, and H. O. Lutz, *J. Phys. B*, to be published.

14. R. J. Fortner, P. H. Woerlee, S. Doorn, Th. P. Hoogkamer, and F. W. Saris, *Phys. Rev. Lett.* **39**, 1322 (1977); Th. P. Hoogkamer, P. H. Woerlee, R. J. Fortner, and F. W. Saris, *J. Phys. B* **10**, 3245 (1977).

15. N. Luz, S. Sackmann, H. O. Lutz, R. v. Reenen, R. McMurray, and I. v. Heerden, *Tenth International Conference on the Physics of Electronic and Atomic Collisions* (ICPEAC), Book of Abstracts, p. 888, Commissariat a l'Energie Atomique, Paris (1977).

16. H. O. Lutz, W. R. McMurray, R. Pretorius, R. J. van Reenen, and I. J. van Heerden, *Phys. Rev. Lett.* **40**, 1133 (1978).

17. C. H. Annett, B. Curnutte, and C. L. Cocke, *Phys. Rev. A* **19**, 1038 (1979).

18. G. B. Schmid and J. D. Garcia, *Phys. Rev. A* **15**, 85 (1977).

19. B. Fastrup, in *Proceedings of the IXth International Conference on the Physics of Electronic and Atomic Collisions* (ICPEAC), J. S. Risley and R. Geballe, eds., p. 361, University of Washington Press, Seattle (1975).

20. J. U. Andersen, E. Laegsgaard, M. Lund, and C. D. Moak, *Nucl. Instrum. Methods* **131**, 341 (1976).

21. W. Brandt and K. W. Jones, *Phys. Lett.* **57A**, 35 (1976).

22. J. S. Briggs, *J. Phys. B* **8**, L485 (1975).

23. W. R. Thorson, *Phys. Rev. A* **12**, 1365 (1975).

24. G. Soff, W. Betz, B. Müller, W. Greiner, and E. Merzbacher, *Phys. Lett.* **65A**, 19 (1978).

25. K. A. Jamison and P. Richard, *Phys. Rev. Lett.* **38**, 484 (1977).

26. W. Mehlhorn, *Phys. Lett.* **26A**, 166 (1968); B. Cleff and W. Mehlhorn, *J. Phys. B* **7**, 593 (1974); E. Döbelin, W. Sandner, and W. Mehlhorn, *Phys. Lett.* **49A**, 7 (1974).

27. W. Sandner and W. Schmitt, *J. Phys. B* **11**, 1833 (1978).

28. A. Schöler and F. Bell, *Z. Phys.* **A286**, 163 (1978).

29. E. G. Berezhko, N. M. Kabachnik, and V. V. Sizov, *J. Phys. B* **11**, 1819 (1978).

30. Q. C. Kessel, R. Morgenstern, B. Müller, A. Niehaus, and U. Thielmann, *Phys. Rev. Lett.* **40**, 645 (1978); and Chapter 19.

31. M. Rødbro, R. DuBois, and V. Schmidt, *J. Phys. B* **11**, L551 (1978).

32. U. Fano and J. H. Macek, *Rev. Mod. Phys.* **45**, 553 (1973).

33. K. Taulbjerg and J. Vaaben, private communication.

34. N. Luz, S. Sackmann, and H. O. Lutz, *J. Phys. B* **9**, L15 (1976).

35. J. Hrdý, A. Henins, and J. A. Bearden, *Phys. Rev. A* **2**, 1708 (1970).

36. S. C. McFarlane, *J. Phys. B* **5**, 1906 (1972).

37. E. G. Berezhko and N. M. Kabachnik, *J. Phys. B* **10**, 2467 (1977).

38. S. Datz, J. L. Duggan, L. C. Feldman, E. Laegsgaard, and J. U. Andersen, *Phys. Rev. A* **9**, 192 (1974); and private communication.

39. W. Jitschin, private communication.

40. B.-H. Choi, E. Merzbacher, and G. S. Khandelwal, *At. Data* **5**, 291 (1973).

41. E. J. McGuire, *Phys. Rev. A* **3**, 587 (1971).

27

Coherent Production of Positrons in Heavy-Ion Collisions

J. REINHARDT, B. MÜLLER, AND W. GREINER

Collisions of very heavy ions are discussed as a means of investigating quantum electro-dynamics in the presence of strong external fields. The action of the combined Coulomb field of the closely approaching nuclei leads to strong binding of the inner electron shells and to large induced transitions. The resulting mechanisms for the production of positrons are discussed in detail and compared with recent experiments.

1. Introduction

Figure 1 shows the energy of the strongest bound electron states in an atom as a function of the nuclear charge Z. As can be seen, the binding is only a small fraction of the total electron rest mass in all known elements up to fermium ($Z = 100$). However, if we go beyond this limit, for the moment in a gedanken experiment, we find that the calculations predict the binding energy of the $1s$ state to reach 511 keV around $Z = 140$. The electron then has zero energy, all its mass being "bound away." Even beyond, we find[1] that the binding energy reaches 1022 keV or twice the electron rest energy at $Z \sim 173$. At this point it is favorable to produce an electron in the $1s$ state and there is enough spare energy left to produce a positron to conserve charge and lepton number. Thus an atom with $Z > 173$ and an empty K shell would spontaneously shield itself by two K electrons and emit two positrons of rather well-

J. REINHARDT, B. MÜLLER, AND W. GREINER • Institut für Theoretische Physik der Johann Wolfgang Goethe-Universität, Robert-Mayer-Strasse 8–10, D-6000 Frankfurt am Main, West Germany.

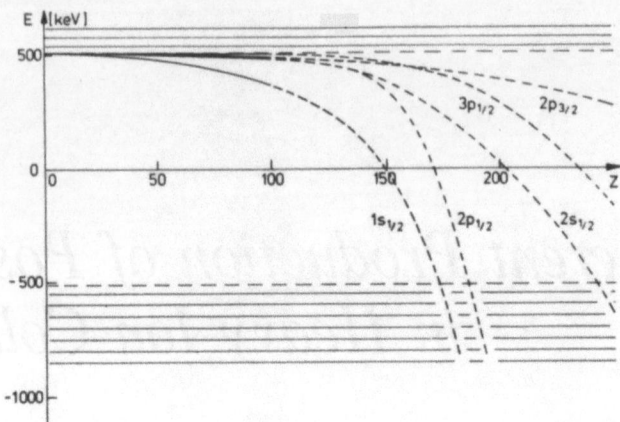

Figure 1. Atomic binding energies as a function of nuclear charge Z, calculated in the Hartree-Fock–Slater approximation.

defined energy. In the terminology of field theory one says that the ground state of the system, the "vacuum state," becomes charged.[2]†

Systems of particles in which the binding energy is a major or even the dominant fraction of the total energy are of great interest in elementary particle physics. If a hadron is a very strongly bound system of several quarks, so that the attempt to ionize one of the quarks leads only to pair production (i.e., meson production), then these ideas should be tested somewhere in a situation where the interactions are known. Very high Z atoms form the unique testing ground, because they involve only electromagnetic interactions and allow precise theoretical predictions.

We must now talk about how to make such atoms with $Z > 130$. Nuclei, of course, with such charge number are not stable. However, the typical time scale for atomic processes to evolve in such an atom is of the order of 10^{-20}–10^{-18} sec. It is therefore sufficient to form a "quasinucleus" for a very short instant of time. This is possible in heavy-ion collisions, where two nuclei can be brought together as close as 20 fm for a typical time of $\sim 10^{-21}$ sec. It has been found that the most suitable way to describe the various stages of the scattering process is the adiabatic picture[4]: the electrons are envisaged as moving very fast (at velocity $v \sim c$) around two slowly moving nuclei (at $v \sim c/10$). Then one takes a snapshot of this "quasimolecule" at every internuclear separation R. The time development is described by the scattering trajectory $R(t)$.

When one solves the (relativistic) Dirac equation for an electron in the electric field of two nuclei,[5] one obtains the quasimolecular correlation diagrams, of which two typical examples are shown in Figure 2: (a) for a Pb–Pb collision and (b) for a U–U collision.[6] The most interesting states are the lowest ones, in particular the one denoted $1s\sigma$. At small distances $R \sim 20$ fm the quasimolecular states approach those of an atom of charge $Z_1 + Z_2$ (i.e., 164 and 184 in our examples). From the

† For the physics of supercritical fields, see the review articles in Reference 3.

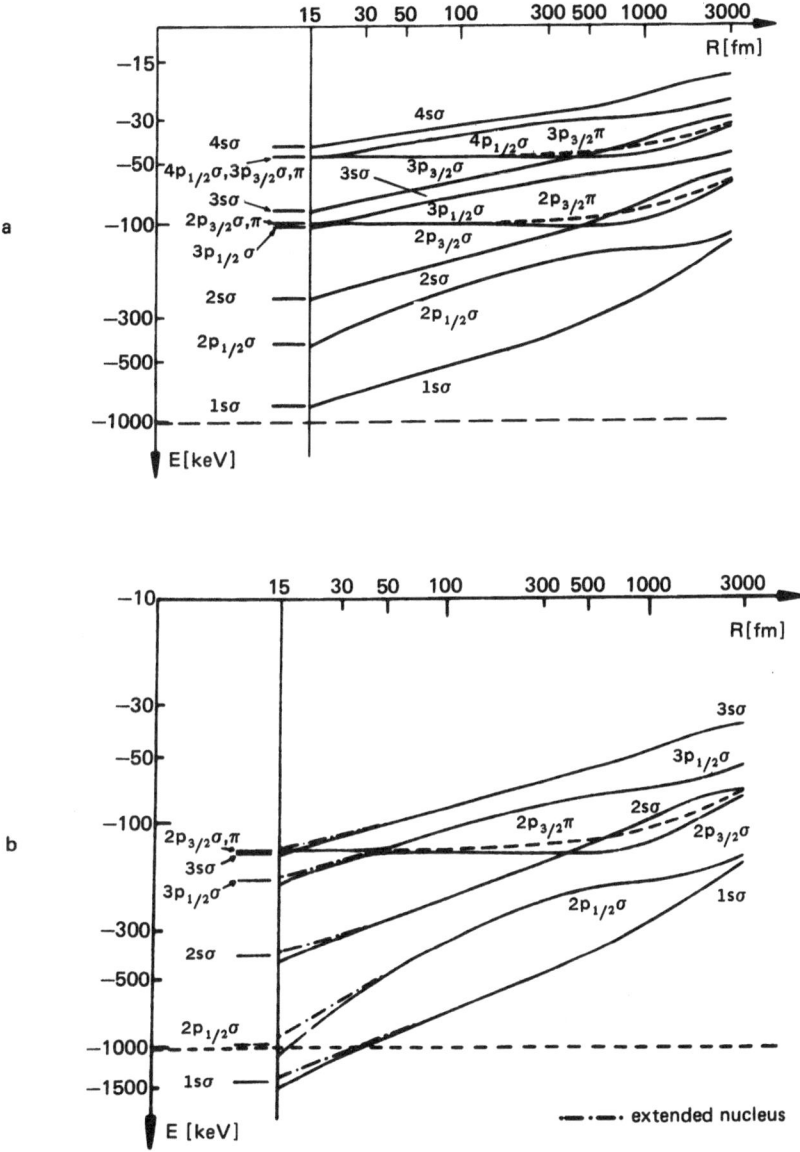

Figure 2. Adiabatic correlation diagram of the quasimolecular systems Pb–Pb (a) and U–U (b). R is the distance between the two nuclei; E is the electronic binding energy. At $R \sim 35$ fm in U–U the $1s\sigma$ state becomes supercritical. The curves are taken from Reference 6.

behavior of the wave functions one finds that in the region $R < 200$ fm the system resembles much more a "quasiatom" with a big blown-up nucleus of radius $R/2$ than a molecule. We conclude that in a heavy-ion scattering process quasiatomic systems of nuclear charge up to 184–190 become available and it is possible to study them.

The major processes allowing for an experimental investigation are the following.

(a) Pair creation and positron emission, when the binding energy of the $1s\sigma$ state approaches 1 MeV (which will be discussed in detail).

(b) Ionization of high energy (delta electrons) out of the low-lying states. By a careful analysis of the numerical calculations and by comparison with analytical models, one finds[7-9] that the ionization probability P of the $1s\sigma$ orbital versus impact parameter b can be directly related to the binding energy $E_B^{1s}(R_0)$ at closest approach. A systematic measurement of $P(b)$ as function of bombarding energy E_{CM} and quasi-atomic charge $Z = Z_1 + Z_2$ could therefore be utilized to determine the binding energy $E_B^{1s}(R, Z)$ right up into the region of supercritical binding ($E_B > 1$ MeV).

These ideas have been successfully tested during the last year, notably in experiments performed at GSI. The first are due to Greenberg, Bokemeyer, and Schwalm,[10] who used a clever Doppler-shift technique to determine the ion scattering angle. Predicted rates in the percent region have been observed. More detailed results can be obtained by ion–x-ray coincidence experiments performed by the Behncke, Liesen, Mokler, Armbruster group at GSI.[11] Just recently a systematic measurement has been used to extract binding energies in the quasimolecular diagram.[12] The results are in rather good agreement, indicating that a spectroscopy of quasiatoms up to $Z \sim 180$ seems to be feasible. Another way to investigate the quasiatomic wave functions is delta-electron spectroscopy. The high-energy part of the spectrum of ionized electrons can be related to the high-momentum components of the quasi-atomic wave functions.[13] It is only because the inner wave functions in atoms beyond $Z \sim 140$ show structures of the order 20–30 fm that very high energy electrons ($E > 1$ MeV) are produced.

(c) X-ray transitions between quasimolecular states when inner-shell vacancies have been formed. Since an inner-shell (e.g., $1s\sigma$) vacancy must be created before radiative decay can occur, ionization is a necessary prerequisite. Unfortunately, the transition energy between two levels varies rapidly with time (see Figure 2) and the overall time dependence introduces an additional broadening of the x-ray energy. Therefore, the quasimolecular x-ray spectra usually have an exponentially falling shape. Very little information can be obtained in this way. However, it was predicted[14] and later experimentally confirmed[15] that the x-ray spectrum exhibits a pronounced directional anisotropy around the transition energy at the distance of closest approach. This phenomenon has to do with the swift rotation of the internuclear axis in the moment of closest approach of the two nuclei. The systematics of the anisotropy peak have been investigated by Wölfli's group at Zürich for light and medium light systems,[16] and up to $Z = 184$ for outer transitions.[17] If plotted versus $Z = Z_1 + Z_2$ the resultant peaks give rise to a spectroscopy of the quasi-atomic states in heavy and superheavy atoms. There is hope that it will be possible to continue it right through $Z = 170$. Calculations for the system Pb–Pb are in progress.

The framework for the description of these phenomena is the same, viz., time-dependent perturbation theory on the adiabatic quasimolecular basis. In the following we will discuss the theory of positron production in detail, keeping in mind that the treatment may be taken as a guideline for the description of ionization and x-ray emission, too.

2. The Question of Coherence

The first attempts to calculate the processes occurring during a heavy-ion collision were based on the quasistatic approximation. A typical example is the calculation of positron production of Peitz et al.[19] Here one assumed that a constant $1s\sigma$ vacancy probability was available during the whole collision. Furthermore, the positron emission was calculated as if the system had infinite time to settle down at every internuclear distance. Thus energy conservation was built in from the beginning and only spontaneous processes were investigated. The stationary phase approximation (see, e.g., Macek and Briggs[20] for x-ray emission) was an improvement only insofar as the coherence between approaching and receding part of the trajectory could be accounted for.

The improved understanding of the collision dynamics showed, however, that the lack of full adiabaticity leads to the occurrence of induced processes, i.e., phenomena where energy is transferred from the nuclear motion into other channels. These effects can be treated by the evaluation of the full Fourier integral for, e.g., the positron amplitudes.[21,22] In general, this procedure led to a broadening of the calculated spectra, which was observed in the MO x-ray experiments and attributed to the time–energy uncertainty relation. Often also a considerable increase in the total cross section was found.

In the following it became clear that also the last approximation, namely, that of constant vacancy amplitudes, had to be dismissed. Especially in heavy systems or at higher bombarding energy, the vacancies are created during the collision and therefore are rapidly oscillating functions of time. Such calculations, where ionization and the subsequent process are treated coherently, could be performed only after the theoretical (and experimental) understanding of the direct excitation process had been achieved. Examples are the calculations of Kirsch et al.[18] for x-ray emission and of Reinhardt et al.[23,24] for positron creation.

In general, the effect of coherence is to reduce the probability. This is exemplified in Figure 3, where we have chosen the process of induced positron emission in Pb–Pb collisions. The solid line accounts for the fully coherent calculation, the expression

$$P_{1s}^{e^+} = 2 \int dE_e \int dE_p \, | \, a^{(2)}_{E_e, E_p; 1s} \, |^2$$

of equation (22) below. The dashed line represents the calculation, where ionization P_{1s} and positron emission P^{e^+} are calculated separately with full intrinsic coherence,

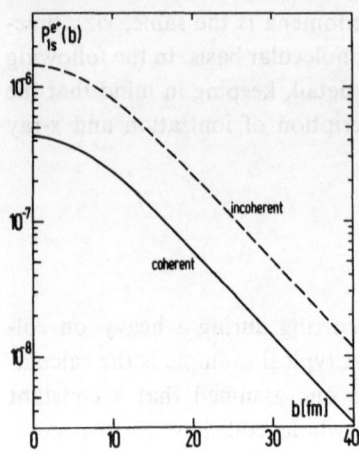

Figure 3. Induced positron emission probability $P_{1s}^{e^+}$ in Pb–Pb collisions versus impact parameter. Ionization (first step) and pair creation (second step) are treated coherently (solid line) and incoherently (dashed line). $2a = 20.9$ fm.

but the product is taken incoherently:

$$\tilde{P}_{1s}^{e^+} = P_{1s} \cdot P^{e^+} = 2 \int dE_e \, | \, a_{E_e,1s} \, |^2 \cdot \int dE_p \, | \, a_{E_p,1s} \, |^2$$

P_{1s} would be observable as the asymptotic $1s\sigma$ ionization probability. The incoherent product is seen to overestimate $P_{1s}^{e^+}$ by a factor of 3 to 4, but to give a reasonable impact-parameter dependence. We conclude that full coherence is necessary to obtain reliable values for processes like induced or spontaneous positron creation.

3. Formalism

To describe the emission of positrons by the various processes qualitatively discussed in the last section it is essential to account for the strong Coulomb force experienced by the electron–positron field at small distances of the colliding nuclei. Since the adiabaticity criterion is satisfied for the motion of inner-shell electrons, it is most natural to adopt the quasimolecular picture. Thus we start from the complete set of solutions of the stationary two-center Dirac equation (5)

$$H_{\text{TCD}}\varphi_q = E_q\varphi_q \tag{1}$$

where

$$H_{\text{TCD}} = \boldsymbol{\alpha} \cdot \mathbf{p} + \beta m + V(\mathbf{r}, \mathbf{R})$$

$$V(\mathbf{r}, \mathbf{R}) = \frac{-Z_1\alpha}{|\mathbf{r} - \mathbf{R}_1|} - \frac{Z_2\alpha}{|\mathbf{r} - \mathbf{R}_2|}$$

and q runs over the set of bound states and the upper and lower continuum. The basis functions $\varphi_q(\mathbf{R}(t))$ depend parametrically on time via the changing internuclear distance $R(t)$. The nuclear motion, therefore, will induce transitions between the adiabatic states φ_q.

To treat the dynamics of the collision and calculate the various possible excitations, in principle one has to solve a problem with infinitely many particles, since according to Dirac's hole picture the negative continuum is occupied with electrons. In addition various bound states may be occupied depending on the charge state of the colliding ions. If one neglects the electron–electron interaction it turns out, however, that it is sufficient to solve the one-electron problem and to include the effect of the Pauli principle only afterwards.

Let us therefore look at the fate of a single electron which initially occupies the level φ_i before the collision. The time development is determined by the time-dependent Dirac equation

$$H_{\text{TCD}}\phi_i = i\frac{\partial}{\partial t}\phi_i \tag{2}$$

with the initial condition $\phi_i(t \to -\infty) = \varphi_i$.

To obtain the functions $\phi_i(t)$ one expands in the adiabatic basis $\varphi_j(\mathbf{R}(t))$:

$$\phi_i(t) = \sum_j a_{ij}(t)\varphi_j e^{-i\chi_j(t)} \tag{3}$$

with the phase

$$\chi_j(t) = \int^t dt' E_j(R(t))$$

Inserting in (2) and using the orthonormality of the basis functions, one obtains a set of coupled differential equations for the expansion parameters a_{ij}:

$$\dot{a}_{ij}(t) = -\sum_k a_{ik}(t)\langle \varphi_j | \frac{\partial}{\partial t} | \varphi_k \rangle e^{-i(\chi_k - \chi_j)} \tag{4}$$

The initial condition is $a_{ik}(-\infty) = \delta_{ik}$. Equation (4) can be solved numerically after the infinite set of basis states φ_j has been suitably truncated. The squared amplitudes $|a_{if}(\infty)|^2$ then give the probability that the electron is excited from state i to state f during the collision.

To discuss the many-particle problem we note that the functions $\phi_i(t)$ form a complete orthonormal set at every instant t. This follows from the hermiticity of the Hamiltonian H_{TCD}, i.e., from the unitarity of the time-development operator:

$$\langle \phi_j(t) | \phi_i(t) \rangle = \langle \varphi_j(-\infty) | U^\dagger(t, -\infty)U(t, -\infty) | \varphi_i(-\infty) \rangle = \delta_{ij} \tag{5}$$

or equivalently

$$\sum_i a^*_{ki}a_{li} = \sum_i a^*_{ik}a_{il} = \delta_{kl} \tag{6}$$

Therefore the set $\phi_i(t)$, already containing the dynamical excitations, can be used as a basis for solving the many-particle problem. To do this, one can expand the total wave function in a basis of many-electron configurations which may be represented by Slater determinants of the single-particle basis functions ϕ_i. The amplitude for exciting a final configuration starting from a given initial configuration turns out to be just the determinant of the corresponding single-particle amplitudes $a_{ij}(t)$.

More formally one can use the language of quantum field theory to construct a state vector $|\psi_H\rangle$ in the Heisenberg picture, i.e.,

$$\frac{\partial}{\partial t}|\psi_H\rangle = 0 \tag{7}$$

The (time-dependent) field operator may be expanded as

$$\hat{\psi} = \sum_q \hat{b}_q \phi_q(t) \quad \text{and} \quad \hat{\psi}^\dagger = \sum_q \hat{b}_q^\dagger \phi_q^\dagger(t) \tag{8}$$

where the creation and destruction operators satisfy the anticommutation relation

$$\{\hat{b}_q, \hat{b}_{q'}^\dagger\} = \delta_{qq'} \tag{9}$$

Now we define a set of initially occupied states (e.g., the negative energy continuum and several bound states) denoting them by $q < F$. The complementary set of empty levels is described by $q > F$. As usual, operators for holes can be introduced by the canonical transformation

$$\left.\begin{array}{l} \hat{d}_q = \hat{b}_q^\dagger \\ \hat{d}_q^\dagger = \hat{b}_q \end{array}\right\} \quad \text{for } q < F \tag{10}$$

leading to the following representation of the field operator:

$$\hat{\psi} = \sum_{q<F} \hat{d}_q^\dagger \phi_q + \sum_{q>F} \hat{b}_q \phi_q \tag{11}$$

The operators $\hat{b}_q, \hat{d}_q$ so defined both annihilate the state vector $|\psi_H\rangle$:

$$\begin{array}{ll} \hat{b}_q |\psi_H\rangle = 0 & \text{for } q > F \\ \hat{d}_q |\psi_H\rangle = 0 & \text{for } q < F \end{array} \tag{12}$$

Note that the number of "particles" and "holes" defined by the basis set ϕ_q always remains zero:

$$\langle \psi_H | \hat{b}_q^\dagger \hat{b}_q | \psi_H \rangle = \langle \psi_H | \hat{d}_q^\dagger \hat{d}_q | \psi_H \rangle = 0$$

The "physical" particles and holes are described by the wave functions φ_q from equation (1). For the corresponding operators we will write $\hat{b}_q, \hat{b}_q^\dagger, \hat{d}_q, \hat{d}_q^\dagger$ so that the field operator reads in this representation

$$\hat{\psi} = \sum_{q<F} \hat{d}_q^\dagger \phi_q e^{-i\chi_q} + \sum_{q>F} \hat{b}_q \phi_q e^{-i\chi_q} \tag{13}$$

Inserting (3) into (11) we obtain the transformation between both sets of creation and destruction operators:

$$\hat{d}_q^{\,+} = \sum_{r<F} \hat{d}_r^{\,+} a_{rq} + \sum_{s>F} \hat{b}_s a_{sq} \qquad \text{for } q < F$$

$$\hat{b}_q = \sum_{r<F} \hat{d}_r^{\,+} a_{rq} + \sum_{s>F} \hat{b}_s a_{sq} \qquad \text{for } q > F \tag{14}$$

To calculate the number of particles (or holes) excited in a particular level one has to take the expectation value of the number operator $\hat{n}_p = \hat{b}_p^{\,+}\hat{b}_p$ or $\hat{n}_q = \hat{d}_q^{\,+}\hat{d}_q$, respectively. This is immediately found using (12) and the anticommutation relations.

Number of particles: $\qquad N_p = \langle \psi_H | \hat{b}_p^{\,+}\hat{b}_p | \psi_H \rangle = \sum_{r<F} | a_{rp} |^2, \qquad p > F$

$$\tag{15}$$

Number of holes: $\qquad N_q = \langle \psi_H | \hat{d}_q^{\,+}\hat{d}_q | \psi_H \rangle = \sum_{s>F} | a_{sq} |^2, \qquad q < F$

Thus, if one is interested only in the number of created particles (holes) it is sufficient to calculate the single-particle transition probabilities for the various initially occupied (empty) states and to sum them incoherently.

If one is interested in the number of correlated particle–hole pairs one has to calculate the expectation value of $\hat{n}_p \cdot \hat{n}_q$. This leads to

$$N_{p,q} = \langle \psi_H | \hat{b}_p^{\,+}\hat{b}_p\hat{d}_q^{\,+}\hat{d}_q | \psi_H \rangle$$

$$= \sum_{\substack{r<F \\ s>F}} [| a_{rp} |^2 | a_{sq} |^2 - a_{rp}^* a_{rq} a_{sq}^* a_{sp}]$$

$$= N_p \cdot N_q + \left| \sum_{r<F} a_{rp}^* a_{rq} \right|^2 \tag{16}$$

where the Pauli exclusion principle enters in the sign of the exchange term on the right-hand side.

To isolate the various contributions discussed in the last section, let us work out (15) and (16) in perturbation theory. The single-particle amplitude for producing a pair consisting of a hole in level q and a particle in level p reads to first order

$$a_{pq}^{(1)}(t) = -\int_{-\infty}^{t} dt' \langle \varphi_q | \frac{\partial}{\partial t'} | \varphi_p \rangle e^{i[\chi_q(t') - \chi_p(t')]} \tag{17a}$$

and to second order

$$a_{pq}^{(2)}(t) = -\sum_r \int_{-\infty}^{t} dt' a_{pr}^{(1)}(t') \langle \varphi_q | \frac{\partial}{\partial t'} | \varphi_r \rangle \cdot e^{i(\chi_q - \chi_r)} \tag{17b}$$

In particular, taking p and q as electron and positron states in the continuum, equation (17a) describes the direct pair creation due to the rapidly varying Coulomb field (process f in Figure 4). In equation (17b) deeply bound inner-shell states, which

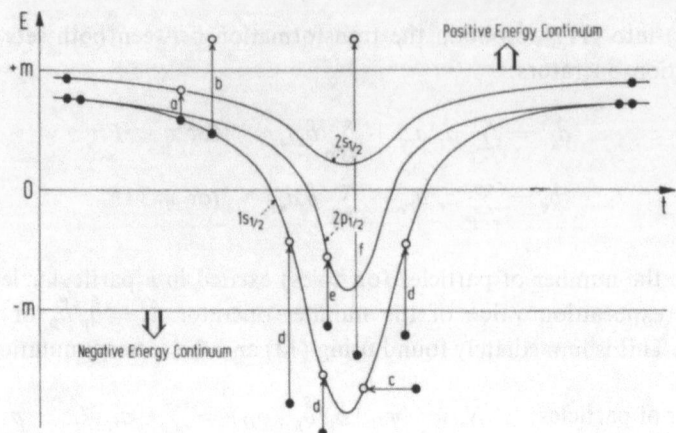

Figure 4. Schematic representation of pair-production processes in heavy-ion collisions. a, b, ionization; c, spontaneous and d, e, induced vacuum decay; f, vacuum polarization shake-off.

are initially occupied, will be the most important intermediate states $r = 1s, 2p_{1/2}, \ldots$ (process b and d in Figure 4). It follows naturally that the one-step process and two-step contributions with various intermediate states r cannot be distinguished and must be added in phase, i.e., $a_{pq} = a_{pq}^{(1)} + a_{pq}^{(2)} + \cdots$. In perturbation theory the number of particle–hole pairs is just

$$N_{p,q} = |a_{qp}|^2 = |a_{pq}|^2 \tag{18}$$

which follows from the second term of equation (16) using

$$a_{rs} \ll 1 \quad \text{if } r \neq s$$
$$a_{rs} \approx 1 \quad \text{if } r = s$$

and

$$a_{rs} = -a_{sr}^*$$

The number of holes is

$$N_q = \sum_{p>F} N_{pq} \tag{19}$$

This is only valid in perturbation theory, where many-particle–many-hole excitations are negligible. The total cross section for the emission of electron–positron pairs is obtained by integrating the squared pair amplitude over electron and positron energy and over impact parameter b and by summing over angular momentum $\varkappa$:

$$\sigma = 2\pi \sum_{\varkappa} (2j_i + 1) \int_0^\infty b \, db \int_{-\infty}^{-m} dE_p \int_m^\infty dE_e \, |a_{E_e, E_p}(\infty)|^2 \tag{20}$$

where $2j_i + 1$ denotes the multiplicity of the initial state.

4. Numerical Results

The actual numerical calculations of positron emission in heavy-ion collisions performed up to now have employed several approximations.

(a) The amplitudes $a_{pq}(t)$ have been calculated in perturbation theory, i.e., the integrals (17a) and (17b) have been evaluated numerically prescribing Rutherford trajectories for the nuclear motion.[24] The full solution of the coupled channel equations (4) is presently attempted. It will improve the perturbative results since the transfer of vacancies between bound states is included. Furthermore the "rescattering" of continuum particles may deform the shape of the emission spectra of δ electrons and positrons (see Note Added in Proof on page 359).

(b) The sum over $\varkappa$ was restricted to angular momentum $j = \frac{1}{2}$ ($\varkappa = \pm 1$), since these states are most severely affected by the strong field. Furthermore, we have taken into account radial coupling ($\dot{R} \, \partial/\partial R$) only since the corresponding matrix elements become large at small internuclear distance. Rotational coupling ($\Omega \cdot \mathbf{j}$) is not expected to be effective in coupling levels which are energetically far apart as must be achieved to produce pairs. It could play a role in the formation of inner-shell vacancies, especially in the $np_{1/2}\sigma$ states.[25] Even then its influence at very small impact parameters must be small ($\Omega \sim vb/R^2$), and only these contribute significantly to pair creation. Vacancy formation on the outgoing branch of the nuclear trajectory, on the other hand, is not relevant for positron production.

(c) All wave functions and matrix elements have been calculated in the monopole approximation,[26] which was proven to be remarkably good for the very heavy systems of interest for positron production. With the availability of full two-center continuum wave functions, however, a considerable improvement in the description of special features (such as angular distributions, etc.) could be achieved. The effect of the finite nuclear size is included and was found to reduce coupling matrix elements by up to 30%.

(d) The influence of the transverse electromagnetic field is not included in the instantaneous Coulomb potential of equation (1), i.e., magnetic and retardation effects are neglected. Furthermore, the energy shifts due to quantum electrodynamics (QED) radiative corrections (vacuum polarization, self-energy) are expected[27−29] to be much too small to influence the positron production rates significantly.

(e) As already implied in the formalism of the last section we neglect the electron–electron interaction. Its main importance lies in the lowering of binding energies which could be included, e.g., by using an effective Thomas–Fermi screening function to modify $V(r)$. The nondiagonal couplings due to electron–electron interaction may be neglected for inner-shell excitation since they lead to relaxation times large compared to the collision time.[30]

Before discussing the resulting positron emission probabilities, let us first take a look at the behavior of the coupling matrix elements. Figure 5 shows the radial matrix element $\langle E_p \, | \, \partial/\partial R \, | \, n \rangle$ (responsible for the induced emission of positrons

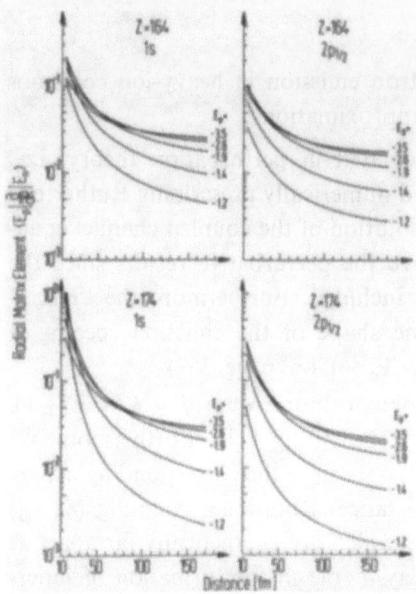

Figure 5. Matrix elements for induced emission from the $1s$ and $2p_{1/2}$ states versus two-center distance R.

with various energies E_p) from the inner-shell bound states $n = 1s, 2p_{1/2}$, for Pb–Pb $(Z = Z_1 + Z_2 = 164)$ and Pb–U $(Z = 174)$ collisions. The calculation made use of the Hellman–Feynman identity $(E_1 \neq E_2)$:

$$\langle \varphi_1 | \frac{\partial}{\partial R} | \varphi_2 \rangle = (E_2 - E_1)^{-1} \langle \varphi_1 | \frac{\partial H}{\partial R} | \varphi_2 \rangle \qquad (21)$$

where φ_1, φ_2 are eigenstates of the same Hamiltonian H.

Most striking is the strong increase of the matrix elements at very small internuclear distances R, clearly demonstrating the contraction of the electron and positron wave functions when the external nuclear charge distribution begins to approach the point charge limit. The slope of the curves is even larger than that for the corresponding ionization matrix elements $\langle n | \partial/\partial R | E_e \rangle$; it is steepest for the lowest values of positron kinetic energy, reflecting the smaller energy denominator in (21). The absolute value of the induced positron matrix element $\langle E_p | \partial/\partial R | n \rangle$ lies about one order of magnitude below the ionization matrix element $\langle n | \partial/\partial R | E_e \rangle$ due to the Coulomb repulsion of the positron wave function. It decreases with the principal quantum number of the bound state.

Figure 6 presents the matrix elements for direct pair creation, $\langle E_p | \partial/\partial R | E_e \rangle$, where no intermediate bound states are involved, for various positron energies E_p. The full lines belong to $\varkappa = -1(s_{1/2})$ continuum waves, the dashed lines to $\varkappa = +1(p_{1/2})$ waves. The magnitude of $\langle E_p | \partial/\partial R | E_e \rangle$ grows with charge $Z = Z_1 + Z_2$: Going from $Z = 164$ (Pb–Pb) to $Z = 174$ (Pb–U), which corresponds to a 6% charge increase, leads to a strong enhancement of the matrix elements (cf. Figure 6b). This, of course, will be reflected in a high Z dependence of the positron emission

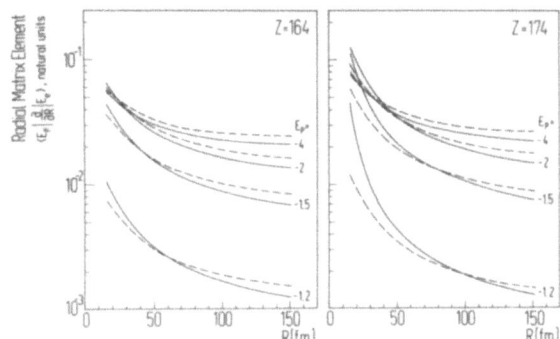

Figure 6. Matrix elements for direct pair creation versus two-center distance R. $E_e = 1.02$. $\varkappa = -1$ (solid line); $\varkappa = +1$ (dashed line).

cross section. In Figure 6 the electron energy is kept fixed at $E_e - m = 10.22$ keV. Variation of E_e would shift the curves but it will not introduce additional structure since it turns out that $\langle \varphi_{E_p} \mid \partial H / \partial R \mid \varphi_{E_e} \rangle$ is nearly independent of E_e. On the other hand, the dependence on positron energy is a result of the Coulomb repulsion factor (see Figure 7).

In perturbation theory [equation (17)] the transition amplitudes for positron emission are Fourier transforms of the coupling matrix elements where the frequency $E_q - E_p$ is governed by the energy matching of initial and final state. We have numerically solved the integrals (17) using the matrix elements and energies of the monopole approximation, prescribing Rutherford trajectories for the nuclear motion.[24] The radial matrix elements were fitted by a power law at distances larger than 200 fm to account for the limited range of validity of the monopole approximation. In addition, they were cut off with a Gaussian factor at $R > 1000$ fm to ensure convergence of the integral. This may be thought to simulate the influence of translational factors, which arise from the transformation between standing and "traveling" molecular orbitals but cannot be determined in the monopole approximation. In the calculation of the shake-off amplitudes $a^{(1)}_{E_e E_p}$ an integration by parts was employed to make the integral convergent.[22]

Before turning to the calculated emission probabilities, let us inspect the time dependence of the transition amplitudes. Their general behavior is quite similar to

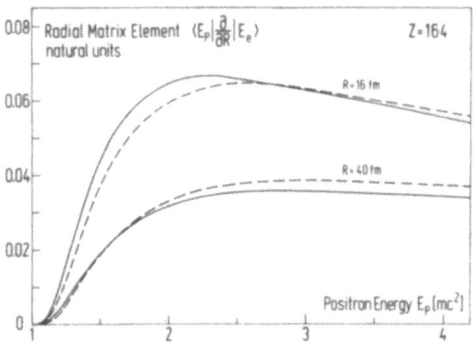

Figure 7. The same as Figure 6, but as a function of positron energy E_p. $E_e = 1.02$. $\varkappa = -1$ (solid line); $\varkappa = +1$ (dashed line).

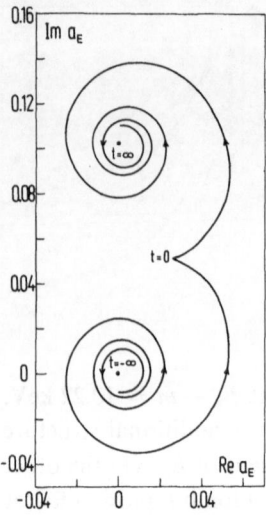

Figure 8. Path described by the $1s\sigma$ ionization amplitude in the complex plane (taken from Reference 13).

the amplitudes for vacancy production (Figure 8). Owing to a steeper increase of the matrix elements, however, the radial scale is contracted and the maximal value of $|a|^2$ is reached at distances R less than 100 fm. The path of the time-dependent positron amplitude in the complex plane is characteristic of the one-step process. With the phase convention $\chi_p(t=0) = 0$ the final amplitude $(t \to \infty)$ is purely imaginary since the integrand is an odd function of time. This is valid both for the direct process $a^{(1)}_{E_e,E_p}$ and for the induced positron emission from bound orbitals brought empty into the collision, $a^{(1)}_{n,E_p}$.

For the two-step process on the other hand

$$a^{(2)}_{E_e,E_p}(t) = -\sum_n \int_{-\infty}^{t} dt'\, a^{(1)}_{E_e,n}(t')\dot{R}(t')e^{i(\chi_{E_p}-\chi_n)}\langle\varphi_{E_p}\,|\,\frac{\partial}{\partial R}\,|\,\varphi_n\rangle$$

$$\equiv \sum_n a^{(2)}_{E_e,E_n;n}(t) \tag{22}$$

the integrand is weighted with the time-dependent ionization amplitude $a^{(1)}_{E_e,n}(t)$, which starts from zero and builds up during the collision. Therefore the shape of the figure of $a^{(2)}_{E_e,E_p;n}(t)$ is distorted and its symmetry is lost. Figure 9 shows that $a^{(2)}_{E_e,E_p;1s}(t)$ at the energies $E_e = 1.02$, $E_p = -1.7$ for the collision Pb–Pb at impact energy $E_{\mathrm{lab}} = 5.85$ MeV/u (distance of closest approach $2a = 15.8$ fm) and impact parameter $b = 0$. The final amplitude $(t \to \infty)$ has a complex value defining a phase angle $\Delta\phi_{1s}$ relative to the imaginary axis. It is important to know the value of this angle since the two-step amplitudes must be added coherently to each other and to the one-step amplitude $a^{(1)}_{E_e,E_p}$. The latter is positive imaginary, therefore the shake-off process and the induced positron emission via the $1s$ state interfere somewhat destructively. Further investigations reveal that the phase angle $\Delta\phi_n$ increases (1) with increasing electron energy, (2) with decreasing (kinetic) positron energy, (3)

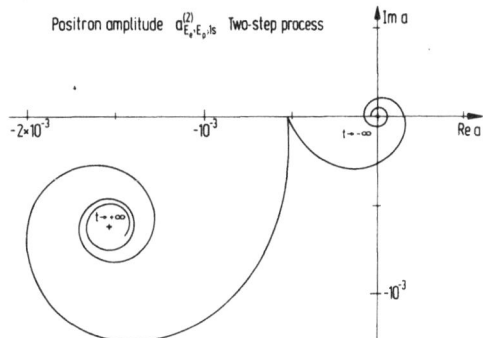

Positron amplitude $a^{(2)}_{E_e,E_s,1s}$ Two-step process

Figure 9. Path described in the complex plane by the amplitude for induced positron emission out of the $1s\sigma$ state. Observe the lack of symmetry characteristic of a second-order process. Pb–Pb; $2a = 15.8$ fm, $E_e = 1.02$, and $E_p = -1.7$.

with increasing binding energy of the intermediate bound state. The last point is demonstrated in Figure 10, where the complex amplitudes $a^{(2)}_{E_e,E_p;n}(\infty)$ for induced emission of an electron–positron pair via the bound state n and the direct amplitude $a^{(1)}_{E_e,E_p}$ for $\varkappa = \pm1$ are shown. The parameters are the same as in Figure 9 except for the slightly higher electron energy $E_e = 1.2$.

Figure 11 explicitly demonstrates the time dependence of the differential probability of induced positron emission from the $1s$ level integrated over electron energy,

$$\frac{dP^{e^+}}{dE_p} = 2\int_m^\infty dE_e \, | \, a^{(2)}_{E_e,E_p;1s\sigma} \, |^2$$

for various energies E_p. As we have already stressed, the excitation happens at very small internuclear distances reflecting the strong deformation of the electronic wave functions. We note that, although there is no way of measuring the excitation during the course of the collision, the curves of Figure 9 and 11 have some physical meaning. In a thought experiment one could imagine a mechanism that stops the relative nuclear motion at an arbitrary distance R_0. Then the radial coupling vanishes and the excitation amplitudes are frozen in at their momentary values. For example, such a mechanism would be provided by the nuclear interactions if the nuclei underwent fusion. The "high Fourier frequencies" of the collision then would show up particularly in a slower decrease of the high-energy part of the emitted particle (electron or positron) spectrum since here the virtual excitations are large (cf. Figure 11).

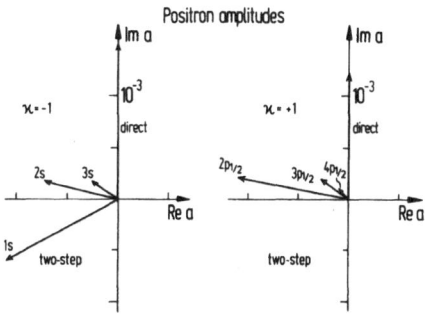

Positron amplitudes

Figure 10. Phase and amplitude of the various second-order processes for s wave ($\varkappa = -1$) and p wave ($\varkappa = +1$). $E_p = -1.5$ and $E_e = 1.2$.

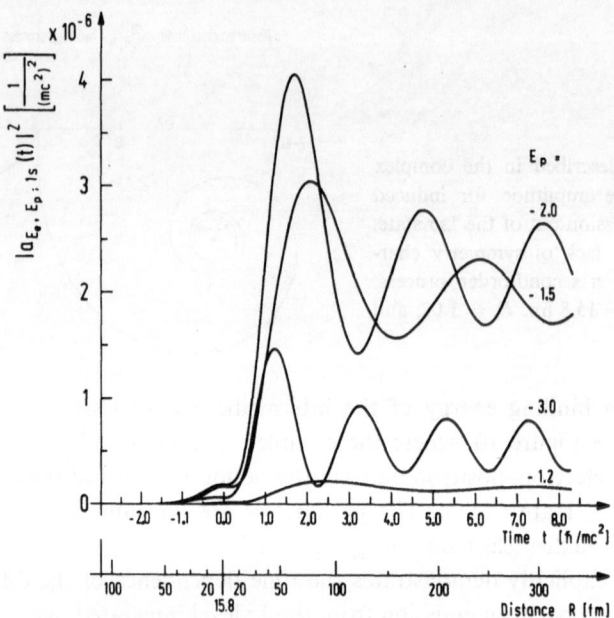

Figure 11. Time evolution of the probability for positron emission. Most oscillations vanish after integration over positron energy. Pb–Pb; $2a = 15.8$ fm, $b = 0$, and $E_e = 1.02$.

In deep inelastic nuclear reactions the nuclei stay together for a prolonged time interval Δt during which the radial velocity R is small. The effect of such a time delay on the electron–positron field is to change the interference between excitations taking place at the incoming and outgoing branches of the trajectory. This is readily demonstrated looking again at the time-dependent path of the amplitude in the complex plane. Figure 12 shows how the amplitude $a^{(2)}_{E_e, E_p; 1s}(t)$ of Figure 9 changes if various delay times Δt are introduced between the branches of an otherwise unchanged

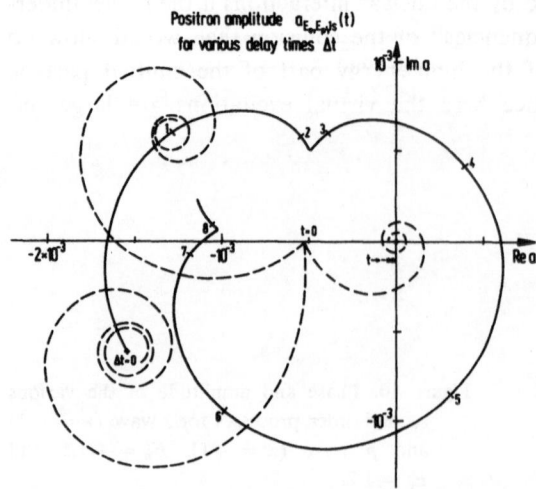

Figure 12. The influence of a time delay Δt at closest approach of the nuclei on the induced positron amplitude $a_{E_e, E_p; 1s}(t)$ (in units of 10^{-21} sec). The path of $a(\infty)$ is shown as a solid line, the amplitude $a(t)$ in two cases as dashed curve.

Rutherford hyperbola. Obviously the spiral path belonging to the outgoing nuclear motion is rotated around the point $a(t=0)$. Knowledge of the usual amplitudes $a(t=0)$ and $a(t \to \infty)$ for $\Delta t = 0$ allows a simple analytic determination of the influence of a delay time. The full curve shows the motion of the final values of the amplitude $(t \to \infty)$ in the complex plane as a function of delay time Δt (noted in multiples of 10^{-21} sec). The epicycloid path depends on the two phases $(E_e - E_p)\Delta t$ and $(E_{1s} - E_p)\Delta t$. For the first-order amplitude we have only one phase angle and the final amplitude moves on a simple circle around $a(0)$. For a fixed time interval Δt the discussed effect leads to marked oscillations in the excitation spectra. Summation over various channels (e.g., intermediate states n or electron energies E_e, if the electron is not detected) will partially smear out this interference pattern.

The investigation of this behavior is partially interesting in connection with the suggestion to use deep inelastic collisions as a means to enhance the rate of spontaneous positron production in supercritical collisions.[31] The amplitude of the latter process grows (to first order) linearly with Δt while the induced contribution gives rise to oscillations. Therefore both processes may be distinguished provided that collisions with Δt large enough do take place.

In the following figures we will show our final results on positron production in subcritical systems. We assume that all inner-shell states are occupied initially and have to be emptied during the collision by direct ionization. The positron emission probability is obtained from the squared coherent sum of one-step and two-step amplitudes $|a^{(1)}_{E_e,E_p} + \sum_n a^{(2)}_{E_e,E_p;n}|^2$ calculated in perturbation theory. The results for the s and p channels are then added and multiplied by the degeneracy factor 2.

As intermediate states we take $n = 1s, 2s, 3s$ ($\varkappa = -1$) and $n = 2p_{1/2}, 3p_{1/2}, 4p_{1/2}$ ($\varkappa = +1$). This basis set should embody most of the relevant Hilbert space, hence only slightly underestimate the cross sections. Another source of possible errors is the neglect of mutual coupling between bound states. This effect can only be treated by a coupled channel calculation.

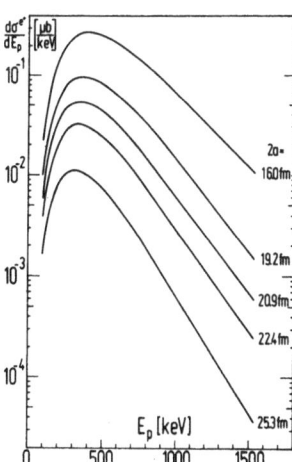

Figure 13. Differential positron cross section in Pb–Pb for various bombarding energies. Shake-off and induced QED positrons (6 states).

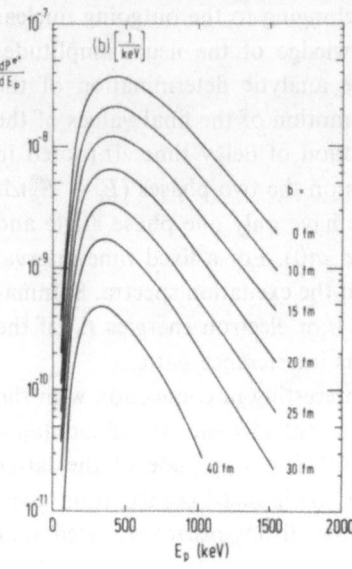

Figure 14. The same as Figure 13, but for various impact parameters.

Figure 13 shows the differential positron spectra with respect to kinetic energy $d\sigma^{e^+}/dE_p$ for Pb–Pb collisions at various bombarding energies ranging from 3.68 to 5.8 MeV/u. The spectra peak between 300 and 400 keV kinetic positron energy and fall off almost exponentially at higher energies. The same behavior is found for the positron energy spectrum $dP^{e^+}/dE_p(b)$ at various impact parameters (Figure 14). The slope of the exponential is determined mainly by the Fourier decomposition of the nuclear motion $R(t)$. Therefore it becomes steeper for distant collisions, i.e., at large impact parameters or at low impact energies.

Figures 15a and 15b show representative QED positron spectra $d\sigma^{e^+}/dE_p$ for the system Pb–Pb and Pb–U at $E = 4.45$ MeV/u in comparison with the background from nuclear Coulomb excitation. The nuclear background in ^{208}Pb–^{208}Pb originates essentially from pair conversion of only two γ lines and therefore looks quite different.

Figure 15. Energy distribution $d\sigma^{e^+}/dE_p$ of positrons produced in (left) ^{208}Pb–^{208}Pb and (right) ^{208}Pb–^{238}U collisions. Left, $2a = 20.9$ fm; right, $2a = 22.0$ fm. Long-dashed line, positrons from all quasimolecular (QED) processes (shake-off and induced, 6 states). Short-dashed lines, background positrons from the conversion of γ rays following nuclear Coulomb excitation. Full curve, total sum.

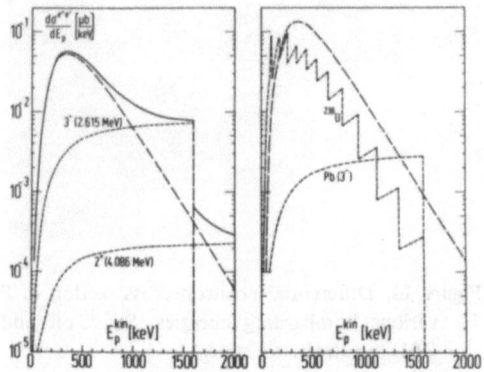

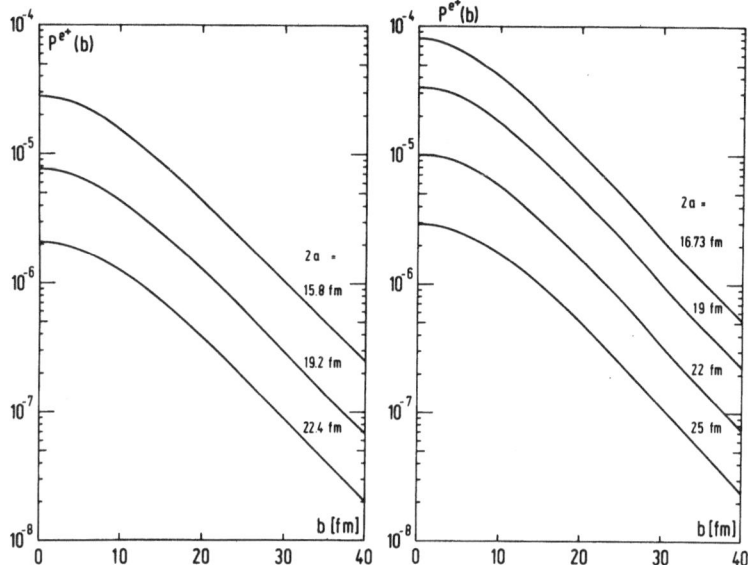

Figure 16. Impact-parameter dependence of QED positron production (shake-off and induced, 6 states) in (left) Pb–Pb and (right) Pb–U collisions for various bombarding energies.

Setting a window at positron energies below 1 MeV will greatly suppress this background. In the ^{208}Pb–^{238}U system, on the other hand, many γ transitions occur between 1 and 3 MeV showing a serrate structure with an overall shape similar to that for QED positrons.

An important piece of information is contained in the impact-parameter dependence $P^{e^+}(b)$ shown in Figures 16a and 16b. The curves decrease monotonically with impact parameter, similar to those obtained for the ionization probability. The slope, however, is even steeper, so that 90% of the cross section comes from collisions with $b < 30$ fm. This is the consequence of the pronounced maximum of the coupling matrix elements at small internuclear distances. The nuclear background (Figure 17) has a different impact parameter dependence and falls off even more steeply with b (note that we are comparing a process on the "atomic" scale with a nuclear process). The appearance of a bump at about $b = 7$ fm in Figure 17a is caused by the reorientation effect of the 3$^-$ (2.615-MeV) level of ^{208}Pb.

Two experiments measuring positrons in collisions of very heavy ions recently have been performed at GSI. Backe et al.[32] employed a solenoidal spectrometer to collect the positrons. They measured total cross sections and differential cross sections with respect to the scattered ion in two angular windows. The nuclear background is well understood in ^{208}Pb–^{208}Pb collisions and can be subtracted. Figure 18 shows the result at $2a = 19.2$ fm (4.85 MeV/u) compared with the theoretical prediction. The two bumps of the calculated curve originate from symmetrization with respect to projectile and target, the left one being due to forward scattering of the projectile

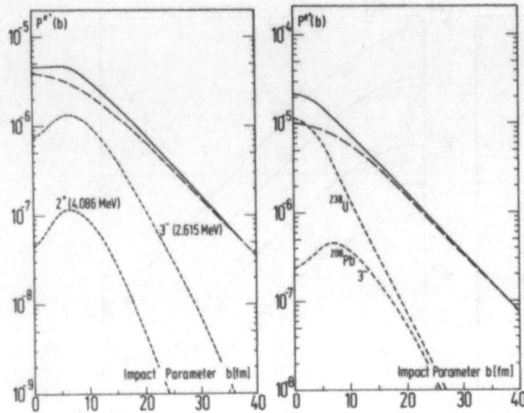

Figure 17. Impact-parameter dependence of pair production in (left) ^{208}Pb–^{208}Pb and (right) ^{208}Pb–^{238}U collisions. Left, $2a = 20.9$ fm; right, $2a = 22$ fm. Nuclear transitions (short-dashed lines), the QED prediction (long-dashed line), and the total curve (solid line) are shown.

nucleus. The measured values are some 30% above theory while the angular dependence agrees very well (the nuclear background looks much different).

With a different experimental setup Kozuharov et al.[33] performed coincidence measurements for several systems at 5.85 MeV/u projectile energy. Using a solenoid spectrometer they singled out an energy window of 100 keV width near the expected peak of the positron spectrum. Figure 19 shows the probability dP^{e^+}/dE_p as a function of c.m. scattering angle. The nuclear background, which has been estimated from the measured γ spectrum using an effective conversion coefficient, has been subtracted. The theoretical curves represent the ratio of the symmetrized positron cross section to the scattering cross section. Theory now lies above experiment but, again, the angular distribution agrees quite well. Figure 20 gives our results for the most comprehensive data, the total positron cross sections σ^{e^+} as a function of projectile energy. The calculated curve is compared with total cross sections measured at GSI.[32] In both collision systems the experimental values lie above theory by up to a factor of 2. The dependence on bombarding energy, however, agrees very well. Most con-

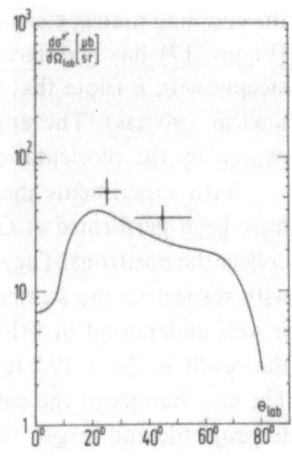

Figure 18. Differential positron cross section in Pb–Pb collisions as function of (averaged) scattering angle. Shake-off and induced QED positrons (6 states); $2a = 19.2$ fm. The experimental points are taken from Reference 32 and are corrected for the nuclear background.

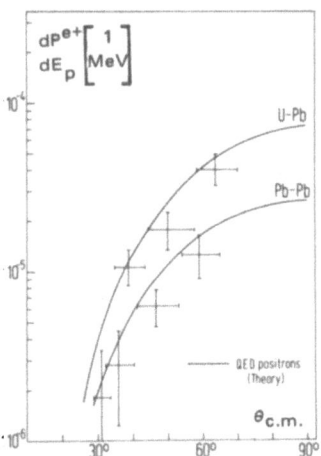

Figure 19. Probability for positron emission as a function of projectile c.m. scattering angle for (5.9 MeV/u) ^{208}Pb–^{208}Pb and ^{238}U–^{208}Pb collisions in the energy window 0.44–0.55 MeV. Theory is compared with measurements from Reference 33, where nuclear background has been subtracted. Both curves are symmetrized with respect to projectile and target nucleus.

vincing in this respect are the Pb–Pb data where the slope of the experimental curve can be explained only by the sum of QED and nuclear positrons, but not by the individual contributions alone. This is also true for the differential cross sections $d\sigma^{e+}/d\Omega_{\mathrm{ion}}$ at scattering angle $\vartheta_{\mathrm{lab}} = 45°$ measured by the same group. The absolute agreement is better, leaving a discrepancy of only about 30%.

Let us now discuss in more detail the various contributions to the positron spectrum. Our results presented so far contained shake-off positrons and those induced from several inner-shell levels up to $4p_{1/2}$. Table 1 lists the induced positron production from various levels separately for the chosen parameters $Z_1 + Z_2 = 164$, $2a = 20.9$ fm. For comparison the right column of Table 1 lists the induced positron cross sections calculated with a constant hole amplitude set equal to unity, i.e., assuming the collision of two naked nuclei. More illustratively, the contributions of the various channels as a function of energy are displayed in Figure 21. In the left part of this figure the cross section for induced positron production via the level $ns\sigma$ (circles) or $np_{1/2}\sigma$ (crosses) have been marked at the corresponding binding energies in the united atom limit. In addition the right half of the figure shows the differential cross section for direct pair creation ($\varkappa = \pm 1$), which is seen to decrease

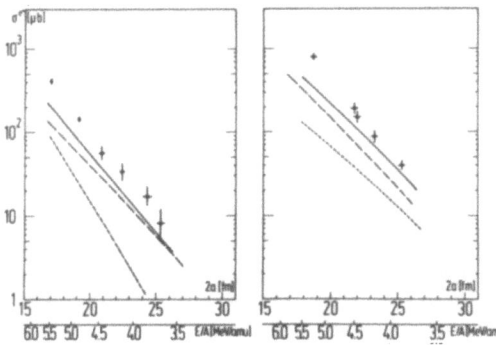

Figure 20. Total positron cross sections for (left) ^{208}Pb–^{208}Pb and (right) ^{208}Pb–^{238}U collisions in dependence of bombarding energy or, equivalently, of distance of closest approach $2a$. The experimental points are taken from Reference 32 (GSI). Short-dashed line, nuclear background; long-dashed line, shake-off and induced QED positrons (6 states); solid line, total positron yield.

Table 1. Total Cross Section for Induced Positron Production from Various Bound States for the System Pb–Pb at $2a = 20.9$ fm

$\varkappa$	State	σ_{ind}^{e+} (2 step) $(\mu b)^a$	σ_{ind}^{e+} (1 step) $(\mu b)^b$
-1	$1s$	2.8	1790
	$2s$	1.26	27
	$3s$	0.28	3.5
	Direct	5.6	5.6
$+1$	$2p_{1/2}$	2.6	121
	$3p_{1/2}$	0.32	4.7
	$4p_{1/2}$	0.10	1.24
	Direct	4.5	4.5

[a] Two-step (no vacancy is present initially).
[b] Calculation assuming totally empty shells.

exponentially with electron energy. The lower set of points has been calculated assuming initially occupied states (two-step process). Although the probability for vacancy production in higher orbitals becomes very large, the largest individual contributions to the positron cross section originate from the deepest bound states. This is even more pronounced in the collision of naked nuclei (upper set of points). Here the contribution from the $1s\sigma$ level is very large and by far dominant, justifying the name "induced decay of the neutral vacuum."[21]

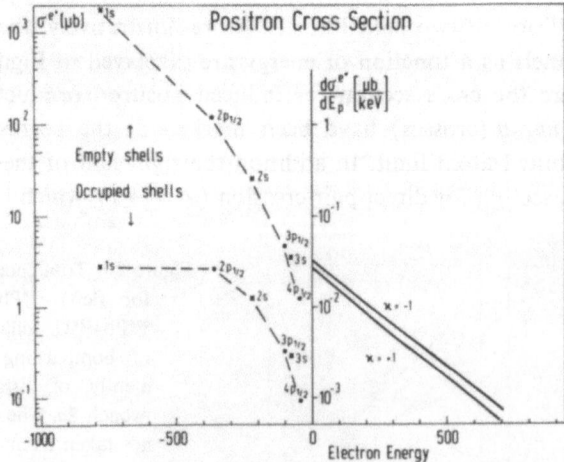

Figure 21. Various contributions to the QED positron production: induced vacuum decay (left side) and direct pair creation (right side). The contributions from the bound states are shown over the maximal binding energy. The dashed lines between the dots are drawn simply to guide the eye. Pb–Pb; $2a = 20.9$ fm.

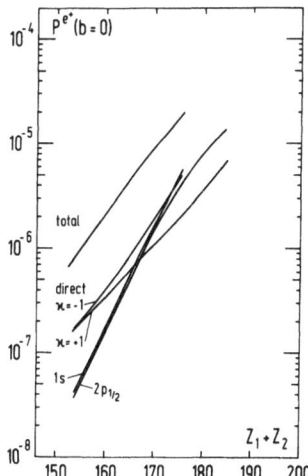

Figure 22. Charge dependence of the various quasi-molecular mechanisms of pair production. The distance of closest approach is kept fixed. $2a = 20.9$ fm.

The contributions to positron production from the various discussed channels cannot be identified in a single experiment since their behavior is quite similar. It is, however, interesting to study the variation of their magnitude and relative proportion with total nuclear charge $Z_1 + Z_2$. Since in the region $Z\alpha > 1$ the binding of inner shells increases strongly and the innermost levels begin to approach the lower continuum one expects a steep rise of the induced positron emission. This is demonstrated in Figure 22, where the positron probability P^{e^+} at impact parameter $b = 0$ is shown for symmetric (or nearly symmetric) collisions with the distance of closest approach $2a = 20.9$ fm. The induced contributions from $1s\sigma$ and $2p_{1/2}\sigma$ grow very fast with charge $Z_1 + Z_2$. The curves for the charge dependence of the total cross section look very similar, their slope being somewhat smaller due to the contribution of higher impact parameters. At $2a = 20.9$ fm we find the scaling behavior Z^n with $n \approx 19$. If the incident velocity is kept fixed the exponent is smaller, e.g., $n \approx 15$ at $E/A = 5.9$ MeV/u.

5. Dynamical Treatment of Supercritical Collisions

The formulation of our theory of positron production in heavy-ion collisions presented before is applicable only in subcritical collisions where no bound states have joined the lower continuum. We have expanded the electron wave function [equation (3)] in the adiabatic molecular basis comprising the electron continuum $|\varphi_{E_e}\rangle$, various bound states $|\varphi_n\rangle$, and the positron continuum $|\varphi_{E_p}\rangle$. This leads to coupled differential equations for the expansion coefficients which have been treated in perturbation theory. In the supercritical situation one, or several, of the bound states has disappeared from the discrete spectrum $|\varphi_n\rangle$. Its wave function is smeared out in the continuum $|\varphi_{E_p}\rangle$ as a Breit–Wigner resonance with a certain width Γ, indicating the possibility of spontaneous decay. Let us discuss three different methods of solving the problem of dynamical transitions involving such a resonance.

(I) Since the continuum wave functions in the vicinity of a resonance differ drastically from their normal shape, the use of the states $|\varphi_{E_p}\rangle$ in the expansion (3) leads to a pathological and nearly singular bound-state–continuum and continuum–continuum coupling which is highly impractical for numerical calculations. Nevertheless, one might artificially discretize the spectrum of the two-center Dirac operator, e.g., by imposing a boundary condition on the wave functions, and solve the coupled channel equations. The motion of the resonance state through the lower continuum leads to a great number of avoided crossings with transition probabilities very close to unity (see Figure 23). Since we are interested in the (very small) fraction of holes which finally remain in the lower continuum, it will be difficult to obtain reliable results. The choice of this basis is obviously not well suited to solve the dynamics.

(II) For an alternative treatment of the problem, one has to change the basis. One simple way to do this is to freeze the basis near the critical internuclear distance R_{cr}, where the bound state joins the continuum. The so-defined subcritical wave functions $|\varphi^0\rangle$ are, of course, no longer eigenstates to the Hamiltonian $H_{TCD}(R)$ if $R < R_{cr}$. Therefore the coupled system (4) is modified to

$$\dot{a}_{ij}(t) = -i \sum_k a_{ik}(t)\langle \varphi_j^0 | \Delta V(R) | \varphi_k^0\rangle$$
$$\times \exp\left[-i \int^t dt'(E_k^0 + \langle \varphi_k^0 | \Delta V | \varphi_k^0\rangle - E_j^0 - \langle \varphi_j^0 | \Delta V | \varphi_j^0\rangle)\right] \quad (23)$$

where $\Delta V(R) = H(R) - H(R_{cr}) = V(R) - V(R_{cr})$. This equation in principle is exact. It turns out, however, that its treatment in first order is not sufficient since the strong distortion of the adiabatic wave functions under the action of ΔV is important. While the true binding energy E_{res} of the "diving" state resonance (as obtained from a phase-shift analysis of $|\varphi_{E_p}\rangle$) increases strongly for closely approaching nuclei, the first-order approximation $E_n^{(1)} - E_n^0 = \langle \varphi_n^0 | \Delta V | \varphi_n^0\rangle$ remains small. However, this method could be applied in the context of a coupled-channel approach. It then has the advantage that it can be tested by applying it to

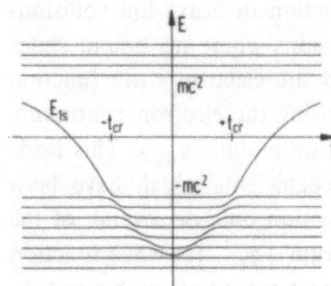

Figure 23. Schematic view of the quasibound overcritical $1s$ state penetrating the discretized negative-frequency continuum. Each avoided crossing contributes to spontaneous positron creation.

subcritical collisions with an arbitrary value of "R_{cr}" and comparing with the result using the normal basis set.

(III) We want to find a dynamical description of the excitations in the presence of a "diving" level which is a smooth extension of the treatment of subcritical collisions. To achieve this, we observe that the resonance wave function, i.e., the continuum solution $|\varphi_{E_{res}}\rangle$, closely resembles a normal bound state except for an oscillating tail of small amplitude indicating the occurrence of tunneling through the particle–antiparticle gap between $E - V(r) + m$ and $E - V(r) - m$. Therefore it may be possible to construct a "reasonable" bound state $|\phi_R\rangle$ by cutting off $|\varphi_{E_{res}}\rangle$ at a carefully chosen distance $r = r_0$ and normalizing the truncated wave function.

The problem then is to construct a new continuum $|\tilde{\varphi}_{E_p}\rangle$ which is orthogonal to the resonance state $|\phi_R\rangle$

$$\langle\tilde{\varphi}_{E_p}|\phi_R\rangle = 0 \tag{24}$$

and complete. Such a problem has been treated in the context of nuclear physics using a projection operator technique (cf. Reference 34; see also Reference 35). In this language we originally have an operator projecting onto the set of free positron states

$$P_0 = \oint |\varphi_{E_p}\rangle\langle\varphi_{E_p}| \tag{25a}$$

and the complement

$$Q_0 = \oint |\varphi_\alpha\rangle\langle\varphi_\alpha| \tag{25b}$$

where $|\varphi_\alpha\rangle$ includes all bound states $|\varphi_n\rangle$ and the upper continuum $|\varphi_{E_e}\rangle$. We have

$$(H - E_p)|\varphi_{E_p}\rangle = 0$$
$$(H - E_\alpha)|\varphi_\alpha\rangle = 0 \tag{26}$$

and

$$P_0^2 = P_0, \quad Q_0^2 = Q_0, \quad P_0Q_0 = 0, \quad P_0 + Q_0 = 1 \tag{27}$$

Our objective is to remove the resonance contained in $|\varphi_{E_p}\rangle$ from P space and to transfer it to Q space:

$$Q = \oint |\varphi_\alpha\rangle\langle\varphi_\alpha| + |\phi_R\rangle\langle\phi_R| \tag{28a}$$
$$P = 1 - Q \tag{28b}$$

Here and in the following derivations we make the important assumption that $|\phi_R\rangle$ is orthogonal to all states $|\varphi_\alpha\rangle$

$$\langle\varphi_\alpha|\phi_R\rangle = 0 \tag{29}$$

so that P and Q are also projection operators. Except for this property the choice of $|\phi_R\rangle$ is arbitrary and must be made on physical, not mathematical, grounds. We will come back to this in a moment.

We now define the modified continuum $|\tilde{\varphi}_{E_p}\rangle$ by the eigenvalue equation of the projected Hamiltonian PHP:

$$(PHP - E_p)|\tilde{\varphi}_{E_p}\rangle = 0 \tag{30}$$

Using (28b), (28a), (26), and (29) this equation can be transformed to

$$(H - E_p)|\tilde{\varphi}_{E_p}\rangle = \langle\phi_R|H|\tilde{\varphi}_{E_p}\rangle|\phi_R\rangle \tag{31}$$

Thus the modified continuum satisfies the original Dirac equation with an additional inhomogeneous term proportional to $|\phi_R\rangle$. Equation (37) may be solved by help of the Green's function $G = (H - E)^{-1}$. The formal solution

$$|\tilde{\varphi}_{E_p}\rangle = \text{const} \times \left(1 - \frac{G|\phi_R\rangle\langle\phi_R|}{\langle\phi_R|G|\phi_R\rangle}\right)|\varphi_{E_p}\rangle \tag{32}$$

may be employed to derive the orthogonality of the modified continuum states:

$$\langle\tilde{\varphi}_{E_p'}|\tilde{\varphi}_{E_p}\rangle = \langle\varphi_{E_p'}|\varphi_{E_p}\rangle = \delta(E_p - E_p')$$

and (33)

$$\langle\tilde{\varphi}_{E_p'}|H|\tilde{\varphi}_{E_p}\rangle = E_p\,\delta(E_p - E_p')$$

For numerical purposes a different approach is more useful. Due to its special form (it has a degenerate kernel) the integrodifferential equation (31) can be solved by integration. We make the ansatz

$$|\tilde{\varphi}_{E_p}\rangle = a|\varphi_{E_p}\rangle + b|\varphi_{E_p}^{(i)}\rangle \tag{34}$$

where $|\varphi_{E_p}\rangle$ is the solution of the homogeneous equation (26a) and $|\varphi_{E_p}\rangle$ solves the inhomogeneous equation with an arbitrary constant $\gamma \neq 0$ (e.g., $\gamma = 1$):

$$(H - E_p)|\varphi_{E_p}^{(i)}\rangle = \gamma|\phi_R\rangle \tag{35}$$

The constants a and b are then uniquely defined by the orthogonality requirement (24)

$$|\tilde{\varphi}_{E_p}\rangle = a\left[|\varphi_{E_p}\rangle - \frac{\langle\phi_R|\varphi_{E_p}\rangle}{\langle\phi_R|\varphi_{E_p}^{(i)}\rangle}|\varphi_{E_p}^{(i)}\rangle\right] \tag{36}$$

and by the asymptotic normalization condition of the wave function $|\tilde{\varphi}_{E_p}\rangle$. The nondiagonal matrix elements of H, i.e., those coupling the P and Q space, are

$$\langle\phi_R|H|\tilde{\varphi}_{E_p}\rangle = -a\frac{\langle\phi_R|\varphi_{E_p}\rangle}{\langle\phi_R|\varphi_{E_p}^{(i)}\rangle}\cdot\gamma \tag{37}$$

The new basis set $|\varphi_{E_e}\rangle$, $|\varphi_n\rangle$, $|\phi_R\rangle$, $|\tilde{\varphi}_{E_p}\rangle$ can be used to expand the time-dependent wave function (3). We obtain the usual set of differential equations with one additional coupling between the resonance state and the modified continuum:

For example, Equation (17a) is modified to

$$a^{(1)}_{1s,E_p}(t) = -\int_{-\infty}^{t} dt' e^{i(\chi_{Ep}-\chi_R)}\left[\langle\tilde{\varphi}_{E_p}|\frac{\partial}{\partial t'}|\phi_R\rangle + i\langle\tilde{\varphi}_{E_p}|H|\phi_R\rangle\right] \quad (38)$$

The second term can be understood to describe a spontaneous transition with the decay width[2]

$$\Gamma(E_p) = 2\pi\,|\,\langle\tilde{\varphi}_{E_p}|H|\phi_R\rangle\,|^2 \quad (39)$$

Apart from the spontaneous decay induced transitions are also present and the theory allows, in principle, a separate investigation of these processes.

The main problem in this formalism is the proper choice of the resonating bound-state wave function $|\phi_R\rangle$. As indicated above, this may be done by truncating a continuum wave function in the resonance region. Numerical investigations have shown that the modified continuum wave functions $|\tilde{\varphi}_{E_p}\rangle$ are quite sensitive to the precise nature of the cutoff. While the decay width $\Gamma(E_{res})$ comes out correctly in various procedures, a sharp cutoff at some value r_0 leads to an irregular R dependence of the matrix elements. Furthermore, if r_0 is kept fixed, the wave function $|\phi_R\rangle$ does not go over into $|\varphi_{1s}\rangle$ at the critical internuclear distance, so that there is no smooth transition between the subcritical and supercritical regime.

A gentler way to truncate the wave function was found to be the use of a modified potential like

$$V(r) = \begin{cases} V(r) & \text{for } r < r_+ \\ V(r_+) & \text{for } r \geq r_+ \end{cases}$$

with

$$E_{res} - V(r_+) + m = 0$$

The value of r_+ is defined as the outer edge of the classically forbidden tunneling region between electron and positron states. At larger distances the potential is kept fixed so that the wave function goes to zero. With this choice discontinuities in $|\phi_R\rangle$ are avoided. Coupled-channel calculations with the matrix elements [equation (38)] are in progress. Their results will be reported elsewhere.

6. Summary

We have discussed the theory of positron production in very heavy ion collisions as an example to show the importance of coherence for quasimolecular processes. The theory agrees well with present experimental data from GSI. It is noteworthy that the agreement could not be achieved without treating the full coherence between ionization and pair creation and between first- and second-order processes. This insight has equal bearing for related phenomena, e.g., multiple excitation processes in inner shells or molecular x-ray transition.

We also note that the agreement between the observed spectra and theoretical predictions constitutes the first evidence for the novel aspects of quantum electrodynamics of strong fields: the nonperturbative character of direct pair production (the "shake-off of the vacuum polarization cloud") and the onset of the change of the vacuum state from neutral to being charged. The identification of the spontaneous vacuum decay, however, must await further calculations and future experiments with the heaviest collision systems, such as U–Cm.

ACKNOWLEDGMENTS

We are much indebted to W. Betz, V. Oberacker, and G. Soff for many helpful discussions and several figures. We would also like to thank H. Backe, H. Bokemeyer, and J. S. Greenberg for valuable conversations about the experiments. This work was supported by the Bundesministerium für Forschung und Technologie (BMFT) and by the Gesellschaft für Schwerionenforschung (GSI).

References

1. W. Pieper and W. Greiner, *Z. Phys.* **218**, 327–340 (1969); S. S. Gershtein and Ya. B. Zeldovich, *Sov. Phys. JETP* **30**, 358 (1970); V. S. Popov, *Sov. J. Nucl. Phys.* **15**, 595 (1972).
2. B. Müller, H. Peitz, J. Rafelski, and W. Greiner, *Phys. Rev. Lett.* **28**, 1235–1238 (1972); B. Müller, J. Rafelski, and W. Greiner, *Z. Phys.* **257**, 62–77, 183–211 (1972).
3. B. Müller, *Ann. Rev. Nucl. Sci.* **26**, 351–383 (1976); J. Reinhardt and W. Greiner, *Rep. Prog. Phys.* **40**, 219–295 (1977); J. Rafelski, L. P. Fulcher, and A. Klein, *Phys. Lett.* **38C**, 228–361 (1978).
4. K. Smith, B. Müller, and W. Greiner, *J. Phys. B* **8**, 75–101 (1975).
5. B. Müller, J. Rafelski, and W. Greiner, *Phys. Lett.* **47B**, 5–7 (1973); B. Müller and W. Greiner, *Z. Naturforsch.* **31a**, 1–30 (1976).
6. W. Betz, Diploma thesis, Frankfurt (1976).
7. G. Soff, B. Müller, and W. Greiner, *Phys. Rev. Lett.* **40**, 540–544 (1978).
8. W. Betz, G. Soff, B. Müller, and W. Greiner, *Phys. Rev. Lett.* **37**, 1046–1049 (1976).
9. B. Müller, G. Soff, W. Greiner, and V. Ceausescu, *Z. Phys.* **A285**, 27–30 (1978).
10. J. S. Greenberg, H. Bokemeyer, H. Emling, E. Grosse, D. Schwalm, and F. Bosch, *Phys. Rev. Lett.* **39**, 1404–1407 (1977).
11. J. S. Macdonald, P. Armbruster, H.-H. Behnke, F. Folkmann, S. Hagmann, D. Liesen, P. Mokler, and A. Warczak, *Z. Phys.* **A284**, 57–59 (1978).
12. H.-H. Behnke, D. Liesen, S. Hagmann, P. H. Mokler, and P. Armbruster, *Z. Phys.* **A288**, 35 (1978).
13. G. Soff, W. Betz, B. Müller, W. Greiner, and E. Merzbacher, *Phys. Lett.* **65A**, 19–22 (1978).
14. B. Müller and W. Greiner, *Phys. Rev. Lett.* **33**, 469–473 (1974).
15. J. S. Greenberg, C. K. Davis, and P. Vincent, *Phys. Rev. Lett.* **33**, 473–476 (1974).
16. Ch. Stoller, W. Wölfli, G. Bonani, M. Stöckli, and M. Suter, *J. Phys. B* **10**, L347–350 (1977).
17. W. Wölfli, E. Morenzoni, Ch. Stoller, G. Bonani, M. Stöckli, *Phys. Lett.* **68A**, 217 (1978).
18. J. Kirsch, W. Betz, J. Reinhardt, G. Soff, B. Müller, and W. Greiner, *Phys. Lett.* **72B**, 298–302 (1978).
19. H. Peitz, B. Müller, J. Rafelski, and W. Greiner, *Lett. Nuovo Cimento* **8**, 37–42 (1973).

20. J. H. Macek and J. S. Briggs, *J. Phys. B* **7**, 1312–1322 (1974).
21. K. Smith, H. Peitz, B. Müller, and W. Greiner, *Phys. Rev. Lett.* **32**, 554–556 (1974).
22. G. Soff, J. Reinhardt, B. Müller, and W. Greiner, *Phys. Rev. Lett.* **38**, 592–595 (1977).
23. B. Müller, V. Oberacker, J. Reinhardt, G. Soff, W. Greiner, and J. Rafelski, *J. Phys. Soc. Jpn.* **44** (Suppl.) 838–846 (1978).
24. J. Reinhardt, V. Oberacker, B. Müller, W. Greiner, and G. Soff, *Phys. Lett.* **78B**, 183–188 (1978).
25. G. Heiligenthal, W. Betz, G. Soff, B. Müller, and W. Greiner, *Z. Phys.* **A285**, 105–106 (1978).
26. G. Soff, W. Betz, J. Reinhardt, and J. Rafelski, *Phys. Scr.* **17**, 417–419 (1978).
27. M. Gyulassy, *Phys. Rev. Lett.* **32**, 921–925 (1974).
28. G. A. Rinker and L. Wilets, *Phys. Rev. A* **12**, 748–762 (1975).
29. K. T. Cheng and W. R. Johnson, *Phys. Rev. A* **14**, 1943–1948 (1976).
30. T. H. Rihan, N. S. Aly, E. Merzbacher, B. Müller, and W. Greiner, *Z. Phys.* **A285**, 397–403 (1978).
31. J. Rafelski, B. Müller, and W. Greiner, *Z. Phys.* **A285**, 49–52 (1978).
32. H. Backe, L. Handschug, F. Hessberger, E. Kankeleit, L. Richter, F. Weik, R. Willwater, H. Bokemeyer, P. Vincent, Y. Nakayama, and J. S. Greenberg, *Phys. Rev. Lett.* **40**, 1443–1446 (1978).
33. Ch. Kozhuharov, P. Kienle, E. Berdermann, H. Bokemeyer, J. S. Greenberg, Y. Nakayama, P. Vincent, L. Handschug, E. Kankeleit, H. Backe, *Phys. Rev. Lett.* **42**, 376–379 (1979).
34. W. L. Wang and C. M. Shakin, *Phys. Lett.* **32B**, 421–424 (1970).
35. M. Micklinghoff, Ph.D. thesis, Hamburg (1977).

Note Added in Proof

Meanwhile, we have performed coupled channel calculations for the amplitudes $a_{pq}(t)$. We find that the ionization probabilities are (almost linearly) increased by a factor of 3 to 5, whereas the pair creation probability is raised only by less than a factor of 2. As a result, in Figure 18 the data now fall right on the curve; the same holds for the Pb–Pb data of Figure 20. For the data of Figure 19, on the other hand, our new curve is too high by a factor of 2. As new, improved experimental results are being obtained, a conclusive comparison with the data will be possible in the near future.

28

Alignment of Core-Excited Autoionizing Be⁺ Projectiles Excited in Single Gas Collisions

P. Bisgaard, R. Bruch, P. Dahl, and M. Rødbro

Angular distributions of Auger electrons emitted from Be II $1s(2s2p\,{}^3P^0)\,{}^2P^0$, $1s(2s2p\,{}^1P^0)\,{}^2P^0$, and $(1s2p^2)\,{}^2D$ states excited in 100–500-keV Be⁺ on He and Be⁺ on CH_4 collisions have been measured. For the ${}^2P^0$ states the source-frame anisotropy coefficients are deduced, and the alignment parameters are calculated. The Be⁺ on He collisions lead to strong alignment where the $M_L = 0$ substates are preferentially populated. Weaker alignment is obtained in Be⁺ on CH_4 collisions.

1. Introduction

Recently the ejected-electron spectra of highly excited autoionizing levels of Li, Be, B, and C have been studied by using the projectile–electron spectroscopy method.[1-5] It has been demonstrated that singly and doubly core-excited states are abundantly populated in 100–500-keV Li⁺, Be⁺, B⁺, and C⁺ single collisions with He and CH_4.

Here, we report observation of strong anisotropies in 100–500-keV projectile Auger-electron emission following ion–atom and ion–molecule[6,7]† collisions. These

---

† In Reference 6 the ⁴P⁰ data are incorrect.

P. Bisgaard, R. Bruch, P. Dahl, and M. Rødbro • Institute of Physics, University of Aarhus, DK-8000 Aarhus C, Denmark.   R. Bruch's permanent address: Fakultät für Physik, Universität Freiburg, D-78 Freiburg, West Germany.

anisotropies can be directly related to the aligned population of magnetic sub-states.[8-9]

2. Experiment

Be$^+$ ion beams were supplied by the 600-keV heavy-ion accelerator at the University of Aarhus. The experimental setup[10] is shown in Figure 1. Electrons ejected from the target region were analyzed by an electrostatic parallel-plate spectrometer, which can be rotated between 0° and 150°. The pressure in the differentially pumped gas cell was kept low enough (typically 10^{-3} Torr) to ensure single-collision conditions.

In continuation of our spectroscopic work,[1,5] angular distributions for the predominant Be II lines were measured for the Be$^+ \rightarrow$ CH$_4$ and Be$^+ \rightarrow$ He collision systems. The methane target was chosen because the projectile K-excitation cross section is large; on the other hand the He target provides a much simpler collision system.

Representative projectile Auger spectra for 500-keV Be$^+ \rightarrow$ He impact are shown in Figure 2. The spectra were fitted with a least-squares fitting program to obtain line intensities for the initial states $1s(2s2p\ ^3P^0)\ ^2P^0$, $1s(2s2p\ ^1P^0)\ ^2P^0$, and $(1s2p^2)\ ^2D$ decaying to $1s^2\ ^1S$ measured relative to the $(1s2s^2)\ ^2S \rightarrow 1s^2\ ^1S$ line. We note that the cross sections must be corrected for Doppler anisotropy.[11] For, e.g., a $^2P^0$ state the correction on the cross-section ratio can be written as follows:

$$\frac{\sigma(\theta_0)\ ^2P^0}{\sigma(\theta_0)\ ^2S} = \frac{[E_0(\cos\psi)/E]\ ^2P^0}{[E_0(\cos\psi)/E]\ ^2S}\ \frac{\sigma(\theta)\ ^2P^0}{\sigma(\theta)\ ^2S}$$

where $\sigma(\theta_0)$ is the cross section at angle θ_0 in the frame of moving Be$^+$ projectile,

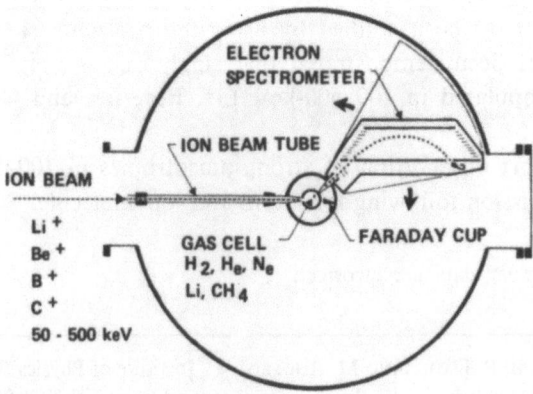

Figure 1. Schematic diagram for projectile–electron angular distribution measurements.

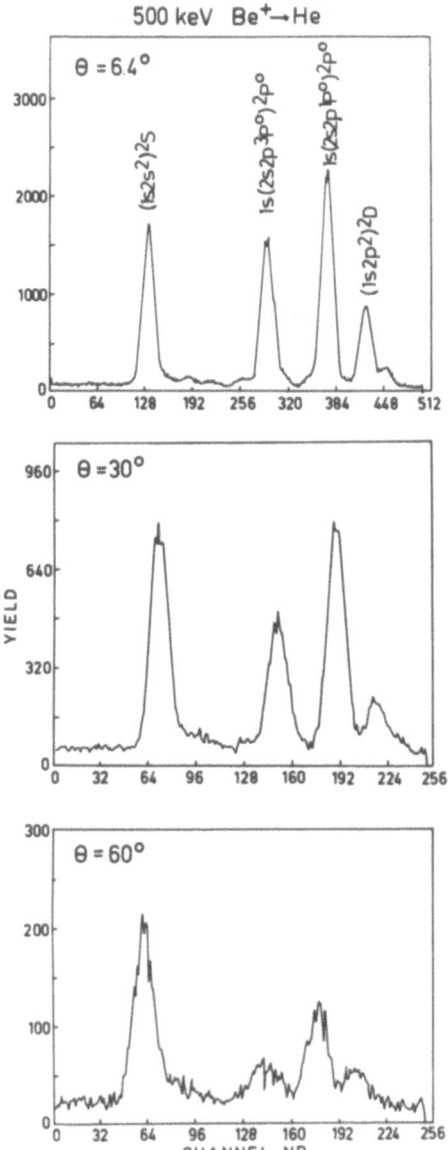

Figure 2. Autoionization spectra produced by beryllium projectiles incident on a helium gas target recorded at $\theta = 6.4°$, $\theta = 30°$, and $\theta = 60°$.

$\sigma(\theta)$ is the cross section at angle θ in the laboratory system, and $E_0(\cos \psi)/E$ is the Doppler anisotropy factor for single differential cross sections. Here, E_0 and E are the source and the laboratory electron energies, respectively, and $\psi = \theta_0 - \theta$. For the $^2P^0$ states the cross-section ratio $\sigma(\theta_0)^2P^0/\sigma(\theta_0)^2S$ is plotted versus θ_0 in Figure 3.

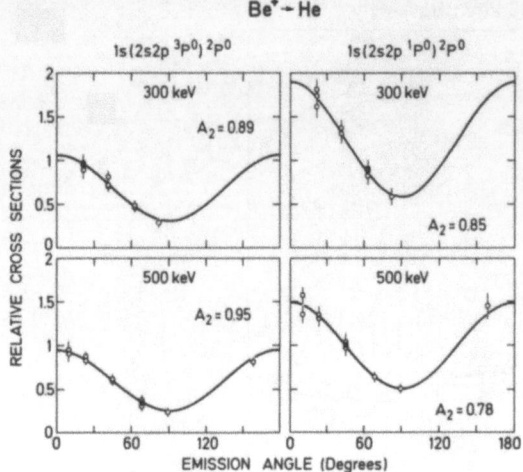

RELATIVE CROSS SECTIONS

Be⁺ → He

1s(2s2p ³P⁰) ²P⁰   1s(2s2p ¹P⁰) ²P⁰

300 keV   300 keV
A₂ = 0.89
A₂ = 0.85

500 keV   500 keV
A₂ = 0.95
A₂ = 0.78

EMISSION ANGLE (Degrees)

Figure 3. Measured angular distribution of ejected autoionization electrons from projectile states in Be II excited in 300 and 500 keV Be⁺ → He collisions.

3. Angular Distribution of Auger Electrons

The angular distribution $I(\theta_0)$ of Auger electrons reflects alignment of the excited projectile state. For the case of beryllium the LSM_LM_S coupling is appropriate for the initial and final projectile states since the natural level widths of the autoionizing states in Be II are of the order of 0.02 eV[12] whereas the fine-structure separations are of the order of a few meV.[13]

Table 1. Angular Distributions and Anisotropy Coefficients Corresponding to the Auger Decay of the $(1s2s^2)$ ^2S, $1s(2s2p$ $^3P^0)$ $^2P^0$, $1s(2s2p$ $^1P^0)$ $^2P^0$, and $(1s2p^2)$ ^2D initial States[a]

Initial state	Emitted electron	Angular distribution	Anisotropy coefficient
$(1s2s^2)$ ^2S	ε_s	isotropic	$A_0 = 1$
$1s(2s2p$ $^3P^0)$ $^2P^0$ $1s(2s2p$ $^1P^0)$ $^2P^0$	ε_P	$1 + A_2 P_2(\cos\theta_0)$	$A_2 = \dfrac{2[\sigma(1,0) - \sigma(1,1)]}{\sigma(1,0) + 2\sigma(1,1)}$ $\dfrac{\sigma(1,0)}{\sigma(1,1)} = \dfrac{1 + A_2}{1 - 0{,}5A_2}$
$(1s2p^2)$ ^2D	ε_d	$1 + A_2 P_2(\cos\theta_0) + A_4 P_4(\cos\theta_0)$	$A_2 = \dfrac{10}{7}\dfrac{\sigma(2,0) + \sigma(2,1) - 2\sigma(2,2)}{\sigma(2,0) + 2\sigma(2,1) + 2\sigma(2,2)}$ $A_4 = \dfrac{6}{7}\dfrac{3\sigma(2,0) - 4\sigma(2,1) + \sigma(2,2)}{\sigma(2,0) + 2\sigma(2,1) + 2\sigma(2,2)}$

$$P_2(\cos\theta_0) = \tfrac{1}{2}[3\cos^2\theta_0 - 1], \qquad P_4(\cos\theta_0) = \tfrac{1}{8}[35\cos^4\theta_0 - 30\cos^2\theta_0 + 3]$$

[a] Here we assume that the fine-structure splitting is small when compared to the natural linewidth.

Owing to the cylindrical symmetry about the ion beam axis the excitation cross sections $\sigma(L, M_L)$ do not depend on the sign of M_L. The angular distribution $I(\theta_0)$ obeys the symmetry condition $I(\theta_0) = I(\pi - \theta_0)$ and may be expanded in even-order Legendre polynomials

$$I(\theta_0) = C \sum_{n=1}^{\infty} A_{2n} P_{2n}(\cos \theta_0), \qquad A_0 = 1$$

When, as in the present case, the final ionic state is a ^1S state, simple relations are found between the A_{2n} coefficients and the $\sigma(LM_L)$ cross sections.[9] These relations are given in Table 1.

4. Results and Discussion

The unknown anisotropy coefficient A_2 and the scale factor C for the $1s(2s2p\,^3P^0)\,^2P^0$ and $1s(2s2p\,^1P^0)\,^2P^0$ states were found by fitting

$$I(\theta_0) = C[1 + A_2 P_2(\cos \theta_0)]$$

to the angular distribution data (Figure 3). The A_2 coefficients and the alignment parameters $\sigma(1, 0)/\sigma(1, 1)$ are presented in Table 2. In Figure 4, these numbers are plotted as a function of the projectile energy. An interesting feature of these data is the strong alignment produced in Be+ → He collisions with a preferential population of the $M_L = 0$ substate. For the Be+–CH$_4$ collision system, the aligned population of magnetic substates is less pronounced, and crossovers of the alignment in the observed energy range might occur.

Table 2. Measured A_2 Coefficients and Alignment Parameters $\sigma(1, 0)/\sigma(1, 1)$

| Collision system | Projectile energy (keV) | Be II $1s(2s2p\,^3P^0)\,^2P^0 \to (1s^2\varepsilon p)\,^2P^0$ | | Be II $1s(2s2p\,^1P^0)\,^2P^0 \to (1s^2\varepsilon p)\,^2P^0$ | |
		Anisotropy coefficient A_2	Cross section ratio $\sigma(1, 0)/\sigma(1, 1)$	Anisotropy coefficient A_2	Cross section ratio $\sigma(1, 0)/\sigma(1, 1)$
Be+ → He	200	0.37 ± 0.10	1.68	0.84 ± 0.10	3.17
	300	0.89 ± 0.05	3.41	0.85 ± 0.05	3.22
	500	0.95 ± 0.05	3.71	0.78 ± 0.05	2.92
Be+ → CH$_4$	100	0.36 ± 0.06	1.66	0.00 ± 0.05	1.00
	200	-0.03 ± 0.03	0.96	0.20 ± 0.03	1.33
	300	-0.03 ± 0.03	0.96	0.16 ± 0.03	1.26
	500	0.13 ± 0.03	1.21	0.32 ± 0.03	1.57

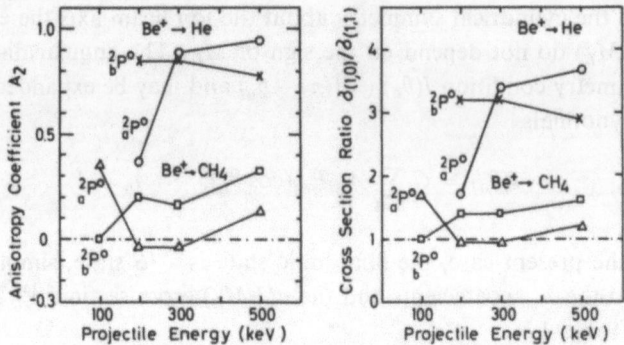

Figure 4. (Left) The A_2 coefficients of the $1s(2s2p\ ^3P^0)\ ^2P^0$ and the $1s(2s2p\ ^1P^0)\ ^2P^0$ states in Be II. (Right) Alignment parameter of the $1s(2s2p\ ^3P^0)\ ^2P^0$ and the $1s(2s2p\ ^1P^0)\ ^2P^0$ states in Be II. $^3_aP^0 = 1s(2s2p\ ^3P^0)\ ^2P^0$, $^1_bP^0 = 1s(2s2p\ ^1P^0)\ ^2P^0$. The straight lines are drawn through the experimental data for guiding the eye.

The impact velocities in this investigation range from 0.7 to 1.5 a.u., i.e., from 0.2 to 0.4 times the Be $1s$ orbital velocity, or 0.5 to 1.1 times the $1s$ velocity in He. The core excitation in $Be^+ \rightarrow$ He collisions may then predominantly be due to molecular promotion of a He $1s$ electron in the $2p\sigma$ orbital. By rotational coupling at small internuclear separations the electron is transferred to the $2p\pi_x$ orbital, which correlates with a $2p$ orbital of Be. Together with a K-vacancy sharing this leads to the $1s2s2p$ configuration of Be II. To explain the preferential population of $M_L = 0$ substates we suggest that the electron in the $2p\pi_x$ orbital by a second rotational coupling is transferred to the $3d\sigma$ orbital, which also correlates with Be $2p$. The $3d\sigma$ and $2p\pi_x$ levels are nearly degenerate except for very small internuclear separations. During the collision the molecular axis rotates through almost $180°$, but with the promoted electron in the $2p\sigma$, $2p\pi_x$, and $3d\sigma$ orbitals in the early, the

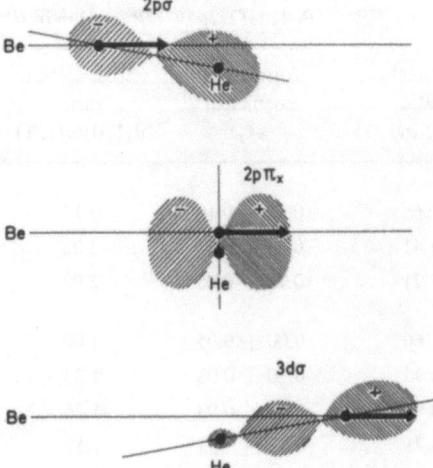

Figure 5. Sketch of the evolution of the promoted electron wave function during the collision.

intermediate, and the late stage, respectively, the electron wave function does not rotate (see Figure 5). This may be quantitatively the expected excitation mechanism for relatively high impact velocities.

In the $Be^+ \rightarrow CH_4$ collision, a Be $1s$ electron is promoted by the $2p\sigma\text{-}2p\pi_x$ coupling at small separations between Be and C. The molecular target structure complicates the final correlations to the outer shell of Be, so that the resulting alignment may be weak. This is in accord with the difference between the $Be^+ \rightarrow He$ and the $Be^+ \rightarrow CH_4$ data, but it is noticed that even for $Be^+ \rightarrow CH_4$ collisions substantial alignments have been observed.

References

1. M. Rødbro, R. Bruch, and P. Bisgaard, *J. Phys. B* **10**, L275–L279 (1977).
2. M. Rødbro, R. Bruch, P. Bisgaard, P. Dahl, and B. Fastrup, *J. Phys. B* **10**, L483–487 (1977).
3. R. Bruch, M. Rødbro, P. Bisgaard, and P. Dahl, *Phys. Rev. Lett.* **39**, 801–804 (1977).
4. P. Bisgaard, R. Bruch, P. Dahl, B. Fastrup, and M. Rødbro, *Phys. Scr.* **17**, 49–52 (1978).
5. M. Rødbro, R. Bruch, and P. Bisgaard, *J. Phys. B* **12**, 2413–2447 (1979).
6. P. Bisgaard, R. Bruch, P. Dahl, B. Fastrup, and M. Rødbro, ICPEAC X, Paris, Abstracts of papers, Vol. 2, pp. 1024–1025 (1978).
7. P. Bisgard, R. Bruch, P. Dahl, and M. Rødbro, *J. Phys. (Paris) Colloq.* **40**, C1-243–245 (1979).
8. B. Cleff and W. Mehlhorn, *J. Phys. B* **7**, 593–604 (1974).
9. W. Mehlhorn, private communication.
10. M. Rødbro, Ph.D. thesis, University of Aarhus (1977).
11. P. Dahl, M. Rødbro, B. Fastrup, and M. E. Rudd, *J. Phys. B* **9**, 1567–1579 (1976).
12. H. P. Kelly, Atomic Inner Shell Processes, B. Craseman, ed., Vol. 1, pp. 331–353, Academic Press, New York (1975).
13. S. Goldsmith, *J. Phys. B* **7**, 2315–2319 (1974).

29

Interference Effects
in K-Shell Vacancy Sharing

R. Schuch, G. Nolte, W. Lichtenberg, and H. Schmidt-Böcking

The sharing of K-shell vacancies between the heavy and light collision partner was investigated for the collision system 32-MeV S on Ar in dependence on impact parameter. The sharing probability was found in agreement with the analytical expression of J. S. Briggs up to impact parameters of double K-shell radius. Strong interference effects were predicted in this calculation if a K-shell vacancy is already present on the incoming part of the collision. This effect was found in the impact-parameter dependence of the vacancy-sharing probability measured with sulfur projectiles with charge 15+ on an Ar gaseous target.

The production and sharing of K-shell vacancies is considerably well described in terms of a molecular orbital model. In this model the vacancies are produced by rotational coupling in the $2p\sigma$ orbital and are then shared by radial coupling between the K shells of heavier and lighter collision partner. This vacancy-sharing process (VS) is well described by the Meyerhof formula.[1] It was derived by applying a formalism of Demkov[2] for the case of producing the $2p\sigma$ vacancies during the collision ("one-passage" VS). This theory of Demkov was developed for charge transfer where the vacancy is already present on the incoming part of the collision ("two-passage" VS) with the approximation of an impact-parameter (b) independent transfer probability—e.g., only collisions of impact parameter zero were assumed. The Meyerhof formula can therefore only be applied to total cross sections.

R. Schuch and G. Nolte • Physikalisches Institut der Universität Heidelberg, 6900 Heidelberg, Federal Republic of Germany.   W. Lichtenberg and H. Schmidt-Böcking • Institut für Kernphysik, 6000 Frankfurt, Federal Republic of Germany.

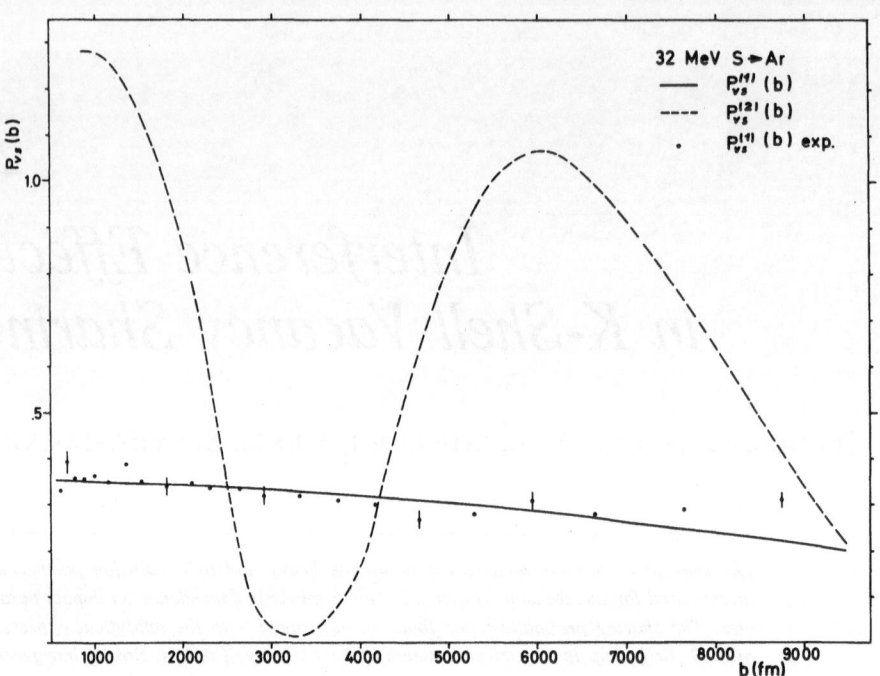

Figure 1. Impact-parameter dependence of K-shell vacancy sharing for the collision system 32-MeV S on Ar. Data points are for S^{5+} on Ar. The solid line represents the "one-passage" and the dashed line the "two-passage" vacancy-sharing process described by the analytical formula of Reference 3.

In a treatment of the K-shell vacancy-sharing process by J. Briggs[3] the impact-parameter dependence of the "one-passage" VS and of the "two-passage" VS was calculated.

An impact-parameter dependence has been found also in the experimental investigation[4] of the vacancy-sharing probability for the collision systems 35-MeV Cl on Ti and on Ni. The measured impact-parameter dependence of VS [$P_{VS}(b)$] was in accord with the analytical formula of Reference 3 for both collision systems.[4]

In Figure 1 the probability $P_{VS}^{(1)}(b)$ from measurements of 32-MeV S^{5+} on Ar is presented. The fluorescence yields of Macdonald *et al.*[5] are used for determining the vacancy-sharing probabilities from the x-ray yields of S and Ar. The error bars only include statistical errors. The measured vacancy-sharing probability $P_{VS}^{(1)}(b)$ is in good agreement with the prediction of the analytical expression for $P_{VS}^{(1)}(b)$[3] at impact parameters below 7000 fm (approximately the double K-shell radius). At large impact parameters, however, the sharing probability remains constant, in contradiction to the analytical expression of Reference 3 (Figure 1, solid line).

A two-state MO calculation was performed[3] for the collision system oxygen on neon. The result of this MO calculation (Figure 3 in Reference 3) also shows, in accord with the experimental data of S^{5+} on Ar presented here, a much weaker decrease than the analytical expression at large impact parameters.

The dashed line in Figure 1 represents the analytical solution[3] of the "two-passage" vacancy-sharing process. The oscillatory structure reflects the interference of the vacancy transfer amplitudes of the incoming and outgoing part of the collision. Experiments with highly ionized sulfur projectiles selected for charge state 15+ and hitting a thin Ar gas target were performed at the MPI für Kernphysik, Heidelberg. This collision system, where a *K* vacancy is already present in the incoming part of the collision, allows the investigation of the "two-passage" vacancy-sharing process. A strong interference structure was found also in the experimental data.[6]

The wavelength of the oscillation is in accord with the prediction of the analytical expression (Figure 1, dashed line) but the oscillations are not in phase. A similar phase difference has also been noticed by Briggs, when he compared the analytical expression with the result of a two-state MO calculation for oxygen on neon.[3] For comparison of the measured "two-passage" VS probability with theoretical predictions a two-state MO calculation for the collision system S^{15+} on Ar has to be performed.

ACKNOWLEDGMENT

This work was supported by Bundesministerium für Forschung und Technologie.

References

1. W. E. Meyerhof, *Phys. Rev. Lett.* **31**, 1341 (1973).
2. Y. u. N. Demkov, *Sov. Phys. JETP* **18**, 138 (1964).
3. J. S. Briggs, Atomic Energy Research Establishment, Technical Report No. 594, unpublished, (1974).
4. R. Schuch, H. Schmidt-Böcking, R. Schule, and I. Tserruya, *Phys. Rev. Lett.* **39**, 79 (1977).
5. J. R. Macdonald, R. Schule, R. Schuch, H. Schmidt-Böcking, and D. Liesen, *Phys. Rev. Lett.* **40**, 1330 (1978).
6. R. Schuch, G. Nolte, H. Schmidt-Böcking, W. Lichtenburg, *Phys. Rev. Lett.* **43**, 1104 (1979).

30

Coherent Excitation of Ionic States by Correlated Collisions in a Crystal Lattice

S. DATZ

Ions moving through crystals at small angles with respect to atomic rows and planes undergo correlated collisions which confine their motion ("channeling"). Well-channeled ions collide only with conduction electrons. For ion velocities $v_i \gg v_0$ ionization is similar to electron bombardment at $v_e = v_i$. The ion passing the atoms spaced at a distance d in the crystal rows experiences a coherent perturbation of frequency v_i/d and may be resonantly excited at $(v_i/d) = \Delta E_{ij}/h$. This is seen in an enhanced ionization probability. The relation of this effect on the states of the penetrating ions and the crystal fields is discussed.

1. Introduction

The subject matter of this chapter is a bit esoteric in the context of this volume in that it involves correlation and coherence in *multiple* atomic collisions in an ordered crystal lattice. Here we will require a rudimentary knowledge of a phenomenon called "channeling," which is caused by correlated collisions of energetic ions with crystal atoms arranged in rows (axial) or sheets (planar) and some of the consequences of the ordered motion of channeled particles on ionization and excitation events taking place while the ion moves through the crystal. We will demonstrate that for certain ionic systems (e.g., hydrogenlike B^{4+}, C^{5+}, N^{5+}, O^{7+}, and F^{8+} in the velocity range $6 \gtrsim v_i \gtrsim 12 v_0$ where $v_0 = e^2/\hbar$) collisional lifetimes are sufficiently long to

S. DATZ • Oak Ridge National Laboratory, Oak Ridge, Tennessee 37830, U.S.A.

give well-defined $n = 1$ states, and finally we shall investigate the effect of spatially (time) coherent excitations on causing resonant transitions in the moving ions.

2. Channeling and Ionic Charge States

The general concept of channeling is illustrated in Figure 1. Consider a projectile ion aimed for a hard collision with an atom contained in a closely packed (i.e., low-index) atomic row in a crystal. Because of the preset arrangement of the atoms with respect to its path the projectile will undergo a set of correlated collisions with the equally spaced row atoms. It will be slightly deflected by the repulsive potential of the first atom, again by the second, and so forth. The final collision will thereby be considerably softened, and, depending upon the angle ψ with respect to the atomic row, it may not "collide" at all with the original target atom. This is the basis of channeling; i.e., within a small critical angle ψ_c with respect to an atomic row or plane, determined by the interatomic potential and spacing d of atoms, the particle will be deflected by a continuum potential made by properly summing the atomic potentials.

In Figure 2 we show some cases of low-index axes and planes in the face-centered cubic crystal. For particles entering in a close-to-axial direction at high velocity, Lindhard[1] showed that the close collisions with the resultant rows or "strings" of atoms are avoided for incidence angle ψ_c if

$$\psi_c \gtrsim (2Z_1Z_2e^2/E \cdot d)^{1/2} \tag{1}$$

For planar channeling, i.e., confinement of the ion's motion between sheets of planar density n_p (atoms cm^{-2}), the critical angle ψ_p is

$$\psi_p = (2\pi n_p Z_1 Z_2 e^2 a_{TF}/E)^{1/2} \tag{2}$$

where a_{TF} is the Thomas–Fermi screening length.

The primary result of channeling is to prevent small-impact-parameter collisions from occurring. Hence, such effects as the elimination of Rutherford scattering and nuclear stopping and the reduction of x-ray and nuclear reaction yields are observed.[2]

Of greatest importance to the subject in question here is that for channeled ions of $v_i \gg v_0$ close electronic interactions are limited to valence and/or conduction electrons, all of which have velocities lower than the penetrating ion. Thus, for electron capture and loss from the ion, we can view the penetrating ion as stationary and being

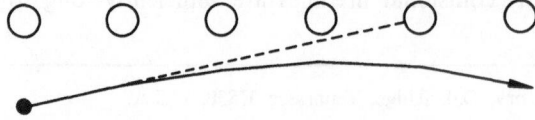

Figure 1. Effect of correlated collisions with a row of atoms in a crystal on penetrating ion trajectories ("channeling").

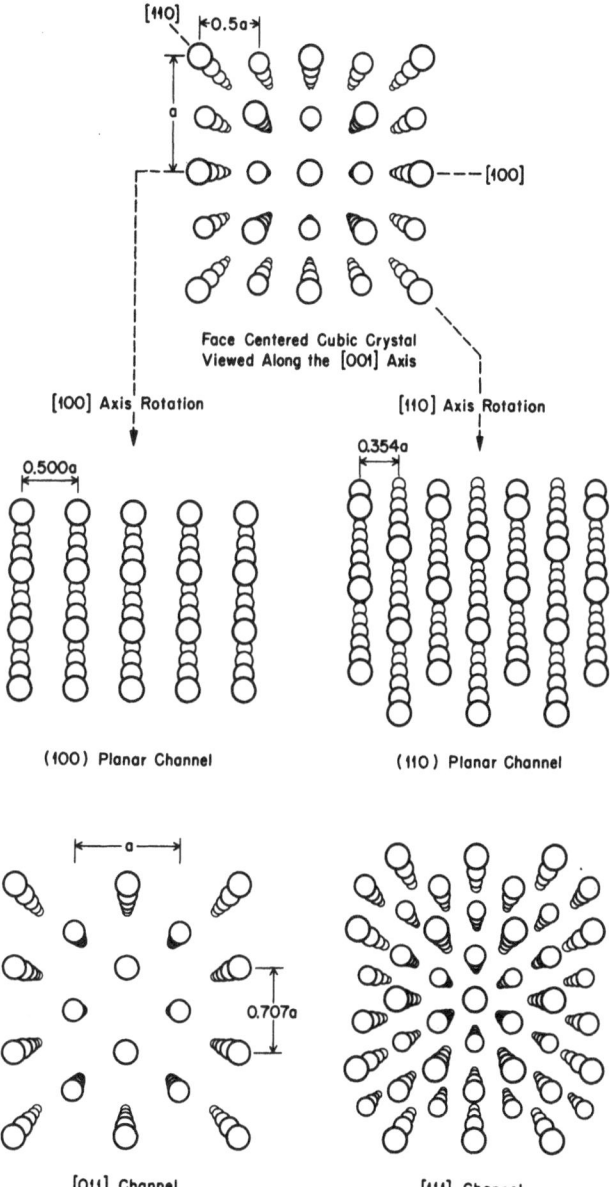

Figure 2. Illustration of some axial and planar channels in a face-centered cubic crystal.

bombarded by electrons of energy $\frac{1}{2}m_e v_i^2$. For example, an ion moving in a channel with an energy of 2 MeV/amu may be viewed as being bombarded with electrons with 1 keV of energy.

The result of this can be seen in Figure 3, where we show the emergent charge state fractions ϕ_i for 40-MeV oxygen ions which have been injected into a thin Au crystal.[3] When the beam is directed in a "random" (nonchanneled) direction the observed emergent distribution is independent of the initial charge on the oxygen ion ($6^+, 7^+$, or 8^+) and also independent of the crystal thickness over the range 0.143–0.663 μm. This is to be anticipated since "normal" electron capture and loss cross sections for ions of this velocity in gases ($Z \lesssim 8$) are on the order of 10^{-17} cm². Solid densities are on the order of $\sim 5 \times 10^{22}$ atoms/cm³. A mean free path for charge exchange is only ~ 0.02 μm so that a steady state ("equilibrium") will be reached independent of the initial charge state.

The same situation does not obtain if we instead inject the beam in a channeled direction, in this case the $\langle 110 \rangle$ axial direction. The emergent charge fractions are a strong function of the input charge. Equilibrium is not attained and the relevant charge-changing cross sections must be much reduced. From the matrix of values of emergent charge distributions for different input charges (or from the variation of charge fraction with thickness[3]) we can derive the relevant electron capture and loss cross sections for channeled ions.

Taking a case in point we have found an electron loss cross section for O^{7+}, $\sigma_{7,8} = 3 \times 10^{-19}$ at 28 MeV in the $\langle 110 \rangle$ channel of Ag.[4] There are ~ 10 valence electrons per Ag atom which then gives a cross section per electron of 3×10^{-20} or about two times as high as one would calculate from Sampson and Golden's[5] method for ~ 900-eV electrons on $O^{7+}(1s)$. An important point here, however, is that in the crystal channel any collision leading to excitation soon leads to ionization, e.g., the ionization cross section for $O^{7+}(2p)$ is ~ 10 times that for $1s$.[5] Insofar as the excitation cross section for the $1s$ electron is equal to its direct ionization cross section, the effective ionization cross section in a channel can be as much as twice that for a single electron collision.

The electron capture (recombination) cross sections in the channel for O^{7+}

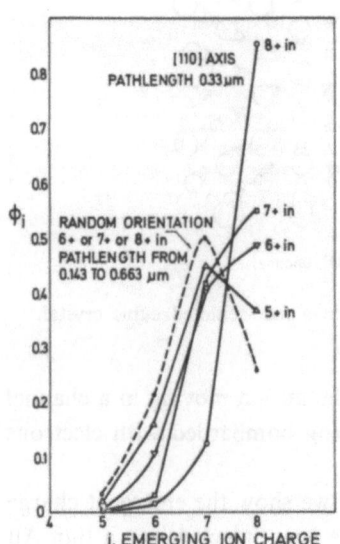

Figure 3. Fractional populations of charge states of 40-MeV oxygen ions emerging from an Au crystal when aligned in a random (dashed line) and [110] channeled (solid lines) direction as a function of incident-ion charge state ($n+$ in).

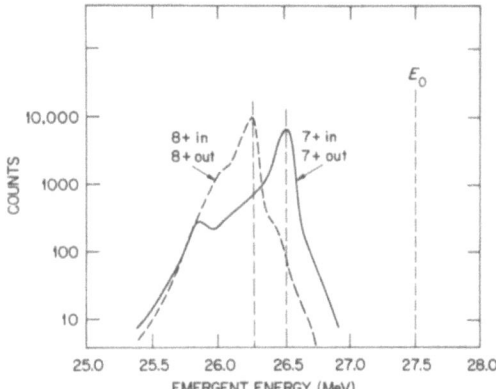

Figure 4. Emergent energy spectra of 27.5-MeV O^{7+} and O^{8+} ions incident on the (111) planar channel of Ag (5417 Å thick). The spectra shown are for the ions having the same charge as the incident beam.

and O^{8+} were even lower[4] ($\sigma_{8,7} \simeq 9 \times 10^{-20}$, $\sigma_{7,6} \simeq 6 \times 10^{-20}$) so that for crystal thicknesses on the order of 0.1–1 μm an appreciable fraction of O^{7+} ions can pass through with no charge-changing collisions and those that are ionized have only a very small chance of recapturing an electron.

Further evidence for the absence of charge-changing collisions was shown in measurement of energy-loss spectra of ions emerging in a given charge state as a function of the input charge. Figure 4 shows such spectra for O ions incident at 27.5 MeV passing through a 5400-Å Ag crystal.[6] The energy loss for fixed-charge ion is closely proportional to q^2.[7] In fact, detailed studies[8] of this type have enabled us to measure the effect of screening by bound $1s$ electrons upon electronic energy loss. For ions of lower Z (e.g., B^{4+} and C^{5+}) the relevant ionization cross sections are, of course, larger ($\sigma_i \propto Z^{-4}$); however, the use of energy-dispersed spectra enables us to pick out those ions that have not undergone charge exchange.[8]

3. Resonant Coherent Excitation

What we have demonstrated thus far is that the concept of states of certain ions moving in solids under certain conditions is a meaningful one, i.e., they are not collision broadened so that discrete eigenstates are not useful concepts in their description. In this case it is also meaningful to discuss transitions between these eigenstates (i.e., to perform a type of spectroscopy to better describe these states). In this the lattice periodicity supplies requisite exciting frequencies and the presence of an excited state is signaled by an increased ionization probability.[9]

Although the potential controlling the detailed trajectory of a channeled ion can be well represented as a continuum, the ion passing between ordered rows of atoms with velocity v_i experiences a coherent periodic perturbation of frequencies $v = K(v_i/d)$, $K = 1, 2, 3, \ldots$, where d is the distance between atoms in the row (Figure 5). When one of these frequencies coincides with $v_r = \Delta E_{ij}/h$, where ΔE_{ij} is the energy difference between states i and j of the ion, a resonant coherent excitation

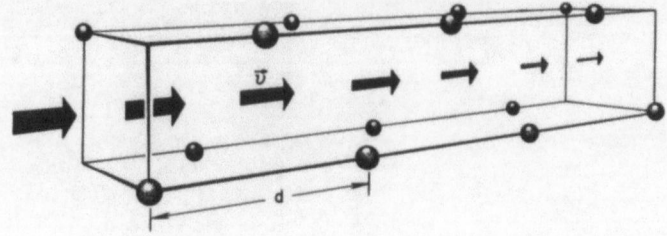

Figure 5. Scheme showing the influence of the symmetry of the perturbations on an ion penetrating in a ⟨100⟩ direction of a face-centered cubic lattice.

might occur. If the excitation does occur an enhanced ionization probability in the transmitted fixed-charge fraction should be observed.

In Figure 6 for incident N^{6+} we display the total population fraction of N^{6+} emerging from an 850-Å-thick Au crystal as a function of the incident energy.[9] The spacing along a ⟨111⟩ row is 7.064 Å. Resonances are seen for $n = 1$ to $n = 2$

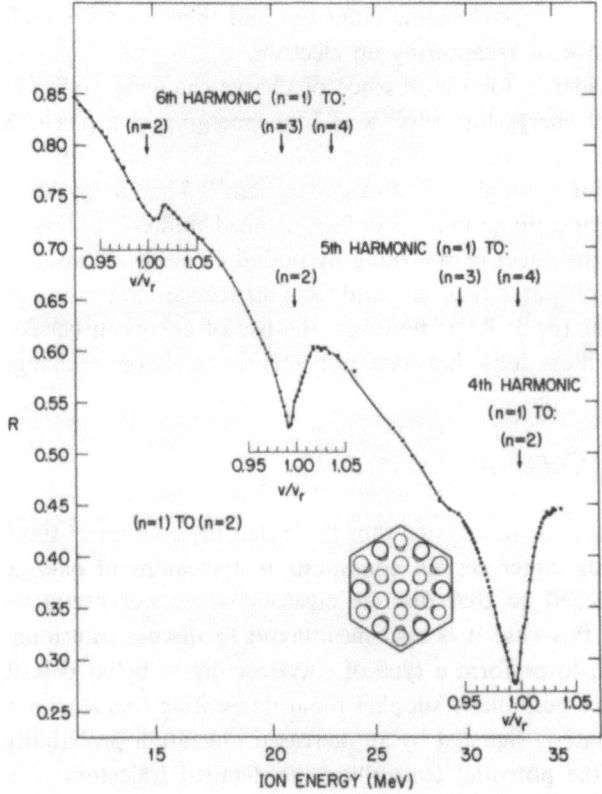

Figure 6. Ratio, R, of N^{6+} counts to the sum of N^{6+} and N^{7+} counts as a function of initial energy for ⟨111⟩ axial channeling in Au. The incident beam is N^{6+}. The additional scales are in terms of the resonant velocity calculated for the vacuum state of N^{6+}. For $(n = 1)$ to $(n = 2)$: 4th harmonic, $E_r = 33.09$ MeV; 5th harmonic, $E_r = 21.19$ MeV; and 6th harmonic, $E_r = 14.72$ MeV.

transitions with $K = 4, 5$, and 6, and for $n = 1$ to $n = 3$, $K = 5$. The additional scales shown are given in terms of v/v_r, where v_r is the resonant velocity calculated for the $n = 1$ to $n = 2$ vacuum states of the ion.

Although the total $(Z - 1)^+$ fraction is plotted, the energy-loss spectra for one-electron ions consist of two parts: (1) fixed-charge-state ions, which appear as a low-energy-loss peak, and (2) partially charge equilibrated ions, which have passed closer to atom rows and appear as a separable peak towards higher energy losses. By deconvolution we can show that the resonance affects *only* the fixed-charge peak. Hence, the actual dips are deeper than shown.

The Oak Ridge group has observed almost 50 $n = 1$ to $n = 2$ resonances (axial and planar) with hydrogenlike ions of $Z = 5$–9 in the energy range $1 \leq E \leq 3.5$ MeV/amu. For all the simpler features of the resonances, some generalizations can be made: (1) narrower channels give stronger resonances, (2) higher harmonic resonances are weaker, (3) there are slight shifts in the resonance peaks to lower (v/v_r), and (4) the peaks are asymmetric towards lower (v/v_r).

To understand these and other features let us concentrate on a single channel, the $\langle 100 \rangle$ axial channel. The configuration of this channel, i.e., alternating pairs of atoms acting on the ion, is illustrated in Figure 5. The electric potential in this axis can be shown to have the following form[9]:

$$V = \sum_{klm}^{even} V_{klm} \cos(2\pi kz/a) \cos(2\pi lx/a) \cos(2\pi my/a)$$
$$+ \sum_{klm}^{odd} V_{klm} \sin(2\pi kz/a) \sin(2\pi lx/a) \sin(2\pi my/a) \tag{3}$$

where the axis is centered on the direction of motion $\langle 100 \rangle$ and x and y are centered on the orthogonal $\langle 010 \rangle$ and $\langle 001 \rangle$ channels. The term a denotes the unit cell length; for the $\langle 100 \rangle$ axis, $a = d$.

Using a Molière potential to describe the Au atom, the Fourier components of the electric field can be calculated[10] and are shown for $K = 2, 3$, and 4 in Figure 7. The value at the centerline is for those ions which pass exactly down the center of the channel. The A and B directions denote the field experienced along paths that take the ion toward atomic rows and in a direction between atomic rows, respectively. The explanations of two of the features mentioned above are immediately obvious; higher harmonics have weaker Fourier components and trajectories constrained closer to atomic rows have stronger Fourier components.

More significantly a qualitative difference between even and odd harmonics is demonstrated. For centerline trajectories even harmonics only the parallel component is present, where in odd harmonics only the perpendicular component is present. Thus, for example, in the $1s$ to $2p$ transition only $1s \rightarrow 2p_z$ excitation is possible in even harmonics, and only $1s$ to $2p_{x,y}$ is possible in odd harmonics.

Other features appear if one considers injecting an ion at a slight angle to the centerline (i.e., the ion has transverse velocity in the B direction of Figure 7. Then in addition to the high frequency (v_i/d) a lower-frequency component corresponding

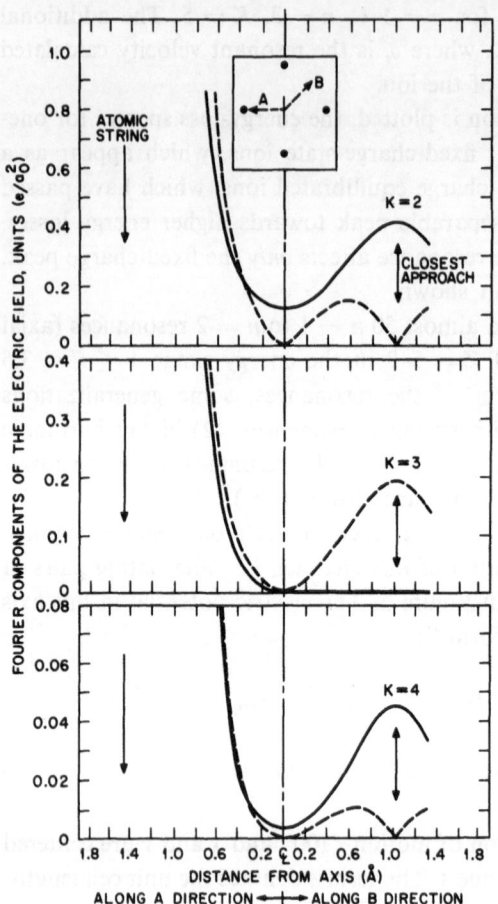

Figure 7. Fourier coefficients of the longitudinal and transverse components of the electric field for trajectories parallel to the $\langle 001 \rangle$ axis of Au vs. displacement from the channel center line; toward an atomic row (A) and between atomic rows (B). Solid line, $|E_\parallel|$; dashed line, $|E_\perp|$.

to the rate at which atomic rows are crossed is introduced and here again differences between odd and even harmonics enter.

For odd harmonics the field contains the frequencies $(v/a)(k \cos \theta \pm l \sin \theta)$; $l = 1, 3, 5, \ldots$. The $2p_x$ and $2p_y$ (degenerate) states are the strongest, and only the sideband frequencies $v_0 \pm v_0/K \tan \theta$ appear. The central frequency v_0 is missing because for odd K the field changes phase one cycle as the trajectory moves along direction B a distance a. For even harmonics the field contains frequencies (v/a) $\times (k \cos \theta \pm l \sin \theta)$; $l = 0, 2, 4, \ldots$. The central component is mainly in the z direction so that the central frequency v_0 is maintained.

The predictions of this model are amply borne out in Figures 8 and 9. Figure 8 shows an odd ($K = 3$) harmonic resonance for O^{7+} ions. For tilts in the B direction the central frequency disappears and only sidebands are seen. For example, a tilt of $0.7°$ causes an ion to step over a distance a once every 80 atoms of its passage in the Z direction (thus once per 240 cycles of the third harmonic) giving sidebands 0.4 above and below the frequency $v_0 = K(v/a)$ ($\pm 0.4\%$ in v/v_r). Figure 9 shows an equivalent tilting experiment but this time for an even ($K = 2$) harmonic for N^{6+} ions. The central feature is seen to be maintained, *but* note that the central feature here is a *doublet*.

At this point we have demonstrated the mechanics of the excitation process well enough. The central question now is what can be learned about the crystal field and the detailed state of the ion in its electronic milieu.

Two principal effects occur which perturb the energy levels of an ion moving in a crystal channel compared to its vacuum state. First is that the ion is contained in a potential gradient (crystal field) formed by bounding atomic rows or planes. Simply put, the potential becomes more positive moving away from the center of the channel so that larger orbits on the channeled ion are more weakly bound than inner ones compared to the vacuum state. The result is, e.g., a reduction in the $n = 1$ to $n = 2$ spacing and accounts for the general observation of lowered resonant velocities.

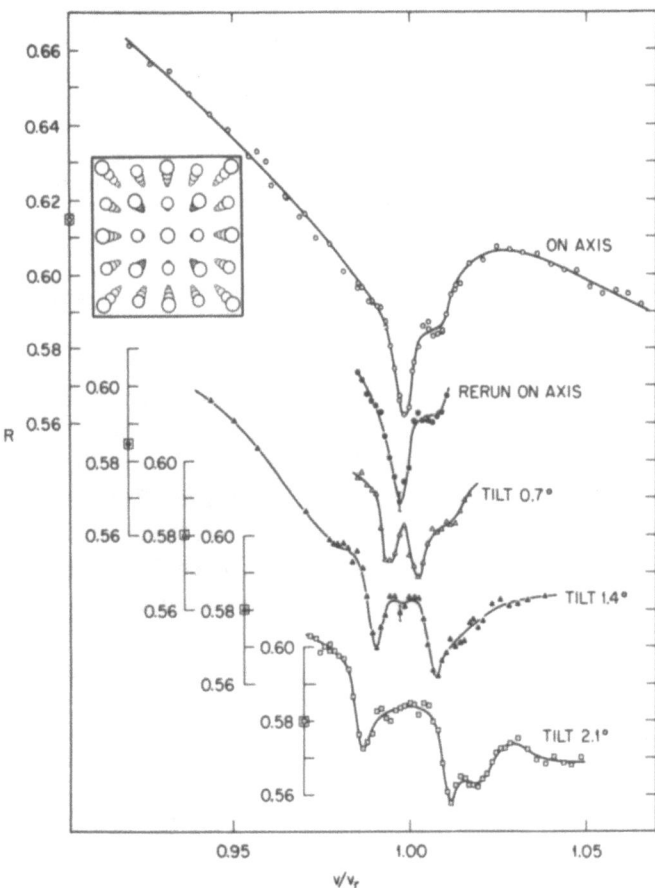

Figure 8. Effect of a small tilt in the (B) direction (into the {100} plane) upon an (odd) third harmonic resonance. O^{7+}, $\langle 100 \rangle$ axial channeling in Au. For ($n = 1$) to ($n = 2$) and the 3rd harmonic, $E_r = 38.22$ MeV.

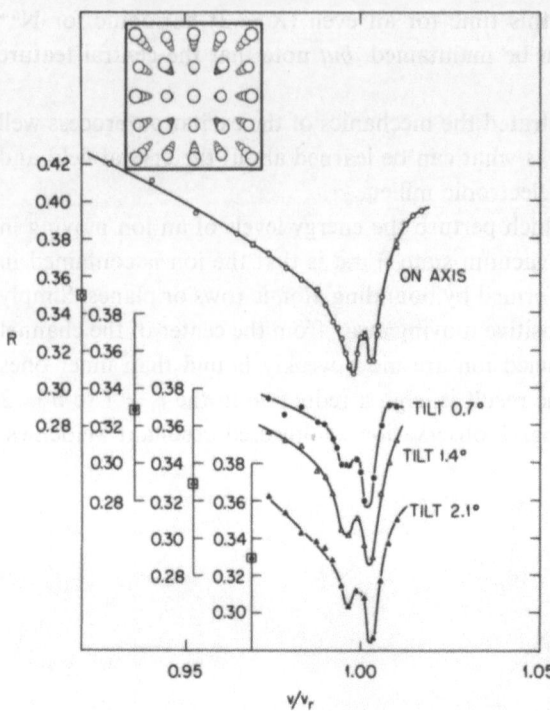

Figure 9. Effect of a small tilt in the (B) direction (into the {100} plane) upon an (even) second harmonic resonance. N^{6+}, $\langle 100 \rangle$ axial channeling in Au. For $(n = 1)$ to $(n = 2)$ and the 2nd harmonic, $E_r = 44.10$ MeV.

Ions moving on paths that are off the center of the channel experience sharper field gradients and probably account for the asymmetric broadening to lower velocities.

The second effect is caused by the presence of an electron "wake"; electrons scattered from the ion $(v_i \gg v_0)$ create an enhanced electron density wave which follows the ion; the distance to the first node in the wave being $\sim 2\pi\hbar v_i/\omega_p$, where ω_p is the plasmon frequency and the integrated enhanced charge density in this part of the wake is equal to $-q$, the charge on the moving ion. Thus the ion experiences a velocity-dependent dc field which can act to Stark mix the ionic states. Since the field acts always in the z direction the mixing occurs between the $2p_z$ and $2s$ states and the doublets observed in even harmonics are due to these splittings.

It follows that the separation between the two minima in Figure 9 is a measure of the wake field in the vicinity of the moving ion. If we assume a uniform field ξ, the first-order Stark splitting is $6e\xi a_0/Z$. The data of Figure 9 give a value of $0.10e^2/a_0$ for this splitting, from which we deduce $\xi = 0.12e/a_0^2$, which is of the expected order of magnitude.[11]

Crawford and Ritchie[12] have calculated the details of the anticipated effects of these perturbations using Hartree–Fock relativistic wave functions for the Au atoms with Wigner–Seitz boundary conditions. The result for 44.1-MeV N^{6+} ions in the $\langle 100 \rangle$ axial channel of Au is shown in Figure 10. Since the $2p_{x,y}$ orbitals extend toward the bounding atomic rows they are more strongly shifted by the static potential than the $2p_z$ or $2s$. The mixing of the $2s$ and $2p_z$ by the polarization from the wake

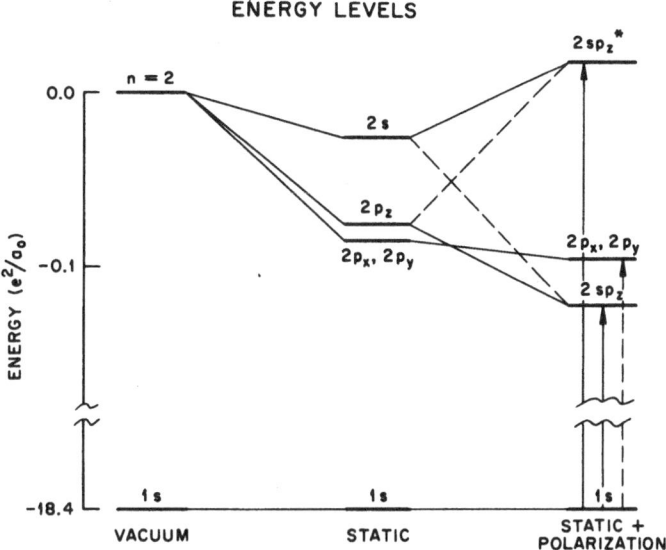

Figure 10. Effect of static crystal field and polarization (wake) field upon the $n = 2$ energy level of N^{6+} (see Reference 12). $\langle 001 \rangle$ axial channeling in Au; velocity $= 11.3v_0$.

field is indicated on the right side of the figure. Transitions exciting $2p_z$ (even) excite the top and bottom states while odd harmonics excite the central line. Details of these calculations will be given in a forthcoming paper.[12] An example of their results is shown in Figure 11, where the energy-level shifts for $2p_{x,y}$ excitation obtained from C^{5+}, N^{6+}, and O^{7+} spectra in axial $\langle 100 \rangle$ and $\langle 111 \rangle$ channels and slight planar tilts into (100) are compared with their theoretical expectations. In passing it is interesting to note that a full Hartree–Fock description of the potential was required to bring the $\langle 111 \rangle$ axial results into agreement with experiment.

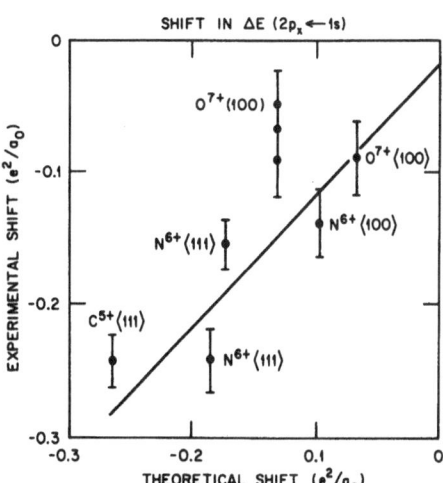

Figure 11. Comparison of calculated crystal field with values obtained from experimental shifts in resonant velocity.

To this point we have been discussing only axial channeling and slight tilts into the (100) plane (i.e., $\cos \theta \simeq 1$); in fact resonance coherent excitation (RCE) should occur everwhere in the plane and should be observable as long as collisional processes do not destroy the integrity of the initial and final states. Here, however, we need two indices to describe the harmonic $K(k, l)$ in question, i.e., in equation (3) even (Stark-split) harmonics should appear at any value of θ for k and l both even. Odd harmonics should appear at k and l both odd. [The field strengths should vary $\sim(k^2 + l^2)^{-1}$.] These expectations are demonstrated in Figure 12 for N^{6+} ions in the (100) plane of Au. The $\langle 100 \rangle$ axial channel is at $\theta = 0°$, the $\langle 110 \rangle$ at $\theta = 45°$, and other higher-order axial channels appear in between. The abscissa gives the energies in MeV/amu, where resonances should appear. Axial $\langle 100 \rangle$ resonances for $K = 2$ and 3 have already been shown (Figures 8 and 9), and the slight θ tilts for $K = 2$ (Figure 9) can now be identified as a (2, 0) resonance and for $K = 3$ (Figure 8) the "side bands" referred to above are really the (3, 1) and (3, −1) planar resonances.

In Figure 13 we show energy scans taken at $\theta = 6°$ and $\theta = 38°$ in the (100) plane for N^{6+}. The 6° tilt demonstrates the continued splitting of the (3, 1) and (3, −1) resonances indicated in Figure 8. The 38° tilt, however, introduces a Stark-split (2, 2) resonance into the picture (cf. Figure 12). Aside from a further demonstration of the generality of RCE, planar measurements provide other parameters which can be used better to determine the crystal and ionic parameters. The Stark

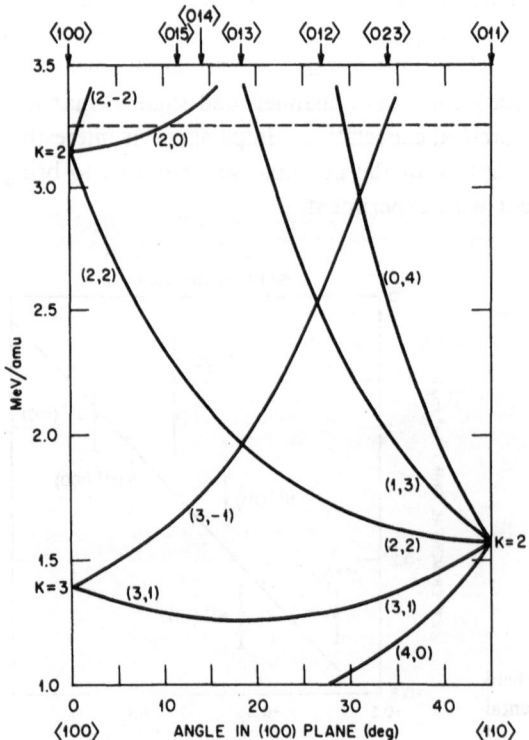

Figure 12. Calculated energy in MeV/amu for coherent $n = 1$ to $n = 2$ excitation in N^{6+} in the (100) plane of Au as a function of angle between the $\langle 100 \rangle$ and $\langle 110 \rangle$ axes.

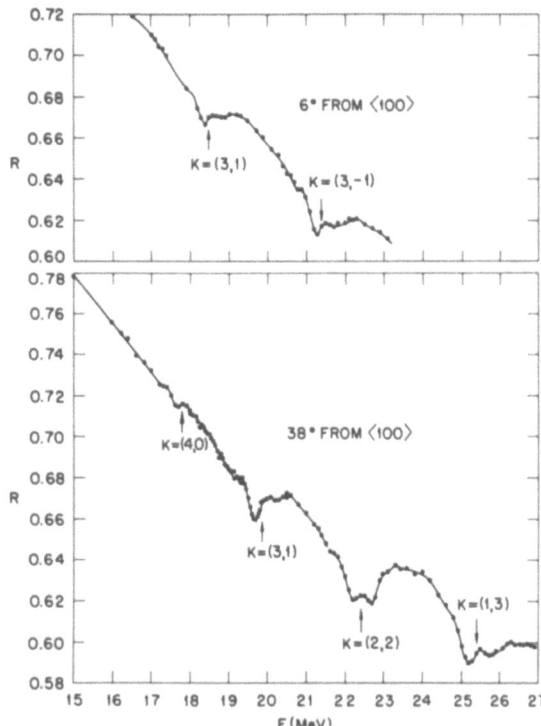

Figure 13. Demonstration of resonances for N^{6+} in the (100) plane. Note Stark-split (2, 2) resonance.

splitting of the line depends to a large extent upon the strength of the wake field at the ionic core which should be velocity dependent. Thus measuring the splitting of a (2, 2) resonance from, e.g., 3 MeV/amu at $\theta = 3°$ to 1.6 MeV/amu at $\theta = 38°$ provides a means for determining the velocity dependence in the same harmonic. Tilting into a plane also involves a change in the projected path length through the crystal so that changes in resonance depths for the same harmonic at different tilt angles can yield information on coherence lengths; e.g., in Figure 13 the (3, 1) harmonic appears at 1.35 MeV/amu at 3° and again at 38°. The resonance depth should also depend on interplanar spacing (i.e., stronger field components accompany narrower spacings) so that the strengths of transverse gradients can be tested. An example of this effect is shown in Figure 14 for the narrower (110) plane (see Figure 2) at a tilt of 10.3° from the $\langle 100 \rangle$ axis towards $\langle 111 \rangle$. The depth of the (3, 0) resonance is $\sim 20\%$ as compared with typical $\sim 4\%$ depths in the wider (100) planes.

Numerous spectra of type shown here have been recorded by the Oak Ridge group and are in the process of analysis and it is hoped that a highly detailed description of the state of the channeled ion and its interaction with the crystal fields will emerge.

Although no unambiguous results on the observation of radiation from resonantly excited states emerging from crystals have yet been reported, analysis of the results of our enhanced ionization experiments should lead to the proper design of experiments on the observation of postfoil radiation.

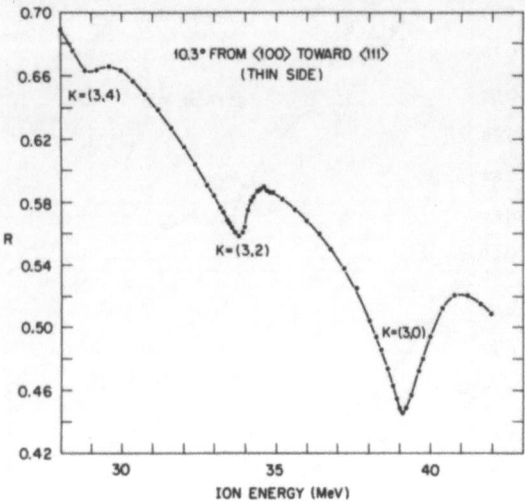

Figure 14. Resonances in the (110) plane of Au for O^{7+} ions.

ACKNOWLEDGMENT

The work reported here is based mostly upon the efforts of the Oak Ridge Accelerator Based Atomic Physics group over approximately the last 8 years. This productive group of co-workers is listed in the references and I am here representing them.

Research sponsored by the Division of Chemical Sciences, Office of Basic Energy Sciences, U. S. Department of Energy under contract No. W-7405-eng-26 with Union Carbide Corporation.

References

1. J. Lindhard, *Phys. Lett.* **12**, 126 (1964).
2. D. S. Gemmell, *Rev. Mod. Phys.* **46**, 129 (1974).
3. S. Datz, F. W. Martin, C. D. Moak, B. R. Appleton, and L. Bridwell, *Radiat. Eff.* **12**, 163 (1972).
4. S. Datz, C. D. Moak, and J. R. Barrett, unpublished results.
5. L. B. Golden and D. H. Sampson, *Astrophys. J.* **170**, 181 (1971); and *J. Phys. B* **10**, 2229 (1977).
6. S. Datz, *Nucl. Instrum. Methods* **132**, 7 (1976).
7. C. D. Moak, S. Datz, B. R. Appleton, J. A. Biggerstaff, M. D. Brown, H. F. Krause, and T. S. Noggle, *Phys. Rev. B* **10**, 2681 (1974).
8. S. Datz, G. Gomez del Campo, P. F. Dittner, P. D. Miller, and J. A. Biggerstaff, *Phys. Rev. Lett.* **38**, 1145 (1977).
9. S. Datz, C. D. Moak, O. H. Crawford, H. F. Krause, P. F. Dittner, J. Gomez del Campo, J. A. Biggerstaff, P. D. Miller, P. Hvelplund, and H. Knudsen, *Phys. Rev. Lett.* **40**, 843 (1978).
10. C. D. Moak, S. Datz, O. H. Crawford, H. F. Krause, P. F. Dittner, J. Gomez del Campo, J. A. Biggerstaff, P. D. Miller, P. Hvelplund, and H. Knudsen, *Phys. Rev. A* **19**, 977 (1979).
11. R. H. Ritchie, W. Brandt, and P. M. Echenique, *Phys. Rev. B* **14**, 4808 (1976).
12. O. H. Crawford and R. H. Ritchie *Phys. Rev. A*, in press.

31

Alignment, Orientation, and the Beam–Foil Interaction

R. M. Schectman, L. J. Curtis, and H. G. Berry

We present the results of a number of recent measurements of alignment and orientation for a variety of atomic systems. The significance of these results in understanding the ion–foil interaction process is discussed.

1. Introduction

The general aim of our study of the ion–foil interaction is twofold: (1) to provide as complete as possible a description of the state of the outgoing beam produced when ions are transmitted through thin foils and (2) to construct a physical model of the interaction process that can explain these results. In constructing such a model it is instructive to consider three distinct classes of interaction, one or all of which may contribute to the phenomena observed: (1) excitation by the bulk, (2) electron capture—both at or near the surface and of secondary electrons traveling with the emerging beam, and (3) interaction with the surface and with surface electric fields. In terms of these processes, one can attempt to assess the relative importance of bulk and surface interactions in determining the properties of the observed outgoing beam, as well as try to determine the relative importance of collision processes vis à vis electron capture. It is also of great importance to discover whether there are significant effects of surface electric fields and—if so—what the strength, range, and

R. M. Schectman and L. J. Curtis • The University of Toledo, Toledo, Ohio 43606, U.S.A.
H. G. Berry • University of Chicago, Chicago, Illinois 60637 and Argonne National Laboratory, Argonne, Illinois 60439, U.S.A.

time-dependent characteristics of these fields are. The results to be presented here furnish much descriptive information concerning the nature of the interaction, but not a complete model of the interaction process. They do, however, suggest an important role for surface effects, and are strongly suggestive of an important role in these processes for electron capture.

2. Phenomenology

The most complete description of the beam that emerges from the foil is contained in the specification of the density matrix of this system, and the experiments described here are designed to measure part of this density matrix. While recent work has shown that present experiments do not *require* the interaction process to be spin-independent,[1] all experiments are, in fact, compatible with such an assumption and —since theoretical arguments generally also lead to this assumption—it has been adopted in the analysis of our results, where the portion to the density matrix studied is presented in the $|LM_L\rangle$ representation. For states of $L \leq 1$, the optical measurements carried out determine the entire density-matrix block as, e.g., was presented in our earliest work describing the orientation produced by transmission of ions through tilted foils.[2] For larger L, field free measurements determine only combinations of density-matrix elements and it is convenient to carry out a spherical tensor expansion of ϱ, in terms of which the expansion coefficients ϱ_q^k with $k \leq 2$ are then uniquely determined by our experiments.[3] An equivalent parametrization of the outgoing beam which can provide a direct physical interpretation has been given by Fano and Macek,[4] who introduce the alignment (A) and orientation (O) parameters. There is a one-to-one correspondence between the alignment/orientation parameters and the ϱ_q^k's introduced earlier, so that measuring the alignment and orientation is equivalent to specifying the accessible part of the density matrix. A generalization of the approach of Fano and Macek to the case of mixed parity coherences and radiation emitted in the presence of electromagnetic field has been carried out by Gabrielse,[5] and is particularly useful in describing hydrogenic systems.

3. Experiments

All experiments to be described here involve detection of radiation emitted by the beam subsequent to traversing the foil. In some cases, quantum beats were measured; in other cases, the detailed polarization state of the emitted light (specified by the three relative Stokes parameters M/I, C/I, and S/I) was determined—sometimes as a function of the azimuthal angle of observation, ϕ. In all cases, determination of the density matrix describing the emergent beam was the aim of the measurements.

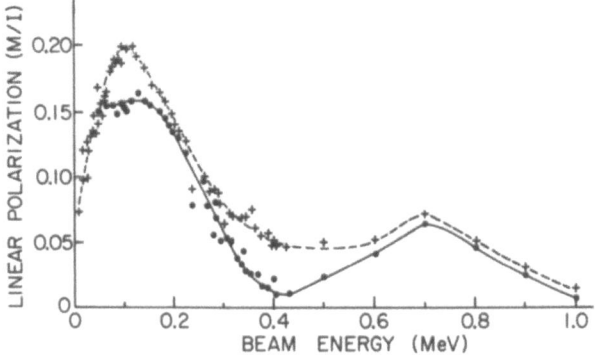

Figure 1. The linear polarization fraction M/I for the $3p\,^1P$ level of He I as a function of energy. $+$, current density 30 μA/cm^2; $\bullet$, zero current density extrapolation. For this case, $A_0^c = -\frac{2}{3}M/I$.

4. Results for the $3p\,^1P$ and $4d\,^1D$ Levels of He I

4.1. Foils Perpendicular to the Incident and Outgoing Beams

In this case, only a single ϱ_q^k, ϱ_0^2 (proportional to a single relative Stokes parameter, M/I) is nonvanishing, and Figures 1 and 2 show the variation of this parameter with energy for the two states studied. Note that ϱ_0^2 is always positive and that, in both cases, it oscillates with energy. A noteworthy aspect of Figures 1 and 2 is the beam current density dependence of the alignment,[6] which occurs in both cases, and itself oscillates with energy as shown in Figure 3.

4.2. Tilted Foils

Here, field free measurements can determine the four ϱ_q^k's with $k \leq 2$ (i.e., the four Fano–Macek parameters). Measurements at one detection position (θ, ϕ)

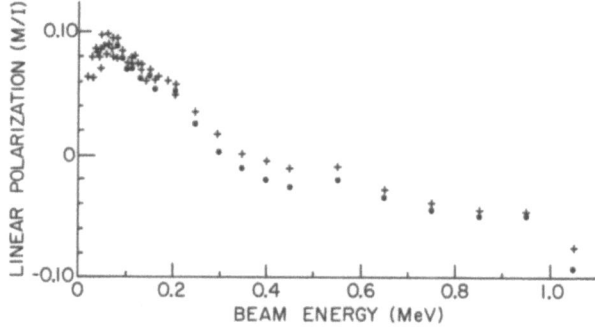

Figure 2. The linear polarization fraction M/I for the $4d\,^1D$ level of He I as a function of energy. $+$ and $\bullet$ as in Figure 1.

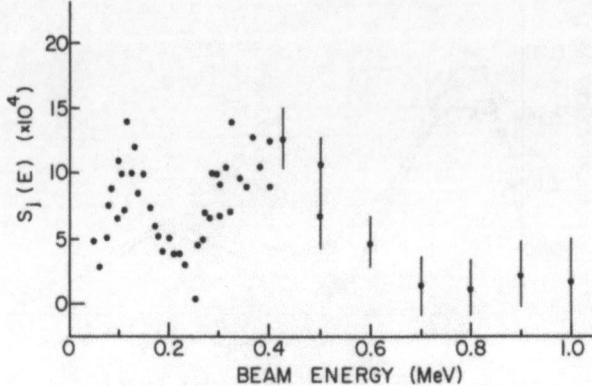

Figure 3. The rate of current density dependence of the linear polarization $S_j = \Delta(M/I)/\Delta(j)$ for the $3p\,^1P$ level of He I as a function of energy.

provide three relative Stokes parameters. For $\theta = \pi/2$, $\phi = 0$, these have been measured between 0° and 60° in 5° increments over the entire energy range 30–1000 keV for both the $2s\,^1S$–$3p\,^1P$ transition at 4016 Å and the $3p\,^1P$–$4d\,^1D$ transition at 4922 Å. The results for the latter transition for a tilt angle $\alpha = 45°$ are shown in Figure 4. From these measurements, the alignment and orientation parameters

$$A_1^c \sim C/I$$

and

$$O_1^c \sim S/I \qquad (1)$$

are directly determined; however, only the combination

$$(A_0^c + A_2^c \cos \phi) \sim M/I$$

is obtained. We have therefore carried out a number of measurements of M/I vs. ϕ

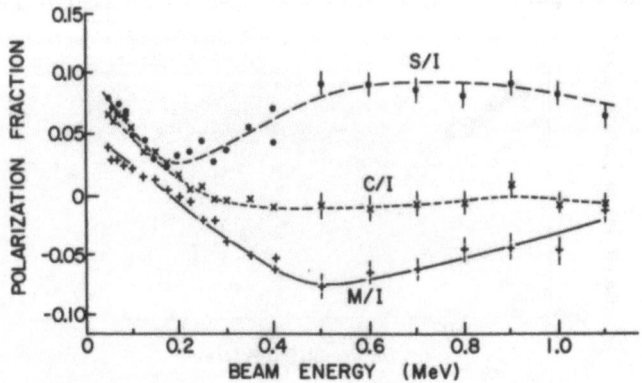

Figure 4. Relative Stokes parameters M/I (+), C/I (×), and S/I (●) for the $4d\,^1D$ level as a function of energy.

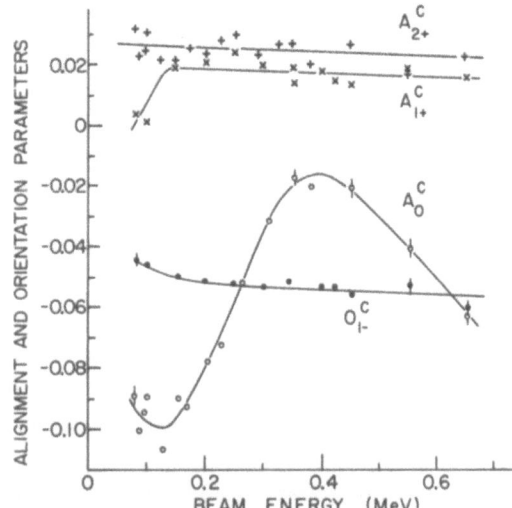

Figure 5. Alignment and orientation parameters for the $3p\,^1P$ level of He I vs. energy: A_2^c (+), A_1^c (×), A_0^c (○), and O_1^c (●).

for the 5016-Å transition, with the results for $\alpha = 45°$ shown in Figure 5. Similar measurements for the 4922-Å transition are in progress.

Comparison of Figures 1 and 5 shows that ϱ_0^2 is essentially unchanged by rotating the foil through 45°; other measurements suggest that the angular dependence of the other ϱ_q^k's is also energy independent. It thus seems likely that, to a good approximation, one can write

$$\varrho_q^k(E, \alpha) = g_q^k(E)f_q^k(\alpha) \tag{2}$$

This is well illustrated, for example, in Figure 6, where all of the measured values of ϱ_1^2 for the $3d\,^1D$ level, measured between 100 and 425 keV, are plotted as a function of the foil tilt angle after factoring out the energy dependence measured for a tilt angle of $\alpha = 45°$ (data for $3p\,^1P$ corresponding to Figure 4). These results agree very well with a single universal—here linear—curve representing the observed

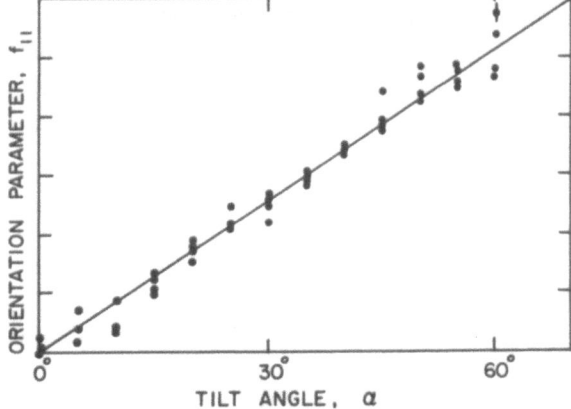

Figure 6. Angular dependence of the orientation, $f_1^1(\alpha)$ for the $3p\,^1P$ level of He I.

angular variation. For all cases measured to date, such an approximation seems valid and the resulting $f_q^k(\alpha)$ are as follows: f_0^2: constant for $3p\ ^1P$; f_1^2: linear for both $3p\ ^1P$ and $4d\ ^1D$; f_2^2: quadratic for $3p\ ^1P$; f_1^1: linear for $3p\ ^1P$, quadratic for $4d\ ^1D$.

5. Interpretation

One feature of the excitation by foils normal to the beam displayed in Figures 1 and 2 is that M/I is everywhere positive (A_0^c everywhere negative). It should be noted that this is, indeed, the sign expected from electron pickup in the simple model that the ion emerges from the foil and captures an electron whose velocity relative to the foil is small compared with that of the ion itself.[7] If one next turns one's attention to the observed oscillations in A_0^c with outgoing ion velocity (energy), it is tempting to try to relate them to the oscillatory electron wake which is set up by the ion's traversal through the foil.[8]† for a plasma frequency $\omega_p \sim 10^{15}\ \mathrm{sec}^{-1}$, the assumption of electron pickup from an oscillating charge density extending some few angstroms beyond the foil can give a reasonable fit to the experimental data. Scattering from an oscillatory potential of similar characteristics also would give rise to such oscillation in A_0^c.

The observation for the $3p\ ^1P$ that A_0^c does not change significantly when the foil is tilted is also consistent with the simple electron pickup model described earlier[7] where the direction of the principal axis for the alignment is determined by the beam velocity. It is also expected if the alignment is produced in the bulk. The variation of the three alignment parameters with foil tilt angle is *not* what would result from alignment produced parallel to the tilted foil normal. Since capture of secondary electrons has been suggested above as a significant contributor to our observations, it is interesting to observe that measurements of the dependences of the yield of such electrons upon foil tilt angle[10] is proportional to $1/\cos \alpha$, due to an increase with tilt angle in the number of electrons that can reach the final surface without absorption. This same mechanism requires that the secondary electron density is asymmetric about the incident beam in exactly the way required to produce orientation of the sense observed in all measurements carried out to date.

Finally, we note that the lack of oscillations with energy in measurements of the orientation suggest that the mechanism for producing it may be different from that producing the alignment.

ACKNOWLEDGMENTS

A number of our colleagues have greatly assisted us in carrying out this program of research. In particular, we wish to acknowledge the contributions of S. T. Chen,

---

† See also Reference 9 for an experimental verification of the effects of this potential in a beam–foil experiment.

D. G. Ellis, G. Gabrielse, T. Gay, R. D. Hight, S. Huldt, and A. E. Livingston. This work was supported in part by the U.S. National Science Foundation and U.S. Department of Energy.

References

1. D. G. Ellis, *J. Phys. B* **10**, 2301 (1977).
2. H. G. Berry, L. J. Curtis, D. G. Ellis, and R. M. Schectman, *Phys. Rev. Lett.* **32**, 751 (1974).
3. D. G. Ellis, *J. Opt. Soc. Am.* **63**, 1322 (1973).
4. U. Fano and J. H. Macek, *Rev. Mod. Phys.* **45**, 553 (1973).
5. G. Gabrielse, Ph.D. thesis, University of Chicago (1978).
6. R. D. Hight, R. M. Schectman, H. G. Berry, G. Gabrielse, and T. Gay, *Phys. Rev. A* **16**, 1805 (1977); also T. J. Gay and H. G. Berry, *Phys. Rev. A* **19**, 952 (1979).
7. D. G. Ellis, *Proceedings of the Fifth International Conference on Beam Foil Spectroscopy*, Lyon, France (1978).
8. J. Neufield and R. H. Ritchie, *Phys. Rev.* **98**, 1932 (1955).
9. D. S. Gemmell, J. Remillieux, J. C. Poizat, M. J. Gaillard, R. E. Holland, and Z. Vager, *Phys. Rev. Lett.* **34**, 1420 (1975).
10. J. Schader, B. Kolb, K. D. Sevier, and K. O. Groeneveld, *Nucl. Instrum. Methods* **151**, 563 (1978).

32

Polarized-Photon, Scattered-Ion Coincidence Study of Mg⁺ Collisions with He, Ne, and Ar

N. ANDERSEN, T. ANDERSEN, C. L. COCKE, AND E. HORSDAL PEDERSEN

The impact-parameter dependence of the Mg II (3s ← 3p) emission has been studied for 3–15 keV Mg⁺–He, Mg⁺–Ne, Mg⁺–Ar collisions by means of photon, scattered-particle coincidence measurements, including polarization analysis. The resonance excitation can be described by means of two mechanisms: At low collision energies, the Mg II (3s → 3p) excitation occurs at molecular curve crossings, whereas direct excitation dominates at higher energies. The direct mechanism is essentially the same for all three collision systems and exhibits a high degree of coherence. The quasimolecular mechanism to a large extent depends on the asymmetry of the collision. The asymmetric Mg⁺–He, Mg⁺–Ar systems show substantial coherent excitation, whereas the resonance level is noncoherently excited in the quasisymmetric Mg⁺–Ne collision.

1. Introduction

Until about 1974, the study of outer-shell excitation in atomic collisions in the lower kiloelectron volt region concentrated on the closed-shell systems represented by the rare-gas or the alkali-ion–rare-gas systems. Collision systems characterized by the presence of an isolated electron outside two closed shells, the so-called quasi-

N. ANDERSEN, T. ANDERSEN, C. L. COCKE, AND E. HORSDAL PEDERSEN • Institute of Physics, University of Aarhus, DK-8000 Aarhus C, Denmark.   Dr. N. Andersen's permanent address is Physics Laboratory II, H. C. Ørsted Institute, DK-2100 Copenhagen Ø, Denmark.   Dr. C. L. Cocke's permanent address: Department of Physics, Kansas State University, Manhattan, Kansas 66506, U.S.A.

one-electron systems, were studied only in specific cases such as the H–rare-gas or He$^+$–rare-gas or the alkali-ion–alkali-atom systems which, in many respects, exhibit special characteristics. During the last few years, the quasi-one-electron systems have been the object of intensive experimental[1–5] and theoretical[6,7] investigations with particular emphasis on the alkali-atom(or alkalilike-ion)–rare-gas systems in the energy region from threshold and up to 100 keV.

So far, most of the experimental work dealing with the quasi-one-electron systems has been concentrated on measurements of the total cross sections by means of optical spectroscopy. These studies have revealed that two types of inelastic processes dominate, (i) excitation of the projectile valence electron from the ground state to the resonance state; (ii) one- and two-electron excitation of the target atom in the quasi-symmetric systems such as Na–Ne, Mg$^+$–Ne, or K–Ar.

The projectile excitation is an important process in all the systems investigated, whereas target excitation plays a significant role only in the quasisymmetric systems. In addition, the light emitted from the projectile resonance levels has been found to be polarized. This polarization is strongly dependent on the impact energy with an energy dependence more or less common for all the systems.[1,6]

The quasi-one-electron systems seem sufficiently simple to justify a more detailed study of the inelastic-collision processes with the emphasis on the projectile resonance-line excitation. Such an investigation should yield information about impact-parameter-dependent excitation probabilities, polarization, and coherence effects for the projectile-resonance level. Since the resonance level for alkali atoms and alkalilike ions of interest is situated only a few electron volts above the ground state, it is necessary to use coincidence technique for this study. We have chosen the ^{24}Mg$^+$ ion as a convenient projectile because in this case the polarization will not be affected by hyperfine-structure effects.

2. Total- and Differential-Emission Cross Sections

For the Mg$^+$–He, Mg$^+$–Ne, Mg$^+$–Ar systems, the Mg II ($3s \leftarrow 3p$) emission cross sections exhibit two maxima, a minor one near 5 keV and a dominant one near 50 keV.[1] These two maxima indicate that the resonance excitation may take place via two mechanisms[2,6,7]:

(i) At energies below 10 keV, the excitation of the Mg$^+$ $3s$ valence electron occurs at the molecular-curve crossings appearing when the electron shells of the two cores interpenetrate. In this mechanism, the rare-gas electrons will play an active role during the formation and breakup of the quasimolecule even though these electrons end up in excited states only for quasisymmetric collision systems. This mechanism predicts excitation probabilities with pronounced threshold values, reflecting the beginning of the interpenetration of the two colliding atoms, and a sharp rise in the excitation probabilities close to threshold values as a result of the core-interpenetration process.

(ii) At energies above 10 keV, the Mg II $(3s \rightarrow 3p)$ excitation predominantly occurs due to the direct electrostatic interaction between the Mg⁺ $3s$ electron and the rare-gas atom. This excitation will reach a maximum in the region of velocities that satisfies the Massey criterion $\Delta E \cdot a/\hbar v = \pi$. Here, ΔE is the energy defect of the excitation process and a is the effective collision length, approximately equal to the size of the orbital for the valence electron. Since the Mg⁺ $3s$ electron will extend over several atomic units, this mechanism is expected to be operative at larger internuclear distances than mechanism (i). Hence, the target electrons are expected to play a rather passive role in the direct excitation mechanism. The delocalized character of this excitation implies that the excitation probabilities should exhibit a slowly varying dependence on impact parameter without definite threshold values.

The two mechanisms are thus expected to exhibit very different impact-parameter dependences of the Mg II $(3s \rightarrow 3p)$ excitation probability, and a direct test can be performed by measuring the angular dependence of the excitation probability. In the present case, coincidence measurements between scattered magnesium ions and Mg II $(3s \leftarrow 3p)$ photons emitted perpendicular to the scattering plane were carried out.

Figure 1 shows an example from the Mg⁺–Ne system of the experimentally determined emission probability P versus the reduced scattering angle τ, which is equal to the product of the beam energy E and the scattering angle θ. The total-emission cross section versus beam energy is also shown, and the arrows indicate the energies at which the differential measurements are performed. The results obtained at lower energies exhibit the properties demanded by mechanism (i), a threshold value followed by a sharp increase in emission probability. At higher energies, a significant tail appears on the emission-probability curve at small τ values. The emission probability is varying slowly over a large impact-parameter range without a well-defined threshold. These features are indicative of mechanism (ii).

For a comparison of the three collision systems, the emission probability versus

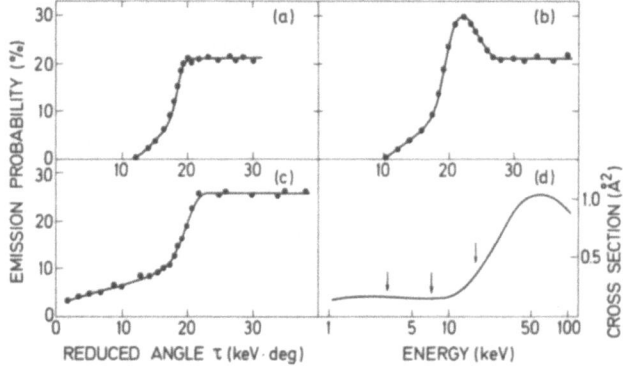

Figure 1. Mg II $(3s \leftarrow 3p)$ emission probability versus reduced scattering angle τ for (a) 3-, (b) 7-, and (c) 15-keV Mg⁺–Ne collisions. The Mg II $(3s \leftarrow 3p)$ total-emission cross section versus impact energy is shown in (d) for Mg⁺–Ne.

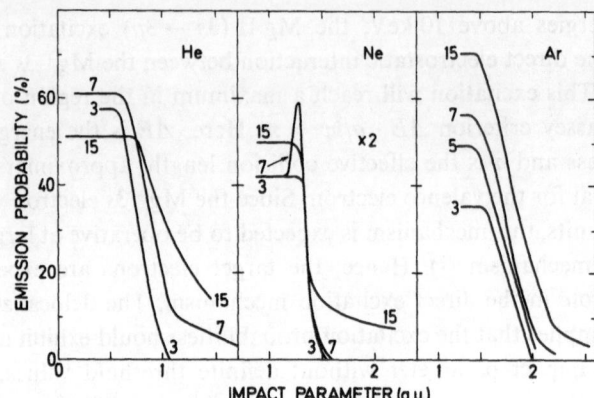

Figure 2. Mg II ($3s \leftarrow 3p$) emission probabilities versus impact parameter (*b*) for Mg⁺–He, Mg⁺–Ne, Mg⁺–Ar collisions. Impact energies are indicated on the curves. Experimental points are omitted for clarity.

impact parameter are shown in Figure 2. The impact parameter *b* was evaluated from the scattering angles by using a Born–Mayer-type scattering potential $V(R) = A \cdot \exp(-\alpha R)$, the parameters of which are calculated from the scaling laws for closed-shell interactions.[8] Figure 2 presents a comparison between the emission probability of Mg II ($3s \leftarrow 3p$) as a function of the impact parameter *b* for the Mg⁺–He, Mg⁺–Ne, Mg⁺–Ar systems. At low energies (~3 keV), the well-localized impact-parameter threshold for excitation shows that only mechanism (i) is operative here. The threshold is followed by a sharp rise in *P* towards smaller *b* values. For the quasisymmetric Mg⁺–Ne system, the emission probability goes up much faster than seen for the asymmetric Mg⁺–He, Ar systems. At 15 keV, the emission probability gradually appears at larger impact parameters, showing the onset of mechanism (ii).

3. Comparison with Theory

Theoretical calculations of the impact-parameter dependence of the two mechanisms have been made by the collision group at Orsay[4,7,9] and by Nielsen and Dahler.[6] Mechanism (i) has been treated for the quasisymmetric Na–Ne, Mg⁺–Ne systems[4,9] by means of *ab initio* molecular calculations. As the inelastic processes in the Mg⁺–Ne system are caused by the $4f\sigma$ promotion, the "frozen-orbital method" is appropriate for the *ab initio* handling of the diabatic MO crossings for this system. The threshold value observed near $R_c = 1.3$ a.u. is in reasonable agreement with *ab initio* calculations.[4] A more detailed comparison has been made with the isoelectronic Na–Ne system[9] for which the experimental and theoretical results are in good agreement.

Mechanism (ii) has been treated in a perturbed atomic description.[6,9] Model potential methods[6] have been successful for a number of systems such as Li–He[10] and Na–He.[11] The qualitative agreement between theory and experiment is good

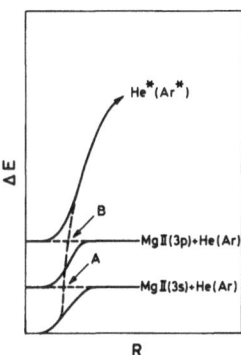 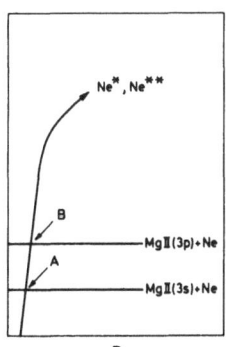

Figure 3. Schematic potential curves for the Mg II $(3s \rightarrow 3p)$ excitation in Mg⁺–He (Ar), and Mg⁺–Ne collisions.

and supports the correctness of the proposed mechanisms even though full quantitative agreement is still lacking.

The direct-excitation mechanism (ii) is essentially the same for all rare-gas targets, as also demonstrated by the common location of the high-energy maximum in the total cross sections of the Mg II $(3s \leftarrow 3p)$ emission. In contrast to this, the quasi-molecular mechanism (i) will depend strongly on the asymmetry of the system. The interaction at the potential curve intersections (Figure 3) will increase, changing from Mg⁺–Ne (quasisymmetric) to Mg⁺–He(Ar) (asymmetric), leading to more adiabatic behavior at the crossings. The consequences will be an overall reduction of inelastic processes at the primary crossing A, and a strongly reduced rare-gas excitation at intersection B.

4. Polarization Analysis of the Mg II (3s → 3p) Excitation

The identification of the two excitation mechanisms responsible for the Mg II $(3s \leftarrow 3p)$ emission is the necessary background for an investigation of the emission polarizations and coherence effects. The experimental investigation has been carried out by means of polarized-photon, scattered-ion coincidence measurements. A detailed description of this experiment will be given elsewhere. The experimental geometry is shown in Figure 4. The incident beam is directed along the Z axis, the particles are scattered and detected at an angle θ with respect to Z, whereas the photons emitted are detected and their polarization analyzed along the positive Y direction. The axes ξ and η of the detector frame are directed parallel to the Z and X axis, respectively, while the third axis ζ is parallel to the Y axis. The Stokes parameters measured are defined as follows:

$$IP_1 = I(0°) - I(90°)$$
$$IP_2 = I(45°) - I(135°)$$
$$IP_3 = I(RHC) - I(LHC)$$

I is the total intensity, $I(0°)$, $I(90°)$, $I(45°)$, and $I(135°)$ are intensities measured with

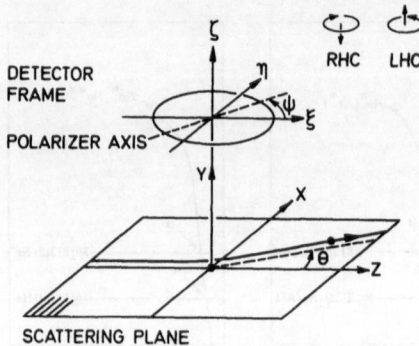

Figure 4. Schematic diagram of the geometry of the scattering events. The axis of the polarizer can be rotated to determine the polarization ellipse. Circular polarization can be measured by insertion of a $\lambda/4$ plate. The orientation of the angular momentum of right- and left-hand circularly polarized photons propagating in the Y direction is indicated.

a linear polarizer set at the angles indicated (Figure 4). I(RHC) and I(LHC) are the right- and left-hand circular polarized components of the photon radiation. The coherency of the photons may be characterized by the degree of polarization,

$$P = (P_1{}^2 + P_2{}^2 + P_3{}^2)^{1/2} \leq 1$$

with $P = 1$ if the light is emitted coherently. In the present case, with a ^2S–^2P transition, it is convenient to correct the measured P_1 and P_2 values for the depolarizing influence of the fine structure, which develops after the collision is completed. P is then given by

$$P = (\tfrac{7}{3}P_1{}^2 + \tfrac{7}{3}P_2{}^2 + P_3{}^2)^{1/2}$$

which is equal to unity if the excitation process is coherent.

The experiments were conducted at 15 keV projectile energy to allow an investigation of both mechanisms. Our results are summarized in Figure 5. There is a clear distinction between the total polarization P obtained for Mg$^+$–He(Ar), and Mg$^+$–Ne collisions. The polarization is large, $0.7 \leq P \leq 1$ for all three systems and impact

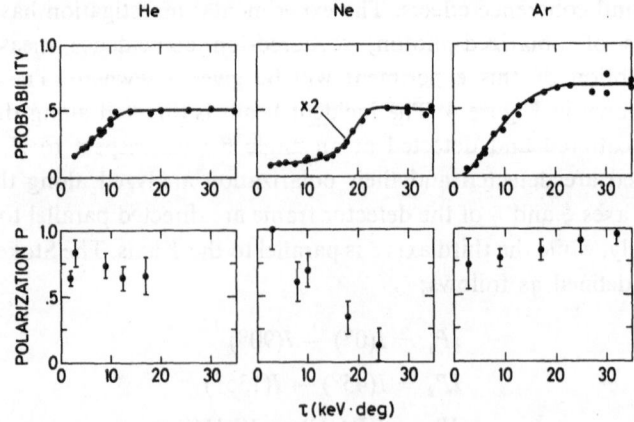

Figure 5. Mg II ($3s \leftarrow 3p$) emission probability and degree of polarization P for 15-keV Mg$^+$–He, Mg$^+$–Ne, Mg$^+$–Ar collisions as function of reduced scattering angle $\tau = E\theta$.

parameters, except for the Mg^+–Ne system, when τ is above 15 keV · deg. The deviation from unity of P can be attributed to cascades and contributions from the Mg II $(3p \leftarrow 3d)$ transition, which is within the wavelength region selected by our interference filter ($\sim$100 Å). The present data indicate that the resonant excitation is highly coherent when the direct mechanism (ii) is operative (small τ values). If the quasimolecular mechanism (i) dominates the excitation (large τ values), the resonant-excitation process is still highly coherent for the asymmetric systems, whereas the excitation is incoherent for the quasisymmetric Mg^+–Ne system. Inside the molecular crossings in the Mg^+–Ne system, many reaction channels, including both one- and two-electron excitation processes, may lead to the $3s \rightarrow 3p$ excitation. This causes a breakdown of the simple one-electron description and accounts for the lack of coherence. Thus the asymmetric systems may be treated correctly as quasi-one-electron systems over the whole impact-parameter range, while rather a 21-electron description is needed for the Mg^+–Ne system in the low-impact-parameter region to account for the resonant excitation.

The orientation of the Mg II 3^2P level can be obtained directly from the circular-polarization data. In all cases, the light is preferentially right-hand circularly polarized. The circular polarization reaches values as high as 0.7 at small τ values. This effect seems intuitively reasonable from Figure 4. Theoretical considerations[11] based upon mechanism (ii) are in agreement with the obtained circular polarization.

5. Conclusion

The present study has revealed that the quasi-one-electron systems can be understood on the basis of the two discussed excitation mechanisms. It seems appropriate to extend these studies to the lighter systems such as Be^+–He, Be^+–Ne[12] to obtain a unified picture of the simple collision systems. Total and differential cross-section measurements suggest that these systems may be relatively simple to describe theoretically. In contrast, total cross-section and polarization measurements for heavier collision systems seem to indicate that more complicated descriptions will be needed to account for the excitation processes in these systems.

References

1. N. Andersen, T. Andersen, and K. Jensen, *J. Phys. B* **9**, 1373 (1976); J. Ø. Olsen and T. Andersen, *J. Phys. B* **8**, L421 (1975); J. Ø. Olsen, N. Andersen, and T. Andersen, *J. Phys. B* **10**, 1723 (1977).
2. W. Mecklenbrauck, J. Schön, E. Speller, and V. Kempter, *J. Phys. B* **10**, 3271 (1977); H. Alber, V. Kempter, and W. Mecklenbrauck, *J. Phys. B* **8**, 913 (1975); V. Kempter, *Proceedings of the 9th International Conference on the Physics of Electronic and Atomic Collisions*, 1975, University of Washington Press, Seattle, Invited Papers, pp. 327–344, and references therein; L. Zehnle, E. Clemens, P. J. Martin, W. Schäuble, and V. Kempter, *J. Phys. B* **11**, 2133 (1978), and references therein.

3. R. Düren, M. Kick, and H. Pauly, *Chem. Phys. Lett.* **27**, 118 (1974).
4. J. Fayeton, N. Andersen, and M. Barat, *J. Phys. B* **9**, L149 (1976); J. Fayeton, thesis, Orsay (1977).
5. V. L. Ovchinnikov, O. B. Shpenik, and I. P. Zapesochnyi, *Sov. Phys. Tech. Phys.* **19**, 382 (1974); O. B. Shpenik, V. L. Ovchinnikov, and I. P. Zapesochnyi, *Ukr. Phys. J.* **19**, 1933 (1974).
6. S. E. Nielsen and J. S. Dahler, *J. Phys. B* **9**, 1383 (1976); *Phys. Rev. A* **16**, 563 (1977); J. Manique, S. E. Nielsen, and J. S. Dahler, *J. Phys. B* **10**, 1703 (1977).
7. C. Gaussorgues, V. Sidis, and M. Barat, *Proceedings of the 10th International Conference on the Physics of Electronic and Atomic Collisions*, Abstracts, pp. 706, Commissariat à l'Energie Atomique, Paris (1977).
8. N. A. Sondergaard and E. A. Mason, *J. Chem. Phys.* **62**, 1299 (1975).
9. E. Horsdal Pedersen, P. Wahnón, C. Gaussorgues, N. Andersen, T. Andersen, K. Bahr, M. Barat, C. L. Cocke, J. Ø. Olsen, J. Pommier, and V. Sidis, *J. Phys. B* **11**, L317 (1978).
10. S. E. Nielsen, N. Andersen, T. Andersen, J. Ø. Olsen, and J. S. Dahler, *J. Phys. B* **11**, 3187 (1978).
11. S. E. Nielsen, private communication (1978).
12. J. Ø. Olsen, K. Vedel, and P. Dahl, *J. Phys. B* **12**, 929 (1979).

33

Coherence Effects
in Heavy-Ion–Atom Collisions

C. BOTTCHER

A new approach to charge capture and ionization by highly stripped projectiles is described and shown to explain cross-section systematics through the Periodic Table. Oscillations in cross section with respect to charge state observed around atomic number 70 are explained as an f-wave resonance in the target-electron–projectile scattering. The ratio of H_2 to H cross sections for both light and heavy projectiles is shown to fit a two-center coherent scattering model; independent scattering by the two centers is not a good assumption for velocities below 4 a.u. Similar coherence effects are predicted in stripping by molecular gases even in multielectron processes where the independent atom model might be thought valid. Recent experiments on the forward peak of electrons ejected from the projectile show interesting structure which can be partly explained without invoking interference effects.

1. Introduction

We shall survey a number of problems in the field of heavy-ion–atom (or molecule) collisions where coherence or interference between quantal scattering amplitudes plays a significant role. In Section 2 we show how interference between short-range and long-range amplitudes explains anomalies in the variation of total cross sections with charge state. The ratio between capture from atomic and molecular targets is discussed in Section 3, and it is shown that a wide range of data is fitted by a two-center scattering model. A similar model also predicts significant effects in stripping. In Section 4 we examine the forward peak in projectile ionization which is potentially a rich source of interference effects, though none have yet been clearly identified. Atomic units are used throughout.

C. BOTTCHER ● Oak Ridge National Laboratory, Oak Ridge, Tennessee 37830, U.S.A.

2. Systematics of Charge Capture by Highly Stripped Ions

We suppose that capture from a hydrogen atom by a highly stripped projectile at intermediate velocities $(1 < v < 3)$

$$A^{+q} + H \rightarrow A^{+(q-1)} + H^+ \tag{1}$$

takes place in two stages. In the first, an amount of energy Δ is transferred to the bound electron to remove it from the field of H^+. Then the electron moves to the saddle point between H and A, whence it may be captured by the heavy ion or escape depending on its acquired kinetic energy. The capture cross section

$$\sigma_C = \sigma_L \exp(-2v/q^{1/2}) \tag{2}$$

where σ_L is the binary encounter loss cross section, related to the $e + A$ elastic scattering amplitude f by

$$\sigma_L = \int_{\theta_0}^{\pi} |f|^2 \sin\theta \, d\theta, \qquad \cos\theta_0 = 1 - \Delta/v^2 \tag{3}$$

The second factor in (2) is calculated from classical transition state theory. If only

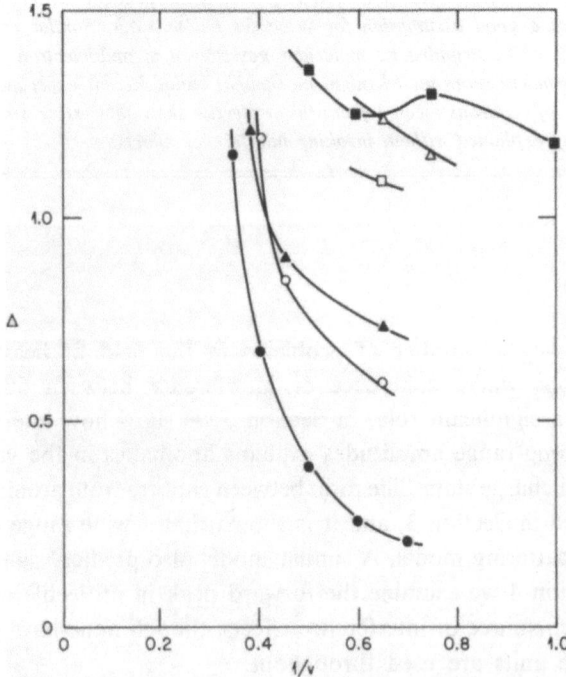

Figure 1. Variation of Δ defined in (4) with $1/v$ for different projectiles: ●, Si; ○, Fe; ▲, Mo; ■, Ta; △, W; □, Au.

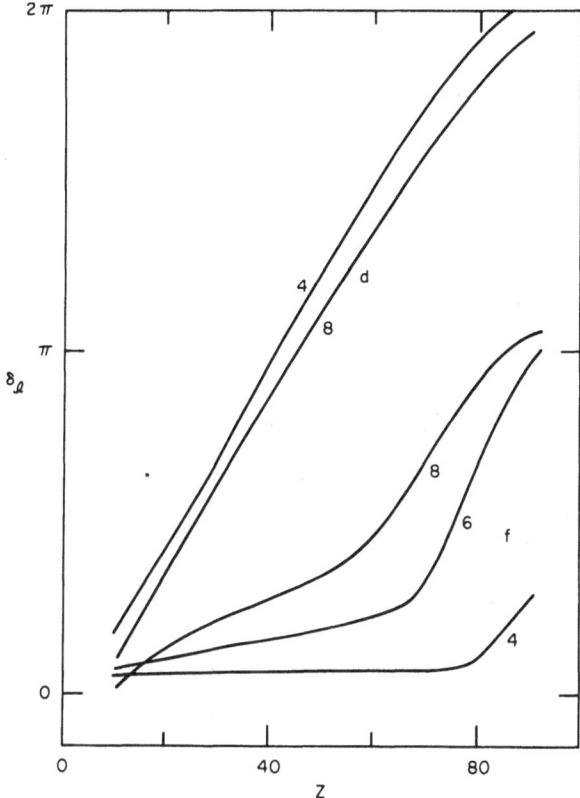

Figure 2. Variation with atomic number Z of d, f phase shifts for scattering of electrons by a Thomas–Fermi field with long-range Coulomb interaction. The numbers beside each curve denote the charge state q. All results for a velocity $v = 1.5$.

the Coulomb field of A^{+q} is considered, we have

$$\sigma_L = \frac{2\pi}{\Delta} \left(\frac{q}{v} \right)^2 \tag{4}$$

To test (2) and (4) empirical values of Δ were calculated from a wide range of recent measurements[1] and plotted against $1/v$ in Figure 1; each point is an average over all available q. Theoretically, one would expect $\Delta \simeq 1$. The apparent discrepancy between light and heavy projectiles is entirely due to projectiles in low stages of ionization, and it can almost certainly be resolved by using a realistic $e + A$ potential.

Improved calculations of f have been undertaken[2] to explain the anomalies seen in the variation of σ_C with q for projectiles of atomic number $Z \simeq 70$. In Figure 2 we plot the Coulomb phase shifts $\delta(l = 2, 3)$ in a Thomas–Fermi potential as a function of Z for $q = 4, 6, 8$ and $v = 1.5$. Phase shifts $\delta(l \geq 4)$ are small, while $\delta(0, 1)$ are similar in behavior to $\delta(2)$; dependence on v, at least for $1 < v < 3$, is slight. The uniquely rapid variation of $\delta(3)$ for $Z \sim 75$, $q \sim 6$ (in fact, a shape

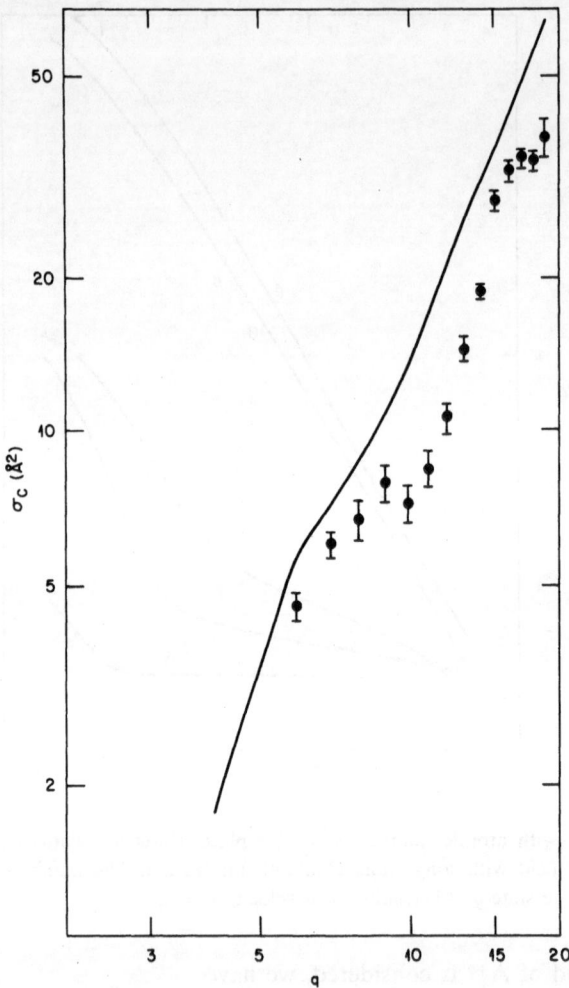

Figure 3. The capture cross sections for Ta ions at $v = 1.55$ are plotted against q. The dots are
measurements and the solid line is the theory (2) with $\Delta = 1$.

resonance) largely accounts for the observed anomaly, as illustrated in Figure 3.
After the leading term (4), the most important contribution to σ_L is the Coulomb
short-range cross term, so that the effect is correctly described as interference.

3. Atomic Versus Molecular Targets

A large body of data has now accumulated on process (1) with both H, H_2 at
the same values of q, v. Thus the ratio

$$R = \sigma(H_2)/2\sigma(H) \tag{5}$$

can be compared with the oft-quoted value of unity, which indeed must be attained at sufficiently high energies. Treating the molecular T matrix as a sum of atomic T matrices with a phase difference appropriate to the equilibrium internuclear separation ϱ we find to a good approximation that

$$R = 1 + C_\perp C_\parallel \tag{6}$$

The transverse coherence factor

$$C_\perp = \langle T_A T_B \rangle / \langle T_A{}^2 \rangle \tag{7}$$

where A, B refer to the two centers and the brackets to an average over impact parameters and molecular orientations. In a simple model

$$T = 2^{-1/2} \exp\left(-\frac{b}{b_0}\right), \qquad \sigma(H) = \pi b_0{}^2 \tag{8}$$

$$C_\perp \simeq \exp\left[-\left(\frac{\varrho}{2.7b_0}\right)^2\right] \tag{9}$$

so that for $b_0 > 1$, $C_0 \simeq 1$. The longitudinal coherence factor

$$C_\parallel = \frac{\sin X}{X}, \qquad X = \frac{\omega\varrho}{v} \tag{10}$$

where ω is a mean energy transfer. The surprising feature of (6)–(10) is that for almost all available measurements $C_\perp \sim 1$ and the scattering by the two centers is highly coherent. The simple argument that $R = 1$ is thus invalid and measurements that find $R = 1$ happen to be in a region where $C_\parallel$ is passing through zero.

In support of this contention, we have plotted R for all Oak Ridge data[1] against $1/v$ in Figure 4 (actually the ratio of Δs, averaged over q as before). A reasonable fit is obtained with $C_\perp = 2$, $\omega = 4.5$; this value of ω corresponds to capture into the state $n = 0.3q$ in reasonable agreement with most theories. The value of $C_\perp$ argues a T matrix that peaks around $b \sim 1$, again not an unreasonable result. Similarly, the recent Belfast data[3] with Li projectiles (Figure 5) are fitted by $C_\perp = 2$, $\omega = 3.7$, suggesting that most capture is into the ground states of the ions. For some velocity > 4, we can expect $b_0 \to 0$, $C_\perp \to 0$ and R to turn over and $\to 1$, but Figures 4 and 5 show that this region has not yet been probed.

Another coherence effect has recently come to light in stripping processes:

$$A^{+q} + B \to A^{+q'} + B. \tag{11}$$

It might be though that for $Q = q' - q > 1$, the cross section $\sigma_{mol}(Q)$ for a molecular stripper would be the sum of $\sigma_{at}(Q)$ for the constituent atoms. In particular, the ratio

$$R(Q) = \sigma(Q)/\sigma(Q - 1) \tag{12}$$

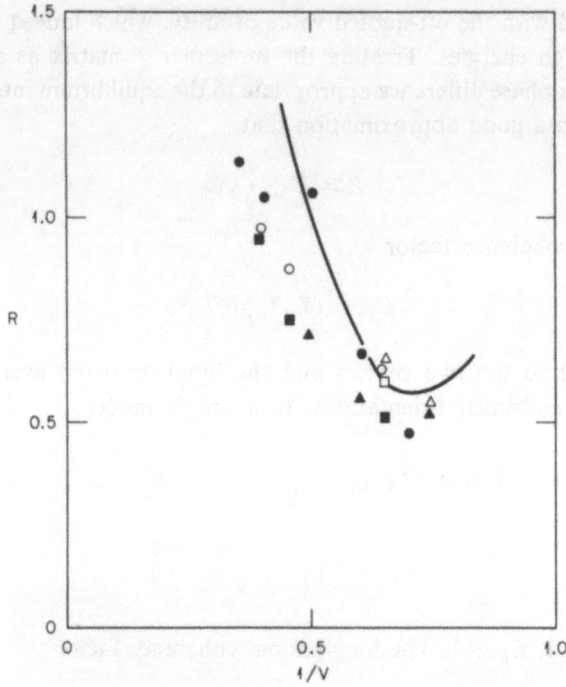

Figure 4. The ratio (5) of the H_2 to twice the H cross sections is plotted versus $1/v$ for all the Oak Ridge data. Symbols have the same meaning as in Figure 1.

representing the probability of removing the last electron should be the same for both atom and molecule. However, we encountered great difficulty fitting N_2 and SF_6 stripping data (with Fe^{+4} at 1.3 MeV/amu) by means of a theory that is quite successful for atomic strippers. This theory (A in Figure 6) is consistently lower than experiment.[4]† A direct calculation in which averaging over impact parameters and molecular orientations was carried out numerically, almost doubles R and even overshoots experiment for $Q < 4$ (M in Figure 6). The numerical calculations are roughly fitted by the expressions

$$R_{at}(Q) = (\tfrac{1}{2})^Q$$
$$R_{mol}(Q) = [\tfrac{1}{2}(1 + C_\perp C_\parallel)]^Q \tag{13}$$

in the earlier notation. It should be borne in mind that the factors $C_\perp$, $C_\parallel$ fall off slowly with velocity, and that the average over molecular orientations weights favorable orientations which enhance $\sigma(Q)$ by several decades. While these predictions have not been confirmed experimentally, a definitive test is not difficult to mount; more refined calculations are in progress.

---

† Older data on I ions in O_2 show similar effects (Betz and Wittkower[5] and Moak et al.[6]). The latter paper plotted $R(Q)$ and noted that it was always ∼0.6.

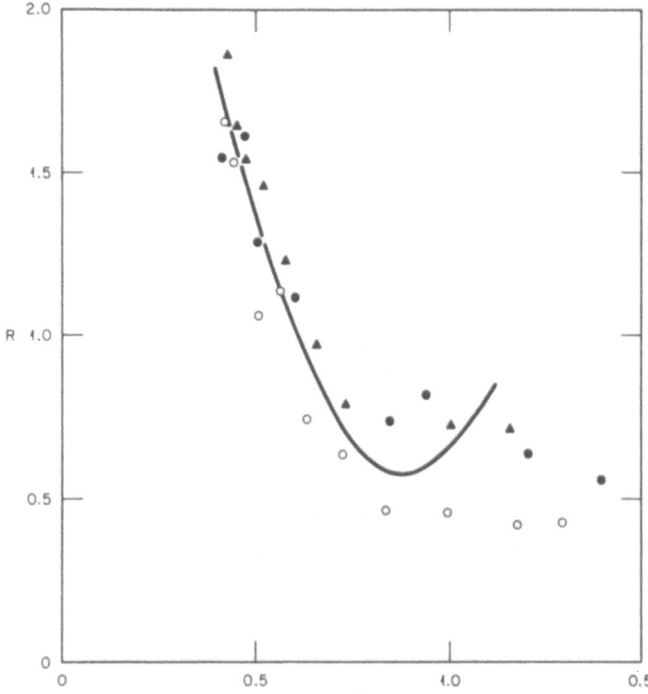

Figure 5. The ratio (5) is plotted versus $1/v$ for all the Belfast data: ●, Li$^+$; ▲, Li^{+2}; ○, Li^{+3}.

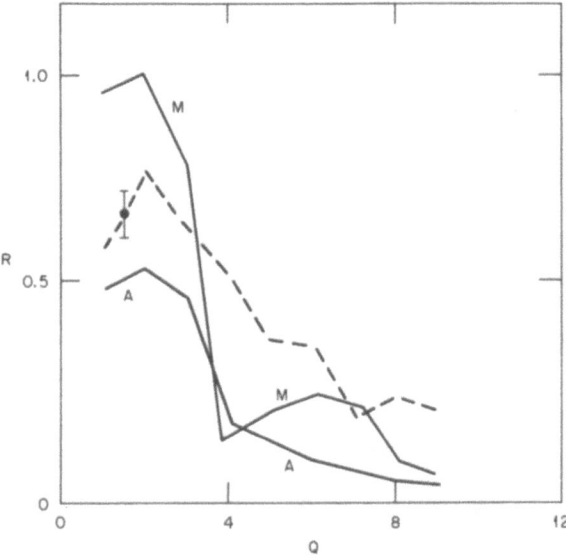

Figure 6. The ratio (12) of successive stripping cross sections versus Q, the number of electrons removed. The dotted line is experiment and the solid lines are the two theories explained in the text.

4. The Forward Peak in Projectile Ionization

In (11) electrons ejected with velocities k close to that of the projectile v (the "forward peak") have two sources: (a) capture into the continuum, which we shall not consider here, and (b) projectile ionization. Recent experiments[7] reveal structures associated with (b) of two types. One type, for which no good explanation has yet been produced, is of small amplitude and rapidly varying with energy; the other consists of broad shoulders symmetric about $k = v$.

Looking for simple explanations to begin with, we apply the impact-parameter Born approximation, making a multipole expansion of the field of B. For practical purposes, the cross section for ejecting an ns electron is

$$\sigma(ns - k') = 2\pi\left(\frac{Z_B}{v}\right)^2 \sum \frac{R_l^2}{(2l+1)}, \qquad R_l = \langle ns \,|\, r^l \,|\, k'l\rangle \qquad (14)$$

where k' is the wave number in the projectile frame. For ions with between 4 and 10 electrons, the dominant process is found to be $2s - k'd$, which is consistent with experiment. Now each partial wave has its own cusp shape F_l, e.g.,

$$F_0 = (1 + x^2)^{1/2} - x$$

$$F_1 = \frac{(1 + 2x^2)}{(1 + x^2)^{1/2}} - 2x$$

$$F_2 = \frac{(4 + 20x^2 + 13x^4)}{4(1 + x^2)^{3/2}} - \frac{13}{4}x$$

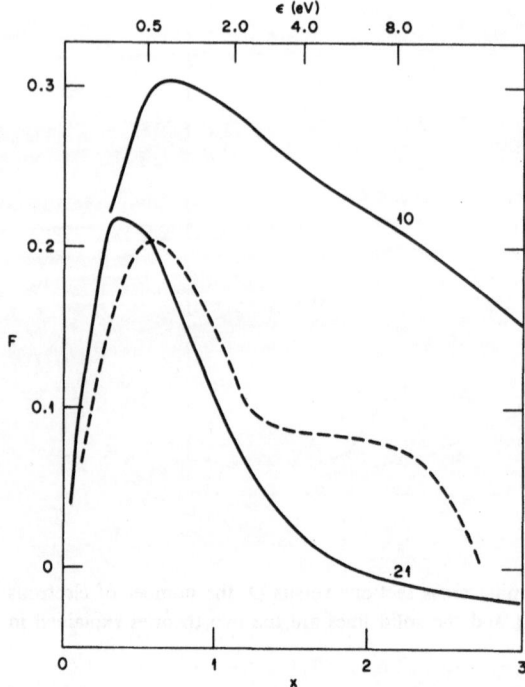

Figure 7. The functions F_1-F_0 and F_2-F_1 defined by (15), labeled 10, 21 respectively. The dotted line roughly follows the experimental $O^{+4}-O^{+5}$ difference, while the upper scale shows the electron energy in the projectile frame for ions of 2.5 MeV/amu and an aperture of $1°$.

where $x = |k - v|/v\theta$ and θ is the aperture of the detector. F_0 is, of course, well known. If each cusp is normalized to unity at $x = 0$ and those for two successive stages of ionization subtracted, the resulting functions must be a linear combination of F_1-F_0, F_2-F_1, etc. For $0^{+4}-0^{+5}$ only F_2-F_1 is possible, and a comparison of theory and experiment (Figure 7) appears to give a satisfactory explanation of the "shoulders" referred to above. More extensive measurements should yield information on the matrix elements R_l, providing a probe of electron scattering by highly stripped ions.

ACKNOWLEDGMENT

Research sponsored by the Division of Physical Research, U.S. Department of Energy, under contract No. W-7405-eng-26 with the Union Carbide Corporation.

References

1. H. J. Kim, P. Hvelplund, F. W. Meyer, R. A. Phaneuf, and P. H. Stelson, *Phys. Rev. A* **15**, 515 (1979).
2. H. J. Kim, P. Hvelplund, F. W. Meyer, R. A. Phaneuf, P. H. Stelson, and C. Bottcher, *Phys. Rev. Lett.* **40**, 1635 (1978); F. W. Meyer, R. A. Phaneuf, and C. Bottcher, to be published.
3. M. B. Shah, T. V. Goffe, and H. B. Gilbody, *J. Phys. B* **11**, L233 (1978).
4. H. Knudsen, C. D. Moak, C. M. Jones, P. D. Miller, R. O. Sayer, G. D. Alton, and M. Bridwell, *Phys. Rev. A* **19**, 1029 (1979).
5. H. D. Betz and A. B. Wittkower, *Phys. Rev. A***6**, 1485 (1972).
6. C. D. Moak, H. O. Lutz, L. D. Bridwell, L. C. Northcliffe, and S. Datz, *Phys. Rev.* **176**, 427 (1968).
7. M. Suter, C. R. Vane, I. A. Sellin, S. B. Elston, G. D. Alton, R. S. Thoe, and R. Laubert, *Phys. Rev. Lett.* **41**, 399 (1978).

34

Excitation of $Cd(5^3P_1)$ and $Cd(5^1P_1)$ in Na^+–Cd Collisions: Optical Polarization and Population of Magnetic Sublevels

VINCENZO AQUILANTI, PIERGIORGIO CASAVECCHIA, AND GAIA GROSSI

Cross sections for excitation of $Cd(5^3P_1)$ and $Cd(5^1P_1)$ in Na^+–Cd collisions have been measured by a combined crossed beam and optical technique. In the investigated energy range, from a few eV to $\sim 3\ keV$, both cross sections show structure, and excitation of $Cd(^1P_1)$ is larger ($\sim 9 \times 10^{-15}\ cm^2$ at $1\ keV$) than excitation of $Cd(^3P_1)$ ($\sim 2.5 \times 10^{-15}\ cm^2$ at $1\ keV$). From the measurement of polarization of the fluorescence lines, cross sections for population of the magnetic sublevels have also been obtained, and considerable anisotropy has been found only for $Cd(^1P_1)$. These results are discussed with reference to the previously investigated Na^+–Hg system.

1. Introduction

In the years elapsed since the appearance of the last editions of Massey's classic books,[1,2] atomic collisions in general and the role played by electronically excited states in particular have been the object of intensive experimental and theoretical studies. Earlier investigations aimed at understanding transport properties, or at clarifying collision-induced phenomena, such as the quenching of resonance radiation, intramultiplet mixing, or the broadening and shift of spectral lines. Presently, interest in this field is being stimulated by the fact that, in order to characterize the properties

VINCENZO AQUILANTI, PIERGIORGIO CASAVECCHIA, AND GAIA GROSSI • Dipartimento di Chimica dell'Università, 06100 Perugia, Italy.

of plasmas for fusion research, the mechanisms of laser action, and a variety of at-
mospheric or astrophysical phenomena, it is necessary to understand elastic and
inelastic atomic collisions more profoundly than allowed by earlier theories and to
obtain for them greater details than within reach of previous investigations, mostly
carried out in bulk experiments. The development of atomic, molecular, and ionic
beam techniques has fostered a significant progress in this field, by allowing the
study of atomic processes under single-collision conditions; further progress is
presently being made by coupling these techniques to optical methods, such as
spectroscopic analysis of emitted radiation and/or laser preparation of selected states.
As a consequence, it is possible by these methods to measure the fundamental
quantities for atomic collisions, namely, the state-to-state cross sections and their
energy dependence. On the other hand, a knowledge of these quantities provides an
experimental basis for further theoretical developments and suggests correlations
between atomic collisions and molecular structure, as emphasized as early as 35 years
ago by Massey and Smith.[3]

In this laboratory, a series of experiments has been carried out on electronic
excitation in ion–atom collisions at impact energies in the range between a few electron
volts and a few kiloelectron volts. The excitation is studied by the measurement of
wavelength, intensity, and polarization of emitted radiation. The purpose of this
paper is to report and discuss results obtained for the excitation of cadmium atoms,
specifically in their 5^3P_1 and 5^1P_1 states, in collisions between sodium ions and cadmium atoms in their ground states. This study is to be considered as a continuation
of our previous work, in particular on the excitation of the 6^3P_1 state of mercury in
collisions between alkali ions and mercury atoms. Since a detailed account of these
latter experiments has recently been published,[4] the present discussion of the
Na^+–Cd system will contain frequent references to the Hg system, in order to stress
analogies and differences between them.

It is beyond the scope of this paper to present a review of related experiments:
a partial list of significant references can be extracted by those quoted in the following
sections and in the paper on Hg.[4] Here we note only that a study of excitation in
collisions between Cd atoms and several ions has been reported by Shpenik et al.[5]
However, for the particular system Na^+–Cd they report only relative excitation of
the 326.1-nm ($5^3P_1 \rightarrow 5^1S_0$) line at energies less than 1 keV. In this paper, we report
absolute cross sections both for the 5^3P_1 and the 5^1P_1 states at energies up to 2.5 keV,
and report also data on optical polarization of emitted light, which contain information on the collisional population of magnetic sublevels.

2. Experimental Conditions

The experiments reported in this paper were carried out by means of an apparatus
previously used in our laboratory for the study of resonance emission in symmetric
and asymmetric alkali ion–atom collisions. The symmetric systems studied were

Na⁺–Na[6] and K⁺–K;[7] the asymmetric systems include Li⁺–Na, Li⁺–K, Na⁺–K, and K⁺–Na,[8] and a study of population of the potassium resonance doublet in collisions of K with all the alkali ions.[9] More recently, collisions of Li⁺, Na⁺, and K⁺ with Hg have also been investigated.[4,10] Since a detailed description of the apparatus can be found in the previous papers, only a brief discussion of the experimental conditions follows.

The alkali ion beams are produced by surface ionization, heating to ∼1200°C porous tungsten disks coated with proper alkali salts, and are accelerated by grids to the desired energy. Up to three different ion emitters can be placed under vacuum and interchanged during an experiment: this ensures the possibility of comparison between different systems. Also, an electron source of the oxide hot-cathode type which has exactly the same dimensions of the ion emitters has occasionally been used, mainly for calibration purposes. The ion beam enters the field-free region collision zone with a circular cross section (∼1.2 cm diameter). Its energy range is typically from a few eV to a few keV (2.5 keV in the present experiments); the energy spread, as measured from retarding potential techniques, is of the order of 0.5 eV.

The cadmium beam is produced by effusion out of a stainless steel oven, operated at a pressure less than 0.1 Torr. It enters the collision zone at a right angle with respect to the ion beam (see Figure 1), and has there a ribbon shape ∼0.03 cm high and ∼2.0 cm wide. The cadmium pressure in the collision zone is typically of the order of 10^{-5} Torr, low enough for preventing pressure depolarization effects.

As illustrated in Figure 1, the collision zone is observed by the optical system at a right angle with respect to the plane defined by the directions of the two beams A 0.25-m optical path monochromator is used for spectral analysis of the emitted radiation: various gratings allow a wavelength range from 200 to 800 nm. The determination of the energy dependence of cross sections and the polarization measurements were carried out using narrow-band interference filters.

A linear polarizing filter is placed along the axis of the optical system (Y axis in Figure 1) in order to measure the polarization of radiation emerging from the beam interesection region. Defining as $I_∥$ and $I_⊥$ the photon intensities measured with the polarizer parallel and perpendicular to the ion beam direction (Z axis in Figure 1),

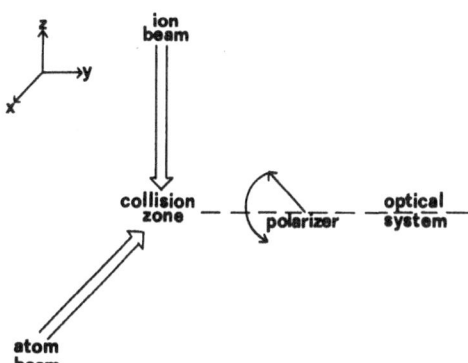

Figure 1. A scheme of the experimental configuration, showing the geometrical arrangements of the two crossed beams and of the optical system.

the polarization fraction P is given as usual by

$$P = (I_\parallel - I_\perp)/(I_\parallel + I_\perp)$$

Further details on experimental conditions can be found in the literature.[4,6–10]

3. Experimental Results

Several atomic lines appear in the spectrum of the emitted light in the 200–800 nm wavelength range at the impact energies considered in this work; an assignment of these lines in terms of transitions between known energy levels of the atomic species involved shows that most of them correspond to radiative decay of some of the lowest excited states of cadmium. Excitation of the sodium D lines was also found to be an important process.

For some of the prominent atomic lines, the dependence of their intensity upon collision energy was studied by means of interference filters. In order to convert the measured light intensities of the various lines into absolute emission cross sections, two independent calibrations were adopted. The first one is based on a measurement of resonance fluorescence from light scattering, and was performed on a sodium beam for the 589-nm line.[11] The second calibration procedure involves the measurement of resonance fluorescence from electron scattering, and was performed on a mercury beam for the 253.7-nm line.[4] Calibration of the optical system for any other line can therefore be obtained by taking into account the dependence on wavelength of both the filter transmissions and of the photomultiplier response. From the uncertainties in the latter factors, and from those due to the two calibration procedures, which however agree within respective limits, it can be estimated that the absolute scale for the cross sections obtained in this work is accurate to within $\sim \pm 20\%$.

In order to obtain the cross sections for collisional excitation of a particular atomic state, some care must be taken in accounting for cascade contribution and anisotropy effects. While the former were shown to be negligible for the states considered here, the latter effects give rise to polarization of the emitted light and can be taken into account when the polarization is measured.

The excitation cross sections from Na^+–Cd collisions for the states $Cd(5^3P_1)$ and $Cd(5^1P_1)$ are shown in Figure 2. They were obtained by the procedure described above from the emission cross sections corresponding to the Cd lines at 324 nm $(5^3P_1-5^1S_0)$ and 228 nm $(5^1P_1-5^1S_0)$. Also shown in Figure 2 is the polarization fraction for the 228-nm line, which was used for the anisotropy correction to the excitation cross section for $Cd(^1P_1)$. No appreciable polarization within the present uncertainty of $\sim \pm 3\%$ was found for the 324-nm line.

As is well known, the anisotropy of impact radiation, as measured from its polarization, is in turn a measure of the deviation from statistical population of

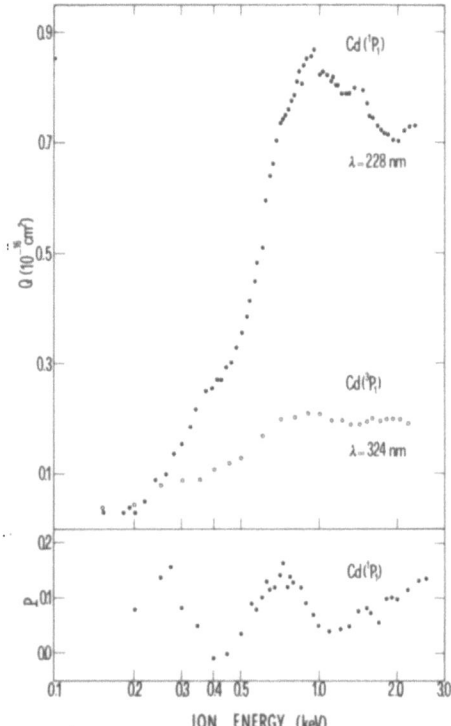

Figure 2. Excitation cross sections for Cd(5^1P_1) and Cd(5^3P_1) in Na+–Cd collisions, as a function of ion energy (systematic uncertainty $\sim \pm 20\%$). In the lower panel, polarization fraction for the 228-nm line (systematic un-

magnetic sublevels m_j. For the two Cd states considered here, $j = 1$ and $m_j = -1$, 0, $+1$. Since, because of cylindrical symmetry, the population of the sublevels having $m_j = +1$ and $m_j = -1$ is the same, the excitation cross section Q for both the singlet and the triplet states can be written as a weighted sum over the two cross sections for population of magnetic sublevels, Q_0 for $m_j = 0$ and Q_1 for $m_j = \pm 1$:

$$Q = Q_0 + 2Q_1$$

The relation between Q_0, Q_1 and the polarization fraction can be extracted from the theory reviewed by Fano and Macek,[12] hyperfine effects being accounted for according to Percival and Seaton.[13] The procedure is similar to that outlined for collisions involving Hg,[4] the only difference being that for natural cadmium we have 75% isotopes with zero nuclear spin, and 25% isotopes with $\frac{1}{2}$ nuclear spin. The final formula is

$$P = 0.75(Q_0 - Q_1)[(Q_0 + Q_1)^{-1} + (7Q_0 + 11Q_1)^{-1}]$$

These two relations between the measured quantities Q and P (Figure 2), and the desired cross sections for population of magnetic sublevels Q_0 and Q_1 allow the extraction of the latter: they are shown in Figure 3 for Cd(5^1P_1). Since for Cd(5^3P_1) the polarization fraction is $P = 0.00 \pm 0.03$, we have in this case the immediate result $Q_0 \approx Q_1$.

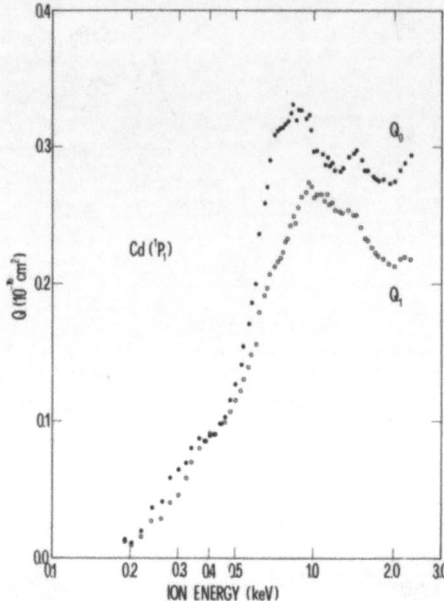

Figure 3. Cross sections Q_0 and Q_1 for the magnetic sublevels $m_j = 0$ and $m_j = \pm 1$ of Cd(5^1P_1) in Na$^+$–Cd collisions obtained from the data in Figure 2 (see text).

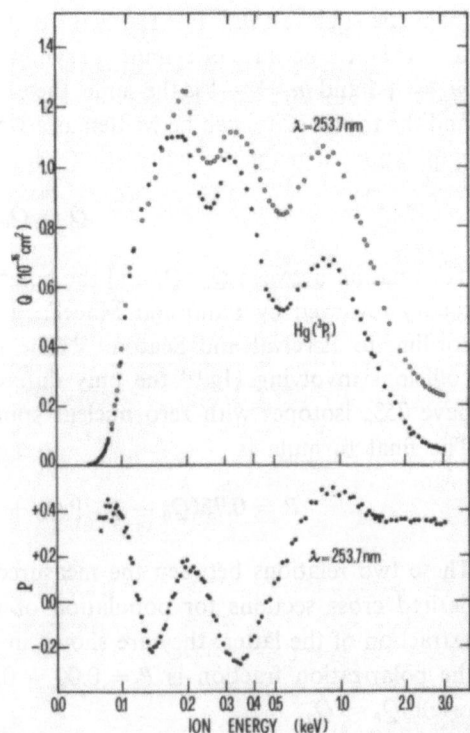

Figure 4. Emission cross section for $\lambda = 253.7$ nm and excitation cross section for Hg(6^3P_1) in Na$^+$–Hg collisions as a function of energy (see Reference 4 for details). In the lower panel, polarization fraction for the 253.7-nm line.

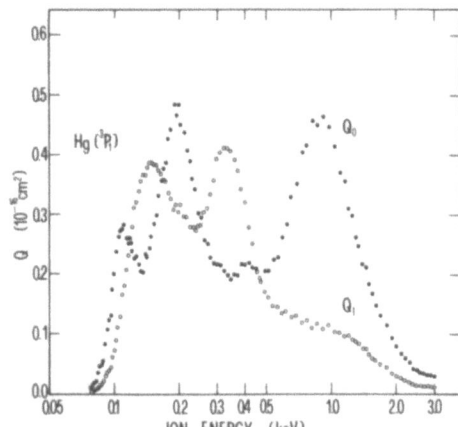

Figure 5. Cross sections Q_0 and Q_1 for the magnetic sublevels $m_j = 0$ and $m_j = \pm 1$ of Hg(6^3P_1) in Na⁺–Hg collisions obtained from the data in Figure 4.

The results that have been obtained for other excitation processes in this Na⁺–Cd system, and the rather similar ones obtained for the Li⁺–Cd and K⁺–Cd systems, will not be reported here. Instead, in order to allow a comparison with a system previously investigated in detail,[4] we report in Figures 4 and 5 the corresponding results obtained for the excitation of Hg(6^3P_1) in Na⁺–Hg collisions.

4. Discussion

The results reported in this paper on excitation of Cd(5^3P_1) and Cd(5^1P_1) in Na⁺–Cd collisions confirm the general observations drawn from our previous work[4] on Hg(6^3P_1) excitation in collisions between Hg and alkali ions. On the other hand, at least one major difference, which must be attributed to the detailed properties of the involved interactions, shows up in these results: the unexpected lack of polarization in the 324-nm line.

The general observations which these results confirm can be summarized as follows.

(i) The excitation cross sections increase as a function of energy and reach values in the 10^{-17}–10^{-16} cm² range (Figure 2). As for Hg,[4] this is an indication that the outer atomic electrons play a dominant role in these processes.

(ii) The cross sections have tails at low energies, which seem to tend to zero at energies much higher than the few electron volts required from endoergicities. The present data do not allow a precise estimate: however, Shpenik et al.[5] give 20 eV for the threshold for Cd(5^3P_1) excitation. It appears altogether that these collisions between closed shell species, Na⁺(3^1S_0) and Cd(5^1S_0) in this case, must possess high energy in order to give rise to excitation. From the similar behavior of Hg excitation (Figure 4), we concluded[4] that the collisions effective for inelastic processes take place in very close encounters, involving short-range repulsive forces.

(iii) Structure appears in the energy dependence of the cross sections in Figure 2. Although not as clearly displayed as for $Hg(6^3P_1)$ (Figure 4), their appearance is an indication that quantum mechanical interference effects take place as the collision partners separate to give rise to the observed products. A discussion of the relevant interactions responsible for this structure should parallel that given elsewhere.[4] Here we stress only that these effects are to be associated with properties of the potential energy curves of the diatomic molecule which is temporarily formed during the collision, and therefore represent a further indication of the quasimolecular nature of atomic collisions.

An interesting feature that emerges from a comparison of the two cross sections displayed in Figure 2 is that excitation of $Cd(5^1P_1)$ is much more prominent than that of $Cd(5^3P_1)$. Although this feature is probably to be attributed to the very complicated nature of the interactions at short range, we note that, assuming that no spin-flip takes place for the unobserved sodium ion, then triplet excitation would be forbidden by the Wigner spin conservation rule.

As anticipated above, the most interesting aspects of this work came from polarization measurements. As shown in Figures 2 and 4, strong polarization effects were observed for $Cd(5^1P_1)$ excitation in Na^+–Cd collisions, and for $Hg(6^3P_1)$ excitation in Na^+–Hg collisions: they indicated a nonstatistical behavior in the cross sections for magnetic sublevels shown in Figures 3 and 5. The lack of polarization in the 324-nm line in Na^+–Cd collisions, and the statistical population of sublevels for $Cd(5^3P_1)$ that this implies, appear therefore rather surprising. It has to be stressed that an entirely similar behavior emerges comparing the results for Li^+ and K^+ on Hg,[4] with those not reported here for Li^+ and K^+ on Cd.

Since it is reasonable to assume that the electrostatic interactions between alkali ions and cadmium atoms should be similar to that between alkali ions and mercury atoms, then the different behavior in the polarization of the $^3P_1$ states is presumably to be explained in terms of a different relative role which the other two relevant interactions, rotational and spin–orbit, play in these systems. In the $Hg(6^3P_1)$ case, the strong polarization effects observed could be rationalized[4] assuming rotational (Coriolis) coupling effects negligible with respect to spin–orbit coupling: in the terminology of molecular spectroscopy, this amounts to describing the process in terms of Hund's cases (a) and (c), which imply the goodness of the quantum number Ω, the projection of j along the internuclear axis. An admittedly tentative explanation of the lack of substantial polarization effects in the $Cd(5^3P_1)$ case may therefore be based on the consideration that the reverse situation holds for cadmium, and namely that in this system rotational coupling is more effective than spin–orbit coupling [as a matter of fact, for the separate atoms the spin–orbit splitting for $Cd(5^3P_1)$ is less than $\frac{1}{3}$ than that for $Hg(6^3P_1)$]. Then an adequate description of these collisions should be formulated in terms of Hund's cases (b) and (d): since Ω is not defined in these cases, the outcome of the collision should be a statistical population of the magnetic sublevels.

Although the above discussion needs to be implemented by further experimental data on similar systems, it can nevertheless be concluded that the present polarization results indicate that particular attention should be paid to the specific role that angular momentum coupling schemes play in atomic collisions.

References

1. N. F. Mott and H. S. W. Massey, *The Theory of Atomic Collisions*, 3rd ed., Clarendon Press, Oxford (1965).
2. H. S. W. Massey and E. H. S. Burhop, *Electronic and Ionic Impact Phenomena*, 2nd ed., Clarendon Press, Oxford (1969).
3. H. S. W. Massey and R. A. Smith, *Proc. R. Soc. London Ser. A* **142**, 142 (1933).
4. V. Aquilanti, P. Casavecchia, and G. Grossi, *J. Chem. Phys.* **68**, 1499 (1978).
5. O. B. Shpenik, A. N. Zavilopulo, and I. P. Zapesochnyi, *Zh. Eksp. Teor. Fiz.* **62**, 879 (1972).
6. V. Aquilanti, *Z. Phys. Chem. (Frankfurt am Main)* **90**, 1 (1974).
7. V. Aquilanti and P. Casavecchia, *J. Chem. Phys.* **64**, 751 (1976).
8. V. Aquilanti and G. P. Bellu, *J. Chem. Phys.* **61**, 1618 (1974).
9. V. Aquilanti, P. Casavecchia, and G. Grossi, *J. Chem. Phys.* **65**, 5518 (1976).
10. V. Aquilanti and P. Casavecchia, *Chem. Phys. Lett.* **47**, 288 (1977).
11. V. Aquilanti, G. Liuti, F. Vecchiocattivi, and G. G. Volpi, *Entropie* **42**, 158 (1971).
12. U. Fano and J. Macek, *Rev. Mod. Phys.* **45**, 553 (1973).
13. I. C. Percival and M. J. Seaton, *Phil. Trans. R. Soc. London Ser. A* **251**, 113 (1958).

35

Spin Polarization in Ion–Atom Collisions Studied with Ion Storage Techniques

HANS A. SCHUESSLER

Spin-dependent charge transfer collisions and spin exchange collisions have been studied for ion–atom systems. The measurements were carried out using a radiofrequency quadrupole ion trap and an atomic beam apparatus. The polarization of the collision partners was changed by radiofrequency and optical pumping techniques. The ion storage exchange collision method and reorientation spectroscopy are described.

1. Introduction

The search to understand the mechanisms governing the transfer of particles in collisions between ions and atoms is still attracting interest from a broadening range of physicists and chemists. So far, the experiments have been carried out with a variety of techniques such as low-pressure mass spectrometry,[1] ion cyclotron resonance,[2] flowing afterglow,[3] and merging beam techniques.[4]

This paper presents the application of an ion storage method to the study of ion–molecule reactions in ultrahigh vacuum and at low relative collision energies. Stored ion collision measurements are just beginning to be recognized as a powerful new tool in atomic collision research and only a few experiments have been carried out so far, for which we refer to two reviews on ion storage techniques by Dehmelt[5] and Schuessler.[6] The purpose of this paper is to report on polarization-dependent

HANS A. SCHUESSLER • Department of Physics, Texas A & M University, College Station, Texas 77843, U.S.A.

aspects of ion–atom collisions, in particular on measurements of spin-dependent charge transfer collis'ons and spin exchange studies made in ion traps at Texas A & M University. All the experiments have been carried out at low energies, where the collision interaction time is either long or comparable to the orbital period of the active electron. This means that the relative speed during the collision is either smaller than, or of the same order as, the average speed of the atomic electron. In this case, coupled molecular state approximations, using impact-parameter methods, are appropriate to describe the collision. Such calculations have been made, for instance, by Bates and McCarrol,[7] by Smith and Olson,[8] and also by Winter and Lane.[9] However, all these calculations do not consider explicitly the dependence of charge transfer on the orientation of the collision partners, an effect which is particularly pronounced for near-resonance charge transfer at low energies.

Following a brief description of the ion motion, the realization of the storage scheme for collision work will be outlined. I shall then describe a charge transfer experiment between polarized collision partners, in which we have studied the electron spin dependence of charge transfer collisions between polarized He+ ions and polarized cesium atoms with the ISEC (ion storage exchange collision) method. Reorientation spectroscopy and its application to spin exchange measurements will then be discussed. Some results of ion storage work investigating simple rearrangement collisions and hydrogen–helium ion–molecule reactions will be mentioned.

2. Ion Storage

There are basically two ion storage devices in use. The Penning ion trap and the radiofrequency (rf) ion trap. Both consist of three electrodes, two end cap electrodes and a ring electrode, which are formed to be complementary hyperboloids of revolution in a quadrupole arrangement. The Penning trap uses a dc electric field for trapping in the axial or z direction and a strong magnetic field for radial or r confinement. The ion orbits are a superposition of a harmonic motion in the z direction, cyclotron orbits in the r–θ plane, and a slow drift motion about the z axis. The work of Walls, Dunn, and colleagues[10,11] on electron recombination has shown that this type of trap can be successfully used to investigate spin-independent processes. However, when spin-dependent aspects are studied the large magnetic field and associated Zeeman splittings restrict its usefulness, and the rf ion trap is better suited for containment. In this device a three-dimensional potential well is generated by electric fields only, and a weak magnetic field can be added at will to define the polarization axis. In such a rf trap the rf driving voltage changes sign before an ion has the chance to escape along a defocusing direction, and ions with the proper initial conditions are always focused back to the center of the quadrupole arrangement by time-varying field gradients. The ions move in a rotationally symmetric potential of the form

$$\phi = (U_0 - V_0 \cos \Omega t)(r^2 - 2z^2)/2r_0^2 \tag{1}$$

Here V_0 is the amplitude of the rf driving voltage at frequency Ω and U_0 is the dc voltage applied in series with it. Figure 1 shows a cross section of the rf ion trap; the potential distribution and the various operating voltages are also indicated. r and z are the cylindrical coordinates and $z_0 = 2^{1/2} r_0$ is the minimum separation of the trap electrodes. The ion motion is described by Mathieu differential equations which may be transformed to have the following normalized form:

$$d^2 x_i / d\xi^2 + (a_i - 2q_i \cos 2\xi) x_i = 0, \qquad i = r, Z \tag{2}$$

with $\xi = \Omega t/2$; $a_z = -2a_r = -8ZeU_0/m\Omega^2 r_0^2$; $q_z = -2qr = -4ZeV/m\Omega^2 r_0^2$, where Z is the charge state of the ion. The a–q plane contains a region in which the solutions of the Mathieu equations are simultaneously stable. This is illustrated in Figure 2. The solution for stable motion can be written as a power series:

$$X_i(\xi) = A \sum_{n=-\infty}^{+\infty} C_{2n} \cos[(2n + \beta_i)\xi] + B \sum_{n=-\infty}^{+\infty} C_{2n} \sin[(2n + \beta_i)\xi] \tag{3}$$

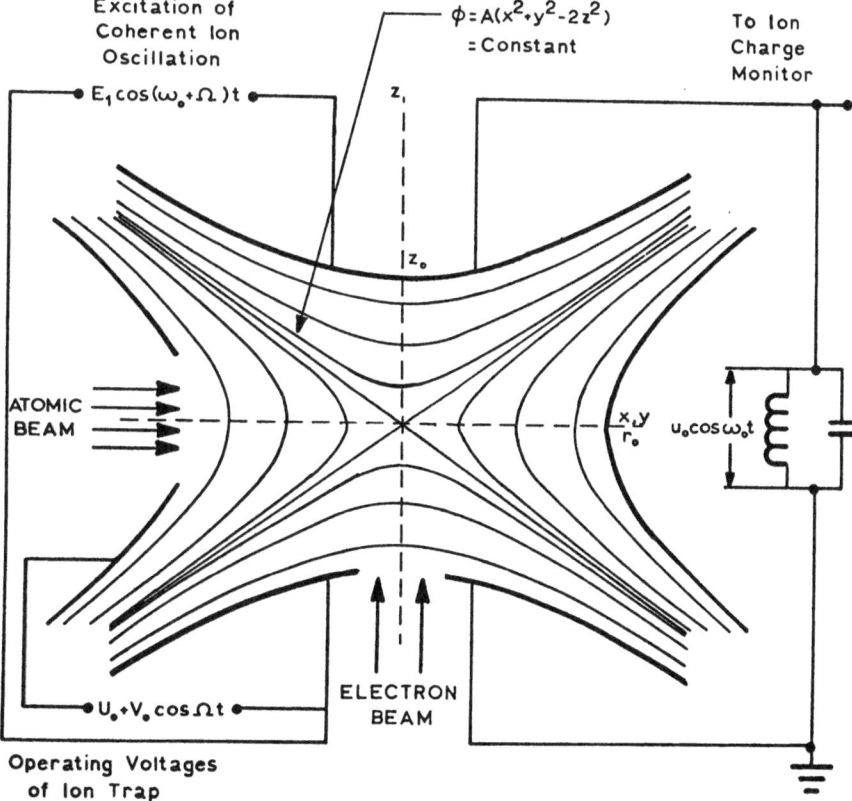

Figure 1. Cross section of the quadrupole ion trap showing the potential distribution of the trapping voltage $U_0 + V_0 \cos \Omega t$. The ion oscillation is excited by a homogeneous rf field at $\Omega + \omega_0$. The voltage $u_0 \cos \omega_0 t$ induced into the tank circuit by the cooperative ion motion is detected as an ion number signal. The atomic beam and electron beam windows are indicated.

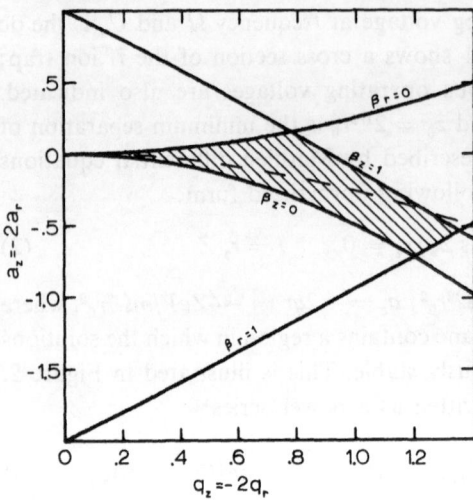

Figure 2. Plot of the a-q diagram for theoretical intensity contours of the ion signal.

The parameters A and B are functions of the initial conditions while β_i and C_{2n} depend on a_i and q_i. The fundamental frequency $\omega = \frac{1}{2}\beta\Omega$ has the largest amplitude. The stability diagram shows the lines with $\beta_i = \text{const}$. An ion is stored when the operating voltages correspond to a_i and q_i values in the stable region and when the amplitude $x_i(\xi)$ of the ion motion is less than the trap dimensions.

It is advantageous in collision studies to operate the rf ion trap in the symmetric potential mode with the same well depth D in the r and z directions or in the symmetric frequency mode where it is required that the axial ion motion frequency ω_Z is equal to the radial ion motion frequency ω_r or $\beta_z = \beta_r$. This latter condition yields the relation

$$a_z = -q_z^2/4 \tag{4}$$

and is shown as a dashed line in Figure 2. The relation is derived for the important case that $a_z \ll q_z$. The ion oscillation frequency for a symmetric trap is given by

$$\omega = q_z\Omega/4 \tag{5}$$

with the depth of the potential well D obtained as

$$D = (V_0/4)(\omega/\Omega) \tag{6}$$

Therefore, in order to operate the trap in the symmetric potential mode, one changes the operating voltages to yield, according to Equation (4), suitable values for q_z. This, of course, changes the frequency of the ion motion, so that the detection circuit has to be broadband.

In ion storage work the ions are usually produced by electron impact ionization of vapor of a selected atom of interest. Storage times ranging from many seconds to several hours have been obtained. The most crucial requirements for long-term storage are ultrahigh vacuum in the 10^{-11}-Torr range and a harmonic trap in which the shape and separation of the electrodes are accurate to about 0.001 in.

An important parameter in collision work is the relative energy or velocity of the collision partners. In our experiment we use a thermal atomic beam and the collision energy is practically the energy of the stored ions. The calculation of the ion energy has to take into account the dynamics of ion cooling during storage and has not yet been done in detail. However, it is known that Coulomb collisions establish a Maxwellian distribution of the ion kinetic energy with a self-collision time t_C given by Spitzer[12] as

$$t_C = 11.4A^{1/2}T^{3/2}/(nZ^4 \ln \Lambda) \tag{7}$$

where A is the molecular weight in amu, T is the temperature in °K, n is the ion density in cm^{-3}, Z is the ion charge, and Λ is the ratio of cutoff distance for the Coulomb potential to the impact parameter for scattering by $\pi/2$. t_C is typically in the millisecond range and is always small compared with the storage time. Ion cooling by evaporation of the ions from the high-energy tail of the Maxwell distribution will also take place until the mean ion energy is about $\frac{1}{10}$ of the well depth. This has been verified experimentally by Werth.[13] In the following we will, therefore, use for the average ion energy $\bar{E}$ and average velocity $\bar{v}$ the relation

$$\bar{E} = 0.1\bar{D} = m\bar{v}^2/2 \tag{8}$$

The number of stored ions and the ion motion were observed with the emission method. The emitted energy of the coherently excited ion cloud is thus measured in a sensitive detection circuit. Normally, the ion motion frequency ω is set slightly below the detection frequency ω_0. To sweep ω through resonance, a linear negative-

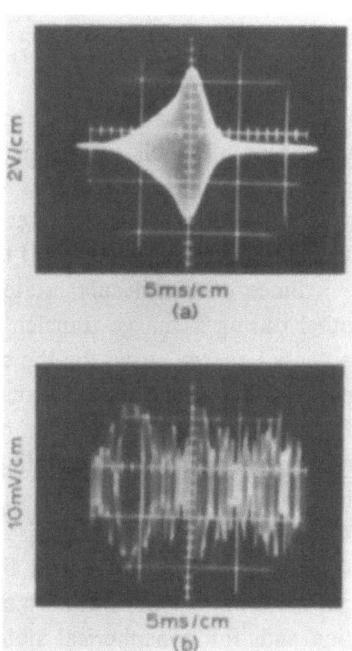

5ms/cm
(a)

5ms/cm
(b)

Figure 3. He$^+$ ion signal oscilloscope traces. (a) Ion number signal detected at ω_0 while being excited at $\Omega + \omega_0$. (b) Noise signal obtained with no ions present by not injecting the ionizing electron beam. The ion trap operating parameters were $V_0 = 200$ V, $U_0 = 7.6$ V dc; excitation amplitude: $E_1 = 1.75$ V, $\Omega = 2\pi$ (1 MHz), $\Omega + \omega_0 = 2$ (1.1312 MHz), $I_e = 1.3$ mA, $V_G = -300$ V dc, $V_s = 4.3$ V, $t_{obs} = 1$ sec, $t_e = 0.15$ sec, $t_D = 0.18$ sec, $t_s = 32$ msec, and $P = 1.5 \times 10^{-9}$ Torr. The signal-to-noise ratio S/N is 175:1. The gain of the detection amplifiers was 1.6×10^4.

going dc sweep is applied to the ring electrode. Simultaneously with the detection sweep, another pulse at the sideband frequency $\Omega + \omega_0$ coherently excites the ion macromotion when $\omega = \omega_0$, finally driving the ions against the electrodes. An emission signal voltage is induced, which is amplified and displayed on a scope. Figure 3 shows such an ion number signal in comparison with the system noise.

3. Spin-Dependent Charge Transfer Collisions

The experiment to be described investigates the near-resonant charge transfer process between stored He$^+$ ions and cesium atoms.[14] The cesium atoms are polarized by optical pumping and impart their polarization to the He$^+$ ions by spin exchange when they pass through the trapping region.

The fact that the spin exchange cross section is large against the charge transfer cross section ensures that the collision partners are highly polarized before the charge transfer becomes effective. The charge transfer process is spin dependent since it proceeds along the following near-resonant channels:

$$Cs(6s^2S_{1/2}) + He^+(1s^2S_{1/2}) \rightarrow \begin{cases} He^*(1s2p^1P_1) + Cs^+({}^1S_0) - 0.52 \text{ eV} \\ He^*(1s2p^3P_{2,1,0}) + Cs^+({}^1S_0) - 0.27 \text{ eV} \\ He^*(1s2s^1S_0) + Cs^+({}^1S_0) + 0.08 \text{ eV} \\ He^*(1s2s^3S_1) + Cs^+({}^1S_0) + 0.88 \text{ eV} \end{cases} \quad (9)$$

A large number of channels of a different nature is also available at the same time:

$$Cs(6s^2S_{1/2}) + He^+(1s^2S_{1/2}) \rightarrow He(1s^2\,{}^1S_0) + Cs^{*+}[5p^5({}^2P)nl] + \Delta E_{n,1} \quad (10)$$

where $n = 6$–10, and $1 = s$, p, and d. In this reaction an electron is transferred from the closed $5p^6$ shell of cesium to the $1s$ shell of helium. An energy level diagram representing these channels is shown in Figure 4. Orbital overlap requires a close collision with a small cross section, but a large number of near-resonant channels and other channels are present that will all contribute to the singlet cross section Q_1.

Since quasimolecular states are formed when the two atoms approach each other during a charge transfer collision, a detailed theory must consider the entire potential energy curve for the collision complex. In one type of process,[15] the important region for charge transfer is between about $5a_0$ and $15a_0$, where a_0 is the Bohr radius. Charge transfer occurs with a large cross section if there are curve crossings or rotational coupling in this region. This type of charge transfer is of importance where the energy separation of reactants and products at $R = \infty$ is large. A second process[16] gives rise to large charge transfer cross sections and occurs when several potential curves lie close together at large separations of the collision partners. This theory has been extended to include long-range forces due to polarization and, for nonspherical states, due to ion–quadrupole interactions.[17]

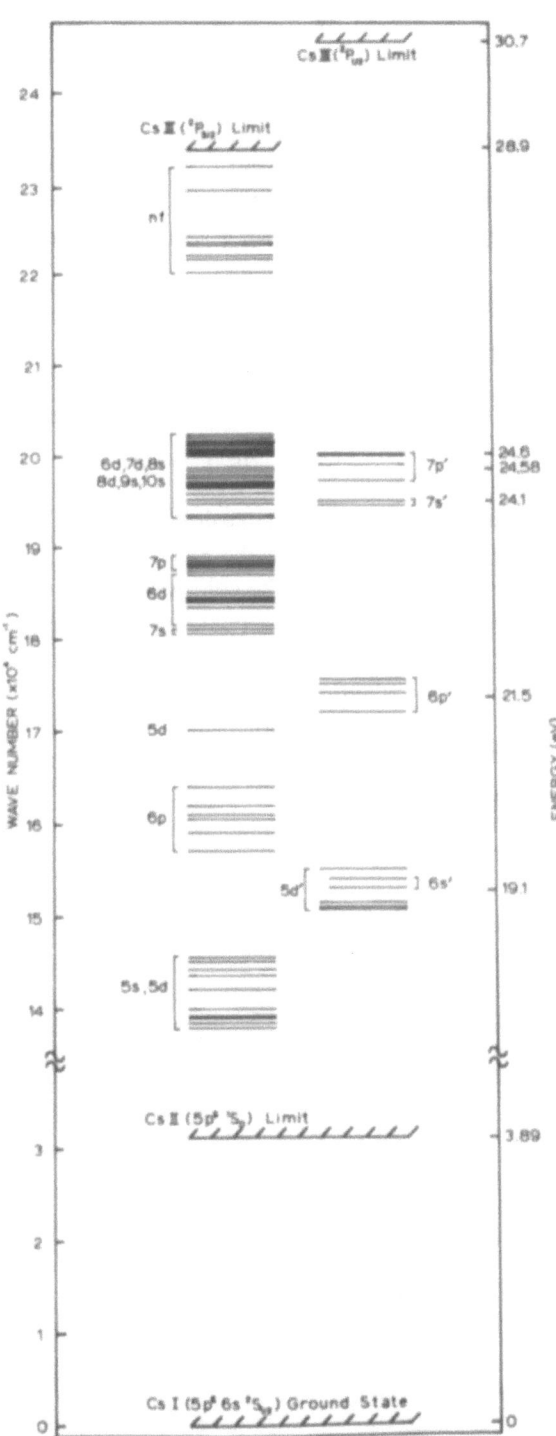

Figure 4. Excited states of the Cs+ ion participating in charge transfer collisions of the Cs–He+ system.

The energy arguments presented so far are necessary but not sufficient for charge transfer. In general, charge transfer should proceed with large cross sections, if also the Wigner spin rule is satisfied. This rule implies that a component of the spin angular momentum is conserved along some quantization axis, which is defined here by the direction of the external magnetic field. We, therefore, treat collisions in which the symmetry of the spin state of the total system is even independently of those for which the spin state is odd. Emphasizing the spin-dependent aspects of the charge transfer process, the following specific reactions occur. A singlet channel

$$Cs\uparrow + He^+\downarrow \rightarrow He^*(\uparrow\downarrow - \downarrow\uparrow) + Cs^+ + \Delta E_1 \tag{11}$$

leads to the singlet excited states described in (9) and (10). Two triplet channels

$$Cs\uparrow + He^+\uparrow \rightarrow He^*\uparrow\uparrow + Cs^+ + \Delta E_3$$
$$Cs\uparrow + He^+\downarrow \rightarrow He(\uparrow\downarrow + \downarrow\uparrow) + Cs^+ + \Delta E_3 \tag{12}$$

lead to the triplet excited states described in (9). The arrows represent the orientation of the electron spin along the external field, just before the collision for the left-hand side, and just after the collision for the right-hand side of the reaction.

In the experiment the relative spin orientation of the collision partners was selected to be predominantly either parallel or antiparallel by suitable application of optical pumping and rf methods. In this way the production of selected excited He* states was favored over others.

The apparatus used for this experiment consists of a rf-quadrupole trap, a cesium atomic beam, and an optical pumping system and is shown in Figure 5. The experiment is submerged in a large homogeneous magnetic field, produced by 150-cm-

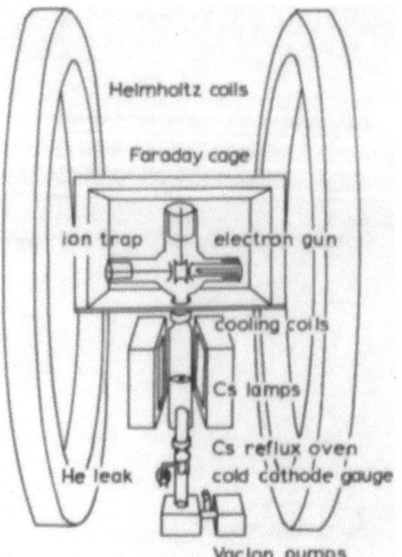

Figure 5. Experimental arrangement for the ion storage experiments on the He$^+$–Cs system.

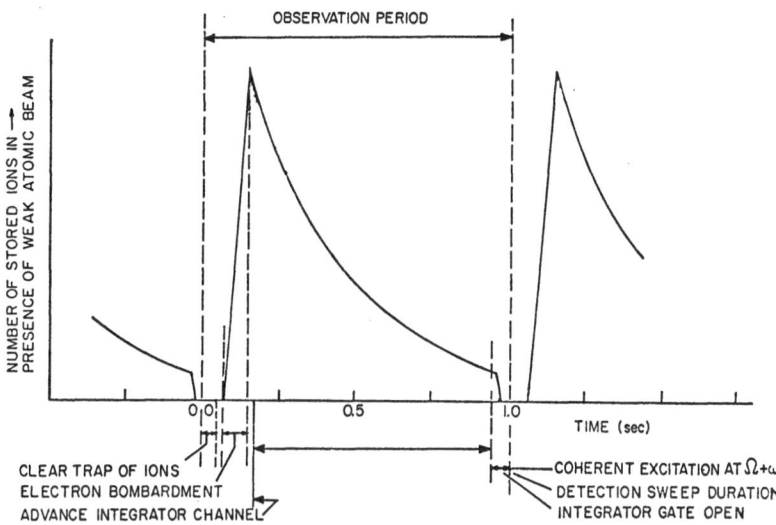

Figure 6. Time evolution of the stored He$^+$ ion number during the periodic sequence of ion observation. The duration of one observation cycle was typically 1 sec.

diameter Helmholtz coils. Helium is leaked into the ultrahigh-vacuum chamber and He$^+$ ions are produced at 10^{-8} mm Hg from the background atoms by a pulse of electrons. The electrons are produced from an electron gun mounted directly behind an array of holes centered in one end cap electrode. After storage, the helium ions are polarized by spin exchange collisions with the optically pumped thermal cesium atomic beam. Subsequently they are slowly lost from the trap by the charge transfer process of interest. Since the spin exchange cross section is larger than the charge transfer cross sections, it follows that the collision partners are polarized before the charge transfer becomes effective. We observed that the rate of neutralization depends on the polarization of the helium ions and cesium atoms, and, to a smaller degree, the relative ion–atom velocity. The polarization of both collision partners can be changed by disorienting the cesium atoms with a suitable rf field at resonance with the hyperfine structure Zeeman splitting. The relative ion–atom velocity is varied by electronically changing the depth of the potential well in which the ions are stored. For each chosen set of parameters the charge transfer process is then studied by counting the number of helium ions left in the trap after a fixed interaction time with the cesium atomic beam. Figure 6 shows the time development of the number of stored ions in an observation cycle.

In a typical measurement the cesium beam polarization is destroyed in alternate cycles by applying an on–off modulated radiofrequency field in resonance with the cesium hyperfine structure Zeeman splitting. The charge transfer occurs then either between polarized (both collision partners have practically the same electronic spin polarization) or unpolarized ion–atom pairs. The effectiveness of the charge transfer process is monitored in each cycle by counting the number of stored He$^+$ ions after a fixed interaction time with the cesium beam. The measurements demonstrate that polarized He^+ ions live longer in the ion trap than unpolarized ones. In other words, the rate at which He^+ ions are lost from the trap by neutralization with cesium atoms depends on the polarization of the collision partners. The rate is faster for the case where the spins of the collision partners are antiparallel and slower when the spins are parallel. We observed the optimum polarization signal $S^*(p)$ after a time $t = 2T_0$, where T_0 is the neutralization time of unpolarized ions. The optimum polarization signal is given by

$$S^* = \tfrac{1}{2}p^2 \, \Delta Q/(\bar{Q} + Q_h) \tag{13}$$

where $Q = Q_3 - Q_1, \bar{Q} = \tfrac{1}{4}(3Q_3 + Q_1)$, and Q_h is the spin-independent cross section due to rf heating, which is constant for a given set of trap parameters. Values for the singlet and triplet charge transfer cross sections were derived from the measurements of S^* at two different ion polarizations p. The singlet cross sections Q_1 and triplet cross sections Q_2 are listed in Table 1 for two different velocities. The ion polarization p is determined from the height ratio of magnetic resonance signals for the double to single quantum transition observed in the ground state of $^3He^+$.

To summarize, we have studied the electron spin dependence of near-resonant charge transfer collisions for the system Cs–He^+. The measurements were made for relative ion–atom energies of 1–8 eV. They demonstrate that in the energy range studied the singlet cross section is about twice as large as the triplet cross section. The cross sections reported in this work are almost an order of magnitude larger than those calculated with the semiclassical impact-parameter method of Rapp and Francis[18] and the spin dependence is less pronounced. Our data show unambiguously that the singlet charge transfer cross section dominates in the Cs–He^+ system and that, therefore, the transferred electron is excited predominantly into the singlet $1s2p\ ^1P_1$ and $1s2s\ ^1S_0$ states of helium, but also into the $^1S_0$ ground state of helium, leaving the cesium ion in one of the many excited states of the Cs^{+*} ion. The importance of the charge transfer channels involving the many Cs^{+*} ion states is presently being investigated further by extending our measurements into the 100-eV range. It should be pointed out that the density of the collision partners does not enter into the calculation of the cross sections and need not be measured. This is an advantage over most other methods for cross-section measurements used in atomic physics.

Table 1. Charge Transfer Cross Sections Q_1 and Q_2 for Helium Ions and Cesium Atoms at Two Relative Energies

Well depth D (eV)	Average relative velocity v (10^6 cm/sec)	Q_1 (10^{-14} cm²)	Q_3 (10^{-14} cm²)
3	0.22	1.0(2)	0.5(1)
8	1.35	1.1(2)	0.5(1)

4. Spin Exchange Collisions

The pulsed spin precession reorientation method[19] was developed to study spin exchange collisions between free ions and atoms. It was applied to stored ^{3}He$^+$ ions which interacted with a beam of polarized cesium atoms. The atomic beam and the ion storage parts of the apparatus were similar to the ones described in Section 3.

For the purpose of explanation, the ideal case of complete polarization with $P = 1$ is assumed. If we assume in addition that the ions have no hyperfine structure, then their electronic spins are all parallel to H_0. This simplified case is easy to discuss. Consider that the constant magnetic field H_0 points in the z direction and a weak rf field of frequency ω is applied in the x–y plane. When the rf field is at resonance with the Zeeman splitting of the ions, the total electronic ion spin is turned out of the z direction. If the H_1 field were applied continuously, the total ion spin would rotate periodically from a parallel to an antiparallel direction and back with respect to H_0. This is possible since, in an ion trap, the ions are ideally isolated from each other and the phase memory time is long.

However, in the pulsed spin precession reorientation scheme, the ions are exposed to a sequence of short resonant rf pulses and not to a continuous rf field. For each pulse, the length T_d and height H_1 are adjusted to produce an inversion of the electronic spin polarization. The pulse parameters then fulfill the conditions $\gamma H_1 T_d = \pi$, $T_d < T_e$, where $\gamma = \mu_B g_s / h$ is the gyromagnetic ratio. The last condition means that the resonant rf pulses, which invert the polarization, are strong enough so that the rotation of the electronic spin is fast compared with the spin exchange pumping of the polarized cesium beam. The polarized cesium beam passes continuously through the trapping region and partly reorients the inverted electronic spins of the ions during the long rf free periods T_p between the short inverting pulses. The average polarization was measured as a function of the pulse separation T_p. Such average polarizations are depicted in Figure 7 for the case of ^{3}He$^+$, including hyperfine structure, by horizontal lines for two different values of T_p. The solid line represents the case in which the pulse separation was equal to the spin exchange time T_e. The dashed line represents the case with T_p half as large. The rate equations for

Figure 7. Pulsed spin-precession reorientation method. (a) Diagram of the hfs Zeeman levels of interest. The magnetic dipole transition, which inverts the occupation number of the (1,1) and (1,0) levels, is indicated. (b) Sequence of the inverting pulses. The frequency of the pulsed rf field is at resonance with one of the hfs Zeeman splittings. T_d and T_p are, respectively, the duration of the inverting pulses and their separations. (c) Time evolution of the instantaneous ion polarization p (sawtoothlike lines) and of the average ion polarization $\bar{p}$ (horizontal lines).

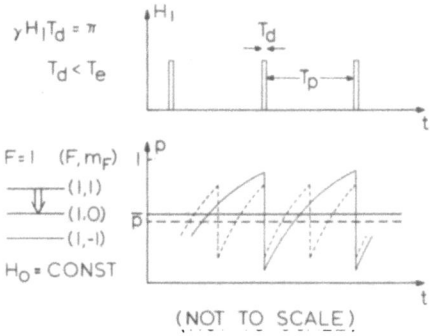

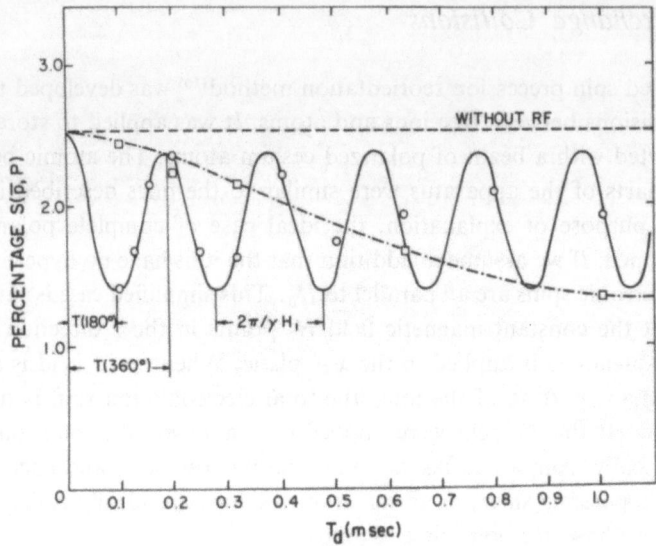

Figure 8. $S(\bar{p}_d, P)$ for various pulse lengths T_d and a constant pulse repetition period $T_p = 80$ msec. Measurements for two different pulse heights H_1 at the d transition frequency. $\bigcirc$, $H_1 \infty 1.8$ V; $\square$, $H_1 \infty 0.15$ V.

spin exchange were solved for each hfs level taking into account that the inversion of the electronic spins by the rf pulses is achieved instantaneously.

The first step in the experiment was to adjust the inverting pulses as shown in Figure 8. For a pulse of the correct length and height the average polarization was at a minimum. Then the effect of spin exchange collisions between $^3\text{He}^+$ and the polarized cesium beam was investigated by observing the average polarization for different intervals T_p between the inverting pulses. From such measurements the spin exchange cross section for the He^+–Cs system was derived and is $Q_e = (1.4 \pm 0.8) \times 10^{-14}$ cm². This ion–atom spin exchange cross section is of similar size to that in atom–atom spin exchange, showing that polarization effects due to the ion charge are small.

5. Other Collision Studies

The present experiments on spin-dependent charge transfer and spin exchange described have been carried out in connection with our spectroscopic work on simple ions and ion molecules, in which suitably chosen collision processes are being employed to polarize and analyze stored ions. We have also studied the production by electron impact[20] and the reaction dynamics of H^+, H_2^+, H_3^+, $^4\text{He}^+$, and HeH^+ in a rf quadrupole ion trap.[21] These collision studies were performed with unpolarized collision partners and are, therefore, only mentioned here.

ACKNOWLEDGMENT

Research supported in part by the National Science Foundation and the Robert A. Welch Foundation.

References

1. D. P. Stevenson and D. O. Schissler, *J. Chem. Phys.* **29**, 282 (1958).
2. R. P. Clow and J. H. Futrell, *Int. J. Mass. Spectrom. Ion Phys.* **8**, 119 (1972).
3. E. E. Ferguson, F. C. Fehsenfeld, and A. L. Schmeltekopt, *Advan. At. Mol. Phys.* **5**, 1 (1969).
4. K. T. Dolder, M. F. A. Harrison, and P. C. Thoneman, *Proc. R. Soc. London* **A264**, 367 (1961).
5. H. G. Dehmelt, *Advan. At. Mol. Phys.* **3**, 53 (1967); **4**, 109 (1969).
6. H. A. Schuessler, *Progress in Atomic Spectroscopy*, p. 999, Plenum Press, New York (1978).
7. D. R. Bates and R. McCarrol, *Proc. R. Soc. London* **A247**, 175 (1958).
8. F. T. Smith and R. E. Olson, *Phys. Rev. A* **7**, 1529 (1973).
9. T. G. Winter and N. F. Lane, *Phys. Rev. A* **17**, 66 (1978).
10. R. A. Heppner, F. L. Walls, W. T. Armstrong, and G. H. Dunn, *Phys. Rev. A* **13**, 1000 (1976).
11. R. D. DuBois, J. B. Jeffries, and H. G. Dunn, *Phys. Rev. A* **17**, 1314 (1978).
12. L. Spitzer, *Physics of Fully Ionized Gases*, p. 76, Interscience, New York (1956).
13. R. Ifflander and H. G. Werth, *Metrologia* **13**, 167 (1977).
14. H. A. Schuessler, *Metrologia* **13**, 109 (1977).
15. J. B. Delos and W. R. Thorson, *Phys. Rev. Lett.* **28**, 647 (1972).
16. N. Y. Demkov, *Sov. Phys. JETP* **18**, 138 (1964).
17. R. E. Olson and F. T. Smith, *Phys. Rev. A* **7**, 1529 (1973).
18. D. Rapp and W. E. Francis, *J. Chem. Phys.* **37**, 2631 (1962).
19. H. A. Schuessler, *Phys. Lett.* **30A**, 350 (1969).
20. H. A. Schuessler, *Proceedings of the Conference on Plasmas in Atomic Physics*, Knoxville, Tennessee (1977).
21. H. A. Schuessler, *Bull. Am. Phys. Soc.* **23**, 532 (1978).

Scattering Experiments with Laser-Excited Atoms

R. DÜREN

Techniques and results of measurements of differential cross sections for laser-excited Na atoms at thermal collision energies are described for mercury and argon as targets. For these examples the evaluation of the experimental data is described, including the determination of the potentials for the interaction in the 3^2P states of Na. In the calculation of the scattering process a close-coupling scheme is employed. The potential matrices for these calculations are obtained in the case of Hg from a pseudopotential calculation incorporated into the evaluation of the measurements, in the case of Ar from a modification of results from CI calculations given previously by other authors.

1. Introductory Survey

The goal of the investigations considered in this report is the determination of the potentials for the heavy-particle interaction in the excited states. More specifically, this goal is achieved by measuring scattering cross sections of an atomic species in an electronically excited state in a beam experiment. This type of investigation is one of the extensions of beam work that has been achieved in recent years and it is a technical achievement—the development of CW-dye lasers—that has made these investigations possible. With these lasers as a light source, a sufficiently high fraction of the atomic beam can be brought into one excited state and the scattering cross sections for this species in that excited state can be studied.

Such scattering experiments have for a long time served as a source of information for the heavy-particle interaction potential in the ground state. In particular,

R. DÜREN • Max-Planck-Institut für Strömungsforschung, D3400 Göttingen, West Germany.

for spherically symmetric interaction pairs without inelastic processes the accuracy of the interaction potentials has been brought below 5% for, e.g., ε, the well depth, and r_m, the internuclear distance of the potential minimum.[1,2] Obviously the hope in the investigations considered here is to do equally well for the excited states. As we will try to demonstrate in this report, this indeed can be reached even though the required effort both in the experimental and the theoretical aspects is considerably higher. The reason is that for the case most investigated, namely, the scattering of Na $2^2P_{3/2}$ or $2^2P_{1/2}$, fine-structure transitions immediately lead to inelastic processes requiring a coupled system in the theoretical treatment.

The same goal—the determination of the potentials for the excited states—is also pursued with other experiments. The most elaborate work is probably the investigation of the far wings of perturbed spectroscopic lines.[3] Most recently the production of van der Waals molecules in jet beams and their spectroscopic analysis has been used.[4] It is obvious that the well-known high accuracy of these spectroscopic methods can hardly be obtained by any other method. On the other hand the difficulties encountered in the evaluation of the data seem to be also great in practice because of the scheme of evaluation and in principle because of the fact that the primary measurement is always a difference between the excited states and the ground-state potential. Finally, this is the goal of many quenching experiments in cells, where the averaging over many atoms poses a serious problem.

The scattering methods may be divided into measurement of integral cross sections and measurement of differential cross sections. Both methods are the same in their basic connection between measured process and potential, but they differ in their particular range of sensitivity: The integral cross-section measurements are most sensitive to the behavior of the potential for large internuclear distances. The differential cross sections, if measured at thermal collision energies, are most sensitive to the region of the potential minimum.

In the integral cross-section measurements the inelastic cross section for transitions from one initial fine-structure level into the other one is usually determined by observing the light from the decay of the exit channel. The light from the initial channel is readily suppressed by filters, and the intensity of light from the exit channel is proportional to the integral inelastic cross section. Experiments have been published for K–He[5] and for the energy dependence of the cross sections for Na interacting with Ne, Ar, and Kr.[6]

In the differential cross-section measurements the scattered particles in the excited state are detected as a function of the scattering angle. A straightforward experiment is the measurement of all the scattered particles without further analysis in the exit channel[7–9] or an energy analysis with a resolution of the order of some vibrational levels of the target molecules.[10,11] A certain disadvantage of this type of experiment is that the fine-structure states are not resolved in the exit channel. But it must be noticed nevertheless that until now the potentials have been determined only from this type of experiment, namely, for Na–Ne,[7] for Na–Hg,[8] and for Na–Ar.[9] Obviously the reason is that the potentials for the fine-structure states are theoretically well understood. Nevertheless the resolution of the fine structure remains an interesting challenge to the experimentalist, and only very recently has this goal been achieved by Kinsey and Pritchard and co-workers[12] for Na–Ar.

Besides the experimental progress in the determination of potentials for the excited states, there is considerable effort in the calculation of these potentials, notably for the "simple" alkali–rare-gas interaction. Beginning with Baylis' work,[13] the pseudopotential methods have been developed.[14,15] Even though some dramatic discrepancies have been detected, it seems that these potentials can be taken as a starting point, and further work is in progress. A new standard of accuracy has been set by an *ab initio* calculation for Na–Ar.[16] In summary, a satisfactory scope of material is available for the comparison of experimental and theoretical results.

Finally, in the survey of the present status it must be noticed that the connection between the scattering experiments and the potentials is also well established[17,18] for the atom–atom interaction considered here.

In the following section we will concentrate our attention on this type of work and will discuss it in detail, referring to examples that we have worked out in our laboratory during the last several years. The experimental work, the potential calculation, and the scattering calculation will be surveyed. In this way the complete path leading from experimental data to the determination of the potentials is described. This may serve as an illustrative example for relatively simple systems that may be extended in future work.

2. Atom–Atom Scattering Experiment

In the excited-state experiments considered here, the ideal case is represented by transitions of the type

$$A(n, l, j, m_j) + B(0) \rightarrow A(n', l', j', m_j') + B(0)$$

Atoms A specified with respect to the quantum number n, l, j, and m_j colliding with target atoms B assumed to be and to stay in the ground state would form the initial state. In the exit channels the atoms A would be observed again completely analyzed with respect to their quantum numbers. The cross sections for such a process would be described as

$$\frac{d\sigma}{d\omega} = \frac{d\sigma}{d\omega} (n'l'j'm_j' \leftarrow nljm_j)$$

In the real experiment two averagings have to be taken into account: First, for the initial state, we have to take into account a distribution, and second we have to sum over various nonresolved quantum numbers of final states. The initial-state distribution will be known in our experiments, and for the final states we will in particular have to sum over j and m_j (note, however, Reference 12).

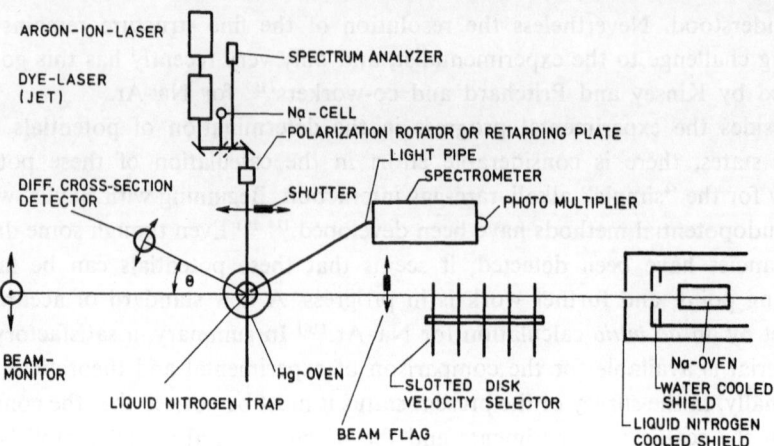

Figure 1. Experimental setup for the measurement of the differential cross section for laser-excited Na.

Figure 1 shows our realization of such an experiment in a survey; a detailed description has been given earlier.[8]

The Na beam is produced in a jet oven, passes then through a mechanical velocity selector of 3% resolution, and enters the scattering area. The primary beam is monitored by a beam monitor. Scattered Na atoms are measured with the differential detector, which moves around the scattering center. The scattering center is defined by the intersection of the Na beam with the target beam. The latter one, directed out of the plane of the drawing, is also produced in a jet oven. Thus we measure the scattered particles in an out-of-plane arrangement. So far the apparatus is a common one often used in ground-state investigations.

The laser part of the apparatus for the excitation consists of a CW-dye laser pumped by an argon-ion laser. The light from the CW-dye laser is directed by mirrors into the scattering volume. For the excitation we use the transition from $F = 2$ in the ground state to $F = 3$ in the $P_{3/2}$ state. To guarantee complete excitation throughout the scattering volume the light passes through a telescope (not shown). The power of the laser necessary to obtain saturation is of the order of 30 mW. The light from the laser is linearly polarized and we rotate the direction of polarization such that it is parallel to the most probable relative velocity in the collision. To monitor the wavelength of the laser so that it matches the resonance line, we measure the fluorescence from the atomic beam. (This together with the spectrum from the spectrum analyzer serves as feedback for a stabilization of the laser as described elsewhere.[19])

The measurement of the differential cross section for the excited state is performed by switching the laser on and off. The observed difference in the scattering signal is then proportional to the difference in the cross sections for the ground-state and the excited-state interaction.

The initial-state distribution is given by the quantum number $j = 3/2$ for the particular transition used and its projection quantum numbers. In the stationary limit (that is, after many spontaneous decay cycles) and for linearly polarized light it is given by

$j, m_j =$	$3/2, -3/2$	$3/2, -1/2$	$3/2, 1/2$	$3/2, 1/2$
$W(j, m_j) =$	$7/84$	$35/84$	$35/84$	$7/84$

These projection quantum numbers refer to the electric field vector as the quantization axis. As mentioned above, we rotate this axis into the most probable relative velocity, which simplifies the theoretical treatment of the scattering process.

For the interaction of Na with Hg the results obtained from the difference signal show a strong oscillatory behavior,[20] which is partly due to the ground-state contribution with its large rainbow oscillations. Above the rainbow angle for the ground state we observe both for Na–Hg and Na–Ar other oscillations which are certainly due to the interaction in the excited state, since in this range the ground-state cross section is monotonic. In our evaluation given below we have for both systems restricted our attention to the excited-state oscillations, and this seems to be sufficient. An example of such a cross section for Na–Hg is given in Figure 3.

Qualitatively the observed oscillations are of somewhat different origin for the two systems Na–Hg and Na–Ar. The coarse structure of the cross section for Na–Hg is given by oscillations appearing for an orbiting case due to the deep potential well for the Π-state interaction.[8] For Na–Ar the well is less pronounced and the oscillations are mainly determined from rainbow scattering.[9]

3. Atom–Atom Scattering Theory

For the scattering experiments considered here, characterized by atom–atom interaction at thermal energies with rare-gas-like targets and involving the lowest P state of the alkali atom, the theoretical treatment is well elaborated[17,18,8] and relatively simple. The main reason for this simplicity is the fact that under these conditions the coupling between the ground state and the excited state is negligible. This same fact allows the potential calculations to be simple and nevertheless comparatively accurate as judged from experiments like the ones discussed here. Another hint of this fact is the relatively high threshold for the excitation of the P states in collisions of ground-state atoms[21,22] in the range of at least several electron volts. Similarly the couplings of the higher states can be safely neglected.

Under these assumptions the qualitative behavior of the potentials for which the scattering process is to be calculated (depicted in Figure 2) is characterized by one or two avoided crossings of the $m_j = \frac{1}{2}$ states. Our examples Na–Hg and Na–Ar are representative of two different cases. In the case of Hg as target, two well-separated avoided crossings (Figure 2b) are clearly visible. In the case of Ar both crossings are also present in principle, but in practice only one broad avoided crossing at a large internuclear distance is found. The pronounced difference in the behavior of the

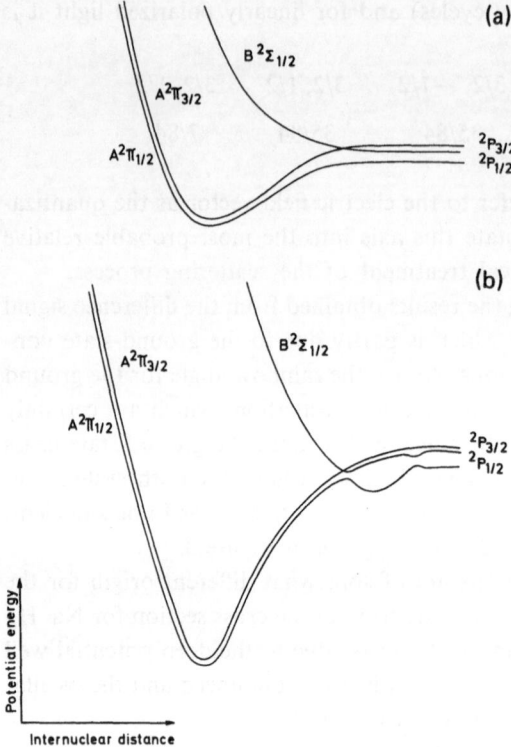

Figure 2. Qualitative potential functions: (a) example for large fine-structure splitting and/or relatively small polarizability of the target (e.g., Na–Ar); (b) example for small fine-structure splitting and/or large polarizability (e.g., Na–Hg).

potential curves can be understood from the values of the fine-structure splitting and the polarizability,[15] the latter being large for Hg and smaller for Ar. Asymptotically the corresponding molecular states connect to the $^2P_{3/2}$ state with the two projections $m_j = \frac{3}{2}$ and $\frac{1}{2}$ and to the $^2P_{1/2}$ state with $m_j = \frac{1}{2}$.

The adiabatic potentials for these states (remember the neglect of all other states) are simply expressed in terms of two radial matrix elements for the P state $V_{11}^{(0)}$ and $V_{11}^{(2)}$:

$$W(^2\Pi_{3/2}) = V_{11}^{(0)} - \tfrac{1}{5}V_{11}^{(2)}$$

$$W(^2\Pi_{1/2}) = -\Delta\varepsilon/2 + V_{11}^{(0)} + \tfrac{1}{10}V_{11}^{(2)} - D \qquad (1)$$

$$W(^2\Sigma_{1/2}) = -\Delta\varepsilon/2 + V_{11}^{(0)} + \tfrac{1}{10}V_{11}^{(2)} + D$$

with

$$D = \tfrac{1}{10}(25\Delta\varepsilon^2 + 10\Delta\varepsilon V_{11}^{(2)} + 9V_{11}^{(2)^2})^{1/2}$$

where $\Delta\varepsilon$ is the fine-structure splitting and where the energy origin is at the $^2P_{3/2}$ level. A spin–orbit interaction independent of the internuclear distance has been assumed. $V_{11}^{(0)}$ and $V_{11}^{(2)}$ are the zeroth- and second-order multipole moments of a multipole expansion of the atomic interaction.

For the scattering calculation[17,18] the expansion of the complete total wave function with respect to the total angular momentum yields the interaction matrix as a product of angular momentum terms and the two radial matrix elements described above. Writing out the respective scattering equations one obtains a set of six coupled channels, which because of parity decouple into two sets of three coupled channels. These equations are treated in standard ways to obtain, e.g., the T matrix and with it the scattering amplitude, and the completely specified cross sections as described above. Weighting for the initial states and summing over the nonresolved quantum number yields the measured quantity.

The principle and practice of the integration are more or less standard and need not be elaborated here. Instead of these details let us consider the implications for the determination of the potential.

First it is noticed that the matrix elements used in this scattering calculation are exactly the same as those obtained in pseudopotential calculations of the potentials. This offers the possibility of combining pseudopotential calculations and scattering experiments in a very direct way. Instead of the variation of the parameters of a phenomenological potential for the heavy-particle interaction, the parameters of the pseudopotentials may be varied to fit the experimental cross section. In some cases, namely, where the parameters refer to the inert target, these will turn out to be invariant[16] for different primary particles.

Second, however, the matrix elements can be obtained from adiabatic potentials as well by inverting equations (1) (under the restrictions mentioned above). To fit the experimental data, a variation of the diabatic matrix elements can then be performed.

Which of these ways is preferred depends among other things upon the quality of the available calculated potentials. In our laboratory we have applied both methods for Na–Hg and for Na–Ar, respectively.

In both cases the final result must be considered from two aspects. One is that a potential function reproducing the experimental data has been established. Thus our main goal is reached and no restrictions are imposed on the way this potential has been established. Second, an attempt is made to differentiate the measurements further by either establishing the underlying pseudopotentials or at least by a modification of known matrix elements.

4. Comparison of Experimental and Theoretical Results

Figure 3 shows the calculated and the exper mental results for Na–Hg at a particular energy where the best fit of the pseudopotential has been obtained. The oscillatory structure of the cross section in both curves is clearly visible and a match of these oscillations throughout the angular range is found. The fit at another energy is less exact; a displacement of the oscillations by $2°$ is observed there.[8] This discrepancy has not been resolved by further variations.

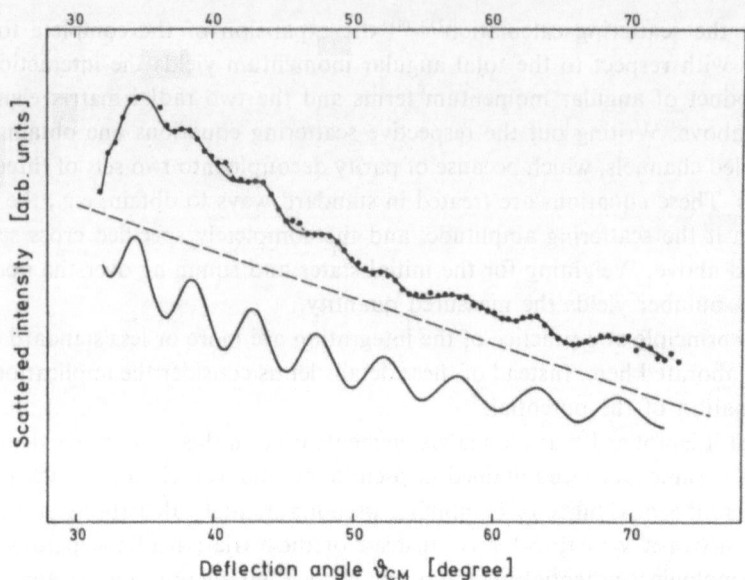

Figure 3. Comparison of experimental and theoretical differential cross sections for Na($^2P_{3/2}$)–Hg. (Arbitrary scale of intensities.) (Top) Experiment: Na($^2S_{1/2}$), Na($^2P_{3/2}$) + Hg; $E = 8.26 \cdot 10^{-3}$ a.u. (Bottom) Theory: total differential cross section; Na($^3P_{3/2}$) + Hg; $E = 8.26 \cdot 10^{-3}$ a.u.; $\varepsilon_* = 1.117 \cdot 10^{-2}$ a.u.; $r_{m*} = 6.70$ a.u.

Table 1 gives the final results, characterized by the well depth of the $^2\Pi_{1/2}$ potential. Since the Na–Hg result is obtained in a pseudopotential fit and since other experiments or calculations are not available, no further comparison can be made. For Na–Ar there are comparable data available and a thorough discussion is presented elsewhere.[9] In summary, the agreement with spectroscopic investigations is very good. The discrepancy with calculated potentials, however, is considerable, especially with respect to the well depth where the differences range from a factor of 2 to 20%. The last value holds for the *ab initio* calculation, where most of the uncertainties inherent in the pseudopotential calculation are removed.

Table 1. Best-Fit Parameters for the Na–Hg and Na–Ar Interatomic Potentials

System	ε [a.u.]	r_m [a.u.]
Na–Hg	1.117×10^{-2}	6.70
Na–Ar	0.255×10^{-2}	5.75

5. Summary

In summary, the molecular beam technique with laser excitation of the beam particles is seen to yield the potential function for the excited states. In particular, at thermal collision energies a high sensitivity to the well depth is noticed, and in this regard the experiments are superior to scattering experiments used to determine integral cross sections, which are most sensitive to the asymptotic behavior. The advantage of these experiments when compared with spectroscopic methods is that only the potentials for the excited states determine the measured quantity.

In this work the complete determination of the potentials, exper mental aspects, scattering calculations, and the role of theoretical potentials have been surveyed. The very few cases where it has been applied may serve as a demonstration of the complete scheme.

In the applications the restrictions at present are basically to be found in the experimental work. The first restriction is that until now the analysis of the exit channel with respect to the fine-structure states has only been realized in one particular example.[12] Even though the potential is well determined without this type of measurement this higher degree of differentiation is clearly desirable.

Second there is a restriction on the types of excitable beam atoms because of the available CW-dye lasers. Further developments in the field of lasers indicates, however, that this restriction will gradually be overcome.

References

1. R. Düren and Ch. Schlier, *J. Chem. Phys.* **46**, 4535 (1967); R. Düren, G.-P. Raabe, and Ch. Schlier, *Z. Phys.* **214**, 410 (1968).
2. U. Buck and H. Pauly, *Z. Phys.* **208**, 390 (1968).
3. G. York, R. Scheps, and A. Gallagher, *J. Chem. Phys.* **63**, 1052 (1975).
4. L. Wharton, D. A. Auerbach, D. H. Levy, and R. E. Smalley, in *Advances in Laser Chemistry*, A. H. Zewail, Ed., Springer, Berlin (1978).
5. R. W. Anderson, T. P. Goddard, C. Porravand, and J. Warner, *J. Chem. Phys.* **64**, 4037 (1976).
6. W. D. Phillips, C. L. Glaser, and D. Kleppner, *Phys. Rev. Lett.* **38**, 1018 (1977).
7. G. M. Carter, D. E. Pritchard, M. Kaplan, and T. W. Ducas, *Phys. Rev. Lett.* **35**, 1144 (1975).
8. R. Düren and H. O. Hoppe, *J. Phys. B* **11**, 2143 (1978).
9. R. Düren and W. Gröger, *Chem. Phys. Lett.* **56**, 67 (1978); *Chem. Phys. Lett.* **61**, 6 (1979).
10. I. V. Hertel, H. Hofmann, and K. J. Rost, *Phys. Rev. Lett.* **38**, 343 (1977).
11. J. A. Silver, N. C. Blais, and G. H. Kwei, *J. Chem. Phys.* **67**, 839 (1977).
12. W. D. Phillips, J. A. Serri, D. J. Ely, D. E. Pritchard, K. R. Way, and J. L. Kinsey, *Phys. Rev. Lett.* **47**, 937 (1978).
13. W. E. Baylis, *J. Chem. Phys.* **51**, 2665 (1969).
14. J. Pascale and J. Vandeplanque, *J. Chem. Phys.* **60**, 2278 (1974).
15. R. Düren, *J. Phys. B* **10**, 3467 (1977).
16. R. P. Saxon, R. E. Olson, and B. Liu, *J. Chem. Phys.* **67**, 2692 (1977).
17. F. H. Mies, *Phys. Rev. A* **7**, 942 (1973).
18. R. H. G. Reid, *J. Phys. B* **6**, 2018 (1973).

19. R. Düren and H. Tischer, Report 4/1978 Max-Planck-Institut für Strömungsforschung, Göttingen (1978).
20. R. Düren, H. O. Hoppe, and H. Pauly, *Phys. Rev. Lett.* **37**, 743 (1976).
21. W. Mecklenbrauck, J. Schön, E. Speller, and V. Kempter, *J. Phys. B* **10**, 3271 (1977).
22. R. Düren, U. Krause, and G. Moritz, *Chem. Phys. Lett.* **56**, 62 (1978).

37

Orbital Angular Momentum State Coherence in Charge Transfer to the n=2 Hydrogen Levels by Fast Protons in Gases

L. LILJEBY, S. MANNERVIK, S. HULTBERG, AND I. A. SELLIN

We find evidence for strong orbital angular momentum state coherence in charge transfer to the $n = 2$ hydrogen levels by protons traversing He, Ar, and O_2 targets at velocities of ~2 a.u. Collision-averaged coherence of the amplitudes f_s, f_{p_0}, describing mixed-parity, oscillating electric dipole moments of the $n = 2$ state electronic charge distributions formed in these single, ion–atom charge transfer collisions are established by a Lyman-alpha quantum-beat difference-signal technique suggested by Eck. Direct comparisons with the earlier solid-target results of Sellin et al. should permit disentanglement of inherently solid-state effects from those due to binary ion–atom encounters. Similarity of the beat amplitudes in solid versus gas targets lends support to the "last-layer" (gas-layer?) hypothesis. Rejection of unwanted, mixed-parity state decay noise in searches for parity violation in fast H beams is discussed. Relevant details of the experimental method are given. Data concerning the relative phases of the difference beats in the three gases are presented. Our most recent results, concerning the beam velocity dependence of the difference beat pattern, are also given.

In the first short accounts of our very recent observations of collision-averaged excitation coherence between $2s$ and $2p_0$ amplitudes (f_s, f_{p_0}) arising from charge transfer to the hydrogen $n = 2$ levels for protons traversing He, Ar, and O_2 targets,

L. LILJEBY, S. MANNERVIK, AND S. HULTBERG • Research Institute of Physics, Stockholm, Sweden.
I. A. SELLIN • University of Tennessee and Oak Ridge National Laboratory, Oak Ridge, Tennessee 37830, U.S.A.

we presented a few preliminary but characteristic results concerning the extent of coherence of these amplitudes.[1] Here space permits the presentation of more details of the experimental method than was previously possible, the inclusion of more data obtained up to the present time, a discussion of results in density-matrix terms, and the presentation of first data concerning beam-velocity dependence of the f_s, f_{p_0} coherence beats observed in Lyman α emission.

Before proceeding to the discussion of methods and previously unpublished results, a summary of earlier work is given.

In Reference 1 we presented evidence for a high degree of excitation coherence in the electron capture to mixed-parity $n = 2$ states of hydrogen by fast protons in gases. To our knowledge there had been no previous experimental observation of such a phenomenon in single-collision charge transfer processes. The many theorists working in the currently very active field[2] of charge transfer seem not to have concentrated on calculating the $s-p$ capture amplitude differences as a test of various competitive theories of charge transfer, presumably because they were unaware that such phase differences can be measured. It had been shown by Sellin *et al.*[3] that such mixed-parity-state excitation coherence is observable in beam–foil excitation; since then it has been of interest to inquire whether such beats are induced by some kind of exit surface capture or electric field effect as suggested by Eck,[4] and thus characteristic of a solid-state effect, or whether such collision-averaged, mixed-parity-state coherence is a prominent feature of charge transfer and perhaps other single ion–atom collision processes. Comparison of the character of electric dipole coherence in foil versus gas targets might then become a promising tool for sorting out intrinsically solid-state from binary ion–atom collision coherence phenomena. We note also that $s-d$ coherence (same parity, different l) had been noted by Burns and Hancock[5] in a solid-target experiment.

The technique suggested by Eck depends on the use of electric probe fields **E**, alternately parallel and antiparallel to the beam, to exploit the fact that an excitation-coherence quantum-beat signal is odd under field reflection, whereas the sum s gnals are even. A simple physical picture of the relationship between true excitation coherence and field reversal is as follows. If there is an initial displacement of the electron charge cloud with respect to the proton, or one develops in time because of an inequality in proton and average electron axial velocity, the displacement will be either enhanced or diminished depending on the direction of **E** relative to that of the charge displacement. Incoherent coupling and quenching effects due to the field, light-intensity anisotropies, etc., depend on $|\mathbf{E}|$, but not on whether **E** is parallel or antiparallel to the quantization axis defined by the common axis ($+z$) of the beam and **E**. Because of axial symmetry Eck also argues that only $m = 0$ states are coupled by the excitation process. Eck's first-order perturbation-theory expression for the Lyman α beat signal is

$$(V/\omega)^2[\tfrac{1}{3}(\sigma_{p_0} + 2\sigma_{p_1}) - \sigma_s]\cos \omega t + 3^{-1/2}(V\omega_0/\omega^2)\langle|f_s|\,|f_{p_0}|\cos \alpha\rangle\cos \omega t$$
$$+ \ 3^{-1/2}(V/\omega)\langle|f_s|\,|f_{p_0}|\sin \alpha\rangle\sin \omega t \qquad\qquad (1)$$

plus terms smaller by a factor of $\Gamma/2\omega$, all exponentially damped with a damping constant $\Gamma/2$. Here Γ is the average of the perturbed s- and p-state decay rates, ω_0 is the Lamb shift, ω is the perturbed $s_{1/2}$–$p_{1/2}$ level splitting ($\equiv \omega_0$ for $|\mathbf{E}| = 0$), V is the dipole matrix element $\langle s_{1/2} | e\mathbf{E} \cdot \mathbf{r} | p_{1/2} \rangle = -3^{1/2}eEa_0/\hbar$, $t = 0$ at the time of excitation, and the excitation amplitudes for s- and p-state excitation are $f_s = |f_s| \exp(i\alpha_s)$ and $f_{p_0} = |f_{p_0}| \exp(i\alpha_p)$, with $\sigma = \langle |f|^2 \rangle$ and $\alpha \equiv \alpha_s - \alpha_p$. The angular brackets refer to collision averages. In our case, where the perturbation is large, this expression is only qualitatively accurate, but can be substantially improved by using a better value for ω obtained by diagonalizing the 3×3 perturbation matrix, incorporating the $p_{3/2}$ state ($\sim 10\omega_0$ away in energy). Hence the sum of the signals yields the incoherent oscillations superposed on nonoscillating perturbed $s_{1/2}$, $p_{1/2}$, and $p_{3/2}$ decays, while the difference yields only the coherent excitation terms, which contain the s–p phase coherence angle α. If $\bar{\alpha} = 0$ (or $\bar{\alpha} = \pi$), for example, the coherent part of the initial wave function would involve $|f_s| u_s \pm |f_{p_0}| u_p$, corresponding to a concentration of the electron charge distribution in the backward (forward for the negative sign) hemisphere. Here u_s and u_p are the field-free spatial eigenfunctions of the $2s$ and $2p$ states. If $\bar{\alpha}$ were near $\pi/2$, however, or $\sin \bar{\alpha} \sim 1$, then the initial wave function would involve $i |f_s| u_s + |f_p| u_p$, corresponding to roughly zero *initial* charge-distribution asymmetry, but one which would reach peak concentration in alternate hemispheres at the subsequent times $t = \pi/2\omega$, $t = 3\pi/2\omega$, etc. Thereafter, the charge-distribution asymmetry would continue to swing periodically between the two limits. For intermediate $\bar{\alpha}$, some combination of initial distortion and initial "velocity" asymmetry would prevail.

The lower curves in Figure 1 display a direct comparison of difference beat signals for thin He targets and C (~ 15 μg/cm^2) targets[3] at the same beam energy (186 keV) and field strength (525 V/cm). A uniform normalization was obtained by normalizing the average intensity of the sum signal from the solid-target data[3] to the present He sum signal just downstream from the cell, facilitating direct comparison of the difference signals on the same vertical scale. The upper curve shows a higher point density signal for $E_{\text{para}} = +525$ V/cm.

It is seen that the excitation coherence beat signal in He is very similar in size, relative to the total intensity, to that in the solid-target case. The raw difference signals have $\sim 7\%$ amplitude relative to the sum signals in the C target case, and $\sim 10\%$ in the He target case. As the cell is only 0.5 mm thick internally, it is difficult to estimate target thickness accurately. It was, however, established by signal-versus-pressure-linearity checks that, to a good approximation, single-collision conditions prevailed for all gases. Hence the results are characteristic of charge transfer to the $n = 2$ states. The He difference beat signal seems to be shifted toward positive values. Part of this shift is certainly due to a beam-induced microdischarge problem creating stray light when the downstream plate is negative, but a real shift cannot be completely ruled out at present.

The similarity of amplitudes in gas versus solid targets lends support to the "last-layer" hypothesis often used in beam–foil excitation models.[6] In fact, the gas

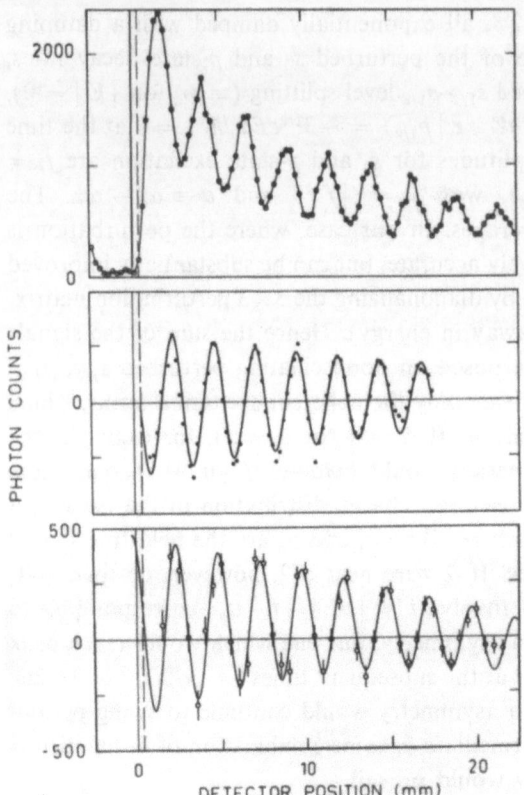

PHOTON COUNTS

DETECTOR POSITION (mm)

Figure 1. Variation of signal strengths with distance downstream at $E = 525$ V/cm and 186 keV proton energy. The lowest curve shows a difference beat signal ($E_{para} - E_{antipara}$) for a ~1 mTorr cm-thick He target. Above it appears a difference signal for the ~15 μg/cm² C target of Reference 4, normalized as explained in the text. The top curve shows an E_{para} signal in the He target, with line segments to guide the eye. The solid line marks the geometric center of the gas cell. The dashed line marks the effective center of the cell arising from the phase delay discussed in the text, and also marks the foil position.

layer adhering to exit surfaces may *cause* the similarity of the single collision and solid target beat amplitudes. Of course the distributed length of a gas cell reduces the apparent modulation amplitude because of the wave form averaging brought about by the spread in place (time) of formation. But as long as the effective length of the cell is short compared to the beat wavelength (~30% in Figure 1), such amplitude reduction is entirely tolerable (~0.8×).

A widespread misunderstanding has been propagated in the beam–foil literature for years.[6] Misinterpretation of the intent of Macek's use of the impulse approximation in his early paper[7] on quantum beats in beam–foil light combined with misunderstanding of how the uncertainty principle applies has long delayed application of the time-of-flight quantum beat method to the gas target case. As clearly demonstrated in Figure 1, the short time of passage through a foil target ($\lesssim 10^{-14}$ sec) represents no fundamental advantage whatever in achieving the impulsive excitation needed to observe superposition states. Since the states are produced in single collisions, if there is a coherence time, it is instead the much shorter binary collision time ($\lesssim 10^{-16}$ sec) that is relevant.

One should then expect coherent excitation of any multipole moment of any excited n state, consistent with symmetry and single-collision excitation times, for many ion–atom collision processes, and even coherent excitation of different n states.

Figure 2 provides a schematic diagram of the apparatus. The experiment was performed by passing a 0.5-mm-diameter, 0.5°-divergence, 186-keV proton beam from the Research Institute of Physics 400-keV isotope separator through a $\lesssim$0.8-mm-long gas cell having 0.5-mm apertures and whose 0.5-mm rear wall formed the upstream boundary of a parallel-plate condenser oriented perpendicular to the beam, and whose symmetry axis coincides with that of the plates. Fields of alternate polarity were established by applying voltages to the downstream plate at 25-mm relative separations. Signals were obtained using a BX762 Bendix encapsulated Channeltron in perpendicular viewing geometry; the Channeltron was sensitive between 1150 and 1900 Å and had a peak sensitivity near the Lyman α line. The detector viewed the radiation through a pair of vertical 0.5-mm straight-edged slits; the slits were 1 cm tall and located 5 cm and 16 cm from the horizontal beam. The finite trapezoidal field of view caused waveform averaging, producing an effective ratio of observed to actual waveform amplitude of about 0.9. Normalization of signal strength per incident particle was accomplished by a Faraday cup mounted behind a 1-mm hole in the downstream plate. The combined cell length and viewing length waveform attenuation factor is thus $\sim\frac{3}{4}$, so that the difference signals in Figure 1 are actually $\sim\frac{4}{3}$ as large as shown. The gas-cell motion was regulated by a stepping motor and an electronic control unit which also supervised the storing of data in a multichannel analyzer. The stepping speed was regulated by beam-current integration.

The solid curve drawn through the He difference oscillations in Figure 1 results

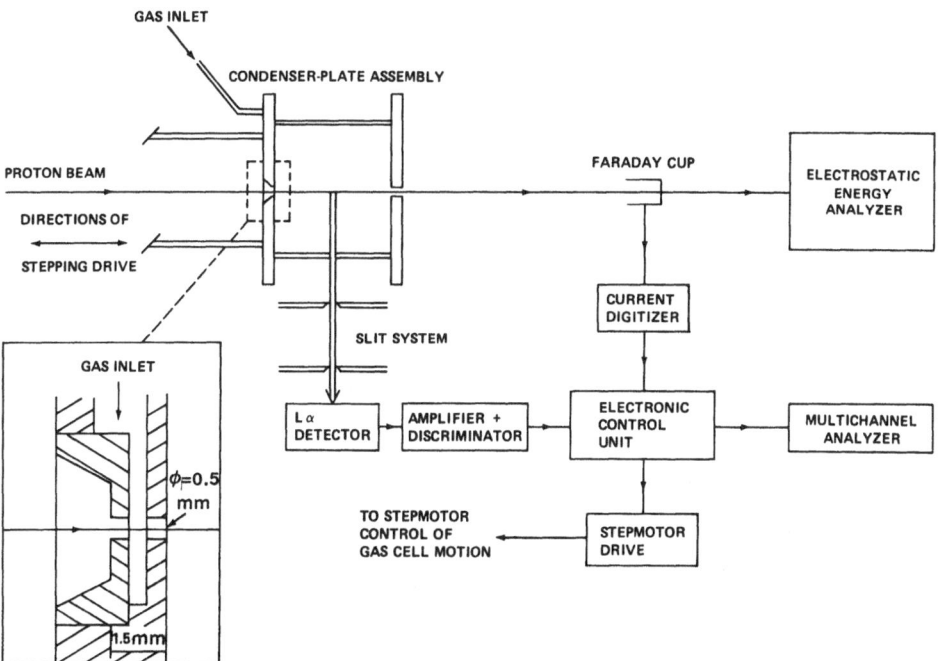

Figure 2. Experimental setup at the 400-keV accelerator.

from least-squares fits to the form $A_{exp}(-\Gamma t)\cos(\omega t - \phi)$, added to a second-degree polynomial background needed to account for the stray light in the 1150–1900-Å region previously mentioned. The fitted frequency and damping constants were $\omega = 15.32 \pm 0.06$ GHz and $\Gamma = 0.33 \pm 0.05$ GHz, compared to calculated values of 15.37 and 0.31 GHz derived from an elementary 3×3 Stark matrix diagonalization. The curve for the solid target data was instead fitted[3] to the coefficients of the difference terms in equation (1), using calculated ω and Γ values.

Caution should be used in interpreting the apparent difference in ϕ for solid and gas targets. Since the field penetrating the gas cell must rise from a small value to full field in a distance of ~ 1 mm, during which time the beat frequency evolves from ω_0 to ω, there is an apparent phase shift which is manifested in the failure of the fitted sum curves to exhibit an extremum at the center of the gas cell. This phase delay is approximately given by the time delay corresponding to half the effective gas cell thickness times $(\omega - \omega_0)$. Using this crude estimate, an apparent gas cell center position can be inferred. This apparent position is indicated by the dashed vertical line in Figure 1. Referred to this position, the phase of the difference oscillations in He—and in fact also in Ar and O_2—versus C targets are qualitatively similar, and the "negative-going" difference oscillations cross the base line near the origin in all cases. Again, a connection between ion–atom collision and the "last-layer" hypothesis—perhaps a *gas* layer—may be implied. As suggested in Reference 1, the initial phase shift due to the field rise time in the present experiment could be eliminated by arranging for either sudden entry into the field at the cell boundary, or by establishing the same electric field within the cell.

Macek and his colleagues have made a more complete analysis[7] of extremely similar-appearing difference signal data from their later, solid-target experiments, performed at overlapping energies. In their data, the negative-going difference-signal fit also crosses the base line at slightly negative times. In contrast to the conjecture in Reference 3 that the observed, negative-going maximum would correspond to peak concentration in the backward hemisphere, Macek has calculated that the electron is ahead of the proton on leaving the foil by $\gtrsim 0.5a_0$. For the case of p in C at 210 keV beam energy, Macek *et al.* have written the complete (trace 1) density matrix in the form

$$
\begin{pmatrix}
0.56 & 0 & 0.22e^{2.06i} & 0 \\
0 & 0.15 & 0 & 0 \\
0.22e^{-2.06i} & 0 & 0.14 & 0 \\
0 & 0 & 0 & 0.15
\end{pmatrix}
$$

Here the rows and columns are labeled according to σ_s, σ_{p_1}, σ_{p_0}, $\sigma_{p_{-1}}$, and the coherence is expressed by the off-diagonal elements σ_{sp_0}, whose maximum possible value is $|f_s||f_{p_0}|$. Since $|f_s||f_{p_0}| = 0.28$ in this case, the coherence is nearly maximal. Since in the present, gas-target data the difference beats are even larger relative to the uniformly normalized sum curves, it is tempting to conjecture that single-collision coherence is even more nearly maximal. A firm conclusion regarding

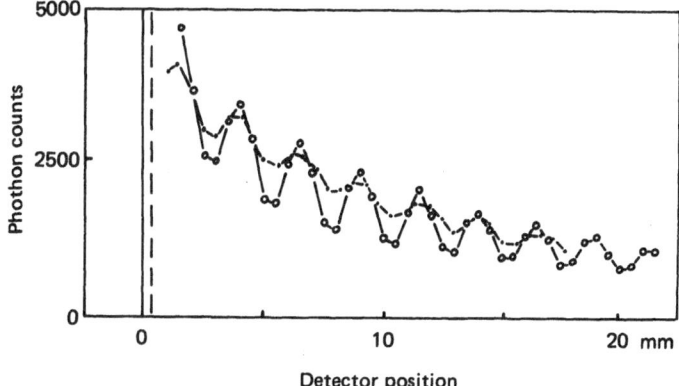

Figure 3. Comparison of sum signals, one from the present measurements with an He target (open circles) and one from the measurements of Reference 3 with a C foil, normalized to the same average intensity (solid points). The two vertical lines mark the position of the gas cell, as discussed in the caption of Figure 1.

this point must await better specification of the diagonal elements at the beam energy in question, since $\sigma_s : \sigma_{p_0} : \sigma_{p_{\pm1}}$ values for this beam energy seem not as yet to be available in the literature.

Figure 3 shows a direct comparison of the sum curves for the situation prevailing in Figure 1, normalized to the same mean intensity. The weaker modulation of the C target curve is undoubtedly related to the distinctly smaller ratio for σ_s/σ_{p_0} found in Reference 7 (4 : 1) than can be estimated from extrapolation of direct measurements of Hughes et al.,[8] up to 110 keV. We point out that the amplitude of modulation in Figure 3 will provide a direct independent measure of $\sigma_s : \sigma_{p_0}$ once the alignment $\sigma_{p_0} : \sigma_{p_{\pm1}}$ is sorted out by suitable polarization measurements. No additional absolute intensity calibrations or 2s quench-field geometry corrections applicable to the direct cross-section measurement technique need be applied.

Additional measurements have been carried out for Ar, O_2, and He target gases at two additional energies in order to determine relative phases and amplitudes. Preliminary results are given in Table 1.

Table 1. Comparison of Phases ϕ for Gases He, Ar, and O_2 at Three Different Energies[a]

Energy (keV)	Relative phase		
	$\Delta\phi(\text{He} - O_2)$	$\Delta\phi(\text{He} - \text{Ar})$	$\Delta\phi(O_2 - \text{Ar})$
110	-0.5 ± 0.3	-0.4 ± 0.2	0.1 ± 0.3
190	0.0 ± 0.3	0.0 ± 0.2	-0.1 ± 0.3
290	0.6 ± 1.0	0.5 ± 0.4	-0.1 ± 1.0

[a] The phase differences are given in radians. Errors are given as one standard deviation.

The remarkable constancy of ϕ for very different gases like O_2 and Ar deserves interpretation. One possible model, arising in a conversation of one of us (IAS) with C. Bottcher, is that the long-range longitudinal Coulomb interactions between a postcollision, excited H atom and a residual, singly charged ion, persisting at times long after charge cloud interpenetration has ceased, leads to the same result independent of target gas simply because the potential at long range is on the average nearly $1/r$ in all cases. Differing phases at different velocities could be understood in terms of the change in time scale for this long Coulomb tail to interact with the emergent excited atom.

Finally, we note that it was current interest in searches for parity violation in atoms that led us to speculate about the role of collisions in producing collision-averaged excitation of mixed-parity excited states. Admixture of unwanted $2p$ state amplitudes when, for example, H atoms in the $2s$ state are prepared by a charge transfer process, might constitute a noxious additional source of background, further complicating attempts to measure weak-interaction-induced admixtures of order 10^{-11} in this atom (or any other atom of interest) excited by collisional means, even if amplified by level-crossing techniques to $\sim 10^{-6}$. Fast-beam experiments of the kind discussed by Williams *et al.*[9] would certainly profit from high rejection of such collision-induced (but obviously parity-conserving) mixed-parity-state decay noise. Of course, in any admixture $\psi = a\psi_{2s} + b\psi_{2p}$, $\psi \to \psi_{2s}$ at sufficiently long times. But regeneration in residual gas can certainly also occur. If one assumes a good, practical vacuum $\sim 3 \times 10^{-8}$ Torr, a regeneration off-diagonal density matrix element $\sim 1 \times 10^{-15}$ cm^2, and a 1-cm path length in a typical microwave region, then $\psi_{2s} \to \psi_{2s} + \varepsilon \psi_{2p_0}$, with $\varepsilon \sim \sigma_{sp_0} \varrho l = \sigma_{sp_0} \pi \sim 1 \times 10^{-6}$. Perhaps the elaborate triple subtraction procedures envisioned in such experiments would then profit from extension to a quadruple subtraction procedure, wherein equivalent data for two beam directions through the apparatus would be needed.

ACKNOWLEDGMENTS

Work supported in part by the Swedish Natural Science Research Council, NSF, DNR, and NASA.

References

1. L. Liljeby, S. Mannervik, S. Hultberg, and I. A. Sellin, *Z. Phys.* **258**, 321 (1978); I. A. Sellin, L. Liljeby, S. Mannervik, and S. Hultberg, *Phys. Rev. Lett.* **42**, 570 (1979).
2. D. N. Tripathy and B. K. Rao, *Phys. Rev. A* **17**, 587 (1978), and numerous references therein; R. A. Mapleton, *Theory of Charge Exchange*, Wiley, Interscience, New York (1972).
3. I. A. Sellin, J. R. Mowat, R. S. Peterson, P. M. Griffin, R. Laubert, and H. H. Haselton, *Phys. Rev. Lett.* **31**, 1335 (1973).
4. T. G. Eck, *Phys. Rev. Lett.* **31**, 270 (1973).
5. D. J. Burns and W. H. Hancock, *Phys. Rev. Lett.* **27**, 370 (1971).

6. *Beam Foil Spectroscopy*, I. Martinson, J. Bromander, and H. Berry, eds., p. 344, North-Holland Publishing Co., Amsterdam (1970); *Beam Foil Spectroscopy*, S. Bashkin, ed., North-Holland Publishing Co., Amsterdam (1973); *Beam Foil Spectroscopy*, I. A. Sellin and D. J. Pegg, eds., Plenum Press, New York (1976).

7. J. Macek, *Phys. Rev. Lett.* **23**, 1 (1969); J. Macek, *Phys. Rev. A* **1**, 618 (1970); A. Gaupp, H. Andrä, J. Macek, *Phys. Rev. Lett.* **32**, 268 (1974).

8. R. H. Hughes, E. D. Stokes, Song-dik Choe, and T. T. King, *Phys. Rev. A* **4**, 1453 (1971).

9. D. J. Miller, *Nature* **274**, 531 (1978); E. N. Fortson, P. Sandars, private communication; E. G. Adelberger, T. A. Trainor, and E. N. Fortson, *Bull. Am. Phys. Soc.* **23**, 546 (1978); C. E. Wieman, E. S. Fry, T. B. Clegg, R. W. Dunford, R. R. Lewis, and W. L. Williams, *Bull. Am. Phys. Soc.* **23**, 39 (1978); R. R. Lewis and W. L. Williams, *Phys. Lett.* **59B**, 70 (1975); W. L. Williams, private communication.

Note Added in Proof

Subsequent to our experiment J. Burgdörter [*Phys. Rev. Lett.* **43**, 505 (1979)] calculated the s–p coherence in hydrogen after electron capture, using the Oppenheimer–Brinkman–Kramers approximation; he obtained results in good agreement with our experimental data.

38

Photon-Scattered Atom Coincidence Study with Vector Polarization Analysis for K–Noble-Gas Collisions

L. Zehnle, E. Clemens, P. J. Martin, W. Schäuble,
and V. Kempter

The linear and circular polarization of the $K(4^2P \rightarrow 4^2S)$ photons detected in delayed coincidence with the inelastically scattered potassium atom is presented for K–He, K–Ne, and K–Ar collisions as a function of the projectile scattering angle. The Stokes parameter analysis reveals that for K–Ar the collisionally excited state is not a pure state. It is concluded that two processes, single excitation of the potassium atom and simultaneous excitation of the potassium and argon atom, lead to $K(4^2P)$ excitation. For K–He and K–Ne the excited state is a pure state; in both cases λ and χ parameters are derived from the measured Stokes parameters. Simple models for the excitation process are discussed in order to explain the dependence of χ on the impact parameter.

1. Introduction

Until very recently the impact-parameter dependence of the electronic excitation process in ion–atom and atom–atom collisions has exclusively been studied by measuring energy-loss spectra of the inelastically scattered projectiles.[1] By measuring

L. Zehnle, E. Clemens, P. J. Martin, W. Schäuble, and V. Kempter • Fakultät für Physik der Universität Freiburg, D 78 Freiburg im Breisgau, Germany.   P. J. Martin's present address: Department of Physics, University of Missouri, Rolla, Missouri 65401, U.S.A.

such spectra as a function of the projectile's deflection angle one obtains the differential cross sections for the occurring inelastic process under study.

For excited states which decay by photon emission the same information can be obtained by studying the time correlation between the scattered projectile and the photon from the decay of the electronically excited atom. The simplest type of such a time correlation measurement uses only the wavelength information contained in the time-correlated emission. By measuring then the number of these coincidence events as a function of the deflection angle one obtains the differential cross section for the inelastic process under study. For atom–atom collisions we have successfully applied this technique to K–noble-gas collisions.[2]

However, with the help of coincidence techniques, even more detailed information on excitation processes can be obtained by studying the polarization properties of the coincident photons. It has been demonstrated by Standage and Kleinpoppen[3] for electron–atom collisions and by Jaecks *et al.*[4] and Vassilev *et al.*[5] for ion–atom collisions that differential magnetic sublevel excitation cross sections can be determined. Of course, such information cannot be obtained from energy-loss measurements. When the polarization state of the coincident photons is completely determined by measuring all Stokes parameters (SP) (see Born and Wolf[6] for an introduction to this subject) it is then possible to characterize the state of the excited atom completely: one is able to examine the coherence properties of the excitation process. In cases where the excited state is a pure state, excitation amplitudes for the inelastic process considered can be obtained. The theory of such measurements is given in References 7 and 8. Hitherto such a complete analysis was only performed for excitation of He(3^1P) in electron–He collisions.[3] For this case it was demonstrated that the excited atom is in a pure state which can then be represented by a wave function.

We report a complete polarization analysis for collisions between potassium atoms and the noble gases He, Ne, and Ar. Linear and circular polarization measurements have been made of the K($4^2P \rightarrow 4^2S$) photons detected in delayed coincidence with the inelastically scattered K atom. For these targets measurements were performed as a function of the deflection angle at 1000 eV potassium beam energy.

2. Apparatus

The apparatus is shown in Figure 1. It consists of the following components: a beam source of fast neutral alkali beams, a scattering cell for the production of the excited atoms, an optical system for the analysis of the wavelength and polarization of the emitted photons, and a neutral beam detector, which is rotatable around the optical system's axis. Also displayed is a block diagram of the coincidence electronic system for studying the time correlation between the photon emission and the arrival of the scattered atom. The following sections describe these components in some more detail.

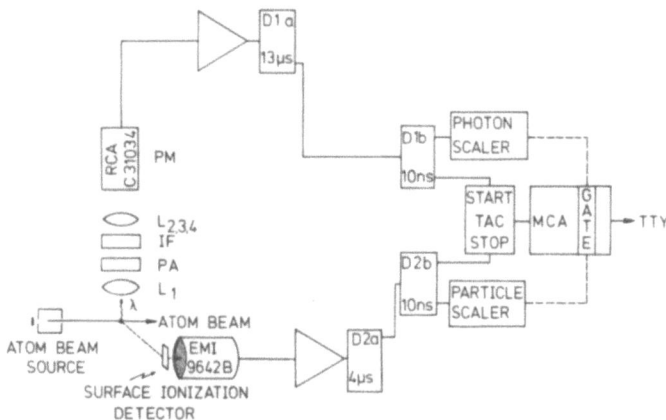

Figure 1. Schematic view of the apparatus. TAC, time-to-amplitude converter; MCA, multichannel analyzer; D1a, D2a, D1b, D2b, discriminators for generating suitable dead times.

2.1. Beam Production and Detection

The fast neutral beam is produced by neutralizing K$^+$ ions obtained by surface ionization on a hot Re ribbon. A detailed description of the beam source is available.[9] The beam energy spread is on the order of or smaller than 2 eV over the entire range of beam energies.

The fast scattered K atoms are detected by surface ionization on a W ribbon kept at about 700 K. The reflection of the fast atoms as ions is a fast process: hence the time relationship to the photon emission is conserved. We do not need to know the absolute detection efficiency; however, we estimate that the overall detection efficiency is on the order of 20% in the beam energy range between 50 and 500 eV. Under the present conditions the detection efficiency is constant for several weeks.

2.2. Scattering Region

Collisionally excited atoms are produced in a collision chamber. The advantage of a chamber as compared to a secondary beam is that the scattering geometry is rather well defined. In particular, the shape of the coincidence peak can be calculated and is in good agreement with the observed peak shape.[10] The length of the chamber is 0.5 cm so that the flight time of the atoms through it is at all energies much longer than the lifetime of the excited state that is studied. Losses due to the finite lifetime are therefore small. The chamber possesses a wide fixed exit slit. Atoms scattered under angles up to about 25° can be detected. The angular resolution of the collision-chamber–particle-detector combination is about 1.3°.[2] Under operating conditions the gas pressure in the chamber was about 1×10^{-3} Torr as measured with a capacitance manometer. We find a linear dependence of the photon count rate on the gas pressure. It was checked experimentally that the circular polarization was indeed pressure independent.

2.3. Optical System

The optical system is rigidly connected to the collision chamber and views the scattering region at 90° to the beam axis. A lens L_1 collects the photons from the collision region; the acceptance solid angle is 0.55 sr. Immediately following is the polarization analyzer PA used for the analysis of the polarization state of the emitted photons in terms of the Stokes parameters. In order to determine the linear polarizations, the light intensity is measured for various orientations of a linear polarizer (equivalent to Polaroid HN 38) relative to the beam axis. The circular polarization of the light is studied by inserting a quarter-wave plate in front of the linear polarizer: the circular polarized light is converted into linearly polarized light, which is then analyzed with the linear polarizer. PA is followed by a second quarter-wave plate, rigidly connected to the linear polarizer, which desensitizes the subsequent parts of the optical system to linear polarization. The nearly parallel light beam passes through the interference filter IF, which is chosen to transmit both fine-structure components of the $K(4^2P \rightarrow 4^2S)$ transition equally well or to select one of the two fine-structure components. Finally, the light is refocused by the lenses $L_{2,3,4}$ onto the cathode of a thermoelectrically cooled photomultiplier PM.

The overall detection sensitivity of the described system for photons of the $K(4^2P \rightarrow 4^2S)$ transition at 7680 Å was determined with a tungsten standard lamp to be 4.6×10^{-5}. By examining the polarization of the unpolarized $K(4^2P_{1/2} \rightarrow 4^2S_{1/2})$ transition, the instrumental linear polarization was found $<2\%$. Linear polarization of the coincident photons of this transition was zero within the error limits. The circular polarization of the noncoincident $K(4^2P \rightarrow 4^2S)$ photons was also $<2\%$. Moreover, zero circular polarization of the coincident $K(4^2P \rightarrow 4^2S)$ photons emitted in $K-N_2$ collisions was observed.

2.4. Coincidence Electronics

Figure 1 shows a block diagram of the electronic system. The time correlation in the output signals of both detectors is determined using a time-to-amplitude converter–multichannel analyzer system. A time spectrum of the coincidence events is recorded: The photons start the time-to-amplitude converter and the pulses produced by the scattered particles stop it. Suitable dead times are inserted in both channels to ensure that a constant background of accidental coincidences is obtained. True events produce a peak whose position is determined by the flight time of the scattered atoms to the detector. When possible, count rates were kept about ten times higher than the background rates at the respective detectors. The scalers monitor the noncoincident count rates in both detection channels. The processing of the time spectra of the coincident events is made by a minicomputer. The shape of the peak in the time-correlated spectrum was compared to that resulting from model calculations;[10] good agreement was found.

3. Theory of the Measurements

The Stokes parameters P_1, P_2, P_3 are defined as

$$IP_1 = I(0) - I(90)$$
$$IP_2 = I(45) - I(135) \tag{1}$$
$$IP_3 = I(RHC) - I(LHC)$$

where I is the total light intensity and RHC and LHC denote right- and left-hand circular polarization, respectively. $I(\alpha)$ is the intensity component linearly polarized at an angle α to the beam direction. P_1 and P_2 are the linear polarizations with respect to the beam axis and under $45°$ to this direction; P_3 is the circular polarization. The SP can be considered as the components of a three-dimensional vector P whose magnitude $|P|$ is the degree of polarization.[6]

From theories treating the collision dynamics of the excitation process under study one should obtain the SP as a function of the particle's scattering angle and of the collision energy. At the present time such results are not available for collisions between heavy particles. In order to gain some insight into the dynamics of the excitation process from the SP that are presented in Section 4, we are forced to use a scheme of data reduction that involves certain assumptions concerning the nature of the collisionally excited state. Some aspects of these assumptions can be checked on the basis of our results (see Section 4).

Our first assumption is that the excitation process is spin independent. This implies that fine and hyperfine interactions play no essential role during the time of the collision. This assumption allows us to express the SP in terms of the parameters O_i^{col} and A_i^{col}, which are expectation value of operators built up from orbital angular momentum operators. Following Fano and Macek[7] and using the fact that the excited state is produced in a collision process possessing reflection symmetry with respect to the collision plane, one obtains the following expressions for P_1, P_2, and P_3 for observation perpendicular to the scattering plane:

$$P_1 = (3/4)h^{(2)}G^{(2)}(A_0^{col} - A_{2+}^{col})/[1 + (1/4)h^{(2)}G^{(2)}(A_0^{col} + 3A_{2+}^{col})]$$
$$P_2 = (3/2)h^{(2)}G^{(2)}A_{1+}^{col}/[1 + (1/4)h^{(2)}G^{(2)}(A_0^{col} + 3A_{2+}^{col})] \tag{2}$$
$$P_3 = (3/2)h^{(1)}G^{(1)}O_{1-}^{col}/[1 + (1/4)h^{(2)}G^{(2)}(A_0^{col} + 3A_{2+}^{col})]$$

$h^{(i)}$ and $G^{(i)}$ $(i = 1, 2)$ are defined by the equations (8) and (40) in Reference 7. For the unresolved $K(4^2P \to 4^2S)$ doublet $h^{(2)} = -2$ and $h^{(1)} = 2$. The values of $G^{(1)}$ and $G^{(2)}$ were calculated using the hyperfine constants and the lifetime of $K(4^2P)$ reported in Reference 11; we obtain $G^{(1)} = 0.459$, and $G^{(2)} = 0.125$ for the unresolved $K(4^2P)$ doublet. The factors $G^{(1)}$ and $G^{(2)}$ appear due to the fact that the excited state develops after the collision under the influence of the weak fine- and hyperfine-structure interactions until the radiative decay occurs.[7]

Our second assumption is that for each spin component the orbital part of the excited state can be represented by a wave function of the form

$$\Psi = a_0 \,|\, 10\rangle + a_1 \,|\, 11\rangle + a_{-1} \,|\, 1 - 1\rangle \tag{3}$$

The incident beam direction has been chosen as quantization axis. Symmetry with respect to reflection in the scattering plane requires that $a_{-1} = -a_{+1}$. Equation (3) fulfilling our first assumption, spin independence of the excitation process, can be applied if only one excitation process leads to production of $K(4^2P)$ atoms. The second assumption is certainly not fulfilled when single excitation of the K atom

$$K(4^2S) + R \rightarrow K(4^2P) + R$$

and simultaneous excitation of the potassium and the noble gas atom

$$K(4^2S) + R \rightarrow K(4^2P) + R^*$$

occur with comparable probability. In this case the orbital part of the excited state for a given spin component is not a pure state. Consequently, this case cannot be represented simply by a wave function in the form of equation (3). In principle the experiment could be conducted using time-of-flight analysis of the scattered atoms allowing one to distinguish between these two processes. Currently our time resolution is not good enough to separate the two processes by their different inelastic energy loss.

In cases where only one excitation process is important we can express the expectation values A_i^{col} and O_{12}^{col} and consequently the SP in terms of $\lambda = |\,a_0\,|^2 / \{|\,a_0\,|^2 + 2\,|\,a_1\,|^2\}$ and the quantum mechanical phase difference χ between a_0 and a_1 (see also Reference 3); from equations (2) and (3) we then obtain

$$P_1 = 3h^{(2)}G^{(2)}(1 - 2\lambda)/(4 - h^{(2)}G^{(2)})$$
$$P_2 = 6h^{(2)}G^{(2)}[\lambda(1 - \lambda)]^{1/2}\cos\chi/(4 - h^{(2)}G^{(2)}) \tag{4}$$
$$P_3 = -6h^{(1)}G^{(1)}[\lambda(1 - \lambda)]^{1/2}\sin\chi/(4 - h^{(2)}G^{(2)})$$

Further quantities that can be used to characterize the polarization state of the coincident radiation are the degree of polarization $|\,P\,|$ and the complex correlation factor μ given by[6]

$$|\,P\,| = (P_1^2 + P_2^2 + P_3^2)^{1/2} \tag{5}$$

and $\mu = |\,\mu\,|\exp(i\beta) = (P_2 + iP_3)/(1 - P_1^2)^{1/2}$.

The importance of these quantities for this experiment is that for coherent excitation in the sense of equation (3) $|\,P\,|$ and $|\,\mu\,|$ are restricted within certain limits which depend on the values of $h^{(i)}$ and $G^{(i)}$. These considerations were first used in Reference 3 to test the coherence properties of the $He(3^1P)$ excitation by electrons. For this special case $(G^{(1)} = G^{(2)})$ one finds that $|P| = |\mu| = 1$ independent of the actual values of the target parameters and, therefore, independent on the electron's deflection angle. In more general cases $(G^{(1)} \neq G^{(2)})$ $|P|$ and $|\mu|$ depend on the actual values of the target parameters, which in turn may be functions of the deflection angle. This implies that, in general, $|P|$ and $|\mu|$ will depend on the deflection angle. For our present experiment $(G^{(1)} \gg G^{(2)})$ we have the limits $0.16 \leq |P| \leq 0.62$ and $0 \leq |\mu| \leq 0.62$. The upper and lower limits reflect the values of $G^{(1)}$ and $G^{(2)}$. Comparing the experimentally derived values for $|P|$ and $|\mu|$ with these predictions, we can decide whether our two assumptions are indeed correct. A further check of our assumptions can be made by comparing the χ values obtained in two different ways from equations (4): χ can either be derived from P_1 and P_2 using P_3 only to determine the sign of χ, or from λ and P_3. If equations (4) are valid, both methods should yield the same results.

4. Experimental Procedure and Results

The result of a coincidence measurement performed at a given detector position and for the polarizer setting α is $I(\alpha)$ in units of the total number of scattered particles. $I(\alpha)$ is derived from the area under the peak in the spectrum of the time-correlated events. The error bars shown with the polarization data result (i) from the uncertainty due to the determination of the number of background events in the time spectrum resulting from accidental coincidences, and (ii) from the statistical error of the channel contents in the channels containing the peak. $I(\alpha)$ is obtained from two independent measurements of approximately 12 h for each polarizer setting. $I(\text{RHC})$ and $I(\text{LHC})$ are obtained from similar measurements at $\alpha = 45°$ and $-45°$ after a quarter-wave plate has been inserted with the slow axis perpendicular to the atomic beam in front of the polarizer.

From auxiliary measurements we found no systematic variation of the Stokes parameters when the collection solid angle of the optical system was varied between 0.06 and 0.5 sr. The change in the SP expected from this variation was calculated from equations (8) and (40) of Reference 7; it is about 5% for our geometry, well within our statistical errors.

Figures 2–4 display P_1, P_2, and P_3 for the transition $K(4^2P \rightarrow 4^2S)$ for K–He, K–Ne, and K–Ar at 1000 eV beam energy. The SP are plotted as a function of the reduced deflection angle $\tau = \theta \cdot E$ (θ and E are the deflection angle and collision energy in the center-of-mass system) and as function of the impact parameter b leading to deflection into θ. The conversion from the detector position into θ and the relation $\theta(b)$ are taken from Reference 2. Some of the values for P_3 were obtained from measurements at negative scattering angles (see Reference 3 for the definition). These values, multiplied by (-1), are also included in Figures 2–4 (see crosses). They coincide with those obtained from measurements at positive scattering angles as required by the parity invariance.[3]

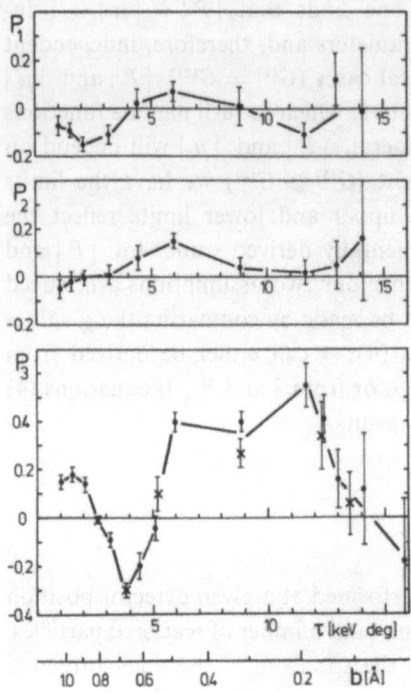

Figure 2. Polarization data for the $K(4^2P \rightarrow 4^2S)$ coincident photons versus the reduced deflection angle τ and the impact parameter b for K–He. $E_{CM} = 93$ eV.

Figures 5–7 show $|\mu|$, β, and $|P|$ as a function of τ and b for K–He and K–Ne. Crosses are shown at τ values where only values of P_3 were measured. For these cases values for P_1 and P_2 were obtained under the assumption that P_1 and P_2 are monotonous functions of τ between two measured points. The absolute upper and lower limits for $|\mu|$ and $|P|$ are indicated by the broken lines. In the cases

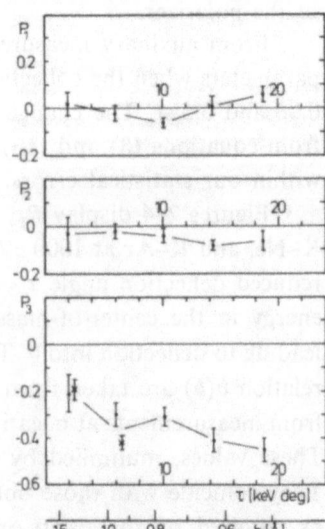

Figure 3. Polarization data for the $K(4^2P \rightarrow 4^2S)$ coincident photons versus the reduced deflection angle τ and the impact parameter b for K–Ne. $E_{CM} = 340$ eV.

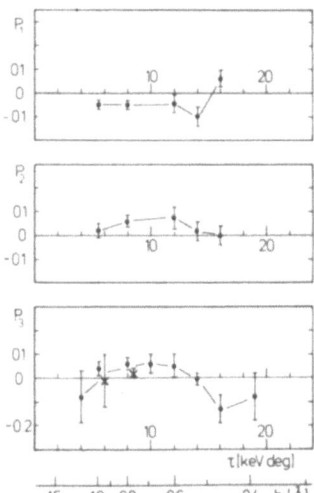

Figure 4. Polarization data for the $K(4^2P \to 4^2S)$ coincident photons versus the reduced deflection angle τ and the impact parameter b for K–Ar. $E_{CM} = 504$ eV.

studied $|\mu|$ and $|P|$ depend on τ as expected from the considerations of Section 3. For He and Ne within the error limits of the experiment, $|P|$ is compatible with coherent excitation in the sense of equations (4). This is not the case for K–Ar. We conclude that for K–Ar one of the two assumptions leading to equations (4) must be violated. Therefore, a data reduction in terms of λ and χ is not useful for K–Ar. On the other hand, such an analysis may be appropriate for K–He and K–Ne.

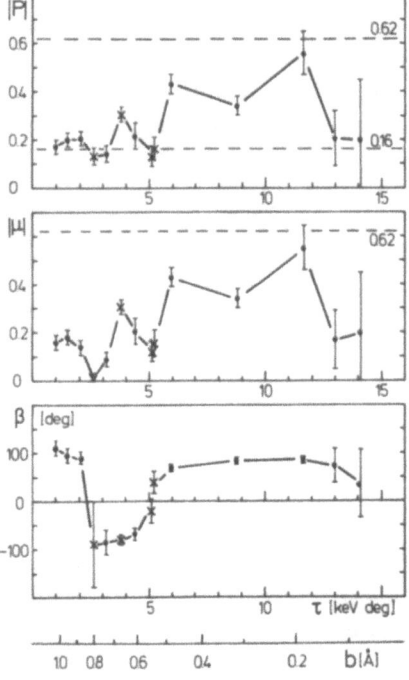

Figure 5. Degree of polarization $|P|$, magnitude $|\mu|$, and phase β of the complex correlation factor μ, and β versus the reduced deflection angle τ and the impact parameter b for K–He. $E_{CM} = 93$ eV.

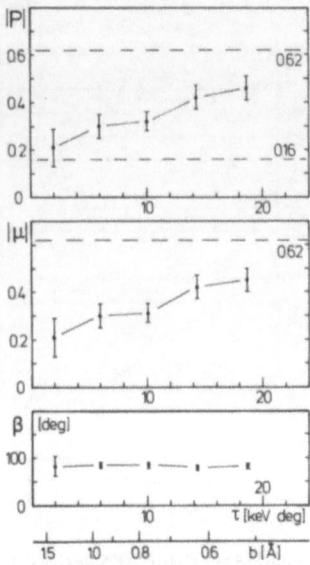

Figure 6. Degree of polarization $|P|$, magnitude $|\mu|$, and phase β of the complex correlation factor μ, and β versus the reduced deflection angle τ and the impact parameter b for K–Ne. $E_{CM} = 340$ eV.

Further support for the use of equations (4) is presented in the following discussion.

We believe that for K–Ar both single and simultaneous excitation (see Section 3) produce $K(4^2P)$ with comparable probability: for such quasisymmetrical systems the probability for target-atom excitation is much higher than for the asymmetrical combinations.[9,12,13] A qualitative and plausible explanation based on diabatic orbital diagrams has been offered in Reference 12. Using this explanation one would predict that also the cross section for simultaneous excitation is large for quasisymmetrical systems. For the asymmetrical combinations K–He, K–Ne integral cross sections for target-atom excitation (and *a fortiori* for simultaneous excitation) were found to be small.[9] Target excitation becomes weaker, compared to projectile

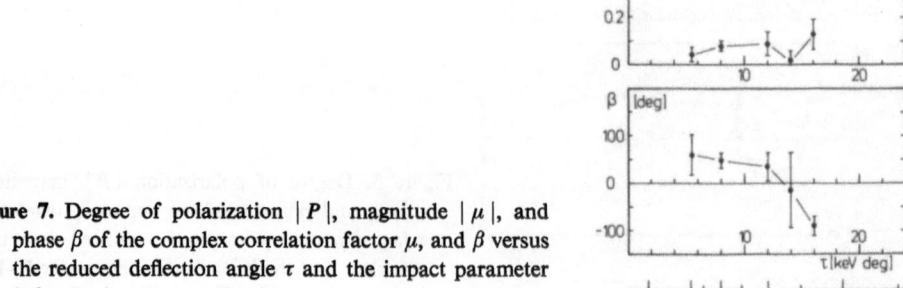

Figure 7. Degree of polarization $|P|$, magnitude $|\mu|$, and phase β of the complex correlation factor μ, and β versus the reduced deflection angle τ and the impact parameter b for K–Ar. $E_{CM} = 504$ eV.

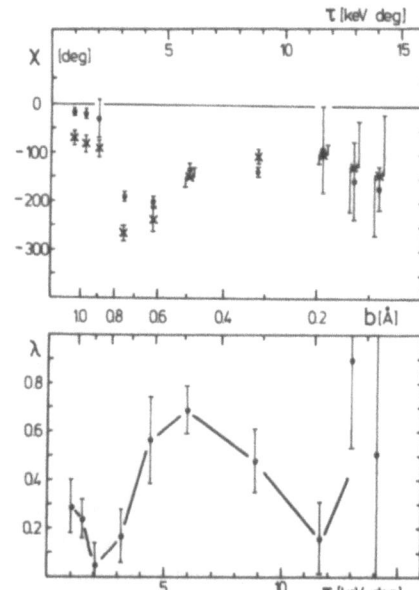

Figure 8. λ and χ data versus τ and b for K–He. χ is calculated from (points) λ and P_3, or (crosses) P_1 and P_2, using P_3 only for the sign of χ. $E_{CM} = 93$ eV.

excitation, the more asymmetrical the studied alkali–noble-gas combination becomes.[13]

Figures 8 and 9 display the parameters λ and χ as a function of τ and b for K–He and K–Ne. They were calculated from equations (4), which were suitably modified to account for the finite solid angle accepted by the optical system. λ is calculated from P_1 and P_2. When equations (4) are valid χ can be calculated in two ways:

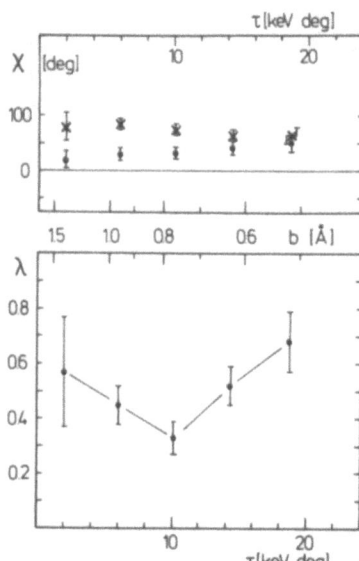

Figure 9. λ and χ data versus τ and b for K–Ne. Symbols as in Figure 8. $E_{CM} = 340$ eV.

(i) The most accurate values (points) are obtained when χ is calculated from λ and P_3.

(ii) Less accurate values (crosses) result when calculating χ from P_1 and P_2 and using P_3 only to obtain the sign of χ. From equations (4) it is clear that χ is only defined modulo π.

The following facts indicate that the assumptions made in deriving λ and χ from the SP are to a good approximation fulfilled:

(i) The χ-values calculated in two different ways are in reasonable agreement.

(iii) Equations similar to those shown in equations (4) can also be obtained for the SP of the two resolved transitions $4^2P_{3/2} \rightarrow 4^2S_{1/2}$ and $4^2P_{1/2} \rightarrow 4^2S_{1/2}$. [17] For both He and Ne the SP have been measured at certain τ values for the unresolved and the resolved transitions. Using the measured SP for the $^2P_{3/2} \rightarrow {}^2S_{1/2}$ line the SP for the $^2P_{1/2} \rightarrow {}^2S_{1/2}$ and for the unresolved doublet have been calculated. These calculated values agree well with the observed ones within the experimental errors. The calculated values are only weakly dependent on the validity of equation (3), however strongly dependent on the values $G^{(1)}$ and $G^{(2)}$. This comparison therefore tests whether equations (2) are valid, e.g., whether the assumption of spin independence is correct. We find that this assumption holds not only for He and Ne, but also for the heavier noble gases Ar (and Kr and Xe which are not discussed here).

5. Interpretation

At the low energies used in our experiments the excitation process is most conveniently described in the quasimolecular picture: Nonadiabatic transitions occur from the quasimolecular ground state $X^2\Sigma$ [emerging from K(4^2S)–He, Ne] to the excited states $A^2\Pi$ and $B^2\Sigma$ [emerging from K(4^2P)–He, Ne]. Two simple models that are currently invoked to explain excitation of alkali atoms in collisions with noble gases are pictured in Figure 10: In case (a) transitions take place only in the relatively well localized region around the crossings. Apart from these regions the system develops adiabatically, implying that the quasimolecule stays in its respective state. Case (a) has been applied to low-energy (100–500 eV) Na–Ne collisions. [14] In case (b) transitions between X and A and B occur over a relatively extended range of internuclear distances with high probability although there are no crossings between the involved states. Moreover, in a wide range of internuclear distances

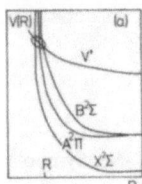 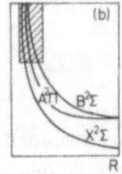

Figure 10. Schematic potential curves for some states of the K–noble-gas systems.

transitions also take place between A and B via rotational coupling. Case (b) has been applied to Li, Na–He, Ne collisions at intermediate energies (1–50 keV).[15] From previous experiments yielding integral[9] and differential[2] cross sections for $K(4^2P)$ excitation no unique answer could be given as to which mechanism describes the excitation process for K–He, K–Ne collisions.

We may ask what new information on the excitation mechanism is obtained from the parameters λ and χ. This question can be discussed qualitatively when the rotating internuclear axis is chosen as quantization axis for the angular momentum states in equation (3). This choice is most useful when the orbital angular momenta remain strongly coupled to the internuclear axis during the collision as in case (a). The wave function of the excited state of the quasimolecule can then be written as

$$\Psi(t, b) = c_\Sigma(b, t) \exp\left[-(i/\hbar) \int_{t_0}^{t} dt V_\Sigma(t)\right] | \Sigma\rangle$$

$$+ c_\Pi(b, t) \exp\left[-(i/\hbar) \int_{t_0}^{t} dt V_\Pi(t)\, dt\right] | \Pi\rangle \qquad (6)$$

where t_0 denotes the time at which the excited molecular states become populated. In case (a) $c_\Sigma(b)$ and $c_\Pi(b)$ would be time independent apart from regions where the crossings occur. For $t \to \infty$ $| \Sigma\rangle$ correlates with $| 10\rangle$ and $| \Pi\rangle$ with $1/2^{1/2} \times [| 11\rangle - | 1 -1\rangle]$. Thus

$$\Psi(b, t = \infty) = a_\Sigma | 10\rangle + a_\Pi 2^{-1/2}[| 11\rangle - | 1 -1\rangle] \qquad (7)$$

where

$$a_\Sigma(b, t = \infty) = c_\Sigma(b, t = \infty) \exp\left[-(i/\hbar) \int_{t_0}^{\infty} dt V_\Sigma(t)\right]$$

and

$$a_\Pi(b, t = \infty) = c_\Pi(b, t = \infty) \exp\left[-(i/\hbar) \int_{t_0}^{\infty} dt V_\Pi(t)\right]$$

From our polarization measurements we obtain in the same manner as in Section 3 $\bar{\lambda} = | a_\Sigma |^2/\{| a_\Sigma |^2 + | a_\Pi |^2\}$ and $\bar{\chi}$, the phase difference between a_Σ and a_Π for $t \to \infty$ (see Figures 11 and 12). $\bar{\lambda}$ gives the probability for populating $B^2\Sigma$ relative to the total excitation probability for $t = \infty$. If, as in case (a), no transitions between the excited states A and B take place, then $\bar{\lambda}$ directly yields the probability for transitions from $X^2\Sigma$ to $B^2\Sigma$ in units of the total excitation probability.

Since the total excitation probability is known for K–He and K–Ne,[2] $| a_\Sigma |^2$ and $| a_\Pi |^2$ can be calculated separately as function of the impact parameter b; the results are shown in Figures 13a and 13b. The structure seen for K–He is not necessarily due to Stueckelberg oscillations. It has been shown[16] that such structure arises in case (b) and is due to transitions between A and B, whose probability is strongly impact-parameter dependent.

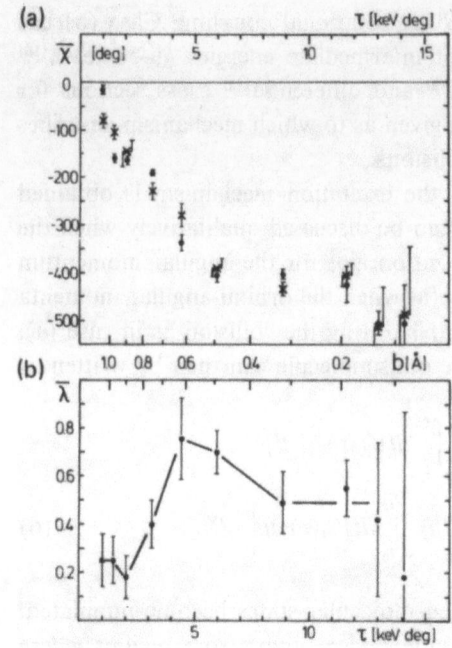

Figure 11. $\bar{\lambda}$ and $\bar{\chi}$ data plotted against τ and b for K–He. $\bar{\chi}$ is calculated from (points) $\bar{\lambda}$ and P_3, or (crosses) P_1 and P_2, using P_3 only for the sign of $\bar{\chi}$. $E_{CM} = 93$ eV.

$\bar{\chi}$ is displayed in Figures 11 and 12 as a function of τ and b. In case (a) $\bar{\chi}$ simply emerges from the different development in time of the excited states $A^2\Pi$ and $B^2\Sigma$. Circular polarization will therefore in general occur if the studied excited atomic state, here $K(4^2P)$, is populated via two different quasimolecular states, here $A^2\Pi$ and $B^2\Sigma$. Denoting t_0 as the time at which transitions to A and B occur at R_0 we find for $\bar{\chi}$ in case (a)

$$\bar{\chi} = -\frac{1}{\hbar}\int_{t_0}^{\infty} dt[V_\Pi - V_\Sigma] \tag{8}$$

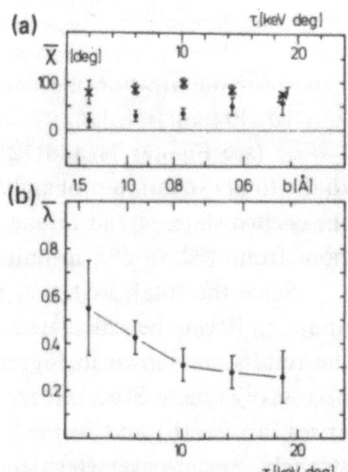

Figure 12. $\bar{\lambda}$ and $\bar{\chi}$ data plotted against τ and b for K–Ne. Symbols as in Figure 11. $E_{CM} = 340$ eV.

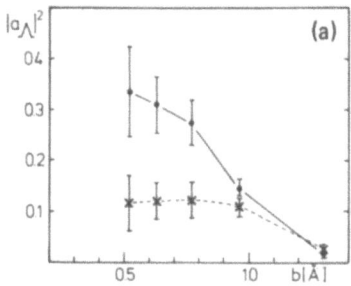 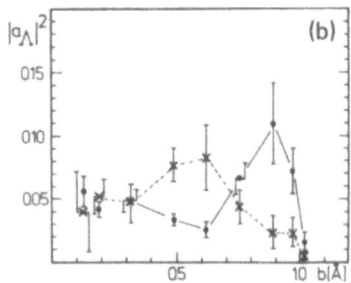

Figure 13. Excitation probabilities $|a_\Sigma|^2$ and $|a_\Pi|^2$ versus b for (a) K–Ne ($E_{CM} = 340$ eV), and (b) K–He ($E_{CM} = 93$ eV). $\times$, $\Lambda = 0$; $\bullet$, $\Lambda = 1$.

where V_Π and V_Σ depend on time via $R(t)$. This simple relation is only obtained if the different time development of the two states between R_0 and the classical turning points can be neglected. More detailed considerations can be found in Reference 17. We have pointed out[17] that mechanisms that are based on curve crossings and neglect interactions between simultaneously excited quasimolecular states [as case (a)] seem to give a weak dependence of $\bar\chi$ on b. Therefore, they would not be able to explain the strong dependence on b that is found for K–He. A strong b dependence of $\bar\chi$ at low collision energies is predicted for Be⁺–He (Nielsen, private communication), where case (b) obviously applies.[16] At the present time it is not clear whether the strong b dependence is due to the nonlocalized primary excitation mechanism or is caused by interactions between the excited states. We believe that the mechanism responsible for K(4^2P) excitation in K–He is similar to case (b). For K–Ne no final decision concerning the excitation mechanism can be made from such qualitative considerations only.

6. Summary

A Stokes parameter analysis of the K($4^2P \rightarrow 4^2S$) photons detected in coincidence with the scattered potassium atoms is presented for collisions of potassium with noble gas atoms. The coherence properties of the excitation process are analyzed. For K–He and K–Ne the parameters λ and χ are derived from the Stokes parameters. The impact-parameter dependence of λ and χ offers a sensitive test for proposed excitation mechanisms.

ACKNOWLEDGMENTS

We thank the Deutsche Forschungsgemeinschaft for financial support of the experiment. One of us (P.J.M.) also gratefully acknowledges a postdoctoral position funded by the Deutsche Forschungsgemeinschaft.

References

1. M. Barat, *Eighth International Conference on the Physics of Electronic and Atomic Collisions*, Invited Lectures and Progress Reports, B. C. Čobič, and M. V. Kurepa, eds., Institute of Physics, Beograd, pp. 43–70 (1973).
2. L. Zehnle, E. Clemens, P. J. Martin, W. Schäuble, and V. Kempter, *J. Phys. B* **11**, 2133–2141 (1978).
3. M. C. Standage and H. Kleinpoppen, *Phys. Rev. Lett.* **36**, 577–580 (1976).
4. D. H. Jaecks, F. J. Eriksen, W. de Rijk, and J. Macek, *Phys. Rev. Lett.* **35**, 723–725 (1975).
5. G. Vassilev, G. Rahmat, J. Slevin, and J. Baudon, *Phys. Rev. Lett.* **34**, 444–447 (1975).
6. M. Born and E. Wolf, *Principles of Optics*, Pergamon Press, New York (1975).
7. U. Fano and J. Macek, *Rev. Mod. Phys.* **45**, 553–573 (1973).
8. K. Blum and H. Kleinpoppen, *J. Phys. B* **8**, 911–925 (1975).
9. W. Mecklenbrauck, Ph.D. thesis, Freiburg (1976).
10. P. J. Martin, G. Riecke, L. Zehnle, and V. Kempter, *Tenth International Conference on the Physics of Electronic and Atomic Collisions*, Abstracts of Papers, Commissariat a l'Energie Atomique, Paris, pp. 1294–1295 (1977).
11. R. W. Schmieder, A. Lurio, W. Happer, and A. Khadjavi, *Phys. Rev. A* **2**, 1216–1228 (1970).
12. J. Fayeton, N. Andersen, and M. Barat, *J. Phys. B* **9**, L149–152 (1976).
13. W. Mecklenbrauck, J. Schön, E. Speller, and V. Kempter, *J. Phys. B* **10**, 3271–3281 (1977).
14. Ch. Gaussorgues, V. Sidis, and M. Barat, *Tenth International Conference on the Physics of Electronic and Atomic Collisions*, Abstracts of Papers, Commissariat a l'Energie Atomique, Paris, pp. 706–707 (1977).
15. J. Manique, S. E. Nielsen, and J. Dahler, *J. Phys. B* **10**, 1703–1722.
16. S. E. Nielsen and J. Dahler, *J. Phys. B* **9**, 1383–1399 (1976).
17. L. Zehnle, E. Clemens, P. J. Martin, W. Schäuble, and V. Kempter, *J. Phys. B* **11**, 2865–2874 (1978).

39

Alignment and Orientation Measurements in He$^+$–He and He$^+$–H$_2$ Collisions

D. H. JAECKS, F. ERIKSEN, AND L. FORNARI

The polarizations of the radiation emitted in excitation processes in ion–atom and ion–molecule collision provide information about the relative populations of magnetic substates. Measurement of this polarized radiation in coincidence with the scattered projectile provides even more information. Specifically, for P states, the quantum mechanical phase differences between magnetic substates can be determined as a function of the collision impact parameter, which in turn provide rigorous tests for various collision models. New insights of the inelastic processes involving ion–molecule collisions is beginning to emerge from this type of polarization measurement.

1. He$^+$–He: A Prototype System

The quantum mechanical collision system of He$^+$–He has become a prototype for the study of a variety of physical processes that occur almost universally in more complicated atomic systems. This is due in part to the relative ease of target and ion preparation, and because of this a variety of rather sophisticated total and differential scattering cross section and probability measurements have been made in the He$^+$–He system: sophisticated in the sense that one puts a great deal of effort into determining the final state of the interacting system. Such measurements have resulted in the observation of a variety of coherence effects, and in the past have led to the discovery

D. H. JAECKS, F. ERIKSEN, AND L. FORNARI • Behlen Laboratory of Physics, University of Nebraska, Lincoln, Nebraska 68588, U.S.A.

of new physical phenomena that are described by relatively simple and tractable theoretical models, thus leading to a better understanding of the collision process. It is important to mention that these theoretical descriptions have been extended to describe similar phenomena in more complicated ion–atom collision systems.

For example, the general concept of electron promotion was useful in describing the pioneering total charge transfer probability measurements of Ziemba, Lockwood, Morgan, and Everhart.[1] These probability results clearly showed that the collision process leading to charge transfer can be thought of as a development of the electronic amplitude along two well-defined, intermediate molecular-electronic energy states, $1s\sigma_g(2p\sigma_u)^2$ and $(1s\sigma_g)^2 2p\sigma_u$.[2,3] Today this concept of electron promotion and the interaction of molecular orbitals at curve crossings is embodied in practically all descriptions of heavy-ion–atom collisions leading to inner-shell vacancies.

"Stueckleberg oscillations" in differential inelastic cross sections, due to electronic curve crossings at intermediate nuclear separation, are also well exhibited in the He^+–He system. Lorents, Aberth, and Hesterman, as early as 1966, reported differential excitation cross sections for the $He(2^3S)$ state showing marked oscillations as a function of scattering angle.[4] These oscillations were interpreted as a coherent summing of two amplitudes caused by a curve crossing at distances on the order of 2 a.u.

Perturbations of the elastic channel amplitude due to inelastic processes are also well exhibited in the He^+–He system, as well as other systems.[5,6] Measurements of the elastic scattering cross section, as a function of scattering angle, show marked deviations from a regular oscillatory pattern at small scattering angles. These angles occur at distances of closest approach of the collisions where a curve crossing occurs between the initial $1s\sigma_g(2p\sigma_u)^2 A\Sigma_g$ state and a molecular state leading to the final state of $He(2^3S) + He^+$.

Oscillations in the total inelastic cross sections (often called Rosenthall oscillations) resulting from couplings among the excited-state channels at large internuclear separations were first found in He^+–He collisions.[7–9] Subsequent studies of this effect with and without polarization analysis of the excited-state radiation in other atomic systems have been numerous.[10] It is the measurement of these various coherence effects that have added greatly to our knowledge and understanding of interaction processes in the He^+–He system, as well as other systems, rather than the precise determination of total cross sections.

As an example of another kind of coherence effect, recent polarized photon-scattered atom coincidence measurements in He^+–He collisions have led to new insights into the coupling mechanism in Σ–Σ, intermediate curve crossings, resulting in the formation of $He(3^3P)$.[11,12] The apparatus used in these experiments is shown in Figure 1. The intersection of the capillary array gas jet forming the helium target, and the ion beam, define the target interaction volume. Slit Sl, being one slit of a parallel-plate analyzer, defines the scattering angle for all observed particles. The intensity of different polarizations of radiation emitted perpendicular to the scattering plane are measured for a variety of scattering angles θ_s.

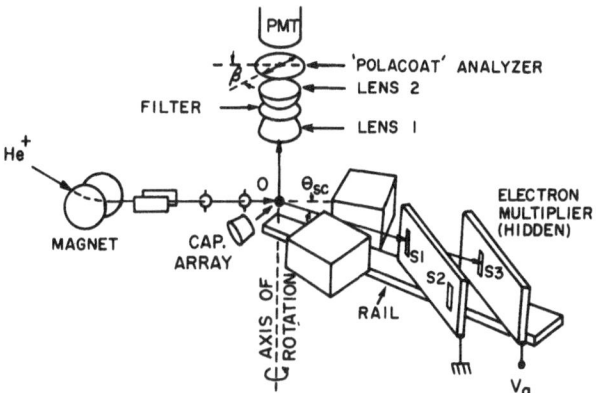

Figure 1. Overall view of apparatus for polarized photon-scattered atom coincidence measurements.

The intensity for any polarization direction β, for a particular final state of the system, characterized by the scattering angle θ_s, can be written[11]

$$I(\beta) = C[28\sigma_0 + 26\sigma_1 + 15(\sigma_0\sigma_1)^{1/2} \cos \Delta\phi \sin 2\beta + (30\sigma_1 - 15\sigma_0) \sin^2 \beta]$$

where σ_0 and σ_1 are the differential cross sections for the magnetic sublevels M_L and $\Delta\phi$ is the phase difference between the scattering amplitudes a_0 and a_1. The use of this form for $I(\beta)$ assumes the He+ target is left only in one final state, thus the term containing $\cos \Delta\phi$ arises from the interference of the two magnetic sublevel amplitudes.

Using acquired data for $I(\beta)$ as shown in Figure 2, the relative values of σ_0 and σ_1 and the relative phase are determined by a least-squares fitting of the data, for a variety of scattering angles. The most striking feature of the analyzed data is

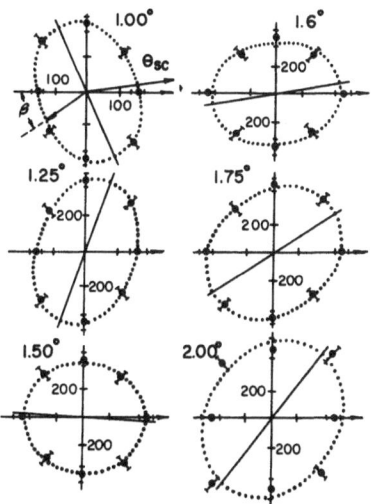

Figure 2. Polar plots of $I(\beta)$ for various scattering angles at 3.05 keV He+ energy.

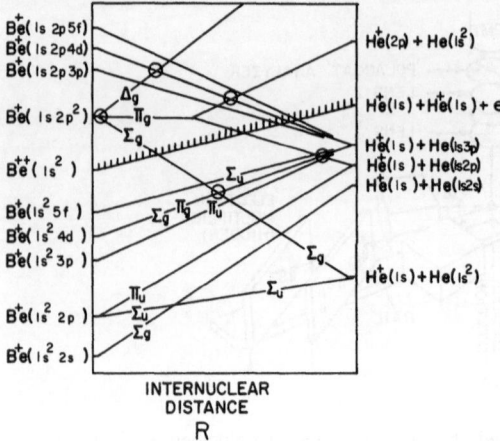

Figure 3. Correlation diagram for He–He+.

the nearly constant $\pm 90°$ phase difference between the two scattering amplitudes for all angles and energies measured.[11]

This 90° phase difference can be interpreted as resulting from the inability of the molecular electron cloud to follow the rapidly rotating internuclear axis; the excited-state molecular wave function is "frozen in space" when it is populated at a curve crossing. (See Figure 3.) This results in a general feature of a 90° phase difference between the amplitudes of the two magnetic sublevels of He(3^3P) when projected onto the laboratory frame.[11] The measured experimental phase values at 1.5 and 3 keV ion energies are consistently near $\pm 90°$. It should be pointed out that this 90° phase factor is rather insensitive to the exact details of the scattering potential and should be present as a general phenomenon in Σ–Σ crossings for other atomic systems.

The theoretical values of σ_0 and σ_1 are sensitive to the exact form of the molecular potentials of the He$_2{}^+$ systems and have not been calculated to date.

2. He+–H₂ System

Triatomic collision studies are now reaching a degree of sophistication that allow new information and insights to be obtained about the collision dynamics from the experimental data.[13–15] One cannot determine the final state of the system completely by a polarized photon coincidence measurement of the type just described; however, new information about the triatomic collision process

$$\text{He}^+ + \text{H}_2 \rightarrow \text{He}(3^3P) + \text{H}_2{}^+ \rightarrow \text{He}(2^3S) + \text{H}_2{}^+ + h\nu$$

can be obtained from such a technique. In this experiment the primary He+ beam and the scattered He(2^3S) determine the scattering plane with the 3889-Å photon being detected perpendicular to the scattering plane. The degree of linear and circular polarization is measured for a specific scattering angle of He(2^3S). The apparatus used is the same as in Figure 1 with the addition of a quarter-wave plate for determining the degree of circular polarization.

Using the polarized-photon–scattered-atom coincidence technique discussed in the He⁺–He measurements, the intensity of radiation $I(\beta)$, for different polarization directions β, was measured for several laboratory scattering angles. These data are shown in Figure 4.

These polarization patterns show the characteristic dipole pattern found in the He⁺–He case and, for the case of 1.5° scattering at 1.5 keV, shows a high degree of alignment. In all cases the dipole pattern is symmetric about the momentum transfer axis of the collision, within experimental error.

The orientation, proportional to $I(45°, 90°) - I(135°, 90°)$, where $I(\beta_1, \beta_2)$ is the photon–particle coincidence rate with polarizer setting β_1, while the fast axis of the quarter-wave plate is at β_2 relative to the beam axis, was also measured. In all cases the orientation was found to be zero, that is, $I(45°, 90°)$ and $I(135°, 90°)$ were found to be equal to each other within one standard deviation.

The zero orientation implies that the radiation associated with the particular collision process is completely incoherent. The summing over all the final possible states of the H_2^+ target after the collision results in an incoherent source, implying the magnetic substates of He(3^3P) are incoherently populated as viewed in the reference frame that has as its symmetry axis the momentum transfer axis. We have thus analyzed the source using this symmetry and obtain two cross sections σ_0 and

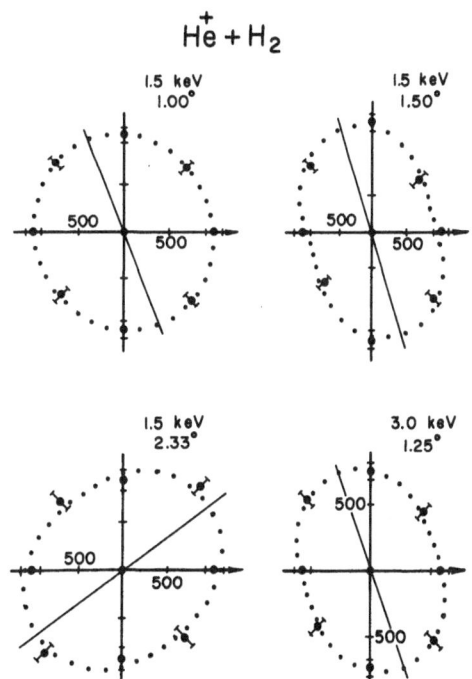

Figure 4. Polar plots of $I(\beta)$ for various scattering angles at 1.5 and 3.0 keV.

σ_1, where σ_0 is the cross section for populating the $M_L = 0$ state and σ_1 is the cross section for populating the $M_L = 1$ state. These cross sections are for averages over the final states of the $H_2{}^+$ target that are populated during the collision and are to be taken in the reference frame where the momentum transfer axis is the polar axis.

Using the incoherent source polarization expressions of Percival and Seaton[16,17] we obtain expressions for the ratio σ_0/σ_1 of 1.50, 6.7, 0.36, for scattering at 1°, 1.5°, and 2.33° scattering at 1.5 keV. We note a rapid change in the relative cross sections as a function of scattering angle. These data are being extended to see if any systematics exist in the changing behavior of σ_0 and σ_1.

References

1. E. P. Ziemba, G. J. Lockwood, G. H. Morgan, and E. Everhart, *Phys. Rev.* **118**, 1552 (1962).
2. E. Everhart, *Phys. Rev.* **132**, 2083 (1965).
3. R. Marchi and F. T. Smith, *Phys. Rev. A* **139**, 1025 (1965).
4. D. C. Lorents, W. Aberth, and V. W. Hesterman, *Phys. Rev. Lett.* **17**, 849 (1966).
5. F. T. Smith, D. C. Lorents, W. Aberth, and R. P. Marchi, *Phys. Rev. Lett.* **14**, 742 (1965).
6. F. T. Smith, R. P. Marchi, W. Aberth, and R. P. Marchi, *Phys. Rev. Lett.* **161**, 31 (1967).
7. H. Rosenthall, *Phys. Rev.* **4**, 1030 (1971).
8. S. Dworestsky, R. Novick, W. W. Smith, and W. Tolk, *Phys. Rev. Lett.* **18**, 939 (1967).
9. V. A. Ankudinov, S. V. Bobashev, and V. I. Perel, *Sov. Phys.-JETP* **33**, 490 (1971).
10. T. Andersen, A. Kirkegaard Nielsen, and K. J. Olsen, *Phys. Rev. A* **10**, 2174 (1974), and references cited therein.
11. D. H. Jaecks, F. J. Eriksen, W. de Rijk, and J. Macek, *Phys. Rev. Lett.* **35**, 723 (1975).
12. F. J. Eriksen, D. H. Jaecks, W. de Rijk, and J. Macek, *Phys. Rev. A* **14**, 119 (1976).
13. F. J. Eriksen and D. H. Jaecks, *Phys. Rev. Lett.* **36**, 1491 (1976).
14. A. V. Bray, D. S. Newman, and E. Pollack, *Phys. Rev. A* **15**, 2261 (1977).
15. V. Kempter, E. Clemens, P. J. Martin, and L. Zehnle, *Tenth International Conference on the Physics of Electronic and Atomic Collisions*, Abstracts of Papers, Commissariat a l'Energie Atomique, Paris, p. 222 (1977).
16. I. C. Percival and M. J. Seaton, *Phil. Trans. R. Soc. London* **251**, 113 (1958).
17. Robert H. McFarland and Edward A. Soltysik, *Phys. Rev.* **127**, 2090 (1962).

40

Simultaneous Electron–Photon Excitation of Atoms

F. H. M. Faisal

Excitations of atoms by simultaneous collisions with electrons and photons are considered theoretically. Their characteristic behavior as functions of laser frequency and polarization direction are discussed in connection with excitations in H atoms. The excitation cross sections show a linear dependence on laser intensity I and are comparable with the ordinary electron excitation cross sections for $I \approx 10^{12}$ W/cm². Such processes are related to but different from the more familiar "on-shell" (or resonant) excitation–collision processes and can permit experimental study of inelastic electron–atom scattering cross sections at "off-shell" electron energies. The differential cross sections show strongly distinguishable dependence on the electron scattering angle as well as on the laser polarization angle. Frequency dependence of the cross sections exhibits both the usual as well as certain "postcollision" resonant excitations (or ionization) at characteristic frequencies less than the first atomic resonance frequency; they are mediated by the simultaneous presence of the electron collision. Besides their basic theoretical interest, off-resonant electron–photon excitations and ionization of H atoms (and their inverse processes) could affect the electron temperature in laser heating of fusion plasma at ultrahigh intensities ($I \gg 10^{12}$ W/cm²), currently in use.

1. Introduction

Since the beginning of this decade considerable progress has been made in the study of properties of electron–atom collisions in the presence of radiation. Broadly speaking there are three kinds of processes that may be distinguished:

F. H. M. Faisal • Fakultät für Physik, Universität Bielefeld, D-4800 Bielefeld 1, Federal Republic of Germany.

(i) selective excitation of one or more target states by resonant-absorption of photons, followed by collision of the excited target with electrons[1-4];

(ii) excitation of a set of target states by collision with electrons followed by emission of photons as time-dependent resonance fluorescence[5-8];

(iii) scattering of electrons with ground-state atoms in the presence of radiation in which a photon is absorbed or emitted by the electron, i.e., stimulated (inverse or direct) bremsstrahlung effects.[9-18]

The scattering process that I intend to discuss has close resemblance to the above processes but at the same time it differs significantly from them. To best clarify the similarities and differences let me briefly consider the energetics of the processes mentioned above.

In type (i) processes the energy is separately conserved between the atom–photon subsystem since the photon is resonantly absorbed; the atom is excited by collision with photons alone. In type (ii) processes the energy is separately conserved between the electron–atom subsystem; the atoms are excited by collisions with electrons only. In type (iii) processes the energy is conserved between the electron–photon subsystem; the target changes no internal energy.

I now consider an electron–photon–atom collision in which all the three pairs of collisional subsystems separately violate energy conservation while maintaining of course the conservation of total energy of the system. In such a collision we have the possibility of observing, for example, inelastic electron–atom differential cross sections at an electron kinetic energy which is off the "energy–shell" from the point of view of the electron–atom subsystem. Consider the energy-level diagrams in Figure 1, where the first two atomic levels $|0\rangle$ and $|1\rangle$ of energy ε_0 and ε_1 of a target atom are shown. Also shown are the incident electron energy E_i and the incident laser photon of energy $\hbar\omega$. As in Figure 1a let the photon frequency be $\hbar\omega < \Delta_{01}$, where $\Delta_{01} = |\varepsilon_0 - \varepsilon_1|$ is the excitation energy of the target and the incident electron energy E_i satisfies the inequality, $\Delta_{01} - \hbar\omega < E_i < \Delta_{01}$. It is clear in this case that neither the electron nor the photon by itself can excite the atomic target. Nevertheless, it is also clear that the total energy of electron plus photon $E_i + \hbar\omega$ is greater than the excitation energy Δ_{01}; the excitation process is therefore energetically completely allowed and hence, in principle, measurable. To be sure the cross section for the "off-shell" excitation under consideration would be (for the same photon flux) much smaller than, for example, the resonant excitation cross sections in type (i) processes. It would, therefore, remain practically unobservable for the small to

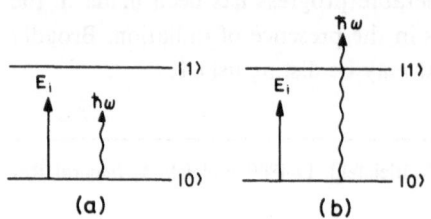

(a)   (b)

Figure 1. A pair of lowest-lying states $|0\rangle$ and $|1\rangle$ of an atomic system are shown along with the incident electron energy E_i and photon energy $\hbar\omega$. 1(a) corresponds to the "red-shifted" scattered electrons and 1(b) corresponds to the "blue-shifted" scattered electrons (final energies $E_f = E_i + \hbar\omega - \Delta_{01}$); $\Delta_{01} = |\varepsilon_i - \varepsilon_f|$ is the excitation energy.

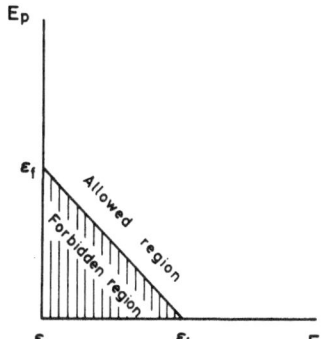

Figure 2. Threshold for simultaneous electron–photon excitation process. Photon energy is E_p and electron energy is E_e. ε_i and ε_f are the energies of the initial and the final states of the target atom. The triangular area is kinematically forbidden. The allowed region is above the line of threshold.

medium photon densities available from more common light sources. However, with very high-power lasers, e.g., Nd^{3+} glass ($\hbar\omega = 1.17$), CO_2 gas ($\hbar\omega = 0.117$), or ruby ($\hbar\omega = 1.78$ eV) lasers (which are becoming increasingly available in atomic collision physics laboratories) already higher photon densities than would be needed for detection of such processes are known to be obtainable.[19] Besides, a new kind of on-resonance excitation (to be referred to as a "postcollision resonance" is predicted to occur with usual cross sections at ordinary intensities.

The threshold condition of simultaneous electron–photon collision with atoms may be indicated simply and generally as in Figure 2. The process is allowed provided $E_e + E_p > E_{if}$ where E_e is the energy of the incident electron, E_p ($= \hbar\omega$) is the energy of the photon, and E_{if} is the energy difference between the initial and the final atomic states. This implies a forbidden triangular region in the energy space separated from the allowed region by a line of threshold (Figure 2).

Referring back to the process depicted in Figure 1a, consider the "energy-loss" spectrum of the scattered electrons. Obviously there would be a large peak due to the elastically scattered electrons at the incident electron energy E_i. There will be, however, a second "red-shifted" peak at $E_f = E_i - (\Delta_0 - \hbar\omega)$ corresponding to the excitation of the atom to $|1\rangle$. Similarly, if $\hbar\omega > \Delta_{01}$, as shown in Figure 1b, there will be a "blue-shifted" peak at $E_f = E_i - (\hbar\omega - \Delta_{01})$. Note also that if $E_i > \Delta_{01}$ then the usual direct electron excitation peak at $(E_i - \Delta_{01})$, and those for the higher excited states of the target, will also arise. In this respect they resemble closely the superelastic peaks observed in group-(i)-type of processes from resonantly prepared states. Furthermore for any combination of E_i and $\hbar\omega$ there will be two (if $E_i > \hbar\omega$) or one (if $E_i < \hbar\omega$) subpeaks at $E_i \pm \hbar\omega$, due to stimulated inverse and direct bremsstrahlung effects.[9–18] Stimulated multiphoton bremsstrahlung spectrum has also been seen recently.[18]

2. Theoretical Considerations

The process of interest to us may be written down as

$$e^-(E_i) + \hbar\omega + A \rightarrow A^* + e^-(E_f) \tag{1}$$

where E_i and E_f ($E_i \neq E_f$) are the incident and final electron energies, A and A^* ($A \neq A^*$) are the incident and final atomic states, and $\hbar\omega$ is the energy of photons from a very high-power laser. This problem could be mathematically formulated somewhat analogously to that of the usual free–free transition by extending it to allow for the final atomic state to be in an excited state (bound or free). However, I shall discuss it from the point of view of the diagrammatic perturbation theory, as has been done recently.[20-24] This formulation is both instructive and general. I shall not discuss the general case but rather consider a simple problem of this kind in some detail. Such a process, for example, is

$$e^-(E_i) + \hbar\omega + H(1s) \to e^-(E_f) + H(2s) \qquad (2)$$

or when the hydrogen atom is left in the final $3s$ state, etc.

It is useful to note that processes of the kind (1) and (2) can occur via off-resonant absorption of not only one photon but also two or more photons. It is nevertheless expected that even at intensities as high as 10^{14} W/cm² the contribution of off-resonant two-photon processes to (1) should not be more than about 1% (the relevant scale of atomic intensity being of order 10^{16} W/cm²) of the single-photon process, whenever the latter is allowed.

The transition amplitude for processes (1) and (2) can be obtained as the sum of the two diagrams shown in Figure 3. The single vertical line represents a free electron while the double lines represent an atom (or the bound electron). The horizontal line with a cross represents collisional interaction of the free electron and the atom and the wavy line represents photon–atom interaction. In principle the class of such diagrams also includes those in which the free electron couples with the external radiation field. However, their contribution can be shown to vanish in the context of transitions between atomic states of the same parity and for high electron collision energy when the plane-wave dipole approximation is applicable.[20] Note the order of interactions in the two diagrams of Figure 3. In Figure 3a the atom interacts first with the photon and then with the electron (assuming later time in the diagram to be vertically higher). In Figure 3b the atom interacts first with the electron and then with the photon. Note also that these diagrams are of first order in the laser field strength but include all orders in the electron–atom interaction through the T operator [see equation (4b) below].

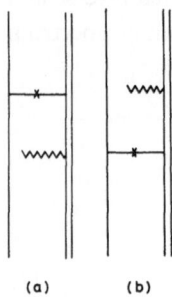

(a)   (b)

Figure 3. Feynman diagrams for the excitation process. The vertical single line stands for the free electron, the vertical double line for the bound electron, and the wavy line for the photon. The horizontal line with a cross represents the collision between the electron and the atom.

   In the resonant absorption as well as in the resonance fluorescence type of processes, essentially one of these two types of diagrams is involved, e.g., Figure 3a for the former and Figure 3b for the latter. In the present off-resonant (off the "energy shell") process the interaction time of photon–atom interaction is so small that it compares in general with that of the usually small electron–atom interaction time. The characteristic time t_p of interaction of photons of frequency ω with a pair of levels of an atom may be given by the inverse of the effective Rabi frequency ω_R associated with the transition:

$$t_p = (\omega_R)^{-1}$$
$$= [(\omega - \omega_{if})^2 + (F\mu)^2 + \gamma_s^2]^{-1/2} \qquad (3)$$

where ω_{if} is the transition frequency between the initial and the final atomic states, F is the electric field amplitude of the photon field, and μ is the dipole transition moment ($\mu \simeq ea_0$ for H). γ_s is the sum of the natural widths of the two levels. For off-resonant transitions, $(\omega - \omega_{if})$ is on the order of electron volts where $|F\mu|$ is (even for high-intensity lasers) several to many orders of magnitude smaller than 1 a.u. Thus only for resonant photons with $\omega = \omega_{if}$, t_p could be much larger than an atomic period. In this case t_p is determined either by the "power width" $|F\mu|$ or by the natural width γ_s, depending on whichever is larger, or by both when the two are comparable. For off-resonant processes, on the other hand, t_p is determined by the large frequency detuning. Hence it is very small and is comparable to the electron collision time (for most electron energies of interest). It follows from uncertainty principle considerations that given a tolerable accuracy of measurement of the energies concerned, the question of which of the two interactions precedes the other is not significant and both orderings in Figure 3 must be taken into account. Explicit calculations confirm this by showing that both diagrams contribute by the same order of magnitude in the off-resonant situation and that only one of them dominates as the frequency is varied to approach an atomic transition frequency. From a simple consideration of the initial conditions it is also possible to tell which of the two diagrams should dominate at resonance. It will be seen later on that the so-called "postcollision resonances" are associated with diagrams of the type shown in Figure 3b.

   The transition amplitude corresponding to the Feynman diagrams in Figures 3a and 3b is given by

$$A_{f \leftarrow i} = -2\pi i\, \delta(E_{ff} - E_{ii} - \hbar\omega)T_{f \leftarrow i} \qquad (4a)$$

where

$$T_{f \leftarrow i} = \sum_j \frac{\langle \phi_f \,|\, F(\boldsymbol{\epsilon} \cdot \boldsymbol{\mu})/2 \,|\, \phi_j \rangle \langle \phi_j \cdot \mathbf{K}_f \,|\, T \,|\, \mathbf{K}_i \phi_i \rangle}{(E_{ii} - E_{jf})}$$
$$+ \sum_j \frac{\langle \phi_f \mathbf{K}_f \,|\, T \,|\, \mathbf{K}_i \phi_j \rangle \langle \phi_j \,|\, F(\boldsymbol{\epsilon} \cdot \boldsymbol{\mu})/2 \,|\, \phi_i \rangle}{(E_{ii} + \hbar\omega - E_{ji})} \qquad (4b)$$

In equation (4b), T is the transition matrix for the ordinary electron–atom collision in the absence of laser radiation and $\sum_j$ implies a summation–integration over the complete set of target atom states. The energy of the electron–atom subsystem is denoted by $E_{jp} = \varepsilon_j + E_p$ where ε_j is the energy of the atomic target in state $|j\rangle$ and $E_p = \hbar^2 K_p^2/2m$ is the energy of the scattering electron, corresponding to the momentum $\mathbf{K}_p$. F is the field amplitude of the laser radiation and $\boldsymbol{\epsilon}$ is the polarization unit vector.

It is clear that in resonant excitation processes only one set of "intermediate states" (which is generally energetically degenerate or almost degenerate) is of importance. A resonant process, therefore, can reveal the geometrical properties (e.g., orientation and alignments of the atom) associated with the given set of "intermediate states." Besides, the member states of the near-degenerate set may be excited in definite phase relations with one another, which can be revealed by, e.g., time-resolved measurements on the excited system. In contradistinction to the resonant case, an "off-shell" excitation process involves the entire atomic spectrum at once and hence is dependent not on a single set of intermediate states but on all possible states of the target, whether they are energetically accessible or not. In other words the process of our interest goes through a whole sequence of virtual transitions from the initial state to the final (energy-conserving) state. The virtual transitions may also be viewed, perhaps more picturesquely, as dressing of the target atom or its polarization, which oscillates rapidly on the time scale of an atomic period. The first term of (4b) may thus be considered to represent (reading from right to left) dressing of the bare initial atomic target state by the field of the incident electron followed by photoabsorption by the dressed atom, leading to the final state. The second term could be similarly viewed as the dressing of the target by the light field followed by collisional excitation of the dressed atom to the final state. It is clear that the associated cross section would provide a more stringent test of the entire set of atomic wave functions (even at high incident electron energy) since it depends on the full information of the target system besides that of its initial and final states. A similar situation arises in off-resonant multiphoton excitation of atoms and molecules.

3. Result for 1s–2s and 1s–3s Transitions in H Atom

Evaluation of the amplitude (4a) is, in general, a formidable computational task. In special circumstances, where further simplifying assumptions hold, it is possible to carry through the calculation to completion and reveal the major aspects of the "off-shell" electron–atom collision process in presence of laser radiation. Considerable theoretical simplification is achieved at high incident electron energy. This allows, according to conventional wisdom, the representation of the incident electron's wave function by plane waves and permits, within tolerable accuracy, the neglect of the exchange effects and use of the dipole approximation (Bethe–Born approximation). Even under the above conditions one is still left with the problem of knowledge of the entire set of target wave functions and that of evaluating the infinite summation–integration in equation (4b). The obvious candidate for the target atom is, therefore, the hydrogen atom or hydrogenic ions. In this case not only the whole set of wave functions is known but powerful representation of the hydrogenic Green's functions have also been constructed[25] which may be used to complete the operation implied by the infinite summation and integration in equation (4b). Details of these techniques and their application are given by Rahman and Faisal.[20–23] Here I shall only discuss the results for the 1s–2s and 1s–3s excitation cross sections obtained under the above approximations for H atoms.

3.1. Total Cross Sections

The total 1s–2s cross section can be written as[22]

$$\sigma^{(\omega)}_{1s \to 2s} = 8\pi\alpha \, \frac{K_f}{K_i} \, A(K_i, K_f) R_{1s \to 2s}(\omega) \tag{5}$$

where

$$A(K_i, K_f) = \frac{\pi}{2K_i} \left[\frac{1}{z} \ln\left(\frac{1+z}{1-z}\right) - 2 \right]$$

$$z = \frac{2K_i K_f}{K_i^2 + K_f^2}$$

$$R_{1s \to 2s}(\omega) = P(\varepsilon_{1s} - \omega) + P(\varepsilon_{2s} - \omega)$$

is a meromorphic function of ω, which describes the response of the atom to the radiation field. The function $P(x)$ is defined by

$$P(x) = \frac{z^8 x^2}{3(2^{1/2})} \sum_{n=2}^{\infty} n(n^2 - 1)\left[n^2 - \left(x + \frac{9}{4x}\right)n + \frac{1}{x^2} + 2\right] \Big/ \left(n - \frac{1}{x}\right)$$

$$\times \left(\frac{1-x}{1+x}\right)^{n-3} \left[\frac{(\frac{1}{2} - x)}{(\frac{1}{2} + x)}\right]^{n-4} (\tfrac{1}{2} + x)^{-8}(1 + x)^{-6} \tag{6}$$

An expression similar to (5) holds for 1s–3s excitation[20]:

$$\sigma^{(\omega)}_{1s \to 3s} = 8\pi\alpha \, \frac{K_f}{K_i} \, A(K_i, K_f) R_{1s \to 3s}(\omega) \tag{7}$$

Here the response function $R_{1s \to 3s}(\omega)$ is defined by

$$R_{1s \to 3s}(\omega) = Q(\varepsilon_{1s} - \omega) + Q(\varepsilon_{3s} - \omega) \tag{8}$$

where

$$Q(x) = \frac{24}{3(3!)^2} x^3 \sum_{m=2}^{\infty} \frac{m(m^2 - 1)}{(m - 1/x)} \frac{4!}{(1+x)^6} \left(\frac{1-x}{1+x}\right)^{m-3} (2 - mx)$$

$$\times \frac{2(3^{1/2})4!}{9} \frac{\alpha^{m-3}}{(\frac{1}{3} + x)^5} \left\{ 1 - \frac{m+2}{4}\beta - \frac{10}{3(\frac{1}{3} + x)\alpha}\left[1 - \frac{m+2}{2}\beta\right] \right.$$

$$+ \frac{(m+2)(m+3)}{20}\beta^2 + \frac{20}{9(\frac{1}{3} + x)^2\alpha^2}\left[1 - \frac{3(m+2)}{4}p\right.$$

$$+ \frac{3}{20}(m+2)(m+3)\beta^2 - \frac{(m+2)(m+3)(m+4)}{120}\beta^3 \Big]\Big\} \tag{9}$$

where

$$\alpha = \frac{\frac{1}{3} - x}{\frac{1}{3} + x} \quad \text{and} \quad \beta = \frac{2x}{\frac{1}{3} + x}$$

The nature of these results as a function of photon frequency can be seen from Figure 4, which shows the 1s–3s electron excitation cross sections scaled to unit light intensity. For illustrative purpose the electron energy is chosen to be 100 eV. It is seen that the optical response of the atom to the photon frequency passes through several resonances. The two resonances on the right-hand side are due to the dipole coupling of the initial 1s state to the higher p states and appear exactly at the atomic level energies (10.2 eV for the 2p level and 12.08 eV for the 3p level) as one would expect. Obviously, resonances further to the right of those shown in Figure 4 arise naturally, corresponding to higher principal quantum numbers $n > 3$, as the photon frequency is further increased. The first resonance structure on the left of Figure 4, however, is of a rather different origin. It occurs at a smaller frequency (1.89 eV) than that of the first excited p state (10.2 eV) and indeed corresponds to the energy difference between the 3s and 2p states of the H atom. In other words, this resonance appears as a result of a coupling of the *final* atomic state to a state lying below it. Clearly such a resonance is not possible for the usual photoabsorption from the

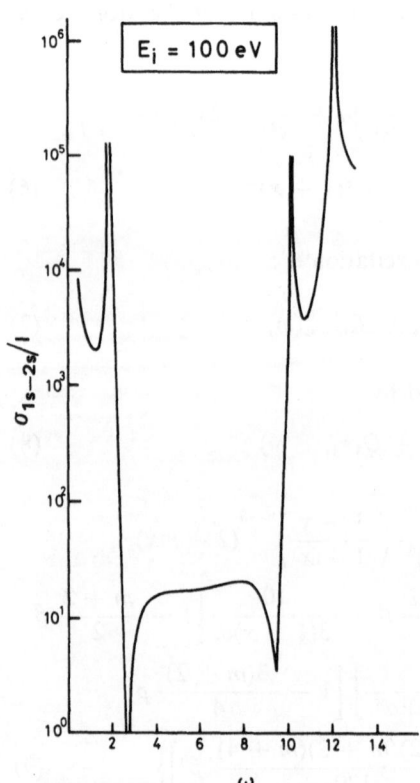

Figure 4. Integrated 1s–3s cross section divided by intensity as a function of photon frequency ω at 100 eV incident electron energy; from Reference 20.

ground-state atom, but it is characteristic of photoabsorption in presence of electron collision with atoms. Diagrammatically these are associated with the contribution of Figure 3b when, in comparison, the contribution from Figure 2a is negligible. It may be visualized as due to the excitation of the atom to an intermediate p state below the final state, by collision with electrons, followed by absorption of the photon to the final s state. The number of such postcollision resonances is necessarily fewer than those of the usual kind associated with the diagrams in Figure 3a. Typical magnitudes of the off-resonant cross sections are associated with the plateau region, e.g., between the first and the second resonance in Figure 4 or with the region of a minimum which is due to the interference between two resonance profiles lying nearest to the minimum. Scattering cross sections are generally several orders of magnitude smaller in these regions than at resonance but their absolute value could be varied directly by varying the intensity of the incident radiation. Thus, for example, for intensities $I \gtrsim 10^{12}$ W/cm² [which is attainable, e.g., with a Nd^{3+}–glass ($\hbar\omega = 1.17$ eV) laser] these cross sections are comparable with the usual electron–atom excitation cross sections.

It is of interest to note that in laser heating of fusion-plasma, intensities as high as 10^{14}–10^{16} W/cm² (from Nd^{3+}–glass and CO_2 lasers) have been in use.[19] At such high intensities "loss" of electron K.E. via such "off-shell" electron–photon excitation channels could be in principle considerable, particularly in the early stage of the process when the plasma is only partially ionized. Besides, the inverse process of stimulated electron–photon recombination will also occur. Contribution of these channels to electron temperature of laser-induced fusion-plasma is not yet known. It should be noted, however, that the present energy exchange mechanism is linearly dependent on the laser intensity. Multiphoton inverse bremsstrahlung effects, on the other hand, predict a nonlinear intensity dependence, contrary to recent experimental findings.[19]

Dependence of $1s$–$2s$ and $1s$–$3s$ cross sections as a function of electron energy is also of interest and they are shown in Figures 5 and 6. It is to be noted that both σ_{1s-2s} and σ_{1s-3s} show a $(\ln E)/E$ behavior at higher energies, which is typical of the present s–s transition process but resembles rather the s–p transition cross section for the usual electron–atom collision. The reason for this reversal of behavior is the requirement of conservation of angular momentum in presence of photons, which leads to a dominant dipole interaction for the scattered electron in the s–s scattering and not in the s–p scattering (for the dipole–photon supplies one unit of angular momentum of the full transition process). Thus the falloff at high energies is slower for the s–s transition and can, in principle, dominate over the simple excitation of the same transition by electrons when (as in the present calculations) the electron energy is above the excitation threshold. In Figures 5 and 6 the fixed photon frequency corresponds to that of Nd^{3+}–glass laser ($\hbar\omega = 1.17$ eV) and cross sections are given in a_0^2, scaled to unit intensity. The final electron energy (E_f), rather than the incident energy (E_i), is shown along the abscissa, where $E_i = E_f + 10.308$ eV and $E_i = E_f + 10.918$ eV for the $1s$–$2s$ and $1s$–$3s$ cross sections, respectively.

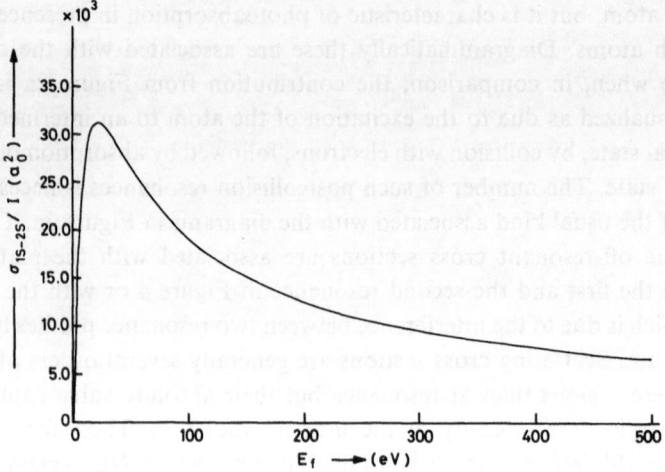

Figure 5. Integrated $1s-2s$ excitation cross section scaled to unit intensity as a function of final electron energy at the photon frequency of Nd^{3+}–glass laser ($\hbar\omega = 1.17$ eV); $E_f = E_i - 10.308$ eV; from Reference 22.

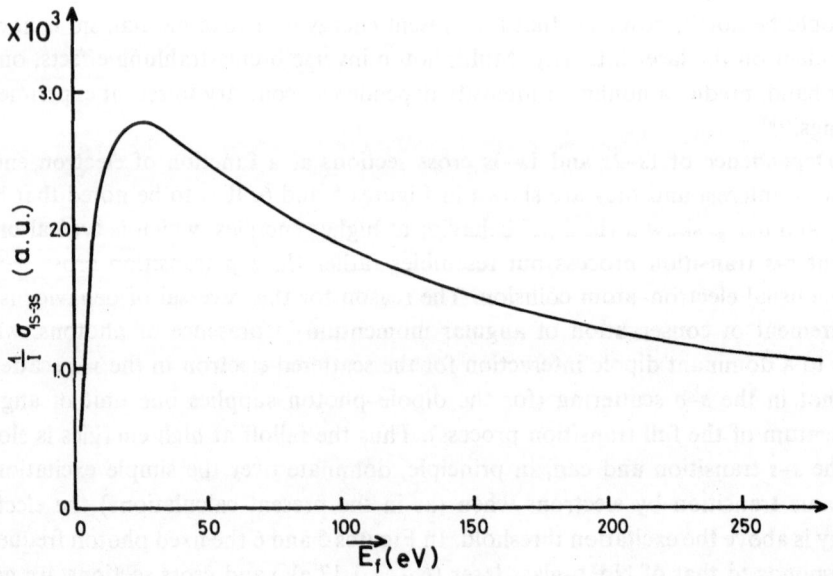

Figure 6. Integrated $1s-3s$ cross section divided by intensity shown as a function of final electron energy at the Nd^{3+}–glass laser frequency ($\hbar\omega = 1.17$ eV); $E_f = E_i - 10.918$ eV; from Reference 20.

3.2. Differential Cross Sections

The angular distribution of the scattered electron for the $1s$–$2s$ and $1s$–$3s$ transitions are similar. They are given by

$$\frac{d\sigma^{(\omega)}_{1s\to2s}}{d\Omega} = 8\pi\alpha I \frac{K_f}{K_i} R_{1s\to2s}(\omega)f(\theta, \varphi, \theta_0, \varphi_0) \tag{10}$$

$$\frac{d\sigma^{(\omega)}_{1s\to3s}}{d\Omega} = 8\pi\alpha I \frac{K_f}{K_i} R_{1s\to3s}(\omega)f(\theta, \varphi, \theta_0, \varphi_0) \tag{11}$$

where $f(\theta, \varphi, \theta_0, \varphi_0)$ is a function of the electron scattering angle $\Omega \doteq (\theta, \varphi)$ and the polarization direction $\epsilon = (\theta_0, \varphi_0)$ of the laser photons. It can be given explicitly as

$$f(\theta, \varphi, \theta_0, \varphi_0) = B_1(\theta) \cos^2 \theta_0 + B_2(\theta) \sin \theta_0 \cos^2(\varphi_0 - \varphi)$$
$$+ B_3(\theta) \sin^2 \theta \cos(\varphi_0 - \varphi) \tag{12}$$

where

$$B_1(\theta) = \left(\frac{K_i - K_f \cos\theta}{q^2} \right)^2$$

$$B_2(\theta) = \left(\frac{K_f \sin\theta}{q^2} \right)^2$$

$$B_3(\theta) = \left(\frac{K_f \sin\theta}{q^2} \right)\left(\frac{K_f \cos\theta - K_i}{q^2} \right)$$

$$q^2 = K_i^2 + K_f^2 - 2K_iK_f \cos\theta$$

Since there is more than one fixed direction in the laboratory corresponding to, for example, the incident electron beam and the incident polarization directions, the excitation cross sections no longer possess any cylindrical symmetry and the distribution is dependent on the azimuth of the scattering angle. Besides, the distribution of the scattered electrons depends parametrically on the direction of polarization of the laser photons. Figure 7 shows the intensity-scaled $1s$–$2s$ differential cross section per steradian for "blue-shifted" scattered electrons at an illustrative incident energy $E_i = 100$ eV and $\hbar\omega = 11.4$ eV, $\theta_0 = 0°$ and $\varphi - \varphi_0 = 90°$. It is seen that the cross sections fall off very rapidly from the forward maximum to zero at $\theta = 6.25°$ but rise again to an almost constant magnitude in the rest of the angular range of θ. This behavior sharply distinguishes them from the angular distribution of the direct electron excitation cross sections for the same transition; at high energy the latter are simply monotonically decreasing functions of θ.[26]

In Figure 8 is shown the $1s$–$2s$ differential cross sections per steradian for "red-shifted" scattered electrons ($E_i = 100$ eV, $\hbar\omega = 1.17$ eV) as function of the angle of polarization θ_0 in the scattering plane. The results shown correspond to a scattering angle of $\theta = 30°$, $\varphi - \varphi_0 = 0°$. The cross section is seen to drop off from a large value at $\theta_0 = 0°$ to a vanishing minimum at $\theta_0 = 20°$, then arches through a maximum at $\theta_0 = 110°$. With different choices of angular parameters other forms of scattering

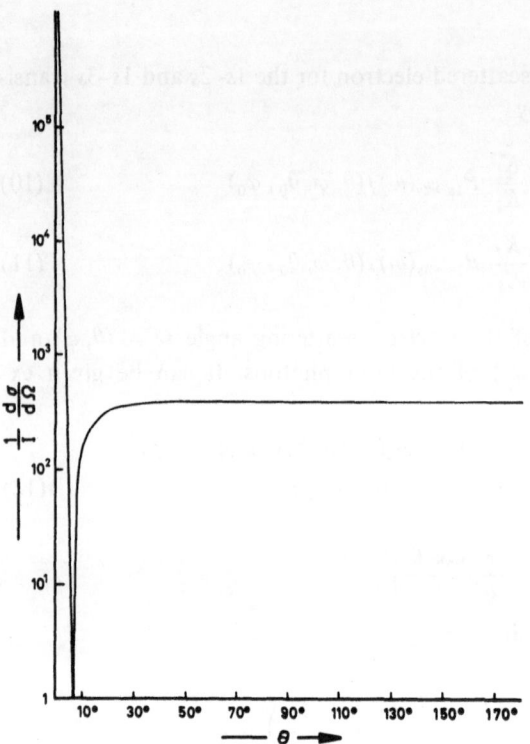

Figure 7. Intensity scaled $1s$–$2s$ differential cross section per steradian for "blue-shifted" electrons; $E_i = 100$ eV, $\hbar\omega = 11.4$ eV, $\theta_0 = 0°$, $\varphi - \varphi_0 = 90°$; from Reference 23.

distribution can be shown to arise and, in general, they are greatly different from their counterparts for atomic excitation by electrons only. Two qualitatively important predictions of the angular distributions (10), are (i) the differential cross section is vanishingly small when the laser polarization direction $\boldsymbol{\epsilon} = (\theta_0\varphi_0)$ is perpendicular to the scattering plane, and (ii) the differential cross section at a fixed angle θ is maximum when the laser polarization direction is parallel to the momentum transfer $\mathbf{q} = \mathbf{K}_i - \mathbf{K}_f$. Such properties of the plane-wave dipole approximation could provide stringent test of its validity or otherwise even at a very high incident energy.

4. Concluding Remarks

Atomic excitations with simultaneous collisions of electrons and photons have certain properties that set them apart from processes initiated by a single beam of projectiles. They are also significantly different from "on-shell" resonant excitation–collision processes and allow the study of inelastic electron–atom scattering amplitude at "off-shell" electron energies. The characteristic electron excitation cross sections depend linearly on laser intensity I and are comparable with direct electron excitation cross sections at $I \geq 10^{12}$ W/cm². Besides their basic theoretical interest, e.g., in permitting possible measurements of "off-shell" electron scattering amplitudes, they

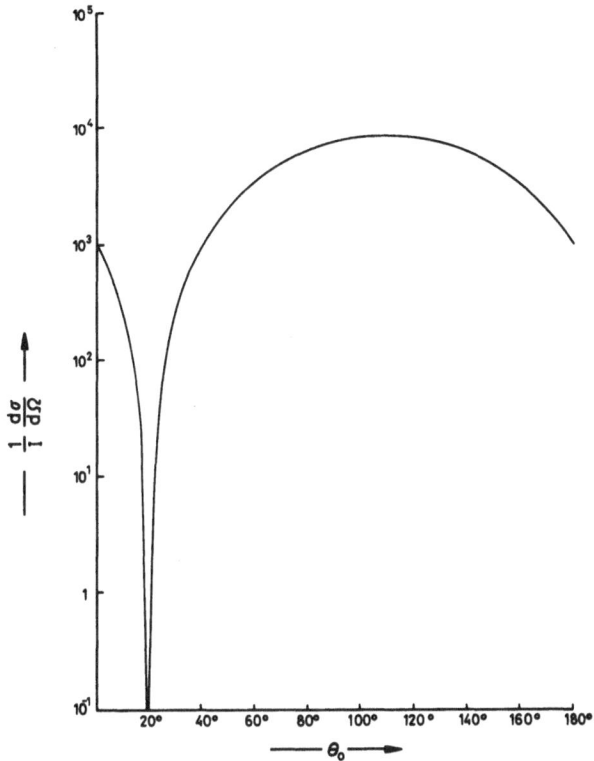

Figure 8. Intensity scaled $1s$–$2s$ differential cross section per steradian for "red-shifted" scattered electrons at $\theta = 30°$, $\varphi - \varphi_0 = 0°$, as a function of the angle of orientation of the polarization vector, θ_0; $E_i = 100$ eV, $\hbar\omega = 1.17$ eV; from Reference 23.

could be relevant in connection with laser heating of fusion plasma experiments which are performed at ultrahigh intensities ($I \gg 10^{12}$ W/cm²). The study of the angular distribution can distinguish a simultaneous electron–photon process from other competing processes when present (because of their characteristic behavior both as a function of scattering angle as well as of laser polarization angle). The frequency response of the system reveals the existence of postcollision resonances at laser frequencies smaller than the first atomic resonance frequency. Since these resonances occur on the energy shell they could be observed even at low laser intensities with commonly available tunable dye lasers. Detection of postcollision absorption could also be made, perhaps more conveniently, by observing the ionization of the target as well. Further theoretical work at low electron energies, for ionization and for other atomic targets, is to be desired.

ACKNOWLEDGMENTS

It is my pleasure to thank Dr. N. K. Rahman for his fruitful collaboration in this field and Professor H. Kleinpoppen for several useful comments.

References

1. U. Fano and J. Macek, *Rev. Mod. Phys.* **45**, 553–573 (1973).
2. J. Macek and I. V. Hertel, *J. Phys. B* **7**, 2173–2183 (1974).
3. I. V. Hertel and W. Stoll, *Adv. At. Mol. Phys.* **13**, 113–228 (1977).
4. H. W. Hermann, I. V. Hertel, W. Heiland, A. Stamatovic, and W. Stoll, *J. Phys. B* **10**, 251–267 (1977).
5. J. Macek and D. H. Jaecks, *Phys. Rev. A* **4**, 228–230 (1971).
6. K. Blum and H. Kleinpoppen, in *Electron and Photon Interaction with Atoms*, H. Kleinpoppen and M. R. C. McDowell, eds., pp. 501–513, Plenum Press, New York (1976).
7. H. Kleinpoppen, in *Comments At. Mol. Phys.* **6**, 35–47 (1976).
8. I. McGregor, K. Blum, and H. Kleinpoppen, in *Physics of Electronic and Atomic Collisions* IPEAC X, p. 511, Commissariat à l'Energie Atomique, Paris (1977).
9. S. Geltman, *J. Quant. Spectrosc. Radiat. Transfer* **13**, 601–613 (1972).
10. F. H. M. Faisal, *J. Phys. B* **6**, L312–315 (1973).
11. N. M. Kroll and K. M. Watson, *Phys. Rev. A* **8**, 804–809 (1973).
12. N. K. Rahman, *Phys. Rev. A* **10**, 440–441 (1974).
13. F. H. M. Faisal, *Phys. Lett.* **50A**, 193–194 (1974).
14. H. Krüger and M. S. Schulz, *J. Phys. B* **9**, 1899–1910 (1976).
15. P. J. K. Langendam, M. Gavrila, J. P. J. Kaandorp, and M. J. Van der Wiel, *J. Phys. B* **9**, L453–457 (1976).
16. D. Andrick and L. Langhans, *J. Phys. B* **9**, L459–461 (1976).
17. K. L. Bell, P. G. Burke, and A. E. Kingston, "Free–free transitions of an electron in the presence of an atomic system," Preprint, Queen's University of Belfast (1977).
18. A. Weingartenshofer, J. K. Holmes, G. Caudle, E. M. Clarke, and H. Krüger, *Phys. Rev. Lett.* **39**, 269–270 (1977).
19. K. Boyer, in *Third Conference on the Laser*, L. Goldman, ed., *Ann. N. Y. Acad. Sci.* **267**, 117–125 (1976).
20. N. K. Rahman and F. H. M. Faisal, *J. Phys. B* **11**, 2003–2014 (1978).
21. F. H. M. Faisal and N. K. Rahman, *Fifth International Conference on Atomic Physics, 1976*, R. Marus, M. H. Prior, and H. A. Shugart, eds., pp. 49–51, Berkeley, University of California, Abstracts.
22. N. K. Rahman and F. H. M. Faisal, *J. Phys. B* **9**, L275–277 (1976).
23. N. K. Rahman and F. H. M. Faisal, *Phys. Lett.* **57A**, 426–428 (1976).
24. V. M. Buimistrov, *Phys. Lett.* **30A**, 136–137 (1969).
25. L. C. Hostler, *J. Math. Phys.* **11**, 2966–2970 (1976).
26. N. F. Mott and H. S. W. Massey, in *The Theory of Atomic Collisions*, 3rd ed., p. 483, Clarendon Press, Oxford (1965).

41

Energy and Angular Momentum Exchanges between Two Electrons in a Coulomb Field

H. G. M. Heideman

A review is presented of recent studies on the phenomenon of "postcollision interaction" (PCI), carried out in our institute. Particular attention is paid to the experimental investigation of orbital angular momentum exchanges and spin effects in the PCI process. A theory is worked out, which is based on the optical-potential method of Feshbach, and which also accounts for postcollisional exchanges of orbital angular momentum.

1. Introduction

During the last few years in atomic collision physics a field has developed that shows some similarity with the study of final-state interactions in elementary particle physics. It concerns the study of the long-range Coulomb interactions between charged particles which have been produced in a primary reaction. Recent studies have shown that such interactions may cause considerable changes in the initial partitioning of energy and angular momentum between the interacting particles and therefore may have a great effect on both the magnitude and energy behavior of various collision cross sections. As these interactions are somewhat difficult to deal with in quantum mechanical scattering theories, it is of importance to obtain experimental information as to their characteristics and effects in various collision processes.

H. G. M. Heideman ● Fysisch Laboratorium der Rijksuniversiteit te Utrecht, Sorbonnelaan 4, The Netherlands.

In electron–atom collisions the first observation of what is now known as the "postcollision interaction" (PCI) were made by Hicks et al.[1] in Manchester. In their study on the electron impact excitation of autoionizing states they observed that the energies of the electrons ejected from autoionizing states shifted to steadily higher values as the incident electron energy was lowered to within a few eV above the threshold of the autoionizing state concerned. The results could be interpreted in terms of a "postcollisional" energy exchange between the ejected and the scattered electron. Shortly after, Smith et al.[2] and Heideman et al.[3] showed that such energy exchanges could also give rise to structures in the excitation curves of singly excited states; if the energy exchange is larger than the initial excess energy of the scattered electron, the latter electron is left with a negative energy and is captured by the residual ion into a bound state of the neutral atom.

Another important example of a collision process where the long-range Coulomb interaction plays an important role is the near-threshold ionization of atoms by electrons. Near the ionization threshold the manifestation and decay of correlations in the motion of the two receding electrons gives rise to a specific threshold law for the ionization process (Wannier,[4] Rau,[5] and Fano[6]) and may cause an additional population of excited states with high orbital angular momentum (Fano[6]). Further examples are near-threshold multiple ionization by electrons or photons or inner-shell ionization followed by Auger emission (Wight et al.[7]).

In this report we will present a selection from the recent investigations that have been carried out in our institute. Special attention will be paid to the study of angular momentum exchanges in postcollision interactions. For a survey of the various other types of experimental investigations that have been performed up till now we refer to the papers by Read,[8] Van Ittersum et al.,[9] and Niehaus.[10]

2. The Near-Threshold Excitation of Autoionizing States

2.1. Indirect Excitation of Bound States via PCI

As mentioned in the Introduction, the energy exchange between the scattered electron and the electron ejected from an autoionizing state may lead to a negative energy for the scattered electron. The latter electron is then captured into an excited state and we have an indirect excitation mechanism according to the scheme

$$e_0 + A \rightarrow A^{**} + e_1 \rightarrow A^+ + e_1 + e_2 \xrightarrow{\text{PCI}} A^* + e_2 \tag{1}$$

By interference with the "background" of the direct excitation this mechanism may give rise to structures on the excitation curves at energies just above the threshold of the autoionizing state concerned. These so-called PCI structures exhibit characteristic features by which they can be distinguished from the well-known negative-ion resonances. They have shapes that differ appreciably from the Fano–Beutler resonance shapes. Further, they appear to shift to higher incident-electron energies when observing states with increasing excitation energies. This is due to the fact that the most probable energy exchange between the scattered and the ejected electron decreases with increase of the excess energy of the scattered electron. Finally, since there must be a maximum value for the energy exchange, it is to be expected that PCI structures do not occur in the excitation of bound states that lie too far below the ionization threshold (see also Reference 9).

In our laboratory we have investigated the above-described mechanism by studying the PCI structures in the optical excitation curves of various helium, neon, and argon states. An optical excitation curve of a particular state is obtained by measuring the light emission following the excitation of that state as a function of the incident electron energy. An advantage of our method[9] is that besides energy exchanges also angular momentum exchanges and possible spin interactions during the PCI process can be investigated. Energy exchanges between the scattered and ejected electron may be studied by measuring the positions of the PCI structures in the excitation of states with different binding energies (Smith et al.,[2] Nienhuis and Heideman[11]) and by determining the energy shifts in ejected electron spectra (Hicks et al.,[1] Roy et al.[12]). Information as to the angular momentum exchange or spin interactions may be obtained by studying the magnitudes of the PCI structures in the excitation of states with different orbital angular momentum or with different spin. In electron excitation experiments, where scattered electrons are detected, bound states with different orbital momentum or spin can usually not be separated (with the possible exception of the lowest-lying states, for which the PCI mechanism is less important or even totally absent). In the measurement of optical excitation curves the various states to be investigated can be separated with optical resolution. In the case of helium, for example, this means that all states with principal quantum number $n \leq 10$ and orbital angular momentum $L \leq 3$ or 4 can readily be resolved.

The Figures 1–4 show a selection of our most recent excitation measurements on a number of helium states in the energy domain from 52 to 72 eV. The figures exhibit a variety of structures occurring in practically every curve. To bring out the structures more clearly we have in most of the curves subtracted the sloping background of the direct excitation in a similar way as was done by Roy et al.[12] Some of the structures can clearly be identified as negative-ion resonances as they occur at the same energy in the different excitation curves. The well-known $(2s^2 2p)^2 P$ and $(2s 2p^2)^2 D$ resonances at 57.2 and 58.3 eV show up very clearly in most of the curves. Near 60 eV and (less clear) near 58.4 eV structures are apparent that exhibit the characteristic PCI features discussed in the beginning of this section. Around 63 and 69 eV additional structures are visible, which are probably caused by negative-ion resonances and PCI mechanisms associated with higher-lying autoionizing states. The latter structures will not be further discussed in the present report. Their interpretation is difficult because of the many possible autoionizing states by which they can be caused. We confine our analysis to the structures around 60 eV, which we interpret as being due to a PCI mechanism via the $(2p^2)^1 D$ autoionizing state, and to

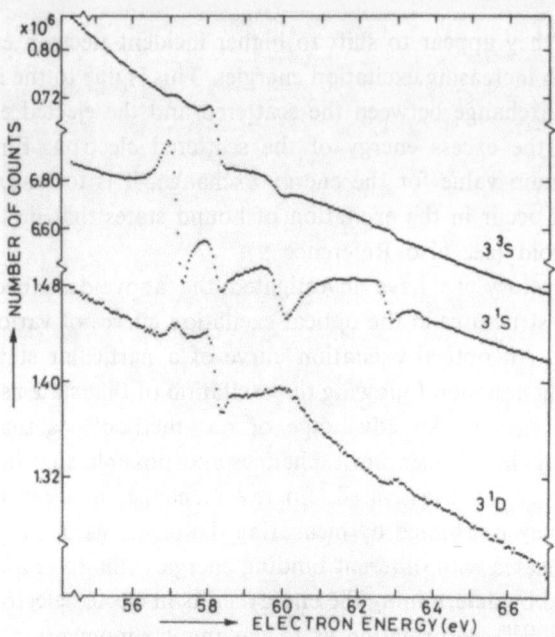

Figure 1. The excitation of three $n = 3$ states of helium by electrons for impact energies between 55 and 67 eV. The vertical scales of the different curves cannot be compared with one another.

the PCI structures near 58.4 eV, which can be caused by the $(2s^2)^1S$ or the $(2s2p)^3P$ state.

In the excitation of the states with $n \geq 4$ (Figures 2–4) the 60-eV structure clearly moves to higher incident energies with decrease of the binding energy of the state observed. For the bound states with $n = 3$ this characteristic behavior is not observed. In the excitation curves of these states the structure appears to be centered around the *same* energy of 59.9 eV [close to the $(2p^2)^1D$ threshold], notwithstanding the fact that some of these states have appreciably different binding energies. Furthermore, the shapes of the structures in the $n = 3$ excitation curves seem to deviate from the shapes of the corresponding structures for larger n. These facts, which were also noticed by King et al.[13] and by Roy et al.,[12] led these authors to suggest that the $n = 3$ structures might not be caused by a PCI mechanism but rather by some three-electron resonance (see King et al.[13]).

There is indeed some experimental (Spence[14]) and theoretical (Ormonde et al.,[15] Nesbet,[16] Kets and Heideman[17]) evidence for the existence of two relatively broad negative-ion resonances around 60 eV; namely, the $(2s2p^2)^2S$ and $(2p^3)^2P$ states (see Nesbet[16]) at 59.3 and 60.4 eV, respectively. It is unlikely, however, that these resonances directly decay to the singly excited states, as do the well-known $(2s^22p)^2P$ and $(2s2p^2)^2D$ resonances at 57.2 and 58.3 eV. The first argument for this statement is the total absence of structure around 60 eV in all four $n = 2$ states (Simpson et al.,[18] King et al.[13]). If a direct decay of a resonance were involved it would be difficult to understand why the resonance does not occur in the excitation of any of the $n = 2$ states. The absence of PCI structures in the $n = 2$ excitation is understandable because these states lie probably too far below the ionization threshold to make their PCI excitation possible (see the beginning of this section). Furthermore, close-coupling calculations by Kets and Heideman[17] have shown that indeed the coupling of the two resonances around 60 eV with the singly excited states is very

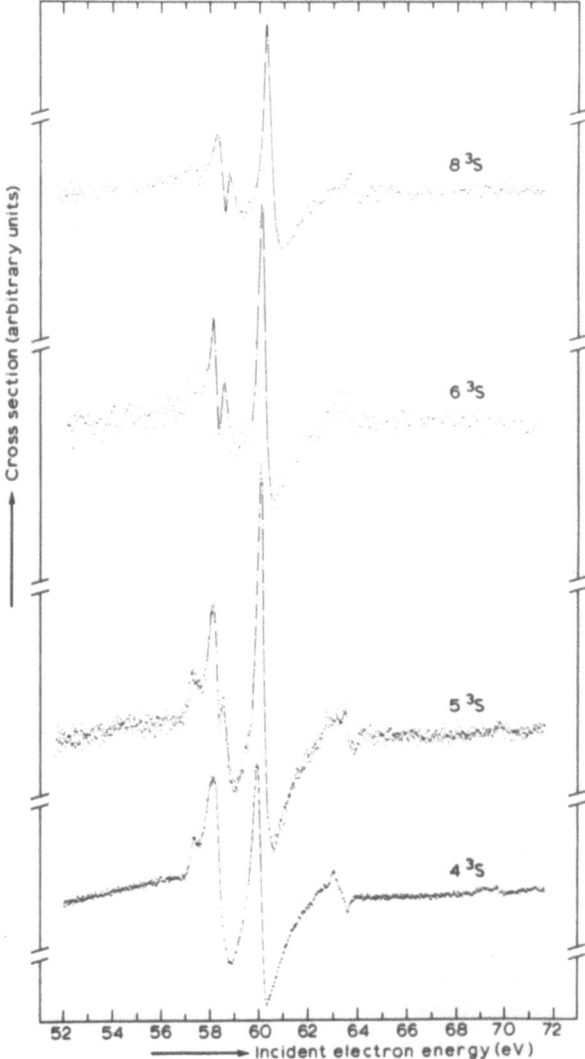

Figure 2. The excitation of the 4^3S, 5^3S, 6^3S, and 8^3S states of helium for impact energies between 52 and 72 eV. In order to display the various structures on a horizontal background the sloping background of the direct excitation has been subtracted from the original data. The vertical scales of the different curves are arbitrary and not comparable. The energy scales have been calibrated against the large $(2s2p^2)^2D$ resonance at 58.30 eV in the 4^3D excitation (see Figure 4).

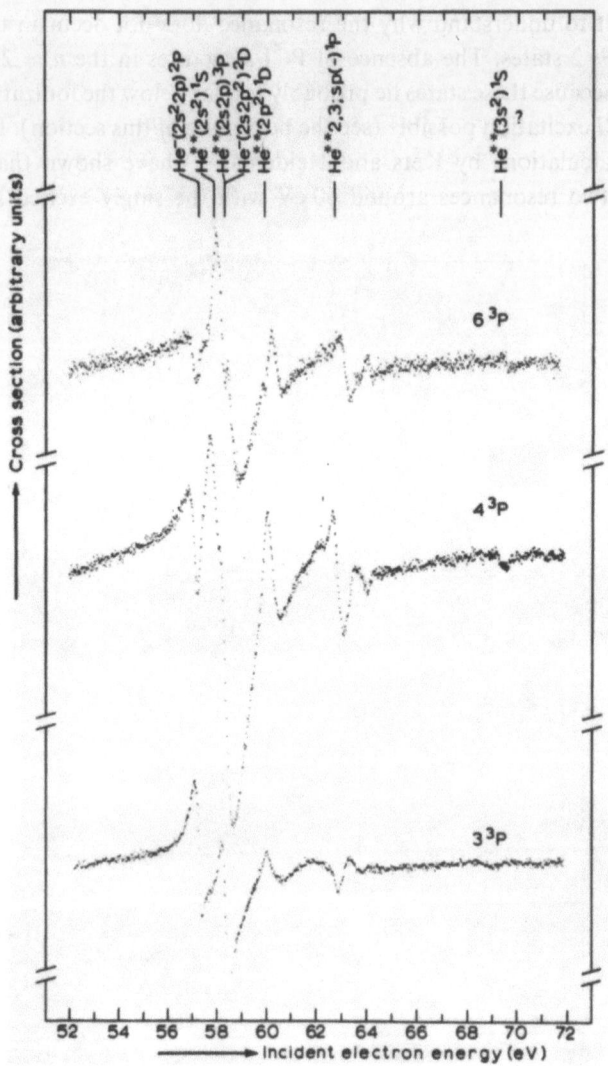

Figure 3. The same as Figure 2, but for the 3^3P, 4^3P, and 6^3P states.

weak. On the other hand, the excitation functions of the doubly excited states appear
to be significantly affected by these resonances. In the next section we will point out
some additional features of the $n = 3$ structures near 60 eV, which strongly support
their interpretation as PCI structures.

In the energy range around 58.3 eV we have a situation where negative-ion
resonances and PCI structures occur at the same energies giving rise to complicated
interference structures. The $(2s^22p)^2P$ and $(2s2p^2)^2D$ resonances are present in
practically every excitation curve. In the $n = 3$ excitation these two resonances are
probably the main contributors to the structure between 57 and 58.5 eV. For higher

principal quantum numbers, however, additional structure appears near 58.3 eV, which more or less moves through the $(2s2p^2)^2D$ resonance when going to larger n. This is in particular evident for the n^3S states, but it can also be observed as a broadening of the lower part of the 58.3 eV structure in the n^3P ($n \geq 4$) excitation and as an extra shoulder on the right flank of the large peak in the 8^3D excitation. This additional structure is most probably caused by a PCI mechanism via the $(2s^2)^1S$ or $(2s2p)^3P$ state (see also van Ittersum et al.[9] and Smith et al.[2]).

By measuring the positions of the PCI structures in the excitation of states with different binding energies and by determining the energy shifts of ejected electrons information can be obtained as to the energy exchange ε in the PCI process. Such

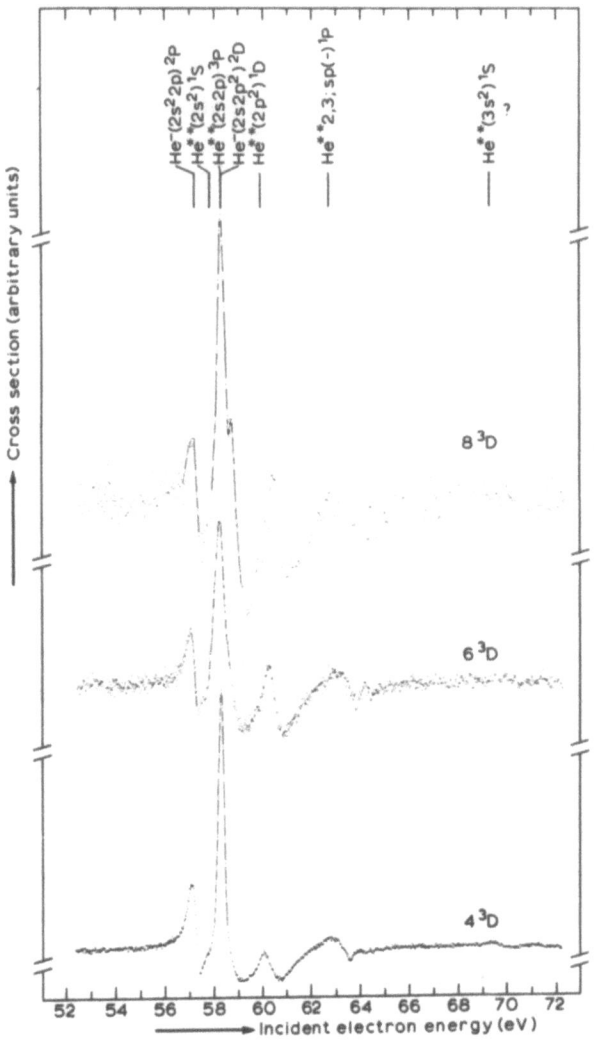

Figure 4. The same as Figure 2, but for the 4^3D, 6^3D, and 8^3D states.

measurements have been performed by among others Hicks et al.,[1] Smith et al.,[2] and Van Ittersum et al.[9] In particular the functional relation between ε and the excess energy E_1 of the scattered electron was studied and compared with various theoretical predictions. It seems questionable afterwards whether such determinations are meaningful at all. The PCI structures in the ejected electron spectra and in particular in the excitation curves appear as relatively broad oscillatory interference structures, so that energy determinations can hardly be made in an unambiguous way (see also King et al.[13] and Morgenstern et al.[19]).

2.2. Orbital Angular Momentum Exchanges and Spin Effects

As discussed in the Introduction, information about the orbital angular momentum exchanges in the PCI process can be obtained by studying the strengths of the PCI structures in the excitation of states with different orbital angular momenta. A first qualitative analysis of the experimental results indicates that the PCI structures are strongest in the excitation of S states. This can be made plausible by a simple estimation. Immediately after the excitation of the autoionizing state (hence, before autoionization) the orbital angular momentum of the scattered electron is most probably zero, because it has very low kinetic energy. After the autoionization the scattered electron may acquire orbital momentum by the postcollisional interaction with the ejected electron and the residual ion. The magnitude of this orbital angular momentum exchange determines the orbital angular momentum of the final excited state after capture of the scattered electron. An upper bound for the angular momentum exchange ΔL may be calculated in a similar way as was done by Fano[6] (see also Heideman et al.[20]). If M ($\leq e^2/r$) is the torque exerted by one electron on the other, Δt ($\approx r/v$) is the postcollision interaction time, r indicates the order of magnitude of the electron–electron and electron–nucleus distances and v is the velocity of the ejected electron, we find

$$\Delta L = M \cdot \Delta t \leq e^2/v \tag{2}$$

In case of PCI via the excitation of the $(2p^2)^1D$ autoionizing state of helium we get $\Delta L \leq 7 \times 10^{-35}$ J s, which is smaller than $1\hbar$. In reality there will, of course, be a probability distribution for the angular momentum exchange with finite probabilities for larger ΔL, but the above estimation suggests that the most probable value for the angular momentum exchange is zero. Therefore, PCI excitation of an S state is more likely than the PCI excitation of a state with larger L.

We will analyze the strengths of the PCI structures caused by the $(2p^2)^1D$ autoionizing state in some more detail. We define a relative strength for the PCI structure in the excitation of a particular state by

$$p_{rel} = \frac{\sigma^r_{max} - \sigma^r_{min}}{\sigma_b{}^r} \tag{3}$$

where σ^r_{max} and σ^r_{min} are the relative cross sections in the main maximum and nearest minimum of the structure, respectively, and $\sigma_b{}^r$ is the relative "background" cross section of the direct excitation in the vicinity of the structure. The p_{rel} values can directly be determined from our relative measurements. The absolute strengths of the structures are then obtained via

$$p = \sigma_b p_{rel} \tag{4}$$

where σ_b is the absolute background cross section near the structure. Values for σ_b were obtained using the absolute measurements of van Raan et al.[21,22] For the states that were not measured by them we estimated the cross sections by applying the n^{-3} scaling rule (n = principal quantum number) for states belonging to the same Rydberg series. In this way we made estimates of the strengths of the PCI excitation mechanisms for the n^1S, n^3S, n^3P, and n^3D states caused by the $(2p^2)^1D$ state. The results are given in Table 1.

In principle the strength of a structure observed in the excitation of a particular state may be influenced by cascading from higher-lying states. We have not corrected our results for this effect. For the n^3S and n^1S excitation the cascade effect is probably very small, because the possible cascade contributions can only come from the higher-lying P states. The magnitudes of the $n^{1,3}P$ structures, however, are appreciably weaker than those of the $n^{1,3}S$ structures. On the other hand the cascade contributions to the n^3P structures, which come for the greater part from the n^3S states, may not be negligible. The effect of the n^3D structures on those in the n^3P excitation and vice versa is probably not so large because the relative intensities of the n^3P and n^3D structures (compared to the direct excitation background) do not differ very much, so that structures and background are affected at roughly the same rate.

Table 1. Experimental Strengths p (in Units of 10^{-22} cm^2) of PCI Structures in the Excitation of Various Helium States[a]

L	n				
	3	4	5	6	8
^3S	47	56	50	36	15
^3P	38	29	16	13	5
^3D	10	5	3.5	3	1
^1S	86	56	35	21	9
Σ	181	146	104.5	73	30
E_0	59.90	59.95	60.05	60.15	60.25

[a] In the fifth row the sums of the p values at each n are given. The last row gives the average energies (in eV) where the PCI structures at each n occur.

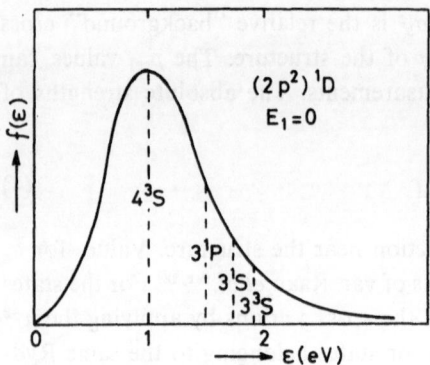

Figure 5. Fictitious energy exchange distribution in case of threshold excitation ($E_1 = 0$) of the $(2p^2)^1D$ state. For finite E_1 values similar distributions may be drawn; their maxima shift to lower ε values and their widths decrease as E_1 increases.

It is clear from Table 1 that for a given principal quantum number n the PCI structures decrease rapidly in strength with increase of the orbital angular momentum L of the state concerned. This indicates that the scattered electron, when it is captured by the residual ion as a result of PCI, most probably has zero orbital angular momentum, in accordance with the estimation based on Equation (2). It is interesting to note that for $n \geq 5$ the ratios of the strengths of the PCI structures in different excitation curves do not vary much with n. For $n = 3$ and 4, however, the relative strength of the structure in the ^3S excitation as compared to those in the ^1S, ^3P, and ^3D excitation is much smaller than the corresponding ratios for $n > 4$. This can be understood[23] if one assumes that the most probable energy exchange between the scattered and the ejected electron reaches a maximum when the autoionizing state is excited at its threshold and hence the scattered electron has an initial energy $E_1 = 0$. The results of Reference 9 suggest that for the $(2p^2)^1D$ state this maximum value is about 1 eV. So it is to be expected that the PCI excitation probabilities will be increasingly suppressed as the binding energies of the states concerned become larger than 1 eV. The situation is illustrated in Figure 5, where for $E_1 = 0$ a fictitious energy exchange distribution[†] with a maximum at 1 eV has been drawn and where the binding energies of some excited states are indicated. That the structure in the 3^3S excitation is still stronger than those in the 3^3P and 3^3D excitation must be due to the fact that zero angular momentum exchange has a much larger probability than exchanges of 1 or $2\hbar$.

By comparing at a given n (with the exception of $n = 3$ and 4; see previous paragraph) the p values for n^3S, n^3P, and n^3D one may get an impression of the relative probabilities for postcollisional angular momentum exchanges of 0, 1, and $2\hbar$; at least within the assumptions and approximations made. In the fifth row of Table 1 the sums are given of the p values for the ^3S, ^3P, ^3D, and ^1S structures, while in the last row the average incident energies are shown, where the PCI structures for each n occur. These numbers suggest a resonancelike behavior of the cross section for excita-

---

[†] In fact a continuous distribution as drawn in Figure 5 cannot exist. The physically accessible points on the curve are those that correspond to ε values that are equal to possible values of the binding energy.

tion of the $(2p^2)^1D$ autoionizing state with a strong peak very close to the threshold. The resonance involved may be the $(2s2p^2)^2S$ or the $(2p^3)^2P$ state (see also the discussion in the fifth paragraph of Section 2.1).

In Figure 5 similar distributions as for $E_1 = 0$ may be drawn for positive E_1 values. Their maxima shift to smaller ε values and their widths decrease as E_1 increases. As a result the probabilities for the right energy exchanges to excite the $n = 3$ states decrease rapidly with increase of E_1 (in particular if the cross section for excitation of the $(2p^2)^1D$ state has a sharp peak at the threshold) and therefore the major contribution to the PCI structures in the $n = 3$ excitation is expected to come from PCI with the $(2p^2)^1D$ state excited very near its threshold. This may explain why the $n = 3$ structures appear at practically the same energy close to threshold of the $(2p^2)^1D$ state (see also the discussion on the $n = 3$ structures in Section 2.1).

Comparison of the p numbers for n^3S excitation with those for n^1S excitation shows that the PCI excitation mechanism is more important for triplet than for singlet states. (The exception for $n = 3$ and 4 has been explained above.) This indicates that also spin effects play an important role in the postcollision interaction. The *relative* importance of the PCI mechanism for triplet excitation can be understood by realizing that triplet states in helium can only be excited by electron exchange (at least for not too large n, where Russell Saunders coupling may break down). As may be seen from the reaction scheme [equation (1)] the PCI excitation process indeed proceeds via an exchange mechanism.

3. Theoretical Models for Postcollision Interactions

The first attempts to describe the observed effects of postcollision interactions in a quantitative way were based upon a classical model introduced by Barker and Berry[24] and modified by Nienhuis and Heideman.[11] A more realistic description of PCI in terms of a sudden approximation has been given by King et al.[25] In this so-called "shakedown" model, which is analogous to the description of Auger processes, the autoionization is viewed as an instantaneous change of the charge of the core, and the probability for the scattered electron to end up in a certain state ψ in the field of the ion is simply taken proportional to the square of the overlap of the outgoing spherical wave with the state ψ. Morgenstern et al.[19] have developed a theory for PCI processes based on a semiclassical description of the decay of an autoionizing state in the time-dependent field of the slowly receding scattered particle. In their theory also possible coherences between excitations of different autoionizing states are discussed. Both the calculations based on the shakedown model as well as those based on the semiclassical theory can very well be fitted to the experimental data by adjusting a number of parameters involving various excitation (or ionization) amplitudes and their relative phases. However, as suggested by Read,[26] there is no doubt that the high quality of the fits is partly due to the large number of adjustable parameters (up to six for the shakedown calculations[26] on ejected electron spectra).

Neither the shakedown model nor the semiclassical theory is able to account for angular momentum exchanges in the PCI process. In the shakedown model no interaction at all is taken into account between the scattered and ejected electron, while in the semiclassical theory the angular dependence of this interaction is neglected. Therefore, neither one of these two theories can be tested against the experimental results on this point. Angular momentum exchanges are in principle included in the formal theory for PCI worked out by Nienhuis[27] and Heideman. Their resulting formalism is a generalization of the standard theory for resonances based on the optical potential method of Feshbach. It gives an expression for the PCI contribution to the amplitude for excitation of a singly excited state. By defining an appropriate projection operator an optical potential is generated which allows a separation of the transition matrix element into two parts; one part T_P, which describes the excitation of the state concerned with the scattering mechanism confined to the subspace of singly excited states, and another part T_{opt}, which contains the effect of the autoionizing states on the excitation process. In the case of excitation of a particular helium state $|i\rangle$ from the ground state $|0\rangle$ with only one autoionizing state $|a\rangle$ contributing to the PCI mechanism T_{opt} can be written as

$$T_{opt} = \sum_{\beta} \left(\frac{1}{2\pi}\right)^3 \int d\mathbf{k}_a \langle \mathbf{k}_i i; \alpha^{P-} | PHQ | \mathbf{k}_a a; \beta^{Q-}\rangle \frac{1}{E_1 - K_a + \frac{1}{2}i\Gamma}$$
$$\times \langle \mathbf{k}_a a; \beta^{Q-} | QHP | \mathbf{k}0; 1^{P+}\rangle \tag{5}$$

where H is the complete Hamiltonian of the system, which can be separated in three different ways in a free Hamiltonian H_α for electron α and a helium atom containing the other two electrons, and the interaction V_α between electron α and the atom:

$$H = H_\alpha + V_\alpha, \qquad \alpha = 1, 2, 3 \tag{6}$$

So we have three different arrangement channels which have to be considered explicitly because exchange processes are essential in the PCI mechanism (see the end of Section 2.2). The arrangement channel in which electron α is free is indicated with the label α. The summation in (5) is over the different arrangement channels.

The projection operator P projects onto the subspace of states with at least one electron in the $1s$ orbital, which is precisely the space of singly excited He states and the ground state of He^+. The complementary operator $Q = I - P$ projects onto the subspace of states with no electron in a $1s$ orbital. So this subspace contains the doubly excited states.

The stationary scattering states $|\mathbf{k}0; 1^{P+}\rangle$ and $\langle \mathbf{k}_i i; \alpha^{P-}|$ are eigenstates of the P-projected Hamiltonian PHP; similarly $|\mathbf{k}_a a; \beta^{Q-}\rangle$ is an eigenstate of QHQ. Finally, E_1 is the excess energy of the incident electron above the threshold of the autoionizing state.

From right to left equation (5) displays the successive steps of the PCI excitation mechanism. First the atom is excited by the incident electron 1 from the ground state $|0\rangle$ to the autoionizing state $|a\rangle$, and the scattered electron moves out with wave vector $\mathbf{k}_a$. The possible decay channels of the autoionizing state give rise to a finite width Γ, which is expressed as an imaginary energy term in the propagator $(E_1 - K_a + \frac{1}{2}i\Gamma)^{-1}$. Finally the scattering state $|\mathbf{k}_a a; \beta^{Q-}\rangle$, in which the outgoing electron β leaves the atom in the state $|a\rangle$, decays to the scattering state in which the ejected electron α (with wave vector $\mathbf{k}_i$) leaves the atom in the singly excited state $|i\rangle$. An integration must be performed over the wave vector $\mathbf{k}_a$ (with corresponding kinetic energy K_a) of the scattered electron immediately after the excitation of the autoionizing state. However, the propagator favors those values of K_a that conserve energy within the width Γ of the autoionizing state.

In view of the physical picture we have of the postcollision interaction the most interesting part of the equation (5) is the first matrix element in the integrand; it contains the interaction between the slow scattered electron β and the fast ejected electron α. To make the notation somewhat more compact (and also more logical) we rewrite this matrix element as follows:

$$\langle \mathbf{k}_\alpha, i_\beta{}^{P-} | PHQ | \mathbf{k}_\beta, a^{Q-}\rangle \tag{7}$$

If we make the assumption that the decay of the autoionizing state occurs when the electron β is essentially free we may replace H in (7) by H_β and $|\mathbf{k}_\beta, a^{Q-}\rangle$ by the free state $|\mathbf{k}_\beta, a\rangle$. We now wish to introduce explicitly in (7) the interaction $V_{\alpha\beta} = |\mathbf{r}_\alpha - \mathbf{r}_\beta|^{-1}$ between the scattered and ejected electron. This can be done by separating the total Hamiltonian as follows:

$$H = H^{\alpha\beta} + V_{\alpha\beta} \tag{8}$$

Next we make use of the Lippman–Schwinger equation, which expresses the scattering state $|\mathbf{k}_\alpha, i_\beta{}^{P-}\rangle$, which is an eigenstate of PHP, in terms of an eigenstate of $PH^{\alpha\beta}P$ at the same eigenvalue E (see Taylor[28]):

$$|\mathbf{k}_\alpha, i_\beta{}^{P-}\rangle = |[\mathbf{k}_\alpha{}^-], i_\beta\rangle + \frac{1}{E^- - PH^{\alpha\beta}P} PV_{\alpha\beta}P |\mathbf{k}_\alpha, i_\beta{}^{P-}\rangle \tag{9}$$

In (9), $[\mathbf{k}_\alpha{}^-]$ is a scattering state, in which electron α interacts only with the ion core and not with electron β. The "free" Hamiltonian $H^{\alpha\beta}$ contains only the interactions between α and the core and between β and the core, but not the interaction between α and β. We now make the approximation of replacing the scattering state $|\mathbf{k}_\alpha, i_\beta{}^{P-}\rangle$ on the right-hand side of (9) by the unperturbed state $|\mathbf{k}_\alpha, i_\beta\rangle$. Substituting in (7) we get

$$\langle \mathbf{k}_\alpha, i_\beta{}^{P-} | PH_\beta Q | \mathbf{k}_\beta, a\rangle \simeq \langle [\mathbf{k}_\alpha{}^-], i_\beta | PH_\beta Q | \mathbf{k}_\beta, a\rangle$$

$$+ \langle \mathbf{k}_\alpha, i_\beta | PV_{\alpha\beta}P \frac{1}{E^+ - PH^{\alpha\beta}P} PH_\beta Q | \mathbf{k}_\beta, a\rangle \tag{10}$$

The first term of (10) can be written as

$$\langle i_\beta | \mathbf{k}_\beta\rangle\langle [\mathbf{k}_\alpha{}^-] | PH_\beta Q | a\rangle \tag{11}$$

Together with (5) this gives the well-known shakedown model. The scattering amplitude is approximated by the overlap of $\langle i_\beta |$ and $| k_\beta \rangle$ multiplied by the decay probability of the autoionizing state and the amplitude for excitation of the auto-ionizing state.

To evaluate the second term of (10) we insert a closure of eigenstates of $H^{\alpha\beta}$:

$$P = \left(\frac{1}{2\pi}\right)^3 \int dk_\alpha \sum_{i_\beta} | i_\beta, [k_\alpha^-] \rangle \langle i_\beta, [k_\alpha^-] |$$

$$+ \left(\frac{1}{2\pi}\right)^6 \int dk_\alpha \, dk_\beta \, | [k_\beta^-], [k_\alpha^-] \rangle \langle [k_\beta^-], [k_\alpha^-] |$$

$$+ \left(\frac{1}{2\pi}\right)^3 \int dk_\beta \sum_{i_\alpha} | [k_\beta^-], i_\alpha \rangle \langle [k_\beta^-], i_\alpha | \qquad (12)$$

where the integrations are over continuum states and summations over bound states. This closure is substituted right after the Green's operator $[E^+ - PH^{\alpha\beta}P]^{-1}$. We get

$$\langle k_\alpha, i_\beta | PV_{\alpha\beta}P \frac{1}{E^+ - PH^{\alpha\beta}P} | i_\beta, [k_\alpha^-] \rangle \langle i_\beta, [k_\alpha^-] | PH_\beta Q | k_\beta, a \rangle \qquad (13a)$$

$$+ \langle k_\alpha, i_\beta | PV_{\alpha\beta}P \frac{1}{E^+ - PH^{\alpha\beta}P} | [k_\beta^-], [k_\alpha^-] \rangle \langle [k_\beta^-], [k_\alpha^-] | PH_\beta Q | k_\beta, a \rangle$$
$$\qquad (13b)$$

$$+ \langle k_\alpha, i_\beta | PV_{\alpha\beta}P \frac{1}{E^+ - PH^{\alpha\beta}P} | [k_\beta^-], i_\alpha \rangle \langle [k_\beta^-], i_\alpha | PH_\beta Q | k_\beta, a \rangle \qquad (13c)$$

The integrations and summations have been omitted for the sake of brevity. The terms (13a) and (13c) do not seem very realistic physically and are probably unimportant as compared to term (13b), which corresponds much better to the intuitive picture we have of the postcollision interaction. Immediately after the decay of the autoionizing state the electron β finds itself in a continuum state in the field of the ion core. The first factor then describes the interaction with the ejected electron α and the capture of β into a bound state $| i_\beta \rangle$. Retaining term (13b) we now can write for (10)

$$\langle k_\alpha, i_\beta{}^{P-} | PH_\beta Q | k_\beta, a \rangle$$

$$\simeq \text{shakedown term} + \int dk_\alpha' \, dk_\beta' \langle k_\alpha, i_\beta | PV_{\alpha\beta}P | [k_\alpha'^-], [k_\beta'^-] \rangle$$

$$\times \left[E^+ - \frac{k_\alpha'^2}{2m} - \frac{k_\beta'^2}{2m} \right]^{-1} \langle [k_\alpha'^-], [k_\beta'^-] | PH_\beta Q | k_\beta, a \rangle \qquad (14)$$

After writing (14) in coordinate representation and separating the angular and radial parts of the wave function, the integration over k_α is performed in a straightforward way by standard contour integration. The radial part of the second term on the right-hand side of (14) can then eventually be written as

$$\sum_{lm} \int dk_\beta T(k_\alpha l_\alpha; nLM; l; k_\alpha^0 L_a; k_\beta l_\beta) \chi(k_\alpha^0 \leftarrow a) \langle lm | l_\alpha m_\alpha L_a M_a \rangle \langle lm | LM l_\beta m_\beta \rangle \qquad (15)$$

where T contains all radial integrals and where $\chi(k_a{}^0 \leftarrow a)$ is the amplitude for decay of the autoionizing state $|a\rangle$ by ejecting electron α with wave vector $\mathbf{k}_\alpha{}^0$. The quantum numbers of the final bound state for electron β are denoted by n, L, and M, whereas k_α, l_α, and m_α are the quantum numbers in the final state of the ejected electron α. The quantum numbers of β and α immediately after the ionization are k_β, l_β, m_β and $k_\alpha{}^0$, L_α, M_α, respectively. The integration over k_β allows (within the width of the autoionizing state) for all possible distributions of the initial energy over electron α and β immediately after the autoionization.

The amplitude for PCI excitation of the state $|i_\beta\rangle$ is now obtained by multiplying equation (15) by the amplitude for excitation of the autoionizing state $|a\rangle$ [see equation (5)]. Practical calculations based on equation (15) are in progress and will be presented in a forthcoming publication.

ACKNOWLEDGMENTS

I would like to gratefully acknowledge the support and advice of my colleagues in Utrecht during the preparation of this paper. In particular I want to thank Jaap van Eck, Willem van de Water, and Gerard Nienhuis; they contributed greatly to the work presented here.

References

1. P. J. Hicks, S. Cvejanovič, J. Comer, F. H. Read, and J. M. Sharp, *Vacuum* **24**, 573 (1974).
2. A. J. Smith, P. J. Hicks, F. H. Read, S. Cvejanovič, G. C. M. King, J. Comer, and J. M. Sharp, *J. Phys. B* **7**, L497 (1974).
3. H. G. M. Heideman, G. Nienhuis, and T. van Ittersum, *J. Phys. B* **7**, L493 (1974).
4. G. H. Wannier, *Phys. Rev.* **90**, 817 (1953).
5. A. R. P. Rau, *Phys. Rev. A* **4**, 207 (1971).
6. U. Fano, *J. Phys. B* **7**, L401 (1974).
7. G. R. Wight and M. J. van der Wiel, *J. Phys. B* **10**, 601 (1977).
8. F. H. Read, Invited Lectures, *Ninth International Conference on the Physics of Electronic and Atomic Collisions*, p. 176, University of Washington Press, Seattle (1975).
9. T. van Ittersum, H. G. M. Heideman, G. Nienhuis, and J. Prins, *J. Phys. B* **9**, 1713 (1976).
10. A. Niehaus, Invited Lectures, *Tenth International Conference on the Physics of Electronic and Atomic Collisions*, Paris, 1977, p. 185, North Holland Publishing Company, Amsterdam (1977).
11. G. Nienhuis and H. G. M. Heideman, *J. Phys. B* **8**, 2225 (1975).
12. D. Roy, A. Delage, and J. D. Carette, *J. Phys. B* **11**, 895 (1978).
13. G. C. King, R. C. Bradford, and F. H. Read, *J. Phys. B* **8**, L477 (1975).
14. D. Spence, *Phys. Rev. A* **12**, 2353 (1975).
15. S. Ormonde, F. B. Kets, and H. G. M. Heideman, *Phys. Lett.* **50A**, 147 (1974).
16. R. K. Nesbet, *Phys. Rev. A* **14**, 1326 (1976).
17. F. B. Kets and H. G. M. Heideman, *J. Phys. B* **10**, L305 (1977).
18. J. A. Simpson, M. G. Menendez, and S. R. Mielczarek, *Phys. Rev.* **150**, 76 (1966).
19. R. Morgenstern, A. Niehaus, and U. Thielmann, *J. Phys. B* **10**, 1039 (1977).
20. H. G. M. Heideman, T. van Ittersum, G. Nienhuis, and V. M. Hol, *J. Phys. B* **8**, L26 (1975).

21. A. F. J. van Raan, J. P. de Jongh, J. van Eck, and H. G. M. Heideman, *Physica* **53**, 45 (1971).
22. A. F. J. van Raan, P. G. Moll, and J. van Eck, *J. Phys. B* **7**, 950 (1974).
23. H. G. M. Heideman, W. van de Water, J. van Eck, and P. H. Peeters, *Proceedings of the Tenth International Conference on the Physics of Electronic and Atomic Collisions*, Paris, 1977, p. 682.
24. R. Barker and H. W. Berry, *Phys. Rev.* **151**, 14 (1966).
25. G. C. King, F. H. Read, and R. C. Bradford, *J. Phys. B* **8**, 2210 (1975).
26. F. H. Read, *J. Phys. B* **10**, L207 (1977).
27. G. Nienhuis and H. G. M. Heideman, *J. Phys. B* **9**, 2053 (1976).
28. J. R. Taylor, *Scattering Theory*, Wiley, New York (1972).

42

Oscillatory Structure
of Excitation Functions
in Atomic Collisions

D. Hasselkamp, A. Scharmann, and K.-H. Schartner

A survey is presented of experiments dealing with the oscillatory structure of outer-shell excitation functions in atomic collisions. Special attention is given to the intermediate and high-energy range. Oscillations in total excitation cross section functions are shown to be a rather widespread phenomenon; they have been observed for a wide variety of collision systems over an energy range extending from threshold to 1 MeV. From a qualitative point of view the oscillatory structure of excitation functions may be explained by a quantum mechanical interference effect as a consequence of the mixing of coherently excited molecular states at large internuclear distance. The applicability of this model to low-energy collisions is generally accepted. New results at intermediate and high impact energies, however, indicate that a refinement of this model or even an entirely new approach is necessary in order to explain the observed effects.

1. Introduction

Quasimolecular processes leading to outer-shell excitation in heavy-particle collisions may lead to the interference of quantum mechanical probability amplitudes of the involved elastic and inelastic channels.[1,2] For example, the double passage of a pseudocrossing of molecular states along two alternative paths during the collision leads to a phase interference effect resulting in the well-known Stückelberg oscillations[3] in the differential cross section as a function of impact energy. The

D. Hasselkamp, A. Scharmann, and K.-H. Schartner ● I. Physikalisches Institut der Justus-Liebig-Universität Giessen, D-6300 Giessen, West Germany.

importance of such effects for the understanding of details of the collision process is evident.

Because the interference patterns in the differential cross sections are very sensitive to the impact parameter, the oscillatory structure is mostly lost when performing the integration over the impact parameter. Thus the corresponding total cross sections are in general a smooth function of the velocity of the incoming particle. However, in the past ten years many examples have been found in which pronounced oscillations are present in total excitation cross sections. This phenomenon may be explained by a model first proposed by Rosenthal and Foley.[4] It assumes the coherent population of two outgoing inelastic molecular potential curves at small internuclear distance and a subsequent interaction of these potential curves at relatively large distance by a term crossing or term approach. Thus the oscillatory structure is explained as a quantum mechanical interference effect of the outgoing wave functions. Because the interaction takes place at large internuclear distances, the mixing of states is relatively insensitive to the impact parameter and the interference pattern is retained in the total excitation cross sections.

The analysis of total cross section oscillations (TCO) on the basis of the Rosenthal model (RM) has been proven to be a powerful tool for "collision spectroscopy" at low impact energies, yielding information about the long-range interaction of molecular potential curves.[5-17] During the last years TCO have also been found at high impact energies corresponding to velocities up to $v = 1$ a.u.[18-23] An *a priori* application of the RM in these cases is not self-evident.

It is the intention of this chapter to review selected data on high-energy TCO, to compare the results with those obtained at low energies, and to discuss the applicability of the RM at high impact energies. The basic ideas underlying the RM will be introduced in Section 2. In Section 3 we will review some of the most interesting data on TCO at low and especially at high impact energies. Main emphasis will be given to the (HeNe)$^+$ system, which has been studied most intensely up to now. A comparison between low-energy and high-energy experiments will be performed.

Oscillations in total charge exchange cross sections into the ground states will be excluded from the discussion.[24,25] Also excluded will be interference effects in connection with optical polarization, which however are closely related to the subject of this article.[26]

2. The Excitation Mechanism for the Appearance of TCO

About 10 years ago Dworetsky et al.[5] measured excitation functions for low-energy He$^+$–He collisions which showed an oscillatory dependence as a function of the projectile energy. Figure 1 shows their results for the excitation to the 3^1S and 3^3S levels. A range of energies can be found where the TCO of the two inelastic channels are in antiphase. Dworetsky et al.[5] suggested that Stückelberg-type oscilla-

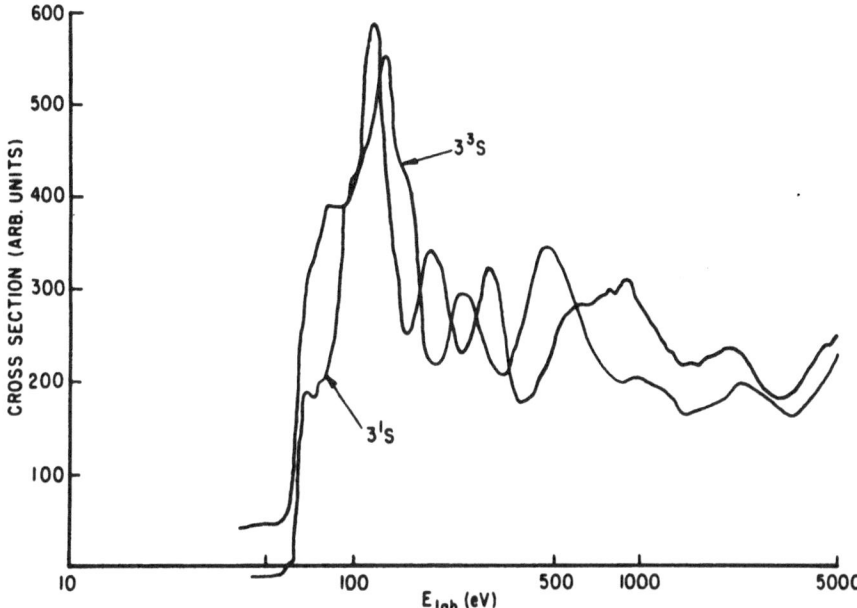

Figure 1. Excitation functions for the $3^{1,3}S$ levels of He measured for He$^+$–He collisions as function of the impact energy.[5] The sum of target and projectile excitation has been measured.

tions might be responsible for the experimental findings. This suggestion was rejected by Rosenthal and Foley,[4] who invoked a three-state model assuming the interference of two inelastic channels at large internuclear distances. The principal idea of the RM is sketched in Figure 2, which is taken from the original paper.[4] The upper row of Figure 2 holds for a two-state system involving a pseudocrossing of the elastic and inelastic channel (a). In this case the double passage of the crossing point along two different paths during the collision will lead to an interference of the associated wave functions giving rise to Stückelberg oscillations in the differential cross sections (b). These oscillations are averaged out in the integration over the impact parameter (c). The lower row of Figure 2 demonstrates the idea of the RM. It assumes a three-level two-crossing scheme with a third crossing of the outgoing inelastic channels at large separations R_2 of the system (d). The excitation probabilities for the two inelastic channels leading to excited states of the separated system oscillate (e). As a result of the mixing of the coherently populated channels at R_2 ($\gg R_1$) interference of the corresponding probability amplitudes takes place. This interference effect at large internuclear distance is only slightly dependent on the impact parameter and TCO arise that are in antiphase for the two inelastic channels (f).

Rosenthal and Foley supported their model by a calculation of the potential curves of the He$^+$–He system,[4] in which the existence of pseudocrossings between the 3^1S and 3^3S states and between the 4^1S and 4^3S states at large distances was demonstrated (Figure 3). These pseudocrossings were found at values of R_2 near 12 a.u. and 40 a.u. The potential curves leading to the 2^1S and 2^3S states showed no

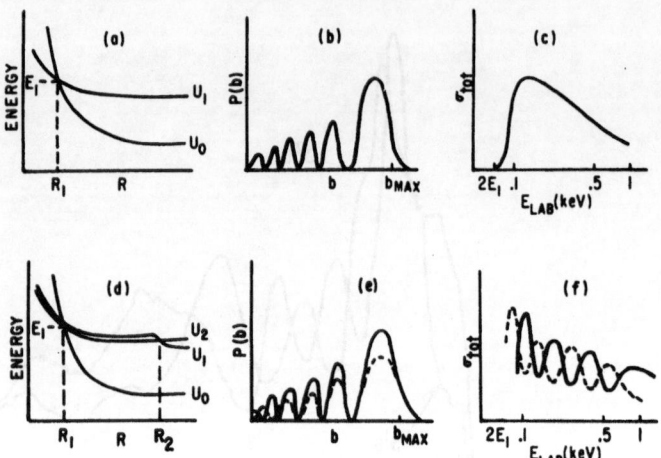

Figure 2. Schematic diagram of potential curves with crossings leading to Stückelberg (a)–(c) and to Rosenthal (d)–(f) oscillations.[4] (a) Two-level crossing scheme. (b) Excitation probability P at fixed impact energy as a function of impact parameter b. (c) Total excitation cross section $\sigma = 2\pi \int P(b)\, db$ as function of impact energy (for a symmetric collision system). (d) Three-level crossing scheme of the Rosenthal model. (e) Excitation probability for levels one and two as a function of impact parameter. Solid line, level one; dashed line, level two. (f) Oscillations of σ for levels one and two. Solid line, level one; dashed line, level two.

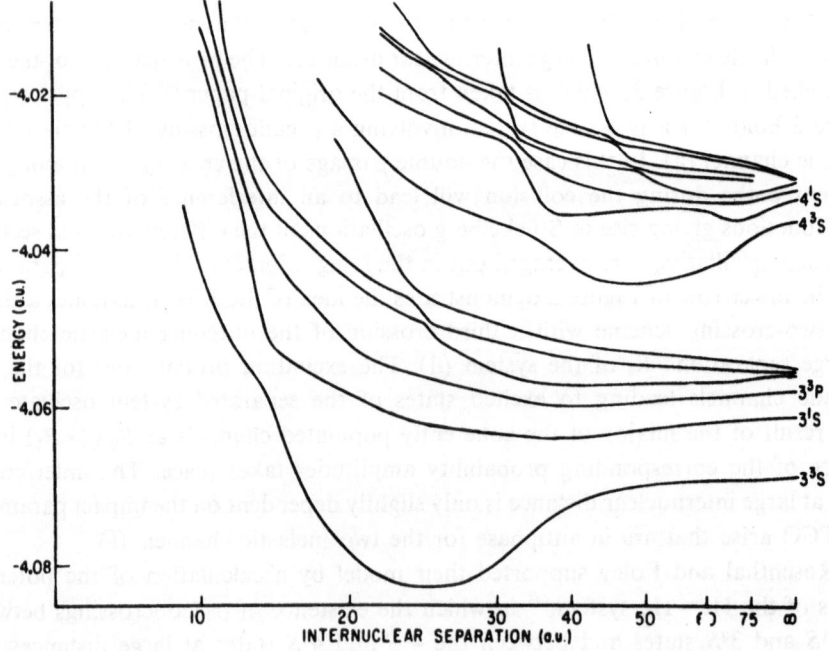

Figure 3. Calculated $n = 3$ and $n = 4$ $^2\Sigma_g$ potential curves of He_2^+ at large internuclear separations.[4]

effective pseudocrossing, and indeed the TCO have been found to be rather weak in this case.[6]

A mathematical treatment of the RM was given by Ankudinov et al.,[27] who derived the total cross sections Q_1 and Q_2 for the excitation of the atomic states 1 and 2. Formally Q_1 and Q_2 can be written as

$$Q_1 = pQ_1' + (1-p)Q_2' + \Delta Q$$
$$Q_2 = (1-p)Q_1' + pQ_2' - \Delta Q$$

(1)

Q_1' and Q_2' are the cross sections for the population of the potential curves U_1 and U_2 at R_1, p gives the probability for a transition between the two terms at R_2 (Figure 2d). ΔQ is the interference term:

$$\Delta Q = 4\pi \int_0^{b_{max}} db \cdot b \operatorname{Re}\left\{ sga_1(R_1)a_2{}^*(R_1) \exp \int_{R_1}^{R_2} \frac{i}{\hbar v} [U_2(R) - U_1(R)]\, dR \right\}$$

(2)

$a_1(R_1)$ and $a_2(R_1)$ are the probability amplitudes for the population of channels U_1 and U_2 at R_1. g and s are mixing coefficients which are determined by the special type of interaction at $R_2 (p = |g|^2 = 1 - |s|^2)$.

The result of the integration in equation (2) over the impact parameter b depends strongly on the phase factors of the probability amplitudes $a_1(R_1)$ and $a_2(R_1)$ whereas the dependence of

$$\langle \Delta U\, \Delta R \rangle = \int_{R_1}^{R_2} [U_2(R) - U_1(R)]\, dR$$

on the impact parameter b will be weak if only $R_2 \gg R_1$. The conditions under which the interference term ΔQ will be present in the total excitation cross section have been discussed by Ankudinov et al.[27] In the case that the integration over the impact parameter b in equation (2) does not result in a cancellation of the interference term they obtained:

$$\Delta Q = [Q_1' Q_2' p(1-p)]^{1/2} \cos \chi_{b=0}$$

(3)

Thus the term ΔQ is an oscillating function with

$$\chi_{b=0} = \frac{1}{\hbar v} \int_{R_1}^{R_2} [U_2(R) - U_1(R)]\, dR + \Delta\phi$$

(4)

Ankudinov et al.[27] also gave expressions for the "initial" phase $\Delta\phi$ for the two cases of a term crossing using the Landau–Zener treatment,[28,29] and of a term approach using the Demkov formula,[30] both applied to the interaction at R_2.

The predictions of the RM concerning TCO as derived by Rosenthal and Foley[4] and by Ankudinov et al.[27] can be summarized as follows: (1) Total cross sections for the excitation of atomic states that are correlated to two interfering molecular potential curves should exhibit oscillations that are in antiphase [equation (1)]. (2) The oscillation should be regular as a function of the inverse relative velocity, i.e., the extrema should be equally spaced. (3) From the oscillation frequency on the inverse velocity scale the effective energy area $\langle \Delta U \, \Delta R \rangle$ between the two interacting potential curves can be determined provided the phase $\Delta \phi$ is constant or can be neglected. (4) Once the effective energy area is known the location of the outer crossing can be approximated by assuming some mean energy difference $\Delta U = U_2 - U_1$. (5) The maximum modulation depth in the total cross section function is predicted to be 50% [equations (1) and (3)] by Ankudinov et al.[27] Maximum modulation depth is achieved if the two molecular potential curves U_1 and U_2 are equally populated (i.e., $Q_1' = Q_2'$) and total mixing occurs at R_2 ($p = \frac{1}{2}$).

When using the RM in connection with experimental data one has to be aware of the presuppositions made: the model assumes the formation of a transient quasi-molecule with well-defined incoming and outgoing potential curves; it is further assumed that the population of the outgoing inelastic channels occurs at a localized crossing. The quantitative treatment of the RM by Ankudinov et al.[27] uses the Landau–Zener formula for the calculation of the transition probability at the inner pseudocrossing. Thus the results of Ankudinov et al.[27] depend on the validity of the Landau–Zener approach. Bates[40] has pointed out that the range of validity of the Landau–Zener formula is rather restricted. It follows from the paper of Bates[40] that especially at high impact energies the assumption of a localized crossing point is no longer valid and that transitions due to rotational coupling become important. These critical remarks have to be remembered when a comparison of experimental data with the results of Ankudinov et al.[27] is attempted.

3. Examples of TCO in Outer-Shell Excitation

We shall start with a general survey of experiments dealing with TCO in Section 3.1. In the following we shall present selected experimental results. It is intended to demonstrate the typical features of TCO with main emphasis given to results measured at impact energies above 10 keV for which only limited experimental data are available up to now. For low-energy collisions the reader is referred to the review of Bobashev.[2]

In Section 3.2 we shall present the work of Andersen et al.[12–14] in which a systematic search for TCO in collisions of a variety of light ions with neon atoms was performed. In Section 3.3 the (HeNe)+ system shall serve as an example that has been intensively studied for a variety of excited states from low velocities up to $v \simeq 1$ a.u. The available data seem to be most complete for this system.

3.1. General Survey

Oscillations in total excitation cross section functions (TCO) have been reported for a wide variety of collision systems over an energy range from threshold up to 1000 keV. In Table 1 we have summarized existing experiments specifying projectile

Table 1. Summary of Collisional Systems for which TCO Have Been Found

Projectile	Target	Atoms or ions with oscillations in excitation functions	References
He$^+$	He, Ne, Ar	Projectile and target atoms	5–7, 15, 18, 22
Ne$^+$	He	Projectile and target atoms	7, 19–23, 34
N$^+$, O$^+$, Na$^+$, Mg$^+$	Ne	Projectile and target atoms	12–14
Na$^+$	He, Ne	Projectile and target atoms	8, 9, 13, 14, 31–33
Li$^+$, Na$^+$, K$^+$, Rb$^+$, Cs$^+$, Mg$^+$, Ca$^+$	Cd	Projectile ions, atoms, target atoms	10
Rb$^+$	Ar	Target atoms	11
K	Ar, Kr, Xe	Projectile atoms	16
K$^+$	Ar	Projectile ions	17
K$^+$	Ar	Autoionizing states	35, 36

and target as well as the excited particle for which TCO have been observed. Most of the available material is concerned with the collisions of rare gas and alkali ions with rare-gas atoms. Material is somewhat limited for other collision systems, maybe due to experimental difficulties. Noteworthy is the experiment of Kempter et al.,[16] which seems to be the only one to report TCO in atom–atom collisions. TCO have been observed mostly in the excitation functions of the target atoms as well as in the projectile atoms after charge exchange of the incident ions. Recently TCO have also been reported for excited states of the incident ions[17] and even for autoionizing states.[35,36]

Thus TCO appear to be a very general phenomenon in excitation studies which may be expected to occur in an even larger variety of collision systems compared to what has been investigated so far. In all cases that are summarized in Table 1 the RM (see Section 2) has been used for an explanation of the results obtained. Therefore we may further conclude that the interference of coherently populated inelastic channels at large internuclear distances is a common feature which must be taken into consideration in heavy-particle excitation studies both experimentally and theoretically.

3.2. The Work of Andersen et al.

Interesting results on TCO in the low to intermediate velocity range have been reported by Andersen et al.[12–14] These authors bombarded neon atoms with ions ranging from Li$^+$ to Al$^+$ in an energy range from 0.1 keV to 15 keV with the intention to find general conditions which have to be met by a collisional system to exhibit TCO in the excitation functions. They studied the excitation of the 3p states ($2p_{1-10}$

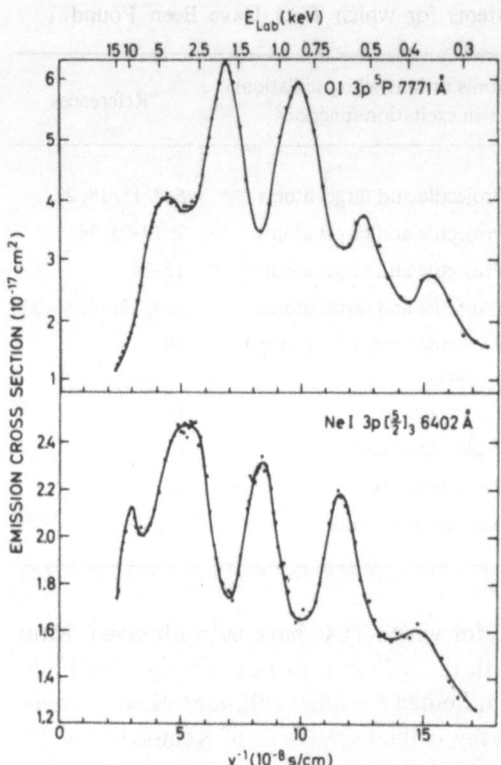

Figure 4. Absolute emission cross sections for the production of 7771-Å O I ($3p^5P$) and 6402-Å Ne I ($3p[\tfrac{5}{2}]_3$) radiation in O+–Ne collisions, as function of inverse velocity. Impact energy given for comparison.[14]

in Paschen notation) of neon together with the excitation resulting from a charge exchange reaction of the ionic projectiles. Only those states of the charge exchange process with simultaneous excitation were investigated for which the energy defect of the reaction was close to that of the direct excitation process of the neon atoms.

Prominent oscillations of the excitation functions of the direct and the charge exchange channels were observed only with the projectiles N+, O+, Na+, and Mg+. The results found in these cases are very much the same. Without going into details the following can be said: all of the excitation functions of the ten $3p$ levels of neon exhibit TCO; in each of the four collision systems a charge exchange reaction was observed which showed an antiphase behavior with respect to the neon excitation functions.

Figure 4 gives an example of TCO in the emission cross sections for radiation resulting from the decay of the $2p_9$ state ($3p[\tfrac{5}{2}]_3$ in Racah notation) of neon and the $3p^5P$ state of oxygen, respectively, as a function of the inverse relative velocity. The two excitation functions are found to oscillate in antiphase. The oscillation maxima (or minima) are equidistant, the mean distance between two maxima is equal to $\Delta(v^{-1}) \simeq 3.2 \times 10^{-8}$ sec/cm. Using equation (4) the quantity $\langle \Delta U \, \Delta R \rangle$ can be calculated from

$$\langle \Delta U \, \Delta R \rangle = 2\pi\hbar/\Delta(v^{-1}) \tag{5}$$

It follows that $\langle \Delta U \, \Delta R \rangle \simeq 13$ eV Å. This value is in general agreement with data extracted from other low-energy experiments.[2]

The reaction scheme leading to the excitation processes as shown in Figure 4 can be written as

$$O^+(^4S_{3/2}) + Ne(^1S_0) \rightarrow O^+ + Ne(2p^53p[\tfrac{5}{2}]_3) + 18.68 \text{ eV} \qquad \text{(direct excitation)}$$
$$O^+(^4S_{3/2}) + Ne(^1S_0) \rightarrow O(2p^33p^5P) + Ne^+ + 18.55 \text{ eV} \quad \text{(charge exchange reaction)}$$

Both reactions show an almost equal energy defect. It was therefore concluded that the corresponding molecular channels interact according to the model developed by Rosenthal[4] (Section 2).

Andersen et al.,[14] however, pointed out that the physical situation must be more complicated than assumed in the two-state RM. They examined the excitation functions of all $3p$ levels that lie close in energy. It was found that all ten excitation functions exhibit TCO. Furthermore it was demonstrated that the emission cross-section curves for eight $3p$ levels ($2p_2$–$2p_9$) exhibit the same number of extrema, which occur at the same velocities with identical spacings between the maxima. The excitation functions for three of these levels are shown in Figure 5. The excitation functions for the remaining two $3p$ levels ($2p_1$ and $2p_{10}$) deviate from the others in that the oscillation pattern is shifted towards lower velocities and in the case of the $2p_1$ level additional peaks are found. Figure 6 shows the excitation functions arising from the decay of the $2p_1$ ($3p'[\tfrac{1}{2}]_0$), $2p_9$ ($3p[\tfrac{5}{2}]_3$), and $2p_{10}$ ($3p[\tfrac{1}{2}]_1$) levels of neon as a function of the inverse velocity of the incoming O$^+$ ions. Andersen et al.[14] gave no explanation

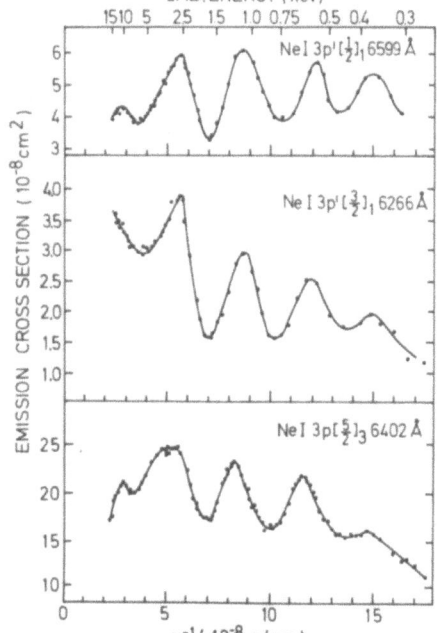

Figure 5. Absolute emission cross sections for the production of 6599-Å Ne I ($3p'[\tfrac{1}{2}]_1$), 6266-Å Ne I ($3p'[\tfrac{3}{2}]_1$), and 6402-Å Ne I ($3p[\tfrac{5}{2}]_3$) radiation in O$^+$–Ne collisions, as a function of inverse velocity. Impact energy given for comparison.[14]

for the deviation found in the case of the $2p_1$ and $2p_{10}$ levels (which are the highest and lowest in energy of the group of the $3p$ levels, see Figure 17), though they were able to identify the charge exchange channel leading to the $3p^3P$ state of oxygen to oscillate in antiphase with the additional peaks found for the $2p_1$ level of neon. Nevertheless it is a notable feature that a group of states shows oscillatory structure in antiphase with a different inelastic channel and furthermore that one of these states $(2p_1)$ seems to interact with at least two charge exchange channels.

Another example of the work of Andersen et al.[14] is presented in Figure 7 for the Na$^+$–Ne system. Oscillatory structure is again found in all ten $3p$ excitation functions as well as in the charge exchange channel leading to the $3p^2P^0$ state of Na. The antiphase behavior of the direct and the charge exchange channel is obvious; a small shift of the maxima for the $2p_1$ $(3p'[\frac{1}{2}]_0)$ and $2p_{10}$ $(3p[\frac{1}{2}]_1)$ levels is noted by the authors. The values of $\langle \Delta U \Delta R \rangle$ are the same for the ten $3p$ levels; however, the initial phases $\Delta\phi$ [equation (4)] were found to differ between the three $3p$ excitation functions shown in Figure 7.

Similar results as noted for the O$^+$–Ne and Na$^+$–Ne system have been found for the N$^+$–Ne and Mg$^+$–Ne systems. A weak tendency towards TCO was reported for the $3p$ levels of neon with the projectiles Li$^+$, Be$^+$, Al$^+$; no structure was present with the projectiles B$^+$, C$^+$, F$^+$, and Ne$^+$.

Andersen et al.[14] compared their results with the available theoretical predictions (see Section 2). They concluded that a qualitative explanation of the observed TCO may be given in terms of the Rosenthal model; however, they also stated that

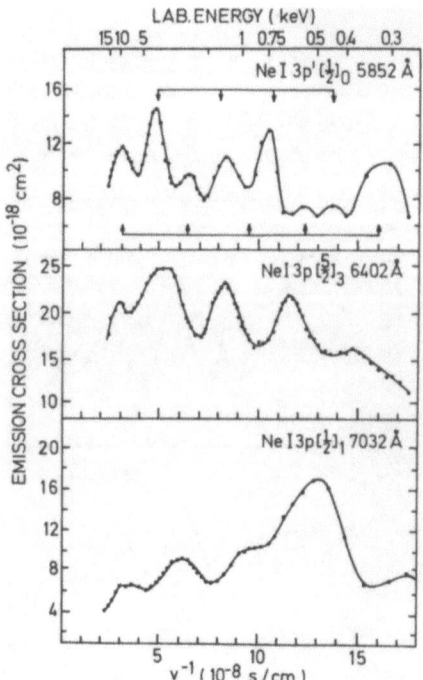

Figure 6. Absolute emission cross sections for the production of 5852-Å Ne I $(3p'[\frac{1}{2}]_0)$, 6402-Å Ne I $[3p[\frac{5}{2}]_3)$, and 7032-Å Ne I $(3p[\frac{1}{2}]_1)$ radiation in O$^+$–Ne collisions, as a function of inverse velocity. Impact energy given for comparison.[14]

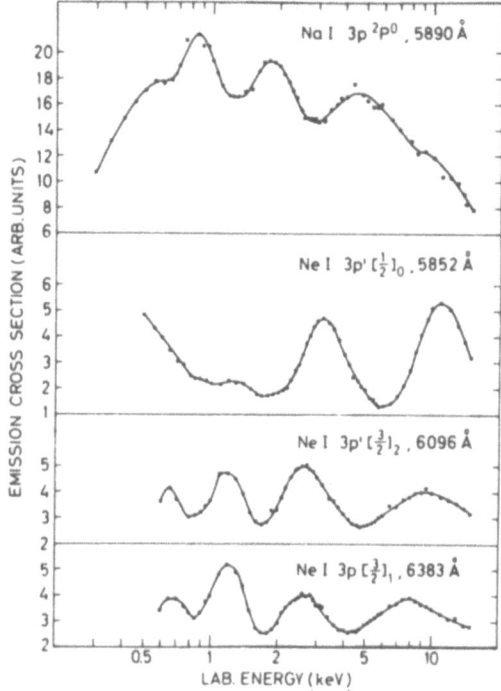

Figure 7. Emission cross section for production of 5890-Å Na I ($3p^2P^0$), 5852-Å Ne I ($3p'[\frac{1}{2}]_0$), 6096-Å Ne I ($3p'[\frac{3}{2}]_2$), and 6383-Å Ne I ($3p[\frac{3}{2}]_1$) radiation in Na+–Ne collisions, as a function of the impact energy. Cross sections are given in arbitrary units.[14]

the more quantitative treatment of the RM as given by Ankudinov *et al.*[27] is not in agreement with the experimental findings. It was suggested that a refined theoretical treatment of the RM is needed. No "selection rules" could be given by Andersen *et al.*[14] from which the occurrence of phase interference phenomena in total excitation cross sections could be predicted from the knowledge of the structure of the collision partners alone. More information about the potential curves of quasimolecules at large internuclear distances is essential. At present such potential diagrams are available only for a limited number of collision systems at small internuclear distances and for low-lying states (with the exception of Rosenthal's calculation for the He+–He system, see Figure 3). Work in this field, however, seems to be in progress, as can be seen from a recent qualitative analysis for the Na+–Ne system[37] which has attempted to give an explanation for the observed TCO in this system.[8,14,31–33] It should be noted that further experimental information about TCO in the Na+–Ne system can also be evaluated from the polarization measurements of Tolk *et al.*[33]

3.3. The He+–Ne System and the Ne+–He System

Excitation functions for a variety of excited states have been measured for both the He+–Ne system and the Ne+–He system over a wide range of energies from the thresholds to 1 MeV. The experimental investigations are summarized in Table 2.

Table 2. Experimental Studies of TCO in the (HeNe)$^+$ Collisional System

Projectile	Target	Excitation of	Energy range (E_{Lab})	Authors
He$^+$	Ne	He*	Threshold to 5 keV	Tolk et al.[7]
Ne$^+$	He	He*, Ne*		
Ne$^+$	He	He*	10–75 keV	Andersen et al.[34]
He$^+$	Ne	He*	0.3–150 keV	Wolterbeek Muller and de Heer[18]
He$^+$	Ne	He*, Ne*	Threshold to 9 keV	Isler[15]
Ne$^+$	He	He*	100–1000 keV	Hasselkamp et al.[19]
Ne$^+$	He	He*	10–75 keV	Veje et al.[20]
Ne$^+$	He	He*	10–75 keV	Larsen[21]
He$^+$	Ne			
Ne$^+$	He	He*, Ne*	10–150 keV	Andresen et al.[22]
Ne$^+$	He	He*, Ne*	50–1000 keV	Hasselkamp et al.[23]

Tolk et al.[7] studied the energy range above the threshold. Their results (Figure 8) show characteristic differences in the excitation functions for the He$^+$–Ne system and the Ne$^+$–He system: TCO have been observed for the He$^+$–Ne system (also investigated by Isler[15] for the He and Ne resonance lines), no structure has been found for the Ne$^+$–He system; the threshold behavior of the excitation functions for the Ne$^+$–He system is gradual while a sharp onset is observed for the He$^+$–Ne system (Figure 8). An explanation of the observed differences for the two systems has been given by Tolk et al.[7] For the He$^+$–Ne system they assume a transition from the ground molecular state to two excited channels at a localized inner crossing and a subsequent phase interference according to the RM (Section 2). For the Ne$^+$–He system they assume a nonlocalized transition range to the excited states. In this case the phase difference between the excitation amplitudes of the outgoing channels is no longer defined and the interference of states will not give rise to TCO. Tolk et al.[7] plotted their results for the He$^+$–Ne system on a $1/v$ scale and demonstrated that the extrema are equally spaced (Figure 9). For the 4^3S–2^3P transition a change of the oscillation frequency is observed (Figure 9). It would be interesting to compare this result with the corresponding antiphase channels. There is the possibility that the molecular curve leading to the 4^3S state of He interferes with two other inelastic channels of the system, similar to the result found by Andersen et al.[14] for the $2p_1$ level of Ne in the O$^+$–Ne system. However, no antiphase channels have been observed so far for the excitation functions shown in Figures 8 and 9.

The pattern of the TCO in Figures 8 and 9 becomes less clear for $1/v \to 0$, i.e., for higher velocities. From inspection of Figure 9 one can expect only one or no further oscillation at higher velocities. The question arises whether excitation functions at higher velocities are smooth or exhibit TCO with higher frequency. An answer to this question can be obtained from a number of investigations which shall be discussed in the following.

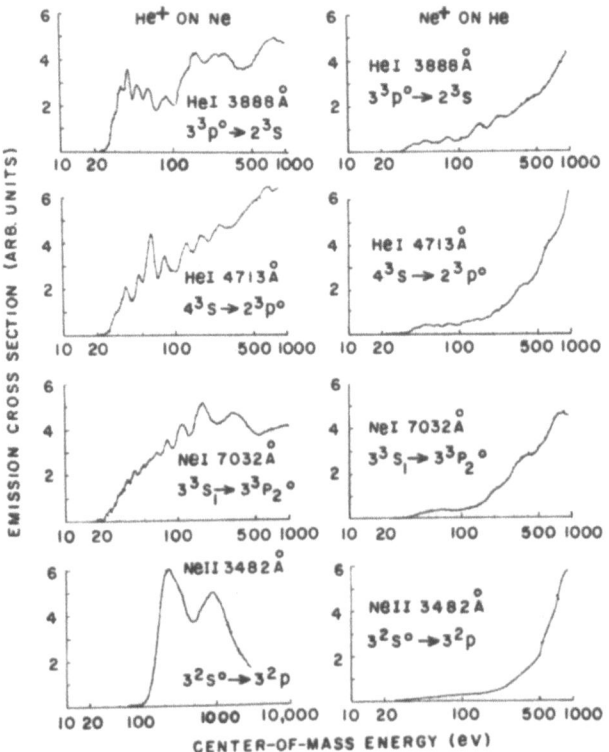

Figure 8. Emission cross sections of prominent optical transitions observed in He+–Ne and Ne+–He collisions as a function of the center-of-mass energy. Structure reproducible within 2%.[7]

Wolterbeek Muller and de Heer[18] measured cross sections for the charge exchange excitation of helium ions in collisions with rare-gas atoms for impact energies up to 150 keV; i.e., in this experiment relative velocities as high as $v = 1$ a.u. are reached. Figure 10 gives an example of their results for the He+–Ne system. Cross sections for the charge exchange into the 3^1P and 3^3P level of helium are shown as a function of the inverse relative velocity. Results of Tolk *et al.*[7] (see Figure 8)

Figure 9. Relative total-emission cross sections plotted as a function of $1/v$ (see Figure 8).[7] The v is proportional to $(E - U)^{1/2}$ with E the center-of-mass energy and U the average excited-state energy above the ground state. The straight lines are least-squares fits of the phase integers n versus locations in $1/v$ of cross-section maxima.

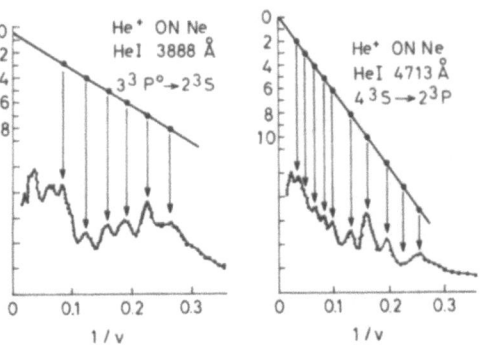

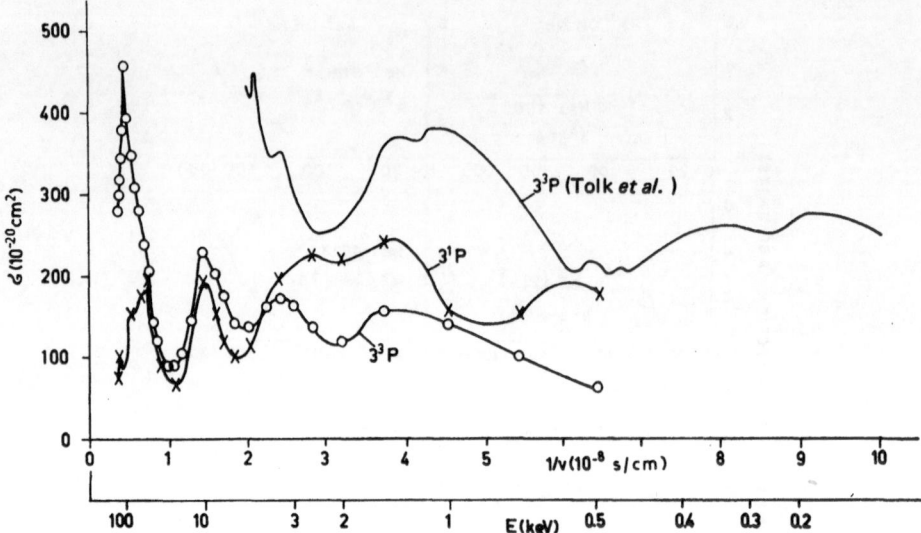

Figure 10. Cross sections for capture into the $3^{1,3}P$ levels of He for $He^+ + Ne$ collisions as a function of the inverse velocity $1/v$, impact energies given for comparison.[18] Values from Tolk et al.[7] for the 3^3P level shown for comparison (see Figure 8).

are included in Figure 10 for comparison. We point out two important features which are characteristic for the He excitation functions in the (He Ne)$^+$ system at high energies: (1) The excitation functions exhibit prominent oscillatory structure at high velocities. (2) There is a range of velocities where oscillations in the singlet and triplet excitation functions for levels with the same n, L quantum number are similar in magnitude, frequency, and phase. Thus there are two points of interest which differ from the situation found at low impact energies. On one hand the appearance of strong oscillations is not expected from low-energy data, as can be concluded from the work of Tolk et al.[7] (see Figure 9). The high-velocity oscillations are found to oscillate with considerable higher frequency. For example, Wolterbeek Muller and de Heer[18] give a value of $\langle \Delta U \Delta R \rangle = 41$ eV Å for the 3^3P excitation function while Tolk et al.[7] derived $\langle \Delta U \Delta R \rangle = 9$ eV Å for the same excitation function above threshold. On the other hand the similar behavior of excitation functions of singlet and triplet levels with the same quantum numbers n, L seems to be a general feature for the (HeNe)$^+$ system at high velocities while at energies just above threshold corresponding singlet and triplet excitation functions rather show the tendency toward antiphase oscillations.[18] This conclusion is supported by the work of Dworetsky et al.[5] on the He$^+$–He system in which the excitation functions of the 3^1S and 3^3S levels were found to oscillate in antiphase at energies just above threshold but clearly show the tendency towards an in-phase behavior at higher velocities, as can be seen from Figure 1.

We are forced to believe that the mechanism leading to TCO must be different at low and high impact velocities. Though it is not unlikely that the interference of states, as postulated by the RM, may involve different molecular potential curves at different velocities (leading to different oscillation patterns), it must be kept in mind that the RM uses a quasimolecular approach which may not be appropriate at higher velocities and which certainly breaks down at $v \simeq 1$ a.u. In view of this fact the appearance of TCO at high velocities is a remarkable effect which is not clearly understood at present. In the following we shall present more results on TCO at high relative velocities. The basic statements of the RM will sometimes be used for a qualitative analysis. The critical remarks given above concerning the applicability of the RM to high-velocity collisions should be kept in mind.

TCO in the helium excitation functions have also been observed for the direct excitation of He by Ne^+ ions at higher energies.[19-23] This is in contrast to the low-energy experiment of Tolk et al.[7] (see Figure 8 and the discussion at the beginning of this section). Figure 11 shows excitation cross section curves for the 4^1S and 4^3S levels of helium[19,23] together with the results of Wolterbeek Muller and de Heer[18] for the He^+–Ne system for the same excited states. The comparison is performed for equal velocities of the He^+ and Ne^+ ions. There is a striking agreement in the observed structure and the absolute magnitude of the cross sections of the excitation functions of the direct (Ne^+–He) and the charge exchange (He^+–Ne) processes with the exception of the triplet levels at the highest velocities. Moreover, the cross-section functions for triplet and singlet excitation in either He^+–Ne of Ne^+–He collisions have similar

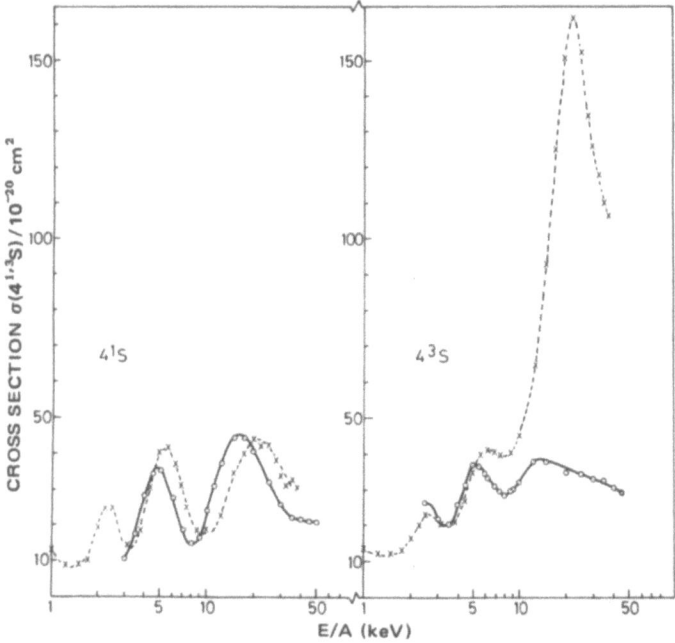

Figure 11. Absolute excitation cross sections for the $4^{1,3}S$ levels of He as a function of the impact energy E divided by the mass number A of the projectile. $\bigcirc$, Ne^+–He(1^1S) → $Ne^+ + He(4^{1,3}S)$;[23] ×, He^+–Ne(2^1S) → He($4^{1,3}S$)–Ne^+.[18]

structure and absolute magnitude with the exception of the charge exchange reaction at the highest velocities where a triplet-to-singlet ratio near the statistical value of 3 is found.

The similarities between the He$^+$–Ne and the Ne$^+$–He system were also reported by Andresen et al.[22] for the excitation to the $3^{1,3}S$, $4^{1,3}D$, and 3^3P levels of helium. Apparent excitation cross sections for the $4^{1,3}D$ levels as a function of the inverse velocity of the projectiles are given in Figure 12 together with the capture cross sections of Wolterbeek Muller and de Heer[18] for the He$^+$–Ne system. The data of Andresen et al.[22] are given in arbitrary units, however corrected for the spectral sensitivity of their optical detection system. In contrast to the $4^{1,3}S$ levels (Figure 11) the oscillation pattern of the $4^{1,3}D$ excitation functions looks more complicated. Andresen et al.[22] explained the structure in the $4^{1,3}D$ excitation functions by two superimposed oscillations (see full lines in Figure 12) of different frequency which result from the interference of three instead of two molecular potential curves at large internuclear distances.

These authors also proposed an explanation for the observed agreement in the singlet and triplet excitation functions for the same n, L quantum numbers and for the similarities found in the He$^+$–Ne and Ne$^+$–He systems. On the basis of a potential curve calculation by Sidis and Lefèbvre-Brion[38] they argue that the quantity $\langle \varDelta U \varDelta R \rangle$, which determines the oscillation frequency in the RM, is almost the same for singlet and triplet excitation and also for both collision systems. This explanation is, however, of qualitative nature because the calculation of Sidis and Lefèbvre-Brion[38] is restricted to the lowest molecular states and to small internuclear distances.

Recently Hasselkamp et al.[23] measured the excitation functions for the n^1S ($n = 3$–6) series of helium using Ne$^+$ ions with energies ranging from 50 to 1000 keV.

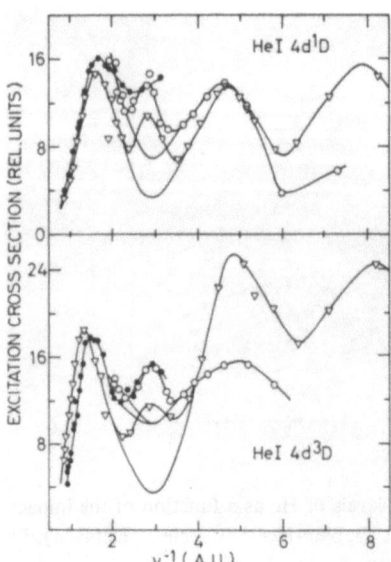

Figure 12. Apparent cross sections for excitation of the $4d^{1,3}D$ levels of He as a function of inverse velocity $1/v$. ●, He$^+$–Ne;[22] ○, Ne$^+$–He;[22] ▽, He$^+$–Ne.[18] Note that the cross sections are given in relative units, so that a relative comparison of the cross sections of the two levels is possible.

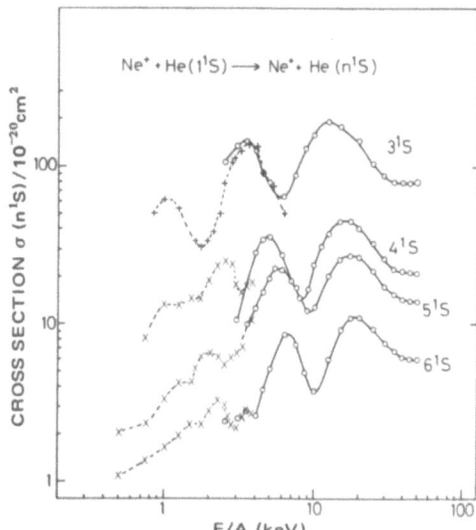

Figure 13. Absolute excitation cross sections for the n^1S ($n = 3, 4, 5, 6$) levels of He excited by Ne$^+$ impact as a function of the energy E divided by the mass number of the projectile A. ×, Larsen;[21] +, Andresen et al.;[22] ○, Hasselkamp et al.[23]

The intention of this work was to extend the energy range of a previous experiment on the 4^1S-level excitation[19] and to examine the excitation functions of high-lying states which are nearly degenerate in energy with several other states of the(HeNe)$^+$ system (see Figure 17). In such a case one might expect that the corresponding molecular potential curves interact with several others so that the interference pattern in the total cross sections, which have been observed for low-lying states, should wash out. The result of this experiment is shown in Figure 13. Strong TCO are also observed for the 5^1S and 6^1S excitation functions. Included in Figure 13 are data reported by Andresen et al.[22] and by Larsen.[21] The relative cross sections of the groups mentioned have been normalized to the absolute data of Hasselkamp et al.[23] It should be noted that the prominent high-energy oscillations observed by Hasselkamp et al.[23] for all four excitation functions tend to fade out or become irregular at lower energies with the exception of the 3^1S function.

By subtraction of a smoothly varying cross-section function $\bar{Q}$ Hasselkamp et al.[23] extracted the oscillating part ΔQ of their excitation functions. They demonstrated that $\Delta Q(n^1S)$ is regular when plotted as a function of $1/v$ by fitting the data points by a cosine function (Figure 14). From inspection of Figure 14 one finds that the oscillation frequency increases with increasing main quantum number of the excited helium level. In Table 3 the calculated values of $\langle \Delta U \Delta R \rangle$ [equations (4) and (5)] are tabulated together with the results of other experiments on the He$^+$–He and Ne$^+$–He systems for relative velocities $v > 0.1$ a.u.

Further interesting information has been extracted by Hasselkamp et al.[23] from their work on the excitation of the n^1S levels of helium by Ne$^+$ ions of high energy. These authors note that the oscillation parameter $\langle \Delta U \Delta R \rangle$ (Table 3) varies linearly with the main quantum number n (Figure 15). This result is hard to understand on the assumption of a transient quasimolecule. Though an increase of the

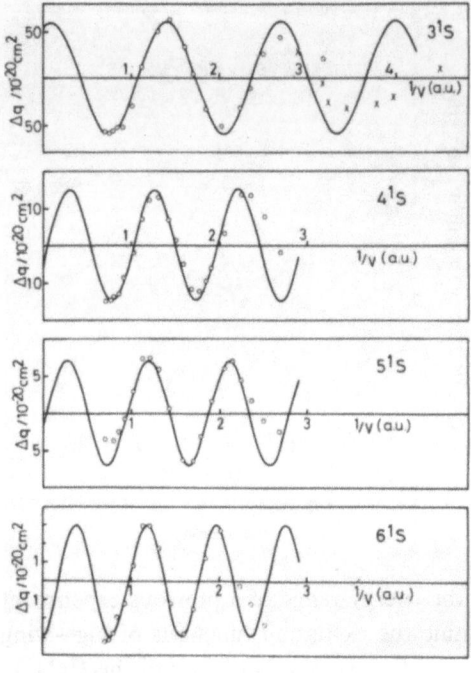

Figure 14. Modulation of the excitation functions of the n^1S levels of He excited by collisions with Ne^+ ions, obtained from the cross sections shown in Figure 13 by subtraction of a mean cross-section value. O, Hasselkamp et al.[23]; ×, Andresen et al.[22]; —, cosine function.

quantity ΔR with n might be reasonable (see for example the calculation of Rosenthal and Foley[4] for the $(HeHe)^+$ system, Figure 3), one would expect a decrease of the quantity ΔU for higher-lying states due to the increasing density of molecular potential curves leading to atomic states near the ionization limit. In a first approximation, based on the term behavior of atomic states,[39] one can estimate the functional dependence $\Delta U \sim n^{-3}$. In order to explain the observed $\langle \Delta U \Delta R \rangle \sim n$ one must assume $\Delta R \sim n^4$. This functional dependence seems unreasonable, because one would have to accept the interaction of molecular states at internuclear distances

Table 3. Experimental Values of $\langle \Delta U \Delta R \rangle$ (eV Å) for the n^1S Excitation Functions of He in He^+–Ne and Ne^+–He Collisions

Level	$v > 0.5$ a.u.		$v < 0.5$ a.u.[a]	
	Ne^+–He (Hasselkamp et al.[23])	He^+–Ne (W. Muller[18])	Ne^+–He (Kopenhagen[21,22])	He^+–Ne (W. Muller[18])
3^1S	71	62	36/59 (42)	62
4^1S	89.5	82	(59)	44
5^1S	100	—	(92)	—
6^1S	115.0	—	(41.5)	—

[a] The data in parentheses are estimated by Hasselkamp et al.[23] from the published data.

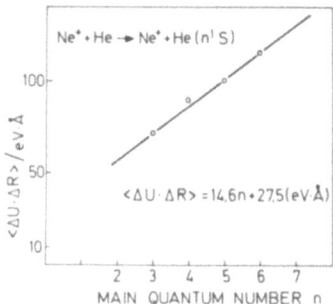

Figure 15. Oscillation frequency of the cosine functions of Figure 14 in units of $\langle \Delta U\, \Delta R \rangle$, as a function of the main quantum number n.[23]

in the order of 100 Å or more for high-lying states. We therefore conclude that the physical meaning of the results shown in Figure 15 is not understood at present.

Another point of interest is the modulation depth defined as the maximum values of the ratio $|\Delta Q|/\bar{Q}$, where $\bar{Q}$ is the smoothly varying part of the excitation function and ΔQ is the oscillatory part. In general, the modulation depth is rather small in the threshold region of the excitation functions (Figures 1, 8, 10). In many cases it is larger at higher energies as can be seen from the work of Andersen et al.[14] (Section 3.2) and from the experiments on the (HeNe)+ system which have been reviewed in this section. Hasselkamp et al.[23] report modulation depths near 0.5 for the excitation of the n^1S levels of helium by high-energy Ne+ ions (Table 4). A maximum value of 0.5 has been predicted by Ankudinov et al.[27] on the basis of the RM (Section 2) for equal population of the two outgoing inelastic channels and for maximal mixing at an outer crossing. The predictions of Ankudinov et al.,[27] however, are restricted to low-energy collisions and can therefore not be applied to the high-energy work of Hasselkamp et al.[23]

At the end of this section on the (HeNe)+ collision system we shall shortly discuss the search for excitation functions which oscillate in antiphase. As has been discussed in the previous sections this antiphase behavior has been found for many collision systems, mostly in the excitation functions due to direct and charge exchange excitation, but not until recently for the (HeNe)+ system. If the interpretation of the oscillatory structure of excitation functions as a quantum mechanical interference effect is allowed (this is most certainly the case for low relative velocities, but is not clear for velocities near 1 a.u.) oscillations that are 180° out of phase should also be present in the excitation functions of the (HeNe)+ system. Andresen et al.[22] measured emission cross sections for several Ne I ($3p$–$3s$) spectral lines for both the He+–Ne and the Ne+–He systems. Figure 16 gives an example of their work.[22]

Table 4. The Modulation Depth $|\Delta Q|/\bar{Q}$ in the n^1S Excitation Functions[23]

He level	3^1S	4^1S	5^1S	6^1S
Modulation depth $\Delta Q/\bar{Q}$	0.49	0.45	0.38	0.45

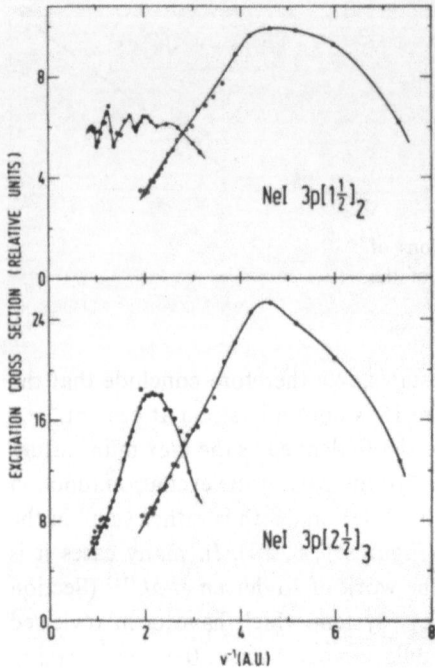

Figure 16. Cross sections for populating the $3p[1\frac{1}{2}]_2$ and $3p[2\frac{1}{2}]_3$ levels of Ne in collisions of He$^+$ with Ne and of Ne$^+$ with He, as a function of the inverse impact velocity $1/v$.[22] ●, He$^+$–Ne; ○, Ne$^+$–He.

No excitation functions have been found which exhibit an antiphase character with respect to the oscillatory structure of the helium excitation functions measured by the same group. Mostly the cross-section curves vary smoothly as a function of the impact energy. Figure 17 gives a diagram of the excited states of the (HeNe)$^+$ system

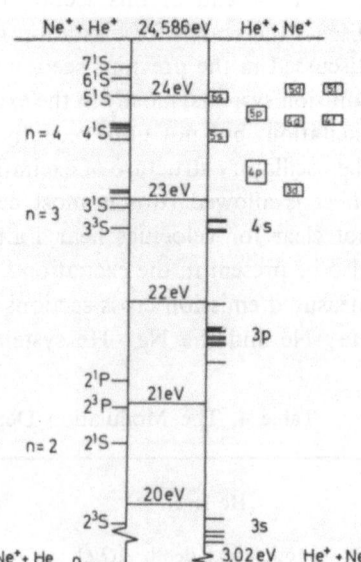

Figure 17. Energy diagram of the (HeNe)$^+$ system for infinite internuclear separations.

for infinite separations of the collision partners. The $3p$ levels of neon lie energetically well separated from all other excited states. If one assumes that this holds also for the corresponding molecular states, no interference of these states with other molecular curves is expected. Therefore the smooth variation of the $3p$ excitation functions with energy is not surprising when an interpretation on the basis of the RM is used.

Oscillatory structure has been found by Hasselkamp[41] in the excitation functions of the $4s$ levels of neon for the Ne$^+$–He system at high energies. The energy defect for the excitation of the Ne $4s$ levels is nearly the same as for the He $n = 3$ levels for the (HeNe)$^+$ system. Indeed the excitation functions for the charge exchange reaction leading to the $4s(^1P_1, {}^3P_1)$ levels of neon were found to oscillate $180°$ out of phase with respect to the excitation functions for the direct excitation reaction leading to the 3^1P state of helium.[19] This experimental result may indicate that the quasimolecular model of Rosenthal and Foley[4] is still valid for the (HeNe)$^+$ system at relative velocities near 1 a.u. However, more experimental and theoretical work is needed in order to come to a definite conclusion about the range of validity of the RM at high velocities.

4. Summary

We have reviewed experiments dealing with the oscillatory structure in outer-shell excitation functions. Selected results have been discussed in detail. Typical features of oscillations in total cross section functions (TCO) can be summarized as follows:

1. The term "oscillation" in excitation functions is used if the oscillatory pattern is regular as a function of the inverse velocity, i.e., if the extrema are equally spaced on a $1/v$ scale.

2. TCO in excitation functions have been observed for a wide variety of collisional systems and excited states. However, no criteria have yet been established for a collision system to exhibit TCO (or not).

3. The appearance of TCO is not restricted to low impact velocities but has been observed up to velocities $v \simeq 1$ a.u.

4. TCO are not restricted to excited levels with low main quantum number. In the case of the He n^1S levels TCO have been observed also for $n = 5, 6$.

5. In many cases a combination of two excitation functions has been found which exhibit an oscillation pattern $180°$ out of phase with respect to each other. In ion–atom collisions this is often true for a direct and for a charge exchange channel with nearly the same energy defect of the reaction.

6. In-phase oscillations have been observed for groups of excited states, e.g., for several $3p$ states of neon excited by N$^+$, O$^+$, Na$^+$, and Mg$^+$. In-phase behavior is also found for singlet and triplet excitation functions of helium for the same n, L quantum numbers at high velocities.

7. The oscillation frequency is not constant over a wide range of velocities. In general frequencies at low velocities are considerable smaller than at high velocities. Sudden changes of the oscillation frequency have been observed.

All phenomena associated with TCO may be explained by a quasimolecular approach which assumes a quantum mechanical interference of coherently excited states at large internuclear distances. This model, however, is only of qualitative nature as long as detailed information about the mechanism of coherent population, about the interfering states, and about the interference process itself is not available. At present this is not the case for most of the collision systems that have been investigated by experiment. The model is unsatisfactory in connection with TCO at velocities near 1 a.u. because the assumption of defined molecular states is invalid in the high velocity range. One may hope for a better understanding of the quantitative aspects of TCO from future experimental and theoretical work.

ACKNOWLEDGMENT

The authors are indebted to Professor N. Grün for numerous valuable discussions and suggestions.

References

1. M. Barat, in *Proceedings of the Eighth International Conference on the Physics of Electronic and Atomic Collisions, Invited Papers and Progress Reports*, B. C. Čobić and M. V. Kurepa, eds., p. 43, Institute of Physics, Belgrade (1973).
2. S. V. Bobashev, in *Proceedings of the Seventh International Conference on the Physics of Electronic and Atomic Collisions, Invited Papers and Progress Reports*, T. R. Govers and F. J. de Heer, eds., p. 38, North Holland, Amsterdam (1972).
3. E. C. G. Stückelberg, *Helv. Phys. Acta* **5**, 369 (1932).
4. H. Rosenthal and H. M. Foley, *Phys. Rev. Lett.* **23**, 1480 (1969).
5. S. Dworetsky, R. Novick, W. W. Smith, and N. Tolk, *Phys. Rev. Lett.* **18**, 939 (1967).
6. S. H. Dworetsky and R. Novick, *Phys. Rev. Lett.* **23**, 1484 (1969).
7. N. H. Tolk, C. W. White, S. H. Dworetsky, and L. A. Farrow, *Phys. Rev. Lett.* **25**, 1251 (1970).
8. S. V. Bobashev, *Zh. Eksp. Teor. Fiz. Pis'ma Red.* **11**, 389 (1970) [English transl. *Sov. Phys.-JETP Lett.* **11**, 260 (1970)].
9. S. V. Bobashev and V. A. Kritskii, *Zh. Eksp. Teor. Fiz., Pis'ma Red.* **12**, 280 (1970) [English transl. *Sov. Phys.-JETP Lett.* **12**, 189 (1970)].
10. O. B. Shpenik, A. N. Zavilopulo, and I. P. Zapesochnyi, *Zh. Eksp. Teor. Fiz.* **62**, 879 (1972) [English transl. *Sov. Phys.-JETP* **35**, 466 (1972)].
11. S. V. Bobashev, V. I. Ogurtsov, and L. A. Razumovskii, *Zh. Eksp. Teor. Fiz.* **62**, 892 (1972) [English transl. *Sov. Phys.-JETP* **35**, 472 (1972)].
12. T. Andersen, A. Kirkegard Nielsen, and K. J. Olsen, in *Proceedings of the Eighth International Conference on the Physics of Electronic and Atomic Collisions, Book of Abstracts*, p. 219, Belgrade (1973).
13. T. Andersen, A. Kirkegard Nielsen, and K. J. Olsen, *Phys. Rev. Lett.* **31**, 739 (1973).
14. T. Andersen, A. Kirkegard Nielsen, and K. J. Olsen, *Phys. Rev. A* **10**, 2174 (1974).

15. R. C. Isler, *Phys. Rev. A* **10**, 2093 (1974).
16. V. Kempter, B. Kübler, and W. Mecklenbrauck, *J. Phys. B* **7**, 149 (1974).
17. S. V. Bobashev, in *Proceedings of the Ninth International Conference on the Physics of Electronic and Atomic Collisions, Book of Abstracts*, p. 735, University of Washington Press, Seattle (1975).
18. L. Wolterbeek Muller and F. J. de Heer, *Physica* **48**, 345 (1970).
19. D. Hasselkamp, A. Scharmann, and K.-H. Schartner, in *Proceedings of the Ninth International Conference on the Physics of Electronic and Atomic Collisions, Book of Abstracts*, p. 733, University of Washington Press, Seattle (1975).
20. E. Veje, B. Andresen, and K. Jensen in *Proceedings of the Ninth International Conference on the Physics of Electronic and Atomic Collisions, Book of Abstracts*, p. 739, University of Washington Press, Seattle (1975).
21. H. B. Larsen, Diplomarbeit, Kopenhagen (1975).
22. B. Andresen, K. Jensen, and E. Veje, *Phys. Rev. A* **16**, 150 (1977).
23. D. Hasselkamp, A. Scharmann, and K.-H. Schartner, to be published.
24. J. Perel, H. L. Daley, and F. J. Smith, *Phys. Rev. A* **1**, 1626 (1970).
25. R. E. Olsen, *Phys. Rev. A* **6**, 1822 (1972).
26. N. H. Tolk and J. Kraas, Chapter 43 in this volume.
27. V. A. Ankudinov, S. V. Bobashev, and V. I. Perel, *Zh. Eksp. Teor. Fiz.* **60**, 906 (1971) [English transl. *Sov. Phys.-JETP* **33**, 490 (1971)].
28. L. D. Landau, *Phys. Z. Sowjetunion* **2**, 46 (1932).
29. C. Zener, *Proc. R. Soc. London Ser. A* **137**, 696 (1934).
30. Yu. N. Demkov, *Zh. Eksp. Teor. Fiz.* **45**, 195 (1963) [English transl. *Sov. Phys.-JETP* **18**, 138 (1964)].
31. W. Maurer and K. Mehnert, *Z. Phys.* **106**, 453 (1937).
32. S. V. Bobashev and V. A. Kharchenko, in *Proceedings of the Tenth International Conference on the Physics of Electronic and Atomic Collisions, Book of Abstracts*, p. 964, Paris (1977).
33. N. H. Tolk, J. C. Tully, C. W. White, J. Kraus, A. A. Monge, D. L. Simms, M. F. Robbins, S. H. Neff, and W. Lichten, *Phys. Rev. A* **13**, 969 (1976).
34. N. Andersen, K. Jensen, C. S. Newton, K. Pedersen, and E. Veje, *Nucl. Instrum. Methods* **90**, 299 (1970).
35. Yu. F. Bydin, V. A. Vol'pyas, and S. S. Godakov, *Phys. Lett.* **50A**, 239 (1974).
36. Yu. F. Bydin, S. S. Godakov, and V. M. Lavrov, in *Proceedings of the Tenth International Conference on the Physics of Electronic and Atomic Collisions, Book of Abstracts*, p. 966, Paris (1977).
37. Ch. Courbin-Gaussorgues, M. Barat, and V. Sidis, in *Proceedings of the Ninth Summer School and Symposium on the Physics of Ionized Gases, Book of Abstracts*, p. 77, Institute of Physics, Belgrade (1978).
38. V. Sidis and H. Lefèbvre-Brion, *J. Phys. B* **4**, 1040 (1971).
39. J. van Eck, F. J. de Heer, and J. Kistemaker, *Physica* **30**, 1171 (1964).
40. D. R. Bates, *Proc. R. Soc. London Ser. A* **257**, 22 (1960).
41. D. Hasselkamp, private communication.

43

Quantum Mechanical Phase Coherence and Polarization in Low-Energy Ion–Atom and Ion–Surface Collisions

Norman Tolk and Joseph Kraus

Recent ion–atom and ion–surface experimental and theoretical studies involving quantum mechanical phase coherence phenomena are reviewed. The coherence effects are manifested in the experiments described here as energy-dependent oscillatory structure in charge or excited state yields and in one case by the degree of ellipticity of emitted radiation. In each case, the physical mechanism may be simply formulated in terms of a two-step model involving (a) the coherent population of two or more states and (b) the differential phase evolution of the states with time.

1. Introduction

We present a review of recent experimental and theoretical studies of low-energy ion–atom and ion–surface inelastic collisions involving quantum mechanical phase coherence effects. Much progress has been made over the past few years in understanding theunderlying physical mechanisms associated with collision-induced electronic outer-shell inelastic phenomena for both gas[1–4] and surface collisions.[5–11] These processes lead in many cases to pronounced oscillatory structure in the energy-dependent inelastic cross sections and to strong optical polarization of the emitted radiation. The results of these studies provide a wealth of information on the detailed nature of the quasimolecular potentials of the collision system and their coherent interactions.

Norman Tolk and Joseph Kraus • Bell Laboratories, Murray Hill, New Jersey 07974, U.S.A.

In this chapter, three cases are treated: Section 2 deals with oscillatory structure and strong optical polarization effects in optical radiation from low-energy Na⁺–Ne ion–atom collisions, Section 3 with oscillatory structure in low-energy ion scattering from surfaces, and Section 4 with pronounced elliptic polarization of Balmer radiation from low-energy grazing-incidence collisions of hydrogen ions on surfaces. The first case arises from a simple biparticle collision event; the second is also due to a biparticle collision interaction, where, however, one of the collision partners resides on the surface; the third case is quite different and may be understood purely as an interaction of the colliding ion with the surface as a whole. Although representing widely dissimilar collision systems, the formalism used to describe the physical model in each case is essentially identical and may be conveniently separated into two parts: (a) the coherent population of two or more states which ultimately contribute to the final atomic state observed, and (b) the subsequent differential phase evolution of the states with time. The major emphasis in this chapter is on identifying and discussing the physical processes responsible for the observed coherent collision phenomena.

2. Quantum Mechanical Phase Interference and Optical Polarization in Low-Energy Na⁺–Ne Inelastic Collisions

We have made a study of collisional excitation in low-energy (100 eV to 6 keV) Na⁺–Ne collisions and have measured absolute emission cross sections for each polarization component as a function of bombarding energy for optical transitions arising from excited $3p$ electrons in NeI and NaI.[4] These measurements show highly regular oscillations in the Na⁺–Ne cross sections obtained as a function of energy. The emission cross sections measured for the 10 levels arising from transitions from the $3p$ levels of NeI exhibit oscillatory structure of the same spacing and phase. The NaI cross sections also show oscillations alike in spacing and phase, but in antiphase with the Ne*($3p$) cross sections.

In addition, we observe strong polarization effects in the optical radiation arising from collisional excitation of NeI($3p$) states. In one case where radiation originates from a $J = 1$ to $J = 0$ transition in NeI, oscillatory structure in the energy dependence of the emission cross section is measured to be due entirely to the component of the emitted radiation polarized perpendicularly to the incident-beam direction. These measurements suggest a simple model which relates the final atomic excited states to the quasimolecular states which participate in the quantum mechanical phase-interference process. We have identified the $^1\Pi(\Omega = \pm 1)$ and $^3\Pi(\Omega = \pm 2)$ states as the major contributors to the quantum mechanical phase-interference processes, and not the simple single-electron diabatic states as believed previously.[2,3] Using this model, we have been able to account quantitatively for the magnitudes of the oscillation amplitudes in the polarization components of the NeI and NaI emission cross sections.

2.1. Apparatus

The Na+–Ne results were obtained using a low-energy ion accelerator capable of producing ion beams of well-defined energy ($\Delta E < 1$ eV) in the range from 10 eV to 6 keV with beam particle currents varying from 10^{12} particles/sec at 10 eV to 10^{14} particles/sec at 6.0 keV. The beam was directed using electrostatic lenses from the source chamber through an intermediate pumping region into the collision chamber. Optical radiation from the collision region was measured at 90° with respect to the beam direction using a 0.3-m $f/5$ monochromator and an S-20Q phototube. Single-photon counting techniques were used to detect and process the photon signals. An EOA L-101 spectral irradiance standard employing a calibrated tungsten coiled-coil quartz–iodine lamp was used to determine the absolute spectral response of the system. Polarization data were acquired with a Polaroid HN-38 polarizer. Gas-pressure information was obtained using a calibrated capacitance manometer. Details of the calculation of the absolute emission cross sections are discussed elsewhere.[4]

2.2. Results

Most of the visible NeI radiation within the wavelength region studied (2000–8000 Å) arises from ten levels which are in the $2p^5(^2P^0_{3/2})3p$ and $2p^5(^2P^0_{1/2})3p$ configurations as shown in Figure 1. In addition, most of the sodium radiation originates from the two NaD(3p) levels also shown in Figure 1. Thus radiation observed in the Na+–Ne experiment arises from the excitation of both Ne and Na into $3p$ electronic states

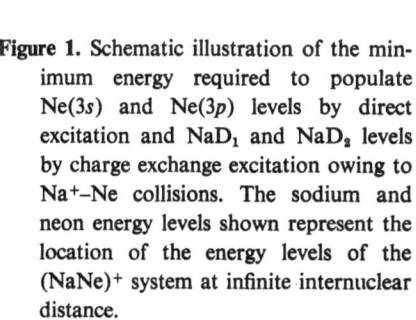

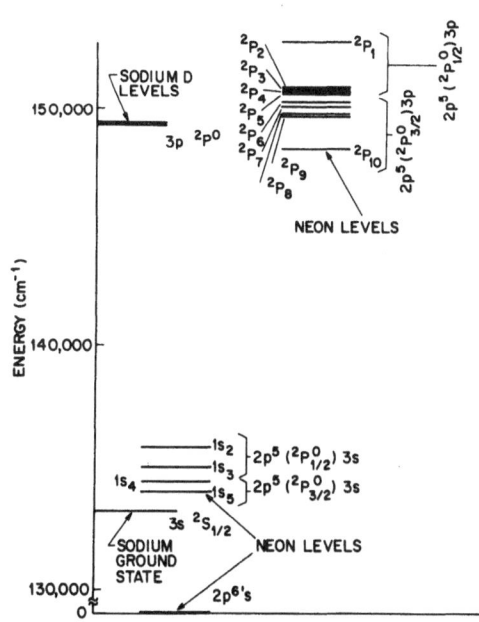

Figure 1. Schematic illustration of the minimum energy required to populate Ne(3s) and Ne(3p) levels by direct excitation and NaD₁ and NaD₂ levels by charge exchange excitation owing to Na+–Ne collisions. The sodium and neon energy levels shown represent the location of the energy levels of the (NaNe)+ system at infinite internuclear distance.

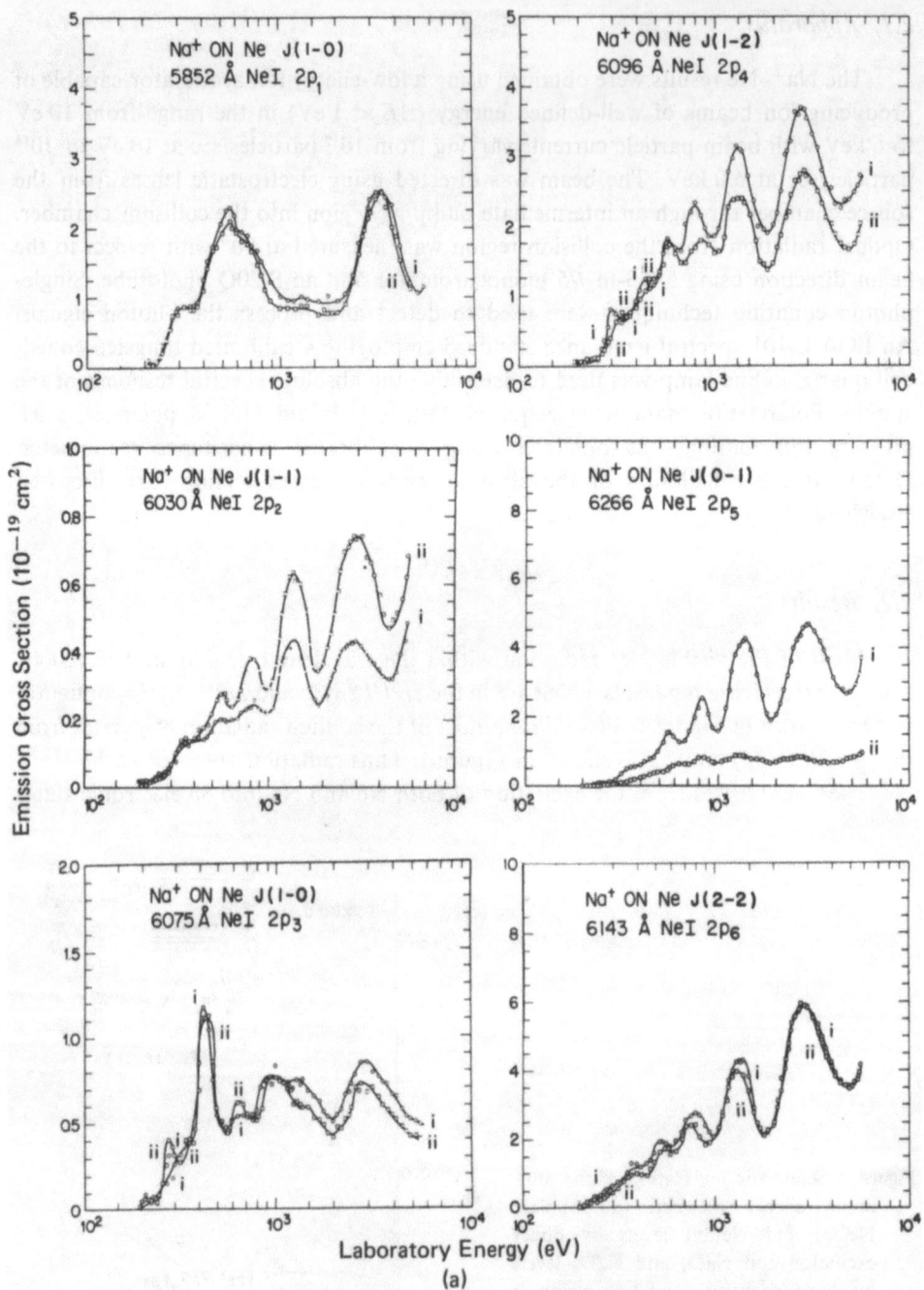

Figure 2. Absolute emission cross sections, plotted as a function of bombarding energy, for the perpendicular (i) and parallel (ii) components of optical radiation emitted at 90° from the beam direction. (a) The radiation comes from optical transitions arising from the decay of $3p$ electrons in NeI excited by Na^+-on-Ne collisions. Results are shown for six of the ten levels in the $2p^5(3p)$ configurations in NeI. (b) The radiation comes from optical transitions arising from

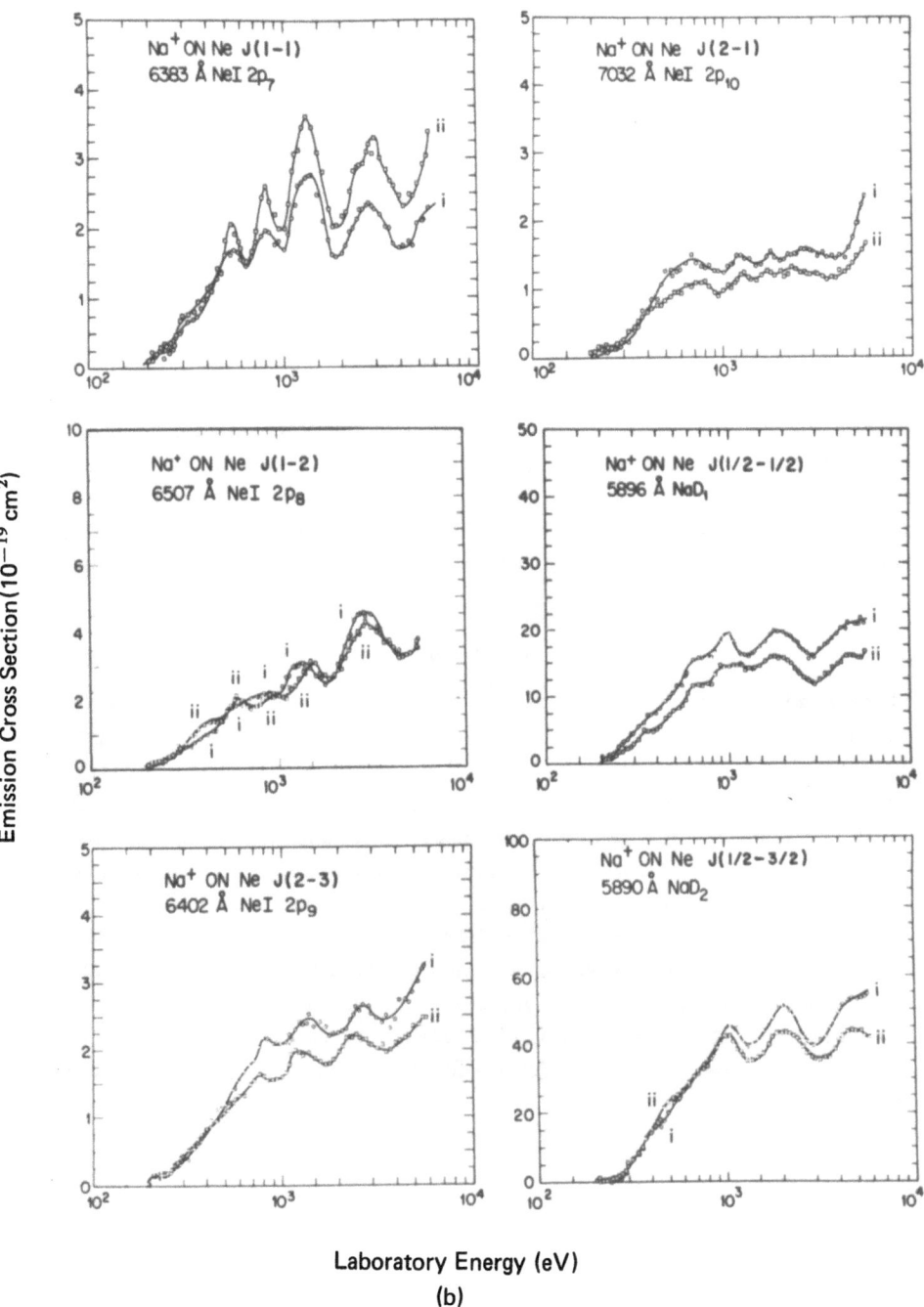

Emission Cross Section $(10^{-19}\ cm^2)$

Laboratory Energy (eV)

(b)

the decay of $3p$ electrons in NeI and NaI excited by Na⁺-on-Ne collisions. Results are shown for four of the ten levels in the $2p^5(3p)$ configurations in NeI and from the two NaD levels. The structure in the cross sections is reproducible to better than 5%, the relative magnitude between each of the two polarization components are uncertain to 5%, and the absolute magnitudes of the total emission cross sections are uncertain to 50%.

as a result of direct and charge-exchange collision processes,

$$Na^+ + Ne \rightarrow \begin{cases} Na^+ + Ne^*(3p) \\ Na^*(3p) + Ne^+ \end{cases} \tag{1}$$

Figures 2a and 2b show 12 sets of absolute emission cross sections plotted as a function of ion-beam energy for the perpendicular and parallel polarization components of NeI($3p \rightarrow 2s$) and NaD($3p \rightarrow 2s$) radiation arising from Na$^+$–Ne collisions. In each case, the data were taken under single-collision conditions in the linear region of the photon-signal-versus-pressure curve. Except for the two $J = 0$ cases, $2p_1$ and $2p_3$ (Paschen notation), the emission cross sections measured for the populations of the NeI($3p$) levels all show regular oscillatory structure of the same spacing and phase. The two NaD emission cross sections also show oscillations alike in spacing and phase, but in antiphase with the NeI structure observed.

Polarization fraction data plotted as a function of laboratory ion-beam energy are shown in Figure 3 along with the perpendicular and parallel components of the absolute emission cross sections for the excitation of the 6266-Å NeI optical emission line owing to Na$^+$–Ne collisions. The polarization fraction Π (equivalent to the Stokes parameter M/I) is defined in the usual manner:

$$\Pi = (I_{\parallel} - I_{\perp})/(I_{\parallel} + I_{\perp}) \tag{2}$$

where $I_{\parallel}$ and $I_{\perp}$ are photon intensities measured with polarizer parallel and perpendicular to the beam direction. Polarization anisotropy has been accounted for in Figure 3 but not in Figure 2. Figure 3 represents a particularly favorable case involving a $J = 1$ to $J = 0$ transition. It should be noted in this case that the absolute value of the polarization fraction is surprisingly large and that the appearance of the data is consistent with the assumption that all of the oscillatory structure arises from the polarization component perpendicular to the ion-beam direction.

Assuming effects due to cascading transitions are negligible, absolute cross sections for populating the ten Ne($3p$) and two Na($3p$) fine-structure states can be obtained from the measured absolute emission cross sections. Since many of the NeI $3p$ levels can radiate to more than one lower state, this requires taking proper account of branching ratios. Total cross sections for formation of each level can be written in the form

$$\sigma_{tot} = \tfrac{2}{3}\sigma_{\parallel} + \tfrac{4}{3}\sigma_{\perp} \tag{3}$$

where

$$\sigma_{\parallel} = \sigma_{\parallel}^{(rad)} B_i \tag{4}$$

$$\sigma_{\perp} = \sigma_{\perp}^{(rad)} B_i \tag{5}$$

and $\sigma_{\parallel}^{(rad)}$ and $\sigma_{\perp}^{(rad)}$ are the measured absolute emission cross sections for parallel and perpendicular polarization, respectively. B_i is the inverse of the branching ratio for the appropriate transition i,

$$B_i = \sum_j \frac{A_j}{A_i} \tag{6}$$

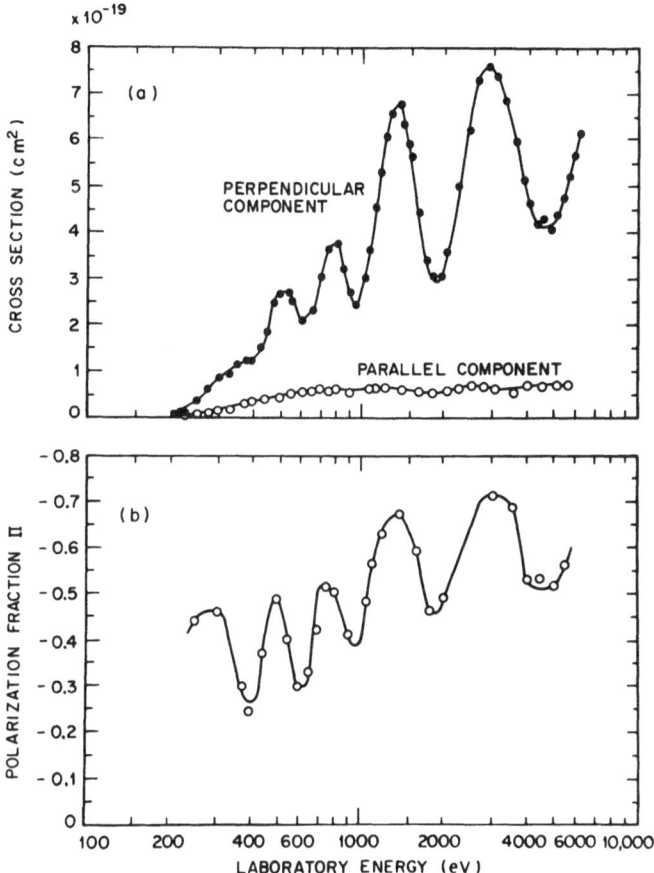

Figure 3. (a) Absolute emission cross sections as a function of ion-beam energy for the perpendicular and parallel components of NeI 6266-Å radiation arising from low-energy Na^+–Ne collisions. These components have been corrected for intensity anisotropy such that the sum will give the total absolute emission cross sections. The structure in the cross sections is reproducible to better than 5%, while the absolute magnitudes of the total emission cross sections are uncertain to 30%. The small amount of structure present in the parallel component may be attributed to an imperfect polarizer, a finite acceptance angle, and the inexactness of hypothesis (d). (b) Polarization fraction $\Pi = (I_\parallel - I_\perp)/(I_\parallel + I_\perp)$ as a function of ion-beam energy for NeI 6266-Å radiation (arising from a $J = 1$ to $J = 0$ transition) as a result of Na^+ + Ne collisions.

where A_j is the radiative transition probability to lower state j. Branching ratios were obtained from transition probabilities reported by Bennett and Kindlman[12] and by Schectman et al.,[13] and are listed in Table 1. The total cross section for formation of Ne(3p), obtained by summing σ_{tot} over all ten Ne(3p) levels, is shown in Figure 4. The total cross section for formation of Na(3p), obtained by summing over the two NaD levels, is also shown in Figure 4. Note that (i) the oscillations in the two cross sections are 180° out of phase,[14–16] (ii) the amplitudes of the oscillations in the two cases are approximately equal, and (iii) the magnitudes of the average

Table 1. Inverse Ratios $\sum_j A_j/A_i$, where A_i are Transition Probabilities Derived from the Work of Bennett and Kindlmann[12] and of Schectman et al.[13]

State	Line (Å)	Number of allowed transitions	$\sum_j A_j/A_i$
$^2p_1$	5852	2	1.01
$^2p_2$	6030	4	11.08
$^2p_3$	6075	2	1.02
$^2p_4$	6096	3	3.35
$^2p_5$	6266	4	2.22
$^2p_6$	6143	3	2.20
$^2p_7$	6383	4	1.66
$^2p_8$	6507	3	1.75
$^2p_9$	6402	1	1.00
$^2p_{10}$	7032	4	1.66

cross sections for production of Na*(3p) and Ne*(3p) are in the ratio of approximately 3:2 at the intermediate and higher energies.

Observations (i) and (ii) support the hypothesis that oscillatory structure is due to interference between quasimolecular states of the $(NeNa)^+$ collision system associated with direct and charge exchange processes. This is consistent with the energy levels of $(NaNe)^+$ at infinite internuclear separation. As shown in Figure 1, the NaD levels fall within the energy range over which the Ne(3p) levels are distributed and are consequently nearly degenerate. Both sets of cross sections are measured to have the same energy thresholds within experimental error at approximately 120 eV in the center-of-mass system which is about 100 eV above that required for populating the levels calculated strictly on the basis of conservation of energy. This is further evidence that the participating levels are coherently populated.

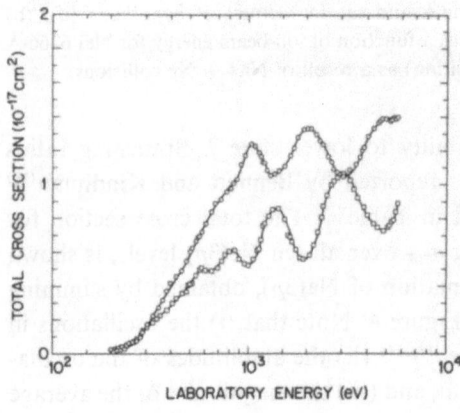

Figure 4. Absolute population cross sections for the Ne*(3p) levels (-□-□-) and Na*(3p) levels (-○-○-) plotted as a function of laboratory energy.

2.3. Na^+–Ne Analysis

It is important to note that although there are many molecular states involved, our results are characterized by their relative simplicity. This is illustrated by the following: only a single oscillatory pattern dominates; the population of the neon levels tend to oscillate in phase with each other and in antiphase with the sodium populations; strong polarization effects are observed; and the amplitudes of the oscillatory structure vary widely from one neon polarization component to another.

In order to explain these observations, we invoke the following four hypotheses.

(a) Oscillations in the energy dependence of the cross sections result from interference between one or more pairs of excited levels which are populated coherently at small internuclear separation and then interact at large separation, as illustrated schematically in Figure 5. This dual-coupling mechanism, first proposed by Rosenthal and Foley,[17] has been established in several related atomic collision processes.[1]

(b) Nonadiabatic interaction at large internuclear separation arises from "quasi-resonant charge exchange" of the type described by Lichten.[18]

(c) The distribution of final atomic level populations, through the quantum mechanical sudden approximation, is a direct reflection of the composition of the precursor molecular electronic states. This hypothesis is stated below in mathematical terms, and it allows us to account quantitatively for the cross sections for formation of various excited Na and Ne levels.

(d) At the outer coupling region, the molecular axis coincides with the laboratory z direction. This hypothesis, approximately valid for low-angle scattering, greatly simplifies the analysis of polarization data.

We discuss each of these hypotheses in detail below, and then examine their consequences.

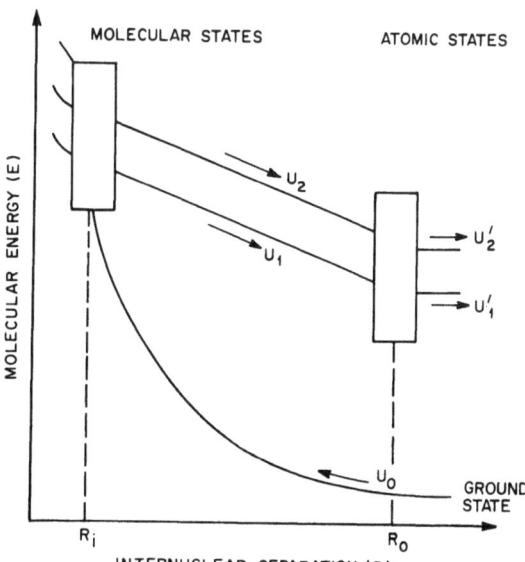

Figure 5. Schematic illustration of the dual-interaction model.

With regard to hypothesis (a), the Rosenthal–Foley (dual-coupling) model can be broken down into three separate parts:

(i) The primary excitation mechanism, in which a transition is made from the ground U_0 state to at least two inelastic channels U_1 and U_2 (at the inner internuclear separation R_i in Figure 5).

(ii) The inelastic channels U_1 and U_2, between which a phase difference develops during the outgoing part of the collision ($R_i \leq R \leq R_0$ in Figure 5).

(iii) A second interaction region (at the outer internuclear separation R_0 in Figure 5), at which interference occurs between the inelastic channels.

It can be easily shown[2] that the phase difference $\Delta\phi$ developed in the interval R_i to R_0 during the outgoing part of the collision can be expressed approximately as

$$\Delta\phi = (1/\hbar)\langle ER \rangle/v + \Delta\phi_0 \tag{7}$$

where $\langle ER \rangle$ is the area between curves bounded by the two interaction regions at R_i and R_0,

$$v^2 = (2/m)(E - U) \simeq (2/m)(E - U_{\mathrm{av}}) \tag{8}$$

and $\Delta\phi$ is the initial phase difference which would result from an infinitely fast collision. It can also be easily shown that the cross section is proportional to $\cos^2[(\frac{1}{2})\Delta\phi]$. Consequently it will exhibit maxima (minima) whenever the following condition is met:

$$\Delta\phi = 2\pi n = (1/\hbar)\langle ER \rangle/v + \Delta\phi_0 \tag{9}$$

The above expression suggests that if cross-section data are plotted as a function of $1/v$, assuming a proper choice of U_{av}, then the maxima should be equally spaced. An integer n may be assigned to each maximum such that, for example, $n = 1$ refers to the first peak which appears as v decreases from infinite velocity. The y intercept of a plot of n vs. $1/v$ will give the initial phase difference to an additive integral multiple of 2π, and the quantity $\langle ER \rangle$, which is important for comparison with calculated potential curves, can be derived from the slope of the line. For this case when U_{av} is 60 eV, $\langle ER \rangle$ equals about 1.27×10^{-7} eV cm. This is similar to values of $\langle ER \rangle$ found for example in the (HeNe)$^+$ collision case.[1]

The above analysis provides a quantitative measure of the average energy and splitting between the interacting excited states responsible for the observed oscillatory behavior, but it does not identify these states. To do this we first need to examine the electronic character of the (NaNe)$^+$ molecular states.

The colliding atoms approach initially along the $^1\Sigma^+$ ground-state potential curve of (NeNa)$^+$. This curve is very repulsive, reflecting the behavior of two rare-gas-like atoms; thus at small internuclear separations it will interact strongly with excited-state curves. We are interested particularly in those excited states which at large internuclear separation correlate with Ne*($2p^5$, $3p$) + Na$^+$ and Na*($3p$) + Ne$^+$($2p^5$) states. At moderate-to-large internuclear separation, these excited states can be accurately described as one-electron states outside of a Na$^+$($2p^6$)–Ne$^+$($2p^5$) core. Configurations corresponding to excitation of the Na$^+$ core or inner-shell excitation of Ne would be associated with too high an energy to contribute significantly in this region.

The symmetry of the $Na^+(2p^6)$–$Ne^+(2p^5)$ core must be either $^2\Sigma^+$ or $^2\Pi$, since it is produced by removal of one electron from $Ne(^1S)$; i.e., the vacant $2p$ orbital may have σ or π symmetry. Similarly, the outer $3p$ electron can occupy molecular orbitals of either σ or π symmetry. The $(NaNe)^+$ molecular states can be derived by combining a core state and outer-electron state according to the rules given by Herzberg, as shown in Table 2.[19] The subscripts on the molecular states indicate the value of $\Omega = \Lambda + \Sigma$ (Ω is the sum of the orbital Λ and spin Σ projections along the internuclear axis). There are two of each of the states listed (Na^+–Ne and Na–Ne^+), resulting in 7 states in all. Primary excitation must occur at small internuclear separations where the energies of the repulsive $NaNe^+$ ground state and the excited states are similar. The excitation is viewed as a one-electron process; i.e., a $2p$ electron of Ne is promoted to one of the excited molecular orbitals listed in Table 2. Both radial and rotational coupling may be important in the inner interaction region; thus the projection of orbital angular momentum along the internuclear axis need not be conserved during the excitation process. We might expect spin to remain unchanged during excitation (the Wigner spin-conservation rule). However, our results are strong evidence that spin is not conserved; both singlet and triplet excited states are produced during the collision. This is consistent with the anticipated magnitude of spin–orbit coupling in the excited states of $(NeNa)^+$.

With regard to hypothesis (b), we note that in the Rosenthal–Foley mode, states are excited coherently during primary excitation, evolve independently in the intermediate region, and are then recombined in an outer interaction region.[17] We hypothesize that in this case nonadiabatic coupling in the outer region is due to quasiresonant charge exchange. As indicated in Figure 1, the $Na^*(3p)$–Ne^+ and Na^+–$Ne^*(3p)$ levels are nearly degenerate. In analogy to the discussion of Lichten,[18] we propose that at moderate internuclear separations (e.g., between roughly 3 and 10 Å), Na^+ and Ne^+ appear very similar to an excited $3p$ electron. Therefore the orbitals of the outer electron can be very roughly characterized as approximately gerade (g) or ungerade (u). In order of increasing energy, these orbitals may be labeled σ_g, π_u, π_g, and σ_u, with σ_g the most strongly bonding and σ_u the most

Table 2. Molecular States Derived from Core- and Outer-Electron States[19]a

Core state	Outer-electron state	Molecular states in $\Lambda\Sigma$ coupling
(a) $^2\Sigma^+$	$^2\sigma$	$^1\Sigma_0^+$, $^3\Sigma_0^+$, ±1
(b) $^2\Pi$	$^2\sigma$	$^1\Pi_{\pm1}$, $^3\Pi_{0,0\pm1}$, ±2
(c) $^2\Sigma^+$	$^2\pi$	$^1\Pi_{\pm1}$, $^3\Pi_{0,0} \pm 1$, ±2
(d) $^2\Pi$	$^2\pi$	$^1\Sigma_0^+$, $^3\Sigma_0^+$, ±1, $^1\Sigma_0^-$, $^3\Sigma_0^-$, ±1, $^1\Delta_{\pm2}$, $^3\Delta_{\pm1,\pm2,\pm3}$

a The subscripts on the molecular states refer to values of Ω.

strongly antibonding. Nonadiabatic transitions are associated with the changeover from the nearly g and u character of the orbitals of the outer electron at intermediate internuclear separations to the asymptotic situation where the outer electron resides either on the Ne$^+$ or on the Na$^+$. This changeover occurs at distances where the splitting between the g and u states becomes comparable to the asymptotic splitting between the Na*(3p) and Ne*(3p) states. We estimate this to be 10–15 Å.

The hypothesis that coupling in the outer interaction region is due to the quasi-resonant charge exchange mechanism [hypothesis (b)] is strongly reinforced by the observations that Na* and Ne* oscillations occur 180° out of phase and the amplitudes of the oscillations are approximately equal (Figure 4). Note that if the g and u symmetry were exact then the average ratio of Na*(3p)-to-Ne*(3p) populations would be unity. The experimentally measured ratio of about 1.5 (see Figure 4) thus provides an indication of the deviation from exact resonance.

We may deal with hypothesis (c) by noting that at relatively large internuclear separations the electronic wave functions of (NaNe)$^+$ molecular states correlating with Na*(3p)–Ne$^+$ and Na$^+$–Ne*(3p) can be expressed accurately as linear combinations of atomic states:

$$\Phi_k \cong C_k \sum_{i=1}^{6} \sum_{j=1}^{6} a_{ijk} \xi_i \phi_j + (1 - C_k^2)^{1/2} \sum_{i=1}^{36} b_{ik} \xi_0^* \phi_i \qquad (10)$$

where $k = 1, \ldots, 72$; Φ_k is an adiabatic molecular wave function, and ξ_i, ξ_0^+, ϕ_i, and ϕ_j^+ are wave functions of atomic Na(3p), Na$^+$(^1S), Ne(3p), and Ne$^+$($^2P_{1/2}$, $^2P_{3/2}$), respectively. The coefficients a_{ijk} and b_{ik} are elements of a unitary transformation relating the spin and orbital angular momentum coupling scheme of the molecular states to that of the isolated atom states. The coefficients C_k describe the relative mixing of the Na$^+$–Ne* and Na*–Ne$^+$ configurations in Φ_k. The rapid variation of C_k with internuclear separation in the outer coupling region, $R \sim R_0$, is responsible for the quasiresonant charge exchange mechanism, hypothesis (b).

Hypothesis (c) can be stated in two parts: (i) the relative populations of excited atomic Ne*(3p) levels arising from a particular (NaNe)$^+$ molecular state are proportional to the intermediate internuclear separation values of $|b_{ik}|^2$ of equation (9); and (ii) the relative populations of excited Na*(3p) levels are proportional to $\sum_j |a_{ijk}|^2$. This hypothesis is applicable at high nuclear velocities where the quantum mechanical sudden approximation is valid. It is difficult to ascertain a priori whether this high-velocity limit is attained in the present case at the collision energies studied. Hypothesis (c) requires that the projections along the internuclear axis of both spin and orbital angular momentum, Σ and Λ, be separately conserved as the atoms recede at large internuclear separation. It should be accurate to assume that the projection of total angular momentum, $\Omega = \Sigma + \Lambda$, be conserved since at large separation nonadiabatic interaction is due almost entirely to radial coupling. Angular velocities are too small to be effective. Separate conservation of Σ and Λ requires, in addition, that spin–orbit coupling be too weak to promote transitions at large internuclear separation. Whether this is true at the nuclear velocities encountered here depends upon the spacial extent of the spin–orbit recoupling region; i.e., the region where molecular interactions are about the same magnitude as the asymptotic separations of the Ne* lines (~ 0.1 eV). We estimate this width to be of order $2a_0$. If this estimate is correct, hypothesis (c) should be valid except perhaps at the lowest collision energies studied. Ultimately, of course, our strongest justification for invoking this hypothesis is the simplicity and accuracy of the resulting interpretation of our experiments, as discussed below.

As a consequence of hypothesis (c), a particular pattern of atomic level populations can serve as a "fingerprint" of the precursor molecular state. As illustrated below, populations of the Ne*($3p$) levels provide information about both the Ne$^+$ core and the outer $3p$ electron; M_1 and M_s of the excited neon are the sum of $\Lambda + \lambda$ and $\Sigma + \sigma$ of the Ne$^+$Na$^+$ core and the excited electron. On the other hand, Na*($3p$) levels reflect solely the properties of the outer $3p$ electron; M_1 and M_s of excited sodium will be equal to λ and σ of the outer electron. Therefore by combining experimental information about both Na* and Ne* excited states we can obtain complete information about the outer electron, the core, and the NeNa$^+$ molecular states.

Hypothesis (d) is closely related to hypothesis (c). Since the processes contributing to the oscillatory phenomena involve predominantly small-angle scattering such that the sodium-beam particles are not significantly deflected from the beam direction (the laboratory z direction), then at the outer coupling region the molecular axis coincides with the laboratory z direction. Therefore by hypothesis (c) the

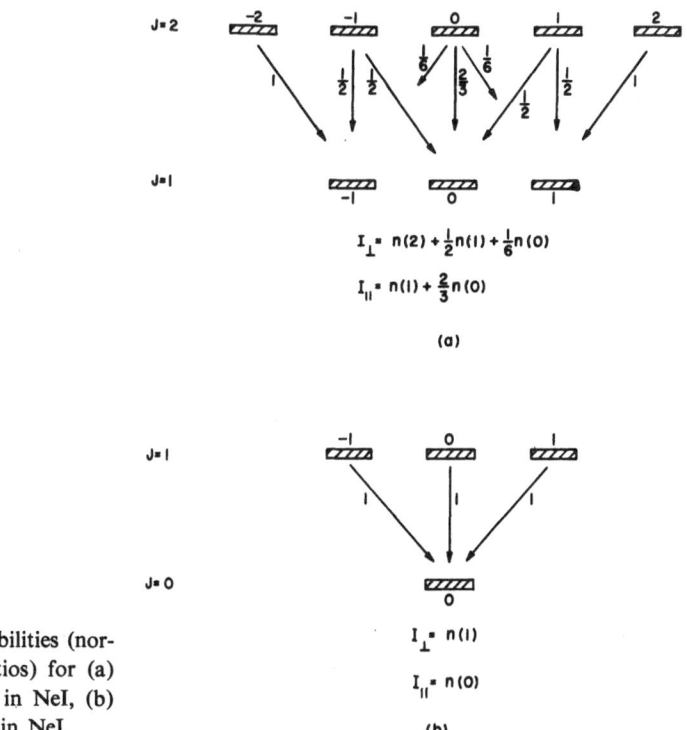

Figure 6. Transition probabilities (normalized branching ratios) for (a) the $1s_3$–$2p_5$ transition in NeI, (b) the $1s_5$–$2p_6$ transition in NeI.

laboratory M_j value of the final atomic state, which governs the polarization of emitted light, is equal to the Ω value of the molecular state from which it was formed. Because the outer coupling region occurs at finite internuclear separation ($\gtrsim 10$ Å) and the collisions occur at nonzero impact parameters, the assumption that the molecular z axis coincides with the laboratory z axis is not exact. This will result in some smearing of observed polarization.

The contributions to the polarization components of each of the 12 optical lines studied from the appropriate atomic M_j states were determined in the standard way from the transformation rules found in Condon and Shortly.[20] Examples are given for the $2p_5$ and $2p_6$ states of neon in Figure 6. A compilation of these results for both the neon and the sodium lines studied are given in Table 3.

Table 3. Theoretically Derived Intensities of the Perpendicular and Parallel Components of Radiation from Ten Lines Arising from $2s$–$3p$ Transitions in Ne I and Two Lines Arising from $2s$–$3p$ Transitions in Na I, Expressed in Terms of the Fractional Contributions of Each of the M_J Upper-State Levels, $n(M_J)^a$

State	Line (Å)	Intensity
$^2P_1$	5852 $J(1\text{–}0)$	$I_\perp = \frac{1}{2}n(0)$ $I_\parallel = \frac{1}{2}n(0)$
$^2P_2$	6030 $J(1\text{–}1)$	$I_\perp = \frac{1}{2}n(1) + \frac{1}{2}n(0)$ $I_\parallel = n(1)$
$^2P_3$	6075 $J(1\text{–}0)$	$I_\perp = \frac{1}{2}n(0)$ $I_\parallel = \frac{1}{2}n(0)$
$^2P_4$	6096 $J(1\text{–}2)$	$I_\perp = n(2) + \frac{1}{2}n(1) + \frac{1}{6}n(0)$ $I_\parallel = n(1) + \frac{2}{3}n(0)$
$^2P_5$	6266 $J(0\text{–}1)$	$I_\perp = n(1)$ $I_\parallel = n(0)$
$^2P_6$	6143 $J(2\text{–}2)$	$I_\perp = \frac{1}{2}n(2) + \frac{5}{8}n(1) + \frac{1}{2}n(0)$ $I_\parallel = \frac{3}{4}n(2) + \frac{1}{4}n(1)$
$^2P_7$	6383 $J(1\text{–}1)$	$I_\perp = \frac{1}{2}n(1) + \frac{1}{2}n(0)$ $I_\parallel = n(1)$
$^2P_8$	6507 $J(1\text{–}2)$	$I_\perp = n(2) + \frac{1}{2}n(1) + \frac{1}{6}n(0)$ $I_\parallel = n(1) + \frac{2}{3}n(0)$
$^2P_9$	6402 $J(2\text{–}3)$	$I_\perp = n(2) + \frac{3}{5}n(2) + \frac{7}{15}n(1) + \frac{1}{5}n(0)$ $I_\parallel = \frac{2}{5}n(2) + \frac{16}{15}n(1) + \frac{3}{5}n(0)$
$^2P_{10}$	7032 $J(2\text{–}1)$	$I_\perp = \frac{7}{10}n(1) + \frac{3}{10}n(0)$ $I_\parallel = \frac{3}{5}n(1) + \frac{2}{5}n(0)$
NaD$_1$	5896 $J(\frac{1}{2}\text{–}\frac{1}{2})$	$I_\perp = \frac{2}{3}n(\frac{1}{2})$ $I_\parallel = \frac{2}{3}n(\frac{1}{2})$
NaD$_2$	5890 $J(\frac{1}{2}\text{–}\frac{3}{2})$	$I_\perp = n(\frac{3}{2}) + \frac{1}{3}n(\frac{1}{2})$ $I_\parallel = \frac{4}{3}n(\frac{1}{2})$

^{a} These expressions were obtained in the standard way from rules found in Condon and Shortley.[20]

Using the above hypotheses we may proceed in our analysis. The experimentally derived cross-section curves were analyzed to obtain the magnitude of the perpendicular ($\perp$) and parallel ($\parallel$) polarization components of radiation from the oscillatory part of the signal. Figure 7 shows four examples of data analyzed in this manner. These results, expressed as the intensity of the oscillation amplitude (ΔI) at 2-keV beam energy and weighted by the proper branching ratios of Table 1, are given in Table 4, column 3.

Each of the 36 molecular state designations [$^1\Sigma^+(0000)$, $^3\Pi(1111)$, etc., where the indices are L, S, Λ, and Σ] was analyzed to determine the mixing coefficients of the asymptotic atomic states, as dictated by hypothesis (c). Molecular states were first transformed into LS coupling-scheme components using standard angular

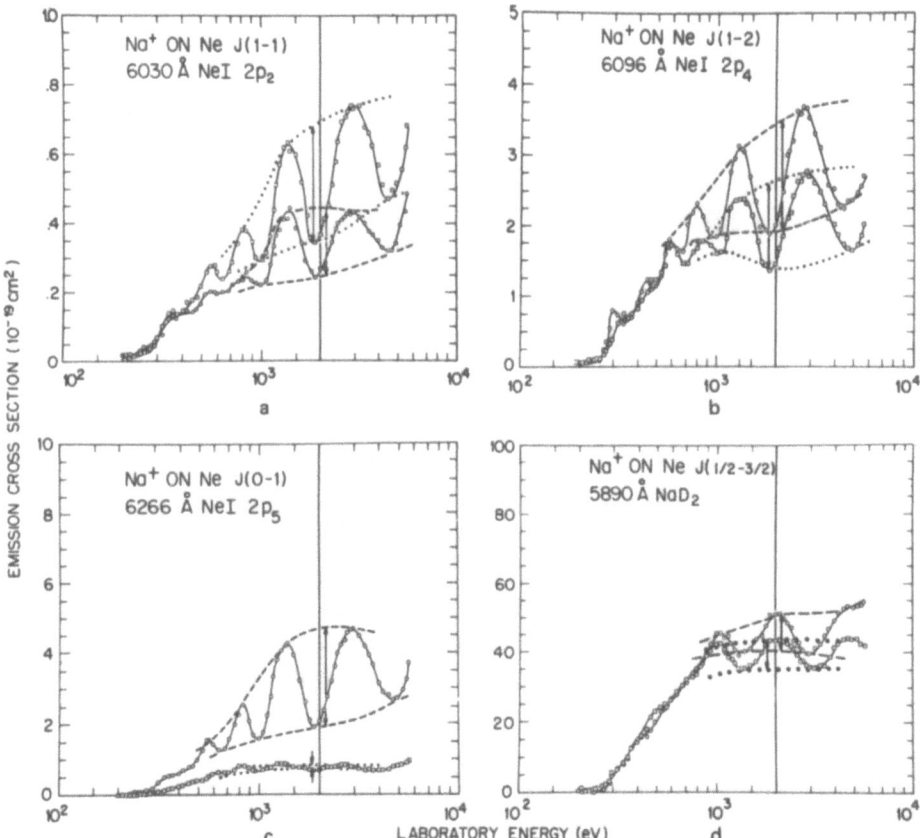

Figure 7. Absolute emission cross sections for the (a) Ne($2p_2$) 6030-Å, (b) Ne($2p_4$) 6096-Å, (c) Ne($2p_5$) 6266-Å, and (d) NaD$_2$ 5890-Å lines showing the intensity amplitudes ΔI of the regular oscillatory structure for each polarization component as a function of laboratory energy. The vertical arrows show the intensity amplitudes at 2 keV. -□-□-, parallel component; -O-O-, perpendicular component. The dots and dashes show the parallel and perpendicular components' intensity amplitudes, respectively.

Table 4. Experimentally and Theoretically Derived Oscillation Amplitude Intensities ΔI

Atomic state[a]	ΔI at 2 keV $(10^{-19}$ cm$^2)$	Experimental ΔI, adjusted[b]	Theoretical ΔI	
			$^1\Pi(\Omega = \pm 1)$	$^1\Pi(\Omega = \pm 1) + {}^3\Pi(\Omega = \pm 2)$
$^2P_1 \perp$	0	0	0	0
$\parallel$	0	0	0	0
$^2P_2 \perp$	0.185	2.05	2.21	2.21
$\parallel$	0.370	4.10	4.42	4.42
$^2P_3 \perp$	0	0	0	0
$\parallel$	0	0	0	0
$^2P_4 \perp$	1.50	5.02	1.95	4.98
$\parallel$	1.25	4.19	3.90	3.90
$^2P_5 \perp$	2.80	6.22	6.50	6.50
$\parallel$	<0.15	<0.33	0	0
$^2P_6 \perp$	2.95	6.49	5.63	6.47
$\parallel$	2.70	5.94	2.25	5.61
$^2P_7 \perp$	0.80	1.33	0.98	0.98
$\parallel$	1.35	2.24	1.95	1.95
$^2P_8 \perp$	1.20	2.10	1.17	1.60
$\parallel$	1.05	1.84	2.34	2.52
$^2P_9 \perp$	0.40	0.40	0	0.54
$\parallel$	0.40	0.40	0	0.54
$^2P_{10} \perp$	0.10	0.17	0.09	0.09
$\parallel$	0.10	0.17	0.08	0.08

[a] Expressed in Paschen notation [C. E. Moore, "Atomic Energy Levels," NBS Circular No. 467 (1949), Vol. 1, p. 76].
[b] Experimental values have been adjusted using branching ratios to reflect population cross-section rather than emission cross-section values.

momentum coupling techniques.[20] Except for $2p_9$ and possibly $2p_1$ and $2p_{10}$, the $2p$ atomic states of NeI may not be described as pure LS coupling states. Schectman et al. have expanded the wave functions of the $2p$ levels in terms of LS functions.[13] These are shown in Table 5. Using these intermediate coupling-scheme results we have developed a matrix which relates the $2pM_j$ states to the molecular states. This is shown in Figure 8 in terms of probabilities. For example, the $^1\Sigma^+(0000)$ state upon dissociation is predicted to produce $2p_1(00)$ with 0.98 probability and $2p_3(00)$ with 0.03 probability, according to hypothesis (c).

By examination of Figure 8, the striking polarization observed in the $2p_5$ case must be due predominantly to the $^1\Pi(1010)$ state. Only three other molecular states could result in the same polarization, and each of these would put far too much intensity into other transitions to be consistent with experiment. Our first attempt,

Table 5. Wave Functions for the $2p$ (Paschen Notation) Levels of Ne I Expanded in Terms of LS Basis Functions (Taken from the Work of Schectman et al.[13])

States		Intermediate coupling coefficients	States		Intermediate coupling coefficients
$2p_1$	$^1S_0$	0.99	$2p_6$	$^3P_2$	−0.65
	$^3P_0$	0.17		$^1D_2$	−0.72
$2p_2$	$^3P_1$	0.80		$^3D_2$	−0.24
	$^1P_1$	0.58	$2p_7$	$^3P_1$	0.28
	$^3D_1$	−0.02		$^1P_1$	−0.39
	$^3S_1$	−0.13		$^3D_1$	0.88
$2p_3$	$^1S_0$	0.17		$^3S_1$	−0.01
	$^3P_0$	−0.99	$2p_3$	$^3P_2$	0.13
$2p_4$	$^3P_2$	0.73		$^1D_2$	−0.43
	$^1D_2$	−0.55		$^3D_2$	0.89
	$^3D_2$	−0.39	$2p_9$	$^3D_3$	1.00
$2p_5$	$^3P_1$	−0.49	$2p_{10}$	$^3P_1$	0.13
	$^1P_1$	0.71		$^1P_1$	0.08
	$^3D_1$	0.49		$^3D_1$	0.00
	$^3S_1$	−0.01		$^3S_1$	0.99

therefore, was to try to explain the oscillatory structure from all of the Ne* states in terms of a single pair of molecular states of symmetry $^1\Pi(\Omega = \pm 1)$. Note that since L is not a good quantum number at finite internuclear separations, both pairs of $^1\Pi$ states listed in Figure 8 possess this symmetry. We are therefore free to choose whatever relative mixing of these two pairs of states will produce the best possible agreement with experiment. The optimal choice of this mixing, $\approx 1 : 1$, results in the predictions of column 4 of Table 4. Agreement is not satisfactory, particularly for the $2p_4$ and $2p_6$ states.

In order to improve agreement with experiment, it is necessary to include at least one state of symmetry different from $^1\Pi$. We can achieve dramatic improvement by including contributions from the $^3\Pi(\Omega = \pm 2)$ states. Column 5 of Table 4 gives the final predictions, assuming contributions in the following proportions: $^1\Pi(10 \pm 10) : ^1\Pi(20 \pm 10) : ^3\Pi(11 \pm 11) : ^3\Pi(21 \pm 11) = 13 : 13 : 6 : 1.2$. Except possibly for $2p_8$, agreement is excellent.

By inclusion of minor amounts of additional states, slight further agreement could be achieved. But if either the $^1\Pi(\Omega = \pm 1)$ or $^3\Pi(\Omega = \pm 2)$ states are omitted, agreement cannot be made satisfactory, regardless of how many other states are included to try to compensate.

We conclude from this analysis of the Ne*($3p$) emission that the oscillatory structure in the cross sections can be attributed primarily to a pair of interacting $^1\Pi(\Omega = \pm 1)$ states. At least one additional pair of states, $^3\Pi(\Omega = \pm 2)$, produces

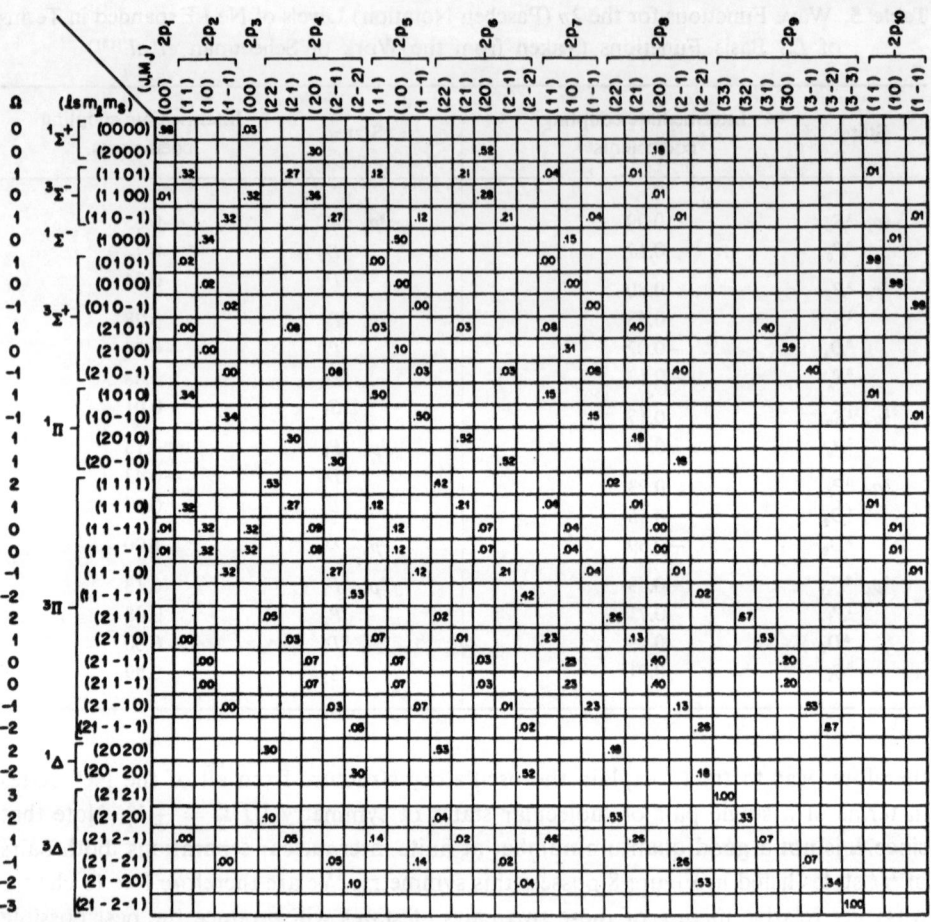

Figure 8. Transformation matrix written in terms of probabilities (obtained by squaring the amplitudes) relating the 36 molecular states ($LSA\Sigma$) to the Ne*($3p$) states (JM_j) expressed in Paschen notation.

similar oscillations. As shown schematically in Figure 9, we view these as completely independent paths. Note that oscillations in the $2p_4$ and $2p_6$ states, those to which $^3\Pi(\Omega = \pm2)$ makes its largest contribution, do not quite line up with those in $2p_5$.

Similarly we may analyze the oscillations from the Na*($3p$) states. The $^1\Pi$ and $^3\Pi(NaNe)^+$ states can be constructed either from a $^2\Pi$ core and $^2\Pi$ outer electron or from a $^2\Sigma^+$ core and $^2\pi$ outer electron, case (b) or (c) of Table 2. We can distinguish between these possibilities by examining the polarization of the Na*($3p^2P_{3/2}$–$3s^2S_{1/2}$) emission. We assume that the symmetry of the Na* excited state is determined solely by the outer electron; the distant Ne$^+$ has no effect. The analysis then parallels that of the Ne* radiation. Invoking part (ii) of hypothesis (c), and making use of the expressions of Table 3 and standard LS-coupling transformation rules,[20] we obtain the following values for $I_\parallel/I_\perp$:

$$I_\parallel/I_\perp = 1.47 \qquad \text{for } \sigma \text{ outer electron}$$
$$I_\parallel/I_\perp = 0.81 \qquad \text{for } \pi \text{ outer electron}$$

(11)

In computing these values it was necessary to take account of the effect of hyperfine structure arising from the nonzero spin of the Na nucleus. This, again, was accomplished using standard angular momentum coupling techniques.

The experimentally observed polarization ratio $I_\parallel/I_\perp$ from the $Na(2p_{3/2})$ state is 1.00 ± 0.05. There are three possible explanations for this discrepancy: (i) Hypothesis (b) is incorrect. (ii) Neither of the molecular orbital diabatic schemes, cases (b) and (c) of Table 2, are present in pure form. (iii) The $^1\Pi(\Omega = \pm 1)$ states are diabatic states of case (c) of Table 2, Σ^+ core and π outer electron, and the $^3\Pi(\Omega = \pm 2)$ states are case (b), Σ^+ core and π outer electron. If this were the case, then the predicted ratio $I_\parallel/I_\perp$ would be 0.93, in closer agreement with the experimental value of 1.00.

Possibility (i) above is extremely unlikely. Our success in describing the Ne* results is a strong indication that part (i) of hypothesis (c), at least, is valid. As shown in Figure 1, the Na* levels are much more closely spaced than most Ne* levels; thus the sudden approximation should be more valid for Na* than Ne*, i.e., part (ii) of hypothesis (c) is almost certainly accurate.

Possibility (iii) above predicts a result that is outside our estimated experimental uncertainty, although not by very much. Therefore it appears that possibility (ii)

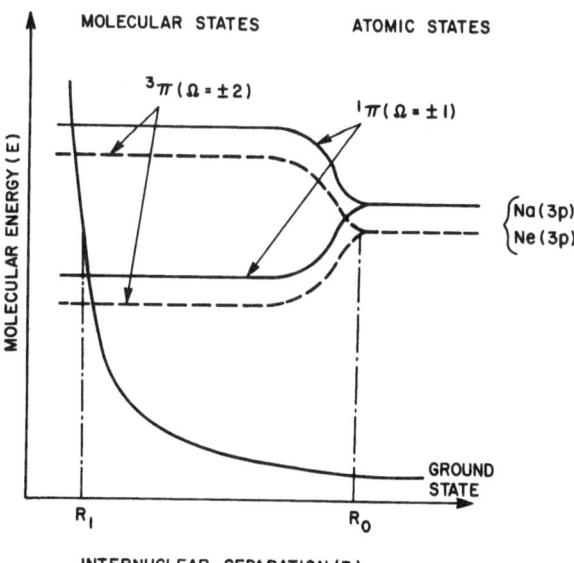

Figure 9. Schematic illustration of two pairs of coherently excited levels which contribute to the observed quantum mechanical phase interference.

is correct, perhaps in combination with possibility (iii). In conclusion, our results suggest strongly that the simple one-electron diabatic models employed previously[2,3] cannot account adequately for the experimental observations.

2.4. Na⁺–Ne Conclusion

Using plausible hypotheses we have been able to account quantitatively for the amplitudes and polarizations of the oscillatory structure in the emission cross sections of $3p$ levels excited in collisions of Na⁺ with Ne. Our analysis of the data provides evidence that the oscillatory behavior arises predominantly from two independent pairs of molecular states, $^1\Pi(\Omega = \pm 1)$ and $^3\Pi(\Omega = \pm 2)$, as illustrated schematically in Figure 9. These states are populated coherently at small internuclear separations, evolve independently as the collision partners recede, and finally interact via a charge transfer mechanism at large ($\sim$10–15 Å) internuclear separations.

Odom, Caddick, and Weiner recently applied this method of analysis successfully to oscillatory structure and polarization in the total emission cross sections arising from K⁺–Ar collisions.[15] They found, in agreement with the Na⁺–Ne work, that the interfering states are the adiabatic states of the quasimolecule and that for this case these states arise from $^3\Pi$ manifolds with no measurable contributions from $^1\Pi$ states. This is further evidence for our sudden approximation hypothesis. Clearly the validity and power of this physical model is supported by its simplicity and by its success in accounting quantitatively for these very detailed and complicated experimental results.

3. Nonadiabatic Neutralization at Surfaces: Angular Dependence of Oscillatory Ion Scattering Intensities

Recent experimental and theoretical studies indicate that collisions of ions with surfaces can provide information about binary ion–atom interactions involving outer-shell electrons that may be difficult to obtain in gas phase experiments.[5-8] In particular, large-angle ion–atom scattering can be observed much more easily from a surface than in the gas phase. Furthermore, impact parameter selection is easily achievable in ion–surface scattering due mostly to the fact that the physical presence of the surface removes from consideration those atoms that are scattered into it. This produces easily observable differential inelastic effects which may be studied in the gas phase only by elaborate coincidence measurements. Although, in many cases, outer-shell processes can be significantly altered by the existence of the surface or bulk, we have nonetheless demonstrated that certain types of ion–surface processes are dominated by specific ion–atom interactions, and that the presence of a surface modifies the results in small and (in some cases) easily accountable ways. By selecting only those backscattered particles whose energy losses correspond to reflection by a particular surface atomic species, it is possible to isolate those events for which the

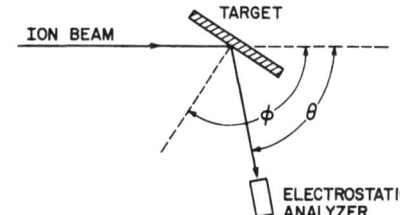

Figure 10. Schematic representation of ion surface scattering experimental configuration.

ion is scattered off a single surface atom from those involving multiple collisions or penetration into the solid. The variation of the intensity of this specular peak with incident energy and angle reveals direct information about the binary ion–atom interactions.

In this section, we report on a recent study[6,8] of the mechanism responsible for the dramatic oscillatory behavior first observed by Erickson and Smith[5] in the yield of He$^+$ ions scattered from surfaces. In particular we have performed measurements of the dependence on scattering angle θ and target orientation ψ of the location of oscillation maxima.[6] The experimental configuration is shown in Figure 10. These experiments provide important information about the ion–surface interaction at low energies (0.5–3 keV) and support the view that the oscillatory behavior arises from quantum mechanical phase interference between near-resonant ionic and neutral levels. This collision phenomenon is similar in many ways to the Na$^+$–Ne biparticle collision case discussed above. We present a semiquantitative model based on this picture which accounts for the positions and spacings of oscillation peaks, and for the distinctive angular dependence observed.

3.1. Apparatus

The apparatus, which is similar to that described in the previous section, consists of an ion source, electrostatic focusing lenses, a Wien filter, a target chamber, and an electrostatic analyzer able to move through laboratory scattering angles ranging from 20° to 135°. The pressure in the target chamber was typically 2×10^{-8} Torr. Ion yields, normalized to integrated beam current, were obtained by sweeping the electrostatic analyzer over the entire energy range of scattered ions and extracting the peak intensity at the energy corresponding to binary collision with the specified surface atom.

3.2. Results

Detailed measurements have been performed of the scattering of 200–3000-eV He$^+$ ions by Pb and GaP targets. We note three major observations: (1) As shown in Figure 11 the intensity of scattered He$^+$ oscillates as a function of incident energy. (2) Our measurements show no observable shifts in the positions of oscillation maxima for both Pb and Ga (in GaP) for target orientation angles ψ ranging from

110° to 180°.[6] The oscillatory structure changes markedly as the scattering angle θ is changed (see Figure 11). Furthermore, these changes depend on the nature of the target species; Ga peaks shift to lower energy as θ is decreased, whereas Pb peaks shift mostly in the opposite direction.

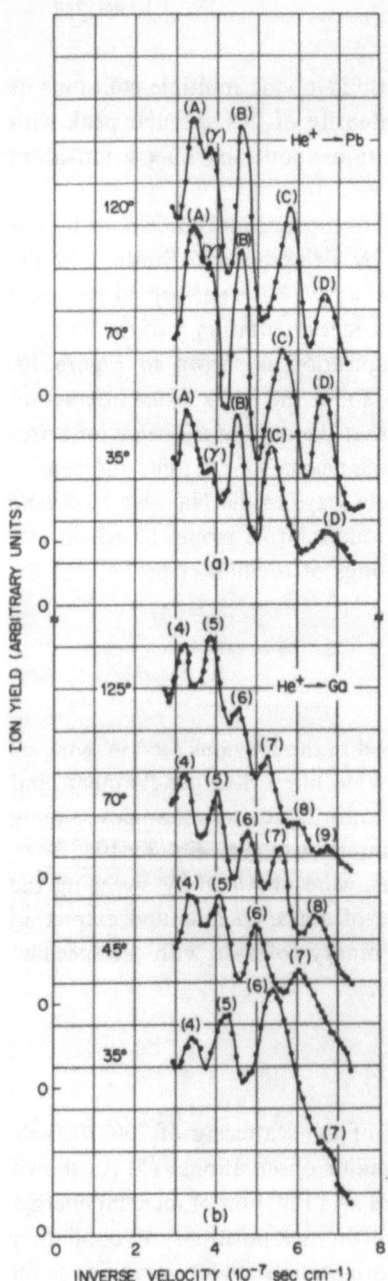

Figure 11. Relative intensity of scattered He+ ions versus the inverse of the initial velocity, for several scattering angles θ. (a) Pb target; (b) GaP target (Ga surface peak).

3.3. Discussion

Most of the observations can be understood within the framework of the model illustrated schematically in Figure 12. Situations where the presence of a pair of closely lying quasimolecular states in interference lead to oscillations in total or differential cross sections are familiar in the field of gas-phase ion–atom collisions.[1-4] In the present case the two states correspond to an He^+ ion in the vicinity of a neutral surface atom S (curve b) and a neutral He atom in the vicinity of surface atom S^+ which is missing an electron from a level nearly resonant with the ionization potential of He (curve a). This picture is strongly supported by the fact that regular oscillatory behavior has been observed only for target species with d-state energy levels lying within 10 eV of the He ionization potential, 24.6 eV.[5]

We hypothesize that at an internuclear separation $R \simeq R_m$ a coherent mixing of the two levels occurs because of the exchange (charge-transfer) interaction. As the collision evolves, a differential phase $\Delta\phi$ develops between the two paths, on both the incoming and outgoing legs, until the mixing region is traversed again. The intensities I_+ and I_0 of scattered ions and neutral atoms are determined by the total accumulation of differential phase[21]:

$$I_+ = \alpha_+ + \beta \cos^2(\Delta\phi/2)$$
$$I_0 = \alpha_0 + \beta \sin^2(\Delta\phi/2)$$

(12)

The coefficients α_+, α_0, and β are slowly varying functions of the incident ion energy which include damping effects due to other neutralization processes and to the finite width of the surface d level. Thus the ion intensity will exhibit maxima whenever the condition $\Delta\phi = 2\pi n$ is met. The phase $\Delta\phi$ is given by

$$\Delta\phi = \frac{1}{h} \int \Delta E(t)\, dt = \frac{1}{h} \int \frac{\Delta E(\mathbf{R})}{v(\mathbf{R})}\, d\mathbf{R}$$

(13)

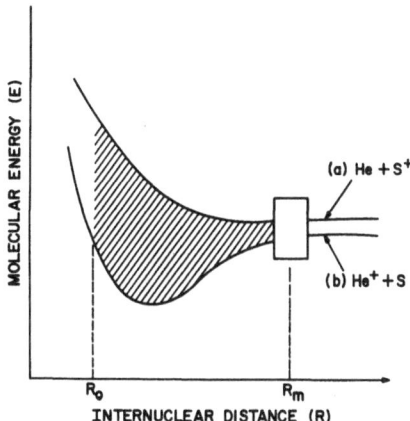

Figure 12. Schematic illustration of near-resonant charge-exchange model.

where the integral is evaluated along the trajectory from the initial to the final crossing of the mixing region at R_m. $v(\mathbf{R})$ is the instantaneous component of velocity tangent to the trajectory, and ΔE is the splitting between curves a and b of Figure 12.

Equations (12) and (13) provide a basis for analysis of our experimental results. Note first that, at least at high energies where velocity is most nearly constant over the path of integration, Equation (13) can be approximated as

$$\Delta\psi \simeq (2/hv) \int_{R_0}^{R_m} \Delta E(R)\, d\mathbf{R} = (2/hv)\langle ER\rangle \qquad (14)$$

R_0 is the turning point, and the factor of 2 arises from inclusion of both incoming and outgoing legs. Thus the oscillation peaks should be approximately equally spaced when plotted versus inverse relative velocity, $1/v$. As can be seen from Figure 11, although there is some deviation, this is nearly the case.

Using Equation (14), we can obtain an experimental measure of the quantity $\langle ER\rangle$. We extract from Figure 1 the values 17.7 and 23.5 eV Å for He$^+$ on Pb and GaP, respectively. These numbers are consistent with the values 22 eV Å for He$^+$–Pb and 19 eV Å for He$^+$–Ga estimated by a simple calculation, and are similar to those found in the previous section for Na$^+$–Ne.

In order to employ the model to account for angular distributions, it is necessary to specify more completely the quantities in equation (13). There are two extremes in which this task is simplified, as shown in Figure 13. The first extreme occurs if the ion–surface interaction is invariant to motion along the surface, and depends only on the distance $R_\perp$ between the ion and the surface (Figure 13a). The velocity v in Equation (13) then refers to the component of velocity normal to the surface, and R_m defines a plane parallel to the surface. In this purely surface effect limit there would be a very strong, predictable shift of oscillation peaks depending both on the scattering angle θ and the target orientation ψ. The fact that we see no variation with ψ whatever and a very different dependence on θ from that predicted effectively rules out this limiting case.

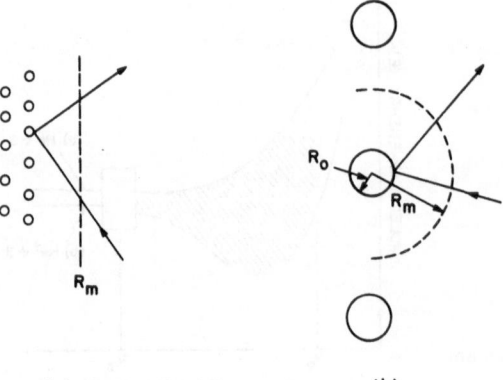

(a)   (b)

Figure 13. Schematic representation of (a) surface interaction and (b) binary interaction.

The second and more realistic extreme is the ion–atom limit in which the ion interacts with only the particular surface atom from which it scatters. The mixing distance R_m then defines a sphere surrounding the surface atom (Figure 13b), and the velocity appropriate to equation (13) is the radial velocity v_r. This model predicts that (a) the results will be independent of target orientation ψ, in agreement with experiment, and (b) the θ dependence is determined both by the amount of time spent in the phase development region and by the amount of phase area $\langle ER \rangle$ swept out during the collision.

We can rewrite equation (13) in the form

$$\Delta\phi = \frac{2m}{h} \int_{R_0}^{R_m} \frac{\Delta E(R)\, dR}{\{2m[E - V(R)] - L^2/R^2\}^{1/2}} \tag{15}$$

where $V(R)$ is the effective ion–atom potential, E is the initial energy, and L and m are the angular momentum and reduced mass of the ion–atom pair. L is related to the final scattering angle θ through the well-known classical deflection function,

$$\theta = \pi - 2L \int_{R_0}^{\infty} \frac{dR}{R^2\{2m[E - V] - L^2/R^2\}^{1/2}} \tag{16}$$

The solid curves of Figure 14b are calculated from equations (15) and (16), assuming ΔE to be constant for $R < R_m$ and $V(R)$ to be of the form Ze^2/R. The best-fit values of ΔE and Z are 8.8 eV and 25. Even using these vastly oversimplified forms for ΔE and V, the model is able to account satisfactorily for the angular dependence and the deviation from equal peak spacing in the He$^+$–GaP results.

It should be noted that the oscillations observed in this ion–surface experiment are of the kind referred to in differential ion–atom collisions as "Stueckleberg" oscillations.[21] In this respect they are a different kind of oscillation than the "Rosenthal" type discussed in the previous section, which appear in the total rather than differential cross sections.

Using the binary ion–atom interaction picture as established experimentally, Tully and Tolk extended the theory of ion neutralization to nonadiabatic processes, which is a necessary step in explaining the oscillatory behavior.[8] Calculations based on this theory showed excellent agreement in fitting the data for the cases of He$^+$–Ga and He$^+$–Pb collisions. Details of this calculation are published elsewhere.[8]

The oscillatory behavior observed in low-energy ion–surface scattering represents a classic example of quantum mechanical phase interference arising from the principle of superposition, similar to the Na$^+$–Ne biparticle collision case treated above. Our measurements of the angular dependence of oscillation maxima constitute strong evidence that the phenomenon arises from near-resonant charge exchange associated with a specific ion–atom interaction. Nevertheless, the electronic potentials involved may be profoundly influenced by the nature and proximity of neighboring surface and bulk atoms.[5-8]

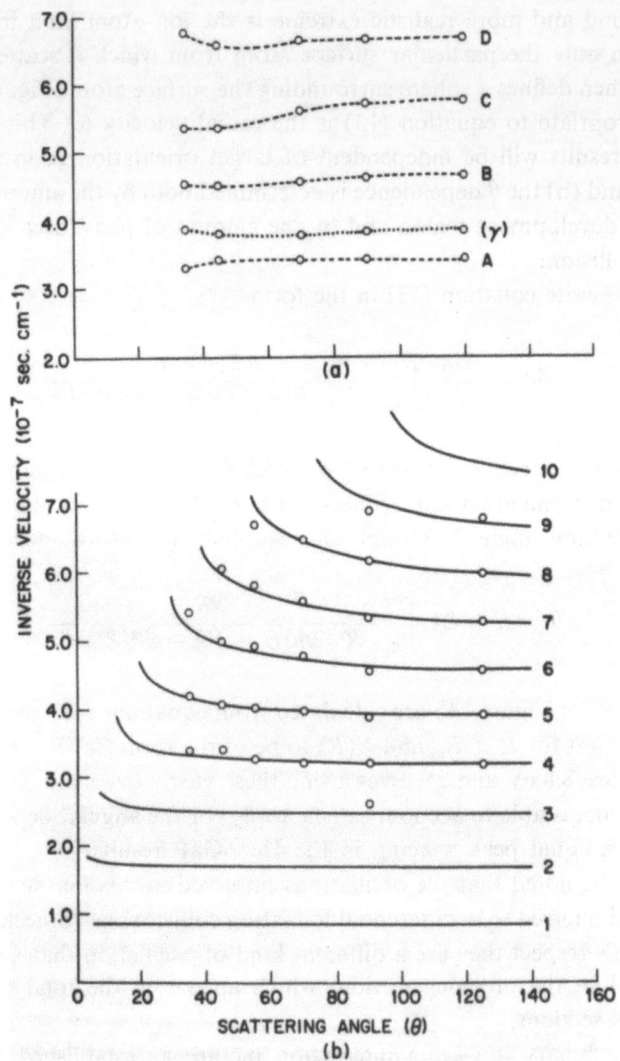

Figure 14. Angular dependence of peak positions. Points are experimental. (a) He$^+$–Pb; peaks are lettered arbitrarily. (b) He$^+$–GaP; peaks are labeled by phase index $n = \Delta\phi/2\pi$. The solid curves are theoretical.

4. Elliptic Polarization of Balmer Radiation from Low-Energy Grazing Incidence Collisions of Hydrogen Ions on Surfaces

Our discussions up to now in this chapter have dealt with coherent inelastic phenomena arising from bipartical collisions even though in Section 3 the bipartical collision occurred at a surface. In this section we discuss a similar coherent phenomenon which may only occur in the presence of a surface.

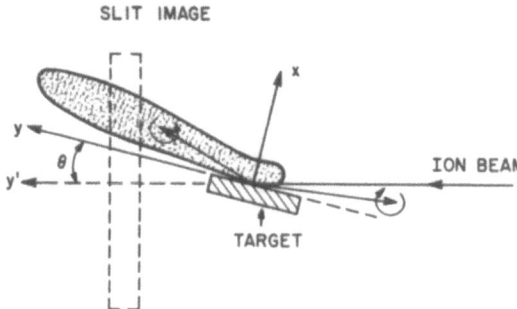

SLIT IMAGE

ION BEAM

TARGET

Figure 15. Schematic representation of
experimental configuration.

We have recently observed strong elliptic polarization arising from the *p*-state component of Balmer radiation emitted following low-energy (0.6–10 keV/nucleon) grazing-incidence collisions of beams of H^-, H^+, H_2^+, and H_3^+ ions on various sputter-cleaned polycrystalline metal surfaces, in particular Pb, maintained under ultrahigh vacuum (10^{-10} Torr).[11] Andrä and co-workers[9,10] and Berry *et al.*[22] have observed elliptic polarization from grazing-incidence experiments at higher energies (40 keV to 2.5 MeV). These measurements are the first to be performed at low energies, with a variety of molecular hydrogen ions, and in a clean environment. Our observations lead us to suggest a physical model involving a direct anisotropic surface electron-pickup mechanism with subsequent modification of polarization due to electrostatic interaction of the atom with the surface causing quantum mechanical phase evolution similar to that described in the previous two sections.

4.1. Apparatus

The apparatus for this experiment is again similar to that described in previous sections with the exception that the grazing incidence collisions occur in an ultrahigh-vacuum chamber capable of sustaining an ambient pressure of 10^{-10} Torr. Mass-separated atomic and molecular ion beams are produced in the energy range from 50 eV to 10 keV. Polarization data are acquired with a Polaroid HN-38 polarizer which removes spectrometer polarization bias, and a retardation plate with quarter-wave retardation at 656.3 nm. Single-photon counting techniques are used to detect and process photons emitted normal to the plane defined by the incoming beam and surface normal (out of the page in Figure 15).

4.2. Results

The optical radiation observed in this experiment arises from the decay of 3*s*, 3*p*, and 3*d* states of hydrogen excited by grazing-incidence collisions of H^-, H^+, H_2^+, or H_3^+ on an approximately 2-mm-wide Pb surface as shown in Figure 15. Ta, Cu, Mo, and C targets were also used; however, the Pb target was found to produce the highest light intensity as well as the highest degree of polarization.

Relatively small amounts of linear polarization were observed. In terms of normalized Stokes parameters,[23] typical values were $M/I \approx 0.01$ and $C/I \approx -0.10$. Consequently the major axis of elliptic polarization was observed to be approximately at 45° from the x axis in Figure 15 in the clockwise direction. The greatest contribution to the fractional polarization in this experiment arose from the normalized Stokes parameter expressing circular polarization given by

$$S/I = (I_{RH} - I_{LH})/(I_{RH} + I_{LH}) \tag{17}$$

where I_{LH} and I_{RH} are the intensities of left and right circularly polarized light, respectively. Strong left-hand polarization with values of S/I of the order of -0.50 were observed. These measurements were found to be only very weakly dependent on grazing incidence angle θ (see Figure 15) in the range of measurement, $\theta = 1$ to 7 deg. Introduction of a small partial pressure of oxygen (10^{-7} Torr) caused severe reduction in the degree of circular polarization, thereby underscoring the importance of clean surfaces.

As shown in Figure 15, measurements as a function of the position y' of the detector slit along the beam direction were acquired by sampling a region of observation defined by a one-to-one image of the monochromator entrance slit of dimensions 2 cm by 0.2 cm. A typical measurement of intensity and degree of circular polarization S/I as a function of y' is shown in Figure 16 for the case of 10 keV H_2^+ grazing-incidence (6°) collision on a 2-mm-wide Pb target centered at $y' = 0$. Note that S/I changes sign for large negative values of y', suggesting that ions that are scattered at large angles, $>90°$, have orientation opposite to those scattered at small glancing angles. Note also that $|S/I|$ decreases markedly as a function of y' downstream from the target, a behavior not observed in previous measurements.[9,10,22] Since projectiles are not slowed appreciably during grazing collisions, we can relate directly this decay with distance to a decay with time. From Figure 16 we extract a decay of circular

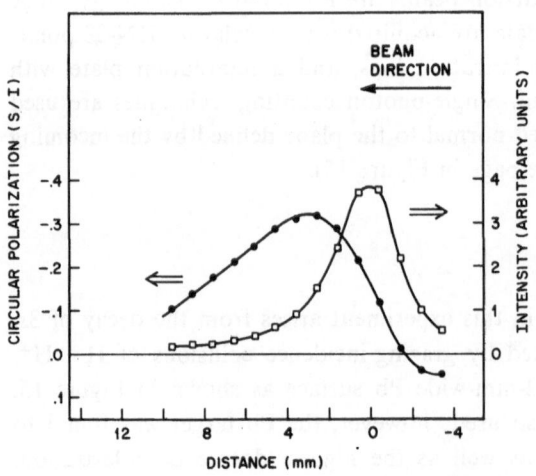

Figure 16. Radiation intensity (open boxes) and circular polarization S/I (solid circles) of H_α radiation (6563 Å) shown as a function of distance along the beam for 10 keV H_2^+ on Pb at a 6° grazing angle. The target region is 2 mm long in the beam direction centered at 0 mm. The acceptance width for radiation detection, determined by the slit width, is 2 mm. $\tau[H(3s)] = 1.6 \times 10^{-7}$ sec (15.4 mm); $\tau[H(3p)] = 5.4 \times 10^{-9}$ sec (5.2 mm); $\tau[H(3d)] = 1.56 \times 10^{-8}$ sec (15.0 mm).

polarization with time of $\sim 5 \times 10^{-9}$ sec. Roughly the same decay time was obtained for a variety of incident energies and with H^+ and H_3^+ as well, and it corresponds very closely to the 5.4-nsec lifetime of the $3p$ state of H. We deduce from this that while all three components, $3s$, $3p$, and $3d$, contribute in undetermined ratios to the total intensity, the principal, if not exclusive, contributor to circularly polarized radiation is the $3p$ state.

The energy dependence of circular polarization S/I in H_α from H^+, H_2^+, and H_3^+ grazing-incidence collisions on Pb is plotted in Figure 17a. Figure 17b portrays the same data as in Figure 17a but plotted as a function of energy per nucleon showing that the resulting $H^*(3p)$ "loses memory" as to whether the projectile began as part of a H^+, H_2^+, or H_3^+ complex and depends only on its initial velocity. Similar results were also obtained for H^- projectiles.

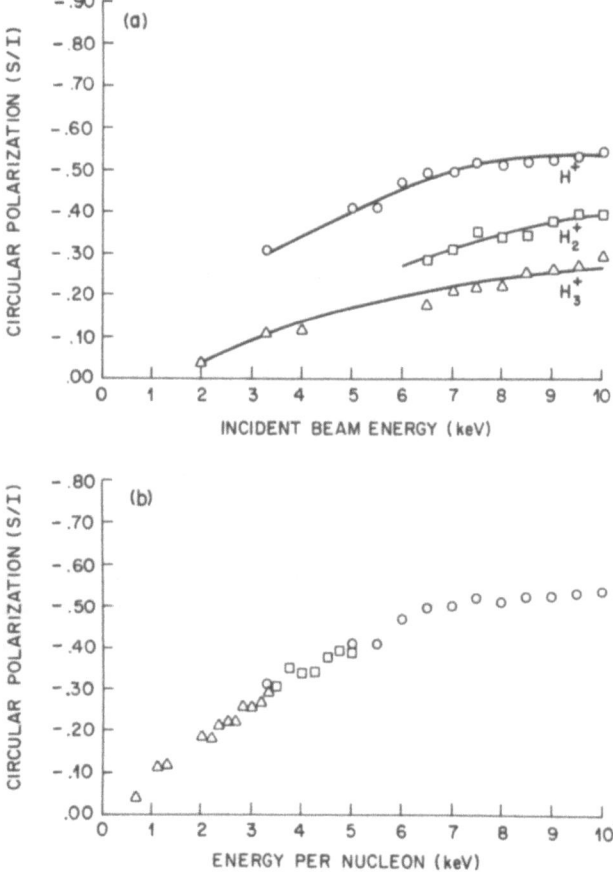

Figure 17. (a) Measurements of circular polarization S/I for beams of H^+ ($\bigcirc$), H_2^+ ($\square$), and H_3^+ ($\triangle$) incident at $6°$ grazing angle on lead as a function of beam energy. (b) Same data as in Figure 17a plotted as a function of energy per nucleon.

4.3. Discussion

We present a model to explain the characteristics of the observed polarization similar to that previously discussed which separates the mechanism into two parts: (a) the initial production of anisotropic states, and (b) their subsequent evolution due to a surface electrostatic interaction.[24] In order to identify the physical mechanism responsible for the production of anisotropic states [part (a)] and thereby responsible for the observed orientation, we consider an initially unpolarized ion or atom moving on a straight-line trajectory in the vicinity of a structureless "jellium" surface. This model, although simple, is sufficient to encompass the major electronic processes thought to be operative.[25] In this model, in order to obtain orientation effects, we must include explicit velocity-dependent interactions. We can do this by attaching a "translational factor" to the electronic wave function of the moving atom to account for the fact that its electrons are moving along with the nucleus.[26] For example, for a hydrogen atom,

$$\phi_{\mathrm{H}} \approx R_{nl}(\gamma) Y_l{}^m(\theta, \phi) \exp(-k_v y) \tag{18}$$

where nk_v is the y component of the atom velocity. If we approximate the electronic wave functions of the jellium by

$$\phi_s \simeq \exp(ik_y y + ik_z z - \xi x) \tag{19}$$

the first-order matrix element governing capture of a surface electron by the ion is of the form

$$\langle \phi_s \mid \tilde{V}(x) \mid \phi_{\mathrm{H}} \rangle \sim \langle \exp(ik_y y + ik_z z) \mid \tilde{V}(x) \mid R_{nl}(\gamma) Y_l{}^m(\theta, \phi)$$
$$\times \exp\{(2\pi/3)^{1/2}\gamma[(\xi + k_v)Y_1{}^1(\theta, \phi) + (\xi - k_v)Y_1{}^{-1}(\theta, \phi)]\}\rangle \tag{20}$$

Since ξ and k_v are positive constants (of roughly the same magnitude in our experiments), $m = -1$ states will be populated preferentially over $m = 1$ states, producing circular polarization, in agreement with our observation.[†] This can be arrived at by an intuitive classical argument: For an ion traversing the surface in the $+y$ direction (see Figure 15), the electron state corresponding to counterclockwise electron orbital motion (which when close to the surface most closely matches surface electron velocities or angular momentum states) will be preferentially populated.

To illustrate part (b), consider an atom prepared in a pure p ($m_l = -1$) state (pure orientation) at a distance x_0 from the surface with constant velocity $v_\perp$, experiencing only static interaction normal to the surface. In our viewing geometry

---

[†] From equation (20) the orientation effects arise because of presence of the spherical harmonics $Y_1{}^1$ and $Y_1{}^{-1}$ that appear within an exponential. If we were to expand this exponential, keeping only the term linear in the Y's, we would obtain an approximate selection rule that orientation would be produced only in $l = 1$, $m = \pm 1$ states, consistent with our experimental observation that the predominant polarization effects arise from the $3p$ state.

the normalized Stokes parameters would be given by

$$S/I = -\cos(\Delta/hv_\perp), \qquad M/I = 0, \qquad C/I = \sin(\Delta/hv_\perp) \qquad (21)$$

where

$$\Delta = \int_{x_0}^{\infty} [E_{p_x}(x) - E_{p_y}(x)] \, dx \qquad (22)$$

$v_\perp$ is the component of velocity perpendicular to the surface, and $E_{p_x}(x)$ and $E_{p_y}(x)$ are the energies of the p_x and p_y states at a distance x from the surface. Thus if the system begins in a pure orientation state, the major axis of elliptic polarization must be at $\pm 45°$ to the surface normal, consistent with our experimental observations. At velocities $v_\perp \ll h^{-1}\Delta$ polarization effects will be washed out. Since they are not washed out in our low-energy experiments [$v_\perp$ ranging from $(3.5$ to $14) \times 10^6$ cm/sec, corresponding to energies 6.5–110 eV assuming specular reflection] Δ must be small. S/I, as shown in Figure 17, increases with increasing energy, i.e., increasing $v_\perp$. In addition, M/I is nearly zero and C/I (not shown) decreases with increasing $v_\perp$ in qualitative agreement with equation (21). By fitting these results with equation (21), we obtain as a crude estimate $\Delta \simeq 0.1$ eV Å; i.e., one to two orders of magnitude smaller than typical values of this quantity obtained previously in Sections 2 and 3. This appears reasonable only if the initial production of polarization occurs at a distance x_0 that is significantly larger than the turning point, so that the integral in equation (22) is greatly reduced. Thus we propose the following picture: The production of orientation occurs as the particle recedes from the surface at distances of at least 2 or 3 Å. This is consistent with our observations that the polarization does not depend upon the initial charge or molecular identity of the projectile and that the sign of the circular polarization is determined by the direction of the outgoing particle as shown in Figure 16.

In summary, we may explain the observed elliptic polarization (with major axis at 45° to the surface normal) arising from grazing incidence collisions as follows. As the incident ion approaches the surface at a grazing angle and collides with one or a few atoms on the surface, it experiences a myriad of competing neutralization, excitation, deexcitation, and momentum-changing processes. If it is initially a molecule, it may be broken apart. Some anisotropically excited hydrogen atoms may be produced by these essentially ion–atom collisions,[27] but as the atoms leave the surface, deexcitation and ionization processes rapidly deplete the excited states.[25] The major contribution to this effect then comes from anisotropic electron pickup at large distances (almost entirely into the $m_l = -1$ angular momentum state). If this state were able to immediately decay the emitted light would be exclusively circularly polarized. However, the inferred presence of an electrostatic interaction immediately splits the $m_l = -1$ level into x and y components (along the surface normal and perpendicular to it). Since the x and y components of the wave function are equal in amplitude then only elliptic polarization with the major axis at 45° can result as differential phase evolves between the two components.

Even though the effect described in this experiment is a purely ion–surface phenomenon and thereby significantly different from the previous two collision systems described in this chapter, the model can still be described in terms of the coherent population of two levels which then evolves phase differentially as the collision proceeds in time.

5. Conclusion

Using a thematic approach, we have examined coherent phenomena arising in three different collision systems representing (a) strictly ion–atom collisions, (b) biparticle collisions on a surface, and (c) ion–surface grazing incidence collisions, a case which appears to be a purely ion–surface interaction. In each case we have been able to satisfactorily explain the basic physical process by assuming the coherent population of two or more states whose relative phase evolves with time.

A major finding in this work is that ion–surface collisions may have much in common with ion–atom collisions. It is particularly gratifying that a simple quantum mechanical formalism of established use in gas phase collisions may be readily applicable to ion–surface interactions. Of course there remain significant differences between gas phase and surface collision phenomena. From these differences however, we may learn about the surface itself and about its effect on the incident ion. Ion–surface collision experiments of these types constitute a new class of differential scattering studies which hold great promise for providing significant insight into the mechanisms responsible for heavy-particle inelastic collision phenomena.

ACKNOWLEDGMENT

The authors gratefully acknowledge helpful discussions with John Tully.

References

1. N. H. Tolk, C. W. White, S. H. Dworetsky, and L. A. Farrow, *Phys. Rev. Lett.* **25**, 1251 (1970); S. Dworetsky, R. Novick, W. W. Smith, and N. Tolk, *Phys. Rev. Lett.* **18**, 939 (1967); R. F. Stebbings, R. A. Young, C. L. Oxley, and E. Ehrhardt, *Phys. Rev.* **138**, A1312 (1965); M. Lipeles, R. Novick, and N. Tolk, *Phys. Rev. Lett.* **15**, 815 (1965); S. V. Bobashev and V. A. Karchenko, in *Electronic and Atomic Collisions, Proceedings of the Tenth International Conference on the Physics of Electronic and Atomic Collisions*, Paris, 21–27 July 1977, G. Watel, ed., p. 445, North Holland, Amsterdam (1978).
2. N. H. Tolk, C. W. White, S. H. Neff, and W. Lichten, *Phys. Rev. Lett.* **31**, 671 (1973); N. Tolk, J. C. Tully, C. W. White, J. Kraus, A. A. Monge, and S. H. Neff, *Phys. Rev. Lett.* **35**, 1175 (1975).
3. T. Andersen, A. Nielsen, and K. J. Olsen, *Phys. Rev. Lett.* **31**, 739 (1973); T. Andersen, A. K. Nielsen, and K. J. Olsen, *Phys. Rev. A* **10**, 2174 (1974).

4. N. H. Tolk, J. C. Tully, C. W. White, J. Kraus, A. A. Monge, D. L. Simms, M. F. Robbins, S. H. Neff, and W. Lichten, *Phys. Rev. A* **13**, 969 (1976).
5. R. L. Erickson and D. P. Smith, *Phys. Rev. Lett.* **34**, 297 (1975); T. W. Rusch and R. L. Erickson, in *Inelastic Ion–Surface Collisions*, N. H. Tolk, J. C. Tully, W. Heiland, and C. W. White, eds., p. 73, Academic Press, New York (1977).
6. N. H. Tolk, J. C. Tully, J. Kraus, C. W. White, and S. H. Neff, *Phys. Rev. Lett.* **36**, 747 (1976).
7. H. H. Brongersma and T. M. Buck, *Nucl. Instrum. Methods* **132**, 559 (1976).
8. J. C. Tully and N. H. Tolk, in *Inelastic Ion–Surface Collisions*, N. H. Tolk, J. C. Tully, W. Heiland, and C. W. White, eds., p. 105, Academic Press, New York (1977).
9. H. J. Andrä, *Phys. Lett.* **54A**, 315 (1975); H. J. Andrä, R. Fröhling, H. J. Plöhn, and J. D. Silver, *Phys. Rev. Lett.* **37**, 1212 (1976).
10. H. J. Andrä, R. Fröhling, and H. J. Plöhn, in *Inelastic Ion-Surface Collisions*, N. H. Tolk, J. C. Tully, W. Heiland, and C. W. White, eds., p. 329, Academic Press, New York (1977).
11. N. H. Tolk, J. C. Tully, J. S. Kraus, W. Heiland, and S. H. Neff, *Phys. Rev. Lett.* **41**, 643 (1978).
12. W. R. Bennett and P. J. Kindlmann, *Phys. Rev.* **149**, 38 (1966).
13. R. M. Schechtman, D. R. Shoffstall, D. G. Ellis, and D. A. Shojnacki, *J. Opt. Soc. Am.* **56**, 1585 (1966).
14. Z. Z. Latypov and A. A. Shopenko, *Zh. Eksp. Teor. Fiz. Piśma Red.* **12**, 177 (1970) [English transl.: *JETP Lett.* **12**, 123 (1970)]; S. V. Bobashev, *Zh. Eksp. Teor. Fiz. Piśma Red.* **11**, 389 (1970) [English transl.: *JETP Lett.* **11**, 260 (1970)].
15. R. Odom, J. Caddick, and J. Weiner, *Phys. Rev. A* **15**, 1414 (1977).
16. V. A. Ankudinov, S. V. Bobashev, and V. I. Perel', *Zh. Eksp. Teor. Fiz.* **60**, 906 (1971) [English transl.: *Sov. Phys.-JETP* **33**, 490 (1971)].
17. H. Rosenthal and H. M. Foley, *Phys. Rev. Lett.* **23**, 1480 (1969); H. Rosenthal, *Phys. Rev. A* **4**, 1030 (1971).
18. W. Lichten, *Phys. Rev.* **139**, A27 (1965).
19. G. Herzberg, *Molecular Spectra and Molecular Structure I. Spectra of Diatomic Molecules* 2nd Ed., p. 318, Van Nostrand, Princeton, New Jersey (1950).
20. E. V. Condon and G. H. Shortley, *The Theory of Atomic Spectra*, pp. 45–78, Cambridge University Press, New York (1959).
21. E. C. G. Stueckleberg, *Helv. Phys. Acta* **5**, 370 (1932).
22. H. G. Berry, G. Gabrielse, A. E. Livingston, R. M. Schectman, and J. Desesquelles, *Phys. Rev. Lett.* **38**, 1473 (1977).
23. N. Born and E. Wolf, *Principles of Optics*, 4th ed., p. 30, Pergamon Press, Oxford (1970).
24. Y. Band, *Phys. Rev. A* **13**, 2061 (1976), and references therein.
25. H. D. Hagstrum, in *Inelastic Ion-Surface Collisions*, N. H. Tolk, J. C. Tully, W. Herland, and C. W. White, eds., p. 1, Academic Press, New York (1977).
26. D. R. Bates and R. McCarrol, *Proc. R. Soc. London Ser. A* **245**, 175 (1958).
27. F. J. Eriksen, D. H. Jaecks, W. deRijk, and J. Macek, *Phys. Rev. A* **14**, 119 (1976).

44

Polarized Electron–Lithium Scattering

W. Raith, G. Baum, D. Caldwell, and E. Kisker

An experiment is discussed that will measure spin-dependent asymmetries in the scattering of polarized electrons on polarized Li atoms. Results are presented for polarizing the atomic beam by the method of optical pumping and for techniques of polarization reversal in both beams which minimize systematic errors.

1. Introduction

This is a status report on an experimental program to study spin effects in scattering of polarized electrons by polarized one-electron atoms. The e–Li scattering is of particular interest because lithium is the simplest alkali-metal atom; in principle, lithium can exhibit all the effects of more complex atoms. The theoretical treatment is therefore a greater challenge than that of hydrogen but is not yet as involved as that of the heavier alkali metals. Relativistic spin–orbit coupling effects are small for lithium and do not complicate the analysis, but short-range correlations (e.g., virtual excitation of core electrons) might contribute to the e–Li scattering. Polarized particle scattering is expected to provide a very sensitive method for studying the important details of the scattering process and to provide a stringent test of the different scattering theories.

Different experiments on spin effects in electron scattering from one-electron

W. Raith, G. Baum, D. Caldwell, and E. Kisker ● Fakultät für Physik, Universität Bielefeld, D-4800 Bielefeld, Federal Republic of Germany.   D. Caldwell's present address: Physics Department, Yale University, New Haven, Connecticut 06520, U.S.A.   E. Kisker's present address: Institut für Festkörperforschung der Kernforschungsanlage Jülich, 5170 Jülich, Federal Republic of Germany.

atoms are being pursued at other laboratories: at Yale University *e*–H scattering[1] is being studied by using a Fano-effect polarized electron source;[2] at the University of Stirling, Scotland, *e*–K scattering[3] is being studied, and the polarized electron source is based on low-energy Mott scattering. In both experiments the atomic beam is polarized by state selection in a permanent six-pole magnet. We employ optical pumping for polarizing the lithium atomic beam, and the polarized electrons are produced by field emission from ferromagnetic EuS on tungsten.[4,5]

2. Theoretical Expectations

The experimental asymmetry Δ, measured with both incoming particles polarized, is defined as

$$\Delta = (C^{\uparrow\downarrow} - C^{\uparrow\uparrow})/(C^{\uparrow\downarrow} + C^{\uparrow\uparrow}) \tag{1}$$

where $C^{\uparrow\downarrow}$ and $C^{\uparrow\uparrow}$ are the count rates obtained with antiparallel and parallel particle polarizations, respectively. The theoretically interesting asymmetry of the differential cross sections $\sigma^{\uparrow\downarrow}$ and $\sigma^{\uparrow\uparrow}$,

$$A = (\sigma^{\uparrow\downarrow} - \sigma^{\uparrow\uparrow})/(\sigma^{\uparrow\downarrow} + \sigma^{\uparrow\uparrow}) \tag{2}$$

is connected to Δ by

$$\Delta = P_e P_{\text{Li}} A \tag{3}$$

where P_e is the electron-beam polarization and P_{Li} is the polarization of the lithium valence electrons.

Our goal is to measure with good energy resolution the asymmetry in elastic scattering and to extend these studies to very low electron energies (1 eV and below). At first, however, we will investigate electron impact ionization, as the groups at Yale and Stirling have done for the *e*–H and *e*–K scattering. The impact ionization is technically easier because the ion can be detected with minimal background problems. Since the Yale results[1] showed remarkable deviations from theoretical predictions, there is great interest in such measurements with other atoms. Also of great interest are experiments on electron impact excitation of resonance transitions (e.g., Li $2s \rightarrow 2p$) with polarized particles.[6]

The motivation for studying elastic *e*–Li scattering is discussed in a recent paper of Temkin and co-workers.[7] It would be very interesting indeed to compare their modified polarized orbital (MPO) calculations with experiment, particularly under conditions for which the MPO results differ significantly from those of close-coupling (CC) calculations.[8] For the differential cross section in the case of unpolarized particles, σ_0, the two theoretical methods yield almost identical results. In Figure 1 the values of σ_0 are plotted versus the cosine of the scattering angle θ for an electron energy of 0.1 Ry = 1.36 eV. The difference between the MPO and CC results is perhaps barely measurable in the vicinity of $\theta \sim 120°$ and negligible elsewhere.

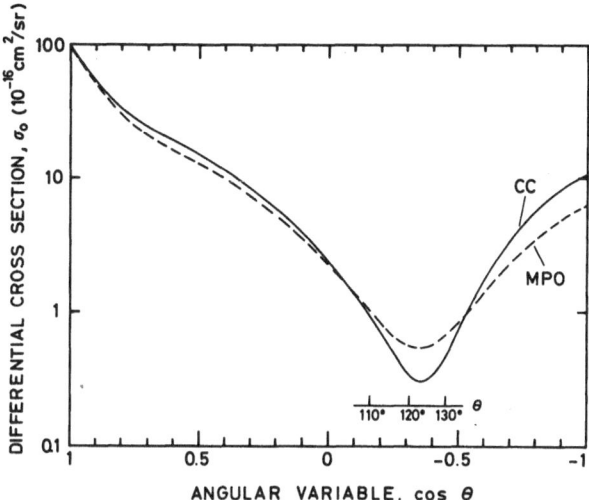

Figure 1. Differential elastic *e*–Li scattering cross sections for an electron energy of 1.36 eV, obtained from close-coupling results (CC)[8] and modified polarized-orbital calculations (MPO).[7]

However, for the asymmetry A the two methods lead to markedly different results, as shown in Figure 2. Thus elastic scattering with polarized particles should help significantly to clarify the problem.

The first measurements shall be made with a fixed scattering angle of $\theta = 120°$ and variable incident electron energy. The electron polarization will be monitored with the Mott detector. As a supplementary measurement we intend to scatter un-polarized electrons (also for $\theta = 120°$) from polarized Li atoms and measure the

Figure 2. Asymmetry in elastic *e*–Li scattering for polarized incoming particles of 1.36 eV electron energy calculated from close-coupling (CC)[8] and modified polarized-orbital (MPO)[7] theory. Note that positivity of equation (5) requires that $A \geq -\frac{1}{3}$.

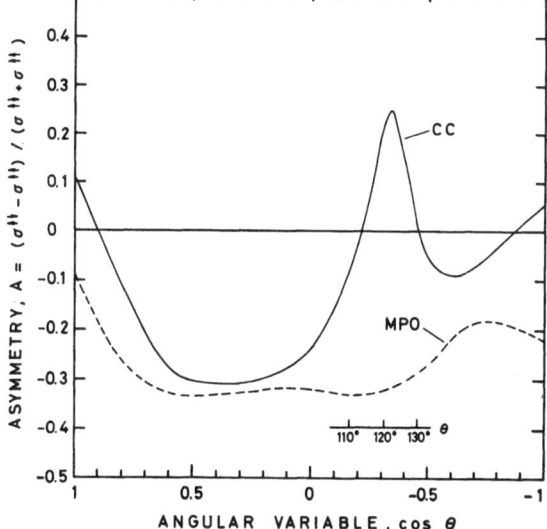

polarization of the scattered electrons

$$P_e' = BP_{Li} \tag{4}$$

where B is another independent parameter which can be used, together with our measurements of A and other measurements of σ_0, to make a very thorough comparison with theoretical parameters. It is customary to describe the elastic scattering in terms of two complex amplitudes, either s and t, the singlet and triplet scattering amplitudes or, $f = (s + t)/2$ and $g = (s - t)/2$, the direct and exchange amplitudes. The amplitudes s and t are related to σ_0, A, and B by

$$|s|^2 = \sigma_0(1 + 3A) \tag{5}$$

$$|t|^2 = \sigma_0(1 - A) \tag{6}$$

$$|s| \cdot |t| \cos \alpha = \sigma_0(1 + 2B - A) \tag{7}$$

where α is the phase angle between s and t.

3. Polarized Electron Source

The polarized electron source provides a highly polarized, monoenergetic electron beam of extremely high brightness. Relevant technical data are listed in Table 1. The most recent work on the source has been concerned with electron polarization reversal. The direction of the electron polarization inside the EuS layer is the same

Table 1. Characteristics of Polarized Electron Source

Emitter—ferromagnetic EuS on tungsten tip	
Temperature	$\sim$10 K
Vacuum	$<10^{-10}$ Torr
Magnetic field	$\sim$50 G longitudinal
Electron beam	
Current, I	$\sim$10^{-8} A
Polarization, P	$\sim$0.85, transverse
Energy width, ΔE	<0.1 eV
Emittance, ε	0.8×10^{-6} rad cm
At energy, E_0	3 eV
Figures of merit	
$M_1 = IP^2$	7×10^{-9} A
$M_2 = M_1 \varepsilon^{-2} E_0^{-1}$	4×10^8 A rad^{-2} cm^{-2} eV^{-1}
$M_3 = M_2 \Delta E^{-1}$	4×10^4 A rad^{-2} cm^{-2} eV^{-2}

as the direction of magnetization of the emitting region. In operating the source with only a small longitudinal magnetic field applied at the tip, the magnetization is still tangential to the EuS layer, leading to a transverse polarization of the extracted field-emission beam. The azimuthal orientation of the polarization vector depends on the direction of magnetization in the emitting EuS layer. (Larmor precession in the applied longitudinal magnetic field is of no concern since it can be controlled by choosing the operating conditions.) We found that the EuS has an axis of easiest magnetization (and sometimes two axes) and that in cooling the tip below the Curie temperature the magnetization vector will orient itself parallel to that axis. By simply raising the temperature of the tip temporarily above the Curie point, and letting it drop again, the polarization will fall back to the former direction or reverse itself, apparently in a random fashion for the majority of our samples. This provides a convenient method for polarization reversal. Most important, this reversal is free of any possible systematic effects connected with changes in the electron optical settings of the source as these are unaltered during the reversal process. A few of our samples showed a preference of the magnetization vector for one of the two directions associated with the axis of easy magnetization. Here a reversal—again without any electron optical changes—can be achieved by applying a transverse magnetic field (2 kG) in the proper direction while the tip temperature is raised temporarily above the Curie point.

To avoid systematic effects in our asymmetry measurements which might be connected with drifts of the experimental parameters, the polarization directions of the incoming particles with respect to each other (antiparallel or parallel) will be reversed frequently. The atomic polarization, which can be switched with ease by simply reversing the circular polarization of the pumping light, will be altered in short intervals, the electron polarization in longer intervals.

4. Polarized Atomic Beam

The simplest method for polarizing an alkali atomic beam is passage through a six-pole magnet which accomplishes a high-field Zeeman-state separation such that atoms with $M_s = +\frac{1}{2}$ are transmitted and those with $M_s = -\frac{1}{2}$ rejected. This method, however, has a great disadvantage as the six-pole magnet cannot be reversed to transmit atoms with $M_s = -\frac{1}{2}$. In order to reverse the atomic polarization with respect to a given external magnetic field one would have to employ more elaborate schemes such as a fast adiabatic passage. Another drawback of the six-pole magnet is the fact that the atoms state-selected in a strong field reduce their electronic spin polarization when they enter a weak magnetic field in which the hfs coupling is reestablished. The achievable polarization depends on the nuclear spin I and is given by

$$P_{\text{Li}} = (1 + 2I)^{-1} \tag{8}$$

For lithium-6 with $I = 1$ it follows that $P_{\text{Li}} \leq \frac{1}{3}$. The reduced polarization consider-

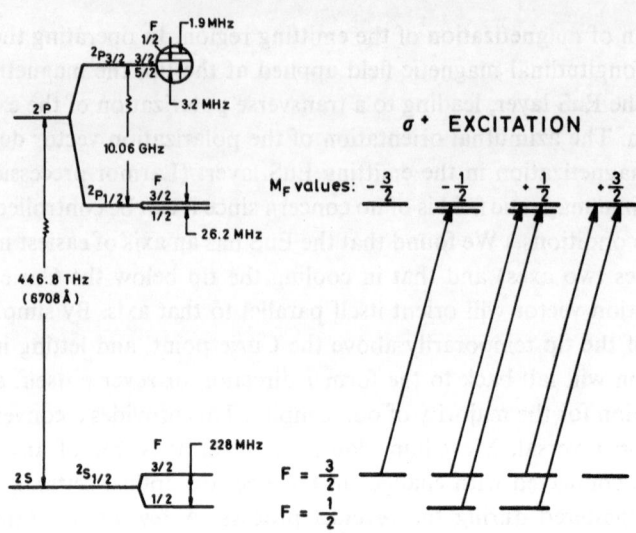

Figure 3. Energy-level diagram of lithium-6 and illustration of the relevant σ^+ excitations in the optical-pumping process which help to transfer the atoms into the ground-state sublevel ($F = \frac{3}{2}$, $M_F = +\frac{3}{2}$). Since the hfs components of the $P_{1/2}$ state are not resolved, excitations to the $F = \frac{1}{2}$ sublevels (not shown in figure) also participate in the pumping process.

ably decreases the experimental sensitivity. Both disadvantages can be avoided by employing the optical pumping method; it can, in principle, yield an easily reversible spin polarization close to unity.

The essence of the optical pumping method is to use the M dependence of the absorption and emission of circularly polarized resonance light to create an orientation in the ground state in which all spins are either parallel or antiparallel to the direction of the incoming radiation. In the case of lithium, the process is complicated by the hyperfine-structure interaction. Nevertheless, the pumping can still be accomplished provided a single-isotope beam is used and pumping transitions from both hyperfine levels can be equally induced.[10] Figure 3 depicts the energy levels of the ground state and first excited state of lithium-6, as well as the absorption processes which produce the orientation. By using a single-mode dye laser, bandwidth ~ 50 MHz, for exciting the lithium D_1 ($2S_{1/2} \rightarrow 2P_{1/2}$) transition we resolve the hyperfine structure of the ground state. We do not resolve the hyperfine splitting of the $2P_{1/2}$ state, but this is not of any disadvantage for the pumping process.

The lithium atomic beam system is similar to that described previously;[9] its relevant characteristics are listed in Table 2. The laser is tuned to the desired transition and maintained there using a microprocessor-controlled stabilization circuit similar to a scheme described by Düren and Tischer.[11] In order to be able to pump both hyperfine levels of the ground state, we split off a part of the laser beam by diffraction from an acousto-optical modulator operating at 228 MHz, thus creating two light beams with exactly the frequency difference corresponding to the hfs separation. In general, the light beam corresponding to the zeroth-order diffraction max-

Table 2. Characteristics of Polarized Atomic Beam System

Li oven

Operating temperature	900°C
Li content	200 g
Running time	60 h
Heating power	1 kW
Orifice diameter	1.5 mm

Laser (Spectra-Physics Model 375 dye laser pumped by Model 164 argon ion laser)

Light power	~20 mW cw
Frequency stabilization	5 MHz
Intensity ratio of light beams corresponding to zeroth and first-order diffraction from the acousto-optical modulator	4 : 1

Atomic beam at scattering region

Cross section	7 mm²
Intensity	>10¹⁴ atoms/sec
Mean atom velocity	2×10⁵ cm/sec

imum is tuned to pumping of the $F = \frac{1}{2}$ level, and the laser is stabilized there; the light beam diffracted in first order by the acousto-optical modulator excites the $F = \frac{3}{2}$ component. This allows us to vary the relative power in each beam by simply varying the power delivered to the crystal. After being split, the two light beams are reflected by mirrors such that they are converging, then they are expanded cylindrically, passed through the quarter-wave plate, and brought to a common focus along the lithium beam (see Figure 4). We adjust the crystal so that both scattered

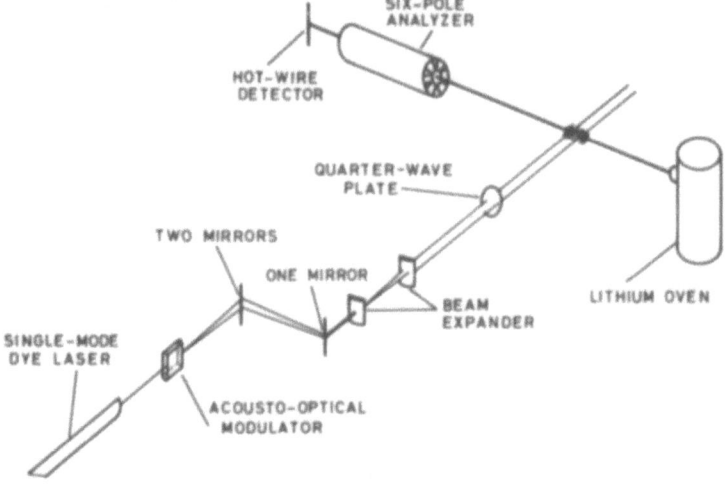

Figure 4. Schematic drawing of the laser light setup used for optical pumping.

and unscattered light beams lie in a plane perpendicular to the lithium beam, thus ensuring that any Doppler shift in the absorption will be in the same direction for both.

The lithium polarization is analyzed by transmitting the beam through a six-pole magnet following the pumping zone. At the exit of the six-pole magnet, the beam intensity is monitored with an oxidized-tungsten hot-wire detector. By using the intensity I_0 of an unpolarized beam as a reference signal, one can determine the polarization achieved by pumping alternatingly with σ^+ and σ^- light and measuring the intensities I^+ and I^-, respectively. Ideally, one would get for complete pumping $I^+ = 2I_0$ and $I^- = 0$. For incomplete pumping the polarization can be estimated according to

$$P_{\min} \leq P \leq P_{\max} \tag{9}$$

with

$$P_{\min} = (I^+ - 3I^- - 2I_0)/6I_0 \tag{10}$$

and

$$P_{\max} = (3I^+ - 2I^-)/6I_0 \tag{11}$$

The degree of polarization that we will ultimately obtain with this method is not yet precisely predictable as there are still a number of factors that have to be optimized. The polarization appears to depend sensitively on parameters such as relative intensity of the two laser beams and amount of laser beam expansion over

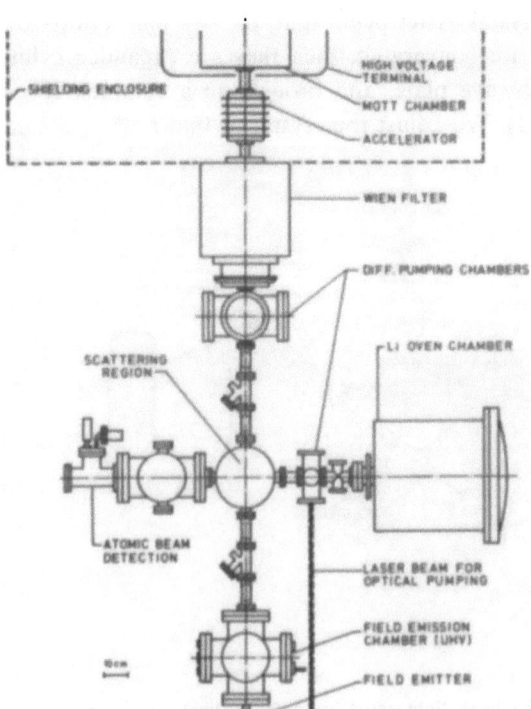

Figure 5. Layout of the apparatus of the e–Li scattering experiment.

the lithium beam, as well as on more common parameters such as laser intensity, retardance of the quarter-wave plate, etc. At this stage, the degree of polarization we obtain lies between $P_{min} = 0.57$ and $P_{max} = 0.69$ for a lithium beam intensity of $\approx 4 \times 10^{13}$ atoms per second.

The layout of the e–Li scattering experiment is shown in Figure 5. The scattering chamber is now being assembled. The crossed beams lie in the horizontal plane. The spin polarization of the ^{6}Li atoms is preserved by a weak magnetic guiding field oriented parallel to the atomic beam. At the scattering region the electron polarization is transverse to the electron beam direction; it lies in the horizontal plane parallel to atom polarization and atomic beam. The electron beam is slightly curved owing to the weak transverse magnetic field of about 20 mG. In the impact ionization experiment the Li$^+$ ions produced will be drawn out and accelerated toward the detector. In the differential scattering experiment the detector will be positioned at various scattering angles. The scattering plane lies in the vertical plane, orthogonal to the atomic beam.

ACKNOWLEDGMENTS

The authors acknowledge the valuable help of W. Schröder and the assistance given by the technical staff of the Bielefeld Physics Department. This research is being supported by the University of Bielefeld under Project No. 2854.

References

1. M. J. Alguard, V. W. Hughes, M. S. Lubell, and P. F. Wainwright, *Phys. Rev. Lett.* **39**, 334–338 (1977).
2. P. F. Wainwright, M. J. Alguard, G. Baum, and M. S. Lubell, *Rev. Sci. Instrum.* **49**, 571–585 (1978).
3. D. Hils and H. Kleinpoppen, *J. Phys. B* **11**, L283–L287 (1978).
4. G. Baum, E. Kisker, A. H. Mahan, W. Raith, and B. Reihl, *Appl. Phys.* **14**, 149–153 (1977).
5. E. Kisker, G. Baum, A. H. Mahan, W. Raith, and B. Reihl, *Phys. Rev. B* **18**, 2256–2275 (1978).
6. H. Kleinpoppen, *Advances in Quantum Chemistry* **10**, 77–141 (1977).
7. A. K. Bhatia, A. Temkin, A. Silver, and E. C. Sullivan, *Phys. Rev. A* **18**, 1935–1948 (1978).
8. P. G. Burke and A. J. Taylor, *J. Phys. B* **2**, 869–877 (1969).
9. M. J. Alguard, J. E. Clendenin, R. D. Ehrlich, V. W. Hughes, J. S. Ladish, M. S. Lubell, K. P. Schüler, G. Baum, W. Raith, R. H. Miller, and W. Lysenko, *Nucl. Instrum. Methods* **163**, 29–59 (1979).
10. G. Baum, C. D. Caldwell, and W. Schröder, *Appl. Phys.* (in press).
11. R. Düren and H. Tischer, Max-Planck-Institüt für Strömungsforschung, Göttingen, Bericht 4/1978.

45

Polarization Effects Caused by Interference in Electron Scattering and Photoionization

JOACHIM KESSLER

Experiments of the third generation in electron scattering and in photoionization are discussed. They aim at a complete determination of the parameters (moduli and phases) by which theory describes these processes. First an electron triple-scattering experiment is described where the change of electron spin polarization due to scattering is observed. Second, an experimental result is given on spin polarization of photoelectrons produced from an unpolarized target by unpolarized light. Finally it is shown that the statement "number of necessary experiments equals number of independent parameters" is misleading.

1. Introduction

Most of the experimental studies discussed in this volume belong to a new generation of experiments, which implies that they often need a rather long time for development. It is therefore well that this volume include a few chapters about experiments that are being done but not yet completed. This may stimulate simultaneous theoretical work and give us the opportunity to hear the advice of colleagues.

Let me first discuss electron scattering.

JOACHIM KESSLER • Physikalisches Institut, Universität Münster, 4400 Münster, Schlossplatz 7, Germany.

2. Electron Scattering

Here the experiments of the first generation are cross-section measurements. They are completely satisfactory only in the simplest case where a single complex amplitude $f = |f| e^{i\gamma}$ suffices for describing the scattered wave. Since the phase factor, according to the principles of quantum mechanics, cannot be observed, a measurement of the differential cross section $d\sigma/d\Omega = |f|^2$ yields all the information that is obtainable.

This situation changes as soon as more than one amplitude is needed to describe the scattering process. If, e.g., spin–orbit interaction or exchange scattering are to be considered, one needs at least two scattering amplitudes

$$f = |f| e^{i\gamma_1}, \qquad g = |g| e^{i\gamma_2} \tag{1}$$

where g is related to the behavior of the electron spins during scattering. In order to dig out all the information hidden in these amplitudes one needs to observe several quantities, not just a cross section.

For the particular case of elastic electron–atom scattering including spin–orbit interaction, a measurement of

$$\frac{d\sigma}{d\Omega} = |f|^2 + |g|^2 \tag{2}$$

yields just $|f|^2 + |g|^2$. Additional independent information is obtained by measuring the spin polarization after scattering of an initially unpolarized electron beam. It is given by

$$P = i \frac{fg^* - f^*g}{|f|^2 + |g|^2} = -2 \frac{|f||g|\sin(\gamma_1 - \gamma_2)}{|f|^2 + |g|^2} \tag{3}$$

This interference term of the two amplitudes makes their relative phase $(\gamma_1 - \gamma_2)$ accessible.

It is obvious that this still does not give all the information hidden in the amplitudes. In order to obtain all the information that can be obtained, one has to take a polarized beam, scatter it, and observe how its polarization changes during scattering. Depending on which polarization component one observes, one measures

$$U = i \frac{fg^* + f^*g}{|f|^2 + |g|^2} = 2 \frac{|f||g|\cos(\gamma_1 - \gamma_2)}{|f|^2 + |g|^2} \tag{4}$$

or

$$T = \frac{|f|^2 - |g|^2}{|f|^2 + |g|^2} \tag{5}$$

Having made these measurements one can determine the three parameters $|f|$

and $|g|$ [from (2) and (5)] and $(\gamma_1 - \gamma_2)$ [from (3) and (4)]. That is all the information that is obtainable since an absolute determination of the phases is impossible. The four measurements discussed are not independent of each other since from (3)–(5) it follows that $P^2 + U^2 + T^2 = 1$. In spite of this one needs all of the four measurements even though only three parameters have to be determined, since otherwise there would remain an ambiguity: a sign would be left undetermined. I will come back to this point at the end of this chapter.

The present knowledge of the four quantities (2)–(5) is quite unbalanced. Cross sections have been measured in single-scattering experiments for more than 60 years; they are well known in comparison to the other quantities.

Measurements of the polarization P belong to the next generation of experiments. Sir Harrie Massey, whose 70th birthday gives the occasion for this volume, pointed out that electron spin plays an important role also at scattering energies typical for atomic physics. Much theoretical and experimental work has been done since the pioneering paper by Massey and Mohr.[1] Despite severe discrepancies between theory and experiment in the beginning, we can now say that the spin polarization caused by spin–orbit interaction in elastic potential scattering is very well understood for energies between 100 eV and a few MeV. Only in rare cases were there persistent discrepancies that were hard to get rid of. One of those rare exceptions is the polarization peak near 110° for 150-eV electrons scattered by xenon (Figure 1). Theoreticians kept predicting a high positive peak whereas we could not help finding a small negative peak.[2] It was only this year that more satisfactory theoretical data could be computed by using McCarthy's optical potential.[4]

Discrepancies between theory and experiment do, however, exist outside the energy range mentioned. When one goes below 100 eV the discrepancies increase as the electron energy decreases, so that the task for second-generation experiments is by no means finished, just as it is not finished for the first-generation experiments.

Let us now have a look at the present knowledge of the quantities U and T. They have to be determined in triple-scattering experiments as will be seen below. This is a new generation of experiments which, in atomic physics, are in their initial stage. Such experiments are certainly difficult, but they are feasible, as has been shown by many analogous experiments that nuclear physicists have been doing for quite some time and as has also been shown by one experiment with electrons in the 200-keV range by van Duinen et al.[5]

At the energies atomic physicists usually use to study electron scattering no such experiment has been made. (I am not discussing exchange scattering, where we did make a triple-scattering experiment.) All one has are theoretical predictions like that shown in Figure 2 for T, which is the factor describing the change due to scattering of the polarization component $\mathbf{P}_p$ parallel to the scattering plane. According to Figure 2, which is valid for a mercury target, there are certain combinations of energy and scattering angle where there should be a strong reduction or even a change of sign of $\mathbf{P}_p$. Not a single point of this theoretical map has been checked by experiment.

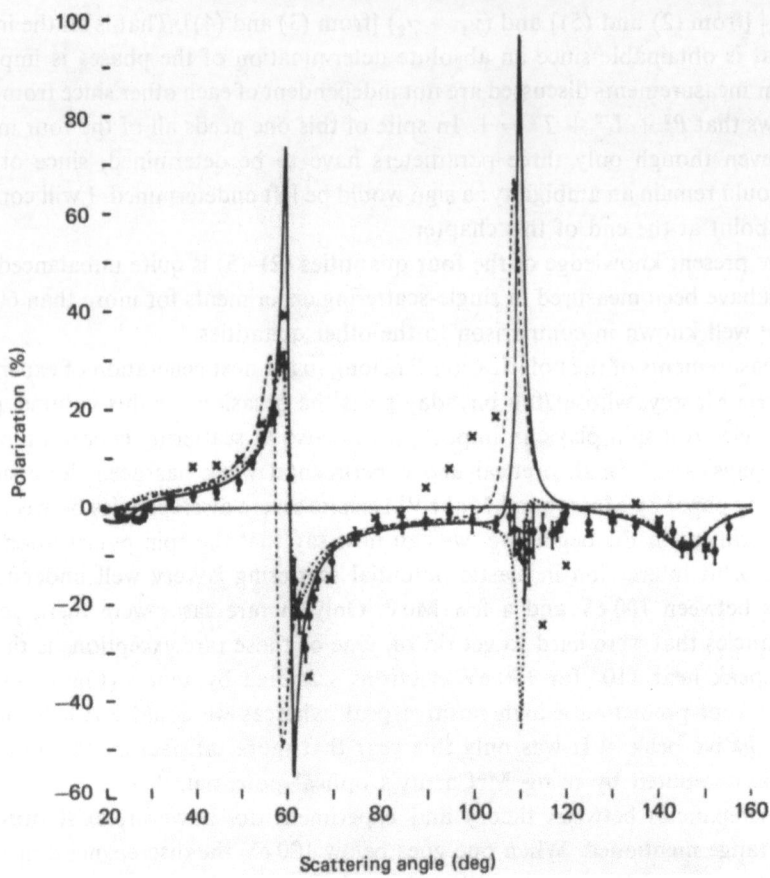

Figure 1. Experimental and theoretical results for polarization of 150-eV electrons scattered elastically by xenon atoms. (Further details in References 2–4.)

We have therefore started such a project: Dr. Kollath and Mr. Wübker have constructed an apparatus for a triple-scattering experiment that looks like the one shown in Figure 3. The first scattering produces a polarization perpendicular to the first and parallel to the second scattering plane. We may replace this part of the apparatus by a Fano-effect source, but I will discuss it as it is shown here. The second scattering process, also on a mercury target, changes the polarization vector. In order to find its transverse component parallel to the second scattering plane we measure the left–right asymmetry by the two counters of the Mott detector which are perpendicular to the second scattering plane. This measurement gives a combination of T and U. Then the longitudinal component of the polarization is rotated by a Wien filter through 90° and measured by the two other counters at large angle in the Mott detector. This gives another combination of T and U, so that these quantities can now be evaluated separately. These measurements will be made at electron energies of a few hundred eV and will hopefully yield

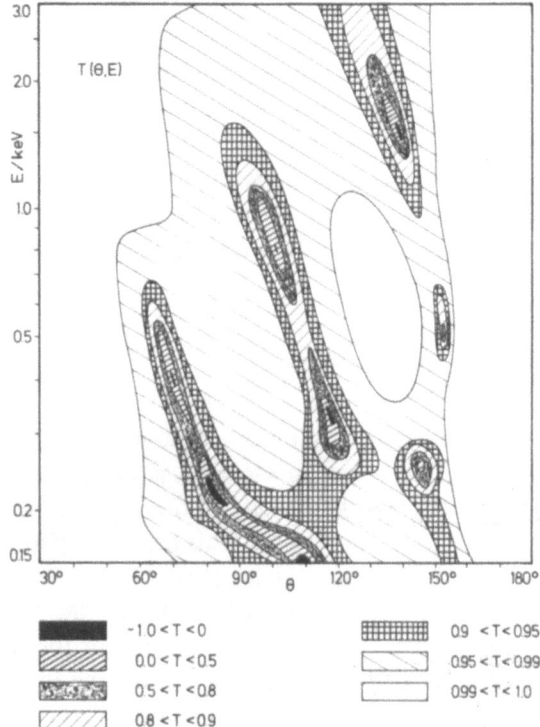

Figure 2. Contours $T(\theta, E) = \text{const}$ for mercury. (The polarization component parallel to the scattering plane is reduced by the factor T due to scattering. Further details in References 6.)

■	$-1.0 < T < 0$	▦	$0.9 < T < 0.95$
▨	$0.0 < T < 0.5$	▱	$0.95 < T < 0.99$
▨	$0.5 < T < 0.8$	□	$0.99 < T < 1.0$
▨	$0.8 < T < 0.9$		

some experimentally verified points in the unexplored landscapes of which I have shown an example.

A new generation of experiments frequently entails a new generation of pitfalls. This was the situation when passing from cross-section measurements to polarization measurements, and it was the reason for the unreliability of most of the data taken

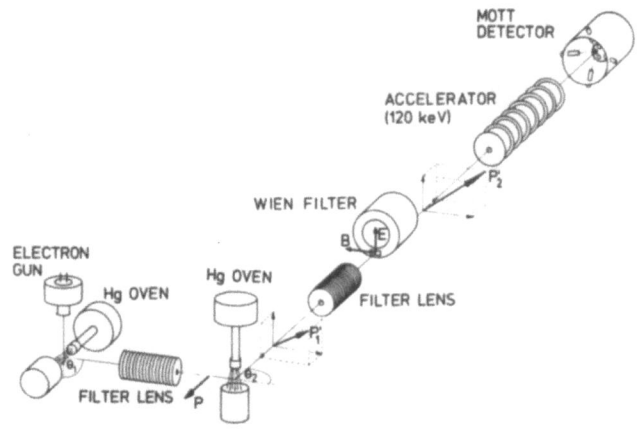

Figure 3. Apparatus for triple-scattering experiment.

in the first 15 years of such studies (in the 1940s and 1950s). We hope to have learned, over the years, enough about electron scattering to perhaps avoid a similar situation in the third-generation experiments.

So far I have discussed a process that is described by two scattering amplitudes. I am not going to discuss one of the many scattering processes such as inelastic exchange scattering that need more than two amplitudes for their description, even though we are preparing such experiments (see Chapter 47 by Dr. Hanne). Let me, instead, switch over to interference effects in photoionization.

3. Photoionization

When you photoionize a certain atomic state you can reach various continuum states. As illustrated in Figure 4, photoionization of a state with quantum numbers $l, j = l + \frac{1}{2}$, leads, because of the selection rules $\Delta l = \pm 1$, $\Delta j = 0, \pm 1$, to three different continuum states. This aspect of photoionization is best known because it is derived from the angular parts of the wave functions, the well-known spherical harmonics. What is not well known are the radial parts of the wave functions and, following from them, the radial parts of the transition matrix elements, which we have denoted by $R_{-1/2}$, $R_{1/2}$, $R_{3/2}$ for the three transitions of Figure 4. The phase shifts of the continuum wave functions describing the photoelectrons are denoted by $\delta_{-1/2}$, $\delta_{1/2}$, $\delta_{3/2}$.

In order to determine all these parameters one has to do quite a few independent experiments. First theoretical studies of how the parameters are related to observable quantities have been made by Cherepkov[7] in Leningrad and by Lee[8] in Fano's group. The quantity where interference among the various continuum waves does not play a role is the integral photoionization cross section Q, which only depends on the squares of the matrix elements. Its measurement has been started in a first

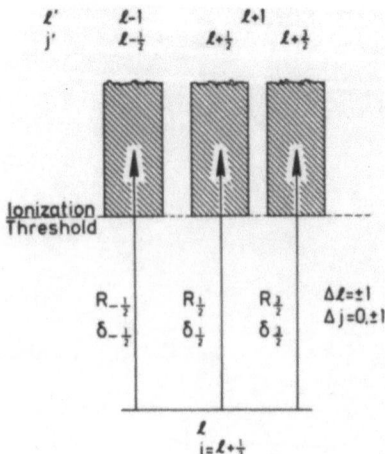

Figure 4. Schematic diagram of a photoionization process.

Table 1. Observable Quantities in Photoionization Experiments

$$
\left.
\begin{array}{c}
Q(R_i{}^2) \\
\beta \\
\bar{P} \\
\gamma_1 \\
\gamma_2
\end{array}
\right\}
\text{ functions of } R_i{}^2,\ R_i R_k \genfrac{}{}{0pt}{}{\sin}{\cos}[\delta_i - \delta_k]
$$

generation of experiments not long after the detection of the photo effect. The experiments of the second generation do concern quantities that are affected by interference among the various continuum waves. They are being made in several laboratories and deal either with the angular distribution of the photoelectrons or with the spin polarization, averaged over all angles, of the photoelectrons that are produced by circularly polarized light. If one measures the angular distribution

$$
\frac{d\sigma}{d\Omega} = \frac{Q}{4\pi}\left[1 - \frac{\beta}{2}P_2(\cos\theta)\right] \tag{6}
$$

where Q is the integrated cross section and P_2 a Legendre polynomial, one obtains the quantity β, which contains interference terms as indicated in Table 1. If one measures the average polarization $\bar{P}$ of the photoelectrons produced by circularly polarized light, one observes another combination of interference terms. The three measurements discussed so far do not suffice to completely unravel the parameters R_i and δ_i. This needs an even higher level of sophistication and brings us to the third generation of experiments. Instead of observing either the angular distribution or the spin polarization of the photoelectrons one has to combine these measurements by observing the angular dependence of the spin polarization. If again circularly polarized light is used for photoionization one has for the photoelectron polarization along the direction of the photon spins (Figure 5)

$$
P_{\parallel}(\theta) = \frac{\bar{P} - \gamma_1 P_2(\cos\theta)}{1 - \tfrac{1}{2}\beta P_2(\cos\theta)} \tag{7}
$$

Measurement of the angular dependence of this polarization component gives γ_1, which again is a function of the matrix elements and the phases. In order to find the last quantity of Table 1 one can use an unpolarized light beam for photoionization

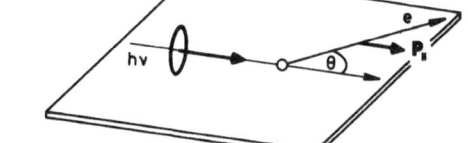

Figure 5. Photoionization experiment as described by equation (7).

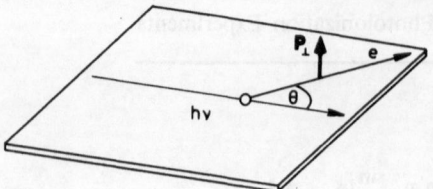

Figure 6. Photoionization experiment as described
by equation (8).

and observe the electron polarization perpendicular to the reaction plane (Figure 6),

$$P_\perp(\theta) = \frac{-\gamma_2 \sin \theta \cos \theta}{1 - \frac{1}{2}\beta P_2(\cos \theta)} \tag{8}$$

This measurement yields γ_2, which also contains interference terms of the parameters R and δ.

There is no reason to be surprised that polarized electrons should be obtained by photoionizing unpolarized atoms by unpolarized light. It is just as in electron scattering where polarized electrons are obtained by scattering unpolarized electrons on unpolarized targets. Axial symmetry is not violated since integration over the angle yields polarization 0, as is seen by equation (8) and illustrated in Figure 7.

We are presently performing an experiment aiming at the measurement of the photoelectron polarization produced by unpolarized light. An encouraging result has been obtained by Dr. Heinzmann, who used the apparatus shown in Figure 8: Unpolarized vacuum ultraviolet radiation in the 150-nm range is used for photoionizing unpolarized lead atoms. The photoelectrons ejected under 45° pass through a lens system and are accelerated into the Mott detector for polarization analysis. The polarization that is perpendicular to the reaction plane is observed by the left–right asymmetry in the backward counters. The forward counters are used for monitoring instrumental asymmetries. The second light source under −45° is also used for eliminating spurious asymmetries.

The observed polarization was

$$P(45°) = 0.17 \pm 0.04$$

The value for γ_2, the phase following from this result, and other details such as energy selection are given by Heinzmann, Schönhense, and Wolcke in Chapter 48 of this volume.

Figure 9 shows a more sophisticated version of the apparatus that we have now completed. The angle of emission studied here is the magic angle $\theta_m = 54°44'$.

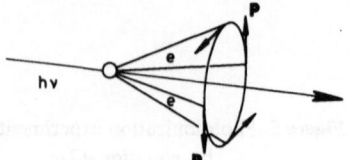

Figure 7. Polarized photoelectrons produced by shining
unpolarized light on unpolarized atoms.

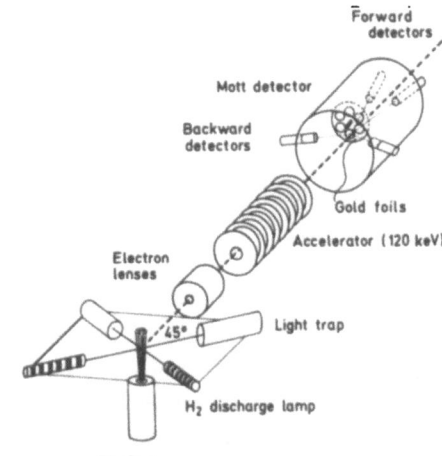

Figure 8. Apparatus for measuring photoelectron polarization produced by shining unpolarized light on unpolarized atoms.

Since $P_2(\cos \theta_m) = 0$, we do not have to worry about the term in the polarization formula equation (8) containing β, which is usually not known. We have also introduced an electron spectrometer so that we can select electrons of a certain energy corresponding to a well-defined energy state of the residual ion. We want to study noble gases in this apparatus, which is in a stage at which we hope to have first results in a few months.

4. Conclusion

Let me conclude with a somewhat provocative remark of a more fundamental nature about such investigations. It goes without saying that one needs n independent measurements if one wants to determine n independent parameters. In the case of

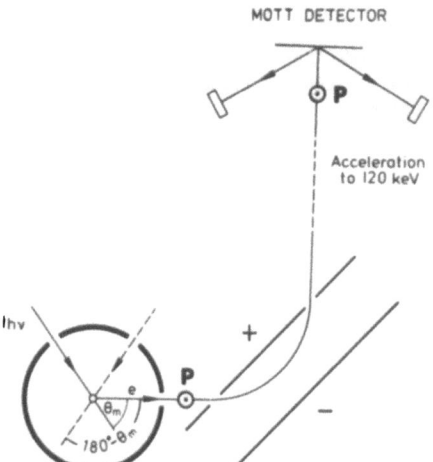

Figure 9. Schematic diagram of the improved version of the apparatus shown in Figure 8.

the two scattering amplitudes discussed before, one wants $|f|$, $|g|$, and $\gamma_2 - \gamma_1$; consequently one might expect to have to do three measurements. I have, however, explained that four experiments are needed, since $d\sigma/d\Omega$, P, T, and U have to be determined. This is because you do not observe $(\gamma_2 - \gamma_1)$ itself, but you measure $\sin(\gamma_2 - \gamma_1)$ and $\cos(\gamma_2 - \gamma_1)$. Both of these measurements are necessary, though not independent, since a measurement of $\sin(\gamma_2 - \gamma_1)$ alone leaves you with two values for $(\gamma_2 - \gamma_1)$; only if you also know $\cos(\gamma_2 - \gamma_1)$ can you eliminate this ambiguity. I cannot think of an experiment that yields $(\gamma_2 - \gamma_1)$ directly. The situation is similar in exchange scattering and in inelastic scattering where more than three parameters are needed: it is always the sines and the cosines of the phase differences that are observed and not the phase differences themselves.

Since one needs two experiments to determine one relative phase unambiguously, the statement "number of necessary experiments equals number of independent parameters to be determined" is misleading. There are quite a few examples in the literature where experimentalists have been misled by this statement, considering their task to be finished when it really was not.

Coming back to the case of photoionization: I have discussed five experiments for determining five parameters, three magnitudes and two relative phase shifts. Even though this seems to sound reasonable at first, I am somewhat suspicious if this is all one has to do, since the interference terms in the β and γ discussed before contain the sines or cosines of the phase differences. A generalization of Cherepkov's explicit results to the more general case I discussed here should clarify the situation. Perhaps the fact that a third-generation experiment has now been started will stimulate such theoretical work.

References

1. H. S. W. Massey and C. B. O. Mohr, *Proc. R. Soc. London Ser. A* **177**, 341 (1941).
2. J. Kessler, C. B. Lucas, and L. Vušković, *J. Phys. B* **10**, 847 (1977).
3. K. Schackert, *Z. Phys.* **213**, 316 (1968).
4. C. B. Lucas and I. E. McCarthy, *J. Phys. B* **11**, L301 (1977).
5. R. J. van Duinen and J. W. G. Aalders, *Nucl. Phys. A* **115**, 353 (1968).
6. J. Kessler, *Polarized Electrons*, Springer, Berlin (1976).
7. N. A. Cherepkov, *Sov. Phys.-JETP* **38**, 463 (1974).
8. C. M. Lee, *Phys. Rev. A* **10**, 1598 (1974).

46

Spin Polarization of Electrons by Resonance Scattering from Mercury

E. Reichert

Hg^- compound ion states formed by electron impact with neutral mercury atoms at collision energies of 4.55, 4.71, 4.94, and 5.51 eV show up in the $^2S_{1/2}$, $^2D_{3/2}$, $^2D_{5/2}$, and $^2D_{5/2}$ scattering states in order of increasing energy. This identification is supported by the observed energy dependence of (i) elastic (e^-, Hg) differential cross section, of (ii) 6^3P_0 excitation of mercury by electron impact, and of (iii) spin polarization of electrons scattered elastically from mercury.

The cross section of electrons scattered from mercury atoms at collision energies around the thresholds of 6^3P excitation is dominated by the formation and subsequent decay of autoionizing states of the Hg^- ion.[1-12]

The left part of Figure 1 from Reference 11 shows the energy dependence of the elastic differential cross section for some fixed scattering angles observed experimentally. Four structures are resolved at collision energies of 4.55, 4.71, 4.94, and 5.51 eV. Most likely all these features are due to negative ion states formed temporarily in the collision process.

The $6^3P_{0,1,2}$ excitation cross sections are influenced strongly by some of these states.[3,7,9,12] Figure 2 from Reference 12 shows an example. In the lower part the current of metastable Hg atoms produced by an electron beam passing through mercury vapor is plotted as a function of electron energy. Only the $^3P_0$ metastable

E. Reichert ● Institut für Physik, Johannes Gutenberg-Universität, Mainz, Germany.

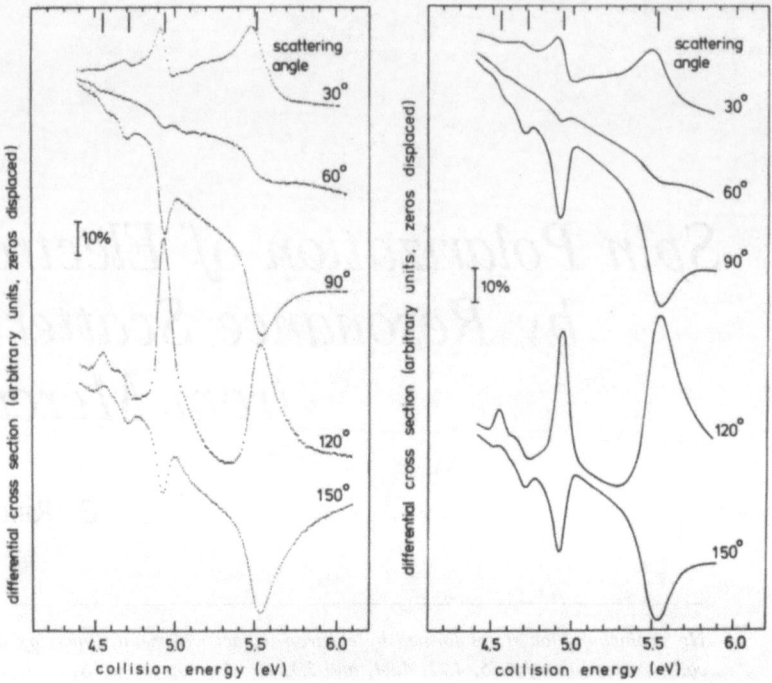

Figure 1. Energy dependence of elastic differential cross section for fixed scattering angles 30°, 60°, 90°, 120°, and 150°. Left-hand part: experiment; right-hand part: calculation. The magnitude of the bars represents 10% of scattered current at a collision energy of 4.40 eV for each scattering angle. The positions of the resonance energies are marked at the top of each diagram.

state of mercury can be excited at the collision energies shown so the curve represents the $^3P_0$ excitation function of mercury. For comparison the energy dependence of electron current scattered elastically through an angle of 40° from the same mercury target is drawn in the upper part of Figure 2. One sees that just above threshold almost all of the $^3P_0$ excitation goes via the negative ion state at 4.71 eV while there is only a weak contribution of this state to the elastic cross section.

Attempts to identify these Hg⁻ states have been made by Fano and Cooper[2] and by Heddle.[8] The present work tries to reexamine these classifications including new experimental data on the differential cross section, part of which is shown in Figure 1, and on the spin polarization of electrons scattered elastically from mercury. Following the arguments of Fano and Cooper and of Heddle, all the negative ion states in question should have $6s6p^2$ configuration. Russel–Saunders terms $^4P_{1/2,3/2,5/2}$, $^2D_{3/2,5/2}$, $^2P_{1/2,3/2}$, $^2S_{1/2}$ can be formed with this configuration. Heddle suggests that the terms $^4P_{1/2}$, $^4P_{3/2}$, $^2D_{3/2}$, $^2D_{5/2}$ correspond to the four structures of Figure 1 in order of increasing collision energy. Another feature observed by Ottley and Kleinpoppen[7] and by Zapesochnyi and Shpenik[3] in the 6^3P_1 excitation function at 5.23 eV, but not seen in Figure 1, is interpreted by Heddle as being due to the $^4P_{5/2}$ compound ion state.

At first sight the classification given by Heddle is consistent with the dependence on scattering angle of the resonant structures seen in the elastic cross section (Figure 1). By conservation of parity only electrons in the entrance channel with an even angular momentum quantum number are able to take part in the formation of compound Hg^- states with configuration $6s6p^2$. In addition conservation of total angular momentum is to be obeyed. So the $^4P_{1/2}$ compound ion state should show up in the ^2S scattering state, the others in ^2D scattering. The curves plotted in Figure 1 confirm this expectation. The resonances at 4.71, 4.94, and 5.51 eV, for example, change their form from peak to dip structure at scattering angles around 60° and 120°. This is what one would expect in case of a resonance in the D-scattering state, because the d-wave contribution in the partial wave expansion of the direct scattering amplitude goes through zeros at angles of 55° and 125°. The feature at 4.55 eV is small in most cases but does not disappear at any scattering angle. This is the behavior one could expect in the case of a resonance in the S-scattering state.

What are the values j of total angular momentum associated with the three negative ion states showing up in D scattering? To answer this question we tried to find out in what way resonance terms have to be incorporated in the d waves of a partial wave expansion of the elastic scattering amplitudes so that the cross section

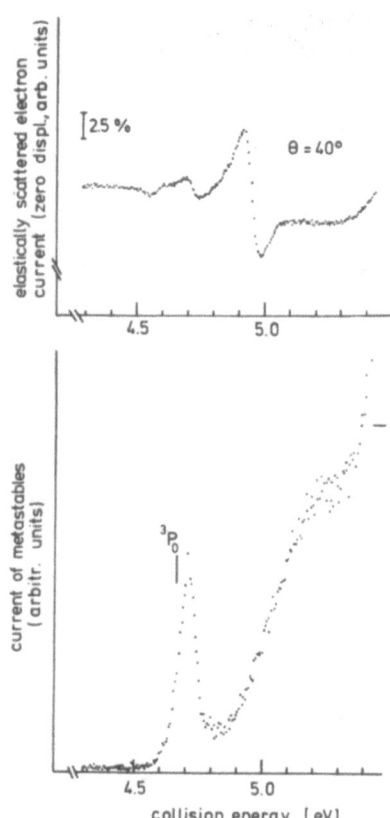

Figure 2. Upper part: Energy dependence of elastic differential cross section for scattering angle 40°. The magnitude of the bar represents 2.5% of scattered current at 4.30 eV. Lower part: Current of metastable Hg atoms produced by electron impact as a function of collision energy. The bar indicates the energy of $^3P_0$ excitation threshold.

calculated with these amplitudes will best match the energy dependence of the elastic differential cross section observed experimentally. We started with the following expansions of scattering amplitudes describing the elastic collision of spin-$\frac{1}{2}$ particles from spin-0 targets[13]:

$$f(E, \theta) = \frac{1}{2ik} \sum_{j=1/2}^{\infty} \sum_{l=j-1/2}^{j+1/2} (j + \tfrac{1}{2})(S_l^j - 1)P_l(\cos \theta) \tag{1}$$

$$g(E, \theta) = \frac{1}{ik} \sum_{j=1/2}^{\infty} \sum_{l=j-1/2}^{j+1/2} (j - l)(S_l^j - 1)P_l^1(\cos \theta) \tag{2}$$

where E is the collision energy, k the wave number, θ the scattering angle, j the quantum number of total angular momentum, l the quantum number of orbital angular momentum, P_l are Legendre polynomials, P_l^1 are associated Legendre polynomials, and S_l^j is the element of the S matrix for elastic scattering with angular momenta j and l. In case of an unpolarized primary electron beam the amplitudes f and g are connected with the differential elastic cross section I by (see, e.g., References 14 and 15)

$$I = |f|^2 + |g|^2 \tag{3}$$

It was assumed that at collision energies in the neighborhood of the Hg$^-$ levels the corresponding S-matrix elements can be described by[16,17]

$$S_l^j = B_l^j + \frac{\Gamma_e}{\Gamma} \frac{2i}{\varepsilon - i} e^{2i\chi} \tag{4}$$

Γ_e is the partial width of the resonance in the elastic channel, Γ is the total width of the resonance, $2\varepsilon = (E_r - E)/\Gamma$, E_r is the resonance energy, and χ a phase shift related to the potential scattering. In the case of only one open scattering channel, χ is identical to the phase shift due to potential scattering. B_l^j is a background term varying slowly with energy.

It is further assumed that all nonresonant terms in the expansions (1) and (2), including all B_l^j in (4), may be computed with sufficiently good approximation using phase shifts extrapolated from phase shifts given by Walker[18,19] in a relativistic one-channel calculation of electron scattering at mercury.

According to Figure 1 four resonance terms have to be included in the expansions (1) and (2). Most of the parameters of these terms are unknown. So by way of trial and error we varied the parameters E_r, Γ, Γ_e/Γ, χ, j, and l of each resonance term until fairly good agreement to the experimental curves plotted in Figure 1 as well as to the data of Burrow and Michejda[6] was obtained. The parameter values giving the best match to experiment are listed in Table 1. The curves in the right-hand part of Figure 1 are calculated with this set of parameters showing the goodness of fit reached. An instrumental energy resolution of 60 meV has been taken into account.

Table 1. Resonance Parameters of Best Fit

E_r (eV)	l	j	Γ (meV)	Γ_e/Γ	χ (rad)
4.55	0	$\frac{1}{2}$	0.5	1.00	1.339[a]
4.71	2	$\frac{3}{2}$	40.0	0.06	0.40
4.94	2	$\frac{5}{2}$	50.0	0.20	0.37
5.51	2	$\frac{5}{2}$	200.0	0.20	−0.10

[a] Walker's potential s-wave phase shift.

Table 1 deviates from the classification proposed by Heddle[8] in the j value of the compound ion state at 4.94 eV with $j = \frac{5}{2}$ instead of $j = \frac{3}{2}$. To decide the question which of the two j values is to be preferred in the classification of the resonance at 4.94 eV we looked for the energy dependence of the spin polarization of the scattered electrons. Applying the well-known formula

$$IP = i(f^*g - g^*f) \tag{5}$$

to the computation of the component P of spin polarization transverse to the scattering plane (see, e.g., References 14 and 15) one obtains the solid-line curves in Figures 3 and 4 if set of parameter values listed in Table 1 is used. One arrives at the other curves if combinations of total angular momentmu quantum numbers different from that in Table 1 are assumed for the negative ion states at 4.94 and 5.51 eV (the other parameter values of Table 1 left unchanged). The experimental values of electron spin polarization in Figures 3 and 4 do not agree quantitatively with any of the computed curves but they clearly show that the classification of Table 1 is to be preferred. The solid-line curves are the only ones that at least roughly show the energy dependence of spin polarization observed in experiment.

The assignment $j = \frac{5}{2}$ to the compound ion state of mercury at 4.94 eV is supported by the fact that this state does not show up in the $^3P_0$ excitation function of Figure 2, so its decay into this channel is only weak. This is the behavior one would

Figure 3. Energy dependence of electron spin polarization, scattering angle 50°. ϕ, experimental values; curves are calculated with different pairs (j (4.94 eV), j (5.51 eV)) assigned to the resonances at 4.94 eV and 5.51 eV, respectively; ——, with ($\frac{5}{2}$, $\frac{5}{2}$); $\cdots$, with ($\frac{3}{2}$, $\frac{5}{2}$); —·—, with ($\frac{5}{2}$, $\frac{3}{2}$); – – –, with ($\frac{3}{2}$, $\frac{3}{2}$).

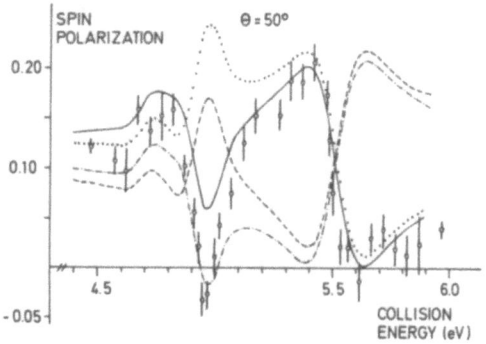

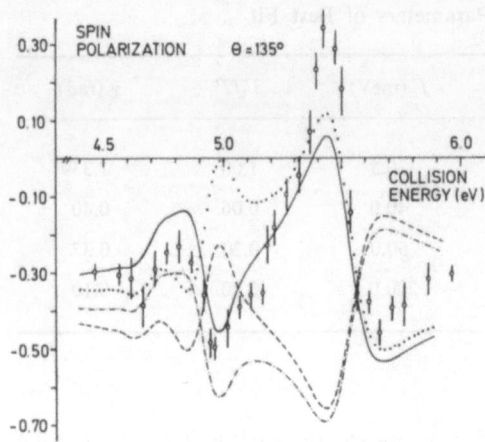

Figure 4. Energy dependence of electron spin polarization, scattering angle 135°; for explanation of symbols used see caption of Figure 3.

expect in case the Hg⁻ state has $j = \frac{5}{2}$. Because of conservation of parity and total angular momentum a $f_{5/2}$ electron would have to be emitted in such a decay, which is highly improbable.

References

1. C. E. Kuyatt, J. A. Simpson, and S. R. Mielczarek, *Phys. Rev.* **138A**, 385–399 (1965).
2. U. Fano and J. W. Cooper, *Phys. Rev.* **138A**, 400–402 (1965).
3. I. P. Zapesochnyi and O. B. Shpenik, *Sov. Phys.-JETP* **23**, 592–596 (1966).
4. A. I. Korotkov, *Opt. Spectrosc.* **28**, 347–349 (1970).
5. M. Düweke, N. Kirchner, E. Reichert, and E. Staudt, *J. Phys. B* **6**, L208–210 (1973).
6. P. D. Burrow and J. A. Michejda, *International Symposium on Electron and Photon Interactions with Atoms*, Stirling (1974).
7. T. W. Ottley and H. Kleinpoppen, *J. Phys. B* **8**, 621–627 (1975).
8. D. W. O. Heddle, *J. Phys. B* **8**, L33–36 (1975).
9. O. B. Shpenik, V. V. Sonter, A. N. Zavilopulo, I. P. Zapesochnyi, and E. E. Kontrosh, *Sov. Phys.-JETP* **42**, 23–28 (1976).
10. M. Düweke, N. Kirchner, E. Reichert and S. Schön, *J. Phys. B* **9**, 1915–1921 (1976).
11. K. Albert, C. Christian, T. Heindorff, E. Reichert, S. Schön, *J. Phys. B* **10**, 3733–3739 (1977).
12. T. Heindorff, L. Koch, and E. Reichert, *Verhandlungen DPG (VI)* **13**, 429–430 (1978).
13. J. M. Blatt and L. C. Biedenharn, *Rev. Mod. Phys.* **24**, 258–272 (1952).
14. N. F. Mott and H. S. W. Massey, *The Theory of Atomic Collisions*, Clarendon Press, Oxford (1965).
15. J. Kessler, *Polarized Electrons*, Springer, Berlin (1976).
16. R. H. Dalitz and R. G. Moorhouse, *Proc. R. Soc. London Ser. A* **318**, 279–298 (1970).
17. J. R. Taylor, *Scattering Theory*, p. 410, Wiley, New York (1972).
18. D. W. Walker, *J. Phys. B* **3**, 788–797 (1970).
19. D. W. Walker, private communication (1975).

47

Spin-Dependent Inelastic Electron–Atom Scattering

G. F. HANNE

A theoretical summary is given so that the observables that are needed for a description of the continuum electrons in spin-dependent inelastic electron–atom scattering can be calculated. Experimental aspects of electron polarization and coincidence photon polarization measurements are discussed. An outline of experiments done so far and feasible future studies is given. Of special interest is a test of the validity of the sudden approximation.

1. Introduction

Spin-dependent electron–atom collisions and coherent excitation of atoms by electron impact are two of the most interesting topics in electron–atom collision studies in the last few years. When these studies are combined a nearly complete determination of all observables should be possible.

There are two approaches to describe inelastic electron–atom scattering. In the first approach the scattered electrons and subsequently emitted photons are observed independently of each other. The probability for finding electrons and photons in certain directions after the scattering is described by various transition (scattering and emission) amplitudes. The atom remains unobserved in this approach. Hence we have to average over all possible atomic states.

In the second approach, which was introduced by Fano and Macek,[1] the angular momentum state of the atoms before and after the scattering is characterized by state multipoles. These can be determined experimentally by electron–photon

G. F. HANNE ● Physikalisches Institut, Westfälische Wilhelms-Universität, Schlossplatz 7, D-4400 Münster, Germany.

coincidence experiments, but for dipole radiation only up to the second rank (intensity, orientation, alignment). Of course, state multipoles can be expressed by transition amplitudes used in the first approach. However, the description in terms of state multipoles gives a more distinct picture when we consider symmetry properties of the collision and use geometrical interpretations of the phenomena, or when we apply first-order theories where the collision is treated as an impulsive transfer of momentum from the incident particle to the atom.

Nevertheless in this work we persist in the description by transition amplitudes, since these are more directly connected with quantum mechanical coherence or polarization phenomena, where we expect first-order theories to fail to give correct results.

The spin dependence of inelastic electron–atom scattering is caused by magnetic interactions, especially spin–orbit interaction, and by electron exchange where the Pauli principle plays the dominant role. If both magnetic interactions and electron exchange can occur in the same collision process we can only distinguish two types of scattering amplitudes, namely, non-spin-flip amplitudes F and spin-flip amplitudes G. If these two types of amplitudes can interfere we have polarization effects even for initially unpolarized electrons. This is a well-known consequence of the spin–orbit interaction for the continuum electron. But also the spin–orbit interaction within the atom, which causes the fine-structure splitting, should lead to polarization effects in inelastic scattering when electron exchange processes are involved.[2] For this effect it can no longer be assumed, however, that the spin–orbit coupling is weak as was done by Percival and Seaton.[3]

2. Theoretical Background

The theoretical background given in this chapter should enable us to calculate all the observables in terms of scattering amplitudes which can be obtained by spin-dependent electron–photon coincidence measurements. We develop the electron–atom state after electron collision and subsequent emission of photons, show how the observables can be expressed by density matrices, and give general amplitude formulas for various experimental conditions.

2.1. Electron–Atom State

2.1.1. Excitation by Electron Impact

The spin–orbit coupling within the atom couples the spin and the angular orbital momentum of the atomic electrons to a total angular momentum F (including hyperfine structure). Hence these are the relevant quantum numbers that describe the excited atomic states. This is the reason why we give here a description in the spin space of the continuum electrons only and why we do not construct a spin space of all colliding electrons, which is commonly used in elastic electron atom scattering.

We define a scattering state of the electron–atom system in the usual way, i.e., the initial (pure) state is $e^{ik_0 z_1}\chi(1)$, $u_0(2, \ldots, N)$, where the quantization axis (z axis) is chosen to be parallel to the momentum of the incident electrons, the scattering plane being the z–x plane.

The asymptotic form of the wave function at a time $t = 0$ after the collision (it is assumed that the collision time is much shorter than the lifetime of the excited states) may be expanded in terms of the exact solutions u_n of the atomic Hamiltonian

$$\phi_{t=0}(1; 2, \ldots, N) \xrightarrow[r_1 \to \infty]{} e^{ik_0 z_1}\chi(1)u_0(2, \ldots, N) + \sum_n \frac{e^{ik_n r_1}}{r_1} \chi_n(1)u_n(2, \ldots, N)$$

where $\chi(1)$ is an arbitrary, but normalized, spin state of the incident plane wave $e^{ik_0 z_1}$ and $u_n(2, \ldots, N)$ are all nondegenerate exact atomic states including continuum states which contain both space and spin coordinates of the atomic electrons. Thus each n is standing for a set of quantum numbers $(\alpha F M_F)$, where α contains all quantum numbers except for the angular momentum quantum numbers $(F M_F)$. Final continuum electron states are described by outgoing spherical waves and spinors χ_n. These are normalized in such a way that $|\chi_n|^2 k_n / k_0$ is the differential cross section for the excitation of an atom from its ground state u_0 to the excited state u_n by electron impact, where k_0 and k_n are the wave numbers of the incident and the scattered electrons, respectively.

To obtain the observables for the continuum electrons we must calculate currents and current densities, which means that the outgoing spherical waves are absent in any expression for observables. Hence with the assumption that an atomic level with definite energy and total angular momentum is separated in an experiment and that we have a pure initial spin state χ, we write this electron excited atom state in a single (spinor) wave function as a coherent sum over all excited magnetic substates [$\alpha F M_F = M$, the summation leaves (αF) unchanged]

$$\phi_{t=0}(\text{electron} + \text{atom}) = \sum_M \chi_M u_M = \sum_M \tilde{A}_M \chi u_M \tag{1}$$

where the amplitude matrix $\tilde{A}_M$ transforms the initial spin state χ into the final one $\chi_M = \tilde{A}_M \chi$.

Note that $\tilde{A}_M$ acts only on the spin states of the continuum electrons, thus $\tilde{A}_M$ is not necessarily invariant under rotation, space reflection, and time reversal as must be the transition matrix operator (often called scattering or collision matrix), which acts on the total electron–atom system. However, the matrix elements of $\tilde{A}_M$ can be expressed by certain matrix elements of the transition matrix operator $\tilde{T}$

$$A_M(m_s \to m_s') = -\frac{4\pi^2 m}{\hbar^2} \Sigma \langle u_M(2, \ldots, N)\chi_{m_s'}(1)e^{ik_n \cdot r_1} \,|\, \tilde{T}(1, \ldots, N)$$
$$\times \,|\, u_0(2, \ldots, N)\chi_{m_s}(1)e^{ik_0 z_1}\rangle$$

where the symbol Σ means correct antisymmetrization and normalization of the wave functions within the amplitude expression, χ_{m_s} and $\chi_{m_s'}$ mean eigenstates of the

Pauli spin matrix σ_z, m is the electron rest mass, and $2\pi\hbar$ is Planck's constant. The matrix elements $A_M(m_s \rightarrow m_s')$ are related to the above introduced non-spin-flip and spin-flip amplitudes F and G by

$$F_M(m_s \rightarrow m_s) = A_M(m_s \rightarrow m_s), \qquad G_M(m_s \rightarrow -m_s) = A_M(m_s \rightarrow -m_s)$$

2.1.2. Subsequent Emission of Dipole Radiation

If inelastically scattered electrons and subsequently emitted photons (dipole transition) are detected in coincidence we can regard this as a transition from the coherently excited atomic state equation (1) to a final electron–atom state. This process may be selected by a photon analyzer. Then, after a sufficiently long time, the electron–atom scattering state may be written as

$$\Psi_{t\rightarrow\infty}(\text{electron} + \text{atom}) \propto \sum_{M_{F'}m_s} \sum \langle \alpha' F' M_{F'}, m_s \,|\, \boldsymbol{\epsilon} \cdot \mathbf{r} \,|\, \phi \rangle \,|\, \alpha' F' M_{F'}, m_s \rangle \qquad (2)$$

where ϕ is the state equation (1), $\boldsymbol{\epsilon}$ the photon polarization vector, $|m_s\rangle$ is a spin eigenstate of the continuum electron and $|\alpha' F' M_{F'}\rangle$ is a final atomic state. The sum is taken over all degenerate final atomic states and the spin states.

From equation (2) we obtain the double differential cross section for this process, i.e., coherent excitation by electron impact into the state ϕ followed by dipole transition into $|\alpha' F' M_{F'}\rangle$

$$\frac{d^2\sigma}{d\Omega_e \, d\Omega_\gamma} \propto \sum_{M_{F'}m_s} \sum |\langle \alpha' F' M_{F'}, m_s \,|\, \boldsymbol{\epsilon} \cdot \mathbf{r} \,|\, \phi \rangle|^2$$

Following Fano and Macek[1] we can rewrite this expression as (we omit in the following the quantum numbers $\alpha' F'$)

$$\sum_{M_{F'}m_s} \sum |\langle \alpha' F' M_{F'}, m_s \,|\, \boldsymbol{\epsilon} \cdot \mathbf{r} \,|\, \phi \rangle|^2$$

$$= \sum_{M_{F'}m_s} \sum \langle M_{F'}, m_s \,|\, \boldsymbol{\epsilon} \cdot \mathbf{r} \,|\, \phi \rangle^* \langle M_{F'}, m_s \,|\, \boldsymbol{\epsilon} \cdot \mathbf{r} \,|\, \phi \rangle$$

$$= \sum_{M_{F'}m_s} \sum \langle \phi \,|\, \boldsymbol{\epsilon}^* \cdot \mathbf{r} \,|\, M_{F'}, m_s \rangle \langle m_s, M_{F'} \,|\, \boldsymbol{\epsilon} \cdot \mathbf{r} \,|\, \phi \rangle$$

$$= \langle \phi \,|\, \boldsymbol{\epsilon}^* \cdot \mathbf{r} \sum_{M_{F'}} |M_{F'}\rangle \langle M_{F'} \,|\, \boldsymbol{\epsilon} \cdot \mathbf{r} \,|\, \phi \rangle \qquad (3)$$

where we have made use of $\sum_{m_s} |m_s\rangle\langle m_s| = 1$.

The last expression of equation (3) can be interpreted as the average value of a tensor operator in the state ϕ, which, by applying the Wigner–Eckart theorem, can be replaced by the more familiar expressions of alignment and orientation as mean values of angular momentum operators[1] or by the most general state multipoles.[4,5]

The exponential decay of the light intensity is omitted in our expressions, which means that we have integrated over the duration of this decay. This is correct if we do not observe within time intervals that are shorter than the mean life time of the excited states. Thus quantum beats or related effects are not considered.

2.2. Calculation of Observables by Density Matrices

2.2.1. Electron Scattering Experiments

Experiments where only scattered electrons are observed may be described by the electron atom state equation (1). From this state, which is a spinor in the space of the continuum electrons, we calculate a density matrix by

$$\tilde{\varrho}' = \left(\sum_M \chi_M u_M\right)\left(\sum_M \chi_M u_M\right)^\dagger = \sum_{M,M'} u_M u_{M'}^* \chi_M \chi_{M'}^\dagger \qquad (4)$$

Since we only observe the scattered electrons we have to integrate over the coordinates of the atomic electrons, hence interferences between different magnetic substates vanish because of the orthonormality of these states. The density matrix (4) then reduces to

$$\tilde{\varrho}' = \sum_M \chi_M \chi_M^\dagger = \sum_M \tilde{A}_M \chi (\tilde{A}_M \chi)^\dagger = \sum_M \tilde{A}_M \chi \chi^\dagger \tilde{A}_M^\dagger = \sum_M \tilde{A}_M \tilde{\varrho} \tilde{A}_M^\dagger$$

where $\tilde{\varrho}$ is the density matrix of the incident electrons and $\tilde{A}_M^\dagger$ is the Hermitian adjoint to the amplitude matrix $\tilde{A}_M$. The relation

$$\tilde{\varrho}' = \sum_M \tilde{A}_M \tilde{\varrho} \tilde{A}_M^\dagger = \sum_M \tilde{\varrho}_{MM}$$

also holds if $\tilde{\varrho}$ describes mixed states of the incident electrons.

Differential cross section $\sigma(\theta)$ and spin polarization $\mathbf{P}_e$ of the scattered electrons are given by

$$\sigma(\theta) = (k_n/k_0) \sum_M \mathrm{Tr}\, \tilde{\varrho}_{MM}, \qquad \mathbf{P}_e = \sum_M \mathrm{Tr}\, \tilde{\varrho}_{MM}\boldsymbol{\sigma} \Big/ \sum_M \mathrm{Tr}\, \tilde{\varrho}_{MM}$$

where $\boldsymbol{\sigma} = (\sigma_x, \sigma_y, \sigma_z)$ are the Pauli matrices, and $\mathrm{Tr}\, \tilde{\varrho}_{MM}$ is the trace of the density matrix $\tilde{\varrho}_{MM}$. Formulas for the flux and polarization of photons emitted from atoms that are excited by electron impact are given in the work of Percival and Seaton.[3]

2.2.2. Electron–Photon Coincidence Experiments

These experiments are characterized by the electron–atom scattering state (2). We write this state in more detail as

$$\sum_{M_{F'},m_s} \langle M_{F'}, m_s \,|\, \boldsymbol{\epsilon} \cdot \mathbf{r} \,|\, \phi\rangle \,|\, m_s, M_{F'}\rangle$$

$$= \sum_{M_{F'},m_s} \langle M_{F'}, m_s \,|\, \boldsymbol{\epsilon} \cdot \mathbf{r} \,|\, \sum_M \tilde{A}_M \chi u_M\rangle \,|\, m_s, M_{F'}\rangle$$

$$= \sum_{M_{F'},M} \langle M_{F'} \,|\, \boldsymbol{\epsilon} \cdot \mathbf{r} \,|\, u_M\rangle \left(\sum_{m_s} \langle m_s \,|\, \tilde{A}_M \,|\, \chi\rangle \,|\, m_s\rangle\right) |\, M_{F'}\rangle$$

$$= \sum_{M_{F'},M} \langle M_{F'} \,|\, \boldsymbol{\epsilon} \cdot \mathbf{r} \,|\, u_M\rangle \chi_M \,|\, M_{F'}\rangle \qquad (5)$$

where we have made use of

$$\sum_{m_s} \langle m_s | \tilde{A}_M | \chi \rangle | m_s \rangle = \sum_{m_s} | m_s \rangle \langle m_s | \tilde{A}_M | \chi \rangle = \tilde{A}_M \chi = \chi_M$$

From equation (5) we construct density matrices for the continuum electrons by

$$\tilde{\varrho} \propto \sum_{M_{F'}} \left(\sum_M \langle M_{F'} | \boldsymbol{\epsilon} \cdot \mathbf{r} | u_M \rangle \chi_M \right) \left(\sum_M \langle M_{F'} | \boldsymbol{\epsilon} \cdot \mathbf{r} | u_M \rangle \chi_M \right)^\dagger$$

$$= \sum_{M_{F'}} \sum_{M,M'} \langle M_{F'} | \boldsymbol{\epsilon} \cdot \mathbf{r} | u_M \rangle \langle M_{F'} | \boldsymbol{\epsilon} \cdot \mathbf{r} | u_{M'} \rangle^* \chi_M \chi_{M'}^\dagger$$

$$= \sum_{M_{F'}} \sum_{M,M'} \langle M_{F'} | \boldsymbol{\epsilon} \cdot \mathbf{r} | u_M \rangle \langle M_{F'} | \boldsymbol{\epsilon} \cdot \mathbf{r} | u_{M'} \rangle^* \tilde{\varrho}_{MM'}$$

This expression is also correct if the incident electrons are only partially polarized. The double differential cross section and the electron spin polarization for an electron–photon coincidence experiment with given photon polarization $\boldsymbol{\epsilon}$ are then

$$\frac{d^2\sigma}{d\Omega_e\, d\Omega_\gamma} \propto \mathrm{Tr}\, \tilde{\varrho}', \qquad \mathbf{P}_e = \mathrm{Tr}\, \tilde{\varrho}' \boldsymbol{\sigma} / \mathrm{Tr}\, \tilde{\varrho}'$$

In a similar way a photon density matrix can be calculated from which the Stokes parameters or the vector polarization $\mathbf{P}_\gamma$ of the emitted light can be obtained.

The photon transition (dipole) operator $\boldsymbol{\epsilon} \cdot \mathbf{r}$ here is defined in the photon frame. A conventional rotation transforms it into the collision frame in which quantization axis and scattering plane (symmetry plane) are defined.

2.3. General Form of Amplitude Terms

Electron scattering experiments are an incoherent average over different final magnetic sublevels and, of course, over initial magnetic sublevels. The latter statement holds for electron–photon coincidence experiments too. The new information that can be obtained from electron–photon coincidence measurements are interferences between transition amplitudes of different final magnetic substates. From these interference terms the magnitude as well as the relative phase of the transition amplitude for a certain magnetic substate can be determined.

Since the matrix $\tilde{A}_M$ contains non-spin-flip amplitudes F_M and spin-flip amplitudes G_M the observables are "products" of such amplitudes. Table 1 shows the general type of the terms that can be obtained but does not contain the exact expressions. The symbols λ and χ in parentheses in this table mean that the expressions in these rows correspond to the pioneer measurements of similar quantities in helium.[6]

In Section 4.1 experiments are described in which the quantity $U(\theta)$ is measured. Spin flips are caused here by electron exchange only. In Section 4.2 the quantity $T(\theta)$ is determined, where again spin flips are caused by electron exchange only.

Table 1. General Form of Amplitude Terms

Primary beam	Electron scattering	Electron–photon coincidence	
Unpolarized incident electrons	$\sigma(\theta) = \sum_{M} (\mid F_M \mid^2 + \mid G_M \mid^2)$	$(\mid F_M \mid^2 + \mid G_M \mid^2)/\sigma$	(λ)
		Re, Im $\mid F_M{}^* F_{M'} + G_M{}^* G_{M'} \mid / \sigma$	(χ)
	$S(\theta) = \text{Im} \sum_{M} F_M{}^* G_M / \sigma$	$S_M = \text{Im } F_M{}^* G_M / \sigma$	
Polarized incident electrons	$T(\theta) = \text{Re} \sum_{M} F_M{}^* G_M / \sigma$	Re, Im $\mid F_M{}^* G_{M'} \pm F_{M'}{}^* G_M \mid / \sigma$	
	$U(\theta) = \sum_{M} (\mid G_M \mid^2 - \mid F_M \mid^2)/\sigma$		
	Incoherent sum over all substates	Interference between different substates	

The experiments described in Section 4.3 lead to an oriented atomic state by polarized electron impact excitation, resulting in a circular polarization of the emitted light or vice versa. These measurements correspond to the quantity $U(\theta)$, where the magnetic sublevels are populated with a different probability for opposite magnetic quantum numbers. In the experiments described in Sections 4.4 and 4.5 the quantities $S(\theta)$ and $S_M(\theta)$ are determined, where spin flips here are caused by spin–orbit coupling and exchange.

This was a rough approach to the formal theory. It is a straightforward procedure to reformulate the expressions when the initial atomic state is not the ground state. Of special interest are so-called superelastic collisions, which mean radiationless deexcitation of excited atoms by electron impact, where the scattered electrons gain energy. The scattering amplitudes for these processes can easily be constructed by time reversal transformations (detailed balance theorem) of the reverse excitation processes.

3. Experimental Aspects

The principal arrangement that is needed for electron spin polarization experiments and electron–photon coincidence experiments is shown in Figure 1. A polarized (or unpolarized) electron beam is produced in an electron source. Part of the electrons are scattered by a target. If the target atoms have unsaturated spin or angular momentum, they may also be polarized and their angular momentum state after the collision may be analyzed. Those electrons that are scattered in a certain direction and that have excited a certain atomic state are selected by an electron energy analyzer. The polarization or the current of these scattered electrons is measured in a Mott detector.

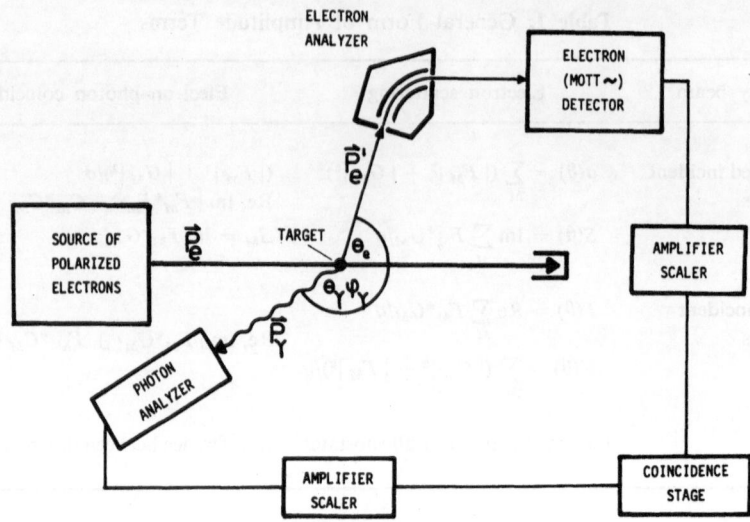

Figure 1. Experimental arrangement that is needed for spin-dependent electron–photon coincidence studies.

The emitted photons are observed by a photon analyzer, which selects a certain transition and measures the polarization or the current of the photons. The incident electron beam and the observed scattered electrons define a scattering plane. The photons, however, should be observed also in a direction out of this plane. Coincidences between inelastically scattered electrons and subsequently emitted photons are measured in a coincidence stage.

Electron-spin and electron–photon polarization coincidence measurements differ with respect to some points. The signal that should be compared may be defined as A^2N, where A is the asymmetry factor of the polarization measurement and N is the fraction of the numbers of detected particles to scattered particles, in our case electrons or photons. For an electron-spin polarization measurement this signal expression is 10^{-4}, which is a value for our Mott detectors. For the photon polarization coincidence measurements the signal expression is 10^{-3}–10^{-4}, which is mainly determined by the solid angle acceptance cone and the efficiency of the photon detector.

A characteristic difference exists between both methods when the error of the polarization measurements is considered. In electron spin polarization analyzers background or error counts are nearly independent of the rate of detected electrons and are typically 7×10^{-2} cps for our Mott detectors. Thus an increase in detected current gives a subsequent decrease of the statistical error. This is different for electron–photon coincidence experiments, where the number of real coincidences is proportional to the total rate of scattered electrons, whereas the number of chance coincidences is proportional to the square of that rate.[7] Hence an optimum value for this rate should not be exceeded in electron–photon coincidence measurements, since the statistical error of the measurement is then mainly caused by chance coincidences and no longer decreases and the signal-to-noise ratio becomes too small.[7] This ratio should be unity or better.

Another point is the energy resolution. The incident energy spread is determined by the electron source and is therefore the same for both polarization measurements. Photon analyzers, however, can easily be made substantially better in resolution than electron analyzers. Hence the level separation in photon polarization measurements can be made orders of magnitude better than in electron scattering experiments. Even simple interference filters often exceed the level separation power of electrostatic electron energy analyzers.

4. Proposed and Working Types of Experiments

In the following we discuss various inelastic electron–atom collision experiments that may be possible with proposed or working (prototype) sources of polarized electrons, for a recent review see Wainwright *et al.*[8] and the electron–photon coincidence method. The corresponding formulas for the observables are discussed in Section 2.3.

4.1. Inelastic Exchange Collisions

Exchange scattering can be directly observed by measuring the change of the polarization of the scattered electrons where it is assumed that only exchange collisions cause spin-flips. This can be achieved by using light atoms as a target or by observing only electrons that are scattered in the forward direction. It is widely established that magnetic interactions are negligible then. However, spin flips due to exchange scattering can only be observed if the atoms have unsaturated spin before scattering ($S \neq 0$) or if there is a change of the atomic multiplicity during the excitation process.

An experiment of the latter type was made in our laboratory.[9] Figure 2 shows the measured ratio P'/P of the final to the initial transversal polarization of electrons scattered in the forward direction which have excited mercury atoms from their 6^1S_0 ground state to the $6^3P_{1,2}$ excited states. Any measured deviation of this ratio

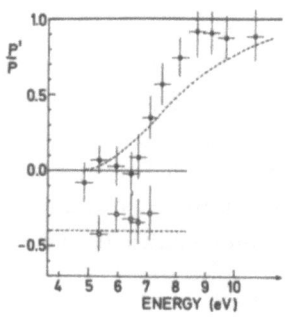

Figure 2. Ratio P'/P measured in the forward direction as a function of incident electron energy for the transitions $6^1S_0 \rightarrow 6^3P_1$ ($\Delta E = 4.89$ eV) (–◆–) and $6^1S_0 \rightarrow 6^3P_2$ ($\Delta E = 5.46$ eV) (–◻–).[9] Theoretical curve (– – –) Born–Ochkur approximation.[10]

from $P'/P = 1$ means a direct observation of electron exchange scattering. It can be seen that for incident energies below 7 eV exchange scattering dominates whereas at energies above 9 eV the measured ratio for the excitation of the 6^3P_1 level tends to $P'/P = 1$. This can be interpreted in terms of intermediate coupling. Because of the breakdown of LS-coupling, there is a certain 6^1P_1 admixture to the wave function of the 6^3P_1 state. This admixture obviously gives the dominant contribution to the cross section above 9 eV incident energy. The excitation then occurs without spin-flip in a direct excitation process, since the atomic multiplicity is not changed during the collision in that case. The theoretical curve is a calculation from Moiseiwitsch[10] in the Born–Ochkur approximation. Though this approximation is very rough it is in quite satisfactory agreement with the experiment.

It is worthwhile noting that the same results can be obtained by an electron–photon coincidence experiment, where we measure the linear polarization of photons emitted by the transition $6^3P_1 \rightarrow 6^1S_0$ in coincidence with those electrons that are scattered inelastically in the forward direction. There is no need for an electron spin polarization measurement. If exchange scattering dominates one would expect, as a consequence of the observation in the forward direction, the threshold value of the light polarization[3,11] for a pure 6^3P_1 state, which is for mercury, including hyperfine interaction, $P_{\text{thr}} = -0.77$. On the other hand, if the 6^1P_1 admixture dominates the excitation, one would expect the threshold value of a pure 6^1P_1 state, which is $P_{\text{thr}} = +0.82$. It seems to be a nice experiment to try to find this remarkable change of the photon polarization.

Scattering of (polarized) electrons on (polarized) atoms with unsaturated spin have been studied for one-electron atoms only. Most of these experiments deal with elastic scattering (see References 8 and 9). Only Lichten and Schultz[12] have reported experimental exchange scattering data for the $1s$–$2s$ inelastic excitation channel for electron–hydrogen collisions, and the New York University group has reported some experimental differential $n^2S \rightarrow n^2P$ cross sections for electron–potassium scattering with spin analysis.[13,14] An analysis of S–P excitations for one-electron atoms in terms of direct and exchange scattering amplitudes has been carried out by Kleinpoppen.[15]

4.2. Ionization of One-Electron Atoms

Though it is not necessary to observe electron exchange directly by measuring a spin flip of electrons or target atoms, there is a spin dependence of the ionization of one-electron atoms by electron impact, which can be obtained by using polarized electrons *and* atoms. Experimental studies exist for the ionization of hydrogen atoms[16] and alkali atoms[17] (see also Chapters 53 and 54 of this book). The ionization cross sections differ when the spins of the colliding electrons are initially parallel or antiparallel. In the experiments an antiparallel–parallel asymmetry

$$A = (\sigma_{\uparrow\downarrow} - \sigma_{\uparrow\uparrow})/(\sigma_{\uparrow\downarrow} + \sigma_{\uparrow\uparrow}) \tag{6}$$

is measured, where the arrows indicate the antiparallel- and parallel-spin configurations. In terms of the direct scattering amplitude $f(\mathbf{k_1}', \mathbf{k_2}')$ and the exchange scattering amplitude $g(\mathbf{k_1}', \mathbf{k_2}')$ (magnetic interactions are not involved in these amplitudes) these cross sections are calculated to be

$$\sigma_{\uparrow\downarrow} = \int \frac{k_1'k_2'}{k_1} (|f|^2 + |g|^2) \, d\mathbf{k_1}' \, d\mathbf{k_2}'$$

$$\sigma_{\uparrow\uparrow} = \int \frac{k_1'k_2'}{k_1} |f - g|^2 \, d\mathbf{k_1}' \, d\mathbf{k_2}'$$

where k_1', k_2', and k_1 mean the wave numbers of scattered, ejected, and incident electrons, respectively. Hence it is the ratio of the interference cross section

$$\frac{1}{2} (\sigma_{\uparrow\downarrow} - \sigma_{\uparrow\uparrow}) = \sigma^{\text{int}} = \int \frac{k_1'k_2'}{k_1} \text{Re}[f^*(\mathbf{k_1}', \mathbf{k_2}')g(\mathbf{k_1}', \mathbf{k_2}')] \, d\mathbf{k_1}' \, d\mathbf{k_2}'$$

to the total cross section $\frac{1}{2}(\sigma_{\downarrow\uparrow} + \sigma_{\uparrow\uparrow})$, which is measured in these experiments [equation (6)].

This type of experiment can of course be used also to measure the corresponding differential cross-section ratios for elastic and inelastic scattering.

4.3. Transfer of Spin Polarization to Atomic Angular Momentum Polarization

Polarized electrons can transfer spin angular momentum to the atom by inelastic exchange collisions. When it is secured that no angular orbital momentum is transferred to the atom, the angular momentum polarization of the atom has the direction of the initial spin polarization. This condition is achieved by observing transitions where the angular orbital momentum of the atoms does not change or at threshold conditions where the electrons do not transfer angular orbital momentum to the atom. After such an excitation the atoms should emit circularly polarized photons in the direction of the spin transfer.

An experiment of this type has been proposed by Farago and Wykes.[18,19] They also discuss this method for use as a spin polarization detector, but no realization has been performed until now.

An interesting variation of this experiment is the time reverse process. Atoms may be excited by circularly polarized light and slow electrons may deexcite the atom by a superelastic collision. If this collision is in a direction where transfer of angular orbital momentum can be neglected, the atomic polarization should be completely transferred to electron spin polarization. The study of superelastic scattering of electrons by laser-excited atoms is feasible now which has been demonstrated by the experiments of Hertel and Stoll.[20] A possible experimental arrangement may be the study of superelastic scattering of electrons in the forward direction. The atoms should be excited by circularly polarized photons which have their momentum parallel to the incident electron beam.

4.4. Spin–Orbit Interaction for the Continuum Electron

In heavier targets the spin–orbit interaction, which is the dominant magnetic interaction, can no longer be excluded during the excitation. Polarization effects due to spin–orbit interaction for the continuum electron have been first found by Eitel and Kessler[21] for the $6^1S_0 \rightarrow 6^1P_1$ transition in mercury and later on have been continued also for a further transition.[22] As an example the result for the $6^1S_0 \rightarrow 6^1P_1$ transition at an incident energy of 180 eV[21] is shown in Figure 3. Theories exist by Madison and Shelton,[23] Bonham,[24] and Yamazaki et al.[25] that are in good agreement with the experimental results.

A simple picture explains this polarization effect as a two-step process, where the electrons are first scattered by the inner atomic potential into a large scattering angle and then, in the outer shell, undergo an inelastic transition without a further significant change of their new direction. Or vice versa, the first inelastically scattered electrons in the forward direction are scattered in the potential of the excited atom which, however, is not very different from the ground state potential in the inner shells. This simple picture explains also the similarity of these polarization curves with those from elastically scattered electrons down to 30 eV incident energy.[22] A theoretical prediction for 300 eV incident electron energy,[25] which contradicts this picture, requires a further experimental study.

We want to emphasize that the measurements done so far are an average over all final atomic magnetic sublevels. The 6^1P_1 state in mercury has three magnetic sublevels (excluding hyperfine interaction) which are degenerate. It is an interesting question whether the polarization curves are different for each sublevel or not. This question can be answered by an electron–photon coincidence experiment where we select a certain sublevel by a photon polarization analyzer. We can either use polarized incident electrons and measure the scattering asymmetry or measure the polarization of initially unpolarized electrons in coincidence with the subsequently emitted photons ($6^1P_1 \rightarrow 6^1S_0$ transition).

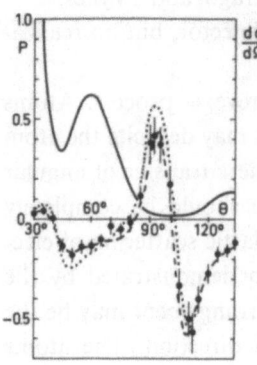

Figure 3. Differential cross section (——, arbitrary units) and electron spin polarization (♦) for inelastic ($6^1S_0 \rightarrow 6^1P_1$) electron mercury scattering at an incident energy of 180 eV measured by Eitel and Kessler.[21] Theoretical curves by Madison and Shelton[23] (– – –, DWBA) and Bonham[24] (···, Special second Born approximation using coupling to the elastic channel). Not shown are various DWB approximations by Yamazaki et al.[25]

4.5. Fine-Structure Interaction

Spin–orbit coupling within the atom is the reason for fine-structure splitting. Fine-structure splitting can only occur if the excited atom has unsaturated spin and angular orbital momentum. It is often assumed that this spin–orbit coupling is so weak that it does not influence the excitation collision process.[3] This approximation is also called the sudden approximation. Speaking in a time-dependent picture it is assumed in this approximation that the collision time is much shorter than the spin–orbit relaxation time.

This cannot be true, on the other side, if fine-structure levels are clearly resolved in an electron scattering experiment. We then must assume that the collision time is longer than the spin–orbit relaxation time, which means a long incident wave packet or a small incident energy spread, which is sufficient to resolve fine-structure levels. The atom is in that case "immediately" excited into a $|J, M_J\rangle$ fine-structure level.

It can be shown[2] that the latter assumption not only changes the polarization formulas for exchange scattering of polarized electrons but leads to polarization effects even for unpolarized electrons. We do not take into account here the spin–orbit interaction for the continuum electron, which of course could give an additional effect.

Let us illustrate this effect for the $6^1S_0 \to 6^3P_0$ transition, where we neglect hyperfine structure ($I = 0$ isotopes). Non-spin-flip and spin-flip amplitudes can interfere if both describe the same transition from a nondegenerate initial state to a nondegenerate final atomic state. This is obviously the case for the transition discussed here, since both states are nondegenerate and the final state can be reached either by non-spin-flip and spin-flip processes, which can interfere.

This can be seen as follows. For simplicity let us assume, which is not essential for this effect, that the total spin quantum number during the collision is conserved

$$m_s + M_S = m_s' + M_{S'}'$$

where m_s, m_s' is the projection of the spin of the continuum electron and M_S, M_S' is the projection of the atomic spin on the quantization axis before and after the scattering, respectively. For $m_s = +\frac{1}{2}$, for example, two processes, without spin flip and with spin flip, are possible by electron exchange[2,9]:

$$(m_s = \tfrac{1}{2}) + (M_S = 0) \to (M_{S'}' = 0) + (m_s' = \tfrac{1}{2}) \qquad \text{non-spin-flip}$$

$$(m_s = \tfrac{1}{2}) + (M_S = 0) \to (M_{S'}' = 1) + (m_s' = -\tfrac{1}{2}) \qquad \text{spin-flip}$$

Since the wave function of the 6^3P_0 level contains both $M_{S'}' = 0$ and $M_{S'}' = +1$ terms, non-spin-flip and spin-flip amplitudes can interfere.

Though it is not possible in principle to separate this polarization effect from the usual spin–orbit interaction for the continuum electron, there may be an observable evidence for this effect: It has been shown[2] that also the interference between the triplet terms and the singlet admixture in the wave function of the 6^3P_1 in mercury should give polarization effects. From the previous measurements[9] it follows that considerable interference terms of this kind should only occur for incident electron energies between approximately 7 and 9 eV. Thus, if polarization effects were observed only at energies below 9 eV and not up to, say, 30 eV, this would be an indication of an effect discussed in this section. An experimental test of these ideas is in progress in our laboratory.

ACKNOWLEDGMENT

The author is indebted to Professor J. Kessler for his interest in this subject and stimulating discussions.

References

1. U. Fano and J. H. Macek, *Rev. Mod. Phys.* **45**, 553–573 (1973).
2. G. F. Hanne, *J. Phys. B* **9**, 805–815 (1976).
3. I. C. Percival and M. J. Seaton, *Phil. Trans. R. Soc. London Ser. A* **251**, 113–138 (1958).
4. U. Fano, *Phys. Rev.* **90**, 577–579 (1953).
5. K. Blum and H. Kleinpoppen, *J. Phys. B* **8**, 922–925 (1975).
6. M. Eminyan, K. B. MacAdam, J. Slevin, and H. Kleinpoppen, *J. Phys. B* **7**, 1519–1542 (1974).
7. R. E. Bell, *Alpha-, Beta- and Gamma-Ray Spectroscopy*, K. Siegbahn, ed., Vol. 2, pp. 905–929, North Holland, Amsterdam (1974).
8. P. F. Wainwright, M. J. Alguard, G. Baum, and M. S. Lubell, *Rev. Sci. Instrum.* **49**, 571–585 (1978).
9. G. F. Hanne and J. Kessler, *J. Phys. B* **9**, 791–804 (1976).
10. B. L. Moiseiwitsch, *J. Phys. B* **9**, L245–247 (1976).
11. J. C. McConnell and B. L. Moiseiwitsch, *J. Phys. B* **1**, 406–413 (1968).
12. W. Lichten and S. Schultz, *Phys. Rev.* **116**, 1132–1139 (1959).
13. K. Rubin, B. Bederson, M. Goldstein, and R. E. Collins, *Phys. Rev.* **182**, 201–214 (1969).
14. M. Goldstein, A. Kasdan, and B. Bederson, *Phys. Rev. A* **5**, 660–668 (1972).
15. H. Kleinpoppen, *Phys. Rev. A* **3**, 2015–2027 (1971).
16. M. J. Alguard, V. W. Hughes, M. S. Lubell, and P. F. Wainwright, *Phys. Rev. Lett.* **39**, 334–338 (1977); see also Chapter 53 of this book.
17. D. Hils and H. Kleinpoppen, *J. Phys. B* **11**, L283–287 (1978); see also Chapter 54 of this book.
18. P. S. Farago and J. S. Wykes, *J. Phys. B* **2**, 747–756 (1969).
19. J. Wykes, *J. Phys. B* **4**, L91–94 (1971).
20. I. Hertel and W. Stoll, *J. Phys. B* **7**, 583–592 (1974).
21. W. Eitel and J. Kessler, *Phys. Rev. Lett.* **24**, 1472–1473 (1970).
22. G. F. Hanne, K. Jost, and J. Kessler, *Z. Phys.* **252**, 141–146 (1972).
23. D. H. Madison and W. N. Shelton, *Phys. Rev. A* **7**, 514–523 (1973).
24. R. A. Bonham, *J. Electron. Spectrosc. Rel. Phen.* **3**, 85–106 (1974).
25. Y. Yamazaki, R. Shimizu, K. Ueda, and H. Hashimoto, *J. Phys. B* **10**, L731–734 (1977).

48

Spin-Polarized Photoelectrons from Unpolarized Lead Atoms Exposed to Unpolarized Radiation

U. Heinzmann, G. Schönhense, and A. Wolcke

The photoelectrons emitted by unpolarized lead atoms exposed to unpolarized vuv radiation are spin polarized. An angular distribution experiment has been performed for two angles: the light beam (spectrum between 145 nm and the photoionization threshold at 167 nm) produced by means of a H_2 discharge lamp is placed at an angle of 45° and 135° with respect to the spin polarization analyzer. A polarization of $17 \pm 4\%$, perpendicular to the reaction plane, has been measured. This result confirms the theories made by Lee and Cherepkov and makes possible calculation of the phase difference of the continuum wave functions of the photoelectrons.

1. Introduction

One of the novel results in photon–atom interactions is the experimental evidence for spin polarization of photoelectrons produced by circularly polarized light. After this Fano effect had been verified for free alkali atoms for which it has originally been predicted,[1] experimental studies were made with other targets and positive results were obtained (for a review see Reference 2). This conspicuous effect of spin-orbit coupling prompted strong interest in the question of whether other experimental methods in the field of photoionization also yield polarized electrons.

In 1974 Cherepkov[3] and Lee[4] predicted an effect (also based on the spin-orbit interaction) that photoelectrons emitted by unpolarized atoms can also be

U. Heinzmann, G. Schönhense, and A. Wolcke ● Physikalisches Institut der Universität Münster, Schlossplatz 7, 4400 Münster, Germany.

polarized even if linearly polarized or unpolarized radiation is used. Because at this photoionization process no photon spin is transferred onto the photoelectrons, the average polarization of these would be zero (contrary to the case of the Fano effect), if they were extracted by an electric field independently of their direction of emission. Therefore using unpolarized radiation an angular distribution experiment has to be performed in order to produce spin-polarized photoelectrons.

The interest in these polarization phenomena comes mainly from the intention to find new experimental methods that complement the measurements of the cross section and angular distribution in order to obtain new detailed and possibly full information about the photoionization process.

It is the purpose of the paper to present for the first time experimental evidence for the theoretical prediction, that spin-polarized photoelectrons can also be obtained by irradiation of unpolarized light at unpolarized atoms. Lead was selected because of its wide fine-structure splitting of the ion, which means that an electron spectrometer did not need to be used. Furthermore lead is a very good example because the photoionization cross section[5] and the Fano effect[6] have also been measured in the same wavelength range as the result described in this paper. This makes possible calculation of the phase difference of the continuum wave functions of the photoelectrons.

2. Experimental Arrangement and Spin-Polarization Results

The experiment has been performed using parts of the apparatus (discharge lamps, oven, vacuum chamber, Mott detector) with which the results of the Fano effect on thallium[7,8] and lead[6] atoms have been obtained. The experimental procedure using the discharge lamps is described in Reference 9, using the oven and the Mott detector in Reference 6. A schematic diagram of the apparatus used is shown in Figure 1. The radiation produced by means of two H_2 discharge lamps, which are placed at an angle of $+45°$ and $-45°$ with respect to the polarization analyzer, passes through the atomic vapor beam into the light trap, where it is absorbed.

The angle $45°$ has been selected because the ports of the vacuum chamber used are only placed at each $45°$. But this fact is no important disadvantage, especially because the product spin polarization times electron intensity has its maximum at the angle of $45°$ (see Section 3).

The photoelectrons produced run into all directions according to the angular distribution rules. Those that pass the grid mounted before the system of electron lenses are accelerated and focused onto the entrance of the 120-keV accelerator. For the measurement of the spin polarization the electrons are scattered by a gold foil in the Mott detector. In order to eliminate instrumental asymmetries of the Mott detector firstly additional detectors in the forward direction (13° scattering angle) have been used and secondly the sign of the electron polarization has been changed

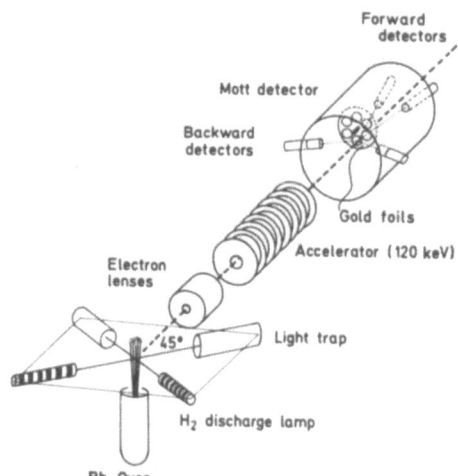

Figure 1. Schematic diagram of the apparatus.

once a minute, changing the sign of the angle between the photon and the electron beam, i.e., alternating the lamp.

Using this procedure a spin polarization of $17 \pm 4\%$ (single statistical error) has been measured. The backward detectors of the Mott analyzer have measured a clear asymmetry of the count rates, which has changed with the change of the lamps, contrary to the count rates in the forward detectors. The direction of the spin polarization vector is shown in Figure 2 with respect to the reaction plane of incoming photons and outgoing electrons. It is interesting to note that it is not important for the polarization vector whether the radiation comes from the left or the right side in Figure 2, as a symmetry consideration shows. Therefore arrows indicating the light momentum have been omitted in Figure 2.

Before the measurement could be performed, a very hard problem had to be solved: (unpolarized) background electrons which came from the inner parts of the oven (1000°C) or have been produced at walls, discharge lamp, atomic beam nozzle, light trap, or other solids by photoionization due to stray light had to be avoided. At first this background was 10^4 times higher than the real count rates expected. At last the intensity of the photoelectrons coming from the atomic beam was twice the background, which could be measured by use of a simple valve above the oven

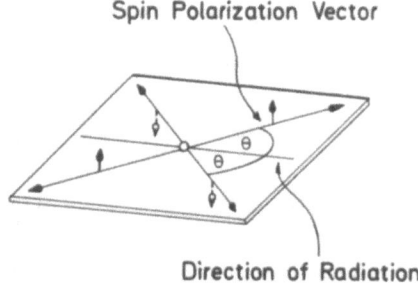

Figure 2. Reaction plane for the photoionization process of lead using unpolarized radiation.

stopping or opening the way of the atomic beam. The background problem has been solved by a complicated system of apertures and capillary tubes between the discharge lamps and the photoionization region and by some apertures at the oven and the light traps (not shown in Figure 1). Some of these apertures have been on different electric potentials, in order to avoid the background electrons reaching the photo-ionization region, but without violating the first principle rule that this region had to be free of any electric or magnetic fields. This behavior was important especially because the photoelectrons produced on lead atoms are very slow (kinetic energy smaller than 1 eV).

3. Discussion

Lead atoms have four valence electrons with the configuration $6s^2p^2(\frac{1}{2}, \frac{1}{2})_0$ in the ground state; this ground state, which can be written as $^3P_0$ in LS characteriza-tion, is the lowest of the three fine-structure levels $^3P_{0,1,2}$. The excited states do not play a role in photoionization processes because the energy difference from the ground state is sufficiently large to make them unoccupied.[10] In the ionization process one of the $p_{1/2}$ electrons makes a transition into either the $s_{1/2}$ or $d_{3/2}$ con-tinuum according to the selection rules. Because the remaining ion with the other p electron in the outer subshell shows a fine-structure splitting (ground state $^2P_{1/2}$ and excited state $^2P_{3/2}$), there are two ionization thresholds, 167.2 and 135.3 nm. Between the first and the second threshold the second Rydberg states can couple with the continua of the first series, which means that autoionization resonances exist in the photoionization cross section.[5]

In the present paper these autoionization resonances probably do not play an important role because the bandwidth of the radiation used was about 20 times larger than the half-width of the resonances. Because the radiation intensity[9] and the photoionization cross section[5] drop to small values above 162 nm and below 150 nm, respectively, one should expect that the measured spin polarization of 17% is an average value for the wavelength range between 162 and 150 nm.

Because no photoelectrons are produced[6] by radiation of a wavelength shorter than 135.3 nm, all ions produced are in the ground state $^2P_{1/2}$. The separation of both fine-structure levels of the ion has thus been performed optically; therefore an electron spectrometer did not need to be used. This resolution of the fine structure induced by the spin–orbit interaction is very important because according to Reference 3 the spin polarization of photoelectrons produced by unpolarized radiation should have opposite sign for both ion channels.

The spin polarization is given by [equations (28), (27), and (2) of Reference 3]

$$P = \frac{-\gamma_2 \cdot s[e \times \varkappa](e\varkappa)}{1 - \frac{1}{2}\beta[\frac{3}{2}(e\varkappa)^2 - \frac{1}{2}]} = \frac{-\gamma_2 \cdot \sin\theta \cos\theta}{1 - \frac{1}{2}\beta P_2(\cos\theta)} \tag{1}$$

where s, e, and $\varkappa$ are the unit vectors of the electron spin, radiation momentum, and electron momentum direction, respectively (the denominator is proportional to the differential photoionization cross section). γ_2 and β are given by

$$\gamma_2 = (-1)^{j-l-1/2} \frac{3l(l+1)}{2j+1} \frac{d_{l+1} \, d_{l-1} \sin(\delta_{l+1} - \delta_{l-1})}{[l \, d_{l-1}^2 + (l+1) \, d_{l+1}^2]} \tag{2}$$

$$\beta = \frac{l(l-1) \, d_{l-1}^2 + (l+1)(l+2) \, d_{l+1}^2 - 6l(l+1) \, d_{l-1} \, d_{l+1} \cos(\delta_{l+1} - \delta_{l-1})}{(2l+1)[l \, d_{l-1}^2 + (l+1) \, d_{l+1}^2]} \tag{3}$$

where j (here $\tfrac{1}{2}$) is the total angular momentum of the ion and l (here 1) is the orbital angular momentum of the photoelectron in the ground state of the atom. $d_{l\pm1}$ are the radial parts of the dipole matrix elements and $\delta_{l\pm1}$ are the phases of the wave functions of the outgoing photoelectrons.

As given by equation (1) the best angle to measure the spin polarization would be the magic angle (about 54°), where the second Legendre polynom $P_2(\cos\theta)$ vanishes, because there the spin polarization does not depend upon the β parameter. But this independence of P from β is only correct if the uncertainty of θ (given by the radiation divergence and the opening aperture of the electron lenses, see Figure 1) can be neglected. In the present case $(\theta = 45°)$ we have an uncertainty of $\pm7.5°$ for θ. Using equation (1) and the measured spin polarization result of 17% one obtains

$$\gamma_2 = -0.34(1 - 0.125\beta)$$

Generally $-1 \le \beta \le 2$, but a complicated consideration using the connection of equations (2) and (3) ($d_{l\pm1}^2$ are known given by the results of the Fano effect[6]) yields $-0.3 < \beta < 1.8$. Performing the β correction one obtains

$$\gamma_2 = -0.32 \pm 0.10$$

where the error bar includes the single statistical error of the polarization result as well as the uncertainties of β and θ.

According to equation (2) one also obtains

$$\gamma_2 = -\frac{3}{2^{1/2}} \frac{2^{1/2} \, d_D/d_S}{1 + 2 \, d_D^2/d_S^2} \cdot \sin(\delta_D - \delta_S) \tag{4}$$

Cherepkov did not include the influence of the spin–orbit interaction on the continuum wave functions of the outgoing electrons: as shown in equations (2) and (3) he does not distinguish between the matrix elements $d_{D_{3/2}}$ and $d_{D_{5/2}}$. But because the transition into the $D_{5/2}$ continuum is forbidden in our case due to the selection rules, equation (4) is also correct for the case of included spin–orbit interaction, as a comparison with the more general theory of Lee[4] shows, if d_D and δ_D are seen as $d_{D_{3/2}}$ and $\delta_{D_{3/2}}$, respectively.

Because the ratio of the partial cross sections, which is independent of the existence and the influence of the autoionization resonances,[6]

$$\frac{Q_{D_{3/2}}}{Q_S} = \frac{2\, d_{D_{3/2}}^2}{d_S^2} = 3.1 \pm 0.5$$

is known from the results of the Fano effect on lead,[6] the phase difference in equation (4) can be calculated as

$$\mathrm{sign}(d_D \cdot d_S) \cdot \sin(\delta_D - \delta_S) = 0.35 \pm 0.11 = |\sin(\delta_D - \delta_S)|$$

The sign of $\sin(\delta_D - \delta_S)$ cannot be specified because the sign of $d_S \cdot d_D$ is unknown. In order to obtain more information one has to measure the β parameter in an angular distribution experiment. According to equation (3) one would obtain $\mathrm{sign}(d_D \cdot d_S) \cdot \cos(\delta_D - \delta_S)$ and together with the result mentioned above $\tan(\delta_D - \delta_S)$ including the sign. But even with the knowledge of β the phase difference could only be determined modulo π.

ACKNOWLEDGMENTS

The authors wish to express their gratitude to Professor J. Kessler for stimulating discussions and gratefully acknowledge support by the Deutsche Forschungsgemeinschaft.

References

1. U. Fano, *Phys. Rev.* **178**, 131 (1969).
2. J. Kessler, *Polarized Electrons*, Springer Verlag, Berlin (1976).
3. N. A. Cherepkov, *Sov. Phys.-JETP* **38**, 463 (1974).
4. C. M. Lee, *Phys. Rev. A* **10**, 1598 (1974).
5. R. Heppinstall and G. V. Marr, *Proc. R. Soc. London Ser. A* **310**, 35 (1969).
6. U. Heinzmann, *J. Phys. B* **11**, 399 (1978).
7. U. Heinzmann, H. Heuer, and J. Kessler, *Phys. Rev. Lett.* **34**, 441, 710 (1975).
8. U. Heinzmann, H. Heuer, and J. Kessler, *Phys. Rev. Lett.* **36**, 1444 (1976).
9. U. Heinzmann, *J. Phys. E* **10**, 1001 (1977).
10. W. R. S. Garton and M. Wilson, *Proc. Phys. Soc.* **87**, 841 (1966).

49

Total Cross Sections for the Scattering of Low-Energy Electrons by Excited Sodium Atoms

B. Jaduszliwer, R. Dang, P. Weiss, and B. Bederson

Preliminary values of the total scattering cross section of electrons on $3P_{3/2}$ sodium atoms have been measured at 1.0, 2.0, and 3.0 eV by an atomic double-recoil method. The atoms are excited by a single-mode tunable dye laser. The atomic recoil due to resonant photon interactions is used to spatially separate the excited atoms and to determine the fraction of excited atoms. Standard recoil techniques are used to determine the total cross sections. The results are compared with close-coupling calculations.

1. Introduction

The single-mode tunable dye laser is being increasingly exploited to perform atom–atom[1,2] and electron–atom[3-5] collision experiments involving excited states of the target beam. Some theoretical aspects of this problem have been discussed by Macek and Hertel.[6] Much of this recent work has been summarized in a review by Hertel and Stoll.[7]

Our efforts have recently been directed towards obtaining reliable, absolute total cross sections for prepared excited states of sodium ($3^2P_{3/2}$) scattered by low-energy electrons, using a "double-recoil" technique.[4] This work is part of a pro-

B. Jaduszliwer, R. Dang, P. Weiss, and B. Bederson • Physics Department, New York University, New York, New York 10003.

gram that includes measurements of state-selected and spin-analyzed total and differential cross sections for elastic, inelastic, and superelastic scattering of $3^2P_{3/2}$ sodium.

2. Excitation of the Atom

Figure 1 shows the hyperfine energy levels for the ground and $3P_{3/2}$ states of sodium in the presence of a magnetic field. At the field strength employed in the interaction region (785 G in the present experiment), the nuclear and atomic magnetic moments are fully decoupled for the excited states, while for the ground state the atoms are in an intermediate field regime, with the dimensionless Rabi parameter $x = g\mu_0 H/\Delta W = 1.24$, where g is the gyromagnetic ratio, μ_0 the Bohr magneton, H the magnetic field intensity, and ΔW the zero-field hyperfine energy separation. It follows that the ground-state level can be conveniently described by the quantum numbers (F, m) while the excited state levels are described in terms of the set (m_I, m_J).

The natural width of these transitions is about 12 MHz, while the laser linewidth is about 50 MHz, so that for mutually perpendicular atom and laser beams the Doppler broadening is generally negligible. The average separation between $3P_{3/2}$ levels differing in one unit of m_J is 1460 MHz, while the splitting of levels of the same m_J and differing by one unit of m_I is 30 MHz (for $m_J = \pm\frac{3}{2}$) or 10 MHz (for $m_J = \pm\frac{1}{2}$). The separation between adjacent ground-state levels varies between 260 and 1220 MHz.

By proper frequency and polarization selection of the laser light, it is possible in principle to resolve each of the different allowed transitions. As is well known, optical pumping depletes most of the ground hyperfine states. The exceptions are caused by the $(2, 2) \rightarrow (\frac{3}{2}, \frac{3}{2})$ and $(2, -2) \rightarrow (-\frac{3}{2}, -\frac{3}{2})$ transitions denoted by arrows in Figure 1, and these are the ones generally employed in laser-excited scattering experiments. Thus, for the conditions of our experiment we have an effective "two-level" system. For laser intensities that are not too high, and linewidths that are not

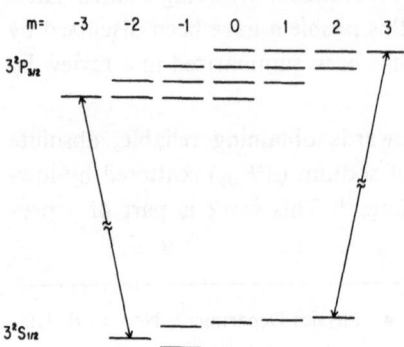

Figure 1. Zeeman structure of the $3S_{1/2}$ and $3P_{1/2}$ states of sodium $(H = 785$ G). The ground-state levels (F, m) are $(1, 1)$, $(1, 0)$, $(1, -1)$, $(2, -2)$, $(2, -1)$, $(2, 0)$, $(2, 1)$, and $(2, 2)$, starting at the lowest one and moving clockwise. The excited-state levels (m_I, m_J) are ordered in columns of constant $m = m_I + m_J$ and rows of constant m_J.

too narrow,[†] perturbation theory is applicable and one can write rate equations to describe the population dynamics of ground and excited states.

An atom will spend about 10^{-5} sec in the region illuminated by the laser. As the time constant of the excitation process will be typically about 10^{-8} sec, each atom will undergo several hundred excitation–decay cycles, and the steady state solution of the rate equations will describe the population of both ground and excited states. The fraction f' of atoms in the excited state will be

$$f' = \varrho/(1 + 2\varrho) \qquad (1)$$

where $\varrho = \lambda^3 P/8\pi ca\varDelta\nu$ is the ratio of stimulated emission rate to spontaneous emission rate. P is the laser power, a is the effective laser beam cross sectional area in the interaction region, and $\varDelta\nu$ is the laser linewidth. It should be noted that f' is the fraction of excited atoms originally in the particular ground-state hyperfine level being pumped by the laser. As there are eight such levels in the sodium ground state, only 12.5% of the atoms are available for excitation.

3. Experimental Method

The natural lifetime of the $3P_{3/2}$ state of sodium is about 1.6×10^{-8} sec. At thermal speeds such an atom will travel about 2×10^{-3} cm before decaying to the ground state. Accordingly, it is necessary for the laser, electron, and atomic beams to overlap. The geometry chosen for our experiment is shown in Figure 2; the three beams intersect each other orthogonally.

In the absence of the laser beam, this geometry is identical to the one used many times in this laboratory to measure ground-state cross sections by the atomic beam recoil technique, described in detail elsewhere.[8,9]

The atomic recoil angles ψ (in the plane defined by the atomic and electron beams) and χ (in the plane defined by the atomic and laser beams), are related to the electron polar and azimuthal scattering angles θ, ϕ by

$$\psi = \alpha - \cos\theta \qquad (2)$$

$$\chi = \beta \sin\theta \sin\phi \qquad (3)$$

where $\alpha = mv/MV$; $\beta = mv'/MV$. mv and mv' are the magnitudes of the electron momentum before and after the collision, and MV is the magnitude of the atomic momentum. If the collision is elastic, $\alpha = \beta$, and it is assumed that α and β are small.

To measure a total ground-state cross section σ one performs a "scattering-out" experiment, measuring the difference between the atom current reaching the detector with the electron beam off, I_0, and on, I. The total cross section is related to the

---

[†] This subject will be discussed in Section 3.

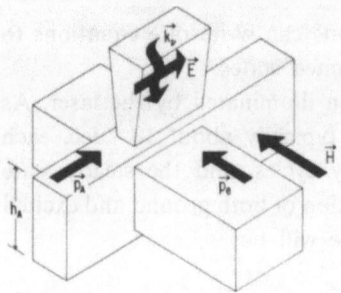

Figure 2. The interaction region. p_A, p_e, and k_ν are the atom, electron, and photon momenta. E gives the laser beam polarization. The magnetic field H is 785 G parallel to the electron beam. h_A is the height of the atomic beam.

experimental parameters by

$$\sigma = (\Delta I/I_0)(V h_A/I_e) \tag{4}$$

where $\Delta I = I_0 - I$, h_A is the atomic beam height, V is the atomic beam velocity, and I_e is the electron current (particles/sec). It is important to stress that in this method it is not necessary to measure absolute atomic beam currents; only the electron current passing through the interaction region has to be measured absolutely.

The applied magnetic field present in the interaction region is parallel to the electron momentum. To observe the $\Delta m = \pm 1$ transitions discussed in the preceding section, the laser light should be linearly polarized with the electric vector parallel to the atomic momentum, as shown in Figure 2. When the laser is tuned to either one of the $(2, \pm 2) \rightarrow (\pm\frac{3}{2}, \pm\frac{3}{2})$ transitions, the atoms will exchange photons with the laser field during their entire passage through the interaction region.

If one assumes that perturbation theory is applicable, then the atomic beam in the interaction region can be represented as an incoherent mixture of ground- and excited-state atoms. On the other hand, for high laser power or narrow band excitation, the target atoms can be represented as a partially coherent superposition of ground and excited states. The description of the electron scattering changes depending on which representation of the target one chooses, as has been discussed in detail by Gersten and Mittleman,[10] Mittleman,[11,12] and Hertel and Stoll.[7] We took the somewhat naive approach of assuming that if the period of the Rabi oscillation of the atom in the radiation field is substantially longer than the lifetime of the excited state against spontaneous decay, there will not be significant coherence between ground and excited states. Given the parameters of our laser, that is indeed the case, and we can represent the atomic target as an incoherent mixture of ground- and excited-state atoms. The measured cross section σ can thus be written in terms of the ground- and excited-state cross sections, σ_0 and σ_e, as

$$\sigma = f\sigma_e + (1 - f)\sigma_0 \tag{5}$$

where f is the actual fraction of atoms in the excited state present in the interaction region. Furthermore, this also means that the rate equation approach is applicable, and equation (1) for f' is also valid. Mittleman[12] obtained equivalent results using the rotating-wave approximation.

One must know f in order to obtain the excited-state cross section σ_e in terms of the measured "effective" cross section σ_0 using equation (5). To determine f we take advantage of the atomic recoil during the resonant photon interactions.

Resonant many-photon recoil is readily observable in an atomic beam experiment.[13] In our geometry the downward transfer per absorbed photon results in a recoil angle $\gamma_0 = h/MV\lambda$, where λ is the photon wavelength. Stimulated emission results in an opposite recoil γ_0 per photon. However, spontaneous decay photons are radiated in essentially random directions. If the atom undergoes many excitation–decay cycles, it will suffer a net downward recoil angle γ

$$\gamma = \gamma_0 f' \tau / \tau_N \tag{6}$$

where $\tau = l/V$ is the passage time through the interaction region of length and τ_N is the natural lifetime of the excited state. In the limit of high laser power the net recoil saturates at $\gamma_0 \tau / 2\tau_N$.

As the detector is at a distance L from the interaction region, the deflection d of the atomic beam due to resonant photon interactions is

$$d = \gamma_0 L f' \tau / \tau_N \tag{7}$$

and f' can be determined from equation (7). Furthermore, in our experiment d is of the order of 0.2 in., so that the atoms that have spent a fraction f' of the passage time in the excited state are spatially separated[†] from those atoms that have not been excited because they are in ground-state hyperfine levels other than the one being pumped by the laser. In these conditions, $f' = f$, the fraction of excited state atoms used in equation (5).

In performing the experiment, therefore, the atom detector is lowered from its axial position by d, so as to measure the current of atoms that have been excited in the interaction region. By turning the electron beam on and off the total electron scattering cross section for excited sodium atoms can be determined using equations (4), (5), and (7).

4. The Apparatus

4.1. The Atomic Beam

Figure 3 shows the experimental arrangement. The sodium beam source is an effusive oven heated up to about 500°C. The beam is polarized by an offset Stern–Gerlach magnet, so that the number of ground-state hyperfine levels in the beam is 4, rather than 8. The magnet also acts as a velocity selector, giving a velocity spread

---

[†] That is to say, we actually have an excited atom beam!

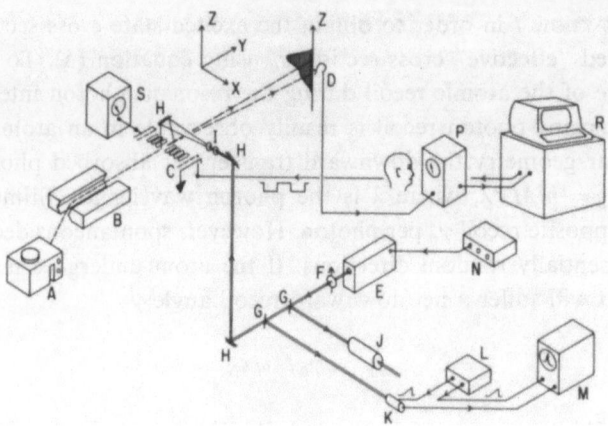

Figure 3. The experimental arrangement. A, sodium oven; B, Stern–Gerlach magnet; C, electron
gun; D, Channeltron electron multiplier; E, tunable dye laser; F, polarization rotator; G,
beam splitters; H, mirrors; I, lenses; J, sodium vapor cell; K, spectrum analyzer; L, spectrum
analyzer ramp generator; M, mode structure monitoring scope; N, tunable dye laser frequency
control; P, atomic beam monitoring electrometer; R, DECLAB-03 computer; S, electron
current monitor.

$\Delta V/V$ of about 0.08. After passing through the interaction region the atoms are
detected by surface ionization on a hot platinum wire. The ions are mass-analyzed
and detected by a Channeltron electron multiplier operated in the current mode,
its output being monitored by an electrometer.

The detector can be moved vertically to scan the beam profile, and it can be
aligned either on-axis to monitor the full atomic beam, or below axis to monitor the
"excited" beam as discussed in Section 3.

4.2. The Electron Beam

The electron gun is similar to the one described by Collins *et al.*[14] The gun
lies between the pole pieces of a permanent magnet, producing a field parallel to the
electron momentum. The electron beam is ribbon-shaped, about 2.5 cm wide and
0.1 cm high (slightly less than the atomic beam). The electron current through the
interaction region is monitored by a digital microammeter, and currents between
200 and 500 µA have been used. The energy width is about 0.40 eV; the mean energy
is corrected for contact potentials and space charge effects.

4.3. Atomic Velocity Determination

Equations (4) and (7) require the explicit measurement of the atomic beam
velocity. This is accomplished by a technique described by Collins *et al.*[15] Electrons
that are inelastically scattered in the forward direction ($\theta = 0$) result in recoiled atoms
for which $\psi = \alpha - \beta$ [equation (2)] and $\chi = 0$ [equation (3)]. Because of the finite

height of the detector, such atoms are detected with greater efficiency than those for which $\theta > 0$, thus giving rise to a peak in the atomic angular distribution as a function of detector position, as shown in Figure 4.

The position x of this forward inelastic scattering peak is related to the electron energy E and atomic velocity by

$$x = \frac{(2m)^{1/2}}{M} L \frac{E^{1/2} - (E - E_e)^{1/2}}{V} \tag{8}$$

where E_e is the excitation energy. Thus, by measuring the displacement of the forward inelastic scattering peak one can determine the mean atomic beam velocity.

4.4. The Laser Beam

A single-mode tunable dye laser using Rhodamine 6G as the lasing medium is pumped by an argon ion laser. The laser frequency can be tuned manually to the desired transition using commercial piezoelectric drives for the cavity back mirror and an intracavity etalon. Coarse tuning is accomplished by monitoring the fluorescence in a sodium vapor cell, and fine tuning by monitoring the atomic beam itself, making use of the atomic recoil during resonant photon interactions. A spectrum analyzer is used to monitor the laser mode structure. A polarization rotator allows us to choose the desired polarization for photons in the interaction region. A cylindrical lens stretches the laser beam along the atomic beam, so as to illuminate the full 2.5 cm length of the interaction region. An auxiliary spherical lens helps to confine the light to the region of interest.

The alignment of all the optical components is best accomplished by monitoring the atomic beam, again taking advantage of the atomic recoil during resonant photon interactions.

4.5. Data Acquisition and Processing

The output of the atomic beam monitor electrometer is fed to the analog-to-digital converter of a DECLAB-03 computer programmed to sample the signal with the electron gun off, turn the electron gun on and sample the signal again,

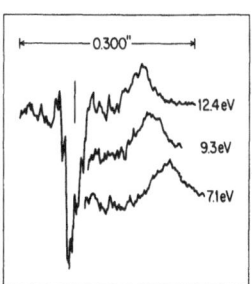

Figure 4. Atomic beam signal with the detector moving parallel to the electron beam, at three different electron energies. The prominent positive peaks are the forward inelastic scattering peaks. The negative peak indicates scattering off the atomic beam. The vertical line above it marks the position of the center of the beam.

and then turn the electron gun off and repeat the cycle. The accurate tuning of the laser to the transition is maintained by constantly monitoring the atomic beam in the electrometer and correcting the laser manually whenever necessary.

After a preset number of cycles, the data accumulated in the computer are processed, and a value for the measured cross section σ in the interaction region is produced. The excited-state cross section is computed from this value using equation (5) and the ground-state cross section values measured by Kasdan et al.[9]

5. Results

In this section we will show first the evidence for the excitation of the atoms and for the validity of the photon recoil technique, and then we will present preliminary values of the total scattering cross section of electrons by $3P_{3/2}$ sodium atoms.

Figure 5 shows the unpolarized atomic beam signal as a function of laser frequency, with the detector on the beam axis. The negative peaks occur at the frequencies at which atoms are recoiled off-axis by resonant photon interactions. If we compare them with the theoretical absorbtion lines on the same figure, we can see that the signal is much more intense for the transitions that do not depopulate their ground-state levels, $(2, -2) \rightarrow (-\frac{3}{2}, -\frac{3}{2})$ and $(2, 2) \rightarrow (\frac{3}{2}, \frac{3}{2})$, than for the other six transitions of comparable strength. This is so because the number of photons ab-

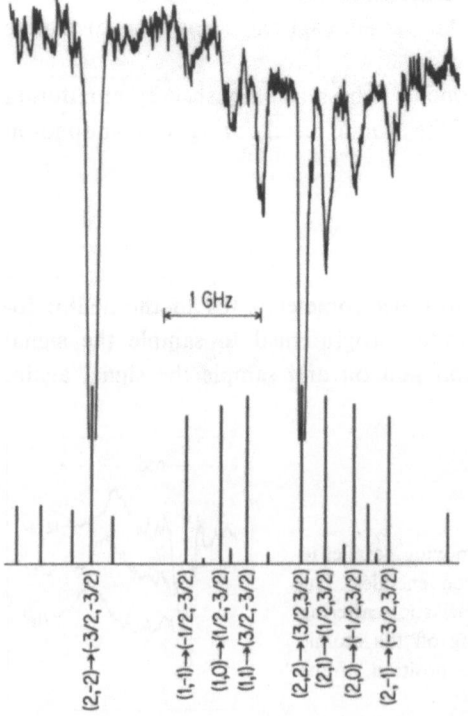

Figure 5. Atomic beam signal as a function of laser frequency. The detector is on-axis, and the beam is unpolarized. The vertical lines at the bottom mark the calculated positions and intensities of all the allowed transitions in the frequency interval being scanned. Only the eight most intense lines have been labeled, in the $(F, m) \rightarrow (m_I, m_J)$ notation.

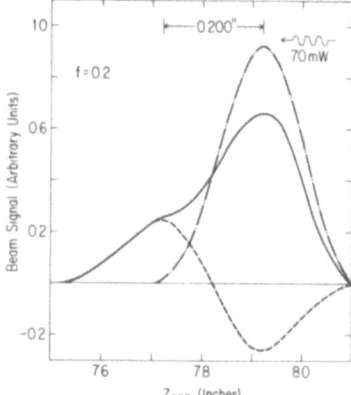

Figure 6. The solid line shows the vertical beam profile with the laser tuned to a nondepopulating transition. The atomic beam is polarized. The laser power (measured at the laser output) was 70 mW. The long-dash line shows the vertical beam profile with the laser off. The short-dash line is the difference signal.

sorbed per atom for the first two transitions is much larger than that for the last six ones, which depopulate their ground-state level.

A typical vertical beam profile with the laser tuned to one of the two useful transitions is shown in Figure 6. The intensity of the difference signal peak ("excited atom beam") is about 25% of the intensity of the beam with the laser off. As the atomic beam is polarized, there are four different ground-state hyperfine levels present in the interaction region, of which only one is pumped by the laser. This means that essentially all atoms available for excitation are actually being excited by the laser.

From the deflection of the "excited atom beam" (0.200 in.) we calculate $f = 0.2$, which is a typical value in our experiments. This value of f corresponds to $\varrho = \frac{1}{3}$. The average number of photons absorbed by each atom when passing through the interaction region is

$$\langle N \rangle = (1 + \varrho)\gamma/\gamma_0 \tag{9}$$

In our case, $\langle N \rangle \cong 370$.

Table 1. Total Cross Sections for the Scattering of Electrons by $3P_{3/2}$ Sodium Atoms

Energy (eV)	Cross section (Å²)
0.84	594 ± 63
1.1	295 ± 56
1.7	234 ± 42
2.3	224 ± 67
3.0	209 ± 33
4.5	127 ± 44
6.0	120 ± 41

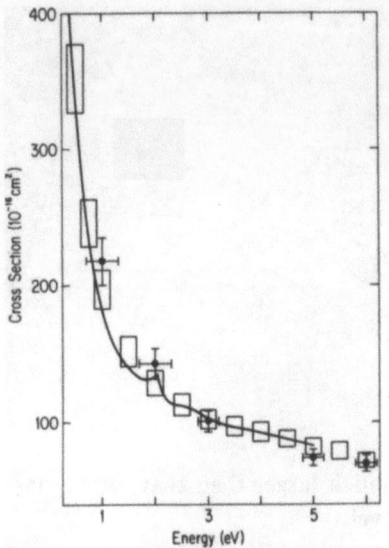

Figure 7. Total cross sections for the scattering of electrons by ground-state sodium atoms. Black circles: this work. The horizontal error bars give the electron energy spread. The vertical bars reflect statistical and possible systematic errors. Rectangles: measurements by Kasdan *et al.*[9] Continuous line: 4-state close coupling calculation by Moores and Norcross.[17]

To take scattering data the detector is set at about 0.200 in. below the beam axis. At this position the contribution to the signal from the 75% of the atoms that are not excited by the laser is negligibly small.

Our total cross section results for the scattering of electrons by sodium atoms in the $3P_{3/2}$ state are presented in Table 1, with the associated standard errors.

The results of the calculations of Moores *et al.*[16] presented in Table 1 of their paper cannot be compared directly to our experimental results since they combine the elements of the T-matrix which describe the scattering of electrons by an ensemble of sodium atoms in which all the $3P_{3/2}$ and $3P_{1/2}$ states are populated, while our experiment deals with scattering by a single hyperfine sublevel of the $3P_{3/2}$ state.

In order to check the possibility of unexpected systematic errors we have performed some measurements of ground-state total cross sections, in conditions as close as possible to the ones used during the excited-state experiments. The results are shown in Figure 7. The error bars were calculated exactly in the same way as for the excited-state cross sections, and it can be seen that there is reasonable agreement with both previous experimental results by Kasdan *et al.*[9] and four-state close coupling calculations by Moores and Norcross.[17]

ACKNOWLEDGMENTS

This work was supported by the National Science Foundation. We wish to thank Mr. Neil Pignatano for his help in programming the DECLAB-03.

References

1. G. M. Carter, D. E. Pritchard, M. Kaplan, and T. W. Ducas, *Phys. Rev. Lett.* **35**, 1144 (1975).
2. R. Düren, H. O. Hoppe, and H. Pauly, *Phys. Rev. Lett.* **37**, 743 (1976).
3. I. V. Hertel and W. Stoll, *J. Phys. B* **7**, 583 (1974).
4. N. D. Bhaskar, B. Jaduszliwer, and B. Bederson, *Phys. Rev. Lett.* **38**, 14 (1976).
5. S. Trajmar, private communication.
6. J. Macek and I. V. Hertel, *J. Phys. B* **7**, 2173 (1974).
7. I. V. Hertel and W. Stoll, in *Advances in Atomic and Molecular Physics*, D. R. Bates and B. Bederson, eds., Vol. 13, pp. 113–228, Academic Press, New York (1977).
8. K. Rubin, B. Bederson, M. Goldstein, and R. E. Collins, *Phys. Rev.* **182**, 201 (1969).
9. A. Kasdan, T. M. Miller, and B. Bederson, *Phys. Rev. A* **8**, 1562 (1973).
10. J. I. Gersten and M. H. Mittleman, *Phys. Rev. A* **13**, 123 (1976).
11. M. H. Mittleman, *Phys. Rev. A* **14**, 1338 (1976).
12. M. H. Mittleman, *Phys. Rev. A* **16**, 1549 (1977).
13. O. Frisch, *Z. Phys.* **86**, 42 (1933).
14. R. E. Collins, B. B. Aubrey, P. N. Eisner, and R. J. Celotta, *Rev. Sci. Instrum.* **41**, 1403 (1970).
15. R. E. Collins, B. Bederson, and M. Goldstein, *Phys. Rev. A* **3**, 1976 (1971).
16. D. L. Moores, D. W. Norcross, and V. B. Sheorey, *J. Phys. B* **7**, 371 (1974).
17. D. L. Moores and D. W. Norcross, *J. Phys. B* **5**, 1482 (1972).

50

Orientation and Alignment in Scattering Processes from Laser-Excited Atoms

H. W. HERMANN AND I. V. HERTEL

We discuss qualitative differences observed in collisions of optically pumped Na(3P) atoms with electrons and diatomic molecules. Varying the scattering geometry and the polarization of the pumping laser prepares well-defined spatial configurations of the colliding Na and allows measurement of details on the ratios and relative phases of the partial transition amplitudes. We compare the electron-collision-induced transitions $P \rightarrow S(I)$, $P \rightarrow D$ in Na (II), and the quenching of Na(3P) by diatomic molecules (III). The three cases exhibit remarkably different responses to orientation (circular) and alignment (linear polarized pumping light) of the Na: I and II depend strongly on the orientation, III not measurably. In contrast, in cases I and III the alignment is strong, while for case II it is weak. Process I is well described by scattering calculations. For cases II and III we present qualitative models to interpret the experimental observations.

Introduction

During the past few years it has become possible to study binary atomic and molecular collision processes in such detail that one may now obtain not only the differential cross sections but also complete information on the state of at least one of the interacting particles after the collision. This has been demonstrated again in the present workshop. Some aspects will be discussed in this paper.

H. W. HERMANN AND I. V. HERTEL ● Sonderforschungsbereich 91, Universität Kaiserslautern.
I. V. Hertel's permanent address: Institüt für Molekülphysik, Freie Universität Berlin.

To be more specific we discuss processes of the type

$$B + A \rightarrow B' + A^* \tag{1}$$

Let B be an electron, atom, ion, or molecule, and A be an atom undergoing an electronic transition during the collision. If the atom A^* is not excited to an S state, it will in general be in a nonisotropic distribution after collision. To describe A^* is the goal of the type of studies reported in this volume. That is to say, one wishes to determine experimentally the density matrix of the atomic system. One has to be aware of the possibility of exciting A^* incoherently if different states B or B' participate in the collision.

It may be useful to emphasize the meaning of coherence and incoherence as applied to stationary states. An atom has been excited coherently if one can find a coordinate frame or a complete orthogonal basis set of atomic states for which the density matrix has only one diagonal term. This is equivalent to $\text{Tr}\,\varrho^2 = 1$. In contrast, the atom is excited incoherently when its density matrix brought into diagonal form contains more than one diagonal element $\text{Tr}\,\varrho^2 < 1$.

In general, a density matrix of an atomic state contains diagonal as well as nondiagonal terms. One may not directly recognize from their values whether the atom is in a pure or mixed state.

The density matrix for atoms prepared in collision processes according to equation (1) is subject to some restrictions due to reflection symmetry with respect to the scattering plane. In consequence, there are fewer independent parameters than density-matrix elements.

In the present paper we specialize furthermore to A^* being in a P state, and we wish to discuss three characteristic examples of such studies. The evaluation of experimental data is, as usual, done by assuming that the electron and nuclear spins can be decoupled from orbital angular momenta during the collision by the appropriate Racah algebra (Percival–Seaton hypothesis). The density matrix will thus be given schematically

$$\varrho\big(A^*(P)\big) = \begin{pmatrix} \varrho_{-1-1} & \varrho_{-10} & \varrho_{-11} \\ \varrho_{0-1} & \varrho_{00} & \varrho_{01} \\ \varrho_{1-1} & \varrho_{10} & \varrho_{11} \end{pmatrix} \tag{2}$$

Reflection symmetry leaves four independent parameters. Following the convention of Reference 3 we adopt

$$\varrho_{00}\,(\text{real}) \rightarrow \quad \lambda = \varrho_{00} = \frac{\sigma_0}{\sigma_0 + 2\sigma_1} \tag{3}$$

$$\varrho_{01}\,(\text{complex}) \quad \nearrow \quad \cos\chi = \frac{\text{Re}\,\varrho_{01}}{(\varrho_{00}\varrho_{11})^{1/2}} \tag{4}$$

$$\searrow \quad \sin\varphi = \frac{\text{Im}\,\varrho_{01}}{(\varrho_{00}\varrho_{11})^{1/2}} \tag{5}$$

$$\varrho_{-11}\,(\text{real}) \rightarrow \cos\delta = \frac{\varrho_{-11}}{\varrho_{11}} \tag{6}$$

with

$$\varrho_{11} = \tfrac{1}{2}(1 - \varrho_{00}), \qquad \varrho_{01} = \varrho_{10}^* = -\varrho_{0-1} = -\varrho_{-10}^*$$

$$\varrho_{-11} = \varrho_{1-1}^*, \qquad \varrho_{-1-1} = \varrho_{11}$$

The new definitions (5) and (6) extend (3) and (4) to the general case of a P-state atom A^*. These four remaining dynamical parameters can be visualized; for instance, by the charge distribution illustrated in Figure 1. The direction of alignment in the x–z scattering plane, indicated by the angle α, is one of these parameters, the ratio of height to thickness and length to thickness of the ellipsoidal shape are two more, and the last parameter gives the angular momentum contained in the electron cloud of the atom with respect to an axis y perpendicular to the collision plane. This is described by dynamical variables constructed from angular momenta in a standard irreducible form.[1] They relate in a simple way to the four density-matrix parameters given in equations (3)–(6) (see, e.g., the review by Hertel and Stoll[2]). The standard choice of the collision frame for presentation of the experimental data is such that its z axis parallels the direction of either the ingoing or the outgoing relative velocity of the interacting particles and the x axis lies in the scattering plane (Figure 2). This is done since calculations are usually given for this particular collision frame. Other choices may be more adapted to the physical problem such as a coordinate system having its z axis normal to the collision plane which is the only axis uniquely defined.

The determination of four dynamical independent parameters necessitates of course four independent experimental data. For any special problem, in principle, two experimental approaches are possible, although their use is restricted by practical reasons:

The photon correlation technique as introduced by Kleinpoppen and co-workers[3] is now most commonly used. The photon emitted from the atom A^* after the reaction

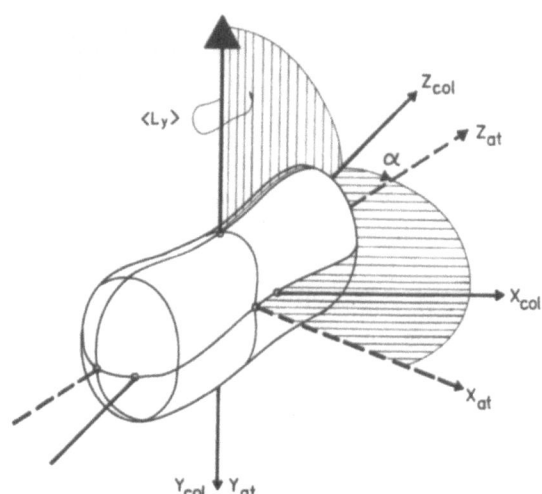

Figure 1. Charge distribution of an atom A^*, excited into a P state by a collision process. The collision plane is indicated by the x and z axis.

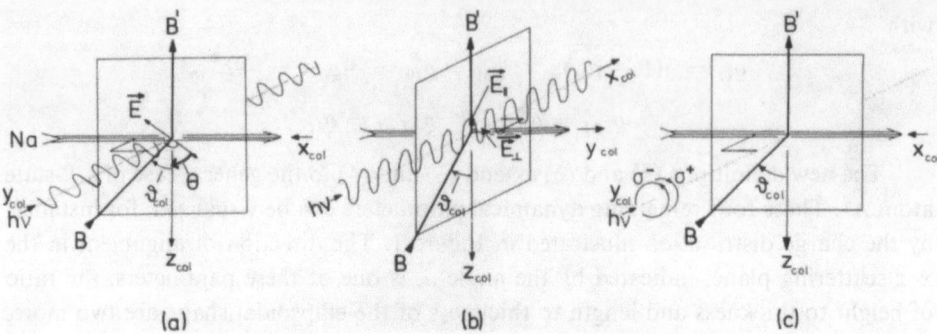

Figure 2. The three types of experiments necessary for complete investigation of excited atoms A^* prepared by collision with particles B. (a) Linear polarized light propagating perpendicular to the $x–z$ scattering plane (two parameters); (b) Linear polarized light propagating within the $x–z$ scattering plane (one parameter); (c) Circular polarized light perpendicular to the $x–z$ scattering plane (one parameter).

(1) is observed in correlation with one of the scattered particles. This can be done either by coincidence techniques or by quasicoincidence, that is, direct photon observation from fast beams (e.g., Reference 4). The four parameters can be extracted by observing for instance the three Stokes parameters for light emitted at right angles to the scattering plane, plus one of the linear Stokes parameters of light emitted parallel to the scattering plane.

The laser excitation technique first demonstrated by our group[5] uses selective magnetic substate preparation of the atom $A^\dagger$ in a collision that is essentially the inverse process to equation (1). By a suitable choice of excitation geometry and state of laser polarization one obtains the parameters of equations (3)–(6) for the collisional excitation process described in equation (1).[†] It should be emphasized that the same density-matrix elements for the original process equation (1) are measured as by photon correlation although the inverse process is actually observed.

The experimental arrangements used in the investigations reported here are illustrated in Figure 2 and differ somewhat from previously used geometries.[7] Three different types of measurements have to be performed to obtain four independent experimental parameters:

(a) *Linear polarized light* is shone *perpendicularly* onto the scattering plane, and the scattering intensity is measured as a function of the angle θ between the electric field vector of the light and the direction of the outgoing scattered particle B' (Figure 2a). This leads to two parameters, one of which may be the ratio of maximum to minimum intensity and the other one the angle $\theta = \alpha$ for which the maximum scattering intensity is observed.

[†] In this comparison with photon coincidence experiments one has also to invert the z axis and thus ends up with the PT inversion for complete equivalence as discussed by Fano.[6]

(b) *The exciting linear polarized light* is *incident within* the scattering plane. Measured is the ratio of the scattering intensities, the electric vector being parallel to the incoming particle beam B' or rectangular to it, respectively (Figure 2b).

(c) *Left and right circular polarized light* is shone *perpendicular* onto the scattering plane (Figure 2c). The fourth parameter is the asymmetry ratio of the scattering intensities.

We shall now present results for the three scattering processes

   I. $Na(3S) + e^- \leftrightarrow Na^*(3P) + e^-$

  II. $Na(3D) + e^- \leftrightarrow Na^*(3P) + e^-$

 III. $Na(3S) + M_2^{\#} \leftrightarrow Na^*(3P) + M_2$

In all cases the present experiments start at the right-hand side of these equations by exciting Na $3^2P_{3/2}(F = 3)$ with a single-mode tunable dye laser.

Example I: Na(3²P) + e → Na(3²S) + e

The $3P \to 3S$ deexcitation process induced by electron scattering is discussed for an incident electron energy of 3 eV (outgoing electron energy 5.1 eV). Since the $3S$ ground state is invariant under reflection on the scattering plane ($3S\ M_L = 0$) the excitation (deexcitation) amplitudes must obey the relation $f_{\pm 1}^{\pm} = -f_1^{\pm}$. Thus, for this particular type of transition only, we have an additional constraint on the density matrix $\varrho_{-11} = -\varrho_{11}$, whereby the set of independent dynamical parameters is reduced to three. Since our experiment allows us to determine four independent parameters, we are fortunately able to obtain the optical alignment parameter.[7]

Since for the $^2P \leftrightarrow {}^2S$ process under discussion direct and exchange amplitudes (or singlet and triplet scattering amplitudes, respectively) participate, the excitation may in general be incoherent. If the collision prepares a pure state (coherent excitation) the set of parameters is further reduced and only two parameters describe the process. In that case the parameters χ and φ given by equations (4) and (5) are equal. Alternatively, the degree of coherence of the excitation may be represented by the expression $\cos^2 \chi + \sin^2 \varphi < 1$.

We shall not indulge here in further discussion of coherence for this particular process; some difficulties that arise will be the subject of a forthcoming publication.[8] The process discussed here has been studied previously[7] and we shall present recently improved results mainly for illustration purposes. Figure 3 shows measurements for linearly polarized excitation of type (a) (see Figure 2) as a function of the angle θ normalized to the scattering intensity ratio of type (b). This measured distribution reflects the charge cloud illustrated in Figure 1, which would be produced in the time inverse process $3S \to 3P$. This qualitative interpretation of the results may be proved explicitly.[2]

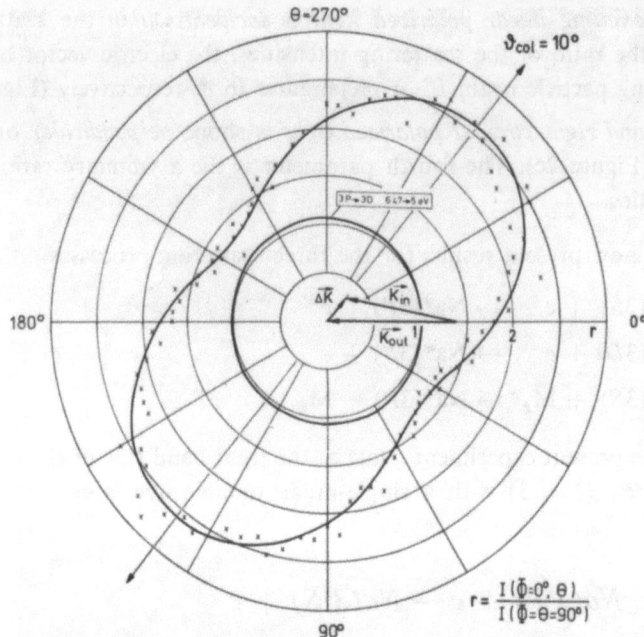

Figure 3. Electron-impact-induced $3^3P \leftrightarrow 3^3S$ transition $(E_{inc} = 3 \text{ eV})$: Scattering intensities (charge distribution) from type (a) measurements, normalized to type (b). A least-squares fit to data is given by the full line. For comparison the $3P \leftrightarrow 3D$ distribution (Figure 7) is also displayed to scale.

Remarkable features of the experimental data are the very large anisotropy and the fact that the charge distribution aligns nearly to the momentum transfer vector ΔK (Figure 3), as would be predicted by the Born approximation. In contrast to the Born approximation the charge distribution has no cylindrical symmetry around ΔK as indicated by the deviation of the minimum scattering intensity ratio from the value 1. The experiments with linearly polarized light give the λ parameter [equation (3)]. It is shown as a function of scattering angle ϑ_{col} in Figure 4. It compares reasonably well with theory; the best agreement is found for a four-state close-coupling approximation.[9]

The observation of left–right circular asymmetry [measurement type (c)] yields additional information. A large asymmetry is shown in Figure 5. It directly relates to the expectation value of angular momentum with respect to an axis y normal to the scattering plane with $\langle L_y \rangle \leq 1$ for a P state. From the asymmetry the phase parameter φ may be extracted by using the previously measured λ parameter.[2] The agreement with the four-state close-coupling calculations[9] is excellent, while a distorted-wave polarized-orbital approximation[10] shows a significant disagreement, as expected for these low energies.

The sign of the left–right circular asymmetry is of significance: According to Fano and Kohmoto[11] it reflects whether the electron observes an effectively repulsive

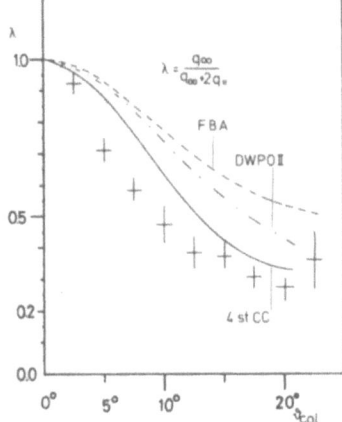

Figure 4. Measured λ parameter for the electron-impact-induced $3P \to 3S$ Na ($3 \to 5.1$ eV) transition as a function of the collision angle ϑ_{col} and comparison with different scattering theories: First Born approximation (- - -); distorted wave polarized orbital (—·—) by J. V. Kennedy;[10] four-states close-coupling (—) calculations by Moores and Norcross.[9]

or attractive force during the collision. A simplified picture is given by Figure 6: As indicated there, after the collision the electrons are detected on the right side of the scattering atomic target, which is oriented either clockwise or counterclockwise. The measured scattering intensities are larger for the first case ($I^+ > I^-$).

Also shown in Figure 6 is a classical model of the electron interacting with the oriented atom in order to illustrate the angular momentum balance of the process. The change of the angular momentum projection quantum number Δm_{el} for the scattered electron and ΔM_L for the atom is indicated with respect to the y axis. In order to reach the detector, the electron has to pass the atom on the left- or right-hand side when the latter is oriented clockwise or counterclockwise, respectively.

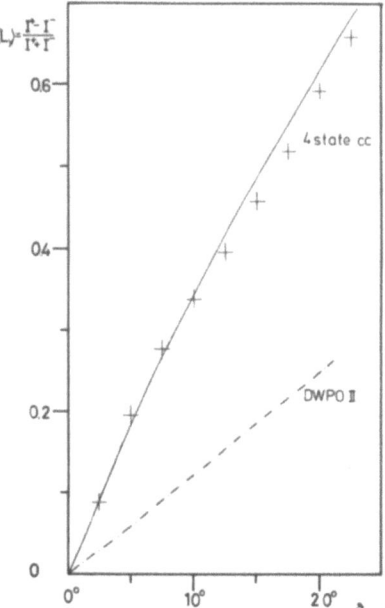

Figure 5. Right–left asymmetry ratio for circular polarized excitation in the electron-impact-induced $3^2P \to 3^2S$ Na ($3 \to 5.1$ eV) transition in sodium and comparison with different theories (same as in Figure 4).

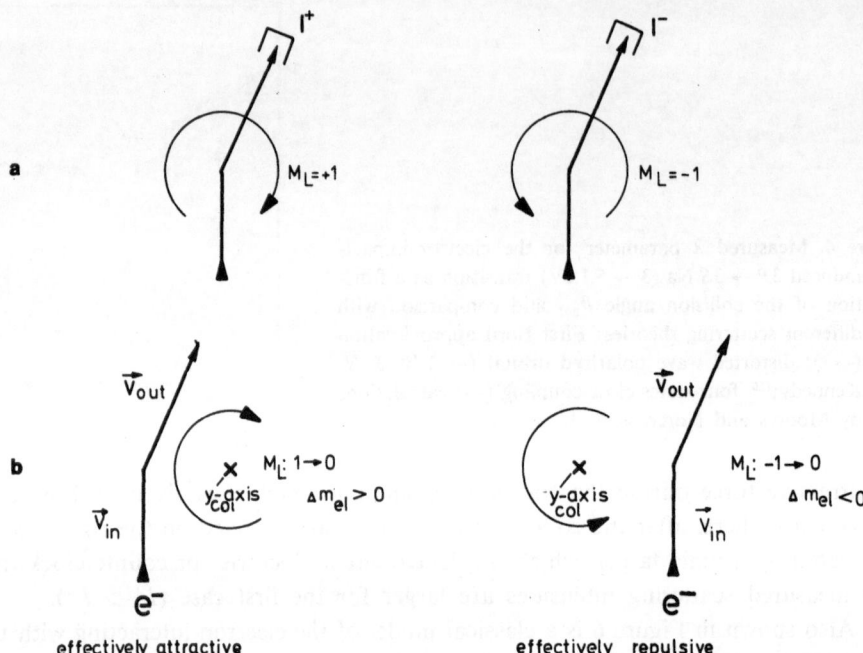

Figure 6. (a) Schematic of the experimental results for type (c) measurement ($e^- + Na^*$ $3P \to 3S$, circularly polarized excitation). Experiment shows that $I^+ > I^-$. (b) Angular momenta of the colliding electron (on the right $m_{el} < 0$, on the left $m_{el} > 0$) and of the laser-excited atom (right: right-hand circularly polarized light $M_L = 1$, left: $M_L = -1$), assuming a constant classical impact parameter. Note the quantization (z) axis points into the page. Comparison of (a) and (b) indicates an effectively attractive force.

This is obvious from angular momentum conservation, since the velocity of the outgoing electron is larger than that of the incoming one. Thereby it is assumed that the classical impact parameter is conserved during the collision.

Comparing this to the experiment, apparently the electron passes the atom on the right-hand side with a higher probability than on the left-hand side. Thus, an effectively attractive force is indicated. We just note this as a phenomenological observation within the framework of this classical model which finds its justification to a certain extent in the quantum mechanical treatment.[11] We summarize the results for the $3^2P \to 3^2S$ transition in sodium induced by electron collisions. We observe (a) a high degree of anisotropy for linearly polarized light excitation and (b) a strong circular asymmetry.

One may understand this qualitatively by recollecting the fact that essentially only two independent amplitudes participate in this process. Their interference easily yields a nonisotropic population (or depopulation) of different magnetic substates. This is even predicted by the Born approximation while, on the other hand, the circular asymmetry cannot be given by such simple axially symmetric theory. It is, however, well described by appropriate close-coupling calculations.

Example II: Na(3P) + e → Na(3D) + e

We now turn to the electron-collision-induced transition $3P \rightarrow 3D$ starting from the laser-excited $3P$ sodium atoms. This case is more general than the previous one, since the five magnetic substates of the final D state partic pate in contrast to only one final magnetic substate for the $P \rightarrow S$ transition. Thus, for the time inverse process starting from an isotropic and incoherently prepared state (which effectively is probed by the laser excitation method) one would not expect a large degree of coherence of the $3P$ atom after the collision. An experimental result for linearly polarized light is shown in Figure 7. Note the enlarged radial scale in this diagram, which clearly shows only a very small but measurable anisotropy. In order to compare with the $3P \rightarrow 3S$ transition, Figure 3 also displays a least-squares fit to the $3P \rightarrow 3D$ data to scale. The anisotropy is hardly noticeable. From Figure 7 we see that, again, the anisotropy shows a maximum approximately in the direction of the momentum transfer vector $\Delta \mathbf{K}$ and again the charge distribution is cylindrically symmetric. The experimental data may be used to determine the λ parameter (F gure 8). Here λ does not depend strongly on the collision angle and is around 0.3, indicating approximately equal population of the $3P$ magnetic substates in the time inverse collision $3D \rightarrow 3P$. This is not surprising.

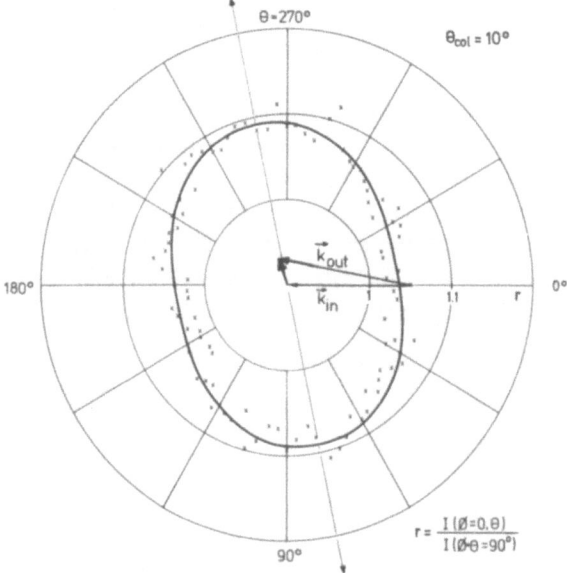

Figure 7. Electron-impact-induced $3^2P \leftrightarrow 3^2D$ Na transition ($E_{inc} = 6.47$ eV, $\vartheta_{col} = 10°$): Scattering intensities (charge distribution) from type (a) measurements, normalized to type (b). Note the magnified radial scale for intensity ratios > 1 in comparison to Figure 3. The solid line gives a least-squares fit to the data and is also shown in Figure 3. The vector diagram indicates the momentum transfer ΔK, which points approximately into the direction of maximum scattering intensity.

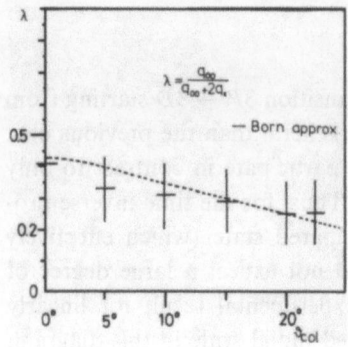

Figure 8. The λ parameter for an electron-impact-induced $3P \leftrightarrow 3D$ Na transition ($6.47 \to 5$ eV) as a function of collision angle ϑ_{col} and comparison with the Born approximation.

In principle eight different transitions between magnetic substates participate. This obviously may lead to a more statistical population. In contrast, for the $3P \to 3S$ case a large anisotropy is expected. For example $\lambda = 1$ for zero angle scattering follows directly from angular momentum conservation rules.

We now turn to the investigation of a possible asymmetry in type (c) measurements. The result is shown in Figure 9. From the previous arguments one might expect that the circular asymmetry is negligible also. To our great surprise, we observe a substantial left–right asymmetry $(I^+ - I^-)/(I^+ + I^-)$ which reaches a maximum at a collision angle of 20°, corresponding to an angular momentum expectation value of 0.3. In a coordinate frame where the z axis is perpendicular to the collision plane, this would correspond, e.g., to a 3:1 population of the ($3P\ M_L = +1$) and ($3P\ M_L = -1$) states, respectively, which indicates a substantially nonisotropic distribution of magnetic substates. This can only be explained by assuming that some transitions among the magnetic substates are predominant, which seems to be in contrast to the observed isotropic distribution, using linearly polarized light.

The measurements with circular polarization of the exciting light, however, select a particular subset of transitions, as indicated schematically in Figure 10. This diagram is consistent with the model discussed in the $3P \to 3S$ case (see Figure 6). The experimental findings are again such that the electron intensity scattered to

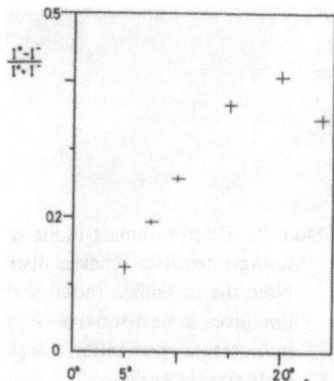

Figure 9. Right–left asymmetry ratio for circularly polarized light excitation in the electron-impact-induced $3P \leftrightarrow 3D$ Na transition ($6.5 \to 5$ eV) in sodium.

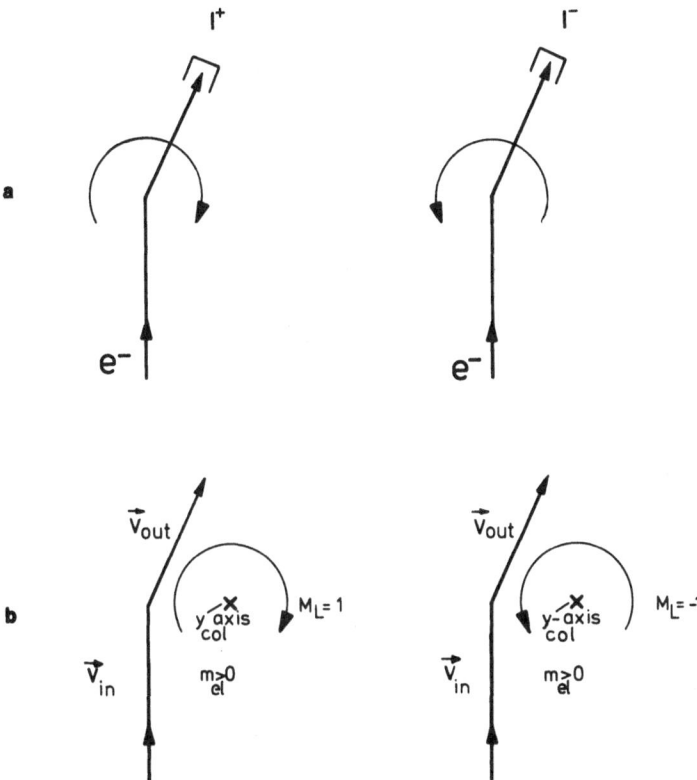

Figure 10. (a) Schematic of the experimental findings for type (c) measurement (e^- + Na $3P \rightarrow 3D$, circularly polarized excitation), otherwise same as Figure 6. (b) $P \rightarrow D$ analog to Figure 6. Experiment shows that $I^+ > I^-$. Note that $v_{\text{in}} > v_{\text{out}}$ (inelastic collision), therefore $\Delta m_{\text{el}} < 0$ assuming effectively attractive force, which leads to the constraint $\Delta M_L > 0$. The transitions $M_L: 1 \rightarrow 2$ are strong, while $M_L: -1 \rightarrow 0, 1, 2$ are weak.

the right from a 3P atom oriented clockwise is larger than the scattering intensity of an atom oriented in the opposite sense. If one assumes that again an effectively attractive force is acting during the collision and adopts the classical model one sees that in this case ($v_{\text{in}} > v_{\text{out}}$) the change in the electron angular momentum Δm_{el} is negative and thus the atomic angular momentum ΔM_L is increased during the collision. (Note that M_L and m_{el} refer here to a quantization axis perpendicular to the scattering plane, i.e., to y_{col}).

Comparing this with the experimental results, we see that the transition $(3P \; M_L = 1) \rightarrow (3D \; M_L = 2)$ is more probable than the transitions $(3P \; M_L = -1) \rightarrow (3D \; M_L = 0, 1, 2)$. This is at least one possibility to interpret the experimental results. The following considerations give a plausible dynamical explanation supporting these arguments.

In Figure 11 we summarize the observations by indicating strong and weak transitions connecting the 3P and 3D substates. From the above arguments (circular

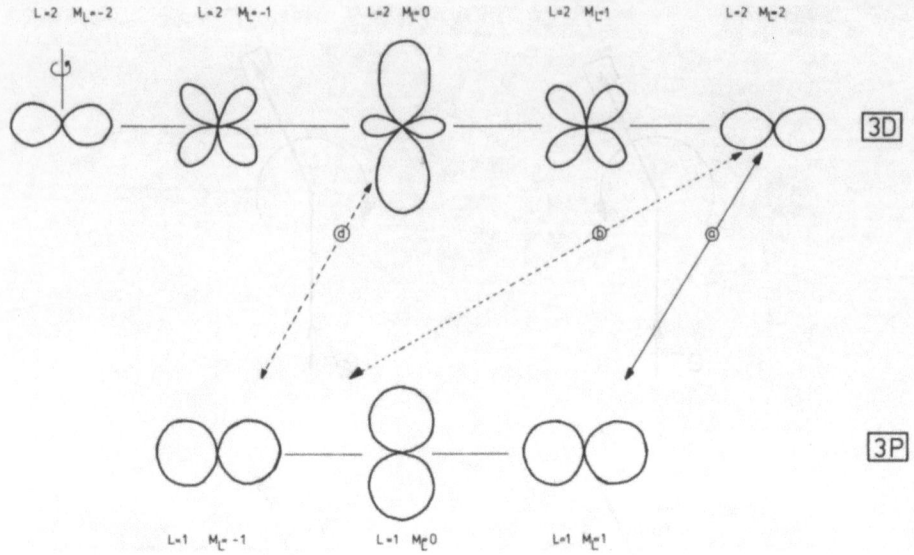

Figure 11. Schematic diagram of the $3P \leftrightarrow 3D$ transition indicating the experimental findings: Strong (—) and weak (- - -) transitions. The shapes of the $3p$ and $3d$ angular wave functions illustrate strong or weak overlap for the transitions ⓐ and ⓓ, respectively.

measurements) we learned that the transition probability ⓐ is large as compared to ⓑ and ⓓ. Again the quantization axis is perpendicular to the collision plane. Figure 11 is suggestive in explaining the differences in the transition probabilities since it is obvious that the suitably weighted overlap between final and initial state is much larger for the transition ⓐ than for ⓓ. In other terms the corresponding Clebsch–Gordan coefficients interconnecting these states in the collision process are large or small, respectively. The weak transition ⓑ in turn involves at least the f partial wave, which may be significantly smaller than the s- and p-wave contribution dominating the other transitions.

We summarize our observations on the electron-collision-induced transition $3P \rightarrow 3D$ in sodium at 5 eV incident energy: A very small linear anisotropy and a rather strong left–right circular asymmetry have been found.

Both observations may be understood by adopting a simple physical picture of the process. However, it would be of great use to have a realistic close-coupling or distorted-wave polarized-orbital calculation.

Example III: $Na(3D) + M_2 \rightarrow Na(3S) + M_2^{\#}$

As a final topic we discuss an interesting and important heavy-particle collision process involving excited sodium and diatomic molecules M_2, the latter being either N_2, H_2, or D_2. In the course of collisions between these reagents the Na atom is quenched into the ground state while the molecule accepts part of the electronic energy to become vibrationally and rotationally excited while the remaing part of initial energy is transferred into relative translational energy of the colliding system. The present knowledge about this process is discussed in a forthcoming review.[12] We wish to discuss some aspects from first observations of polarization effects for this process. Again the sodium has been excited by laser light, this time however only with type (a) and (c) measurements (see Figure 2). A discussion of the possible parameters and the expected reflection symmetries has not yet been given, but a few interesting experimental results have already emerged for small center of mass scattering angles.[13]

We observe a small but significant anisotropy of the differential quenching cross sections for linearly polarized light excitation. This is indicated in Figure 12 together with a least-squares fit to the experimental data. Also given in Figure 12 is the scattering geometry indicating the relative velocity (CMS system) before the collision. Within the limits of experimental error the maximum of the scattering intensity distribution aligns with this CMS direction. That is to say, the preparation of the atomic electron cloud in a $(3p\sigma)_{CMS}$ state gives the largest quenching cross section. We observe a decrease in the anisotropy with increasing temperature of the molecular beam, which probably indicates an influence of the rotational state population of the molecule prior to collision. Thirdly, we were unable to observe a circular asymmetry for small center of mass scattering angles.

In the light of recent quantum mechanical calculations of $M_2 + Na$ potential surfaces we are able to understand the experimental findings.[14,15] Briefly the quenching process may be understood by the potential energy scheme indicated in Figure 13. The atom preparation in the $3P$ state leads partially to a population of the $^2B_2$ surface of the triatomic system. At nonequilibrium distances r for the diatomic molecule a surface crossing between the excited state and the ground state $^2A_1$ is accessible for

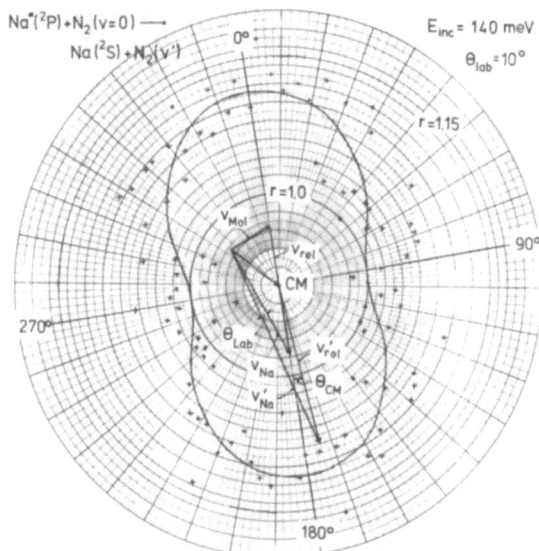

Figure 12. Na–N_2 collisions: Intensity distribution of type (a) measurement in an enlarged radial scale. The solid line gives a least-squares fit to the data. Also shown is the Newton diagram indicating that the maximum of scattering intensity coincides with the incident CMS direction.

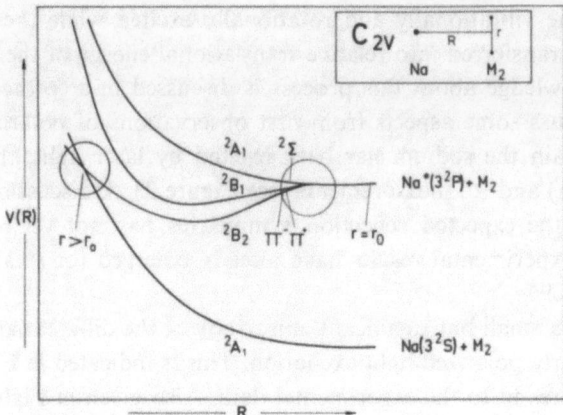

Figure 13. Na* + M_2 quenching schematic potential diagram. For smaller R, the bond length r has been stretched. This leads to a surface crossing between the $^2B_2$ excited and $^2A_1$ ground-state potentials just above the asymptotic 3^2P energy.

thermal incident energies. The transition will thus lead to vibrationally excited molecules.

Tentatively we try to explain the observations with linearly polarized light excitation: Obviously different directions of polarization populate the $^2B_2$ state to a different fraction and thus yield different quenching cross sections. However, the $^2B_2$, $^2B_1$, and $^2A_1$ $(3P)$ states are coupled for larger internuclear distances, as indicated by the circle. Because of a coupling of the orbital electron cloud and the nuclear motion of the three particles involved, transitions are possible by a Demkov mechanism and anisotropies are reduced. An originally populated $^2B_2$ state may be depopulated again during the collision. This apparently is enhanced when the molecule originally rotates. Thus the anisotropy is destroyed with increasing rotational temperature of the molecule as observed.

Why is the $(3p\sigma)_{\text{CMS}}$ state preparation the most favorable for quenching?

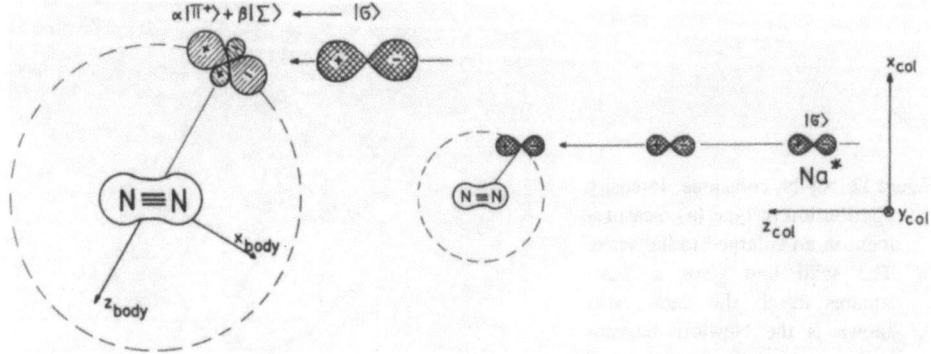

Figure 14. Model illustrating the transition from the CMS system to the body-fixed system for large impact parameters. The σ_{CMS} configuration thereby becomes predominantly a Π_{body} state.

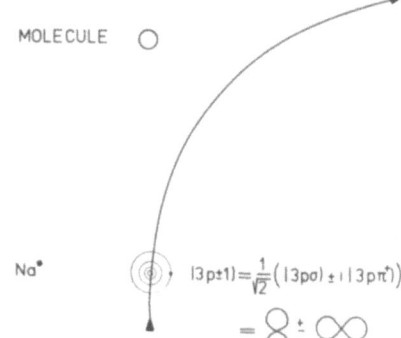

MOLECULE

Na*

$|3 p \pm 1\rangle = \frac{1}{\sqrt{2}} (|3 p \sigma\rangle \pm i |3 p \pi\rangle)$

Figure 15. Model to explain vanishing circular asymmetry. Laser excitation prepares a superposition of σ and π^+ states (see text).

Figure 14 gives a possible explanation: Since we deal with large impact parameters (small scattering angles) an atom originally prepared in a σ_{CMS} state remains in such a state until it actually interacts with the molecule. There, as illustrated in Figure 14, the σ_{CMS} state becomes predominantly a Π^+_{body} state with respect to the Z_{body} axis. This $^2\Pi^+_{body}$ state will partially reach the $^2B_2$ surface and thus lead to quenching, while the Σ_{body} (originally in a $\pi^\pm_{CMS}$ configuration) predominantly will not. So the apparent discrepancy between σ_{CMS} preparation and $^2B_2$ quenching is removed.

It remains to explain why no circular asymmetry is observed. Circular asymmetry necessitates a defined phase relation between different magnetic substates participating in the collision. Obviously in the present case no such phase relation exists: Figure 15 illustrates again the symmetry of the collision experiment. If we choose the quantization axis perpendicular to the collision plane, circular light prepares a $(3P\, M_L = \pm 1)_{Ph}$ state, where the plus or minus sign refers to left or right circular polarized light excitation. These states may also be written as a linear superposition of magnetic substates with respect to the z_{body} axis. Then

$$| 3PM_L = \pm 1)_{Ph} = 2^{-1/2}[| 3P\Sigma) \pm i | 3P\Pi^+)]_{body} \tag{7}$$

The phase between the Σ and the Π state, however, is quite insignificant for the quenching process, since only the population of the $^2B_2$ state is of importance and not the phase relation between the $^2B_2$ and the other two excited states $^2B_1$ and $^2A_1$. Thus, the quenching cross section will not depend on the plus or minus sign in the above equation (7) and in consequence it cannot depend on the sense of orientation.

Summary

We recall that we have shown quite a different behavior of linear anisotropy and circular asymmetry for different collision processes:

I. In the case of the electron-collision-induced deexcitation processes Na($3P$) $+ e^- \rightarrow$ Na($3S$) $+ e^-$ we have found a large linear anisotropy as well as a strong circular asymmetry.

II. For the electron-collision-induced excitation process $Na(3P) + e^- \rightarrow$ $Na(3D) + e^-$ we have found a very small linear anisotropy but a strong circular asymmetry.

III. For the heavy-particle quenching process of excited sodium by diatomic molecules $Na(3P) + M_2 \rightarrow Na(3S) + M_2^\#$ we have observed a significant linear anisotropy but no circular asymmetry for small scattering angles and thermal incident kinetic energies.

All three cases may be qualitatively well understood. A good quantitative agreement with existing quantum mechanical scattering calculations is obtained for the process I. Furthermore we have been able to show that polarization studies for the scattering from laser-excited atoms may lead to useful qualitative models for the dynamics of atomic and molecular collision processes, even if a detailed theoretical treatment is not available.

References

1. U. Fano and J. H. Macek, *Rev. Mod. Phys.* **45**, 533 (1973).
2. I. V. Hertel and W. Stoll in *Adv. At. Mol. Phys.* **13**, 162 (1978).
3. M. Eminyan, K. B. McAdam, J. Slevin, and H. Kleinpoppen, *Phys. Rev. Lett.* **31**, 576 (1973).
4. W. Wittmann and H. J. Andrä, *Z. Physik* **A288**, 335 (1978).
5. I. V. Hertel and W. Stoll, *J. Phys. B* **7**, 570–593 (1974).
6. U. Fano, Chapter 18 of this volume.
7. H. W. Hermann, I. V. Hertel, W. Reiland, A. Stamatović, and W. Stoll, *J. Phys. B* **10**, 251 (1977).
8. H. W. Hermann, I. V. Hertel, and M. H. Kelley, *J. Phys. B*, to be published.
9. D. L. Moores, and D. W. Norcross, *J. Phys. B* **5**, 1482 (1972).
10. J. V. Kennedy, V. P. Myerscough, and M. R. C. McDowell, *J. Phys. B* **10**, 357 (1977).
11. U. Fano and M. Kohmoto, private communication.
12. I. V. Hertel, in *The Excited State in Chemical Physics II*, J. Wm. McGowan, ed., Advances in Chemical Physics, J. Wiley and Sons, New York, 1980.
13. I. V. Hertel, H. Hofmann, and K. A. Rost, *Phys. Rev. Lett.* **38**, 343 (1977).
14. P. Botschwina, W. Meyer, and I. V. Hertel, to be published.
15. P. Habitz, to be published.

51

Photon Polarization Dependence of Superelastic Electron Scattering by Laser-Excited Barium

D. F. Register, S. Trajmar, S. W. Jensen, and R. T. Poe

Superelastic scattering of 30- and 100-eV electrons from the 6s6p ^1P state of Ba 138 prepared by plane-polarized dye laser pumping has been observed for the first time. A Coherent Radiation CR 599 dye laser operating in a single-frequency mode using Rh 110 dye was used to pump up to 30% of the collimated barium beams into the ^1P state. Conventional double hemispherical electron optics was used to achieve the high-energy resolution, incident electron beam collimation, and background discrimination necessary for this experiment. Preliminary measurements indicate that a simple dipole approximation is sufficient to explain the photon polarization dependence of the observed superelastic electron scattering intensity.

1. Introduction

The pioneering research of Hertel and co-workers on electron impact studies of laser-excited sodium[1-4] has provided a new and fruitful tool for investigating electron collision processes. Although this technique can be considered as a time-reversed electron–photon coincidence experiment,[5,6] the laser excitation scheme can, in principle, be applied to a far wider class of atomic problems and can access higher multipoles than are available to coincidence measurements. With the advent of

D. F. Register and S. Trajmar • Jet Propulsion Laboratory, California Institute of Technology, Pasadena, California 91103.   S. W. Jensen and R. T. Poe • Department of Physics, University of California, Riverside, California 92521.   S. W. Jensen's permanent address: National Bureau of Standards, Washington, D.C.

greater frequency ranges and narrower bandwidths in actively stabilized commercial CW dye lasers, a large number of atomic species are now amenable to laser-excited studies.

For most metals, the combination of fine structure, isotope and hyperfine splittings generates a mixture of L, J, or F quantum-numbered states which cannot be resolved by any conventional electron spectrometer. Even if the appropriate energy resolution were available ($\sim 10^{-8}$ eV), it would still not be possible to selectively probe a specific magnetic sublevel of the target. In many cases, however, a 1-MHz bandwidth CW dye laser can selectively prepare a particular fine and/or hyperfine isotopic magnetic sublevel for electron scattering studies. Such is the case with the alkali earth elements and, in particular, atomic barium, which is the subject of the present study.

2. Experiment

2.1. Atomic Barium

Naturally occurring barium has a rich isotopic structure consisting of five plentiful species ($>1\%$ abundance). Three of these are even isotopes (138, 136, 134) while two are odd (137, 135). The principal constituent is Ba 138 (71.7%). The even isotopes have nuclear spin of 0 while the odd I spin is $\frac{3}{2}$. Therefore, a characteristic signature of this element is a well-defined hyperfine structure for its odd isotopes but none for the even ones. Further, the hfs and isotope separations are large (~ 50 MHz) and are easily resolvable with available spectroscopic equipment. (See Figure 1.)

The ground state of Barium is $6s^2$: $^1S_0$. The first optically allowed excited state is $6s6p$: $^1P_1$ at 2.239 eV (5535 Å). This excited state decays primarily to either

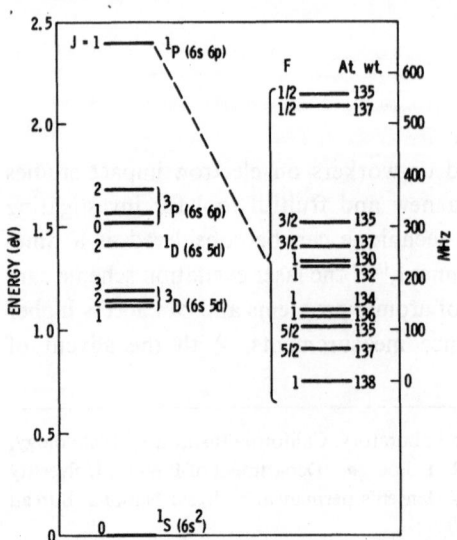

Figure 1. Barium energy level diagram.

the ground state or the $6s5d$: $^1D_2$ level at 1.412 eV. (Decay of the ^1P state to the lower lying ^3P and ^3D states has been observed but with a very low transition probability.[7] For the present discussion these processes may be neglected.) The $^1D_2$ state is meta-stable and has a low transition probability to the ground state. The branching ratio for the $^1P_1$ decay has been given as 23.5:1.[8] However, from the present measurements[9] and other indications,[10] this value is uncertain and may be as high as 200:1.

2.2. Pumping Equations

Restricting the analysis to the Ba 138 isotope reduces the optical pumping scheme to a simple three-level system. For this problem, the optical pumping equations are

$$dN_1(t) = -B_{12}\varrho(\gamma_{12})N_1(t)\,dt + [B_{21}\varrho(\gamma_{12}) + A_{21}]N_2(t)\,dt$$

$$dN_2(t) = -[B_{21}\varrho(\gamma_{12}) + A_{21}]N_2(t)\,dt + B_{12}\varrho(\gamma_{12})N_1(t)\,dt - A_{23}N_2(t)\,dt$$

$$dN_3(t) = N_2(t)A_{23}\,dt$$

where N_1, N_2, N_3 are the populations of the $^1S_0$, $^1P_1$, and $^1D_2$ levels and B_{ij} and A_{ij}

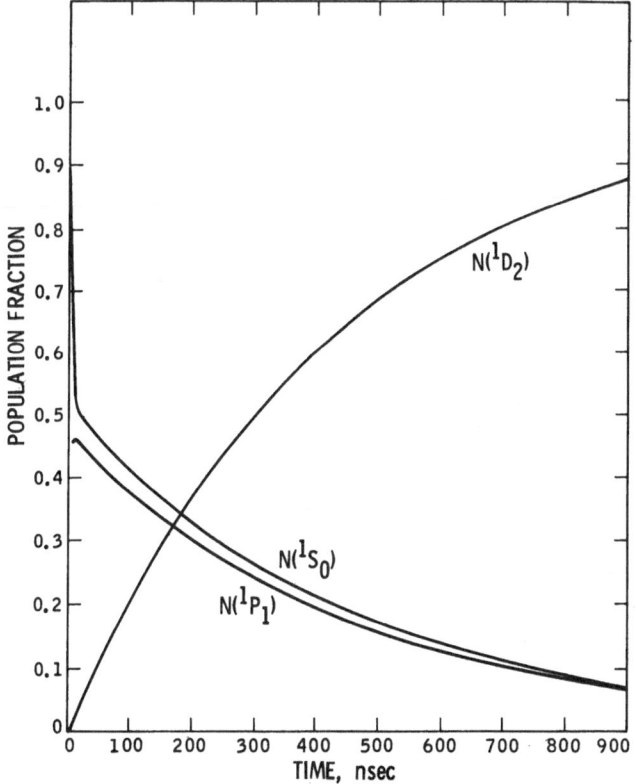

Figure 2. Population fraction of the ^1D, ^1P, and ^1S states as a function of pumping time (23.5 : 1 branching ratio).

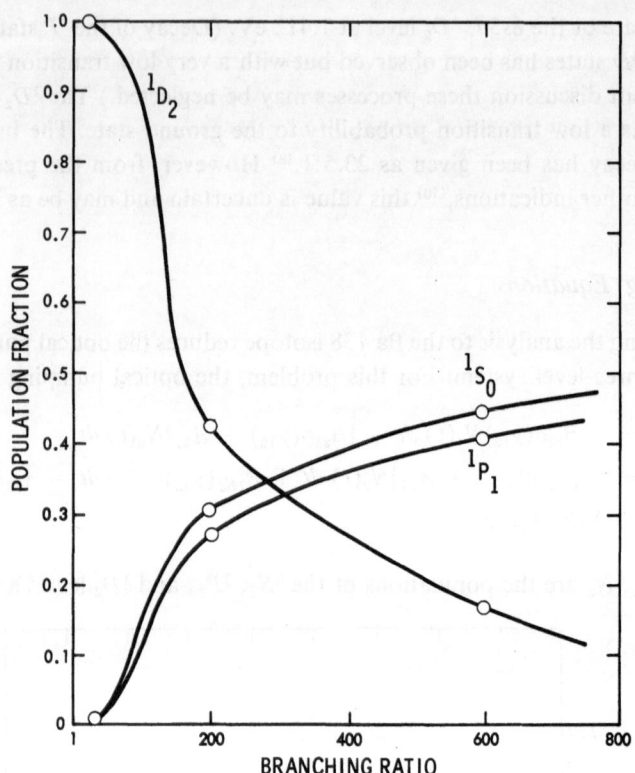

Figure 3. Population fraction of the ^1D, ^1P, and ^1S states as a function of branching ratio (2 μsec illumination time).

are the induced and spontaneous transition rates, respectively, for the i to j transition. These equations may be solved to yield the populations as a function of spontaneous transition probability and $\varrho(\gamma)$ (energy density/Hz). For a 30-mW, 1-MHz dye laser focused into a 2-mm-radius spot, $\varrho(\gamma) \simeq 10^{-11}$ erg/cm³ Hz. Using this value and a branching ratio of 23.5:1, the pumping equations have been solved to yield the populations of the $^1S_0$, $^1P_1$, and $^1D_2$ states as a function of illumination time. For the geometry of this experiment, a reasonable estimate for the illumination time is 1–2 μsec. Using this as a typical value, the pumping equations have also been solved as a function of branching ratio. The results for these two calculations are shown in Figures 2 and 3.

2.3. Dye Laser

The dye laser used throughout these experiments is a Coherent Radiation model 599-21 scanning single frequency CW dye laser. A portion of the laser output is coupled into both a reference cavity and power detector to provide active, closed-loop frequency stabilization and output power compensation. Using Rhodamine 110 dye with glycol solvent, 100 mW of broadband power is available at 5535 Å. In single-frequency operation, the usable output is typically 20 to 30 mW. The dye laser may be scanned in single-frequency mode over a 30 GHz range. With the system properly aligned and thermally stable, typical operating times between laser mode hops are from 5 to 40 min. Mode hopping is primarily due to thermal drifting in the argon laser pump (CR-53G, operated at 2.5 W at 5145 Å). Therefore, all experiments commenced with several hours running time on the argon laser before data acquisition was begun. Typical lifetime of the Rh 110 dye at 2.5 W pumping power is 8 to 10 h. This required a fresh charge of dye before each experimental run.

2.4. Experimental Geometry

The electron spectrometer used in this measurement has been described elsewhere.[9] For the present measurements, the laser crossed the Ba beam at an angle of either 45 or 90 degrees with respect to the electron gun (in the scattering plane). Linearly polarized light from the laser tuned to 5535 Å was passed through a Ba line polarization rotator and excited the $^1S_0-^1P_1$ transition. Energy loss spectra obtained under these conditions clearly show a large number of features that are due to electron collisions with the excited species. Figure 4 demonstrates a typical laser-on–laser-off

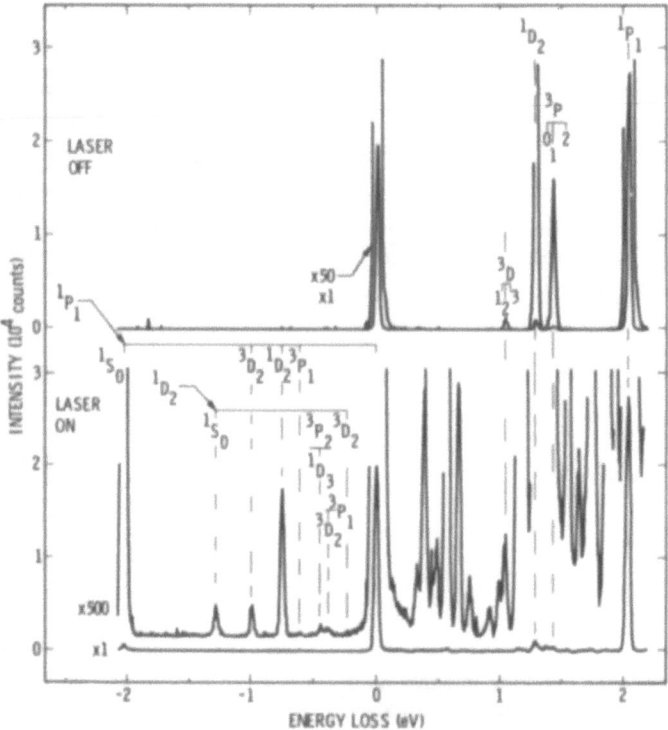

Figure 4. Laser-on, laser-off energy loss spectra of Ba.

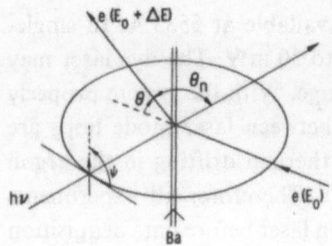

Figure 5. Scattering geometry: θ is the scattering angle, θ_n is the angle between the laser and the outgoing electron, and ψ is the angle between the laser polarization angle and the scattering plane.

measurement; in this figure some 28 features are shown and 23 of these are identified as transitions originating from excited states.[7]

With the detector tuned to superelastic scattering at -2.24 eV energy loss (peak #1 in Figure 4), superelastic scattering as a function of incident laser polarization angle was obtained. The scattering geometry is shown in Figure 5. In the early experiments, an atomic beam collimation of 6:1 was used. These data (Figure 6) exhibited some radiation trapping and odd isotope effects but were sufficient to demonstrate the feasibility of the experiment. Later measurements at 30:1 collimation reduced these effects ($\sim$30-MHz doppler width with no isotope overlap).

3. Data Analysis

Although the multipole moment analysis of Macek and Hertel[11] can be used to advantage in this problem, the basic simplicity of the Ba 138 system lends itself to a more transparent approach. In this analysis we consider the normal scattering plane as the proper reference for the problem with the constraint that the z axis is defined along the outgoing electron direction. The laser, incident in the scattering plane, then defines an angle θ_n with respect to the outgoing electron, and the linear polarization vector of the photon is oriented at an angle ψ with respect to the scattering (x–z) plane (Figure 5).

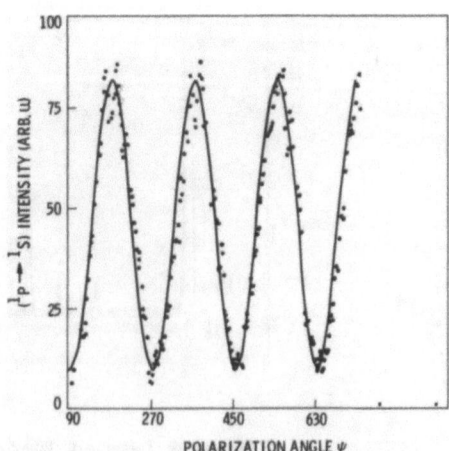

Figure 6. Superelastic scattering intensity as a function of laser polarization angle, 100 eV impact energy, 5° scattering angle, $\theta_n = 40°$. Atomic beam collimation of 6 : 1.

In the photon reference frame, the state Y_1^0 is prepared with unit probability. The problem at hand is to represent this state as a projection onto the Y_1^0 and $Y_1^{\pm 1}$ representation in the outgoing electron reference system. This may be accomplished by use of the complex rotation operators,[12] and the resulting wave function has the form

$$|\Psi\rangle = \sum_m D_{m0}^1(\alpha, \beta) \,|\, 1m\rangle$$

where

$$D_{\pm 1,0}^1(\alpha, \beta) = -m e^{-im\alpha}(\sin \beta)/2^{1/2}$$

$$D_{0,0}^1(\alpha, \beta) = \cos \beta$$

and α, β are the Euler angles which project the photon system onto the outgoing electron coordinates.

The Euler angles are related to the experimental angles θ_n, ψ as follows[4]:

$$\cos \beta = -\cos \psi \sin \theta_n$$

$$\sin \alpha = \sin \psi/\sin \beta$$

The detected superelastic electron scattering intensity may then be written as

$$I = Ak_f/k_0 \left| \sum_m D_{m,0}^1(\alpha, \beta) f_m(\Omega) \right|^2$$

where k_f, k_0 are the final and initial electron momentum and A is a proportionality factor which accounts for detector efficiency and energy dependence, photon, electron, and atomic flux, and the geometric overlap of the electron, laser, and atomic beams.[7] f_m is the scattering amplitude for the $m = \pm 1, 0$ ($^1P_1$) to $m' = 0$ ($^1S_0$) transition. Using the property of reflection symmetry in the scattering plane ($f_1 = -f_{-1}$) and evaluating $D_{m,0}^1$ from the Euler angle relations, the superelastic intensity becomes

$$I = Ak_f/k_0 \cos^2 \psi [2f_1^2 \cos^2 \theta_n + f_0^2 \sin^2 \theta_n + 2 \cos \theta_n \sin \theta_n * (f_0 f_1^* + f_0^* f_1)]$$

Using Kleinpoppen's[13] definition of λ and χ

$$\lambda = |f_0|^2/(|f_0|^2 + 2|f_1|^2)$$

$$f_1 = |f_1| e^{i\chi}$$

the superelastic scattering intensity is proportional to

$$I\alpha \cos^2 \psi \{\lambda \sin^2 \theta_n + (1 - \lambda) \cos^2 \theta_n + 2[\lambda(1 - \lambda)]^{1/2} \cos \chi \cos \theta_n \sin \theta_n\}$$

Finally, if the dependence on laser polarization is neglected ($\cos^2 \psi$), the superelastic scattering intensity may be related to the coincidence rate of Kleinpoppen

Table 1. Maximum/Minimum Superelastic Intensity Ratios at 30 eV, 6:1 Collimation Ratio[a]

θ_s at $E_0 = 30$ eV	R	
	$\theta_y = 45°$	$\theta_y = 90°$
−15	13	—
−10	13	3.1
−5	15	5.2
−2.5	15	9.0
0	18	10.6
2.5	19	8.3
5	14	5.5
10	8.5	3.9
15	9	7.2

[a] θ_s = scattering angle; θ_y = laser–incident-electron angle. The statistical uncertainty in the ratios is $\simeq \pm 3$ at $\theta_y = 45°$ and $\simeq \pm 2$ at $90°$.

and co-workers[13] by making the substitution $\theta_n = \pi - \theta_y$. In this case, the expression is identical to the coincidence measurements which detect photons without regard to polarization with the definition that θ_n is measured to the outgoing electron while θ_y is measured to the incident electron.

A feature of this analysis that is markedly different from the results of Hertel and Stoll for Na is the depth of modulation of the superelastic electron intensity as a function of polarization rotation. In the sodium experiments, the presence of higher-order multipoles does not allow as high a degree of cancellation among the scattering amplitudes. For the case of a simple dipole excitation, the analysis indicates that for $\psi = 90$ and $270°$, this cancellation is complete. For these orientations, the Y_1^0 wave function in the electron reference frame is 0 while the Y_1^1 and Y_1^{-1} components are exactly out of phase. This conclusion is well supported by the data thus far acquired (Figure 6). A summary of the observed maximum to minimum ratios in the superelastic scattering intensity is given in Table 1. Owing to residual radiation trapping and counting statistics, the minimum in the superelastic intensity is not a true zero. In most cases, however, this minimum is less than 10% of the maximum value.

4. Summary

Superelastic electron scattering from laser-excited barium has been observed for the first time. The behavior of the electron scattering as a function of laser polarization angle was found to be consistent with a simple dipole approximation to the problem. As the minimum in the scattered intensity is zero, no information on the

λ and χ values can be deduced from a polarization rotation experiment at a fixed incident photon angle. In order to complete the analysis, work is underway to configure the scattering apparatus to obtain this information.

ACKNOWLEDGMENT

This work was supported in part by NASA contract No. NAS7-100 and in part by the Caltech President's Discretionary Fund.

References

1. I. V. Hertel and W. Stoll, *J. Phys. B* **7**, 570 (1974).
2. I. V. Hertel and W. Stoll, *J. Phys. B* **7**, 583 (1974).
3. H. W. Hermann, I. V. Hertel, W. Reiland, A. Stamatovic, and W. Stoll, *J. Phys. B* **10**, 251 (1977).
4. I. V. Hertel and W. Stoll, *Advances in Atomic and Molecular Physics*, Vol. 13, p. 113, Academic Press, New York (1977).
5. M. Eminyan, K. B. MacAdam, J. Slevin, and H. Kleinpoppen, *Phys. Rev. Lett.* **31**, 576 (1972).
6. M. Eminyan, K. B. MacAdam, J. Slevin, and H. Kleinpoppen, *J. Phys. B* **7**, 1519 (1974).
7. D. F. Register, S. Trajmar, S. W. Jensen, and R. T. Poe, *Phys. Rev. Lett.* **41**, 749 (1978).
8. B. M. Miles and W. L. Wiese, *Atomic Transition Probabilities*, NSRDS-NBS-22, Vol. 3, U.S. Government Printing Office, Washington, D.C. (1969).
9. S. Jensen, D. Register, and S. Trajmar, *J. Phys. B* **11**, 2367 (1978).
10. A. Gallagher, private communication.
11. J. Macek and I. V. Hertel, *J. Phys. B* **7**, 2173 (1974).
12. M. E. Rose, *Elementary Theory of Angular Momentum*, John Wiley and Sons, New York (1957).
13. H. Kleinpoppen, *Atomic Physics*, Vol. 4, p. 449, Plenum Press, New York (1975).

52

The Effect of the Angular Coherence of Each Beam, in a Crossed Beam Experiment, on Observations of the Small-Angle Elastic Scattering

K. Rubin and I. Efremov

We present the results of a crossed electron–potassium beam scattering experiment which, we believe, demonstrates that at very small angles a plane-wave treatment of the scattering event is incorrect if the unscattered particles are not themselves in pure plane-wave states. We suggest instead that one must treat the particles as coherent packets and that this will result in interference effects similar to those which occur in the forward direction in a plane-wave treatment but in this case will occur in a range of angles about zero degrees. The scattering amplitude will now be an integral of the plane-wave amplitude over a range of angles. This presents the possibility of making observations of the phase of the scattering amplitude at angles other than zero degrees (i.e., at zero degrees one could determine the phase of the scattering amplitude from a measurement of the total cross section and the differential elastic cross section at zero degrees by applying the optical theorem).

1. Introduction

We have recently completed a series of measurements in a crossed beam scattering experiment which, we believe, demonstrate that there is sufficient coherence in each of the beams to enable one to observe, at very small scattering angles, effects

K. Rubin and I. Efremov ● Department of Physics, City College of the City University of New York, New York, New York 10031.

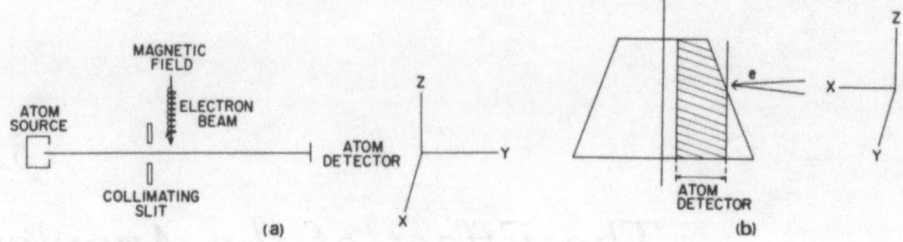

Figure 1. (a) Schematic diagram of the experiment. (b) Atomic beam profile in the plane of the detector.

that can be attributed to the fact that the colliding beams are not plane waves but are in fact superpositions of coherent momentum states. Under such conditions the scattering signal is related to an integral of the scattering amplitude over some range of angles. This offers the possibility of making measurements that are directly related to the scattering amplitude rather than its square.[†] We began thinking about this problem a number of years ago while observing, in a crossed atom–electron beam experiment, the ratio of the number of atoms scattered away from the detector to the total number of atoms falling on the detector as a function of the position of the detector. Figure 1 shows a sketch of the experimental arrangement. In Figure 1b we are looking down the apparatus along the direction of the undeflected atom beam trajectory. We show in the figure the geometrical profile of the atom beam, as determined by the collimating and source slits, and the portion of this profile overlapped by the detector. The region overlapped is of course a function of the detector position. The electrons are not collimated but are confined to magnetic field lines that are at right angles to the atom beam. This causes the electrons to hit the atom beam in a small cone of angles. Our purpose was to use these observations both to determine the effect of the finite resolution of the apparatus on the total cross-section measurements and to make a determination of the elastic differential cross section in the forward direction.[2] As can be seen from Figure 1b, as the detector is moved towards the electron source (to the right in the figure) the resolution of the apparatus improves. This can be understood as follows: The scattering signal is the change in the number of atoms reaching the detector when the electron beam is turned on. This in turn consists of the total number of atoms in the region overlapped by the detector that are scattered minus the number of atoms scattered into this region from the right and minus those atoms in the region of the detector that are scattered through angles that are too small to enable them to miss the left edge of the detector. Both of the effects that detract from the scattering signal decrease as the detector is moved to the right. We would expect then that the ratio of the observed scattering signal to the total number of atoms in the region of the detector would increase as the detector

---

[†] While this work was in progress M. C. Li proposed an experiment to measure the scattering amplitude directly using coherent beams scattering from a potential. His analysis however does not take into account the recoil of the target which is of major importance here. See Reference 1.

is moved to the right. This does in fact happen until we approach very close to the right edge of the beam, at which point the ratio starts to exhibit a behavior that we cannot explain using a standard plane-wave treatment of the scattering (see Figure 9). We will show, however, that if one assumes that the colliding beams have an appropriate set of coherent momentum states, which then allow for interference effects, it might be possible to explain this anomalous behavior.

2. Theory

The scattering of particles using a wave packet formulation has been treated by many authors;[†] however, the possibility of interference effects due to an overlap of the incoming packet with the scattered wave at angles other than the forward direction seems to have been overlooked. This sort of interference should be observable if each colliding packet has an appropriate set of coherent momentum states. By appropriate set of states we mean that if the beams initially contain the states k_a and k_e as shown in Figure 2, then interference could be observed if the states k_a' and k_e' were also present initially in the beams. Note also that k_a' and k_e' will scatter back into k_a and k_e.

To see in detail how this interference comes about consider the collision between two wave packets. The incident wave function has the form

$$\psi = \int A(\mathbf{k}_a)e^{i\mathbf{k}_a \cdot \mathbf{r}_a}B(\mathbf{k}_e)e^{i\mathbf{k}_e \cdot \mathbf{r}_e}\, d^3k_a\, d^3k_e \tag{1}$$

This can be written in terms of the total and relative momentum as

$$\psi = \int C(\mathbf{K}, \mathbf{k})e^{i\mathbf{K}\cdot\mathbf{R}}e^{i\mathbf{k}\cdot\mathbf{r}}\, d^3K\, d^3k \tag{2}$$

where

$$C(\mathbf{K}, \mathbf{k}) = A[(m_a/M)\mathbf{K} - \mathbf{k}]B[\mathbf{k} - (m_e/M)\mathbf{K}] \tag{3}$$

We calculate the scattering in the following way. Consider first the scattering of those states with a fixed total momentum $\mathbf{K}$ and fixed magnitude for the relative momentum $|\mathbf{k}|$. This will correspond to the scattering in the CM of a beam with some initial angular spread given by $C(\mathbf{K}, \mathbf{k})$. It is easy to show that the scattered current per unit solid angle in some direction in the CM is given by

$$j_s(\mathbf{K}, |\mathbf{k}| \hat{\mathbf{r}}) \approx \left[4\pi IM\left\{C(\mathbf{K}, |\mathbf{k}| \hat{\mathbf{r}}) \int \left[C^*(\mathbf{K}, \mathbf{k})\frac{f(\hat{\mathbf{k}} \cdot \hat{\mathbf{r}})}{k}\right] d\Omega_k\right\}\right.$$
$$\left. + \left|\int [C^*(\mathbf{K}, \mathbf{k})f(\hat{\mathbf{k}} \cdot \hat{\mathbf{r}})]\, d\Omega_k\right|^2\right] \tag{4}$$

where the first term is due to the interference between the incoming and scattered

---

[†] See, for example, Reference 3.

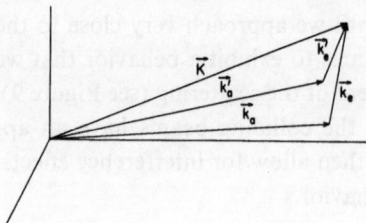

Figure 2. The states k_a and k_e scatter into k_a' and k_e'. The initial presence of both sets of states in the colliding beams could result in interference effects.

wave and represents a decrease in the forward beam due to the presence of the scatterer. When this term is integrated over all angles it gives the total scattering signal. If the incoming wave is a plane wave this will reduce to the usual optical theorem. The scattering current can be transformed into the lab system using

$$j_s(\mathbf{K}, |\mathbf{k}|\hat{\mathbf{f}})\, d\Omega_k = j_e(\mathbf{K}, \mathbf{k}_e)\, d\Omega_{k_e} = j_a(\mathbf{K}, \mathbf{k}_a)\, d\Omega_{k_a} \tag{5}$$

In principle it is possible to observe j_e and j_a by doing a coincidence experiment (i.e., simultaneously observe the momentum of the scattered particles) since $\mathbf{K}$ can then be determined. If either the scattered atom or scattered electron current alone is observed one must sum over all values of the momenta of the particle that is not being observed for a fixed value of the observed momentum.

In an actual experiment the observed scattering currents are averages taken over many collisions. We must therefore do a time average of equation (4). Thus

$$\overline{j_s(\mathbf{K}, |\mathbf{k}|\hat{\mathbf{f}})} \approx \Big\{ 4\pi I M\Big[\int \overline{C(\mathbf{K}, |\mathbf{k}|\hat{\mathbf{f}})C^*(\mathbf{K}, \mathbf{k})}\, \frac{f(\hat{\mathbf{k}}\cdot\hat{\mathbf{f}})}{k}\, d\Omega_k\Big]$$
$$+ \iint \overline{C(\mathbf{K}, \mathbf{k})C^*(\mathbf{K}, \mathbf{k}')} f(\hat{\mathbf{k}}\cdot\hat{\mathbf{f}}) f(\hat{\mathbf{k}}'\cdot\hat{\mathbf{f}})\, d\Omega_k\, d\Omega_{k'}\Big\} \tag{6}$$

The observed signals depend upon the time average of the quantity $C(\mathbf{K}, \mathbf{k})C^*(\mathbf{K}, \mathbf{k}')$. This time average is a measure of the coherence properties of the colliding beams. If for example

$$\overline{C(\mathbf{K}, \mathbf{k})C^*(\mathbf{K}, \mathbf{k}')} = |C(\mathbf{K}, \mathbf{k})|^2\, \delta(\mathbf{k} - \mathbf{k}')$$

Equation (5) reduces to

$$\overline{j_s(\mathbf{K}, |\mathbf{k}|\hat{\mathbf{f}})} \approx \Big\{ 4\pi I M\Big[|C(\mathbf{K}, |\mathbf{k}|\hat{\mathbf{f}})|^2\, \frac{f(0)}{k}\, d\Omega_k\Big] + \int |C(\mathbf{K}, \mathbf{k})|^2\, |f(\hat{\mathbf{k}}\cdot\hat{\mathbf{f}})|^2\, d\Omega_k\Big\}$$

which is the usual plane-wave solution.

The problem becomes relatively simple when treating the scattering of an electron from a heavy atom. Here the relative momentum is approximately equal to the electron's momentum. We can also assume that, for collisions in the range of angles of interest in the present experiment, the magnitudes of the electron's and atom's momenta are not changed by the collision. (The change in the magnitude of the electron's momentum for the range of angles considered here is less than one part in 10^8.)

In this approximation equation (5) becomes

$$j_e(\mathbf{K}, |\mathbf{k}|\hat{\mathbf{r}}_e)$$

$$= \left\{ 4\pi IM \left[\int \overline{A(\mathbf{K} - |\mathbf{k}_e|\hat{\mathbf{r}}_e)A^*(\mathbf{K} - \mathbf{k}_e)B(|\mathbf{k}_e|\hat{\mathbf{r}}_e)B^*(\mathbf{k}_e)[f(\hat{\mathbf{k}}_e \cdot \hat{\mathbf{r}}_e)/k_e]\,d\Omega_{k_e}} \right] \right.$$

$$\left. + \int\int A(\mathbf{K}-\mathbf{k}_e)A^*(\mathbf{K}-\mathbf{k}_e')B(\mathbf{k}_e)B^*(\mathbf{k}_e')f(\hat{\mathbf{k}}_e \cdot \hat{\mathbf{r}}_e)f^*(\hat{\mathbf{k}}_e' \cdot \hat{\mathbf{r}}_e)\,d\Omega_{k_e}\,d\Omega_{k_e'} \right\} \quad (7)$$

Since we are observing the scattered atom beam we want $j_a(\mathbf{K}, \mathbf{k}_a)$, which we obtain from the relationship

$$j_e(\mathbf{K}, |\mathbf{k}_e|\hat{\mathbf{r}}_e)\,d\Omega_{k_e} = j_a(\mathbf{K}, \mathbf{k}_a)\,d\Omega k_a$$

It is convenient at this point to describe the atomic momentum $\mathbf{k}_a$ in terms of the angles (ψ_a, χ_a) as shown in Figure 3 and a similar set of angles (ψ_K, χ_K) to describe $\mathbf{K}$. In our approximation (i.e., $|\mathbf{k}_a| \gg |\mathbf{k}_e|$) we have

$$\psi_K = \psi_a + RX_e \quad (8)$$

$$\chi_K = \chi_a + R\sin\theta_e \sin\varphi_e \quad (9)$$

where $R = |\mathbf{k}_e|/|\mathbf{k}_a|$, $X_e = \cos\theta_e$, and θ_e and φ_e are the polar and azimuthal angles of the electron. Thus

$$j_e(\mathbf{K}, |\mathbf{k}_e|\hat{\mathbf{r}}_e)\,dX_e\,d\varphi_e = j_a(\mathbf{K}, \mathbf{k}_a)\,d\psi_a\,d\chi_a$$

From equations (7) and (8), we see that for a fixed $\mathbf{K}$ the atomic scattering angle ψ_a is directly related to the scattered electron's polar angle, while for a given ψ_a the atomic angle χ_a is related to the electrons's azimuthal angle. If our observations are made as a function of ψ_a (i.e., the detector has sufficient length in the z direction (see Figure 3) to intercept all the scattered atoms) we can write

$$\mathcal{J}_a(\mathbf{K}, |\mathbf{k}_a|, \psi_a)\,d\psi_a = \left[\int j_a(\mathbf{K}, \mathbf{k}_a)\,d\chi_a \right] d\psi_a$$

$$= \left[\int_0^{2\pi} j_e(\mathbf{K}, |\mathbf{k}_e|\hat{\mathbf{r}}_e)\,d\varphi_e \right] dX_e \quad (10)$$

and

$$\mathcal{J}_a(\mathbf{K}, |\mathbf{k}_a|, \psi_a) = -\frac{1}{R} \left[\int_0^{2\pi} j_e(\mathbf{K}, |\mathbf{k}_e|\hat{\mathbf{r}}_e)\,d\varphi_e \right] \quad (11)$$

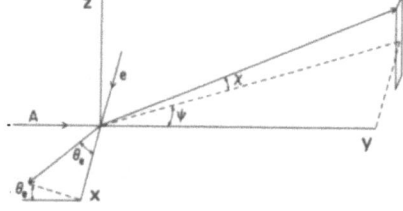

Figure 3. The angles ψ and χ are used to describe the direction of the total momentum $\mathbf{K}$ and the atomic momentum $\mathbf{k}_a$, while θ_e and φ_e are the polar and azimuthal angles of the electron.

Finally the total atom current for a given ψ_a in the range $d\psi_a$ is obtained by integrating equation (10) over the magnitude of $\mathbf{k}_e$ and over the appropriate range of total momenta. Since in general this can be quite complicated we give a specific example below in which we have made a number of simplifying assumptions. This example will serve to illustrate the technique used and demonstrate in this simple case that we can expect to find the ratio decreasing as we approach the edge of the atom beam.

3. Outline of Calculation

We shall assume that both the atom and electron beams are monoenergetic, completely coherent, and that the amplitude of the atomic wave function is constant over the range of angles in the atom beam. The action of the magnetic field in the electron gun causes the electrons to spiral about the field line and strike to atom beam in a small cone of angles. We shall assume that the amplitude of the electron's wave function is independent of φ_e. We must first calculate

$$\int_0^{2\pi} j_e(\mathbf{K}, |\mathbf{k}_e| \hat{\mathbf{r}}) \, d\varphi_e$$

which we can write as

$$\int_0^{2\pi} j_e(\psi_K, \chi_K, X_e) \, d\varphi_e$$

The total atomic scattering current will be given by

$$I_a(\psi_a) \, d\psi_a = -\frac{1}{R} \left[\int dw_K \, d\chi_K \int_0^{2\pi} j_e(\psi_K, \chi_K, X_e) \, d\varphi_e \right] d\psi_a \qquad (12)$$

where X_e is a function of ψ_a and ψ_K. The role of the atomic amplitude in the above is just to determine the limits in the various integrations. We show in Figure 4 the regions of integration assuming that the limits of the angles ψ_a and χ_a are $\pm\alpha$ and $\pm\beta$, and that the maximum electron angle is θ_m. In Figure 4a the points A and X_e^m are the limits for the integral over X_e in equation (6) while the points C and D are the limits of the ψ_K integration in equation (12). The limits for the φ_e integration are determined from Figure 4b. We see, for example, that if $\chi_K < \beta - R \sin \theta_e$ the limits on the φ_e integration are 0 to 2π while if χ_K lies along the solid line the limits are 0 to E and F to 2π. If $\beta \gg R \sin \theta_M$ the limits will be 0 to 2π for most of the range of χ_K. We shall assume that this is the case. Under this condition

$$I_a \sim \int J_e(\psi_K, X_e) \, d\psi_K \qquad (13)$$

We shall only concern ourselves with the first term in equation (6) since the second term represents the differential scattering signal and will be small compared to the first term in the region of the incoming atom beam. We have then

$$I_a(\psi_a) \sim \int d\psi_K IM \left[B(X_e) \int [B^*(X_e')f(\hat{\mathbf{k}}_e' \cdot \hat{\mathbf{r}}_e)/|\mathbf{k}_e|] \, dX_e' \right] \qquad (14)$$

Figure 4. (a) ψ_K vs. X_e for $\psi_a = \pm\alpha$ and $1 \geq X_e \geq X_e^m$. Also shown is the intersection of a line $\psi_K = $ const with the lines representing the boundary of the region of integration in equation (15). For the value of ψ_K shown the limits of the integration over X_e' are X_e^m and A. In addition, the intersection of 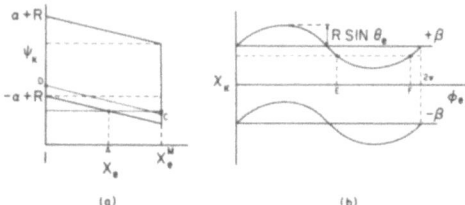 a line ψ_K vs. X_e for $-\alpha \leq \psi_a \leq \alpha$ with the lines $X_e = 1$ and $X_e = X_e^m$ determine the limits of the integral over ψ_K (see for example the points C and D above). (b) χ_K vs. φ_e for $\chi_a = \pm\beta$ and $0 \leq \varphi_e \leq 2\pi$ and a given value of θ_e. Also shown is the intersection of a line $\psi_K = $ const with the lines representing the boundary of the region of integration over φ_e. For example for the value of ψ_K shown the φ_e integration goes from 0 to E and from F to 2π.

where $X_e = (\psi_K - \psi_a)/R$. Finally if the angular range of the beams are small enough we can assume that in the region of integration $f(\hat{\mathbf{k}}_e \cdot \hat{\mathbf{r}}_e) \approx f(0)$. Thus

$$I_a(\psi_a) \sim IM\left\{\int d\psi_K B[(\psi_K - \psi_a)/R] \int B^*(X_e')\, dX_e'\, f(0)\right\} \tag{15}$$

If B is constant I_a will depend only on the area of integration. In Figure 5 we show this area for two values of I_a, one at the center of the beam (region A) and the other at one edge of the beam (region B). Note that the area of integration at the edge is $\frac{1}{2}$ of that at the center. We would expect then that the fraction of atoms scattered from the edge of the atom beam will be $\frac{1}{2}$ of the value at the center. This of course is only true for the assumptions made and for the proper choice of the parameters α and θ_M.

A more realistic calculation has been made which takes into account first the variation of the amplitude of the electron's wave function $B(\mathbf{k}_e)$ with θ_e, due to the fact that the angular distribution of the electron beam is compressed in the forward direction, and secondly the fact that the scattering does not take place at one point but occurs throughout the region of overlap of the electron and atom beams. It should be noted that we do not assume any interference between the scattering from different points in the overlap region but simply sum the scattered currents produced at each point. We have also included in the calculation the variation of the scattering amplitude $f(\hat{\mathbf{k}} \cdot \hat{\mathbf{r}})$ over the range of angles in the beams and calculated both the interference and differential terms in equation (6). We made two calculations of the scattered current, one assuming complete coherence and the other complete incoherence

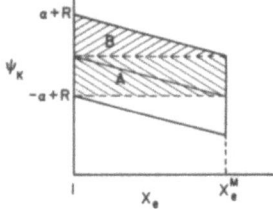

Figure 5. A case in which the fraction of atoms scattered at the edge of the atomic beam will be one-half the fraction scattered at the center of the beam.

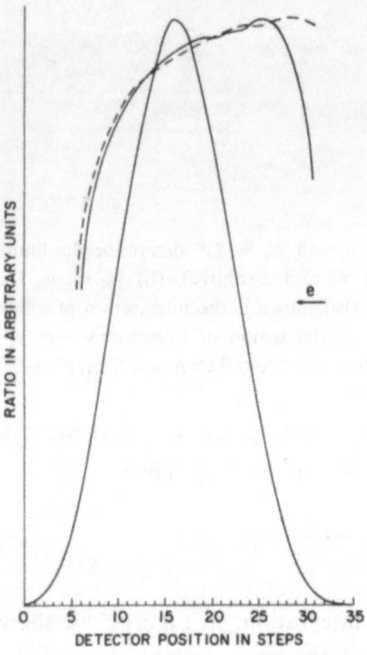

RATIO IN ARBITRARY UNITS

0   5   10   15   20   25   30   35
DETECTOR POSITION IN STEPS

Figure 6. Calculated ratio of the observed atom scattering signal to the total atom beam current as a function of detector position for the coherent case (solid line) and the incoherent case (dashed line) assuming a constant atomic amplitude $A(k_a)$ for the range of angles in the atomic beam. Also shown is the atom beam profile. Each detector step corresponds to an angular displacement of the detector about the center of the scattering region of 3×10^{-5} rad. The maximum electron angle is $5°$.

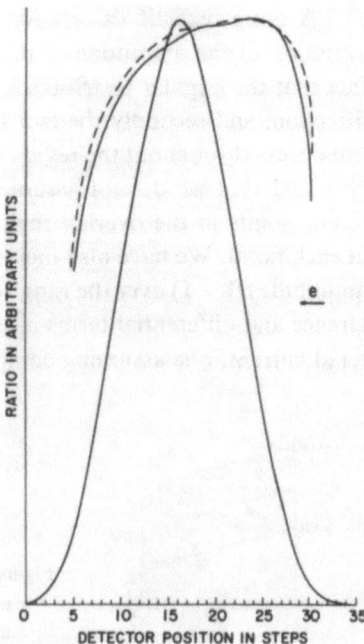

RATIO IN ARBITRARY UNITS

0   5   10   15   20   25   30   35
DETECTOR POSITION IN STEPS

Figure 7. Same as Figure 6 with a maximum of electron angle of $7°$.

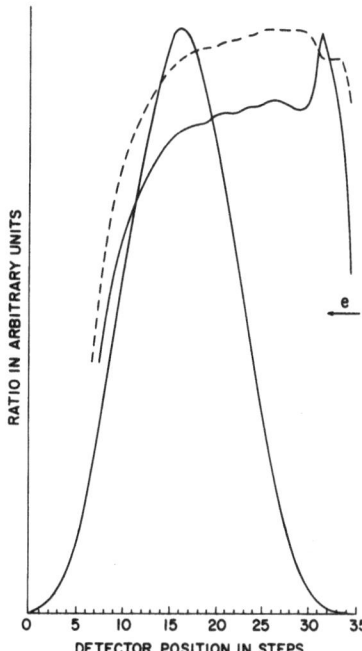

Figure 8. Same as Figure 6 with a tail added to the atomic amplitude $A(k_a)$.

(i.e., the standard plane-wave treatment). The results of these calculations are shown in Figures 6–8. In each figure we show the atom beam profile (integrated over the width of the detector) together with the ratio of the net current scattered out of the region of the detector to the total current intercepted by the detector for both the case of no coherence (dotted line) and complete coherence (solid line). In Figures 6 and 7 the amplitude of the atomic wave function $A(k_a)$ is assumed to be constant. Figure 6 corresponds to a maximum electron angle of 5° while the maximum electron angle in Figure 7 is 7°. In Figure 8 we show the effect of adding a small tail to the atomic amplitude.

4. Discussion of Results

The important features illustrated in the examples discussed above are as follows. If the electron beam has a small enough divergence then the coherent and incoherent results are sufficiently different to make the degree of coherence easily observable experimentally. If, on the other hand, the angular spread is too large the interference effect will be overshadowed by the backscattering of the atoms (i.e., atoms to the left of the detector in Figure 1b scattering back into the detector) due to the fact that not all of the electrons hit the atom beam at right angles. The second thing to notice is that in the coherent calculation in Figure 8, where we have added a tail to the atom beam amplitude, there is a sharp peak near the edge of the atom beam profile. This peak is due to interference between the large amplitude at the center of the profile and the much smaller amplitude near the edge. This causes a large fractional change in the scattering near the edge of the beam. This sort of profile is more realistic than one in which it is assumed that the amplitude $A(k_a)$ is constant. An examination of the experimental beam profile always reveals a tail on the profile that would not be present if the amplitude were constant. The tail in the present experimental profile (Figure 9) is, however, quite long, and a good part of it is probably due to small-angle scattering of the atom beam as it travels from the source to the entrance of the electron gun. We expect that this sort of random scattering produces an incoherent spreading of the beam, and in the experimental results presented here we have subtracted this background from both the incoming atom beam profile and the scattering signal. We show the resultant atom beam profile and ratio curves in Figures 10 and 11. Also shown are the calculated ratios for the coherent case appropriate to the atom beam geometry. The incoherent calculation gives a result similar to the one illustrated in Figure 7 and it is easy to see that the experimental results are in much closer agreement with the coherent than the incoherent calculations and indicate that there is enough coherence in the beams to observe interference effects in the scattering.

We are presently improving the pressure in the region between the electron gun and atom source so as to reduce as much as possible the incoherent tail in the beam profile. This will eliminate the necessity of subtracting a large base line from the experimental results. We will obtain more data, varying parameters such as the angular widths of both the atom and electron beams and the scattering energy, to make certain that the results are in agreement with the coherent calculations. If the results are positive we will then design an experiment that will enable us to extract from the scattering data the phase of the scattering amplitude as a function of the scattering angle.

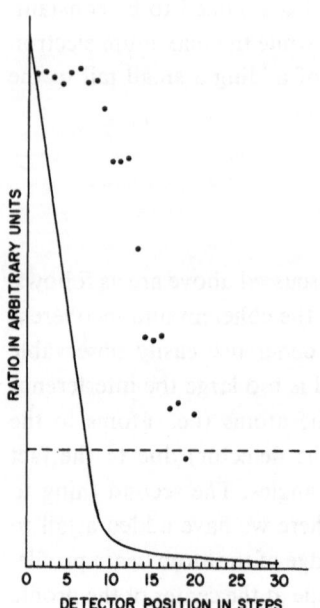

Figure 9. Raw experimental data. The solid line is the observed atom beam profile and the points are the measured values of the ratios.

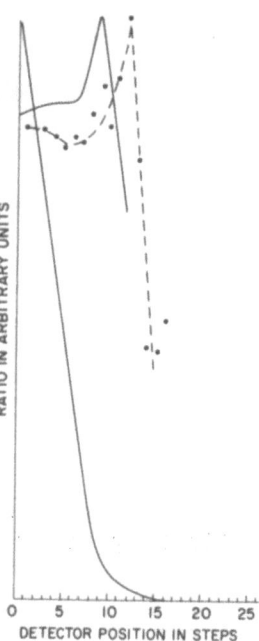

Figure 10. Experimental data with the background subtracted. Also shown are the calculated values of the ratio for the coherent case (solid line). The dotted line through the data points is intended to indicate the trend of the experimental results. The atomic source slit is 0.003 in. and the collimating slit 0.005 in.

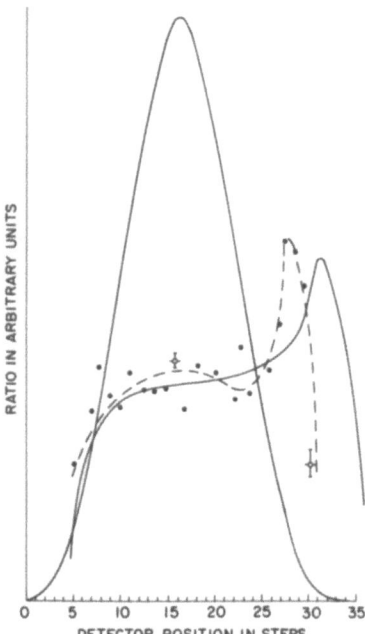

Figure 11. Same as Figure 10 with an atomic source slit of 0.01 in. and a collimating slit width of 0.008 in.

ACKNOWLEDGMENTS

   This work was supported by the National Science Foundation and by the PSC-BHE Research Award Program of the City University of New York.

References

1. Ming Chiang Li, *Phys. Rev. A* **9**, 1635 (1974).
2. K. Rubin, *Proceedings of the Eighth International Conference on the Physics of Electronic and Atomic Collisions*, p. 91. North Holland, Amsterdam (1971).
3. M. L. Goldberger and K. M. Watson, *Collision Theory*, Chapter 3, Wiley, New York (1964).

53

Polarized-Beams Studies of Spin Exchange in Electron–Hydrogen Collisions

M. S. LUBELL

The application of polarized beams to the study of interference phenomena in electron–hydrogen collisions is reviewed. In particular, the results of the first experiment using polarized electrons to study the effects of spin exchange in electron–atom collisions are discussed. Finally, the future of spin exchange studies using polarized electrons is examined.

1. Introduction

1.1. Polarization Experiments

The theoretical construction (see, for example, Reference 1) for the nonrelativistic solution of the electron–hydrogen scattering problem incorporates two scattering amplitudes: the direct and exchange amplitudes, f and g, respectively, or alternatively the singlet and triplet amplitudes, $f + g$ and $f - g$, respectively. Thus the spin-averaged differential cross section is given by

$$\frac{d\bar{\sigma}}{d\Omega} = \frac{1}{4} |f + g|^2 + \frac{3}{4} |f - g|^2 \tag{1}$$

or alternatively by

$$\frac{d\bar{\sigma}}{d\Omega} = |f|^2 + |g|^2 - |f||g| \cos \vartheta \tag{2}$$

M. S. LUBELL ● J. W. Gibbs Laboratory, Yale University, New Haven, Connecticut 06520 U.S.A.

where ϑ is the relative phase between f and g. Since each of the amplitudes is a complex quantity, three independent parameters, such as $|f|$, $|g|$, and ϑ, are needed to describe the scattering into any single channel.

Experiments with unpolarized particles, sometimes called "first-generation" experiments, generally result in the measurement of the spin-averaged cross section, thereby determining a linear combination of the three scattering parameters, as illustrated by equations (1) and (2). Only in the region of resonances can information be gleaned separately about the singlet and triplet amplitudes.[2] Elsewhere, polarization experiments must be performed.

Until recently, the limitations of polarized sources and polarimeters restricted such experiments to two polarization devices (sources or polarimeters) each. If the electron and atom polarization vectors, $\mathbf{P}_e$ and $\mathbf{P}_a$, are defined as the ensemble averages of the expectation values of the respective Pauli spin-operators, then six such "second-generation" experiments are possible with $\mathbf{P}_e$, $\mathbf{P}_a$, $\mathbf{P}_e'$, and $\mathbf{P}_a'$ parallel or antiparallel to each other. Here, as elsewhere in this paper, unprimed quantities refer to incident particles and primed quantities to scattered particles. Of the six possible experiments, shown schematically in Figure 1, however, the two in each row provide identical information.

This information can be expressed in the form of asymmetry parameters, denoted by A, A', and A'' for rows 1, 2, and 3, respectively. In the case of experiment (1), for example, where the scattered electron intensity is measured for the electron and atom spins antiparallel ($\uparrow\downarrow$) and parallel ($\uparrow\uparrow$) the cross-section asymmetry A, given by

$$A = (\sigma_{\uparrow\downarrow} - \sigma_{\uparrow\uparrow})/(\sigma_{\uparrow\downarrow} + \sigma_{\uparrow\uparrow})$$

can be expressed as

$$A = \mathrm{Re}(f^*g)\left(\frac{d\bar\sigma}{d\Omega}\right)^{-1} = |f||g|\cos\vartheta\left(\frac{d\bar\sigma}{d\Omega}\right)^{-1} \tag{4}$$

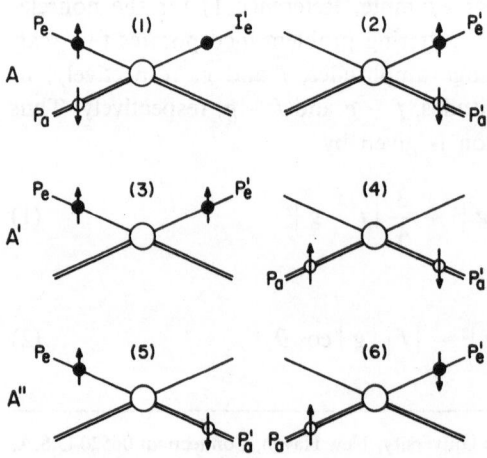

Figure 1. Nonreactive "second-generation" scattering experiments. Particles are incident from the left with electrons denoted by single lines and atoms by double lines. The circles, solid for electrons and open for atoms, indicate measured beams with the measured quantities denoted by P_e, P_a, and I_e for the electron polarization, atom polarization, and electron intensity, respectively. The primes indicate quantities measured in the outgoing channels. The asymmetries measured in a given row are specified by A, A', and A'' in accordance with equations (4), (7), and (12).

Table 1. Polarization Processes[a] and Cross Sections

Process	Cross section
$e\downarrow + A\uparrow \rightarrow e\downarrow + A\uparrow$	$\lvert f \rvert^2$
$e\downarrow + A\uparrow \rightarrow e\uparrow + A\downarrow$	$\lvert g \rvert^2$
$e\uparrow + A\uparrow \rightarrow e\uparrow + A\uparrow$	$\lvert f - g \rvert^2$
$e\downarrow + A\downarrow \rightarrow e\uparrow + A\downarrow$	$\lvert g \rvert^2$
$e\uparrow + A\downarrow \rightarrow e\downarrow + A\uparrow$	$\lvert g \rvert^2$
$e\downarrow + A\downarrow \rightarrow e\downarrow + A\downarrow$	$\lvert f - g \rvert^2$

[a] The electron is denoted by e, the atom by A, and the relative spin orientations by the arrows.

with the aid of the cross-section summaries provided in Table 1. For electron and atom beams with nonunity polarizations, the experimentally measured asymmetry $\Delta^{(1)}$ is related to A by

$$\Delta^{(1)} = P_e P_a A \tag{5}$$

where P_e and P_a are the degrees of polarization of the incident projectile electron and the atomic electron, respectively. In the case of experiment (2), the measured asymmetry $\Delta^{(2)}$ and the physical asymmetry A are similarly related by

$$\Delta^{(2)} = S_e S_a A \tag{6}$$

where S_e and S_a are the analyzing powers of the electron and atom polarimeters.

Experiments (3) and (4) both result in the determination of the asymmetry parameter A' given by

$$A' = \lvert g \rvert^2 \left(\frac{d\bar{\sigma}}{d\Omega} \right)^{-1} - 1 \tag{7}$$

where A' is found from

$$P_e' = A' P_e \tag{8}$$

in experiment (3) and from

$$P_a' = A' P_a \tag{9}$$

in experiment (4), with P_e' and P_a' denoting the degrees of polarization of the scattered electrons and atoms, respectively. If the electron and atom polarimeters again have the analyzing powers S_e and S_a the asymmetries measured in experiments (3) and (4) are given by

$$\Delta'^{(3)} = S_e P_e' = A' S_e P_e \tag{10}$$

and

$$\Delta'^{(4)} = S_a P_a' = A' S_a P_a \tag{11}$$

An analogous set of equations describes the results obtained in experiments (5) and (6) with A'' given by

$$A'' = |f|^2 \left(\frac{d\bar{\sigma}}{d\Omega}\right)^{-1} - 1 \tag{12}$$

In experiment (5) A'' is found from

$$P_a' = A''P_e \tag{13}$$

and in experiment (6), from

$$P_e' = A''P_a \tag{14}$$

The corresponding measured asymmetries are given by

$$\Delta'^{(5)} = S_a P_a' = A'' S_a P_e \tag{15}$$

and

$$\Delta'^{(6)} = S_e P_e' = A'' S_e P_a \tag{16}$$

From equations (2), (4), (7), and (12) it can be seen that A, A' and A'' are related by the expression

$$A - A' - A'' = 1 \tag{17}$$

Thus of the six second-generation experiments shown in Figure 1, only two are actually independent. The predicted values of A_E, A_E', and A_E'' from a number of theoretical calculations for 90° elastic scattering are shown in Figure 2, illustrating the linear relationship among the three asymmetry parameters expressed by equation (17). (The location of the singlet-S resonance is shown by the arrows.)

It might appear from the foregoing discussion that the performance of any two independent second-generation experiments together with the measurement of the spin-averaged cross section $d\bar{\sigma}/d\Omega$ is sufficient to determine the three scattering parameters $|f|$, $|g|$, and ϑ. In fact, however, an ambiguity in the sign of ϑ still remains, since only $\cos\vartheta$ is determined. A further experiment is required in which three polarization devices are employed. Such an experiment might be called a "third-generation" experiment.

Consider, for example, the case of a *longitudinally* polarized electron beam along the x axis incident upon a *longitudinally* polarized hydrogen beam along the y axis with an electron polarimeter employed to determine the z component of the polarization of the scattered electron. The quantity $(P_e')_z$ so measured is related to the incident electron and atom polarizations, $P_e = (P_e)_x$ and $P_a = (P_a)_y$, respectively, according to

$$(P_e')_z = P_e P_a \, \mathrm{Im}(f^*g)\left(\frac{d\bar{\sigma}}{d\Omega}\right)^{-1} = P_e P_a |f||g| \sin\vartheta \left(\frac{d\bar{\sigma}}{d\Omega}\right)^{-1} \tag{18}$$

The determination of $\sin\vartheta$ from such a measurement together with the previous

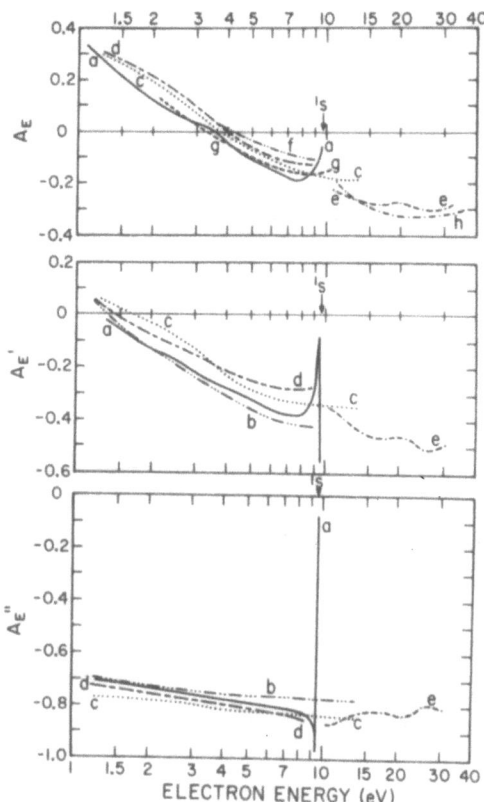

Figure 2. Some theoretical predictions for 90° elastic scattering asymmetries. The curves shown are based upon calculated quantities given in the following references: (a) close-coupling approximation, Reference 3; (b) exchange approximation, Reference 4; (c) polarized orbital calculation, Reference 4; (d) pseudostate close-coupling approximation, Reference 5; (e) pseudostate close-coupling calculation, Reference 6; (f) Kohn variation calculation, Reference 7; (g) virtual 2s variational calculation, Reference 8; and (h) Glauber exchange approximation, Reference 9.

determination of $\cos \vartheta$ serves to specify ϑ completely. The necessity of performing four experiments to determine three scattering parameters in the case of spin exchange collisions is analogous to the similar requirement in the case of spin–orbit collisions.

Since third-generation experiments have only recently become feasible with the newly developed GaAs photoemission polarized electron source,[10] the remainder of this discussion will be limited to second-generation experiments, and the ambiguity in the sign of ϑ will be implicitly understood. Although the need for measuring only two asymmetry parameters is established by equation (17), a prescription must still be developed for selecting those two parameters that can be measured to best advantage. One natural criterion is the asymmetry sensitivity defined for A by $A/\delta A$, where δA is the uncertainty in the measurement of A. In the case of experiment (1), δA is given by

$$\delta A = (P_e P_a)^{-1}(I_e')^{-1/2} = K(P_e P_a)^{-1}(I_e^{(1)} \varrho_a^{(1)})^{-1/2} \qquad (19)$$

where $I_e^{(1)}$ is the incident electron intensity, $\varrho_a^{(1)}$ is the atom density in the interaction region, and K is a proportionality constant which depends upon the scattering cross section and the acceptance of the electron detector. Therefore the sensitivity for experiment (1) is given by

$$A/\delta A = K^{-1}P_eP_a(I_e^{(1)}\varrho_a^{(1)})^{1/2}A \tag{20}$$

If the polarization experiments are limited to those involving measurements of the scattered electron, the sensitivities $A'/\delta A'$ and $A''/\delta A''$ must be calculated for experiments (3) and (6), which use an electron polarimeter. With η denoting the efficiency of the electron polarimeter relative to the electron detector of experiment (1), $A'/\delta A'$ and $A''/\delta A''$ are given by

$$A'/\delta A' = K^{-1}P_eS_e(I_e^{(3)}\varrho_a^{(3)}\eta)^{1/2}A' \tag{21}$$

and

$$A''/\delta A'' = K^{-1}P_aS_e(I_e^{(6)}\varrho_a^{(6)}\eta)^{1/2}A'' \tag{22}$$

A comparison of the sensitivities depends upon the values assigned to P_e, P_a, S_e, η, and the respective electron intensities and atom densities. These values, to a certain extent, in turn depend upon the level of technological development of various experimental techniques. Prior to the development of the GaAs photoemission polarized electron source,[10] reasonable optimal values of I_e and P_e for low-energy polarized electron beams, were $\sim$3 nA and $\sim$0.65, respectively,[11] based upon the Fano effect in Cs.[12-15] (Polarized electron production by means of the Fano effect in Cs will be discussed in succeeding sections.) For a Mott scattering[16-18] electron polarimeter, which is still the most reliable and best understood device applicable to the measurement of electron polarization in electron–atom scattering experiments, appropriate values of S_e and η are $\sim$0.4 and $\sim$10^{-5}, respectively.[19,20] Finally, for readily produced atomic hydrogen beams, values of $\sim$10^9 atoms/cm^3 for a polarized atom density, $\sim$10^{11} atoms/cm^3 for an unpolarized atom density, and $\sim$0.5 for P_a are reasonable estimates. (A value of $P_a \sim$ 1.0 is easily achieved at high magnetic field, but at low field the influence of the hyperfine structure reduces P_a by a factor of 2.[20] Only with the selection of pure hyperfine levels can a unity value of P_a be maintained at low magnetic fields, but the consequent loss of intensity diminishes the effectiveness of the technique.)

If a value of $\sim$3 μA is chosen for I_e for an unpolarized electron beam of modest energy spread ($\sim$100 meV full width at half-maximum), the three sensitivities can be calculated according to equations (20)–(22) with unpolarized beam values used for $I_e^{(6)}$ and $\varrho_a^{(3)}$ and polarized beam values used elsewhere. For elastic scattering, the relative sensitivities, generated with the aid of the values of A_E, A_E', and A_E'' given in Figure 2, are shown in Figure 3. As shown, the sensitivity $A_E/\delta A_E$ compares favorably with the other two sensitivities over a reasonable energy band. Reinforced by this observation and motivated by the desire to carry out collision studies with polarized electrons, a crossed-beams experiment involving polarized electrons and polarized hydrogen atoms was undertaken to determine spin-exchange effects in electron–hydrogen scattering. Reports of the first measurements in this experimental program have been published recently.[21,22]

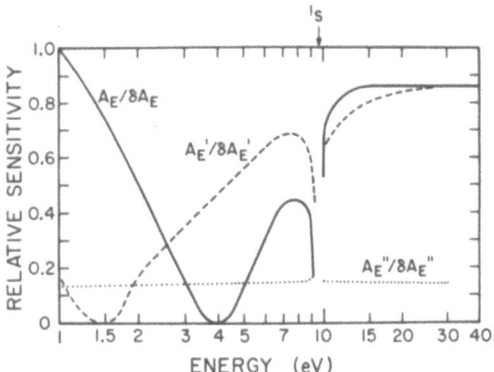

Figure 3. Estimated relative sensitivities for elastic asymmetries A_E, A_E', and A_E'' in electron experiments, based upon the characteristics of a Fano-effect[12-15] polarized electron source and the values of A_E, A_E', and A_E'' given in Figure 2. All curves are normalized to the sensitivity $A_E/\delta A_E$ at an incident electron energy of 1 eV.

1.2. Polarized Electron Sources

The performance characteristics and the status of polarized electron sources have been reviewed a number of times in the last few years.[10,11,20,22-25] As has been stressed in many of these reviews, sources cannot be judged outside the context of their application. For crossed-beams electron–atom scattering experiments, a polarized electron source ideally should possess the following characteristics: high figure of merit $\zeta = P_e I_e^{1/2}$, since the uncertainty δA in an asymmetry measurement is proportional to $1/\zeta$; small generalized emittance[11,20,22,25] ε^* and high brightness I_e/ε^{*2}, since the acceptance of the interaction region is generally quite limited and the magnetic field at the interaction region must be kept small for low-energy scattering; small energy spread, since the intrinsic energy scale of structures such as resonances and threshold effects is ~ 0.05 eV; and capability of frequent polarization reversal in a manner that minimally affects the electron optics, since the systematic uncertainties in an asymmetry measurement are minimized if the modulation of the polarization is rapid compared to drifts in the apparatus provided the modulation itself does not introduce systematic effects.

Naturally the choice of any particular source involves some compromise of the ideal performance characteristics. For the purpose of the first electron–hydrogen measurements, which were to be exploratory in nature, the characteristic of small energy spread was sacrificed. The requirements of small generalized emittance, rapid polarization reversal, and freedom from systematic effects under reversal then dictated the choice of a Fano-effect source, which was free of magnetic fields and was capable of optical rather then electron-optical polarization reversal. The photoemission GaAs source, which not only possesses the necessary reversal characteristics but also has a smaller emittance, higher figure of merit, and smaller energy spread, it must be remembered, had not yet been developed.

A Fano-effect polarized electron source relies upon the spin–orbit interaction in continuum states to transfer the helicity of an incident ionizing photon beam to that of an extracted photoelectron beam. The precise dependence of the photoelectron beam polarization on the energy of a completely circularly polarized incident photon

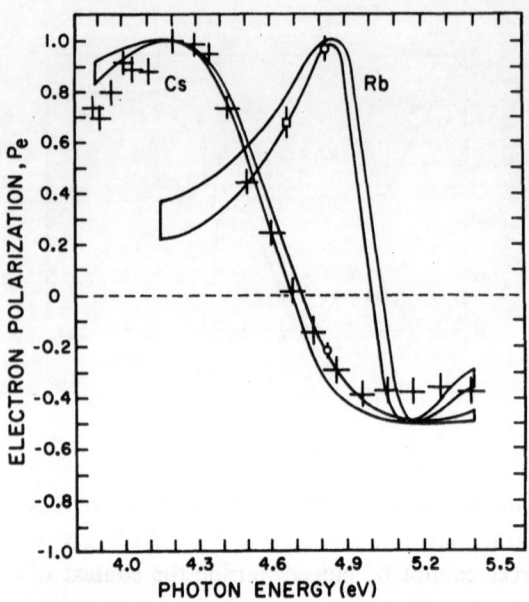

Figure 4. Fano-effect electron polarization for unity light polarization. Curves are taken from the results given by Baum et al.[15] with bands representing one standard deviation uncertainties. Data points: (+) Heinzmann et al.[26]; (○) Drachenfels et al.[27]; (□) Grannemann.[28]

beam is shown in Figure 4. For the case of Cs, it is readily apparent that high polarizations are achievable for incident light of fairly broad bandwidth, starting, for example, from threshold at 3.894 eV = 318.4 nm and extending up for ∼0.5 eV to ∼4.4 eV ≈ 280 nm. The photoionization cross section[29] for Cs over this wavelength band, moreover, is relatively high, starting at 0.225 Mb at threshold and falling to 0.075 Mb at 280 nm, thus permitting the easy generation of photoelectron beams with intensities ≳10 nA.

2. Method

2.1. General Concept

The experimental method that was employed is illustrated by the schematic diagram of the experimental layout shown in Figure 5. The experiment comprises four principal sections—the polarized electron source, a Mott-scattering electron polarimeter, a polarized atomic hydrogen source, and an interaction region with associated electron, atom, and ion detection devices. The electron and atomic hydrogen beams intersect at 90° in a crossed-beams configuration.

2.2. Polarized Electron Production and Electron Polarimetry

In the Fano-effect electron source, polarized uv light photoionizes an atomic beam of Cs in a region maintained at −1 kV electrostatic potential. The photoelectrons which are highly polarized along the light axis are extracted from the ionization region by a potential gradient of ~1 V/cm and then accelerated to ground potential and formed into a beam by means of electron-optical focusing elements. The incident uv light is circularly polarized by a dichroic film polarizer and zeroth-order quartz quarter-wave retardation plate. Rotation of either the polaroid or the quarter-wave plate by 90° reverses the photon helicity and hence the electron polarization.

The bending magnet shown in Figure 5 can be adjusted to deflect the 1-keV electron beam into either the interaction region or the Mott-scattering polarimeter, without affecting the longitudinal nature of the electron polarization. In the Mott branch, the electron beam is first accelerated to 7 keV and then passed through a Wien filter,[18] having crossed transverse electric and magnetic fields, which rotates the electron polarization to the transverse orientation required for Mott analysis. Finally, the transversely polarized electron beam is accelerated to 100 keV and 120° backscattered[17] from thin gold foil targets ranging in thickness from 27 to 53 μg/cm², each having a Formvar backing ~20 μg/cm² thick.

2.3. Polarized Atomic Hydrogen Production and Hydrogen Beam Transport

The atomic hydrogen beam is produced by thermal dissociation of molecular hydrogen gas in a tungsten oven. After collimation, the $M_J = +\frac{1}{2}$ states of the atoms in the partially dissociated beam are selected at high field in a long sextupole magnet. A computer analysis, based upon an optical model of the sextupole focusing properties,[20] shows that the high-field state-selection parameter s of the atomic hydrogen reaching the interaction region is >0.99. Since the scattering experiment is carried

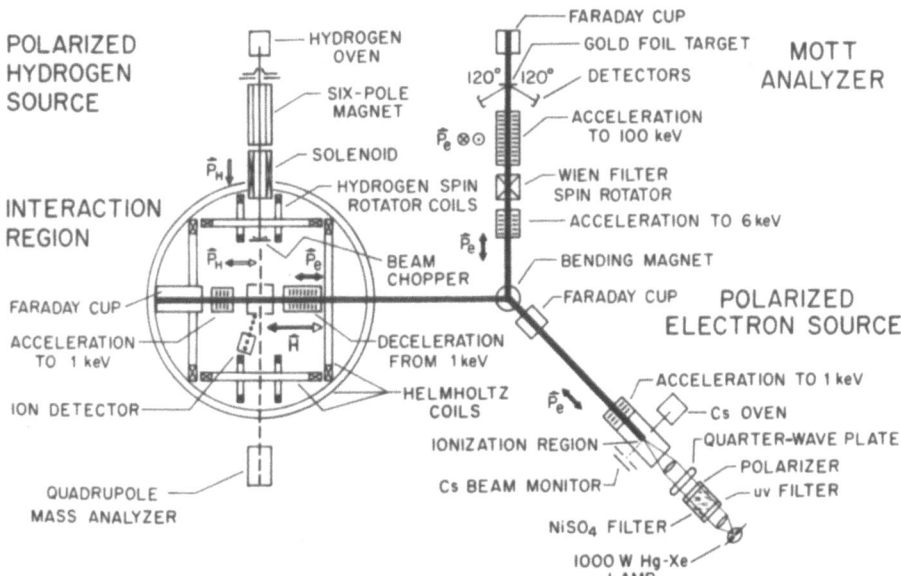

Figure 5. Schematic diagram of the experimental layout.

out in a near-zero-field region, however, the ground-state hyperfine interaction reduces the effective electronic polarization of the atom P_H to $s/2$.[20] The magnetic fields along the hydrogen beam line (see Section 3) are shaped in such a fashion that Majorana transitions[30] are avoided with the atomic hydrogen spins adiabatically following the field lines. Thus, in the interaction region, where there is a 100–200 mG magnetic field coaxial with the electron beam, the atomic hydrogen polarization is transverse to the hydrogen beam, with the direction of the polarization determined by the direction of the magnetic field.

Before entering the interaction chamber, the hydrogen beam undergoes a 100-Hz modulation produced by a tuning fork beam chopper. After leaving the interaction region, the beam enters a separate chamber containing a quadrupole mass analyzer which monitors the relative amounts of atomic and molecular hydrogen. In the interaction region itself, the atomic hydrogen density is estimated to be $(1-2) \times 10^9$ atoms/cm³. The residual gas pressure in the interaction chamber, on the other hand, is maintained at $\sim 10^{-9}$ Torr facilitated by the use of several stages of differential pumping along the hydrogen beam line.

2.4. Polarized Electron Beam Transport

The electron beam is transported from the source to the interaction chamber at an energy of 1 keV. Prior to entering the interaction region, the beam is decelerated to an energy of several eV and is collimated. Then the beam is reaccelerated to the desired potential relative to the electron source, the absolute energy scale being established by the onset of ionization at 13.6 eV. After intersection with the hydrogen beam, the unscattered electrons are once again accelerated to 1 keV and monitored in a Faraday cup.

2.5. Data Acquisition

The electron current arriving in the Faraday cup is digitized by means of an electrometer coupled to a voltage to frequency converter (VFC). The output of the VFC is then counted in a preset scaler which halts data taking after a preset charge has arrived at the cup. For each data interval, lasting 5 to 15 sec, the number of counts registered by an ion detector, located downstream from the interaction region along the hydrogen beam, and by an electron detector, located directly beneath the interaction region to observe 90° scattering, are totaled on two pairs of blind scalers. The "on" scalers record the electron and ion events during the beam-on portion of the 10-msec hydrogen beam chopper cycle, and the "off" scalers record the events during the beam-off portion of the cycle. Of the full cycle, 4 msec are devoted to beam-on measurements and 2 msec to beam-off measurements, with the remaining 4 msec representing a dead time during which the chopper is only partially open.

The quadrupole signals obtained for both the atomic and molecular hydrogen are likewise totaled on pairs of "on" and "off" scalers. Also recorded for each data interval is the accumulated charge in the Faraday cup and the elapsed time of the interval. At the conclusion of a data interval, a PDP-15 computer reads and clears the blind scalers, advances the quarter-wave plate of the polarized electron source by 90°, and reinitiates data accumulation. After 20 to 50 complete revolutions of the quarter-wave plate the run is halted, and the accumulated totals for each quarter-wave plate position are written onto magnetic tape. A complete measurement at a given energy normally comprises eight runs corresponding to the four orientations (90° apart) of the linear polarizer of the electron source for each of the two directions of the magnetic field in the interaction region.

Events obtained from the ion detector, representing total cross-section data for impact ionization, are combined to form the "real" asymmetry Δ_I^R according to

$$\Delta_I^R = (N_I^+ - N_I^- - B_I^+ + B_I^-)/(N_I^+ + N_I^- - B_I^+ - B_I^-) \tag{23}$$

where N_I^+ is the sum of the hydrogen-beam-on ion counts for quarter-wave-plate positions 0 and 2 (0° and 180°), N_I^- is the sum of the beam-on ion counts for quarter-wave plate positions 1 and 3 (90° and 270°), and B_I^+ and B_I^- are the corresponding sums of beam-off ion counts. In addition to the real asymmetry defined by equation (23), two "false" asymmetries Δ_I^{F1} and Δ_I^{F2} can be constructed analogously by combining data from the quarter-wave-plate positions according to the following prescriptions:

$$\Delta_I^{F1}: (0) + (1) - (2) - (3) \tag{24}$$

and

$$\Delta_I^{F2}: (0) + (3) - (1) - (2) \tag{25}$$

For both the real and false asymmetries, the values of Δ_I measured for each run are combined according to their statistical weights to give a final asymmetry for a given energy.

The physical asymmetry for total impact ionization A_I is related to Δ_I^R by the expression

$$\Delta_I^R = P_e P_H (1 - F_I) A_I \tag{26}$$

where P_H is the polarization of the atomic hydrogen and F_I is the fraction of ion counts originating from molecular hydrogen. Measurement of F_I is accomplished by lowering the hydrogen oven temperature from its normal operating value of ~2800 K to 1000 K, where it can be assumed that the beam is entirely molecular in composition. The ratio Δ_I^{cold} of the beam-related ion counting rate $R_I^{cold} = (N_I^{cold} - B_I^{cold})/I_e$ to the beam-on-minus-beam-off quadrupole H_2 molecular signal Q_2^{cold} at 1000 K indicates the sensitivity of the ionization measurement to molecular contamination of the beam. The product of this ratio and the H_2 quadrupole signal at full temperature Q_2^{hot} gives the number of counts attributable to molecular hydrogen at full temperature. Thus F_I is given by

$$F_I = \frac{\Delta_I^{cold}}{\Delta_I^{hot}} = \frac{R_I^{cold}/Q_2^{cold}}{R_I^{hot}/Q_2^{hot}} \tag{27}$$

Measurements of Λ_I^{cold} are interspersed in the data acquisition process at ~6-intervals. As a check on systematic effects, the linearity of the dependence of Λ_I^{hot} on the ratio $\Gamma^{\text{hot}} = Q_1^{\text{hot}}/Q_2^{\text{hot}}$ is monitored constantly, where Q_1^{hot} is the H_1 beam-on-minus-beam-off quadrupole signal at full temperature. Typical values of F_I range from a few percent for electrons incident at the threshold energy of 13.6 eV to ~20% for electrons incident at energies of 100–200 eV. Data acquisition rates for ionization measurements are typified by the accumulation of ~60000 beam-on and 10000 beam-off events in a 10–15-min period for an incident electron energy of 27 eV.

In the case of 90° elastic differential scattering, the much lower event rate (~4 counts/sec at ~8 eV) necessitates a slightly different treatment of the data. Real and false asymmetries, Λ_E^R, Λ_E^{F1}, and Λ_E^{F2}, are formed from the events obtained by the electron detector in a manner analogous to the ionization asymmetries given by equations (23)–(25), except that background events are not subtracted. Thus Λ_E^R is given by

$$\Lambda_E^R = (N_E^+ - N_E^-)/(N_E^+ + N_E^-) \tag{28}$$

where N_E^+ and N_E^- are the sums of beam-on elastic-electron counts for quarter-wave-plate positions 0, 2 and 1, 3, respectively. The background events are taken into account by expressing the physical asymmetry A_E in terms of Λ_E^R according to

$$\Lambda_E^R = P_e P_H (1 - f_E)(1 - F_E)A_E \tag{29}$$

with f_E the ratio of the beam-off rate to beam-on rate averaged over all runs at a given energy. The fraction of elastic events originating from molecules, F_E, is determined in a manner analogous to that used in the determination of F_I. As in the case of ionization, the value of Λ_E for a given energy is obtained from a weighted average of the values measured in individual runs.

At irregular intervals during the data acquisition process, the electron beam is directed into the Mott branch for polarization analysis. A typical pulse height spectrum from one of the two Si surface-barrier electron detectors is shown in Figure 6.

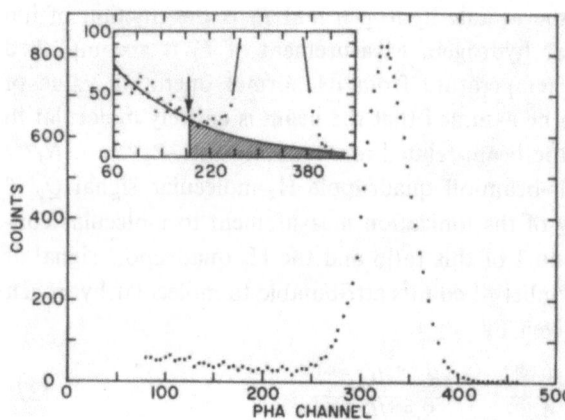

Figure 6. Mott pulse height spectrum. Shaded area represents extrapolation of "inelastic" tail. Arrow indicates the setting of the discriminator threshold.

Discriminator thresholds are set as indicated by the arrow in the figure. For the two detectors, events occurring above threshold are gated according to the quarter-wave-plate position at the electron source and scaled on one of two pairs of scalers, one pair for positive helicity ionizing light at the source and the other pair for negative helicity light. Counts are accumulated for 5-sec intervals, with the quarter-wave plate rotated by 90° between intervals. After an integral number of complete quarter-wave-plate rotations, the scaler totals are recorded. This procedure is repeated for each of the four orientations of the linear polarizer. A complete polarization measurement, consisting of $\sim 4 \times 10^5$ events, requires ~ 4 min, and results in a statistical uncertainty of $\sim 1.6 \times 10^{-3}$ in the measured Mott asymmetry Δ_M.

In order to compensate for differences in acceptance and efficiency of the two detectors, the Mott asymmetry is calculated as

$$\Delta_M = (1 - \xi)/(1 + \xi) \tag{30}$$

where ξ is given by

$$\xi = \left(\frac{N_M^{1+}}{N_M^{2+}} \frac{N_M^{2-}}{N_M^{1-}} \right)^{1/2} \tag{31}$$

with N_M^{1+} and N_M^{2+} the number of counts coming from detectors 1 and 2, respectively, for the case of positive-helicity light, and N_M^{1-} and N_M^{2-} the corresponding counts for negative-helicity light. The quantities $N_M^{1\pm}$ and $N_M^{2\pm}$ are corrected for inelastic events, estimated by extrapolating under the elastic peak the background counts fitted by an exponential of the form $e^{-\beta n}$ where n is the channel number in the pulse height spectrum shown in Figure 6. The inelastic events, indicated by the shaded area in the figure, are typically found to be $\sim 6\%$ of the total counts within the elastic peak. The effects of multiple and plural scattering in the target foil are taken into account by making measurements of Δ_M with several foils of different thicknesses and performing an extrapolation to zero thickness.

Once a value of Δ_M is established, the corresponding electron polarization is calculated from the relation

$$\Delta_M = S_M P_e \tag{32}$$

where S_M is the Mott analyzing power (often called the Sherman function). For the case of 120° Mott scattering from Au at 100 keV, S_M is known to be 0.391 $\pm$ 0.008.[19,20]

3. Apparatus

A scale drawing of the Fano-effect polarized electron source is shown in Figure 7. A complete discussion of the source has been published previously,[22] and the reader is referred to that publication for added detail. Briefly, the source contains a 1000-W uv high-pressure Hg–Xe lamp with a Suprasil quartz envelope. Light from the lamp

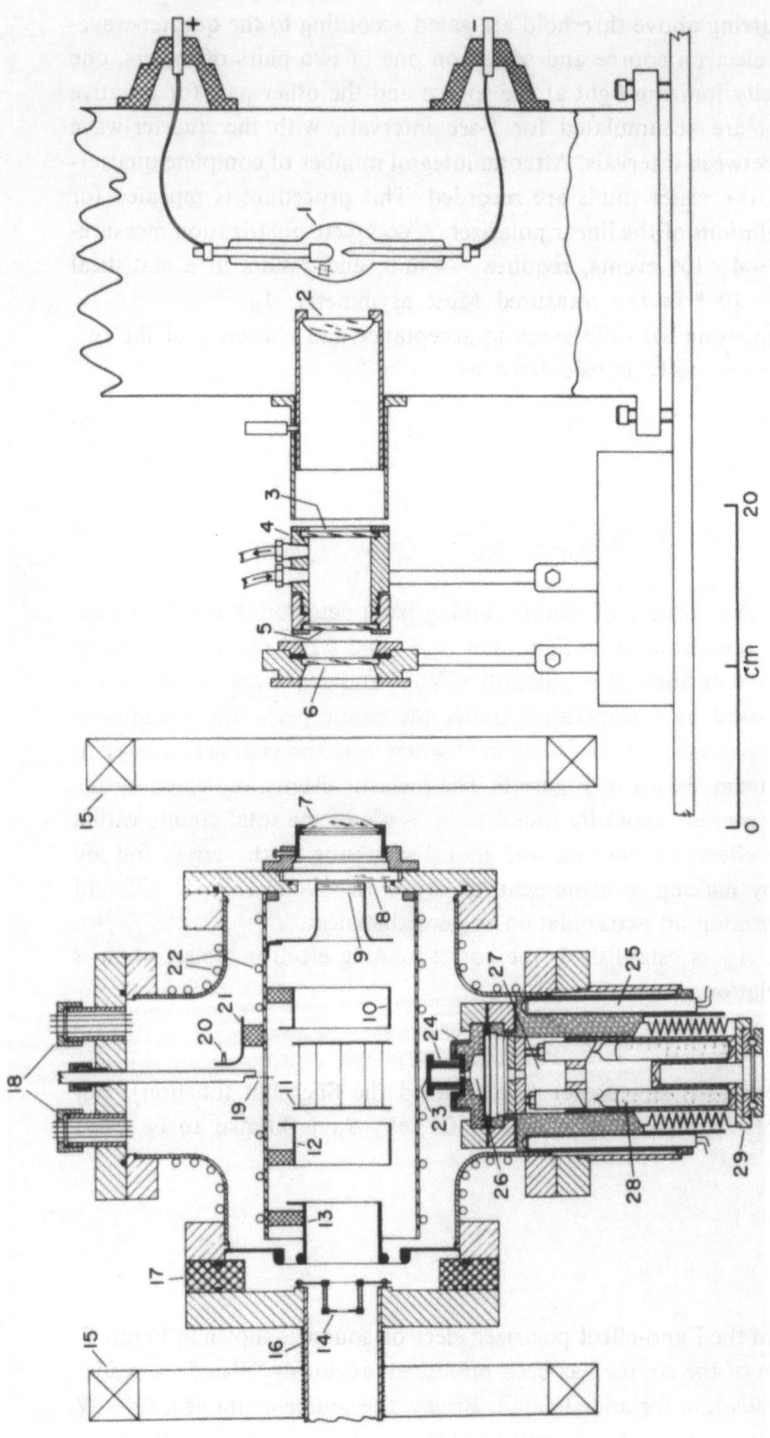

Figure 7. Side-view scale drawing of the Fano-effect polarized electron source showing the following components: (1) 1000-W Hg–Xe cw arc lamp with Suprasil envelope; (2) f/1.5 quartz lens; (3) Corning CS0-56 filter; (4) NiSO₄ absorption cell; (5) dichroic linear polarizer on a 5-cm-diam rotatable quartz disk; (6) 3000-Å zeroth-order rotatable quartz quarter-wave retardation plate; (7) f/3.0 quartz lens; (8) Suprasil vacuum window; (9) heated quartz disk; (10) repeller electrode; (11) ionization region; (12) extractor electrode; (13) focusing electrode; (14) electron collimator; (15) Helmholtz coils; (16) beam pipe with solenoid; (17) Lucite insulating flange; (18) electrical feedthroughs; (19) stainless steel mesh; (20) and (21) hot-wire and ion collector of Cs-beam surface-ionization detector; (22) freon cooling pipes; (23) stainless steel multicapillary orifice; (24) Thermocoax heating coils; (25) twelve 300-W heaters; (26) oven lower chamber; (27) oven upper chamber; (28) six 10-g Pyrex ampoules of Cs metal; and (29) bellows mechanism for breaking ampoules. All vacuum parts were fabricated from stainless steel unless otherwise noted. Not shown are thermocouple oven temperature monitors and an externally operable Cs beam-blocking flag.

passes through a cell 5 cm in length containing an aqueous solution of $NiSO_4$ (100 g/liter), which removes radiation of wavelength longer than 320 nm. The entrance window of the cell is a Corning CSO-56 filter which has a spectral transmission that falls from 0.8 at 310 nm to 0.4 at 280 nm and 0.2 at 270 nm. Thus the wavelength band of the light emerging from the cell is well matched to the requirements for high electron polarization, as illustrated in Figure 4. The exit window of the cell is a manually rotatable quartz plate coated externally with a dichroic linear-polarizing film.

After leaving the absorption cell, the linearly polarized uv light passes through a 300-nm zeroth-order quartz quarter-wave retardation plate to produce circularly polarized light. The quarter-wave plate is automatically rotated by a stepping motor controlled by the PDP-15 computer. In the ionization region the circularly polarized uv light intersects a beam of Cs atoms emanating from a multicapillary orifice of a stainless steel oven. The lower chamber of the oven is operated at a temperature of 595 K, while an upper chamber immediately below the orifice (see Figure 7) is operated at ~625 K to reduce the dimer (Cs_2) content of the beam.[20] Under these conditions, a density of ~10^{12} atoms/cm^3 is produced in the ionization region.

Three orthogonal pairs of magnetic field coils are used to create a weak (<200-mG) magnetic field in the ionization region coaxial with the incident light beam. The shape of the electrodes and their potentials were established with the aid of an electron-trajectory computer program. For the purpose of initial electron-optics tuning, a stainless steel mesh can be inserted into the ionization region at the image of the lamp to produce a ~1-μA photoemission beam of unpolarized "monoenergetic" (~0.2 eV full width at half maximum) electrons.

A scale drawing of the Mott scattering region is shown in Figure 8. The electrons scattered by ±120° in the plane normal to the incident polarization vector are detected by a pair of Si surface barrier detectors. Since the Mott scattering apparatus is maintained at ~100 kV relative to ground, the amplified pulses from each detector are first converted to analog optical signals using light-emitting diodes and then transmitted through Lucite light pipes to ground potential where the optical signals are reconverted to electrical signals by photomultipliers.

The atomic hydrogen beam line is shown in detail in Figure 9. Of particular

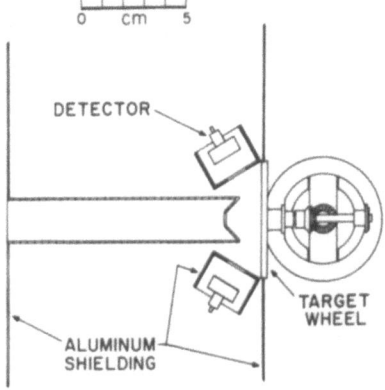

Figure 8. Mott scattering region. The 100-keV transversely polarized electrons enter from the left and are scattered by one of six targets in the target wheel, which can be rotated while the system is under vacuum and at high voltage. Aluminum is used for shielding and chamber construction to maximize the energy loss of surface-scattered electrons which enter the detectors, thereby permitting pulse height discrimination of such background electrons. Ortec model SBEE1000 surface barrier detectors are used for the detection of electrons.

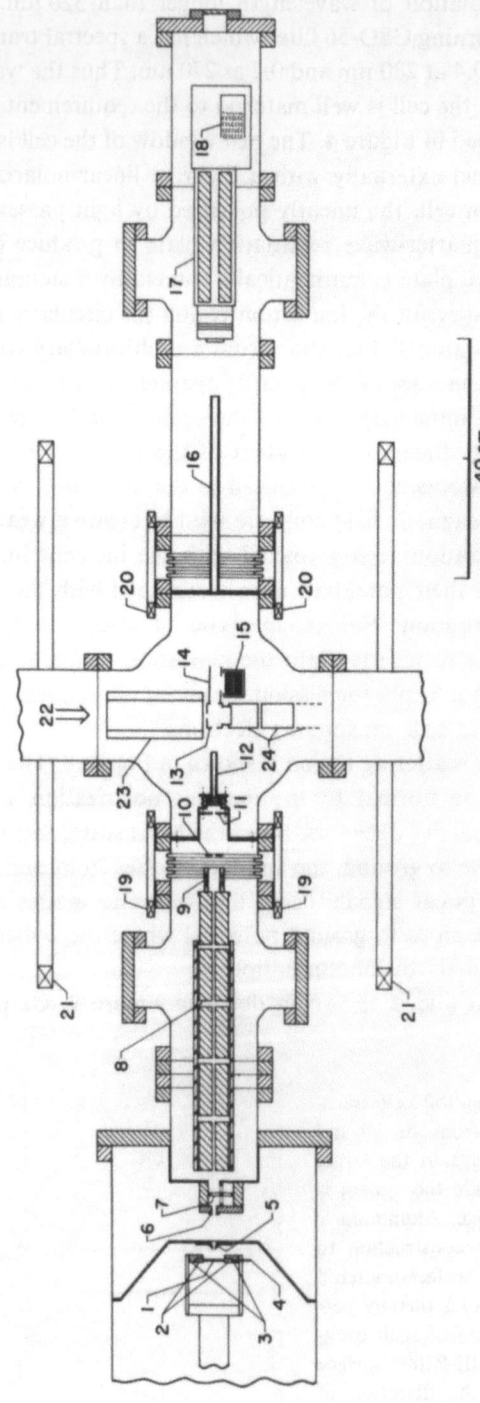

Figure 9. Top-view scale drawing of hydrogen beam line showing the following elements: (1) tungsten oven; (2) molybdenum support blocks; (3) water-cooled copper terminals; (4) hydrogen gas inlet; (5) and (6) hydrogen beam collimators; (7) butterfly valve; (8) sextupole magnet; (9) solenoid; (10) tuning fork beam chopper; (11) hydrogen beam-defining aperture; (12) tapered differential pumping tube; (13) interaction region; (14) electric field shield; (15) ion detector (Johnston model MM1-1S-FDB electron multiplier); (16) differential pumping tube; (17) quadrupole mass analyzer (Extranuclear Laboratories model 270-9); (18) ion detector (EMI model 9603/2B electron multiplier); (19) and (20) hydrogen spin-rotation coils; (21) interaction region magnetic field defining coils (two orthogonal pairs not shown); (22) direction of incident electrons; (23) preinteraction electron deceleration optics; and (24) postinteraction electron acceleration optics.

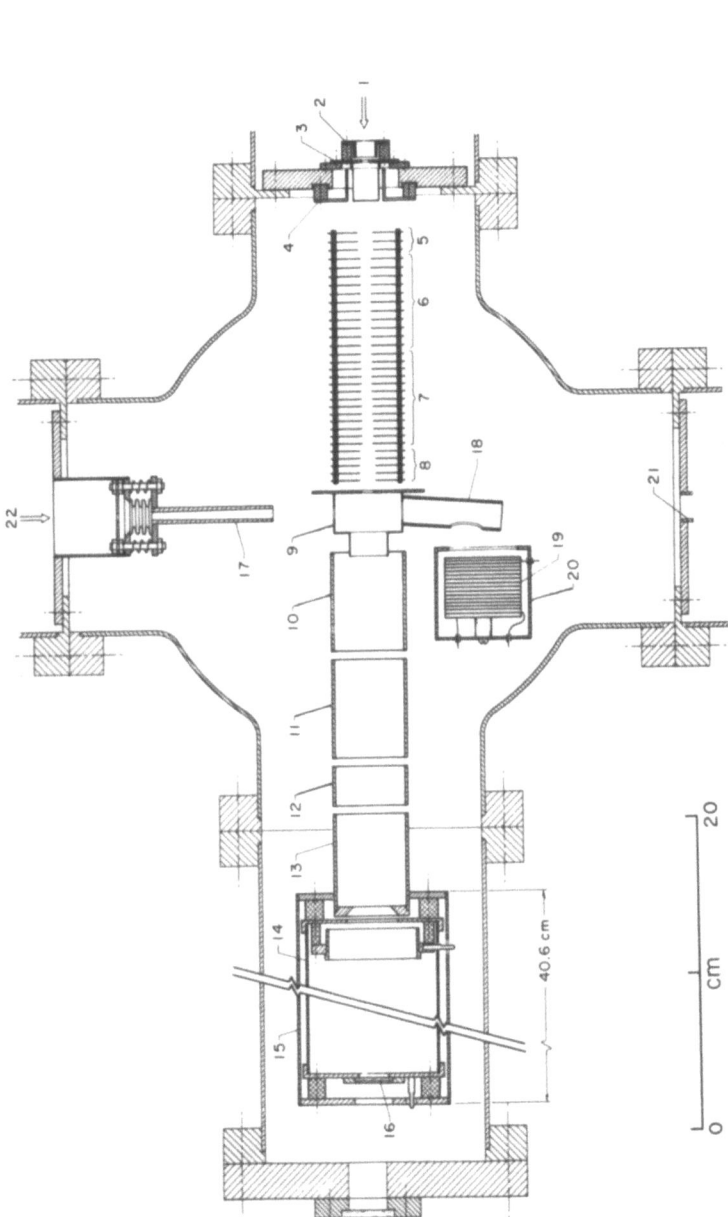

Figure 10. Top-view scale drawing of the opened interaction chamber showing the following elements: (1) direction of incident electrons; (2) secondary electron repeller for (3) electron current sensor; (4) predeceleration electron beam steering plates; (5) predeceleration electron beam focusing optics; (6) first deceleration filter lens;[81] (7) second deceleration filter lens;[81] (8) postdeceleration electron beam focusing optics; (9) interaction region, (10)–(13) postinteraction acceleration optics; (14) Faraday cup with (15) shield; (16) conducting-glass window for electron beam alignment; (17) hydrogen-beam-line tapered differential pumping tube; (18) electric field shield; (19) ion detector (Johnston model MM1-1S-FDB electron multiplier with (20) shield; (21) hydrogen-beam-line differential pumping tube; and (22) direction of incident hydrogen beam.

note are the hydrogen oven and the five-section sextupole magnet. The oven is constructed from a piece of tungsten foil 4.5 cm long by 4.9 cm wide by 0.0025 cm thick, rolled to form a tube 4.5 cm long and ~0.4 cm in diameter. A ~0.15-cm hole in the wall of the tube at its center serves as an orifice. The oven mount allows the oven to be aligned while under vacuum and at its operating temperature of ~2800 K. Each section of the sextupole magnet is 7.62 cm long with a 0.31-cm-diam gap and has a pole tip field strength of 8500 G.

The magnetic field coils used for avoiding the Majorana depolarization effects mentioned in Section 2.3 are also shown in Figure 9 as elements 9 and 19–21. The atoms emerging from the sextupole are aligned in a ~200-G axial field of the small solenoid (9) and then adiabatically rotated into a orientation parallel (or antiparallel) to the electron beam while within the ~5-G field of the small Helmholtz pair (19). The atoms then experience a field produced by the large Helmholtz pair (21) which always remains parallel (or antiparallel) to the electron beam as it monotonically decreases to a value of 100–200 mG at the interaction region. The second small Helmholtz pair (20) serves to maintain a symmetric field configuration. Not shown in Figure 9 are two additional pairs of field coils mutually orthogonal to the large Helmholtz pair (21) which act as Earth's field compensators.

Details of the interaction chamber are shown in top view in Figure 10 and in side view in Figure 11, both drawn to scale. All materials used are compatible with the requirements of ultrahigh vacuum and can withstand bakeout temperatures up to 450°C. The filter lens[31] shown in Figure 11 (element 25) is used to select elastically scattered electrons. The filter lenses shown in Figure 10 serve not only to decelerate the incident electron beam and appropriately trim its phase space but also to analyze the beam energy and width.[22]

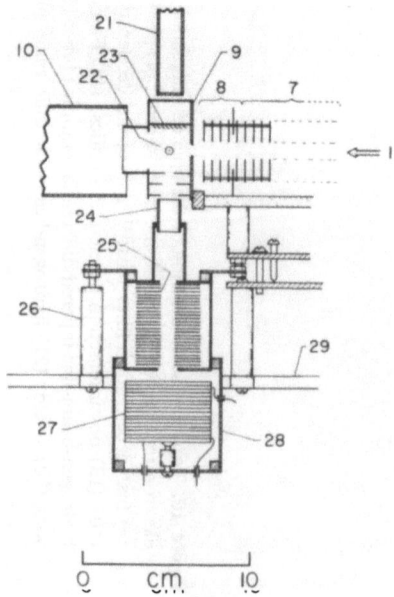

Figure 11. Side-view scale drawing of interaction region looking upstream along hydrogen beam with the following elements shown: (1) direction of incident electron beam; (7)–(10) electron-beam-line components shown in Figure 9; (22) direction of incident hydrogen beam; (23) CuBe baffle for suppression of surface-scattered electrons; (24) scattered-electron collector tube; (25) scattered-electron filter-lens[31] energy analyzer; (26) Steatite ceramic insulator; (27) electron detector [Johnston model MM1-1S-FDB electron multiplier with (28) shield]; (29) support frame, and (30) magnetic-field-probe vacuum insertion tube.

4. Results

The operating characteristics of the Fano-effect polarized electron source are summarized in Table 2. The electron currents given are based upon measurements made with a removable Faraday cup placed ~30 cm downstream from the exit of the source. The transport efficiency from this point to the crossed-beams interaction region ranged from 10% to 30%, so that under normal operating conditions, when, for the purpose of overall efficiency, the output current of the source was maintained at ~10 nA, only about 1–3 nA were available at the interaction region.

The electron polarization was measured by Mott scattering according to the procedure detailed in Section 2.5. Several checks were made to ensure the integrity of the measurement. First, the Wien filter spin rotator (see Section 2.2) was scanned on both sides of its calculated operating point to ascertain that the spin rotation from longitudinal to transverse was 90°. Then the quarter-wave-plate zero position was rotated by 90° in steps of 3° in order to determine the relative orientation of the linear polarizer and quarter-wave plate that produced maximum polarization. The typical electron-polarization dependence upon quarter-wave-plate orientation ϕ is shown in Figure 12 together with a plot of the function $\cos 2\phi$. Finally, the measured polarizations at each of the four polaroid positions were compared for consistency. Within statistical uncertainties, no deviations were found.

The electron polarization was, however, found to depend upon the extraction voltage gradient in the ionization region as shown in Figure 13. The degradation of the polarization with decreasing extraction voltage ΔV is thought to be attributable to e–Cs spin-exchange collisions.[22] This observed degradation necessitated a compromise between energy spread and polarization. Since maximum polarization was desired, ΔV (the voltage between electrodes 10 and 12 in Figure 7) was set at 7.0 V with the consequence that the energy spread of the extracted polarized beam was enlarged to 3 eV full width at half-maximum. Under these operating conditions, the electron polarization P_e was measured to be 0.63 ± 0.03. As a consistency check on

Table 2. Polarized Electron Source Characteristics

Electron polarization, P_e	0.63 ± 0.03
Electron current	
Maximum.	25 nA
Average	10 nA
Energy spread (FWHM)	3.0 eV
Emittance at 1 keV	20 mrad cm
Oven capacity	60 g
Oven lifetime (at 10 nA)	75 h
Atomic beam density	10^{12} atoms/cm³
Polarization reversal	Optical

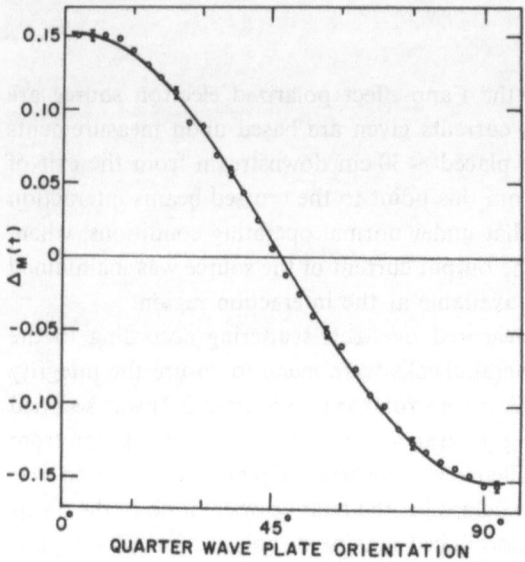

Figure 12. Mott asymmetry Δ_M as a function of Fano source quarter-wave plate orientation ϕ.

the measurement of P_e, the circular polarization of the uv ionizing light was measured as a function of wavelength, and the predicted value of P_e was calculated based upon the known photoionization cross section of Cs[29] and the measured wavelength characteristics of the optical components. A predicted value of 0.64 ± 0.06 was calculated for P_e in this manner, in excellent agreement with the measured value.

Electron polarization monitoring during data acquisition revealed that P_e slowly decreased with time as energy absorption in the dichroic linear polarizer steadily degraded its polarization capabilities. The temporal dependence of P_e was taken into account in the analysis of the collision data.

Several initial measurements of A_E for 90° elastic scattering are shown in Figure 14 along with a representative sample of theoretical curves. The value of A_E obtained from an angular distribution analysis[2] of resonance region data is also shown. Within the context of the preliminary nature of the polarization data, the agreement between theory and experiment is satisfactory.

In the case of total impact ionization, more extensive polarization data have been obtained over an incident energy range from 15 to 197 eV. The measured values

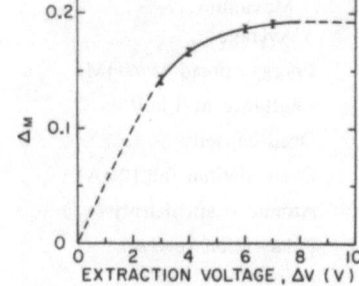

Figure 13. Mott asymmetry Δ_M as a function of Fano source extraction voltage ΔV between electrodes (10) and (12) in Figure 6.

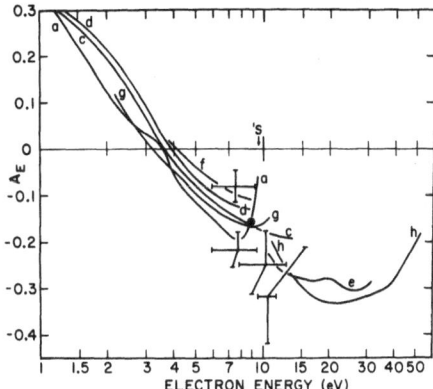

Figure 14. Experimental values of A_E and representative theoretical calculations, taken from the same references as those given in Figure 2. Resonance region measurement from Reference 2 is shown as solid dot.

of A_I are given in Table 3 together with the two false asymmetries, Δ_I^{F1} and Δ_I^{F2}, for each energy. As can be seen, the false asymmetries are all small and their respective chi-squares per degree of freedom calculated for assumed values of zero are generally close to unity, demonstrating the absence of significant systematic effects. In fact, Δ_I^{F1} and Δ_I^{F2}, summed over all energies, are consistent with zero at the statistical limit of 4×10^{-4}. (In the case of $90°$ elastic scattering, the corresponding limit presently stands at 2×10^{-3}.) These small values of the false asymmetries attest to the absence of systematic effects under the optical reversal of the electron polarization. The systematic uncertainties in A_I, included in the overall uncertainties given in Table 3, arise from the following uncertainties: P_e: $\pm 4\%$; P_H: $\pm 2\%$; $(1 - F)$: $\pm 1\%$.

Table 3. Results of Impact Ionization Data Analysis[a]

E (eV)	Δ_I^{F1} (10^{-4})	Δ_I^{F2} (10^{-4})	$\chi^2(0)$/deg freedom Δ_1	Δ_2	A
15	−56(30)	13(30)	16/13	10/13	0.515(32)
17	15(20)	−15(20)	7/10	11/10	0.479(28)
19	−12(27)	−55(27)	5/10	15/10	0.472(35)
23	38(25)	17(25)	15/9	13/9	0.465(35)
27	3(5)	3(5)	71/68	67/68	0.415(20)
34	−00(22)	02(22)	3/5	3/5	0.343(33)
42	05(11)	18(11)	9/10	16/10	0.338(17)
57	−04(15)	01(15)	15/9	6/9	0.256(15)
77	−11(22)	13(22)	10/9	13/9	0.201(20)
107	27(13)	05(13)	21/12	5/12	0.147(11)
147	04(12)	08(12)	12/8	8/8	0.128(13)
197	−20(16)	−00(16)	8/8	7/8	0.074(17)
All runs	3(4)	4(4)	192/171	174/171	

[a] Uncertainties are one standard deviation; those for A include systematic as well as statistical effects.

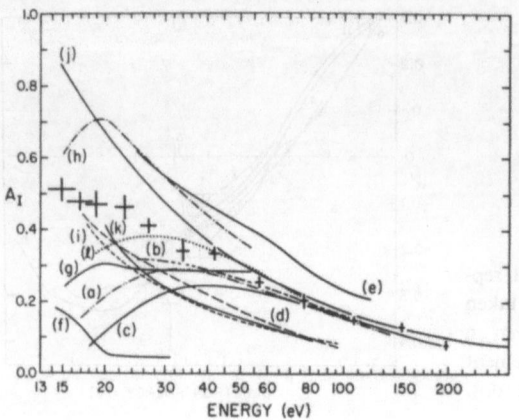

Figure 15. Experimental values of A_I and theoretical curves based upon the following approximations: (a), (b), (c), and (d) Born-exchange (BE) calculations taken from References 32–35 in order; (e) BE calculation with maximum interference taken from Reference 33; (f) BE calculation with angle-dependent potential taken from Reference 36; (g) and (h) spherical-average-exchange calculations, the latter allowing for maximum interference, both from Reference 32; (i), (j), and (k) Glauber-exchange, modified Born–Oppenheimer, and close-coupling calculations, respectively, taken from References 35, 37, and 38; (l) BE calculation taken from Reference 39.

The measured values of A_I are compared in Figure 15 with predicted values from a number of theoretical calculations. In reducing the published theoretical singlet and triplet cross sections to the predictions for A_I shown by curves (j) and (k), equation (4) was converted to the form[21]

$$A = (1 - r)/(1 + 3r) \tag{33}$$

where r is the ratio of the total triplet to total single ionization cross section. For all other curves, the total interference cross section σ_I^{int} obtained by integrating $\text{Re}(f^*g)$ over angle and outgoing energy,[21] was found from the relation[33]

$$\sigma_I^{\text{int}} = \sigma_I - \bar{\sigma}_I \tag{34}$$

where σ_I is the total ionization cross section calculated in the absence of exchange effects and $\bar{\sigma}_I$ is the total spin-averaged cross section with exchange included. The asymmetry A_I was then calculated according to

$$A_I = \sigma_I^{\text{int}}/\bar{\sigma}_I \tag{35}$$

Although a number of theoretical calculations are in good agreement with measured values of $\bar{\sigma}_I$,[41] it can be seen from Figure 15 that none are in agreement with the measured values of A_I below $\sim$50 eV. Since a discrepancy between theory and experiment is not unexpected at low energies, given the nature of the theoretical methods, the appearance of the discrepancy in A_I and not in $\bar{\sigma}_I$ signifies that new information is indeed contained in polarization measurements.

5. Future Developments

While the measurements described in this paper represent a substantial advance in the field of electron–atom collisions with polarized particles, much work remains untouched. Even with considerations of future work restricted to the case of electron–hydrogen collisions it is evident that only the surface has been scratched. In addition to the performance of elastic scattering measurements over a wide range of angles and energies, inelastic scattering measurements are well within the realm of experimental capability. In particular, the measurement of the $2s$ metastable excitation cross section is of interest, since $\bar{\sigma}$ and A' have already been measured.[40] Thus, within the framework of the discussion of Section 1.1, a complete determination of the scattering parameters will have been made once A is measured.

Future asymmetry measurements will be facilitated greatly by the use of improved polarized electron sources. At the present time, photoemission from single-crystal GaAs with circularly polarized light appears to be the most suitable source for low-energy electron–atom scattering experiments.[10] Several such sources have already been developed and used in solid state[42] and high-energy physics experiments.[43] The high-current capabilities ($>10\ \mu A$ under dc operation) and intrinsically narrow energy spread (~100-meV full width at half-maximum) make these sources ideal for the study of narrow-energy features such as resonances and threshold effects. In the case of ionization, for example, the asymmetry parameter A_I is predicted to reach a value of unity precisely at threshold.[44] The large energy spread of the Fano-effect source used in the present electron–hydrogen studies precluded such a threshold measurement, but a GaAs source used in conjunction with an electron monochromator should make the measurement possible.

Although the polarization of a GaAs source is only ~0.4, its high current capabilities and small emittance enable it to have an effective figure of merit, ζ, 20 to 50 times that of the Fano-effect source used in the electron–hydrogen measurements. Thus the sensitivity $A_E/\delta A_E$ shown in Figure 3 for elastic scattering would be increased by a factor of 20–50, making the determination of A_E much more favorable relative to A_E' and A_E'' at virtually all energies. In addition, the substantially increased value of ζ will make ever more feasible the performance of third-generation scattering experiments, thereby eliminating the extant sign ambiguity in ϑ.

In addition to all of these advantages, the optical reversibility of the electron polarization, used with such effectiveness in the Fano-effect source, is still preserved with a GaAs source. Indeed, asymmetries have already been measured to a precision of $\sim10^{-5}$ in a high-energy physics study of parity-violating neutral weak currents.[43] With the advent of new polarized-electron-source technology already upon us, the future of coherence measurements in electron–atom collisions with polarized beams appears almost boundless.

ACKNOWLEDGMENTS

The author is grateful to his collaborators, Dr. M. J. Alguard and Dr. P. F. Wainwright, for their devotion to the experimental program and for their insight and ingenuity during many difficult hours. The author also wishes to thank Dr. G. Baum, Dr. J. Ladish, and Dr. W. Raith for their assistance in the early phases of the program and Dr. C. Tu for a careful reading of the manuscript. Finally the author wishes to acknowledge the support of the U.S. National Science Foundation under Grant No. PHY76-84469.

References

1. P. G. Burke and K. Smith, *Rev. Mod. Phys.* **34**, 458–502 (1962); P. G. Burke, D. F. Gallaher, and S. Geltman, *J. Phys. B* **2**, 1142–1154 (1969); P. G. Burke, in *Atomic Physics*, V. W. Hughes, B. Bederson, V. W. Cohen, and F. M. J. Pichanick, eds., pp. 265–294, Plenum Press, New York (1969); R. J. Drachman and A. Temkin, in *Case Studies in Atomic Collision Physics II*, E. W. McDaniel and M. R. C. McDowell, eds., pp. 399–481, North Holland Publishing Co., Amsterdam (1972); J. Callaway and J. F. Williams, *Phys. Rev. A* **12**, 2312–2318 (1975); and citations in these references.
2. J. F. Williams, in *Electron and Photon Interactions with Atoms*, H. Kleinpoppen and M. R. C. McDowell, eds., pp. 309–338, Plenum Press, New York (1976).
3. P. G. Burke and H. M. Schey, *Phys. Rev.* **126**, 147–162 (1962).
4. A. Temkin and J. C. Lamkin, *Phys. Rev.* **121**, 788–794 (1961).
5. P. G. Burke, D. F. Gallaher, and S. Geltman, *J. Phys. B* **2**, 1142–1154 (1969).
6. J. Callaway and J. F. Williams, *Phys. Rev. A* **12**, 2312–2318 (1975).
7. C. Schwartz, *Phys. Rev.* **124**, 1468–1471 (1961).
8. K. Smith, R. P. McEachran, and P. A. Fraser, *Phys. Rev.* **125**, 553–558 (1962).
9. G. Khayrallah, *Phys. Rev. A* **14**, 2064–2070 (1976).
10. D. T. Pierce and F. Meier, *Phys. Rev. B* **13**, 5484–5500 (1974).
11. M. S. Lubell, in *Atomic Physics 5*, R. Marrus, M. Prior, and H. Shugart, eds., pp. 325–373, Plenum Press, New York (1977).
12. U. Fano, *Phys. Rev.* **178**, 131–136 (1969).
13. M. S. Lubell and W. Raith, *Phys. Rev. Lett.* **23**, 211–214 (1969).
14. J. Kessler and J. Lorenz, *Phys. Rev. Lett.* **24**, 87–88 (1970).
15. G. Baum, M. S. Lubell, and W. Raith, *Phys. Rev. A* **5**, 1073–1087 (1972).
16. N. F. Mott, *Proc. R. Soc. London Ser. A* **124**, 425–442 (1929).
17. L. Mikaelyan, A. Borovoi, and E. Denisov, *Nucl. Phys.* **47**, 328–337 (1963).
18. H. Frauenfelder and A. Rossi, in *Methods of Experimental Physics*, L. C. L. Yuan and C. S. Wu, eds., Vol. 5, Pt. B, pp. 214–274, Academic Press, New York (1963).
19. G. Holzwarth and H. J. Meister, *Nucl. Phys.* **59**, 56–64 (1964); *Tables of Asymmetry, Cross-section and Related Functions for Mott Scattering of Electrons by Screened Gold and Mercury Nuclei*, G. Holzwarth and H. J. Meister, eds., Institut für Theoretische Physik der Universität München, München, Germany (1964).
20. V. W. Hughes, R. L. Long, Jr., M. S. Lubell, M. Posner, and W. Raith, *Phys. Rev. A* **5**, 195–222 (1972).
21. M. J. Alguard, V. W. Hughes, M. S. Lubell, and P. F. Wainwright, *Phys. Rev. Lett.* **39**, 334–338 (1977).
22. P. F. Wainwright, M. J. Alguard, G. Baum, and M. S. Lubell, *Rev. Sci. Instrum.* **49**, 571–585 (1978).
23. J. Kessler, in *Atomic Physics 3*, S. J. Smith and G. K. Walters, eds., pp. 523–541, Plenum Press, New York (1973).
24. G. Baum, E. Kisker, W. Raith, and B. Reihl, *Phys. Rev. B* **18**, 2256–2275 (1978).
25. M. J. Alguard, J. E. Clendenin, V. W. Hughes, M. S. Lubell, K. P. Schüler, G. Baum, W. Raith, and R. H. Miller, *Nucl. Instrum. Meth.* **163**, 29–59 (1979).
26. U. Heinzmann, J. Kessler, and J. Lorenz, *Z. Phys.* **240**, 42–61 (1970).

27. W. von Drachenfels, U. T. Koch, R. T. Lepper, T. M. Müller, and W. Paul, *Z. Phys.* **269**, 387–397 (1974).
28. E. H. A. Granneman, *Polarization Effects in One- and Two-Photon Ionization of Cesium and Rubidium*, academisch proefschrift, University of Amsterdam, 1976 (unpublished).
29. G. V. Marr and D. M. Creek, *Proc. R. Soc. London Ser. A* **304**, 233–244 (1968).
30. N. F. Ramsey, *Molecular Beams*, p. 149, Oxford, London (1969).
31. H. D. Zeeman, K. Jost, and S. Gilad, *Rev. Sci. Instrum.* **42**, 485–489 (1971).
32. M. R. H. Rudge and M. J. Seaton, *Proc. R. Soc. London Ser. A* **283**, 262–287 (1965).
33. R. J. Peterkop, *Zh. Eksp. Teor. Fiz.* **41**, 1938–1939 (1961) [*Sov. Phys.-JETP* **14**, 1377–1378 (1962)].
34. S. Geltman, M. R. H. Rudge, and M. J. Seaton, *Proc. Phys. Soc. London* **81**, 375–378 (1963).
35. J. E. Goldin and J. H. McGuire, *Phys. Rev. Lett.* **32**, 1218–1221 (1974).
36. M. R. H. Rudge and S. B. Schwartz, *Proc. Phys. Soc. London* **88**, 563–578 (1966).
37. V. I. Ochkur, *Zh. Eksp. Teor. Fiz.* **47**, 1746–1750 (1964) [*Sov. Phys.-JETP* **20**, 1175–1178 (1964)].
38. D. F. Gallaher, *J. Phys. B* **7**, 362–370 (1974).
39. M. R. H. Rudge, *J. Phys. B* **11**, L149–L150 (1978).
40. W. L. Lichten and S. Schultz, *Phys. Rev.* **116**, 1132–1139 (1959).
41. W. L. Fite and R. T. Brackman, *Phys. Rev.* **111**, 1141–1151 (1958); R. L. F. Boyd and A. Boksenberg, in *Proceedings of the Fourth International Conference on Ionization Phenomena in Gases, Uppsala, Sweden, 1959*, R. N. Nilsson, ed., Vol. I, pp. 529–536, North Holland Publishing Co., Amsterdam (1959); E. W. Rothe *et al.*, *Phys. Rev.* **125**, 582–583 (1962).
42. G.-C. Wang *et al.*, *Phys. Rev. Lett.* **42**, 1349–1352 (1979).
43. C. Y. Prescott *et al.*, *Phys. Lett.* **77B**, 347–352 (1978); **84B**, 524–528 (1979).
44. H. Klar and W. Schlecht, *J. Phys. B* **9**, 1699–1711 (1976).

54

Ionization of Polarized Alkali Atoms by Polarized Electrons

D. HILS, K. RUBIN, AND H. KLEINPOPPEN

Crossed beams of polarized electrons and polarized alkali atoms have been used to measure the ratio of the interference cross section to the total cross section for electron impact ionization of K and Na. For K our results extend from 6 to 80 eV. For Na we have at present only one preliminary result at 9.3 eV.

1. Introduction

Valuable new information about electron–atom scattering processes can be obtained from experiments with polarized particle beams.[1–3] In this paper we report on measurements to determine the ratio of the interference cross section to the total ionization cross section for electron impact ionization of potassium and sodium. The only other measurements of the interference cross section are those by Alguard *et al.*[4] for hydrogen (see also Chapters 44 and 53 of this book by W. Raith *et al.* and by M. S. Lubell). Our measurements for potassium extend over the energy range 6–80 eV, while for sodium we have at present only one preliminary result at 9.3 eV. The interference cross section is important at the lower energies where its effect is to reduce the ionization cross section from its nonexchange value.[5] In our experi-

D. HILS AND H. KLEINPOPPEN ● Institute of Atomic Physics, University of Stirling, Stirling, Scotland.   K. RUBIN ● City College of the City University of New York, New York.   Dr. Kleinpoppen's temporary address is Fakultät für Physik and Zentrum für interdisziplinäre Forschung of Universität Bielefeld, 48 Bielefeld, West Germany.

ments both the neutral alkali beam and the electron beam are polarized. We therefore do not require a knowledge of the absolute value of quantities like beam intensities, atomic beam density, particle detection efficiency, etc. The major difficulty with low-energy polarized electron experiments which has so far limited the number of investigations[6,7] is the problem of small signals due to the limitations of present polarized electron sources. In the case under study, the situation is more favorable than, for instance, in elastic scattering experiments, because the resulting ion can be detected with a high signal-to-noise ratio.

In the future, incorporation of novel polarized electron sources now under development[6] should see further progress in this area. A preliminary report on this work was given at the Satellite Conference of the Xth ICPEAC, Paris, France, 1977.

2. Experiment

The details of the experiment for the measurement of the interference cross section in K have been described already.[8] We therefore limit ourselves in the following to a brief summary of the experimental method and then concentrate on the different techniques used in the sodium experiment.

The apparatus is shown in Figure 1. A polarized alkali beam is cross-fired with a polarized electron beam (see Hils and Kleinpoppen[8]). Ionized atoms are extracted by a lens system and detected with a channeltron particle multiplier. The ionization signal is measured for the two cases where electron and atom spins are antiparallel and parallel. The ratio χ of these two numbers defines the ionization asymmetry $A = (\chi - 1)/(\chi + 1)$ which determines the ratio of the interference cross section to the total ionization cross section, i.e.,

$$A = P_e P_A \left(\frac{Q_{\text{int}}}{Q_0} \right) \tag{1}$$

where P_e is the electron beam polarization, P_A is the atomic beam polarization, Q_{int} the interference cross section, and Q_0 the total ionization cross section. In our earlier measurements for potassium, the relative orientation of the two spin vectors was controlled by a longitudinal magnetic field between hexapole magnet and interaction region which adiabatically rotated the atomic spin in a direction either parallel or antiparallel to the electron spin direction. For our measurements in sodium, an argon-ion-pumped dye laser was used to periodically depolarize the sodium beam as described below. In this case one measures the ionization difference between a polarized and unpolarized sodium beam when cross-fired with a polarized electron beam. The apparatus used a Rabi-type inhomogeneous magnet to spin-analyze the sodium beam and to verify the destruction of the polarization when the sodium atoms are exposed to the laser light.

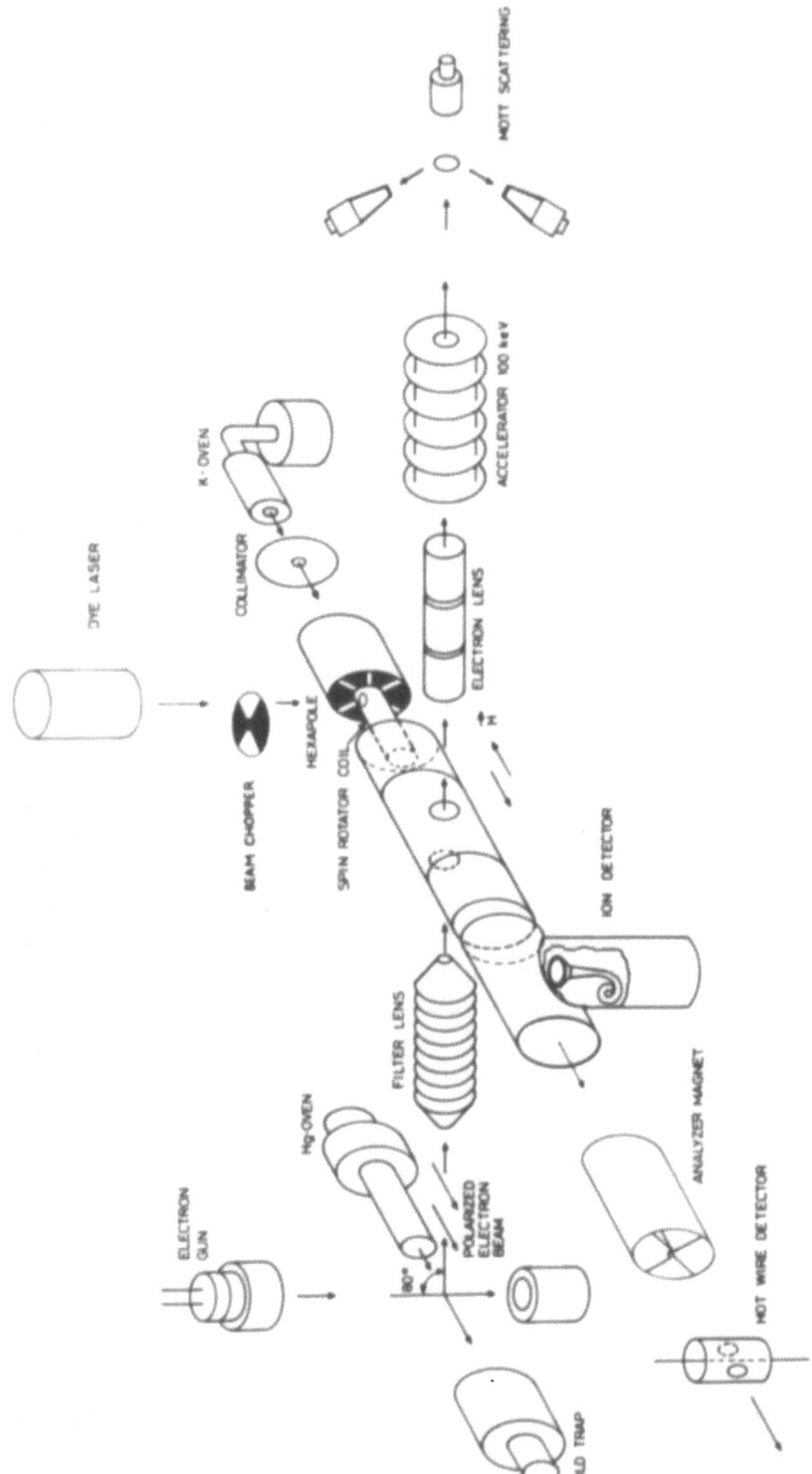

Figure 1. Schematic diagram of the apparatus.

Laser Depolarizer

In the following we discuss laser excitation of a polarized sodium beam with a multimode cw dye laser (Spectra-Physics model 375) pumped with an argon ion laser (Spectra-Physics model 170). This laser has a linewidth of 40 GHz and a mode spacing of 420 MHz. Figure 2 shows the hyperfine splitting of sodium in the ground and first excited state. In our case, tuning to a particular hyperfine transition is not possible because of the presence of many modes within the laser linewidth. Let us consider laser excitation to the $3^2P_{1/2}$ state for simplicity (excitation to the $3^2P_{3/2}$ state works as well). This state splits into the two states $F_u = 2$ and $F_u = 1$, with a hyperfine splitting of $\Delta\nu_u = 192$ MHz. Atoms passing through the hexapole magnet are predominantly in the $3^2S_{1/2}$ $F = 2$ state. To begin the pumping process, therefore, one has to tune the laser to the frequency ν_{21} or ν_{22} (Figure 3). Atoms excited to these states then decay back to either $F = 2$ or $F = 1$ of the $3^2S_{1/2}$ ground state, separated by $\Delta\nu_1 = 1772$ MHz. To avoid trapping of atoms in the $F = 1$ ground state, one has to pump at a frequency ν_{12} as well. In fact Gerritsen and Nienhuis[9] showed that one requires at least three different frequencies, which are illustrated in Figure 3. In our experiment laser excitation at these frequencies is achieved by a combination of Doppler tuning and the multimode nature of the laser. If we denote the angle between sodium beam and the laser by α, then the Doppler shift is given by

$$\nu_D = -\nu\left(\frac{v}{c}\right)\cos\alpha \tag{2}$$

where v is the velocity of the atom. If one takes into account now that there are laser lines every 420 MHz, it is easy to verify that one has to pump at angles $\alpha = 88°$, $84°$, and $96°$ (this refers to an oven temperature of 500°C). To cover these angles with one laser only we irradiate the sodium beam at $\alpha \simeq 90°$. The laser beam after traversing the sodium beam is then reflected from an aluminum foil which lines the inside walls of a cylindrical cell through which the atoms travel. The reflected laser light is sufficiently diffuse to cover the required angles. As the atoms decay back into the ground states we eliminate optical pumping into magnetic sublevels by the application of a small magnetic field roughly perpendicular to the axis of quantization

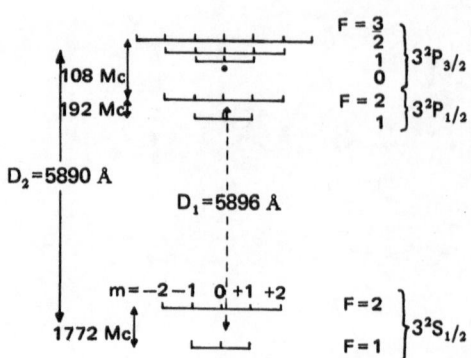

Figure 2. Sodium fine structure and hyperfine structure of the 3^2S ground and 3^2P excited state.

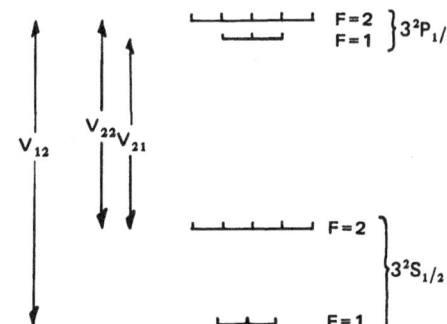

Figure 3. Sodium hyperfine structure of the $3^2S_{1/2}$ ground and $3^2P_{1/2}$ excited state. Shown are the required pumping frequencies ν_{22}, ν_{21}, and ν_{12} if one wants to avoid trapping in the $F = 1$ ground state.

established by the laser polarization. This field mixes the various m_F states and prevents accumulation of atoms in any particular m_F state (Zeeman level).[10] Figure 4 shows the beam splitting with the Rabi analyzer. The bottom part demonstrates the splitting into the two spin components ($^2S_{1/2}$ ground state) with the laser beam off. Here we see the effect of the polarizing hexapole magnet as we observe a strong spin-up peak and a small spin-down peak. The top part of Figure 4 shows the effect of laser irradiation upon beam polarization. Now both spin states are more or less equally populated. For a measurement of the ionization asymmetry the laser beam is mechanically chopped at 30 cps. The corresponding ionization rates are synchronously accumulated by two scalers, one of which registers when the atomic beam polarization is P_A, the other when the beam is unpolarized.

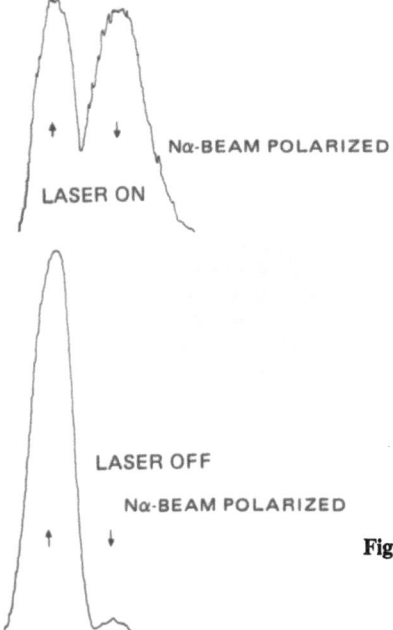

Figure 4. Alkali beam splitting measured with a Rabi-type inhomogeneous magnetic analyzer. $\uparrow = |+\frac{1}{2}\rangle$ state. $\downarrow = |-\frac{1}{2}\rangle$ state. Top part demonstrates the depolarizing effect of laser irradiation. Bottom part shows the effect of the polarizing action of the hexapole magnet.

One of the unwanted side effects of laser irradiation is the production of a small ion signal due to the interaction of the laser beam with the atomic beam. This ion signal originates from laser-induced associative ionization, i.e.,

$$\mathrm{Na^*(3p) + Na^*(3p) + \hbar\omega \rightarrow Na^+ + Na(3s) + e^-}$$
$$\mathrm{Na^*(3p) + Na^*(3p) + \hbar\omega \rightarrow Na_2^+ + e^-}$$
$$(3)$$

and has been observed in other experiments.[11] This background ion signal which is synchronous with the laser pulse is eliminated by biasing the laser cell with a small negative voltage which traps the ions.

To check our ionization asymmetry measurements we studied A as a function of electron beam polarization P_e. According to equation (1) $A \propto P_e$ and we expect to observe the same angular dependence in both parameters. P_e is varied by a rotation of the electron gun in the mercury polarizer. Figure 5 shows the results of such a measurement. The variation of the electron beam polarization $P_e(\theta)$ is shown by the crosses, whereas points with error bars indicate the ionization asymmetry A. Notice the change in sign in both curves and the disappearance of A and P_e for small angles.

We are at present engaged in a program to modify our multimode Spectra-Physics model 375 to single-mode operation. A single-mode laser opens up the possibility of polarizing the atomic beam without using a hexapole magnet. In addition the direction of polarization of the atom beam can easily be changed from $+P_a$ to $-P_a$ by reversing the sense of circular polarization of the laser light.

Let us look again at Figure 2. Consider excitation with circularly polarized light (σ^+) tuned to the $F = 2$ to $F_u = 3$ transition. Since the dipole selection rule for optical transitions is $\Delta F = 0, \pm 1$, this means that the excited atoms can decay back to the original $F = 2$ state only. Transitions to the $F = 1$ ground state are forbidden. Furthermore, due to the selection rule $\Delta M_F = +1$ (σ^+) for excitation, after a few pumping cycles, five out of eight atoms end up in the $F = 2$ $m_F = +2$ state. Since this is a pure spin-up state, the atomic beam polarization is $P_A = \frac{5}{8}$. Reversing the sense

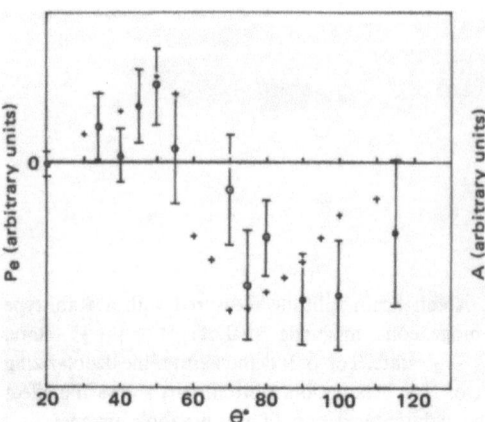

Figure 5. Measured electron polarization and ionization asymmetry as a function of scattering angle in the polarized electron source. $\bar{\phi}$, Ionization asymmetry at 9.3 eV for sodium. $+$, Electron polarization. Arbitrary units are used.

of circular polarized light from right to left (σ^-) also reverses the polarization to $P_A = -\frac{5}{8}$. Such a value compares favorably with polarizations obtained with hexapole magnets, which for Na²³ in a zero-field region is $P_A \leq 0.25$.

3. Results

Figure 6 summarizes all our results at present for the ratio of interference cross section to total ionization cross section. Circular dotted points with error bars are our results for potassium. The crossed point with error bar is our first result for sodium. Also shown in this figure are the results of a theoretical calculation based on the Born approximation by Peach.[12,13] Peach's results are for lithium and sodium. There is little variation in the theoretical results between lithium and sodium and it can therefore be expected that potassium calculations should be similar. While there is fair agreement between theory and experiment at the higher energies, we disagree for small energies. This is expected because the Born approximation is not valid at low energies and also does not include correlations of the two electrons in the final state.

The most interesting region to explore in future measurements is the threshold region. In a recent calculation, Klar and Schlecht[14] demonstrated that the triplet cross section near threshold vanishes as $E^{3.88}$ whereas the singlet cross section for ionization follows Wannier's law[15] and is proportional to $E^{1.127}$. This implies a threshold value $Q_{int}/Q_0 = 1$ [see equations (7) and (8) in Reference 8].

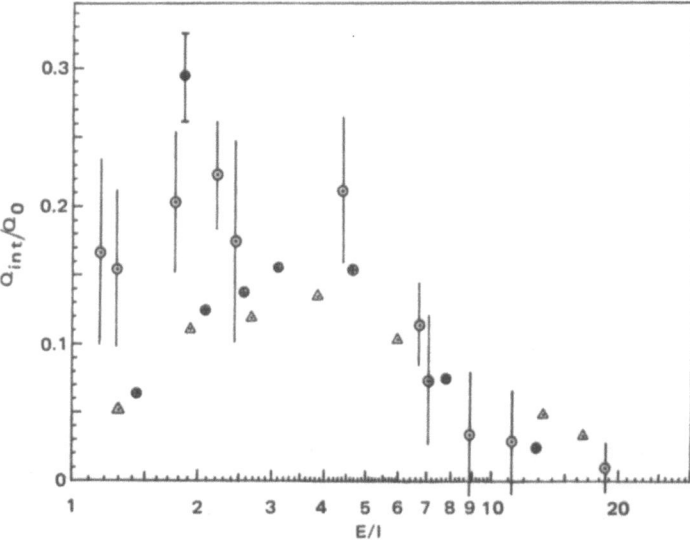

Figure 6. Experimental results and theoretical calculations for the ratio of interference cross section to total ionization cross section. Experimental results: ⌀ K; ⌀ Na. Born approximation by Peach: ⊕ Li; △ Na.

The energy range over which this is valid is, however, not known precisely and more measurements will be necessary to check this prediction.

ACKNOWLEDGMENT

We are grateful to the U.K. Science Research Council for supporting this research project and providing a Senior Visiting Fellowship for one of us (K.R.).

References

1. B. Bederson, *Comments At. Mol. Phys.* **1**, 41–44 (1969).
2. B. Bederson, *Comments At. Mol. Phys.* **1**, 65–69 (1969).
3. H. Kleinpoppen, *Phys. Rev. A* **3**, 2015–2027 (1971).
4. M. J. Alguard, V. W. Hughes, M. S. Lubell, and P. F. Wainwright, *Phys. Rev. Lett.* **39**, 334–338 (1977); see also Chapter 53 of this book.
5. R. Peterkop, *Proc. Phys. Soc.* **77**, 1220–1222 (1961).
6. M. S. Lubell, *At. Phys.* **5**, 325–373 (1977).
7. H. Kleinpoppen, *Adv. Quant. Chem.* **10**, 77–141 (1977).
8. D. Hils and H. Kleinpoppen, *J. Phys. B* **11**, L283–287 (1978).
9. H. J. Gerritsen and G. Nienhuis, *Appl. Phys. Lett.* **26**, 347–349 (1975).
10. G. M. Carter, D. E. Pritchard, and T. W. Ducas, *Appl. Phys. Lett.* **27**, 498–499 (1975).
11. A. v. Hellfeld, J. Caddick, and J. Weiner, *Phys. Rev. Lett.* **40**, 1369–1373 (1978).
12. G. Peach, *Proc. Phys. Soc.* **85**, 709–718 (1965).
13. G. Peach, *Proc. Phys. Soc.* **87**, 381–391 (1966).
14. H. Klar and W. Schlecht, *J. Phys. B* **9**, 1699–1711 (1976).
15. G. H. Wannier, *Phys. Rev.* **90**, 817–825 (1953).

55

Total Cross Section of Interference between Direct and Exchange Interaction in Polarized Electron Impact Excitation of Polarized Atoms

H. KLEINPOPPEN

An interference effect between the direct and exchange interactions should occur and should be observable in the total cross section for electron impact excitation of the resonance transitions $^2S_{1/2} \rightarrow {}^2P_{1/2,3/2}$ of alkali atoms. In order to detect this interference effect both polarized electrons and polarized atoms have to be used. By applying the four-state exchange close-coupling theory of Moores and Norcross, predictions of the interference effect are made for the total and the partial magnetic sublevel cross sections of the first resonance transition of sodium.

Recent progress in technology has enabled experimental physicists to study electron–atom collisions by means of both polarized electrons and polarized atoms.[1-4] Based on the analysis of such experiments, interference effects could be detected in the total ionization cross section of atomic hydrogen[1-2] and alkali atoms.[3-4] Such experiments proved the assumption[5] that the interference between direct and exchange ionization amplitudes may result in an "integral" interference effect in the total ionization cross section.

H. KLEINPOPPEN ● Fakultät für Physik and Zentrum für interdisziplinäre Forschung of Universität Bielefeld, 4800 Bielefeld 1, West Germany. Permanent address: Institute of Atomic Physics, University of Stirling, Stirling, Scotland.

In analogy to this effect, an "integral" interference effect in the excitation of the ^2P states of one-electron atoms should be observable in the total excitation cross section for experiments with partially polarized electrons (degree of polarization P_e) and partially polarized atoms (degree of polarization P_A).

The total cross section of excitation for the transition $^2S_{1/2} \to {}^2P_{1/2,3/2}$ of alkali atoms can be expressed[6] in terms of partial cross sections for exciting the magnetic sublevels with $m_l = 0$ and $m_l = \pm 1$:

$$Q(^2S_{1/2} \to {}^2P_{1/2,3/2}) = Q_{m_l=0} + 2Q_{m_l=\pm 1}$$

$$= Q_0 + 2Q_1 \tag{1}$$

Introducing direct and exchange excitation amplitudes f_0, g_0, f_1 and g_1 for the excitation of the sublevels $m_l = 0$, $m_l = \pm 1$, we obtain[7,8] for Q_0 and Q_1

$$Q_0 = \frac{k_f}{k_i} \iint \frac{1}{2} \{|f_0|^2 + |g_0|^2 + |f_0 - g_0|^2\} \, d\Omega$$

$$= \frac{k_f}{k_i} \iint \{|f_0|^2 + |g_0|^2 - \text{Re}(f_0 g_0^*)\} \, d\Omega$$

$$\tag{2}$$

$$Q_1 = \frac{k_f}{k_i} \iint \frac{1}{2} \{|f_1|^2 + |g_1|^2 + |f_1 - g_1|^2\} \, d\Omega$$

$$= \frac{k_f}{k_i} \iint \{|f_1|^2 + |g_1|^2 - \text{Re}(f_1 g_1^*)\} \, d\Omega$$

(k_i and k_f are wave vectors for in and outgoing electrons).

An interference cross section Q^{int} is determined by the last term $\text{Re}(fg^*)$ in these integrals $[Q^{\text{int}} = (k_f/k_i) \iint \text{Re}(fg^*) \, d\Omega]$ and the total cross sections can be split as follows:

$$Q_0 = Q_0' - Q_0^{\text{int}}, \qquad Q_1 = Q_1' - Q_1^{\text{int}}, \qquad Q = Q' - Q^{\text{int}} \tag{3}$$

where the primed Q's contribute to the sums of the terms $|f|^2$ and $|g|^2$ in the integrals of equation (2).

By using partially polarized electrons and partially polarized atoms for the above excitation, it follows from equation (25) in Reference 7 (by including integration over all scattering directions and introducing f_0, f_1, g_0, and g_1) for the modified total cross sections Q^S, Q_0^S, Q_1^S with polarized electrons and polarized atoms

$$Q^S = Q - P_e P_A \, Q^{\text{int}}$$

$$Q_0^S = Q_0 - P_e P_A \, Q_0^{\text{int}} \tag{4}$$

$$Q_1^S = Q_1 - P_e P_A \, Q_0^{\text{int}}$$

In other words the modified cross sections Q^S for excitation with both spin polarized electrons and polarized atoms can be used to determine the ratios Q^{int}/Q, Q_0^{int}/Q_0, and Q_1^{int}/Q_1:

$$\frac{Q^{int}}{Q} = \frac{Q_0^{int} + 2Q_1^{int}}{Q} = -\frac{1}{P_e P_A}\left(\frac{Q^S}{Q} - 1\right)$$

$$\frac{Q_0^{int}}{Q_0} = -\frac{1}{P_e P_A}\left(\frac{Q_0^S}{Q_0} - 1\right) \qquad (5)$$

$$\frac{Q_1^{int}}{Q_1} = -\frac{1}{P_e P_A}\left(\frac{Q_1^S}{Q_1} - 1\right)$$

Predictions for the ratios of these cross sections can be made, for instance, by applying amplitudes from the four-state exchange close-coupling theory[8] at low electron energies. Figure 1 displays such predictions, which reveal large ratios and even structure in the near threshold region. No experimental data are available for comparison though it should be possible to apply present sources for polarized electrons and polarized atoms to detect the above interference effects. Furthermore, special anisotropy effects in the radiation or the polarization of the emission line are expected if polarized electrons and polarized one-electron atoms are used.

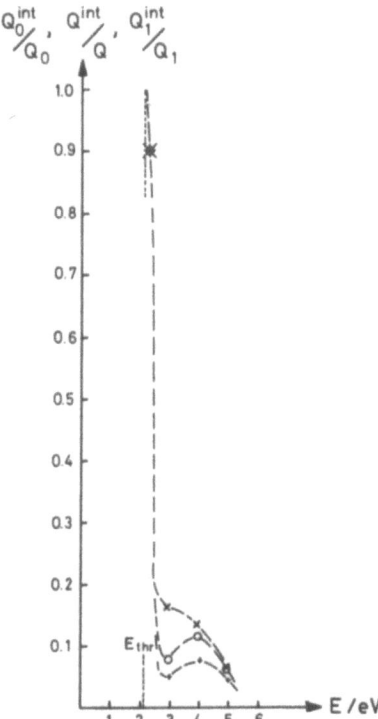

Figure 1. Relative total interference cross sections for sodium $3^2S_{1/2} \rightarrow {}^2P_{1/2,3/2}$ excitation: —○—, Q^{int}/Q, —×—, Q_0^{int}/Q_0, —+—, Q_1^{int}/Q calculated from four state exchange close-coupling theory of Moores and Norcross.[8] E, electron impact energy, E_{thr}, threshold energy.

References

1. M. J. Alguard, V. W. Hughes, M. S. Lubell, and P. F. Wainwright, *Phys. Rev. Lett.* **39**, 334–338 (1977).
2. M. S. Lubell, Chapter 53 of this volume.
3. D. Hils and H. Kleinpoppen, *J. Phys. B* **11**, L283 (1978).
4. D. Hils, K. Rubin, and H. Kleinpoppen, Chapter 54 of this volume.
5. R. Peterkop, *Proc. Phys. Soc.* **77**, 1220 (1961).
6. I. C. Percival and M. J. Seaton, *Phil. Trans. R. Soc. London Ser. A* **251**, 113 (1958).
7. H. Kleinpoppen, *Phys. Rev. A* **3**, 2015 (1971).
8. D. L. Moores and D. W. Norcross, *J. Phys. B* **5**, 1482 (1972).

Author Index

Subject Index

A

Alignment, 116, 121, 124, 143, 148, 151, 181, 215, 219, 220, 225, 318, 327, 361, 364, 366, 367, 389, 390, 391, 594, 625, 629
Alignment tensor, 101, 104, 110
Angular coherence, 651
Angular correlation, 97, 98, 110, 113, 115, 167, 174, 181, 231, 237
Angular correlation parameters λ and χ, 172, 174, 175, 177, 462, 467, 468, 471
Angular distributions of Auger electrons, 215, 361
Anisotropy, 98, 159, 318, 334, 539, 638, 639, 699
Anisotropy coefficients, 364, 365
Atomic-field bremsstrahlung, 189, 197
Auger decay, 302
  electron, 211, 245, 260, 262, 264, 302, 310, 311, 319, 328
  process, 302
  spectrum, 218, 362
  transition, 260
Autoionization, 240, 246, 261, 264, 265, 297, 300, 306, 309, 610, 612
Autoionization rate, 259
Autoionization state, 283, 290, 293, 294, 299, 300, 301, 494, 505, 507

B

Bethe–Heitler calculation, 194
Bethe–Heitler formula, 190, 193, 198
Binary collision, 41, 553
Binary encounter theory, 44, 316
Binary peak, 45
Born approximation, 23, 24, 44, 93, 99, 104, 152, 294, 316, 695
Born–Mayer-type scattering, 398
Born–Oppehmeimer approximation, 136
Born T matrix, 182
Breit–Wigner resonance, 353

Bremsstrahlung, 187, 188, 189, 190, 195, 200, 202, 205, 206, 207, 211, 212, 213

C

Carambole collision, 266, 267
Channeling, 373, 374
Characteristic x-ray line radiation, 205, 315
Charge capture, 404
Charge exchange reaction, 517
Charge transfer collisions, 423, 430, 432
Clementi Wave function, 24, 26
Close-coupling calculations, 167, 271
Closed-channel, 272
Coherence factor, 407
Coherent excitation, 116, 377, 384
Coherent momentum states, 653
Coherent multipoles, 116
Coincidence-time-of-flight technique, 43
Collision spectroscopy, 510
Compton polarimeter, 199
Constant-q geometry, 25
Correlation diagram 333
Coster–Kronig yields, 316
Coulomb ionization, 315, 316, 317
Coulomb-projected Born approximation, 107
Coupled-channel calculations, 357

D

Degree of coherence, 95, 248, 250, 251, 253, 633
Demkov formula, 513
Density matrix, 121, 123, 130, 131, 157
Density-matrix formalism, 108
Dipole polarizability, 271
Dissociative ionization, 42
Distorted-wave approximation, 103, 177
Distorted-wave Born approximation, 47
Distorted-wave calculation, 93, 94
Distorted-wave impulse approximation, 1, 5, 38, 63